Experimental Brain Research Series 25

Springer-Verlag Berlin Heidelberg GmbH

P. Thier H.-O. Karnath (Eds.)

Parietal Lobe Contributions to Orientation in 3D Space

With 197 Figures

Springer

Peter Thier
Sektion für Visuelle Sensomotorik
Neurologische Klinik
Universität Tübingen
Hoppe-Seyler-Str. 3
72076 Tübingen, Germany

Hans-Otto Karnath
Neurologische Klinik
Universität Tübingen
Hoppe-Seyler-Str. 3
72076 Tübingen, Germany

Cover illustration:
Giorgio de Chirico. The Disturbing Muses. 1925. Galleria Nazionale d'Arte Moderna, Rome

---

ISBN 978-3-642-64498-6 ISBN 978-3-642-60661-8 (eBook)
DOI 10.1007/978-3-642-60661-8

Originally published by Springer-Verlag Berlin Heidelberg New York in 1997
Softcover reprint of the hardcover 1st edition 1997

Cover: Design & Production, Heidelberg
Production Editor: Renate Münzenmayer
Typesetting: Camera-ready by authors

SPIN 10527177 3135-5 4 3 2 1 0 – Printed on acid-free paper

Preface

The function of the parietal lobe has been a topic of great interest, its study stimulated by the profound and intriguing perceptual and motor deficits resulting from parietal lobe lesions in humans. The specific role of the parietal cortex has always been a matter of great controversy, with different laboratories emphasizing seemingly exclusive interpretations of parietal lobe functions arranged around a line separating sensory input and motor output, both possibly modulated by attention. Recent work based on awake, behaving monkeys and the study of patients with parietal lobe lesions have unmasked the sensory versus motor dichotomy of parietal lobe function as being both arbitrary and simplistic.

The present book conveys the current view of parietal lobe functions, centering around the idea that parietal lobe areas act as true sensorimotor interfaces contributing to the sensory guidance of movement and to the perception of space by offering non-sensory, mental representations of space suited to the needs of the specific task. It is largely based on a conference on parietal lobe functions held in Tübingen, Germany, in the early summer of 1995. The major goal of this meeting was to further the exchange between neurophysiologists and neuropsychologists interested in this part of the brain. This book aims to cast the productive discussions of this conference into a state-of-the-art overview of present thinking on the role of the parietal lobes and their specific contributions to eye movements, reaching and grasping, attention, perception, and the representation of space.

This book would not have been possible without the encouragement of Johannes Dichgans and the commitment of Dagmar Heller-Schmerold, who tirelessly maintained the communication between the authors, the editors, and the publisher.

We hope that readers from various fields, not only specialists working on the parietal lobe, will find this book a useful and stimulating source of information on this fascinating part of the brain.

Peter Thier
Hans-Otto Karnath

Tübingen, 1996

List of Contributors

R. A. Andersen
California Institute of Technology
216-76
Pasadena, CA 91125, USA

C. Baleydier
INSERM Unité 94
Vision et Motricité
16 Ave du Doyen Lépine
69500 Bron, France

S. Barash
Dept. Neurobiology
Weizmann Institute of Science
Rehovot 76100, Israel

A. Battaglia-Mayer
Istituto di Fisiologia Umana
Università di Roma „La Sapienza"
Piazzale Aldo Moro 5
00185 Rome, Italy

P.-P. Battaglini
Istituto di Fisiologia
Università di Trieste
Via Fleming 22
34127 Trieste, Italy

S. Ben Hamed
CNRS, Laboratoire de Physiologie Neurosensorielle
15 rue de l'Ecole de Médecine
75270 Paris Cedex 06, France

A. Berthoz
CNRS - Collège de France
LPPA
15 rue de l'Ecole de Médecine
75270 Paris Cedex 06, France

E. Bisiach
Dipartimento di Psicologia
Università di Torino
Via Lagrange 3
10123 Torino, Italy

F. Bremmer
Allg. Zoologie & Neurobiologie
Ruhr-Universität Bochum
44780 Bochum, Germany

Y. Burnod
INSERM CREARE
Université Pierre et Marie Curie
9 Quai St. Bernard
75005 Paris, France

L. Buxbaum
Moss Rehabilitation Institute
1200 Tabor Road
Philadelphia, PA 19141, USA

C. M. Butter
Department of Psychology
University of Michigan
East Hall
Ann Arbor, MI 48109-1109, USA

R. Caminiti
Istituto di Fisiologia Umana
Università di Roma „La Sapienza"
Piazzale Aldo Moro 5
00185 Rome, Italy

E. Daprati
Istituto di Fisiologia Umana
Università di Parma
Via Gramsci 14
43100 Parma, Italy

G. Davis
Department of Experimental Psychology
University of Cambridge
Downing Street
Cambridge C2 3EB, UK

J. Driver
Department of Psychology
Birkbeck College
University of London
Malet Street
London WC1E 7HX, UK

J.-R. Duhamel
CNRS, Laboratoire de Physiologie Neurosensorielle
15, rue de l'Ecole de Médecine
75270 Paris Cedex 06, France

L. Ercolani
Istituto di Fisiologia Umana
Università di Roma „La Sapienza"
Piazzale Aldo Moro 5
00185 Rome, Italy

L. Fadiga
Istituto di Fisiologia Umana
Università degli Studi di Parma
Via Gramsci 14
43100 Parma, Italy

A. Fanini
Dipartimento di Scienze Neurologiche e della Visione
Università degli Studi di Verona
Strada Le Grazie
37134 Verona, Italy

M. Farah
Department of Psychology
University of Pennsylvania
3815 Walnut Street
Philadelphia, PA 19104-6196
USA

P. Fattori
Department of Human and General Physiology
University of Bologna
Piazza di Porta S. Donato 2
40127 Bologna, Italy

S. Faugier-Grimaud
INSERM Unité 94
Vision et Motricité
16 Ave du Doyen Lépine
69500 Bron, France

S. Ferraina
Istituto di Fisiologia Umana
Università di Roma „La Sapienza"
Piazzale Aldo Moro 5
00185 Rome, Italy

J. M. Findlay
Department of Psychology
University of Durham
Durham DH1 3LE, UK

L. Fogassi
Istituto di Fisiologia Umana
Università degli Studi di Parma
Via Gramsci 14
43100 Parma, Italy

V. Gallese
Istituto di Fisiologia Umana
Università degli Studi di Parma
Via Gramsci 14
43100 Parma, Italy

C. Galletti
Department of Human and
General Physiology
University of Bologna
Piazza di Porta S. Donato 2
40127 Bologna, Italy

M. R. Garasto
Istituto di Fisiologia Umana
Università di Roma „La
Sapienza"
Piazzale Aldo Moro 5
00185 Rome, Italy

M. Gentilucci
Istituto di Fisiologia Umana
Università di Parma
Via Gramsci 14
43100 Parma, Italy

M. Girelli
Dipartimento di Scienze
Neurologiche e della Visione
Università degli Studi di Verona
Strada Le Grazie
37134 Verona, Italy

M. Glickstein
Department of Anatomy &
Developmental Biology
University College London
Gower Street
London WC1E 6BT, UK

M. A. Goodale
Department of Psychology
University of Western Ontario
London, Ontario N6A 5C2
Canada

W. Graf
CNRS, Laboratoire de
Physiologie Neurosensorielle
15, rue de l'Ecole de Médecine
75270 Paris Cedex 06, France

C. Guariglia
Dipartimento di Psicologia
Università di Roma „La
Sapienza"
Via dei Marsi 78
00185 Rome, Italy

M. Hasselbach
Department of Psychology
University of Michigan
East Hall
Ann Arbor, MI 48109-1109,
USA

W. Heide
Klinik für Neurologie
Medizinische Universität Lübeck
Ratzeburger Allee 160
23538 Lübeck, Germany

U. J. Ilg
Sektion Visuelle Sensomotorik
Universität Tübingen
Auf der Morgenstelle 15
72076 Tübingen, Germany

Y. Inoue
Neuroscience Section
Electrotechnical Laboratory
1-1-4 Umezono Tsukubashi
Ibaraki 305, Japan

A. E. Ipata
Dipartimento di Scienze
Neurologiche e della Visione
Università degli Studi di Verona
Strada Le Grazie
37134 Verona, Italy

M. Jeannerod
INSERM Unité 94
Vision et Motricité
16 Ave du Doyen Lépine
69500 Bron, France

S. Kakei
Department of Physiology
Tokyo Medical & Dental
University, School of Medicine
1-5-45 Yushima, Bunkyo-ku
Tokyo 113, Japan

H.-O. Karnath
Neurologische Klinik
Universität Tübingen
Hoppe-Seyler-Str. 3
72076 Tübingen, Germany

K. Kawano
Neuroscience Section
Electrotechnical Laboratory
1-1-4 Umezono Tsukubashi
Ibaraki 305, Japan

O. Kazennikov
Neurol. Universitätsklinik
Inselspital
Motoriklabor, BHH M-130
3010 Bern, Switzerland

T. Kitama
Neuroscience Section
Electrotechnical Laboratory
1-1-4 Umezono Tsukubashi
Ibaraki 305, Japan

D. Kömpf
Klinik für Neurologie
Medizinische Universität Lübeck
Ratzeburger Allee 160
23538 Lübeck, Germany

M. Kusunoki
Department of Physiology
Nihon University
School of Medicine
30-1 Oyaguchi-Kamamuchi
Itabashi-ku, Tokyo 173, Japan

M. Lappe
Allg. Zoologie & Neurobiologie
Ruhr-Universität Bochum
44780 Bochum, Germany

G. Luppino
Istituto di Fisiologia Umana
Università degli Studi di Parma
Via Gramsci 14
43100 Parma, Italy

M. Magnin
INSERM Unité 371
Cerveau et Vision
18 Ave du Doyen Lépine
69500 Bron, France

C. A. Marzi
Dipartimento di Scienze
Neurologiche e della Visione
Università degli Studi di Verona
Strada Le Grazie
37134 Verona, Italy

J. B. Mattingley
Department of Experimental
Psychology
University of Cambridge
Downing Street
Cambridge C2 3EB, UK

J. H. R. Maunsell
Division of Neuroscience
S-603
Baylor College of Medicine
Houston, TX 77030, USA

B. Mazoyer
Groupe d'Imagerie Neuro-
fonctionnelle
Service Hospitalier Frédéric
Joliot
4 pl du Général Leclerc
91406 Orsay Cedex, France

F. A. Miles
Laboratory of Sensorimotor
Research
National Eye Institute, NIH
Bethesda, MA 20892, USA

A. D. Milner
School of Psychology
University of St. Andrews
St. Andrews, Fife KY16 9JU,
UK

C. Miniussi
Dipartimento di Scienze
Neurologiche e della Visione
Università degli Studi di Verona
Strada Le Grazie
37134 Verona, Italy

R. Müri
Neurol. Universitätsklinik
Inselspital
3010 Bern, Switzerland

A. Murata
Department of Physiology
Nihon University
School of Medicine
30-1 Oyaguchi-Kamamuchi,
Itabashi-ku, Tokyo 173, Japan

K. J. Murphy
Department of Psychology
University of Western Ontario
London, Ontario N6A 5C2
Canada

C. Orssaud
Groupe d'Imagerie Neuro-
fonctionnelle
Service Hospitalier Frédéric
Joliot
4 pl du Général Leclerc
91406 Orsay Cedex, France

M.-T. Perenin
INSERM Unité 94
Vision et Motricité
16 ave du Doyen Lépine
69500 Bron, France

S. Perrig
Institute of Physiology
University of Fribourg
Rue du Musée 5
1700 Fribourg, Switzerland

L. Petit
Section of Functional Brain Imaging
NIMH, Laboratory of Psychology & Psychopathology
Bldg. 10, Rm 4C110
9000 Rockville Pike
Bethesda, MA 20892, USA

C. Pierrot-Deseilligny
Hôpital Pitié-Salpêtrière
Clinique Paul Castaigne
Service de Neurologie 1
47-83, blvd. de l'Hôpital
75651 Paris Cedex 13, France

A. Pouget
Institute for Cognitive & Computational Sciences
Georgetown University
Washington, D.C. 20007-2197
USA

M. Prior
Department of General Psychology
University of Padua
Via Venezia 8
35131 Padova, Italy

C. Rorden
Department of Experimental Psychology
University of Cambridge
Downing Street
Cambridge C2 3EB, UK

E. Rouiller
Department of Neurology
University of Geneva
Hôpital cantonal
Rue Micheli-du-Crest 24
1205 Geneva, Switzerland

M. L. Rusconi
Dipartimento di Psicologia Generale
Università di Padova
Via Venezia 8
35131 Padova, Italy

M. C. Saetti
Istituto di Fisiologia Umana
Università di Parma
Via Gramsci 14
43100 Parma, Italy

H. Sakata
Department of Physiology
Nihon University
School of Medicine
30-1 Oyaguchi-Kamamuchi
Itabashi-ku, Tokyo 173, Japan

T. J. Sejnowski
Howard Hughes Medical Institute
The Salk Institute for Biological Studies
La Jolla, CA 92037, USA

E. Shikata
Department of Physiology
Nihon University
School of Medicine
30-1 Oyaguchi-Kamamuchi
Itabashi-ku, Tokyo 173, Japan

Y. Shinoda
Department of Physiology
Tokyo Medical & Dental
University, School of Medicine
1-5-45 Yushima, Bunkyo-ku
Tokyo 113, Japan

N. Smania
Rehabilitation Center
Ospedale Policlinico Borgo
Roma
37134 Verona, Italy

M. Taira
Department of Physiology
Nihon University
School of Medicine
30-1 Oyaguchi-Kamamuchi
Itabashi-ku, Tokyo 173, Japan

A. Takemura
Neuroscience Section
Electrotechnical Laboratory
1-1-4 Umezono Tsukubashi
Ibaraki 305, Japan

Y. Tanaka
Department of Physiology
Nihon University
School of Medicine
30-1 Oyaguchi-Kamamuchi
Itabashi-ku, Tokyo 173, Japan

P. Thier
Sektion Visuelle Sensomotorik
Universität Tübingen
Hoppe-Seyler-Str. 3
72076 Tübingen, Germany

I. Toni
Istituto di Fisiologia Umana
Università di Parma
Via Gramsci 14
43100 Parma, Italy

S. Treue
Labor für Cognitive
Neurowissenschaften
Sektion Visuelle Sensomotorik
Universität Tübingen
Auf der Morgenstelle 15
72076 Tübingen, Germany

N. Tzourio
Groupe d'Imagerie Neuro-
fonctionnelle
Service Hospitalier Frédéric
Joliot
4 pl du Général Leclerc
91406 Orsay Cedex, France

G. Vallar
Dipartimento di Psicologia
Università di Roma „La
Sapienza"
Via dei Marsi 78
00185 Rome, Italy

R. Walker
Department of Clinical
Neuroscience
Charing Cross Hospital Medical
School
London W6 8RF, UK

T. Wannier
Department of Physiology
University of Bern
Bühlplatz 5
3012 Bern, Switzerland

M. Wiesendanger
Motoriklabor, BHH M-130
Neurol. Universitätsklinik
Inselspital
3010 Bern, Switzerland

Contents

Functional Anatomy of the Parietal Lobe and its Connections

Neglect, Extinction, and the Cortical Streams of Visual Processing

A. D. MILNER

School of Psychology, University of St Andrews, St Andrews, UK

It is often assumed that the symptoms of visuospatial neglect can be mapped on to the known properties of neurones within the primate posterior parietal lobe, especially in what may still be collectively called area 7 (Brodmann 1905). A major plank in the argument is that the visual responses of many of these neurones are enhanced when the receptive-field stimulus is concurrently the target for a saccadic or manual movement (Bushnell et al. 1981). In other words, his neuronal "enhancement" is associated with spatially selective attention to the visual target stimulus. Therefore, given the general belief that the essence of neglect is a lateralized *inattention*, a plausible candidate for the neural substrate of the disorder would seem to be provided by the human homologue of monkey posterior parietal cortex. This putative correspondence is also given superficial support by the fact that area 7 lies in the inferior parietal lobule in the monkey, just as the focus for lesions that cause neglect lies largely in the inferior parietal lobule (IPL) in humans. Indeed, the inferior parietal lobule in both species lies ventrolateral to what is known as the intraparietal sulcus in both species (see Fig. 1).

I will argue that both of these propositions may be incorrect. I will suggest that, while the neuronal enhancement seen in area 7 of the monkey may well underlie attentional phenomena, a lesion of this area is likely to cause *visual extinction* rather than visuospatial neglect. And I will propose that the inferior parietal lobule in humans is probably not homologous to that in the monkey.

The Neglect Syndrome

No two patients with visuospatial neglect are alike. Nevertheless, there are certain symptoms that seem to hang together as a syndrome, providing the term "syndrome" is not understood to imply an obligatory association. While these symptoms often go under different names, we may conveniently use the terminology of Heilman et al. (1983) and call them *extinction*, *hemi-akinesia*,

In: Parietal Lobe Contributions to Orientation in 3D Space (1997). P. Thier and H.-O. Karnath (eds). Springer-Verlag, Heidelberg.

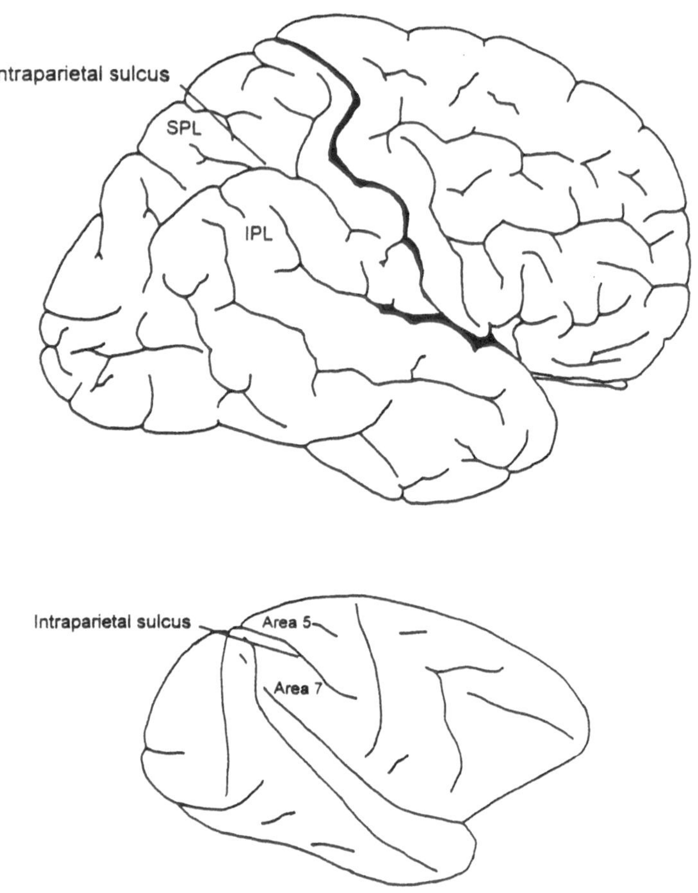

Fig. 1. Diagrammatic views of the lateral surface of the human and macaque cerebral hemispheres, showing the location of the intraparietal sulcus in each case. (*SPL*, superior parietal lobule; *IPL*, inferior parietal lobule)

hemi-inattention, hemispatial neglect, and *allaesthesia*. Of these, the most important for the present discussion are extinction, hemi-inattention, and hemispatial neglect. These three symptoms are defined ostensively, by the tasks used to reveal them, rather than by any understanding of their nature. Thus, extinction can be demonstrated by presenting two visual stimuli briefly, side by side, and is defined as an inability to report the contralesional stimulus in a patient who is able to report either stimulus when presented singly. Hemi-inattention can be demonstrated when this latter condition is *not* fulfilled: namely when the patient fails to detect a contralesional stimulus even when it is presented alone. Visuospatial neglect is not so simply described. It is apparent in many different circumstances, such as when a patient is asked to cancel each item in a symmetrical visual array, say of letters or symbols, or when the patient is asked to copy a complex pattern or picture by drawing, or when the patient attempts to read words or lines of text. In all of these cases, a relative failure to acknowledge items on the left would indicate hemispatial neglect. Another example of a task that reveals hemispatial neglect is the line bisection test, in which the patient is asked to mark a horizontal line at its midpoint, but in fact responds by marking it at some distance to the right of that point. It is axiomatic, of course, that patients only qualify as having one of these various symptoms where their behavior cannot be explained as due to hemianopia or other primary sensory or motor disorder.

It is important to note that not all patients who show one of the symptoms of the neglect "syndrome" necessarily show the others. Thus, there are many patients who show extinction but not neglect, and there are some who show neglect but not extinction (e.g., Barbieri and De Renzi 1989; Liu et al. 1992). [Indeed, the situation is much more complicated than that, since there are also many dissociations among tests of hemispatial neglect itself; patients may, for example, show neglect on a cancellation task but not on line bisection, and vice versa (Halligan and Marshall 1992).] It is a reasonable assumption that if two patients can be found who show a double dissociation – that is, one shows a deficit on task A but not on task B, while the other shows the reverse pattern – then the two tasks are measuring different fundamental processes. I wish to argue for the view that extinction and hemispatial neglect provide such an example, and therefore that one cannot use the same explanation for both disorders.

As the (apparently) simpler of the two disorders, extinction should be easier to explain than neglect. I suggest, as a first approximation, that most cases of extinction can be understood as a disorder of attentional shifting in egocentric space. The disorder is manifest, for example, in the well-known task devised by Posner and his colleagues which is designed to measure a subject's ability to detect visual target stimuli when primed with a cue either on the same side (a valid cue) or the opposite side (an invalid cue). In a study in which the cue preceded the target by between 50 and 1000 ms, patients with parietal lesions and extinction were found to perform catastrophically badly at detecting contralesional signals on "invalid" trials (Posner et al. 1984). And they also performed equally badly when the task was changed such that the cue took the form of a central arrow pointing to left or right: in this case the cue directed or misdirected attention symbolically, without the presence of a competing stimulus on the other ("good") side. These data make it plausible to argue that the essence of extinction is a profound difficulty in switching attention from one part of space to another, in a contralesional direction. But if extinction does result from such an attentional disorder, then it is difficult to argue that neglect is *also* caused by the same attentional disorder, given the double dissociations that have been observed.

Of course, this is not to deny that most patients with neglect do have an attentional disorder, but I would suggest that this is because most of them show extinction as well as neglect. Therefore, the observation that many neglect patients also show the extinction-like effect described by Posner and his colleagues (Morrow and Ratcliff 1988) does not contradict the position I am arguing here. My argument is that although a neglect patient may have an attentional disorder, this does not imply that the neglect is intrinsically *caused* by that attentional disorder.

The difficulty comes, of course, in trying to specify what the underlying disorder in neglect *is* if it is not a form of lateralized inattention. Some recent evidence has accrued from a series of experiments in which Monika Harvey and I have concentrated on one particular neglect phenomenon – specifically we have tried to discover why patients with left-sided neglect (i.e. due to right hemisphere damage) bisect lines to the right of centre. One possibility that some investigators (e.g. Heilman and Valenstein 1979) have suggested is that patients have an orienting imbalance, which causes their manual responses to err towards the right. An alternative idea, however, is that the patients might bisect rightwards because they actually perceive the left half of the line as *shorter* than the right half; if so, the subjective midpoint would be shifted to the right, causing them to make bisection responses to that side.

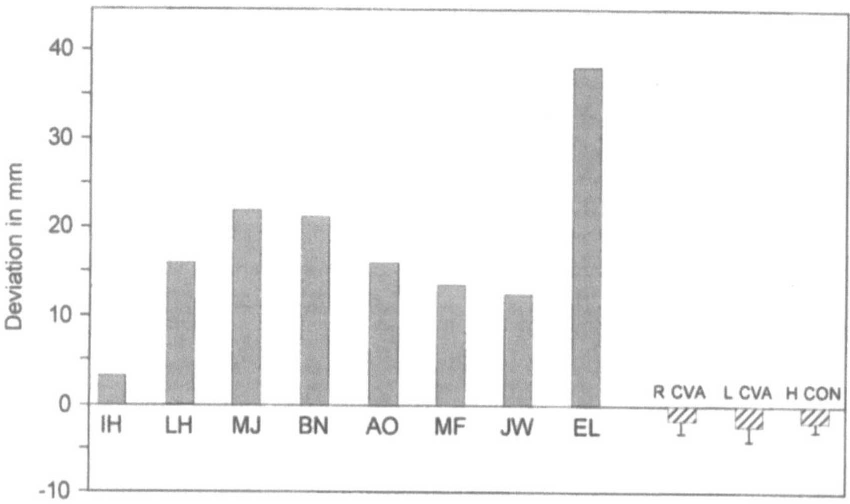

Fig. 2. Bisection errors in a group of eight patients with left visuospatial neglect, along with mean data from control groups of healthy subjects (*H CON*) and patients with right or left hemisphere stroke but no neglect. *R CVA*, right hemisphere stroke patients; *L CVA*, left hemisphere stroke patients. Rightward errors are coded as positive, leftward errors as negative. *Error bars* indicate *SEs* within the control groups. The scores are mean deviations averaged over different spatial positions and cueing conditions. (From Harvey et al. 1995, from which clinical details of the patients may be obtained)

The test we devised to try to decide between these two broad alternatives resembles the "landmark task" (Sayner and Davis 1972; Pohl 1973), in which monkeys are trained to choose one of two foodwells on the basis of the relative proximity of a spatial cue placed between the two foodwells, but closer to one than to the other. Borrowing this task, we presented our patients with a series of lines pre-bisected at various points along their length. Their task was to point to the end of the line that was nearer to the bisection mark. We reasoned that if we presented a line bisected at the midpoint, then the two theories should give opposite predictions: If the patients saw the line as shorter on the left, they should point left; but if they had a rightward motor bias, then, if anything, they should point right. We have now obtained extensive data on this landmark task from eight neglect patients. Fig. 2 shows that all of these patients made substantial rightward errors in bisecting lines, as one would expect.

On the landmark task itself, seven of the patients, when presented with mid-transected lines, pointed predominantly to the left, despite the fact that according to Heilman and Valenstein's (1979) model they should have had a profound reluctance to act leftwards in space. But not all of our patients behaved in this way (see Fig. 3). The exception, patient EL, did behave as Heilman and Valenstein would predict. This supports the idea that line bisection errors can arise from both perceptual *and* directional/response factors in different patients, though the former pattern is more typical of neglect caused by parieto-temporal lesions (Bisiach et al. 1990). The directional/response bias factor may be more typical of frontal and basal ganglia lesioned patients.

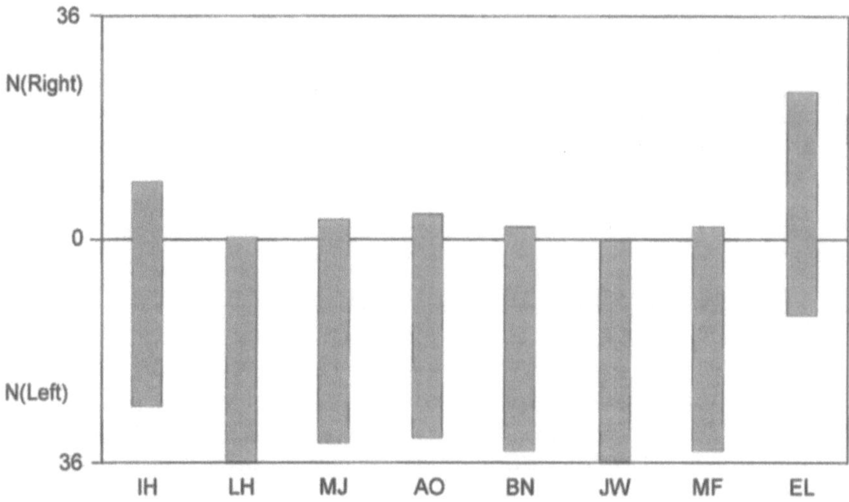

Fig. 3. Total landmark judgements made by each of the eight neglect patients, collapsed across spatial position and cueing conditions. The mean frequency (N) of pointing right (*upward*) and pointing left (*downward*) is shown (out of nine). (From Harvey et al. 1995)

We went on to test directly the idea that most of our neglect patients might have a distorted perception of horizontal lines, by using a psychophysical comparison task. We predicted that they would see two identical lines, presented side by side, as different in length, the left one looking shorter than the right. We constructed a series of perceptual matching tasks using computer graphics, and asked our subjects, in free vision, to compare the two stimuli and to respond using the arrow keys on the computer keyboard to indicate which stimulus was longer. (In a separate series of trials, we asked subjects to indicate which was *shorter*, this balancing allowing us to separate a bias in a patient's responses from one in the patient's perception.) By using a sequence of pairs of these computer-generated stimuli taken from a graded series, we were able to arrive at the point of subjective equality (PSE): that is the point at which a stimulus on the left appeared to the patient to be equal to one on the right. Unfortunately, we were only able to obtain data from three of the eight neglect patients shown in Figs. 2 and 3. All three of them were prone to make bisection errors to the right (see Fig. 2), and they all pointed consistently leftwards in the landmark task (Fig. 3). As shown in Fig. 4,

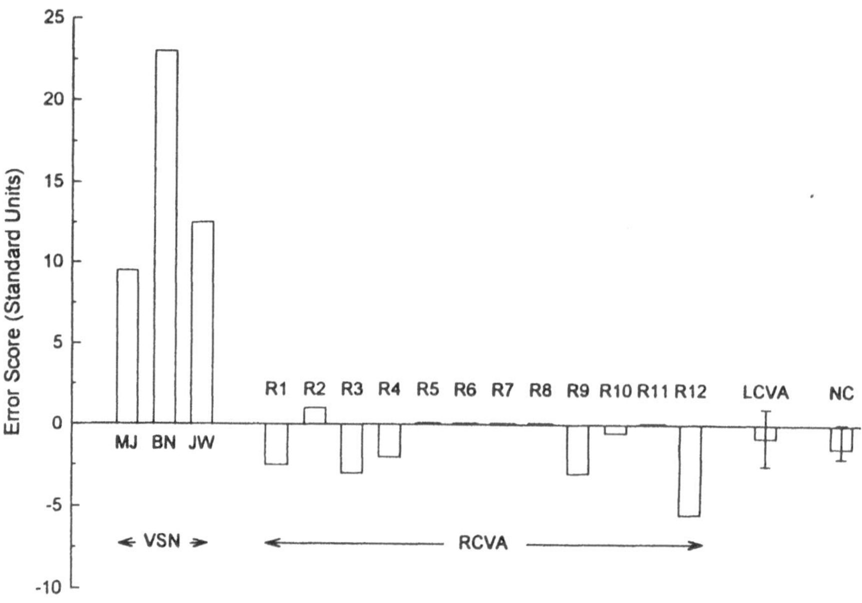

Fig. 4. 'Constant errors' in matching horizontal rectangles presented one on the left and one on the right of a computer screen. The *ordinate* represents the extent to which the left rectangle had to be longer than the right rectangle in order to appear of equal length to the subjects tested. The three patients with visuospatial neglect (*VSN*: *MJ*, *BN*, and *JW*) are each shown individually, as are 12 control patients (*R1* - *R12*) with right hemisphere stroke but no neglect (*RCVA*). The mean results (+/- range) for the control groups of subjects with left-hemisphere stroke (*LCVA*) or without neurological illness (*NC*) are shown on the right. (From Milner and Harvey 1995)

we found that all three had a pronounced perceptual bias such that a horizontal rectangle on the left had to be made much longer in order to appear as long as one on the right (Milner and Harvey 1995). Similarly, comparisons of a random nonsense shape also revealed this leftward underestimation. Yet the same patients made no errors at all when comparing *vertical* rectangles.

As Fig. 4 also shows, our three neglect patients behaved quite differently from controls with left or right hemisphere lesions who did not show symptoms of neglect, and from age-matched healthy controls (in most cases spouses or friends of our patients). None of these control subjects showed the perceptual distortions apparent in our neglect patients.

In summary, we consider that the combined evidence from the landmark task and the matching task strongly suggest that many neglect patients perceptually devalue the linear extent of stimuli on the left side of their subjective space. Such a conclusion may also explain a recent finding by Bisiach et al. (1994). They found that two neglect patients, when asked to set the ends of a "virtual line" equidistantly from a fixed "bisection mark", did so by placing the left end at a much greater distance from the mark than the right end.

The next question to answer, of course, is *why* this perceptual distortion occurs. It might be argued that the distortion effect could simply result from a chronic attentional bias in our patients. We initially thought so ourselves, because as others had found before us, the rightward line bisection errors made by our neglect patients could be reduced by cueing the left end of the line. In our experiments, each cue consisted simply of a visually-confusable letter which had to be reported prior to bisection (Milner et al. 1993). This beneficial effect of cueing in neglect patients superficially resembles the finding that line bisection in normal subjects can be "pulled" in the direction of the cue when the identical cueing procedure is used. In our own study of this phenomenon in normal subjects (Milner et al. 1992), we also examined the effect of these cueing procedures on the landmark task. We found that our letter cues resulted in more frequent choice responses to the uncued end of a centrally-prebisected line (Milner et al. 1992), indicating that the cued half appeared longer. This allowed us to conclude that the effect of cueing on line *bisection* could be attributed to this same distortion in subjects' perception of the line.

But is this also the case for neglect patients? That is, does cueing affect the disordered length perception that seems to be present in neglect, in the same way as it seems to affect length perception in normal subjects? We tested this by providing attentional cues on the left or right to neglect patients while they performed the landmark task.

We discovered, somewhat to our surprise, that our neglect patients behaved quite differently from control subjects. Instead of cueing causing landmark judgements that would indicate a subjective expansion of the cued half of the line, the *opposite* occurred, with the neglect group tending to point significantly more often to the *cued* end of the line (see Fig. 5). As we saw in Fig. 3, our neglect

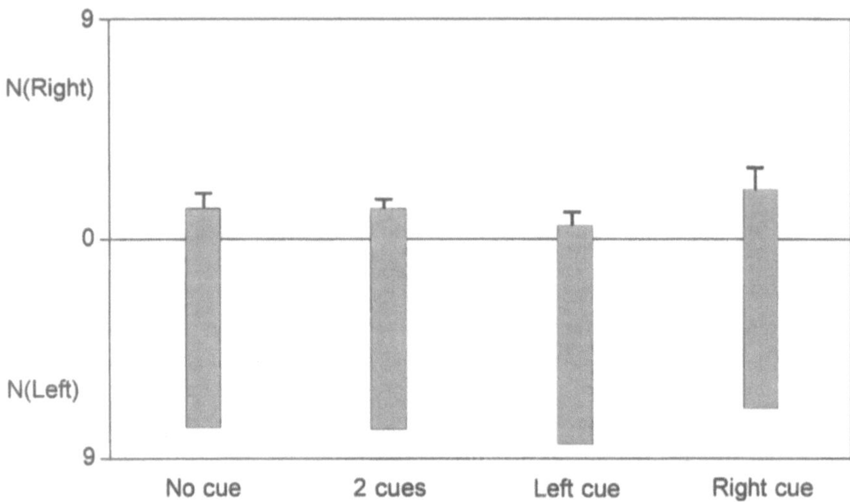

Fig. 5. Cueing effects on the landmark judgements of the eight neglect patients. Frequencies are coded as in Fig. 3. *Errors bars* indicate *SEs*. (From Harvey et al. 1995)

patients normally pointed left in the landmark task, indicating that they under-perceived the length of the left half of the line. If a left-side cue therefore caused an amelioration of this leftward underestimation, it should cause *fewer* pointing responses to the left – yet instead there were significantly *more* leftward responses.

It is implausible to suppose that a cue would affect perception of line length in neglect patients in a way opposite to its effect on perception in everyone else. Instead, it seems more likely that cueing produces its effects in our neglect patients via their orienting behavior, rather than via their perception. Ocular responses are necessarily drawn to the cued end of a line in the landmark task (and similarly, indeed, in the line bisection task), in order for the patient to be able to report the letter cue. This turning tendency might "leak" into manual responding, e.g. through activation of visuomotor systems in the hemisphere contralateral to the cue. This manual bias could cause an increased likelihood of the patients responding to the cued end of the line in the landmark task, which is what we found (Fig. 5). We have suggested elsewhere (Harvey et al. 1995) that this directional response bias may be the cause of cueing effects in both the bisection and landmark behavior of neglect patients.

My conclusion is that a left-side cue does *not* restore perception when it ameliorates bisection errors in neglect. Thus, we have found no support for the idea that a chronic attentional bias is the cause of size distortions, and thereby line bisection errors, in neglect. Yet similar cueing procedures do affect length perception in normal subjects (Milner et al. 1992). Presumably, therefore, that demonstrable effect on perception was overshadowed in our neglect patients by the putative motor biassing induced by the cueing. Such overshadowing would not be surprising if the cortical substrate of the perceptual effect of cueing had been largely destroyed by the neglect-causing lesions.

Perhaps a more likely explanation of the size distortions we have found in neglect is that there is a disruption of some kind of representational network, whose normal role is to mediate the perception of spatial relationships both within and between objects. This idea has been frequently proposed by others (e.g. Bisiach and Vallar 1988) in response to the findings of a range of previous studies (see below) which are difficult to explain at lower levels of perceptual processing. Our own data, for instance, suggest that neglect affects the processing of visual stimuli at a stage after the extraction of object characteristics such as size. I shall return to this argument in a later section.

Lesions That Cause Neglect and Extinction

Despite the existence of theoretically important exceptions, many patients with neglect also do show extinction, or do so at some stage in the course of their illness. This may be attributable to the fact that most of the lesions that cause neglect are large (e.g. Bisiach et al. 1981), and are therefore very likely to disrupt more than one functional system. This reasoning is borne out by comparisons of CT and MRI data from groups of patients with neglect and extinction within the visual modality, which allow us to distinguish quite distinct patterns of lesion foci for the two disorders.

While extinction is relatively common after lesions of either side of the brain, neglect is much more often seen after right-hemisphere lesions. Within the right hemisphere, the lesions causing neglect frequently include the inferior parietal lobule (areas 39 and 40), though the region that is common to most of these posterior lesions is actually parieto-temporal rather than just parietal (Bisiach et al. 1981; Heilman et al. 1983; Vallar and Perani 1986). In fact, Vallar and Perani observed that lesions located more superiorly in the parietal lobe, and not overlapping this more ventral region, were typically *not* associated with neglect. Other CT-scan studies, for example by Kertesz and Dobrowolski (1981), also support this conclusion. They found that vascular lesions restricted to the parietal lobe – that is, not extending into the temporal lobe – resulted in only mild, if any, neglect. Thus, it seems clear that superior parietal areas *per se* are not part of the focus associated with neglect.

Unlike neglect, visual extinction can result from lesions in many different areas of the brain, including the right inferior parietal lobule. But in a recent CT study of extinction, Vallar et al. (1994) noted that although extinction and neglect *together* are typically associated with damage to the right parietotemporal junction, extinction alone is specifically *not* associated with damage to this area.

Instead, such "pure" visual extinction is linked with damage to a range of other structures. That these structures probably include the superior parietal region, is indicated by the previously-mentioned study by Posner et al. (1984). These investigators studied 13 patients with parietal lesions (some right, some left) using their cued visual reaction-time task. They found that the patients who showed the contralesional extinction-like deficit most strongly were those whose lesions (on CT evidence) extended the most into superior parts of the parietal lobe. Consistently with the evidence of Vallar et al., there was no relationship between this extinction-like behavior and the extent of damage in the *inferior* parietal lobule, and indeed little indication that hemispatial neglect was associated with the extinction seen in these patients.

The Functions of Area 7 in Monkeys

Extinction, I have suggested, reflects a pathological difficulty in shifting attention towards the contralesional side of space. The cause of this problem may be an overpowering orienting bias, such that the patient is simply unable to resist turning and maintaining his or her attention to visual stimuli on the *ipsi*lesional side. This kind of picture is readily recognizable from the animal literature (for review see Milner 1987). It is seen whenever an animal is lesioned unilaterally in areas concerned with visuomotor orienting, such as the superior colliculus, substantia nigra, the dorsomedial nucleus (Pdm) of the pulvinar nucleus, and the frontal eye field, as well as the posterior parietal cortex. And it is surely no coincidence that these areas where damage causes extinction and hemi-inattention are the same ones where the neuronal property of enhancement, referred to earlier, has been observed in monkeys (see Desimone and Duncan 1995). We may assume that a finely-tuned balance normally exists between these various paired sensorimotor structures on the two sides of the brain, and that damage on either side will disrupt that balance.

The monkey's area 7 in particular contains many neurones that show enhancement (especially in sub-areas 7a and LIP; Colby et al. 1993). And although no truly neglect-like phenomena have been demonstrated after unilateral damage to posterior parietal cortex, several experimenters have described the occurrence of visual extinction after such lesions (e.g. Rizzolatti et al. 1985; Lynch and McLaren 1989). It seems reasonable to assume that such unilateral damage causes an imbalance between visuospatial orienting or attentional systems within the cortex.

The major constellation of deficits caused by lesions of the monkey's area 7, however, is not obviously attentional in nature, but rather relates to the visuomotor control of behavior. For example, there are disorders of visually guided reaching and grasping by the hand, as well as of visually guided eye and whole-body movements (for reviews see Jeannerod 1988; Milner and Goodale

1995). But again, these deficits fit well with the known neurophysiological properties of neurones in the monkey's area 7. As was discovered over 20 years ago (Hyvärinen and Poranen 1974; Mountcastle et al. 1975), many posterior parietal neurones will not respond well when the monkey is simply exposed to an appropriate visual stimulus. Instead, they become optimally active when the monkey executes a relevant act towards that stimulus. For example, cells of one type become very active when the monkey *reaches* towards the stimulus object, while other cells respond when the monkey *saccades* to the target stimulus, and yet others when the monkey *grasps* it. In short, enhancement is not the only salient property of neurones in area 7: these networks are associated with visuomotor processing as well as with selective attention.

As is now well known, a "dorsal stream" has been identified which passes from the primary visual cortex through a series of prestriate areas, to terminate within area 7 in the posterior parietal cortex (Ungerleider and Mishkin 1982; Morel and Bullier 1990; Young 1992). Of course, "area 7" is now known to include several distinct and interconnecting areas that have a complex pattern of subcortical connectivity, notably with motor and sensorimotor structures in the midbrain and the caudate-putamen. This connectional anatomy is fully consistent with a major role of area 7 in visuomotor control, and a number of authors have argued that this is the primary function of the dorsal stream in the monkey (Glickstein et al. 1985; Jeannerod and Rossetti 1993; Milner and Goodale 1995). The dorsal stream appears to process not only information about spatial location but also about geometric object features (see Sakata et al., this volume), thus providing egocentric visual guidance for a range of motor acts which include hand shaping while reaching for objects and (probably) locomotion through a cluttered environment. In contrast, neurones in the occipito-temporal "ventral stream" are not modulated by the concurrent activities of the animal, although some of them are modulated by the previous training experience of the animal. Instead, the role of the ventral stream seems primarily to be in the mediation of our perception and recognition of the world, to ensure the effective implementation of behavioral strategies over longer periods of time.

Thus, the posterior parietal area of the monkey, generally regarded as the termination point of the dorsal stream of visual cortical projections, seems to do double duty. It is involved both in the visual guidance of action and in the control of selective visual attention. One way of conceptualizing this duality is to argue that shifts of visual selective attention are most parsimoniously seen as a preparatory activation of the very networks concerned with spatially directed action (Rizzolatti et al. 1985, 1994). If this "premotor theory" is correct, visual extinction might be expected to occur following the unilateral disruption of dorsal stream circuits, as well as after unilateral damage to visuomotor structures elsewhere. But, of course, it is also possible that although the two major visual functions of area 7 may be partially related, they are functionally separable. And, indeed, there is anecdotal evidence for dissociations between visuomotor and attentional disorders after parietal lesions in humans.

The Human Parietal Lobe

In trying to establish homologies between monkeys and humans in regard to particular parts of the cortex, cytoarchitecture has been the traditionally favoured approach. But, although this methodology doubtless has a part to play, ultimately the over-riding criteria must be functional and connectional. In the case of the human parietal lobe, there is widespead disagreement among anatomists as to the parcellation of its subareas and their homologies with the monkey brain (Brodmann 1907; von Bonin and Bailey 1947; Eidelberg and Galaburda 1984). There is also very little knowledge of its intrinsic and extrinsic neural connections. At the present time, therefore, one has to be guided primarily by functional comparisons. We know the principal effects of damaging area 7 in monkeys, and we know a good deal about the physiological properties of neurones in this region. In order to infer which parts of the human parietal lobe most resemble area 7, therefore, we should examine the human lesion evidence; and although we do not have single-cell recording data for the human parietal cortex, we do now have some functional-imaging evidence, which to some degree can give us parallel information.

As we have noted, the evidence strongly indicates that one of the major functions of area 7 in monkeys is in visuomotor control. The first step, therefore, is to ask which parts of the parietal lobe in humans play a similar role. The answer seems to be clear. The lesions that cause optic ataxia – that is, disordered visual control of the arm, hand and fingers – are centred in the intraparietal sulcus and adjacent superior parietal lobule (Ratcliff and Davies-Jones 1972; Perenin and Vighetto 1988). These superior lesions in man cause not only visually guided misreaching – as in Bálint's (1909) classic description of optic ataxia – but they also cause disorders of grasp formation (Jeannerod 1986; Jakobson et al. 1991). Similarly, lesions causing slowing of saccadic eye movements to visual targets are centred in posterior parts of the intraparietal sulcus (Pierrot-Deseilligny et al. 1987, 1991). This set of deficits is strikingly similar to that seen in monkeys. Functional imaging studies tell much the same story. Thus Grafton et al. (1996) have recently demonstrated a metabolic activation of posterior parts of the superior parietal lobule (SPL) during performance of both visually-guided reaching and visually-guided grasping tasks. A PET focus in the human SPL has also been observed during visually driven saccadic eye movements (Anderson et al. 1994), and this focus has more recently been narrowed down to the posterior part of the human intraparietal sulcus using functional MRI (Pierrot-Deseilligny and Müri, this volume).

We also noted, however, that a second major function of area 7 in the monkey appears to be attentional: neurones there show selective enhancement, while lesions cause visual extinction. The second set of comparisons we can make, therefore, is to ask whether there are parallel data in humans. The answer is again in the affirmative: PET studies support the proposed homology, in that they reveal a focus of activation in the human superior parietal lobule bilaterally during the exercise of peripheral visuospatial attention (Corbetta et al. 1993). These demonstrations of the active participation of superior areas in the human parietal lobe in selective visual attention agree well with the earlier argument that unilateral *damage* to these areas (among other areas, most of which are themselves connected with dorsal stream areas) is liable to result in visual extinction.

In other words, lesion studies and imaging studies are in rather good agreement with each other in supporting the idea that the human superior parietal lobule and adjacent parts of the intraparietal sulcus are concerned with *both* visuomotor control *and* with selective visuospatial attention. It is therefore arguable that Brodmann (1907) was roughly correct in his proposed homology, i.e. that it is this superior parietal region that contains the human homologue of the monkey's area 7 (along with area 5). The prediction must follow that these superior areas will be the primary recipients of visual inputs from "earlier" human dorsal stream areas, such as MT and MST (which appear to be located more posteroventrally, in the border territory between areas 19, 37, and 39 in the human brain; Zeki et al 1991; Morrow and Sharpe 1990, 1993). Empirical examination of these predictions must of course await detailed connectional studies, which at present are only in their infancy.

The assignment of area 7 to superior parts of the human parietal lobe does of course leave a great deal of the parietal lobe unaccounted for. As we have seen, neglect is typically associated with large lesions that include parts of the right *inferior* parietal lobule, which presumably lies outside the dorsal visual stream. Consistent with this presumption, there is as yet little convincing evidence that unilateral lesions of area 7 or other parts of the dorsal stream in the monkey cause deficits resembling human hemispatial neglect; though, as noted earlier, visual extinction and hemi-inattention have been repeatedly described following such damage. We therefore have little or no functional information as to the evolutionary antecedents of the human inferior parietal lobule, and indeed Brodmann (1905, 1907) found no evidence that the areas he identified there (areas 39 and 40) existed at all in the monkey brain.

But there are scattered clues as to the possible origins of these areas. One such clue may be discernible in some recent data from Karnath (this volume), in which he has demonstrated convincingly that the egocentric spatial coding frameworks underlying the phenomena of neglect and of extinction are very different. Karnath and his colleagues first showed that neglect is influenced both by the orientation of the trunk with respect to the visual array and by manipulations such as neck vibration and vestibular stimulation, which probably operate upon the same proprioceptive system (Karnath et al. 1993). But his more recent work shows that visual extinction phenomena are quite unaffected by these same manipulations, and therefore may be inferred to depend on a simpler form of visuospatial coding. Karnath's data provide confirmatory evidence that we are dealing with two different disorders in visuospatial neglect and visual extinction, but, more significantly, they also indicate that the two brain systems concerned are furnished with differently coded visual information. While extinction can be regarded as a disruption of networks provided primarily with *retinal* information, it is clear that neglect cannot be seen in this way. Instead, neglect evidently involves a disruption of visually coded information within a spatial framework that is dependent on the integration of inputs from other sensory modalities as well as vision.

There are well-known multimodal networks within the monkey's cortex that could constitute the forerunner of such a system. For example, the region in anterior parts of the superior temporal sulcus that we may loosely refer to as STP falls into this category (Bruce et al. 1981; Jones and Powell 1970). Anatomical studies have shown that parts of this region receive visual inputs from the inferior temporal cortex (Seltzer and Pandya 1994), as well as from dorsal stream areas like MT (Boussaoud et al. 1990). Moreover, physiological studies show that STP neurones depend on the striate cortex for their form and motion selectivity (Bruce et al. 1986), just like neurones in the inferior temporal complex (Rocha-Miranda et al. 1975) and area V4 (Girard et al. 1991) in the ventral stream. The present evidence therefore suggests that the major source of specific visual information to STP is likely to be found in the ventral stream rather than in the dorsal stream, though the latter may, for example, provide attentional modulation within the system.

I would argue that such a dependence on the ventral stream would fit well with our current knowledge of the neuropsychology of hemispatial neglect. Neglect seems to affect complex spatial representations of visual scenes and patterns (see review by Bisiach and Vallar 1988), and therefore is not readily understood in terms of damage to visuomotor systems. For example, neglect may affect only one class of patterns, such as faces (Young et al. 1990) or words (Costello and Warrington 1987). Similarly, neglect may be associated with the left side of *objects*, even when they are rotated or located well into the right side of egocentric space (see review by Walker 1995). These clearly documented cases of domain-specific neglect and object-based neglect, respectively, are inexplicable in terms of inattention to contralesional parts of egocentric visual space *per se*. Furthermore, it is well established that neglect can affect visual images reconstructed from long-term memory (Bisiach et al. 1981) as well as patterns reconstructed within visual short-term memory (Bisiach et al. 1979). And as we saw earlier, patients with neglect often show perceptual size distortions (Gainotti and Tiacci 1971; Milner and Harvey 1995).

In other words, the information that is "neglected" in neglect is often not coded in purely egocentric coordinates with respect to the retina, head or body, but instead is frequently, if not always, based on higher-order – including sometimes figural – frames of reference. Indeed some of these disorders can only be explained by assuming that neglect affects processing beyond the point where the perceptual system distinguishes between different classes of visual patterns (e.g. faces). These considerations would all be consistent with the idea that the root cause of neglect may be a disruption of a representational network located in the right inferior parietal lobule. This system would receive highly processed information relating to object identity, and accordingly be in large part dependent on visual inputs from the ventral stream. In humans, it appears that the right hemisphere's ventral stream dominates visual perception and recognition, at least at a global perceptual level, since disorders such as topographic agnosia and prosopagnosia are associated much more with right ventral occipito-temporal lesions than with left (see Grüsser and Landis 1991 for review). It should therefore not be surprising that a system for representing spatial relationships that depended on the ventral stream for important visual inputs would also be more heavily represented in the right hemisphere.

While our knowledge of the functions of the right inferior parietal lobule remains sketchy, there is now some functional imaging data which supports the idea that it has a role in visuospatial recognition. Thus inferior right parietal sites (within Brodmann's area 40) have been found to be activated in PET experiments during the performance of short-term (Jonides et al. 1993) and long-term (Moscovitch et al. 1995) spatial location memory tasks.

In apparent conflict with these conclusions, a series of studies by Haxby et al. (1991, 1994) have reported bilateral activation of the superior parietal lobule during performance of certain visuospatial matching tasks involving mental rotation. However, the reason for these contrary findings could be that the eye movements of their subjects were unconstrained, and indeed were probably more extensive in the "spatial" task condition than in their control no-task condition or their face discrimination condition. If so, then one would expect the areas in the superior parietal lobule and the medial bank of the intraparietal sulcus that are activated during visually guided saccadic eye movements (Anderson et al. 1994; Pierrot-Deseilligny and Müri, this volume) to be more active during this spatial task. In contrast, Jonides et al. (1993), maintained the ocular fixation of their subjects, and perhaps as a consequence observed no evidence a selective superior parietal involvement in their spatial recognition task. It may also explain why the parietal activation was bilateral during the spatial tasks of Haxby and his colleagues.

The evidence summarized in this section supports the idea that there are *at least* two quite distinct regions in the human posterior parietal lobe, each concerned with quite different aspects of visuospatial processing. A superior region, as well as guiding action within an egocentric spatial framework, seems to be involved in spatial attention within that same framework. In contrast, an inferior region may be involved in mediating spatial representation within a more flexible allocentric framework. It is tentatively proposed that lesions of this inferior region in the right hemisphere may interfere with the patient's ability to manipulate and cognitively interrogate the spatial representation system there, and hence cause neglect. It has to be conceded, however, that the regions of area 40 activated in the studies of Jonides et al. (1993) and Moscovitch et al. (1995) probably include the inferior bank of the intraparietal sulcus, and thus may be distinct from the more ventral (parieto-temporal) regions that appear to be implicated in visuospatial neglect. Those PET studies may thus signal the presence of a third system.

Conclusions: Streams of Visual Processing in the Human Brain

We do not have the detailed connectional data that would allow us to map unequivocally the dorsal and ventral streams within the human brain. Yet wherever functional and anatomical homologies have been explored between monkey and man elsewhere within the visual domain, the parallels have been remarkably strong. It is therefore very likely that the human brain does have separate ventral and dorsal visual streams. I have argued from the available functional data that the human dorsal stream terminates in the superior parietal lobule and adjacent intraparietal sulcus. I have further argued that area 7 and the rest of the dorsal stream may have little directly to do with visuospatial neglect *per se*, but that they may be among the several areas associated with visual extinction. It is not denied, of course, that visual extinction will follow damage not only to neural systems within the dorsal stream, but also to various premotor and subcortical systems associated with that stream.

In contrast, the egocentric spatial coordinates used by the dorsal stream for guiding action are not at all consistent with many of the phenomena seen in spatial neglect. Neglect operates within a set of relative spatial frameworks that can rotate with an object and even be domain-specific. It is my contention, in fact, that neglect cannot be fully understood without reference to the perception and representation of objects and their interrelationships. I have suggested that we owe the existence of neglect to the evolution of areas in the right hemisphere that are strictly a part of neither visual processing stream. These areas probably receive visual inputs from both streams, but their inputs from ventral visual areas may be especially critical, not only for their spatial representational functions, but also for the fact that these representations are experienced consciously (Milner 1995).

Acknowledgements. The author is grateful to David Carey, Chris Dijkerman, Monika Harvey and Catrin Pritchard for their help in the preparation of this paper, and to the Wellcome Trust for their financial support.

References

Anderson TJ, Jenkins IH, Brooks DJ, Hawken MB, Frackowiak RSJ, Kennard C (1994) Cortical control of saccades and fixation in man: a PET study. Brain 117:1073–1084

Bálint R (1909) Seelenlähmung des "Schauens", optische Ataxie, räumliche Störung der Aufmerksamkeit. Monatsschr Psychiat Neurol, 25, 51–81. [English translation: Harvey M (1995) Cogn Neuropsychol 12:261–282.]

Barbieri C, De Renzi E (1989) Patterns of neglect dissociation. Behav Neurol 2:13–24

Bisiach E, Vallar G (1988) Hemineglect in humans. In: Boller F, Grafman J (eds) Handbook of neuropsychology, vol. 1. Elsevier, Amsterdam, pp 195–222

Bisiach E, Luzzatti C, Perani D (1979) Unilateral neglect, representational schema and consciousness. Brain 102:609–618

Bisiach E, Capitani E, Luzzatti C, Perani D (1981) Brain and conscious representation of outside reality. Neuropsychologia 19:543–551

Bisiach E, Geminiani G, Berti A, Rusconi ML (1990) Perceptual and premotor factors in unilateral neglect. Neurology 40:1278–1281

Bisiach E, Rusconi ML, Peretti VA, Vallar G (1994) Challenging current accounts of unilateral neglect. Neuropsychologia 32:1431–1434

Bonin G von, Bailey P (1947) The neocortex of macaca mulatta. University of Illinois Press, Urbana.

Boussaoud D, Ungerleider LG, Desimone R (1990) Pathways for motion analysis: cortical connections of the medial superior temporal and fundus of the superior temporal visual areas in the macaque. J Comp Neurol 296:462–495

Brodmann K (1905) Beiträge zur histologischen Lokalisation der Grosshirnrinde. Dritte Mitteilung: die Rindenfelder der niederen Affen. J Psychol Neurol Lpz 4:177–226

Brodmann K (1907) Beiträge zur histologischen Lokalisation der Grosshirnrinde. Sechste Mitteilung: die Cortexgliederung des Menschen. J Psychol Neurol Lpz 10:231–246

Bruce C, Desimone R, Gross CG (1981) Visual properties of neurones in a polysensory area in superior temporal sulcus of the macaque. J Neurophysiol 46:369–384

Bruce CJ, Desimone R, Gross CG (1986) Both striate cortex and superior colliculus contribute to visual properties of neurones in superior temporal polysensory area of macaque monkey. J Neurophysiol 55:1057–1075

Bushnell MC, Goldberg ME, Robinson DL (1981) Behavioral enhancement of visual responses in monkey cerebral cortex. I. Modulation in posterior parietal cortex related to selective visual attention. J Neurophysiol 46:755–772

Colby CL, Duhamel J-R, Goldberg ME (1993) The analysis of visual space by the lateral intraparietal area of the monkey: the role of extraretinal signals. In: Hicks TP, Molotchnikoff S, Ono T (eds) Progress in brain research, vol 95. Elsevier, Amsterdam, pp 307–316

Corbetta M, Miezin FM, Shulman GL, Petersen SE (1993) A PET study of visuospatial attention. J Neurosci 13:1202–1226

Costello ADL, Warrington EK (1987) The dissociation of visuospatial neglect and neglect dyslexia. J Neurol 50:1110–1116

Desimone R, Duncan J (1995) Neural mechanisms of selective visual attention. Ann Rev Neurosci 18:193–222

Eidelberg D, Galaburda AM (1984) Inferior parietal lobule: divergent architectonic asymmetries in the human brain. Arch Neurol 41:843–852

Gainotti G, Tiacci C (1971) The relationship between disorders of visual perception and unilateral spatial neglect. Neuropsychologia 9:451–458
Girard P, Salin PA, Bullier J (1991) Visual activity in macaque area V4 depends on area 17 input. Neuroreport 2:81–84
Glickstein M, May JG, Mercier BE (1985) Corticopontine projection in the macaque: the distribution of labelled cortical cells after large injections of horseradish peroxidase in the pontine nuclei. J Comp Neurol 235:343–359
Grafton ST, Fagg AH, Woods RP, Arbib MA (1996) Functional anatomy of pointing and grasping in humans. Cerebral Cortex 6:226–237
Grüsser O-J, Landis T (1991) Vision and visual dysfunction, vol 12: visual agnosias and other disturbances of visual perception and cognition. Macmillan, London
Halligan PW, Marshall JC (1992) Left visuo-spatial neglect: a meaningless entity? Cortex 28:525–535
Harvey M, Milner AD, Roberts RC (1995) An investigation of hemispatial neglect using the landmark task. Brain Cogn 27:59–78
Haxby JV, Grady CL, Horwitz B, Ungerleider LG, Mishkin M, Carson RE, Herscovitch P, Schapiro MB, Rapoport SI (1991) Dissociation of object and spatial visual processing in human extrastriate cortex. Proc Natl Acad Sci USA 88:1621–1625
Haxby JV, Horwitz B, Ungerleider LG, Maisog JM, Pietrini P, Grady CL (1994) The functional organization of human extrastriate cortex: a PET-rCBF study of selective attention to faces and locations. J Neurosci 14:6336–6353
Heilman KM, Valenstein E (1979) Mechanisms underlying hemispatial neglect. Ann Neurol 5:166–170
Heilman KM, Watson RT, Valenstein E, Damasio AR (1983) Localization of lesions in neglect. In: Kertesz A (ed) Localization in neuropsychology. Academic, New York, pp 471–492
Hyvärinen J, Poranen A (1974) Function of the parietal associative area 7 as revealed from cellular discharges in alert monkeys. Brain 97:673–692
Jakobson LS, Archibald YM, Carey DP, Goodale MA (1991) A kinematic analysis of reaching and grasping movements in a patient recovering from optic ataxia. Neuropsychologia 29:803–809
Jeannerod M (1986) The formation of finger grip during prehension: a cortically mediated visuomotor pattern. Behav Brain Res 19:99–116
Jeannerod M (1988) The neural and behavioral organization of goal-directed movements. Oxford University Press, Oxford
Jeannerod M, Rossetti Y (1993) Visuomotor coordination as a dissociable visual function: experimental and clinical evidence. In: Kennard C (ed) Visual perceptual defects. Bailliere Tindall, London, pp 439–460 (Bailliere's clinical neurology, vol 2, no 2)
Jones EG, Powell TPS (1970) An anatomical study of converging sensory pathways within the cerebral cortex of the monkey. Brain 93:793–820
Jonides J, Smith EE, Koeppe RA, Awh E, Minoshima S, Mintun MA (1993) Spatial working memory in humans as revealed by PET. Nature 363:623–625
Karnath H-O, Christ K, Hartje W (1993) Decrease of contralateral neglect by neck muscle vibration and spatial orientation of trunk midline. Brain 116:383–396
Kertesz A, Dobrowolski S (1981) Right-hemisphere deficits, lesion size and location. J Clin Neuropsychol 3:283–299
Liu GT, Bolton AK, Price BH, Weintraub S (1992) Dissociated perceptual-sensory and exploratory-motor neglect. J Neurol Neurosurg Psychiatry 55:701–706

Lynch JC, McLaren JW (1989) Deficits of visual attention and saccadic eye movements after lesions of parietooccipital cortex in monkeys. J Neurophysiol 61:74–90

Milner AD (1987) Animal models for the syndrome of spatial neglect. In: Jeannerod M (ed) Neurophysiological and neuropsychological aspects of spatial neglect. Elsevier, Amsterdam, pp 259–288

Milner AD (1995) Cerebral correlates of visual awareness. Neuropsychologia 33:1117–1130

Milner AD, Goodale MA (1995) The visual brain in action. Oxford University Press, Oxford

Milner AD, Harvey M (1995) Distortion of size perception in visuospatial neglect. Curr Biol 5:85–89

Milner AD, Brechmann M, Pagliarini L (1992) To halve and to halve not: an analysis of line bisection judgements in normal subjects. Neuropsychologia 30:515–526

Milner AD, Harvey M, Roberts RC, Forster SV (1993) Line bisection errors in visual neglect: misguided action or size distortion? Neuropsychologia 31:39–49

Morel A, Bullier J (1990) Anatomical segregation of two cortical visual pathways in the macaque monkey. Vis Neurosci 4:555–578

Morrow LA, Ratcliff G (1988) The disengagement of covert attention and the neglect syndrome. Psychobiol 3:261–269

Morrow MJ, Sharpe JA (1990) Cerebral hemisphere localization of smooth pursuit asymmetry. Neurology 40:284–292

Morrow MJ, Sharpe JA (1993) Retinotopic and directional deficits of smooth pursuit initiation after posterior cerebral hemispheric lesions. Neurology 43:595–603

Moscovitch M, Kapur S, Köhler S, Houle S (1995) Distinct neural correlates of visual long-term memory for spatial location and object identity: A positron emission tomography study in humans. Proc Natl Acad Sci USA 92:3721–3725

Mountcastle VB, Lynch JC, Georgopoulos AP, Sakata H, Acuña C (1975) Posterior parietal association cortex of the monkey: command function of operations within extrapersonal space. J Neurophysiol 38:871–908

Perenin M-T, Vighetto A (1988) Optic ataxia: a specific disruption in visuomotor mechanisms. I. Different aspects of the deficit in reaching for objects. Brain 111:643–674

Pierrot-Deseilligny C, Rivaud S, Penet C, Rigolet M-H (1987) Latencies of visually guided saccades in unilateral hemispheric cerebral lesions. Ann Neurol 21:138–148

Pierrot-Deseilligny C, Rivaud S, Gaymard B, Agid Y (1991) Cortical control of reflexive visually-guided saccades. Brain 114:1473–1485

Pohl W (1973) Dissociation of spatial discrimination deficits following frontal and parietal lesions in monkeys. J Comp Physiol Psychol 82:227–239

Posner MI, Walker JA, Friedrich FJ, Rafal RD (1984) Effects of parietal lobe injury on covert orienting of attention. J Neurosci 4:1863–1874

Ratcliff G, Davies-Jones GAB (1972) Defective visual localization in focal brain wounds. Brain 95:49–60

Rizzolatti G, Gentilucci M, Matelli M (1985) Selective spatial attention: one center, one circuit, or many circuits? In: Posner MI, Marin OSM (eds) Attention and performance XI. Erlbaum, Hillsdale, pp 251–265

Rizzolatti G, Riggio L, Sheliga BM (1994) Space and selective attention. In: Umiltà C, Moscovitch M (eds) Attention and performance XV. Conscious and nonconscious information processing. MIT Press, Cambridge, pp 231–265

Rocha-Miranda CE, Bender DB, Gross CG, Mishkin M (1975) Visual activation of neurons in inferotemporal cortex depends on striate cortex and forebrain commissures. J Neurophysiol 38:475–491

Sayner RB, Davis RT (1972) Significance of sign in an S-R separation problem. Percept Mot Skills 34:671–676

Seltzer B, Pandya DN (1994) Parietal, temporal, and occipital projections to cortex of the superior temporal sulcus in the rhesus monkey: a retrograde tracer study. J Comp Neurol 343:445–463

Ungerleider LG, Mishkin M (1982) Two cortical visual systems. In: Ingle DJ, Goodale MA, Mansfield RJW (eds) Analysis of visual behavior. MIT Press, Cambridge, pp 549–586

Vallar G, Perani D (1986) The anatomy of unilateral neglect after right-hemisphere stroke lesions. A clinical / CT-scan correlation study in man. Neuropsychologia 24:609–622

Vallar G, Rusconi ML, Bignamini L, Geminiani G, Perani D (1994) Anatomical correlates of visual and tactile extinction in humans: a clinical CT scan study. J Neurol Neurosurg Psychiatry 57:464–470

Walker R (1995) Spatial and object-based neglect. Neurocase 1:371–383.

Young AW, De Haan EHF, Newcombe F, Hay DC (1990) Facial neglect. Neuropsychologia 28:391–415

Young MP (1992) Objective analysis of the topological organization of the primate cortical visual system. Nature 358:152–155

Zeki S, Watson JDG, Lueck CJ, Friston KJ, Kennard C, Frackowiak RSJ (1991) A direct demonstration of functional specialization in human visual cortex. J Neurosci 11:641–649

Parietal Lobe and the Visual Control of Movement

M. GLICKSTEIN

Department of Anatomy and Developmental Biology, University College London, London, UK

One of the most important circuits in the mammalian brain, which links sensory with motor cortical areas for controlling movement, descends subcortically to involve the cerebellum. Most sensory areas of the cerebral cortex project to the pontine nuclei, and the neurons of the pontine nuclei send their axons to the cerebellar cortex where they terminate as mossy fibers (Fig. 1). The cerebellar outflow has a powerful influence on all of the descending motor pathways. In this paper I will:

1. Describe briefly the experiments that led to the discovery of the motor and visual areas of the cerebral cortex, and to the gradual recognition of the large number of visual areas outside the primary visual cortex. In primates, one group of extrastriate cortical visual areas is centered in the parietal lobe. This region of the monkey's cerebral cortex was at first wrongly identified as the primary visual cortex because lesions in it caused visuo-motor deficits which were so severe they seemed to indicate that the animals were blind.
2. Present evidence that the effects of lesions of parietal lobe visual areas in man are virtually identical to those seen in monkeys.
3. Discuss the circuits by which the parietal lobe visual areas are connected to motor centres, and argue that cortico-cortical circuits are not the most important pathways for the visual control of movement.
4. Outline the pathways from cortical visual areas to the pontine nuclei with a brief description of the response properties of pontine visual cells. Although the cerebellum receives visual climbing fibers as well as mossy fibers from other structures, the pathway from parietal lobe visual areas is by far the largest source of visual input to the cerebellum in monkeys and man.
5. Discuss the targets of pontine visual cells in the cerebellar cortex and the projections from those target areas to the cerebellar nuclei.

In: Parietal Lobe Contributions to Orientation in 3D Space (1997). P. Thier and H.-O. Karnath (eds). Springer-Verlag, Heidelberg.

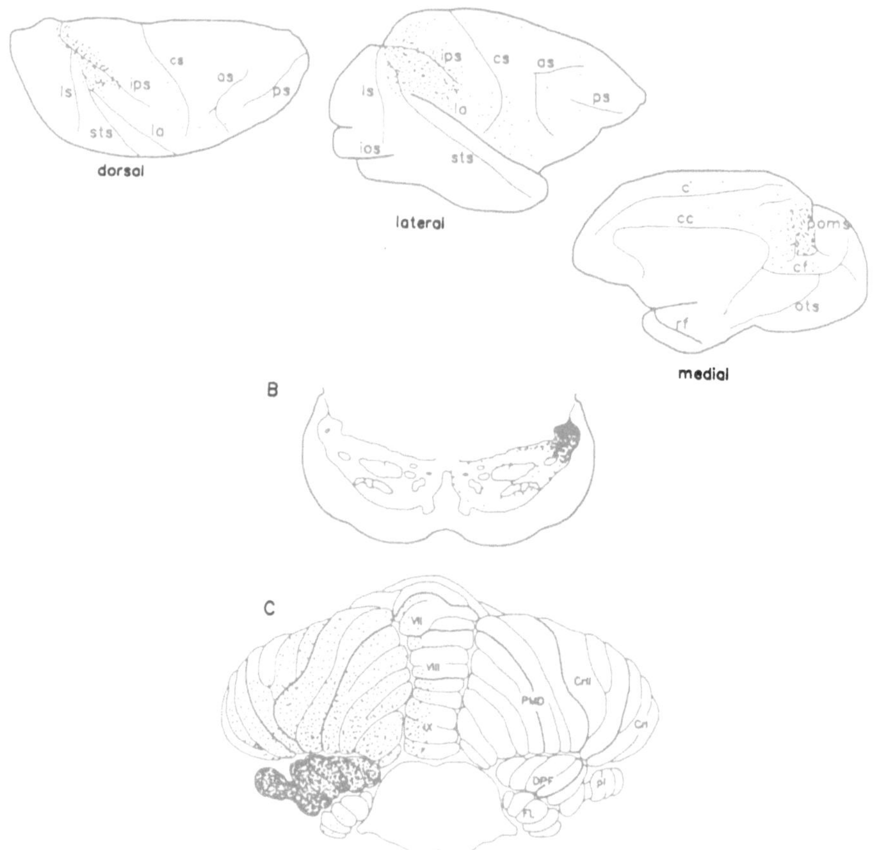

Fig. 1 A-C. The major pathway which transmits visual information from the cerebral cortex to the cerebellum. **A** Extrastriate cortical areas that are labelled after pontine injections. The same cortical visual regions are labelled if the label is confined to the dorsolateral pons or fills the entire pontine nuclei. (From Glickstein et al. 1985, 1990, 1994). **B** Principal target of cortical visual areas in the pontine nuclei. (From Glickstein et al. 1980, 1990, 1994). **C** Projection from the dorsolateral pontine cells onto the cerebellar cortex. Only the contralateral projection is illustrated. There is a roughly symmetrical projection to the ipsilateral cerebellum, but the number of terminals is much smaller. (From Glickstein et al. 1994). *as*, arcuate sulcus; *c*, cingulate sulcus; *cc*, corpus callosum; *cf*, calcarine fissures; *CrI*, Crus I; *CrII*, Crus II; *cs*, central sulcus; *DPF*, dorsal paraflocculus; *Fl*, flocculus; *ips*, intraparietal sulcus; *la*, lateral (Sylvian) fissure; *ls*, lunate sulcus; *ots*, occipito-temporal sulcus; *pl*, petrosal lobule; *PMD*, paramedian lobule; *poms*, parietooccipital medial sulcus; *ps*, principal sulcus; *rf*, rhinal fissure; *sts*, superior temporal sulcus; *VII-IX*, cerebellar vermian lobules

Cerebral Localization: the Discovery of the Motor and Visual Cortex

In the first half of the nineteenth century there was no evidence for regional specialization in the mammalian cerebral cortex. The claims of Gall and the phrenologists (1810-1819) for localization of function in the brain were made on the basis of the weakest of evidence. Flourens (1824) made lesions in different regions of the cerebral hemispheres in various animals and found no evidence that specific deficits were associated with particular cortical areas. Prudent opinion supported Flourens' view that there is no functional localization in the cerebral cortex.

Two developments in the second half of the nineteenth century reversed this conclusion and initiated modern understanding of cortical localization of function. One of the developments was the discovery by Broca (1861) that damage to a particular region of the human frontal lobe is associated with loss of the power of speech. The other was the discovery by Fritsch and Hitzig (1870) in dogs that localized, low-voltage electrical stimulation of a specific region of the frontal lobe, the motor cortex, caused movement on the opposite side of the body. Stimulation outside that area did not have this effect. Whereas stimulation of the appropriate part of the motor cortex produced movement of the forepaw, ablating the same area caused that limb to become clumsy.

Fritsch and Hitzig's discovery prompted a search for possible localization of other functions. The first claim to having identified the cortical visual area was made by David Ferrier (1876; Glickstein 1985). Working first in Yorkshire, and later in London, Ferrier found that electrical stimulation of the angular gyrus of the parietal lobe in monkeys caused the eyes to deviate away from the side which had been stimulated. Ferrier suggested that the electrical stimulation might have mimicked natural optical stimulation of the visual system. Perhaps the monkey was turning his eyes as if to look at a visual stimulus evoked by the electrical stimulation. So, Ferrier reasoned, the angular gyrus might be the site of the primary visual cortex. When he removed the angular gyrus of a monkey bilaterally, this seemed to make the animal blind. Therefore Ferrier concluded that the angular gyrus of the parietal lobe is the primary visual cortex.

Ferrier's observations were limited by the short time that the animals were allowed to survive postoperatively. In his early experiments he routinely killed monkeys three days after he had made a brain lesion, since, if they had been left alive any longer, they would inevitably have died of infection. In later experiments however, Ferrier adopted sterile surgical procedures. His animals could now live for weeks, months, or years after an operation, so he could test them over a much longer postoperative period. With Gerald Yeo, Ferrier (1884) repeated his earlier studies of the effect of angular gyrus lesions. They found that thier animals were not in fact blind; instead, they were markedly impaired at guiding their movement under visual control. Ferrier and Yeo wrote:

> The cherry was laid on the floor in front of it but it was unable to find it though looking eagerly for it. The animal enjoyed its food which it found by groping about with its hand in its cage.
> On the fourth day there was some indication of returning vision. A piece of orange was held before it whereupon it came forward in a groping manner and tried to lay hold but missed repeatedly. When the piece of orange was laid on the floor it stretched out its hand over it, short of it, and round about it before it succeeded in securing it.
> On the fifth day the animal came out of its cage spontaneously and walked about. It never now knocked its head. It was evidently able to see its food, but constantly missed laying hold of it at first putting its hand beyond it or short of it.
> On the sixth day the animal walked about freely avoiding obstacles, but vision was evidently defective, and on several occasions it was seen as if about to climb before it had come sufficiently near the ledge on which it wished to mount. It was however, able to pick-up grains of rice scattered on the floor, but always with uncertainty as to the exact position.

Hermann Munk (1881) completely disagreed with Ferrier's conclusion about the localization of the primary visual cortex in the parietal lobe. He showed that unilateral lesions in the occipital lobe produced hemianopia, and that bilateral lesions produced complete blindness. Munk's discovery of the primary visual cortex in the occipital lobe was confirmed by other investigators and soon demonstrated in the human brain. Salomon Henschen (1890) reviewed post-mortem studies of people who had been hemianopic following brain damage. He confirmed that the human primary visual cortex, like that of monkeys, is in the occipital lobe; in man it is largely contained within the calcarine fissure.

Ideas about how the visual image is processed by the brain beyond the primary visual cortex were at first rather simplistic. Munk (1881) envisioned a single "visual psychic area" surrounding the primary visual cortex in which visual memories were stored. There was little anatomical or physiological evidence of visual inputs to areas beyond the primary visual cortex. However, in 1942 Talbot set out to map the extent of the cortical representation of vision in cats by recording the potentials which were evoked by small spots of light. He found not just one orderly representation of the visual field, but two. The vertical meridian was represented along a diagonal line across the lateral gyrus. The primary retinotopic visual field was presented medial to this line and extended onto the medial face of the hemisphere. Lateral to the diagonal line a second, mirror-image representation of the contralateral visual field was found. Talbot's discovery of a second visual area, visual area II (V2), was confirmed with single-cell recording (Hubel and Wiesel 1965), and over the past 30 years many more visual areas have been identified in various mammals. A recent count by Van Essen et al. (1992) listed a total of 31 extrastriate visual areas in macaque monkeys, all of which receive direct or indirect connections from the primary visual cortex.

On the basis of lesion effects, anatomical connections, and receptive field properties the monkey extrastriate visual areas have been subdivided into two groups. The ventral group projects towards the inferotemporal cortex and is thought to be involved in visual recognition. The dorsal group culminates in the posterior parietal lobe and is thought to play an important part in localizing targets with respect to the observer (Ungerleider and Mishkin 1982). Ferrier's angular gyrus lesions would have destroyed nearly all the dorsal group but they probably spared the ventral set. The dominant input to the dorsal group is provided by the magnocellular component of the visual system; so the extrastriate visual areas of the dorsal group are particularly responsive to visual motion. In addition to their role in locating objects, they probably play a major role in the visual guidance of movement (Glickstein and May 1982; Glickstein et al. in press; Stein and Glickstein 1992).

On Parietal Lobe Function in Visuo-Motor Control

Ferrier and Yeo's results demonstrated that bilateral lesions of the monkey angular gyrus can impair the visual guidance of movement. Balint (1909) described similar symptoms in a patient with bilateral parietal lobe lesions. In addition to visual neglect and oculomotor deficits, Balint's patient could not direct his limbs accurately under visual guidance. In Balint's case, as in most human clinical studies, the lesions were caused by vascular damage and were rather diffuse. However, small calibre high-speed bullets produce much more restricted brain lesions. Hence, studying brain-injured soldiers provided more accurate information about the effects of focal damage to the parietal lobes. Gordon Holmes worked in a military hospital during the First World War, where he examined numerous patients with gunshot wounds. He found that patients with bilateral parietal lobe lesions had characteristic impairments of visually guided movements (Holmes 1918). He described the symptoms of one of his patients as follows:

> Throughout the whole time he was in hospital his most striking symptom was his inability to take hold of or touch any object with accuracy, even when it was placed in the line of vision. When a pencil was held up in front of him he would often project his arm in a totally wrong direction, as though by chance rather than by deliberate decision, or more frequently he would bring his hand to one or other side of it, above or below it, or he would attempt to seize the pencil before he had reached it, or after his hand had passed it. When he failed to touch the object at once he continued groping for it until his hand or arm came into contact with it, in a manner more or less like a man searching for a small object in the dark.
>
> ... but this diminished after some weeks. When asked to touch or grasp my hand or a pencil held in front of his eyes, he groped wildly for it, and, as a rule, brought his hand beyond it when it was within his reach, but he made errors in every direction in the judgement of its position. If his hand, however, came in contact with my arm he moved his fingers promptly along this until they reached my hand or the object it held. In all these attempts he used the left hand in preference to the right, but the errors were equally great with both. That these symptoms cannot be attributed to disturbances in the movements of his arms was shown by the fact that he could always bring his finger accurately and promptly to any point on his own body that was touched. He was extremely slow and awkward in taking food with a spoon; often striking it too heavily on

> the plate, or searching portions of this on which there was no food, but he always brought the spoon quickly and correctly in his mouth.
> When he attempted to touch any object he generally stared at it with widely open eyes, then brought his hand slowly forward from the neighbourhood of his mouth or chest, and continued groping and searching for it with his fingers till he reached it, or even after he had passed it; he has even hit my face with his hand when attempting to seize a pencil I held considerable distance to one side of it. Sometimes he leaned forward in bed and searched for the object at full arm's length when it was quite close to his eyes, but often underestimated its distance, too, and tried to seize it before his hand had reached out.
> Similarly when requested to pick up coins placed on a board in front of him he searched for them with his hand, employing touch rather than vision, though he could obviously see them, and he generally failed to bring his fingers down to any one directly.

The symptoms caused by parietal lobe damage in Holmes' patients resemble those of Ferrier's monkeys. Lesions of the parietal lobe visual areas impair severely the visual guidance of movement in man and monkey.

What Are the Routes Whereby Parietal Lobe Visual Areas Are Connected to Motor Areas?

For vision to guide movement, there must be connections from visual to motor areas of the brain. The assumption is often made that the most important link is made by association fibers linking areas of cortex to each other. Short U-fibers join adjacent gyri of the cerebral cortex, and long association bundles connect more remote cortical areas with one another. Also, the two hemispheres are connected together by the corpus callosum or anterior commissure. But cortico-cortical circuits are probably not the most important pathways linking visual to motor areas. Monkeys in which all of the cortico-cortical fibers between visual and motor areas have been cut can still reach accurately for small morcels of food even when they are moving (Myers et al. 1962; Glickstein and Sperry 1963), although finger guidance may be somewhat impaired (Gazzaniga 1964; Brinkman and Kuypers 1973).

In split-brain monkeys each hemisphere is supplied with visual information by the ipsilateral eye, but it controls the contralateral arm. When only one eye is open, therefore, the arm on that side is controlled by the contralateral hemisphere and that hemisphere receives no visual input. Yet the monkeys can pick up food with great accuracy using that arm. Two kinds of explanation have been put forward to account for the preservation of accurate visually guided arm movements under these conditions. Gazzaniga (1964) suggested that the monkeys may solve the problem using a ‘cross-cuing’ strategy. He proposed that they orient towards a target using the sighted hemisphere, and then reach directly forward using proprioceptive cues to guide the arm controlled by the blind hemisphere.

Brinkman and Kuypers (1973) put forward an alternative hypothesis. They suggested that the monkey might use surviving ipsilateral descending pathways originating from the sighted hemisphere to steer the arm at the shoulder. Final completion of the task would be accomplished using tactile cues when the fingers touched the target.

There is, however, a third possibility. The largest subcortical pathway linking visual with motor areas involves the cerebellum. The dorsal extrastriate visual areas project to the ipsilateral pontine nuclei. These nuclei project in turn bilaterally to the cerebellum (Rosina and Provini 1984). The motor cortex on the side of the brain which is deprived of a cortico-cortical visual input could therefore be guided by visual information routed through the cerebellum. (Glickstein and May 1982; Glickstein et al. in press; Stein and Glickstein 1992). Thus, the most important pathway for the visual guidance of movement may be the cortico-ponto-cerebellar route.

The Pathway from Cerebral Cortex to the Pontine Nuclei

In order to study the visual cortico-pontine projections, we made lesions or injected tracers into selected cortical visual areas to study their efferent targets in the pontine nuclei (Glickstein et al. 1980). In parallel retrograde tracing experiments we filled the pontine nuclei of monkeys (Glickstein et al. 1985) and rats (Legg et al. 1989) with wheat-germ agglutinin horseradish peroxidase (WGA/HRP), and mapped the distribution of retrogradely labelled cells in the cerebral cortex.

The main projection to the pontine nuclei arises from cells in layer V. In some regions layer V is clearly divided into sublaminae Va and Vb. In the rat somatosensory cortex stained to reveal the distribution of the enzyme cytochrome oxidase, the superficial sublamina, Va, stains palely, and the deeper sublamina, Vb, stains darkly. The pyramidal cells found in layer Va project to the basal ganglia, whereas those in layer Vb project to the pontine nuclei. A narrow tier of cells between the two sablaminae projects to both (Mercier et al. 1990).

In rats the whole cerebral cortex, including motor, sensory, and "association" areas, projects to the pontine nuclei (Legg et al. 1989). But in monkeys only about half the cortex does (Glickstein et al. 1980, 1985). The dorsal group of prestriate visual areas provides nearly all the visual corticopontine visual projection in monkeys, whereas the ventral group and the striate cortex itself send virtually no fibers to the pontine nuclei.

Corticofugal fibers travel to the pontine nuclei by way of the internal capsule and the basis pedunculi. We therefore injected wheat-germ agglutinin horseradish peroxidase (WGA/HRP) into selected sites in the rat cerebral cortex and followed the course of labelled fibers through the peduncles en route to the pontine nuclei (Glickstein et al. 1991). Fibers arising from cells in the frontal pole were found to occupy the most ventromedial portion of the peduncle; cells in the occipital and temporal cortex occupy the dorsolateral corner. Somatosensory cortex fibers ran between them.

Probably the same basic arrangement of corticopontine fibers is found in man and monkeys. Just as parietal lobe lesions can impair visuomotor function, so can destruction of corticopontine fibers within the internal capsule or in the basis pedunculi produce similar visuo-motor deficits. Classen et al. (1995) described a patient with symptoms very similar to those following posterior parietal lesions, namely severe visuo-motor apraxia. But the deficit was caused by a subcortical lesion which destroyed the caudal fibers of the internal capsule. The parietal coretx and its cortico-cortical connections were completely spared.

Cortical visual areas in monkeys and rats project to an area which is centered in the dorsolateral region of the pontine nuclei. In cats the cortical visual areas project to a more centrally located group of cells in the pons (Brodal 1972a, b; Glickstein et al. 1972). Cat visual pontine cells which receive this visual input can be activated by visual targets (Glickstein et al. 1972; Baker et al. 1976). They are sensitive to the direction and velocity of moving visual targets; but they are far less influenced by the precise shape or orientation of the stimuli. Similar receptive field properties have been described for visual pontine cells in monkeys (Suzuki and Keller 1984; Thier et al. 1988) The response properties of antidromically activated visual corticopontine cells are similar to those of pontine visual cells (Gibson et al. 1978). Their receptive fields are entirely different from those of cells within the same regions of cortex which do not project to the pons.

To Where do Pontine Visual Cells Project in the Cerebellar Cortex?

Snider and Stowell (1944) were the first to show that visual signals reach the cerebellum, by recording gross potentials on the surface of lobules VI and VII of the cerebellar vermis of cats which were evoked by flashing lights. Because the potentials survived complete decortication, they concluded that the visual input was probably relayed by way of the superior colliculus. Although in all mammals studied the colliculus does provide a visual input to the cerebellum via the pontine nuclei (e.g., Harting 1977; Kawamura and Brodal 1973; Mower et al. 1979), in monkeys, and probably in man also, the input from visual cortical areas is much greater (Glickstein et al. 1990).

Several years after Snider and Stowell's report, Fadiga and Pupilli (1964) showed that visually evoked potentials could be elicited over a much wider area of the posterior lobe of the cat cerebellum than just the vermis.

The potentials recorded by Snider and Stowell and by Fadiga and Pupilli were probably relayed by mossy fibers arising from the pontine nuclei and the nucleus reticularis tegmenti pontis. In addition to these mossy-fiber inputs, climbing fibers also carry visual signals. These originate from cells in the dorsal cap of the inferior olive (Maekawa and Simpson 1973; Maekawa and Takeda 1977). Visual climbing fibers are distributed over a much smaller territory on the cerebellar cortex, and olivary visual cells respond best to targets moving at much slower speeds than pontine visual cells.

Most of the cortical visual input to the cerebellum of monkeys is relayed from parietal lobe visual areas to the pontine nuclei, and thence to the cerebellar cortex. In order to study the cerebellar targets of pontine visual cells, we injected WGA/HRP amongst visually activated cells in the monkey dorsolateral pons (Glickstein et al. 1994). We confirmed that the projection to this region of the pons comes from the parietal lobe visual areas; and we mapped the distribution of orthogradely labelled fibers projecting to the cerebellar cortex. The heaviest projection of pontine visual cells is to the dorsal paraflocculus. As Burne and Woodward (1983) had shown in rats, and we had described in cats (Robinson et al. 1984), in monkeys also the dorsal paraflocculus is a major target for visual-afferent mossy fibers. In addition there is a moderately dense projection to the rostral folia of the uvula, to the paramedian lobule and to Crus II. There were only sparse projections to lobule VI, VII or the ventral paraflocculus. We also injected WGA/HRP into these target areas on the cerebellar cortex and mapped the distributions of retrogradely labelled cells in the pons. These experiments confirmed that the input to these sites on the cerebellar cortex comes from the rostral dorsolateral pons.

We have recently begun to analyze the nuclear targets of cerebellar cortical regions which receive the largest visual input. We injected WGA/HRP into selected locations on the cerebellar cortex and mapped the distribution of orthogradely labelled fibers in the cerebellar nuclei. The dorsal projects to a region in the caudal-ventral region of the posterior interpositus and the adjacent medial dentate nucleus. This projection is of particular interest since it is the same region as that in which Van Kann et al. (1993) found cells which were active before eye movements. The specificity of the projection raises the possibility that there may be a previously unrecognized somatotopic organization in this part of the posterior lobe.

In summary, there is a massive visual projection from the parietal lobes to the cerebellum by way of the pontine nuclei. This pathway serves as one of the major routes through the brain for the visual guidance of movement.

References

Baker J, Gibson A, Glickstein M, Stein J (1976) Visual cells in the pontine nuclei of the cat. J Physiol 255:414–433

Balint R (1909) Seelenlähmung des "Schauens", optische Ataxie, räumliche Störung der Aufmerksamkeit. Monatschr Psychiat Neurolog 25:51–81

Brinkman J, Kuypers HGJM (1973) Cerebral control of contralateral and ipsilateral arm, hand, and finger movements in the split-brain Rhesus monkey. Brain 96:653–674

Broca PP(1861) Remarque sur la siège de la faculté du langage articulé suivie d'une observation d'aphemie (perte de la parole). Bull Soc Anat Paris 36:330–357

Brodal P(1972a) The corticopontine projection from the visual cortex in the cat I. The total projection from Area 17. Brain Res 39:297–317

Brodal P (1972b) The corticopontine projection from the visual cortex in the cat II. The projection from Areas 18 and 19. Brain Res 39:319–335

Burne RA, Woodward DJ (1983) Visual cortical projections to the paraflocculus in the rat. Exp Brain Res 49:55–67

Classen J, Kunesch E, Binkofski F, Hilperath F, Schlaug G, Seitz R, Glickstein M, Freund H-J (1995) Subcortical origin of visuomotor apraxia. Brain 118:1365–1374

Fadiga E, Pupilli GC (1964) Teleceptive components of the cerebellar function. Physiol Rev 44:432–486

Ferrier D (1876) The functions of the brain, 1st edn. Smith-Elder, London

Ferrier D (1886) The functions of the brain, 2nd edn. Smith-Elder, London

Ferrier D, Yeo GF (1884) On the effects of brain lesions in monkeys. Philos Trans R Soc Lond ii, 494

Flourens P (1824) Recherches expérimentales sur les propriétés et les fonctions du système nerveux dans les animaux vertébrés. Crevot, Paris

Fritsch G, Hitzig E (1870) Über die elektrische Erregbarkeit des Großhirns. Arch Anat Physiol wissenschaftl Med:300–332

Gall FJ, Spurzheim JC (1810-1819) Anatomie et physiologie du système nerveux en géneral et du cerveau en particulier avec des observations sur la possibilité de reconnôitre plusieurs dispositions intellectuelles et morales de l'homme et des animaux par la configuration de leur têtes: four volumes and atlas. Schoell, Paris (vols 1, 2); Libraire Greque-Latine-Allemande, Paris (vol 3); Maze, Paris (vol 4)

Gazzaniga MS (1964) Cross-cuing mechanisms and ipsilateral eye-hand control in split-brain monkeys. Exp Neurol 20:11–17

Gibson A, Baker J, Mower G, Glickstein M(1978) Corticopontine cells in Area 18 of the cat. J Neurophysiol 41:484–495

Glickstein M (1985) Ferrier's mistake. Trends Neurosci 8:341–344

Glickstein M, May J (1982) Visual control of movement: the circuits which link visual to motor areas of the brain with special reference to the visual input to the pons and cerebellum. In: Neff WD (ed) Contributions to sensory physiology. Academic, New York

Glickstein M, Sperry RW (1963) Visuo-motor coordination in monkeys after optic tract section and commissurotomy. Fed Proc 22:456

Glickstein M, King R, Stein J (1972) Visual input to the pontine nuclei. Science 178:1110–1111

Glickstein M, Cohen J, Dixon B, Gibson A, Hollins M, La Bossiere E, Robinson F (1980) Corticopontine visual projections in Macaque monkeys. J Comp Neurol 190:209–229

Glickstein M, May J, Mercier B (1985) Corticopontine projections in the Macaque: the distribution of labelled cells after large injections of horseradish peroxidase in the pontine nuclei. J Comp Neurol 235:343–359

Glickstein M, May J, Mercier B (1990) Visual corticopontine and tectopontine projections in the Macaque. Arch Ital Biol 128:273–293

Glickstein M, Kralj-Hans I, Legg C, Mercier B, Ramna-Rayan M, Vaudano E (1991) The organisation of fibers within the rat basis pedunculi. Neurosci Lett 135:75–79

Glickstein M, Gerrits N, Kralj-Hans I, Mercier B, Stein J, Voogd J (1994) Visual pontocerebellar projections in the Macaque. J Comp Neurol 349:51–72

Glickstein M, Buchbinder S, May JL (in press) Visual control of the arm, the wrist, and the fingers; pathways through the brain. Neuropsychologia

Harting JK (1977) Descending pathways from the superior colliculus: an autoradiographic analysis in the Rhesus monkey (Macacca mulatta). J Comp Neurol 173:583–612

Henschen SE (1890) Klinische und anatomische Beiträge zur Pathologie des Gehirns, part 1. Almquist and Wiksell, Upsala

Holmes G (1918) Disturbance of visual orientation. Br J Ophthalmol 2:449–468

Hubel D, Wiesel T (1965) Receptive fields and functional architecture in two non-striate visual areas (18 and 19) of the cat. J Neurophysiol 28:229–289

Kawamura J, Brodal A (1973) The tectopontine projection in the cat: an experimental anatomical study with comments on pathways for teleceptive impulses to the cerebellum. J Comp Neurol 149:371–390

Legg C, Mercier B, Glickstein M (1989) Corticopontine projection in the rat: the distribution of labelled cortical cells after large injections of horseradish peroxidase. J Comp Neurol 286:427–436

Maekawa K, Simpson J (1973) Climbing fiber responses evoked in the vestibulo-cerebellum of rabbit from visual system. J Neurophysiol 36:649–666

Maekawa K, Takeda T (1977) Afferent pathways from the visual system to the cerebellar flocculus in the rabbit. In: Baker R, Berthoz A (eds) Control of gaze by brainstem neurons. Elsevier, Amsterdam, pp 187–195

Mercier B, Legg C, Glickstein M (1990) Basal ganglia and cerebellum receive different somatosensory information in rats. Proc Natl Acad Sci USA 87:4388–4392

Mower G, Gibson A, Glickstein M ((1979) Tectopontine pathways in the cat: laminar distribution of cells of origin and visual properties of target cells in dorsolateral pontine nucleus. J Neurophysiol 42:1–15

Munk H (1881) Über die Functionen der Grosshirnrinde. Hirschwald, Berlin

Myers RE, Sperry RW, Mc Curdy N (1962) Neural mechanisms in visual guidance of limb movement. Arch Neurol 7:195–202

Robinson F, Cohen JL, May J, Sestokas AK, Glickstein M (1984) Cerebellar targets of visual pontine cells in the cat. J Comp Neurol 223:471–482

Rosina A, Provini L (1984) Pontocerebellar system linking the two hemispheres by intracerebellar branching. Brain Res 289:45–63

Snider RS, Stowell A (1944) Receiving areas of the tactile, auditory, and visual systems in the cerebellum. J Neurophysiol 7:331–357

Stein JF, Glickstein M (1992) The role of the cerebellum in the visual guidance of movement. Physiol Rev 72:967–1017

Suzuki D, Keller EL (1984) Visual signals in the dorsolateral pontine nucleus of the alert monkey: their relationship to smooth pursuit eye movements. Exp Brain Res 53:473–478

Talbot SA (1942) A lateral localization in cats' visual cortex. Fed Proc 1:84

Thier P, Koehler W, Buettner UW (1988) Neuronal activity in the dorsolateral pontine nucleus of the alert monkey modulated by visual stimuli and eye movements. Exp Brain Res 70:496–512

Ungerleider LG, Mishkin M (1982) Two cortical visual systems. In: Ingle DJ, Goodale MA, Mansfield JW (eds) Analysis of visual behavior. MIT Press, Cambridge, pp 459–486

Van Essen DC, Anderson CH, Felleman DJ (1992) Information processing in the primate visual system: an integrated systems perspective. Science 255:419–423

Van Kan PLE, Houk JG, Gibson AR (1993) Output organization of intermediate cerebellum of the monkey. J Neurophysiol 69:57–73

Input from the Cerebellum and Motor Cortical Areas to the Parietal Association Cortex

S. KAKEI[1], T. WANNIER[2] and Y. SHINODA[1]

[1] Department of Physiology, School of Medicine, Tokyo Medical and Dental University, Tokyo, Japan
[2] Department of Physiology, University of Bern, Bern, Switzerland

Introduction

The parietal association cortex (Px, areas 5 and 7) is thought to play an essential role in fitting desired movements into extrapersonal space (Hyvärinen 1982). This popular notion about the Px assumes a stream of information that the Px "gathers," "integrates" and "transforms" information about the self and external space. The Px then "sends" the results to the motor systems or higher centers to help these systems generate spatiotemporal muscle activities and determine proper movement trajectories. In this context, corticocortical inputs to the Px from various sensory systems (for example, Dubner and Rutledge 1964; Jones and Powell 1968; Seltzer and Pandya 1980; Avendaño et al. 1988; Felleman and Van Essen 1991) and corticocortical outputs from the Px to the motor cortical areas (for example, Heath and Jones 1971; Kawamura 1973; Strick and Kim 1978; Cavada and Goldman-Rakic 1989; Dum and Strick 1991) as well as corticofugal outputs from the Px to the subcortical motor systems (for example, Mizuno et al. 1973; Brodal 1978; Schmahmann and Pandya 1989, Glickstein, this volume) have been highlighted and investigated extensively by many researchers. However, little attention has been paid to the flow of information in reverse direction, such as inputs from the cerebellum (Fanardzhyan 1964; Sasaki et al. 1972a,b, 1976; Stanton and Orr 1985; Kakei and Shinoda 1990; Wannier et al. 1992; Yamamoto et al. 1992; Kakei et al. 1995).

Using a combination of electrophysiological and morphological methods, we identified two regions of the Px in the cat that receive different types of cerebellar inputs (Wannier et al. 1992) relayed by the ventroanterior-ventrolateral (VA-VL) complex of the thalamus. The present study summarizes the origin of the cerebellar inputs to the Px and the destination of the Px output in order to gain some insight into the functional significance of this input. The results indicate that in the Px, corticocortical neurons projecting to the motor cortex (Mx) and corticofugal neurons reaching the pontine nuclei both receive strong cerebellar input, and furthermore these neurons receive another strong excitatory input from the Mx.

In: Parietal Lobe Contributions to Orientation in 3D Space (1997). P. Thier and H.-O. Karnath (eds). Springer-Verlag, Heidelberg.

Methods

Experiments were performed on cats that were anesthetized with sodium pentobarbital (35 mg/kg, supplemented as required). Cerebellar- or thalamic-evoked field potentials were recorded from the cortical surface on the left side with a silver ball electrode and from within the cortex with a tungsten wire in a glass micropipette (tip diameter 10-20 μm; tip length 15-20 μm; Stoney et al. 1968). Intracellular recordings were made in parietal areas 5 and 7 on the left side with glass micropipettes filled with 3 M KCl, or K-acetate (15-25 MΩ). For morphological identification of recorded neurons, intracellular staining with horseradish peroxidase (HRP) was carried out in eight animals (for details, see Shinoda et al. 1992).

For stimulation, concentric electrodes (inner diameter 0.1 mm; outer diameter 0.3 mm; interpolar distance 1.0 mm) were used. To stimulate the cerebello-thalamic pathway, stimulating electrodes were placed stereotaxically in the brachium conjunctivum (BC) or in the cerebellar nuclei on the right side. In experiments with intracellular recording, eight stimulating electrodes were placed in the left pericruciate gyrus (areas 4γ and 6aβ), and three stimulating electrodes were placed in the pontine nuclei (PN) for antidromic identification of recorded neurons. For stimulation of the Mx, the PN and the BC, rectangular negative pulses (0.2 ms duration, $\leq$500 μA) were applied bipolarly through individual concentric electrodes. The stimulation sites were verified histologically in each experiment.

In two cats, multiple injections of a 50% (w/v) solution of HRP (0.1 μl per injection) were performed in the crown of the suprasylvian gyrus (Ssyl. G.) on the right side and in the caudal bank of the ansate sulcus (Ans. S.) on the left side. Injection sites were identified as receiving cerebellar input by recording cerebellar-evoked field potentials. After a survival period of 48 h, the animals were perfused under deep pentobarbital anesthesia (50 mg/kg). Each section of the brain (75 μm thick) was treated according to the method of Mesulam (1978).

In four cats, a double-barreled electrode was placed in the thalamus. This electrode consisted of two glass micropipettes glued together (Wannier et al. 1992). One micropipette contained a tungsten wire and was used for electrical stimulation and recording field potentials in the thalamus. The other was used for injecting *Phaseolus vulgaris* leucoagglutinin (PHA-L). In two cats, PHA-L was injected into layers III and V of the Mx around the lateral end of the cruciate sulcus to substantiate the projection from the Mx to the Px. The details of PHA-L injection and staining, surgery, and care of the animals have been reported elsewhere (Shinoda and Kakei 1989; Wannier et al. 1992).

Results

Two Regions of the Px That Receive Different Modes of Cerebellar Inputs

Our previous studies (Kakei and Shinoda 1990; Wannier et al. 1992; Kakei et al. 1995) have revealed two regions of the Px which receive different modes of cerebellar input in the cat. These studies have shown that two separate regions in the Px, one area in the crown of the Ssyl. G. and the other in the caudal bank of the Ans. S., receive cerebellar inputs from the interpositus nucleus (IN) and the dentate nucleus (DN) through different regions in the VA-VL complex of the thalamus. They also showed that thalamocortical (TC) fibers terminate at different cortical layers corresponding with the depth of the current sink revealed by laminar field analysis of the cerebellar and thalamic-evoked potential in the two regions.

To relate the depth profile of thalamic-evoked potentials to the laminar distribution of TC axon terminals, laminar field analysis was carried out in the Px during stimulation of the thalamic site that receives cerebellar input, and then PHA-L was injected iontophoretically into the thalamic stimulating site through a double-barreled electrode (Wannier et al. 1992). The thalamic injection site was electrophysiologically determined as follows. In the thalamic position shown in Fig. 1*E*, a large cerebellar-evoked potential was recorded. Electrical stimulation of this site induced surface negative-deep positive potentials in the Ssyl. G. (Fig. 1A, VA-VL), and surface positive-deep negative potentials followed by surface negative-deep positive potentials in the Mx (not illustrated). These responses in the Px were similar to those elicited by stimulation of the brachium conjunctivum (BC, Fig. 1A, BC) except for their latencies. The PHA-L injected in this stimulating site spread about 1 mm in diameter in the VA-VL complex (Fig. 1E), and resulted in orthograde labeling of TC fibers in both the Mx and the Px. In the Px, axon terminals were exclusively distributed in the upper third of layer I in the anterior Ssyl. G. (Fig. 1B, C), whereas in the Mx, axon terminals were observed in both layers I and III (not illustrated).

Figure 2 illustrates an analogous experiment with PHA-L injection in the ventrolateral part of the VA-VL complex (Fig. 2C). Electrical stimulation of this site induced the characteristic surface positive-deep negative potentials followed by surface negative-deep positive potentials in the caudal bank of the Ans. S. and the Mx. Cortical field potentials evoked from this thalamic site (Fig. 2D, VA-VL) and the cerebellar nuclei (Fig. 2D, CN) had similar depth profiles at identical depths of this part of the Px (Fig 2A, E), although the onset of the surface positive-deep negative waves was about 1 ms earlier for the thalamic-evoked potentials (Fig. 2D). PHA-L injection in this thalamic stimulating site (Fig. 2C) resulted in anterograde labeling of TC fibers in both the Mx and the Px. The axon

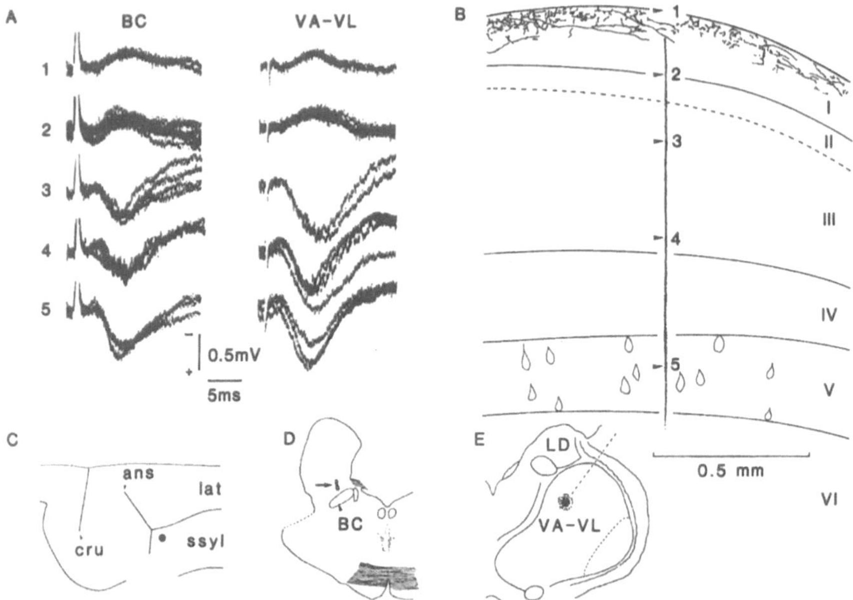

Fig. 1 A-E. Relationship between depth profiles of cerebellar- and thalamic-evoked potentials and laminar distribution of thalamocortical fibers and terminals in the Ssyl. G. **A** Depth profiles of cerebellar-evoked (*BC*, brachium conjunctivum; *left column*) and thalamic-evoked (*VA-VL*, ventroanterior-ventrolateral complex of the thalamus; *right column*) potentials along the recording track shown in *B*. *Numbers* on the left correspond to those along the track in *B*. After recording these potentials, *Phaseolus vulgaris* leucoagglutinin (PHA-L) was injected into the stimulating site in the VA-VL complex through one electrode of the double-barreled electrode. **B** Drawing of PHA-L labeled fibers in layer I of the anterior suprasylvian gyrus (area 5b) based on a single section 100 μm thick. The section is almost parallel to the recording electrode track (*arrow*). **C** Recording site (a *closed circle* in the Ssyl. G.) where laminar field analysis shown in A was made. *cru*, cruciate sulcus; *ans*, ansate sulcus; *lat*, lateral gyrus; *ssyl*, suprasylvian gyrus. **D** Stimulation site of the *BC*. An *arrow head* indicates a stimulation site and an *arrow* indicates a stimulating electrode track. **E** Stimulation and injection site of PHA-L in the VA-VL complex of the thalamus. *LD*, lateral dorsal nucleus. (Modified from Wannier et al. 1992)

terminals found in the Px were limited to the bottom of the bank of the Ans. S. and mainly distributed in layers III and IV with some additional projection to layers I and V (Fig. 2E). The axon terminals in the Mx were found mainly in layer III and less in layers I, V, VI and II. Analysis of the injection sites of the PHA-L revealed that TC neurons projecting to the Ssyl. G. were located more dorsomedially than those projecting to the caudal bank of the Ans. S. (see Figs. 1C, 2C).

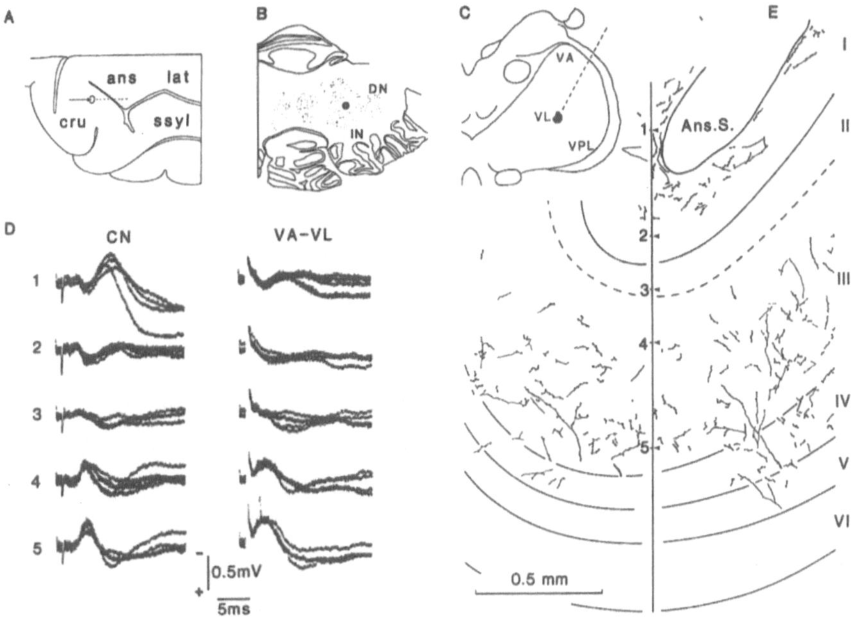

Fig. 2 A-E. Relationship of depth profiles of cerebellar-evoked and thalamic-evoked potentials with laminar distribution of thalamocortical fibers and terminals in the caudal bank of the ansate sulcus (Ans. S.). **A** Position and orientation of the electrode track for laminar field analysis in the Ans. S. seen from the surface of the left hemisphere. *cru*, cruciate sulcus; *ans*, ansate sulcus; *lat*, lateral gyrus; *ssyl*, suprasylvian gyrus. **B** Stimulation site in the cerebellar nucleus. *IN*, interpositus nucleus; *DN*, dentate nucleus. **C** Injection site of PHA-L in the VA-VL complex of the thalamus. *VA*, ventroanterior nucleus; *VL*, ventrolateral nucleus; *VPL*, ventroposterolateral nucleus. **D** Depth profiles of cerebellar-evoked (*left column*) and thalamic-evoked (*right column*) potentials recorded along the electrode track shown in *E*. *Numbers* on the left correspond to those along the track in *E*. PHA-L was injected into the thalamic stimulating site where a large cerebellar-evoked negative field potential was recorded. *CN*, cerebellar nuclei. **E** PHA-L labeled thalamocortical fibers and terminals distributed mainly in layers I and III-IV in the bottom of the bank of the ansate sulcus (area 5a, reconstructed on two consecutive sections 100 μm thick). *Ans. S.*, ansate sulcus. (Modified from Wannier et al. 1992)

Thalamic Distribution of TC Neurons Projecting to the Two Regions of the Px Revealed by Retrograde Labeling with HRP

To further confirm the thalamic relay sites of the cerebellar input to the two regions of the Px, the thalamic distribution of TC neurons was analyzed by

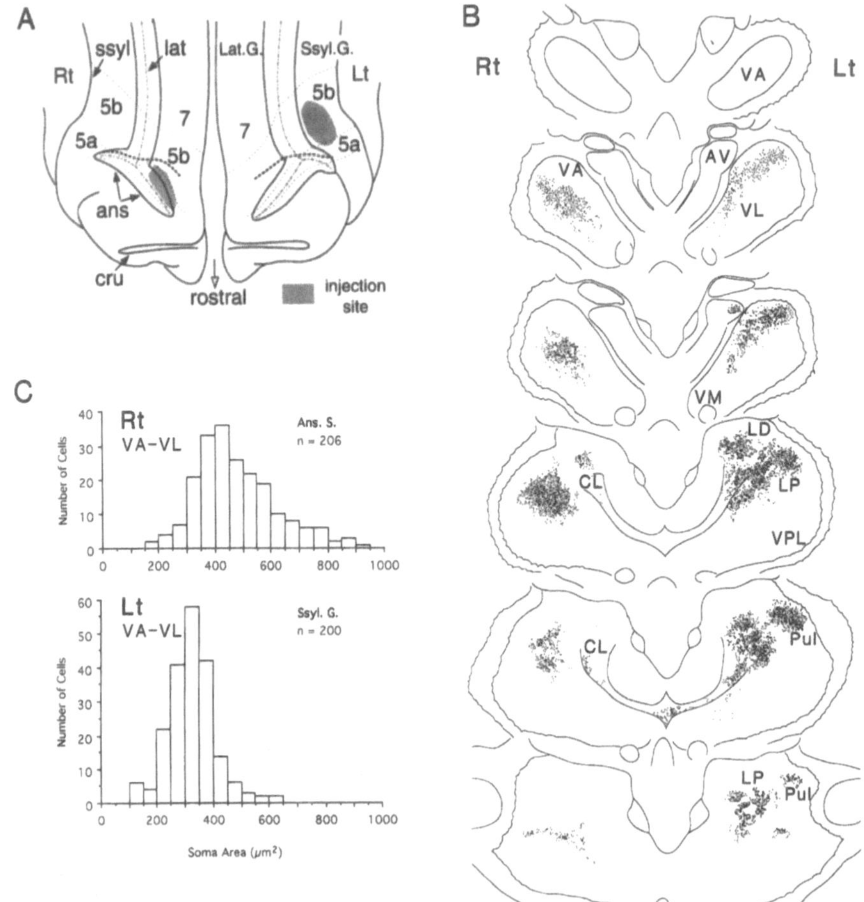

Fig. 3 A-C. Distribution of retrogradely labeled thalamocortical (TC) neurons after injection of horseradish peroxidase (HRP) into the caudal bank of the right ansate sulcus (Ans. S., *Rt*) and into the crown of the left suprasylvian gyrus (Ssyl. G., *Lt*). **A** Dorsal view of rostral half of the cerebral hemispheres of the cat showing the injection sites. After identifying cerebellar-projection areas by recording cerebellar-evoked field potentials, HRP was injected into these cortical sites. Banks of the lateral sulcus and the ansate sulcus are shown "*opened up*". A *thick broken line* on each side of the Px indicates approximate border of the two regions of the Px which receive different modes of cerebellar inputs. The *rostral* region which corresponds to the region shown in Fig. 2 receives cerebellar input to both *deep* and *superficial* layers, while the *caudal* region which corresponds to the region shown in Fig. 1 receives cerebellar input only to *superficial* layers. These border lines were determined by charting distribution of two types of cerebellar-evoked field potentials in the Px. *ans*, ansate sulcus; *cru*, cruciate sulcus; *lat*, lateral sulcus; *ssyl*, suprasylvian sulcus; *Lat.G.*, lateral gyrus; *Ssyl.G.*, suprasylvian gyrus; *5a*, area 5a; *5b*, area 5b; *7*, area 7; *Rt*, right parietal association cortex; *Lt*, left parietal association cortex. **B** Distribution of retrogradely-labeled neurons in the thalamus. Each *dot* represents one neuron. Representative coronal sections (100 μm thick) are separated by 900

μm. *AV*, anteroventral nucleus; *CL*, central lateral nucleus; *LD*, lateral dorsal nucleus; *LP*, lateral posterior nucleus; *Pul*, pulvinar nucleus; *VA*, ventroanterior nucleus; *VL*, ventrolateral nucleus; *VM*, ventromedial nucleus; *VPL*, ventroposterolateral nucleus. **C** Histograms showing the distribution of soma areas of HRP-labeled neurons in the right (*Rt, upper histogram*) and left ventroanterior-ventrolateral (*VA-VL*) complex (*Lt, lower histogram*). (Modified from Kakei et al. 1995)

injecting HRP into the two regions of the Px determined by recording the characteristic cerebellar-evoked field potentials. Figure 3A shows injection sites in the Px. In the right Px (Fig. 3A, Rt), injections were restricted to within the caudal bank of the Ans. S. (areas 5a and 5b), whereas in the left Px (Fig. 3A, Lt), injections were confined within the crown of the Ssyl. G. (mostly area 5b). In the thalamus, many retrogradely labeled neurons were present in the VA-VL complex, the lateralis posterior (LP) -pulvinar (Pul) complex, the central lateral (CL), the central medial (CM) and the lateral dorsal (LD) nuclei on both sides (Fig. 3B). On either side, the labeled neurons formed a dense localized cluster in the VA-VL complex. The cluster in the VA-VL complex continued further caudally, into the LP-pulvinar complex and the CL nucleus. Each cluster shared different parts of each nucleus. The injections into the crown of the Ssyl. G. (Fig. 3A, Lt) labeled cells in the dorsomedial part of the VA-VL complex (Fig. 3B, Lt), while the injections into the caudal bank of the Ans. S. (Fig. 3A, Rt) labeled cells in a more ventrolateral part than the contralateral side (Fig. 3B, Rt). This topographical difference confirms the electrophysiological and morphological data described in Figs. 1 and 2.

Soma areas of labeled TC neurons in the VA-VL complex were measured on each side. Soma areas of TC neurons projecting to the caudal bank of the Ans. S. ranged from 157 to 912 μm^2 (mean ± S.D., 477 ± 142 μm^2, $n = 206$; Fig. 3C, Rt, upper histogram), whereas TC neurons projecting to the crown of the Ssyl. G. had soma areas ranging from 100 to 612 μm^2 (326 ± 86 μm^2, $n = 200$; Fig. 3C, Lt, lower histogram). Thus, TC neurons projecting to the Ans. S. contained much larger neurons than neurons projecting to the Ssyl. G. (Fig. 3C), and this difference was statistically significant ($p<0.01$, Mann-Whitney test).

Cerebellar Input to Px Neurons: Intracellular Recording

To examine the destination of the cerebellar input to the Px, intracellular recordings were made from 176 neurons in the caudal bank of the Ans. S. (areas 5a and 5b) and in the crown of the Ssyl. G. (areas 5b and 7). The experimental setup is shown in Fig. 5C. These Px neurons were classified into four groups according to their antidromic responses. The first group consisted of 72 corticocortical neurons with antidromic responses to stimulation of the Mx. The second group consisted of 48 corticofugal neurons that were activated antidromically by stimulation of the PN. The third group contained one neuron activated antidromically by both the Mx and the PN. The fourth group contained 55 neurons which did not show antidromic responses by stimulation of the Mx and PN. For morphological identification of penetrated neurons, intracellular staining with HRP was performed in 57 neurons after electrophysiological identification of the neurons. Successful staining was achieved in 17 of 31 neurons of the first group, in 17 of 25 neurons of the second group and in the one neuron of the third group. Morphologically, all of the neurons in the first group were layer III pyramidal neurons, and all of the neurons in the second and third groups were layer V pyramidal neurons. Various morphological features of these stained neurons will be described in a separate paper (Kakei et al., in preparation).

Figure 4A shows examples of intracellular records from two corticocortical neurons projecting to the Mx. In A-1, the neuron was located in the caudal bank of the Ans. S. (A-1 in Fig. 4D, top inset), whereas in A-2, the neuron was located in the crown of the Ssyl. G. (A-2 in Fig. 4D, top inset). Action potentials were evoked by stimulation of the Mx at latencies of 0.8 ms (Fig. 4A-1-a) and 2.0 ms (Fig. 4A-2-a), respectively. These spikes were antidromic, since they were evoked without synaptic potentials, and the latencies of the spikes were constant even at the threshold. The latencies of Mx-evoked antidromic spikes ranged from 0.3 to 1.6 ms (mean ± S.D.= 0.8 ± 0.3 ms, n = 54; Fig. 4E, upper histogram) in the caudal bank of the Ans. S., and from 0.4 to 1.6 ms (0.7 ± 0.3 ms, n = 18; Fig. 4E, lower histogram) in the crown of the Ssyl. G. Out of eight stimulating electrodes in the Mx, only one to three sites were effective in producing antidromic spikes in each neuron. There seemed to be some topographical relation between the location of the cell in the Px and the position of the electrodes that were effective in evoking antidromic responses in the Mx. Neurons in the caudal bank of the Ans. S. frequently responded to antidromic stimulation of the lateral or caudal part of the cruciate sulcus, whereas neurons in the crown of the Ssyl. G. responded, although with considerable overlap, to stimulation of the more rostral and medial part of the pericruciate gyrus (Fig. 4D, top inset). Stimulation of the BC produced excitatory postsynaptic potentials (EPSPs) in 47 corticocortical neurons (65%) projecting to the Mx (Fig. 4A-1-b, A-2-b). In the caudal bank of the Ans. S., the latencies of BC-evoked EPSPs ranged from 1.8 to 5.4 ms (mean ± S.D.= 3.0 ± 0.8 ms, n = 33; Fig. 4G, upper histogram), whereas in the crown of the Ssyl. G., those latencies ranged from 2.6 to 5.2 ms (3.2 ± 0.7 ms, n = 14; Fig. 4G, lower histogram). The difference in the mean latencies of the BC-evoked EPSPs was not statistically significant between the two Px regions (p>0.05, Mann-Whitney test). Although the shortest latency of the EPSPs in the Ssyl. G. was 2.6 ms, the latencies of a considerable number of EPSPs in the caudal bank of the Ans. S. were shorter than 2.6 ms.

Figure 4B shows examples of intracellular records from two corticofugal neurons activated by antidromic stimulation of the PN (Fig. 4B-1, 2). In a neuron located in the caudal bank of the Ans. S. (B-1 in Fig. 4D, top inset), antidromic

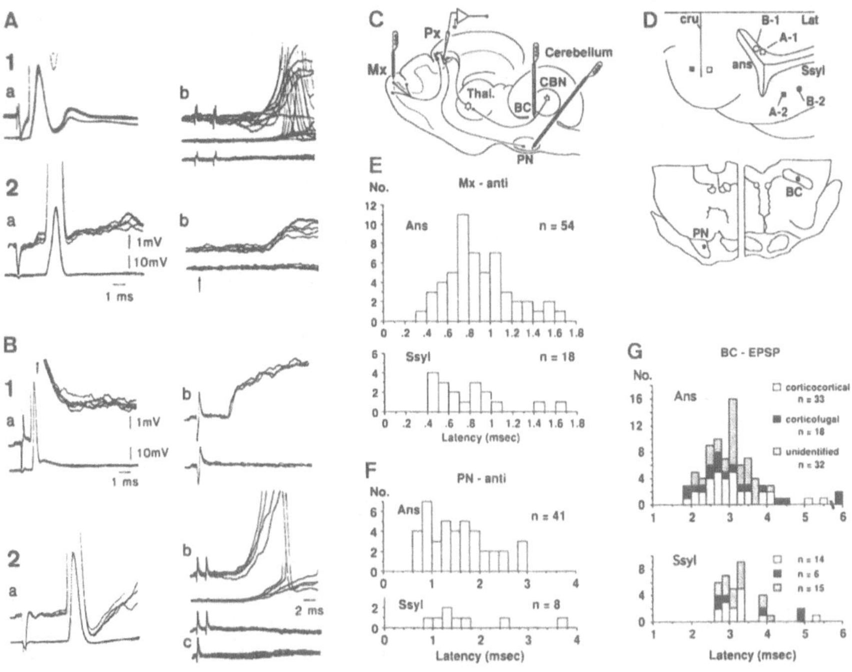

Fig. 4 A-G. Cerebellar input to corticocortical neurons activated antidromically from the motor cortex (Mx) and corticofugal neurons activated antidromically from the pontine nuclei (PN) in the parietal association cortex (Px). **A-1** *and* **-2** Intracellular recordings from corticocortical neurons in the caudal bank of the ansate sulcus (Ans. S., indicated as *A-1* in *top inset* in *D*) and the crown of the suprasylvian gyrus (Ssyl. G., indicated as *A-2* in *top inset* in *D*). **A-1-a** and **A-2-a** Antidromic spikes evoked by stimulation of the Mx. **A-1-b** *and* **A-2-b** Brachium-conjunctivum (BC)-evoked excitatory postsynaptic potentials (EPSPs, *upper traces*) and extracellular field potentials (*lower traces*). **B-1** *and* **-2** Intracellular recordings from corticofugal neurons in the caudal bank of the Ans. S. (indicated as *B-1* in *top inset* in *D*) and the crown of the Ssyl. G. (indicated as *B-2* in *top inset* in *D*). **B-1-a** *and* **B-2-a** Antidromic activation by the PN. **B-1-b** BC-evoked EPSPs (*upper traces*) and extracellular field potentials (*lower traces*). **B-2-b** EPSPs evoked by double-shock stimulation of the BC (*upper traces*) and extracellular field potentials (*lower traces*). **B-2-c** No response to single-shock stimulation of the BC. *Time calibration* in *A-2-a* is applied to all traces except in *B-2-b* and *B-2-c*. **C** The experimental setup. *Mx*, motor cortex; *Px*, parietal association cortex; *Thal.*, thalamus; *PN*, pontine nuclei; *BC*, brachium conjunctivum; *CBN*, cerebellar nuclei. **D**, *top inset* Recording sites of the neurons in A and B in the *Px*, and stimulation sites in the *Mx* for antidromic activation of corticocortical neurons shown in A. *Different symbols* at the recording sites in the *Px* correspond to those at the stimulation sites in the *Mx*. *cru*, cruciate sulcus; *ans*, ansate sulcus; *Lat*, lateral gyrus; *Ssyl*, suprasylvian gyrus. **D**, *bottom left inset* A stimulation site in the ipsilateral *PN* (a *closed circle*). **D**, *bottom right inset* A stimulation site in the contralateral *BC* (a closed circle). **E** Latency histograms of antidromic spikes evoked by stimulation of the *Mx* in neurons in the caudal bank of the *Ans. S.* (*upper histogram*), and in the crown of the Ssyl. G. (*lower histogram*). **F** Latency histograms of antidromic spikes evoked by stimulation

of the *PN* in neurons in the caudal bank of the Ans. S. (*upper histogram*), and in the crown of the Ssyl. G. (*lower histogram*). **G** Latency histograms of *BC*-evoked *EPSPs* in corticocortical, corticofugal and unidentified neurons in the caudal bank of the Ans. S. (*upper histogram*), and in the crown of the Ssyl. G. (*lower histogram*). (Modified from Kakei et al. 1995)

spikes were evoked by stimulation of the PN (Fig. 4D, bottom left inset) at a latency of 0.8 ms (Fig. 4B-1-a). In a neuron located in the crown of the Ssyl. G. (B-2 in Fig. 4D, top inset), antidromic spikes from the PN were also evoked with a much longer latency of 3.5 ms (Fig. 4B-2-a). The latencies of the PN-evoked antidromic spikes ranged from 0.6 to 2.9 ms (mean ± S.D.= 1.5 ± 0.6 ms, $n = 41$) in the caudal bank of the Ans. S. (Fig. 4F, upper histogram), and from 0.9 to 3.7 ms (1.4 ± 0.5 ms, $n = 8$) in the crown of the Ssyl. G. (Fig. 4F, lower histogram). Stimulation of the BC (Fig. 4D, bottom right inset) produced EPSPs in these neurons (24/49, 49%; Fig. 4B-1-b, 2-b). In the caudal bank of the Ans. S., the latencies of BC-evoked EPSPs ranged from 1.8 to 7.5 ms (mean ± S.D.= 3.4 ± 1.6 ms, $n = 18$; Fig. 4G, upper histogram), whereas in the crown of the Ssyl. G., they ranged from 2.6 to 4.8 ms (3.6 ± 1.0 ms, $n = 6$; Fig. 4G, lower histogram). Although the mean latencies of the BC-evoked EPSPs were not significantly different between the two regions ($p>0.05$), a considerable number of the EPSPs were evoked in the caudal bank of the Ans. S. at latencies shorter than 2.6 ms.

In addition to some latency differences of BC-evoked EPSPs, shapes of the BC-evoked EPSPs were systematically different in the two regions of the Px for both corticocortical and corticofugal neurons. In the caudal bank of the Ans. S., rising phases of the EPSPs tended to be steeper (Fig. 4A-1-b, B-1-b) than those in the crown of the Ssyl. G (Fig. 4A-2-b, B-2-b). Similar differences in BC-evoked EPSPs were also observed between neurons with unidentified destinations in these two regions of the Px.

Input from the Mx to Px Neurons

Stimulation of the Mx frequently produced EPSPs and/or inhibitory postsynaptic potentials (IPSPs) in Px neurons (109/159, 69%; Fig. 5, left). Out of 72 corticocortical neurons recorded in the Px, 42 (58%) had Mx-evoked EPSPs, and 15 out of 32 corticofugal neurons (47%) had Mx-evoked EPSPs. These values are probably underestimations, since a buried part of the Mx in the cruciate sulcus was not stimulated in these experiments. In contrast, in the case of unidentified Px neurons, 91% (50/55) had Mx-evoked EPSPs. This extremely high incidence of Mx-induced EPSPs in the unidentified neurons is most likely due to sampling bias. These neurons without antidromic responses would have escaped from our sampling if they had not had orthodromic inputs from either the BC or the Mx. Although the extremely high incidence of Mx-evoked EPSPs in these unidentified

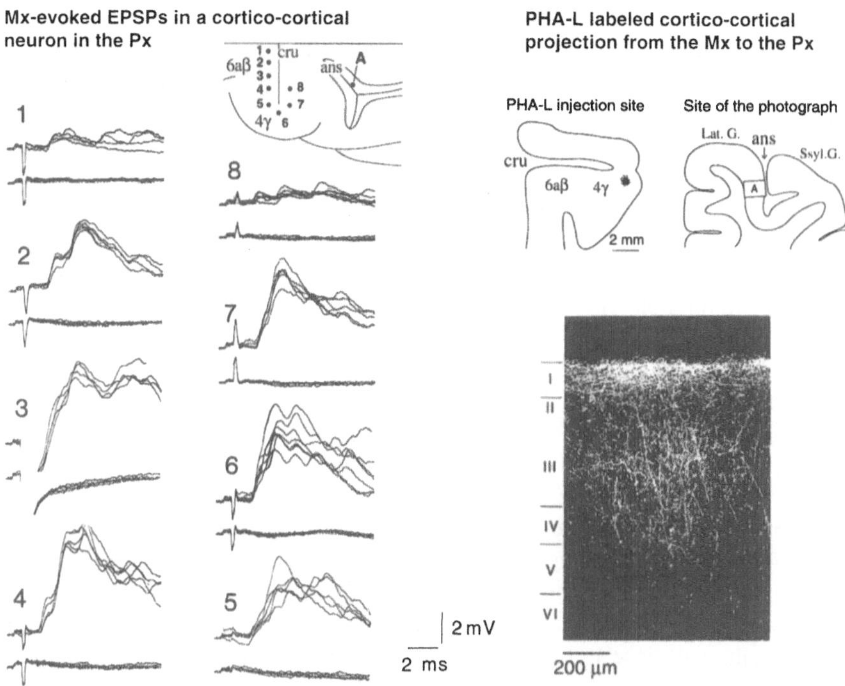

Fig. 5. *Left:* Motor-cortex (Mx)-evoked excitatory postsynaptic potentials (EPSPs) recorded from a corticocortical neuron in the parietal association cortex (Px). The location of the neuron is indicated as *A* in the *inset* on the *top right*. This neuron was antidromically activated from the Mx (electrode No. 3, not shown). The stimulation sites in the Mx are indicated by the *numbered dots* in the inset. Each record (*1-8*) shows the response evoked by stimulation of the site in the Mx indicated by the numbered dot in the inset. Each site was stimulated with a constant current of 500 μA. The lower traces in each record are extracellular field potentials. *Right:* Photomicrograph showing dense corticocortical projection from the Mx to the caudal bank of the ansate sulcus (Ans. S., area 5b) revealed by focal injection of *Phaseolus vulgaris* leucoagglutinin (*PHA-L*) into the forelimb region of the Mx (*top left inset*). Numbers *I-VI* indicate cortical layers. Scale 200 μm. The location of the photograph is indicated by a rectangular (*A*) in the *top right inset. ans*, ansate sulcus; *cru*, cruciate sulcus; *Lat. G.*, lateral gyrus; *Ssyl. G.*, suprasylvian gyrus; *4γ*, area *4γ*; *6aβ*, area *6aβ*. (Modified from Kakei et al. 1995)

neurons may be artificial, it is important that these neurons received strong synaptic input from the Mx. Stimulation of the Mx evoked EPSPs in Px neurons at almost fixed latencies, even though the stimulus intensities were decreased close to the threshold. The distribution of latencies of Mx-evoked EPSPs in the caudal bank of the Ans. S. (0.8-4.4 ms, mean ± S.D.= 1.7 ± 0.6 ms, n = 254) shows a spread similar to that in the crown of the Ssyl. G. (0.8-4.5 ms, 1.7 ± 0.5 ms, n = 163). IPSPs were also evoked by stimulation of the Mx, and their latencies ranged from 2.0 to 4.5 ms (2.8 ± 0.9 ms, n = 29). As shown in the typical potentials in Fig. 5, EPSPs often had more than one peak and were frequently strong enough to produce an action potential (see Fig. 6 in Kakei et al. 1995). The latencies of Mx-evoked second EPSPs ranged from 2.0 to 4.4 ms (mean ± S.D., 2.5 ± 0.6 ms, n = 32) in the caudal bank of the Ans. S., whereas in the crown of the Ssyl. G., they ranged from 2.0 to 4.0 ms (3.0 ± 0.7 ms, n = 19). Moreover, the rising phase of the EPSPs was very steep, which suggests that the synaptic sites were located close to the recording site in each neuron. In many corticocortical and corticofugal neurons, thresholds for evoking monosynaptic EPSPs were as low as 50 μA, and effective stimulation sites for evoking EPSPs were distributed very widely throughout the Mx (Fig. 5, left 1-8 and inset). EPSPs were evoked in single Px neurons by stimulation of stimulation sites 2-8 (1.5-2.0 mm apart) in the Mx and sometimes by stimulation of both areas 4γ and 6aβ (Fig. 5, inset). However, these findings do not necessarily mean that this synaptic input actually originates from the Mx, since any collateral projections of single neurons to both the Px and the Mx can also produce these EPSPs by axon reflex in Px neurons.

To confirm the existence of strong corticocortical input from the Mx to the Px, *Phaseolus vulgaris* leucoagglutinin (PHA-L) was injected into the Mx, and anterograde terminal labeling in the Px was examined. Figure 5 (right) shows a photomicrograph of terminal labeling in the caudal bank of the ansate sulcus (Fig. 5 right, right inset) after injection of PHA-L into the Mx just rostral to the lateral end of the cruciate sulcus (Fig. 5 right, left inset). Numerous terminal branches bore swellings of both en passant and terminal types and clustered to form column-like aggregations. These column-like aggregations ranged in diameter from 0.5 to 1.6 mm (mean ± S.D.= 1.1 ± 0.3 mm, n = 14). No retrogradely labeled neurons were found in the Px, which indicates that recurrent collaterals of corticocortical neurons of the Px play little, if any, role in labeling. Axonal swellings were distributed from layer I to layer VI with a peak of density in layer I. Similar depth distribution of axonal swellings was observed in the other 13 clusters. This strong projection from the Mx to the Px supports the electrophysiological finding of strong monosynaptic excitatory input from the Mx to Px neurons.

Discussion

Our study has identified two pathways from the cerebellum to the Mx and to the PN through the Px in the cat. Furthermore, this part of the Px receives strong excitatory input from the Mx. The following discussion will focus on five points: (1) cerebellar input to the two regions of the Px, (2) cerebellar input to the Mx and the PN through the Px, (3) excitatory input from the Mx to the Px, (4) the functional significance of tight reciprocal connections between the Mx and the Px, and (5) comparison of the present data with organization of the monkey cerebellothalamo-parietal projection.

Multiple Projections from the Cerebellar Nuclei to the Cerebral Cortex

Our previous report has shown that the two regions of the Px receive cerebellar input both from the IN and the DN (Kakei et al. 1995). In areas 4γ and 6aβ, Shinoda et al. (1985a) have identified convergent inputs from both the IN and DN to different corticofugal neurons in the cat. They have also determined that this convergence occurs at the level of the VL nucleus in the thalamus (Shinoda et al. 1985b). Our results suggested a similar organization of the cerebellar input to the Px (see Fig. 1 in Kakei et al. 1995). The inputs from the IN and the DN converge onto the two regions of the Px. There was a tendency that DN stimulation was more effective in the crown of the Ssyl. G. than in the caudal bank of the Ans. S., although this difference was not significant. Stimulation of individual sites in the CN evoked potentials in both the Mx and the Px, and a very focal injection of PHA-L into the VA-VL complex resulted in labeling in both the Mx and the Px (Kakei and Shinoda 1990; Wannier et al. 1992). Therefore, both the Mx and Px receive cerebellar input from a particular site in the CN by way of a common area in the VA-VL complex. However, the question remains as to whether single cerebellar nucleus neurons project to the two regions in the VA-VL complex which contain TC neurons projecting to the two different regions in the Px.

Two Regions of the VA-VL Complex

Our former studies (Kakei and Shinoda 1990; Wannier et al. 1992; Kakei et al. 1995) identified two distinct regions in the VA-VL complex as mediating cerebellar input to the Px; one in the dorsomedial part of the VA-VL complex adjacent to the internal medullary lamina, and the other in the ventrolateral region of the VA-VL complex. In the dorsomedial region, medium- or small-sized neurons project to the crown of the Ssyl. G. and terminate only in layer I. In the ventrolateral region, large-, medium-, and small-sized neurons project to the caudal bank of the Ans. S. and terminate in both layer I and layers III-V. These two regions are also different in terms of their projection to the Mx. Injection of PHA-L into the ventrolateral region produced strong terminal labeling mostly in area 4γ, whereas injection into the dorsomedial region produced moderate or sparse terminal labeling in areas 6aβ and 4γ (Kakei and Shinoda 1990). Furthermore, in the case of injections into the dorsomedial part, labeling in the Px is always stronger than in the Mx. These differences in the efferent organization of the two regions of the VA-VL complex cannot be explained by simple topographical organization, and strongly suggest the existence of a functional differentiation between them.

The two thalamic regions were also different in terms of afferent organization. Nakano et al. (1980) reported that the lateral or ventrolateral portions of the VA-VL complex receive input mainly from the IN and less from the DN, whereas the medial or dorsomedial portions of the VA-VL complex receive input from the DN and the IP. In accordance with this report, Stanton and Orr (1985) have suggested a pathway from the ventral part of the DN and the IP to the middle Ssyl. G. through the dorsal part of the VA-VL complex. In addition, some authors have reported the existence of corticothalamic input from several visual (areas 19 and 20) and putative visuomotor (areas 5b and 7) cortical areas to the dorsomedial portion of the VA nucleus (Robertson and Rinvik 1973; Updyke 1983) that probably corresponds to the dorsomedial portion of the VA-VL complex in our studies. These differences in input-output organization of the two thalamic regions must have some relevance to the functional difference of the two regions of the Px that are connected specifically to the respective thalamic regions.

Cerebellar Input to Corticocortical and Corticofugal Neurons in the Px

In most corticocortical neurons recorded in this study, antidromic spikes were evoked from a rather localized area in the Mx. The relationship between the location of corticocortical neurons in the Px and the stimulation sites which evoked antidromic spikes in these neurons suggests that the two regions of the Px project to the Mx topographically with some overlap. This finding is consistent with previous anatomical data (Kawamura 1973; Waters et al. 1982). All these data suggest that the cerebellar inputs to each region of the Px are distributed widely to the Mx with some overlap. Thus, cerebellar inputs to the Mx through the VA-VL complex have at least three routes: (1) direct projection to the Mx, (2) projection through the caudal bank of the Ans. S., and (3) projection through the crown of the Ssyl. G.

Corticofugal neurons in the Px also received cerebellar input. This result confirmed findings of Yamamoto and his colleagues (Yamamoto et al. 1987; Yamamoto and Oka 1993). Mizuno et al. (1973) demonstrated projection from the Px to the PN in the cat. Oka et al. (1975) revealed a parietopontocerebellar pathway. Together with these other findings, the present results support the idea of a closed circuit between the Px and the cerebellum (Sasaki et al. 1975). However, as long as comprehensive knowledge about the link between the cerebro-pontocerebellar and cerebellothalamoparietal projections is not available, the question as to how the cerebrocerebellocerebral system is really organized remains unanswered. This knowledge is essential to understanding the functional significance of the cerebellar input to the Px.

Input from the Mx to Neurons in the Px

Stimulation of the Mx evoked large EPSPs in the majority of Px neurons (107/159, 67%; Kakei et al. 1995). This result is in agreement with a report by Fanardjian and Papoyan (1995). In addition, most of these neurons received highly convergent inputs from wide regions of the Mx including both areas 4γ and 6aβ (Fig. 5, left). Based on the discussions in our previous report (Kakei et al. 1995), most of these EPSPs are considered to be monosynaptically evoked by stimulation of corticocortical neurons originating from the Mx. Furthermore, we confirmed these physiological data by a morphological method (Kakei et al. 1995, Fig. 5, right). Our results have shown that the stream of information between the Mx and the Px should be considered bidirectional with considerable diversified-convergence in both directions rather than a one-way stream from the Px to the Mx. In other words, the Px and the Mx are under concurrent mutual interaction.

Strong interactions between the Px, Mx, and VA-VL complex may be involved in the generation of cortical "beta" rhythms in unrestrained alert cats. Bouyer et al. (1987) found that "beta" electrocorticographic rhythms (40 Hz) developed in two distinct cortical foci during *motionless focused attention* in cats. One focus was located in motor areas 4γ and 6aβ, and the other focus was located in area 5a along the bank of the ansate sulcus. The two foci were separated by areas 1, 2 and 3, where "beta" rhythms were never recorded. Although functional significance and the generation mechanism of "beta" rhythm have not been fully understood, the tight reciprocal connection between the Px and the Mx, and the cerebellar input to both the Px and the Mx, might have some relevance to the localization and generation of this rhythm.

In addition, we should note that areas 5 and 7 of the cat have reciprocal connections not only with the motor or premotor areas but also with auditory or visual association areas such as suprasylvian fringe or area 20 (Heath and Jones 1971; Reinoso-Suárez 1984). It would not be surprising at all if corticocortical neurons in the Px that project to these sensory association areas were also to receive strong inputs from the Mx and the cerebellum. Consequently, the Px may be a center which enables influence from the motor systems to the sensory systems.

Functional Considerations of the Inputs from the Cerebellum and the Mx

Mountcastle et al. (1975) found in the monkey that most area 5 neurons activated by passive joint rotation were much more active during active movement of the joint. This movement-enhanced activity of the active joint neurons of area 5 results from a change that is wholly central. Such a change in the Px could be evoked by reafferent neuronal activity occurring parallel to efferent commands for movement, e.g., those from the Mx or the cerebellum. This corollary or parallel activity would facilitate the particular topographical regions of the Px (Mountcastle et al. 1975). In this way, internal feedback signals from the cerebellum and the Mx to the Px may function as a kind of filter that selectively enhances or depresses concomitant feedback signals resulting from ongoing movement. This idea reminds us of an experiment performed by Held and Hein (1963). They found that newborn kittens require self-produced movement, with its concurrent visual feedback, to develop visually guided paw placement. This observation suggests that motor commands and resultant sensory feedback signals are combined somewhere in the brain to develop visually guided paw placement. The Px may be a candidate for the site of such integration, since the input-output organization of the Px discussed above suggests that it is in an advantageous position to correlate internal signals from the Mx and the cerebellum with external sensory feedback signals.

Comparison with the Cerebellothalamoparietal Organization of the Monkey

Sasaki et al. (1976) first demonstrated cerebellar input to the Px in the monkey. They found that cerebellar-evoked field potentials in area 5 were elicited from the fastigial nucleus (FN), while stimulation of the DN and the IN was ineffective. Recent morphological studies have demonstrated reciprocal connections between the ventral lateral nucleus pars caudalis and pars postrema (VLc and VLps) of the thalamus and the Px (Kasdon and Jacobson 1978; Miyata and Sasaki 1983; Yeterian and Pandya 1985; Schmahmann and Pandya 1990). These thalamic regions receive strong cerebellar input (Stanton 1980; Asanuma et al. 1983; Anderson and Turner 1991; Yamamoto et al. 1992) not only from the FN but also from the DN and the IN. Although there has been no direct demonstration of the cerebellar input from the IN and the DN to the Px, we strongly suggest the existence of this input in the monkey for the following reasons. The VLc of the monkey has several features suggesting that this nucleus is a homologue of the dorsomedial part of the cat VA-VL complex: (1) it occupies the dorsal or dorsomedial corner of the ventrolateral nuclear mass of the thalamus, and it is bordered by the intralaminar nuclei medially and by the LP nucleus posteriorly; (2) it receives cerebellar inputs from the DN, IN and FN, although FN input to the dorsomedial part of the VA-VL complex of the cat is very weak, if at all (Kyuhou and Kawaguchi 1987); (3) it has abundant projection to the Px; (4) it also has strong projections to the premotor, motor and prefrontal cortices (Kievit and Kuypers 1977; Schell and Strick 1984; Wiesendanger and Wiesendanger 1985; Matelli et al. 1989; Nakano et al. 1992); and (5) it lacks projections to the primary somatosensory areas between the Px and the Mx (Jones et al. 1979; Darian-Smith and Darian-Smith 1993). On the other hand, a possible homologue of the ventrolateral part of the cat VA-VL complex, nucleus ventralis posterolateralis pars oralis (VPLo), shares some features with the ventrolateral part of the cat VA-VL complex, such as its location in the ventrolateral nuclear mass, massive cerebellar input, and strong projection to the motor cortex. Yet, it has only very minor projections to the Px (Miyata and Sasaki 1983; Schmahmann and Pandya 1990). In spite of some differences, all these similarities mentioned above suggest to us the hypothesis that a homologue of the cerebellothalamoparietal projection of the cat must exist in the monkey. To confirm this hypothesis, further direct experimental evidence has to be provided.

Cerebellar Inputs to the Human Posterior Parietal Cortex?

Recent studies, mainly based on clinical observations, suggested the idea of cerebellar contribution to *higher* brain functions (Keele and Ivry 1990; Schmahmann 1991; Akshoomoff and Courchesne 1992). Based on clinical observations of impaired shift of attention in autistic and cerebellar patients, Akshoomoff and Courchesne (1992) proposed that the neocerebellum is necessary for rapid shift in attention between sensory modalities. On the other hand, with the use of positron emission tomography or functional magnetic resonance imaging, it has been suggested that the cerebellum is activated not only in motor control functions (Grafton et al. 1992) but also in *cognitive* brain functions (Petersen et al. 1989; Kim et al. 1994; Griffiths et al. 1994). These observations might be related to the idea of cerebellar contribution to the function of the Px or other association cortices. Recent detailed studies of the human thalamus using various histochemical and immunohistochemical staining techniques have revealed remarkable similarity between the human and the monkey thalamus (Hirai and Jones 1989a,b). These authors identified regions in the human thalamus, dorsal parts of *the ventral lateral posterior nucleus* (VLp) that supposedly corresponded to VLc and VLps of the monkey. The question of this similarity also reflecting similarities of connectivity and function remains unanswered.

Acknowledgements. This research was supported by a grant from the Japanese Ministry of Education, Science and Culture for Scientific Research. We wish to thank Mr. M. Takada for his invaluable technical assistance.

References

Akshoomoff NA, Courchesne E (1992) A new role for the cerebellum in cognitive operations. Behav Neurosci 106:731–738

Anderson ME, Turner RS (1991) Activity of neurons in cerebellar-receiving and pallidal-receiving areas of the thalamus of the behaving monkey. J Neurophysiol 66:879–893

Asanuma C, Thach WT, Jones EG (1983) Anatomical evidence for segregated focal groupings of cells and their terminal ramifications in the cerebellothalamic pathway of the monkey. Brain Res Rev 5:267–297

Avendaño C, Rausell E, Perez-Aguilar D, Isorna S (1988) Organization of the association cortical afferent connections of area 5: a retrograde tracer study in the cat. J Comp Neurol 278:1–33

Bouyer JJ, Montaron MF, Vahnée JM, Albert MP, Rougeul A (1987) Anatomical localization of cortical beta rhythms in cat. Neuroscience 22:863–869

Brodal P (1978) The corticopontine projection in the rhesus monkey. Origin and principles of organization. Brain 101:251–283

Cavada C, Goldman-Rakic PS (1989) Posterior parietal cortex in rhesus monkey: II. Evidence for segregated corticocortical networks linking sensory and limbic areas with the frontal lobe. J Comp Neurol 287:422–445

Darian-Smith C, Darian-Smith I (1993) Thalamic projections to areas 3a, 3b, and 4 in the sensorimotor cortex of the mature and infant macaque monkey. J Comp Neurol 335:173–199

Dubner R, Rutledge LT (1964) Recording and analysis of converging input upon neurons in cat association cortex. J Neurophysiol 27:620–634

Dum RP, Strick PL (1991) Premotor areas: nodal points for parallel efferent systems involved in the central control of movement. In: Humphrey DR, Freund HJ (eds) Motor control: concepts and issues. Wiley, New York, pp 383–397

Fanardjian VV, Papoyan EV (1995) Electrophysiological evidence for a direct neuronal connection from the motor cortex to the parietal association cortex of the cat. Neurosci Lett 184:201–203

Fanardzhyan VV (1964) Recruiting reaction in cerebral cortex in relation to cerebellar stimulation. Fed Proc Transl Suppl 23:1156–1160

Felleman DJ, Van Essen DC (1991) Distributed hierarchical processing in the primate cerebral cortex. Cereb Cortex 1:1–47

Grafton ST, Mazziotta JC, Woods RP, Phelps ME (1992) Human functional anatomy of visually guided finger movements. Brain 115:565–587

Griffiths TD, Bench CJ, Frackowiak RSJ (1994) Human cortical areas selectively activated by apparent sound movement. Curr Biol 4:892–895

Heath CJ, Jones EG (1971) The anatomical organization of the suprasylvian gyrus of the cat. Springer, Berlin Heidelberg New York

Held R, Hein A (1963) Movement-produced stimulation in the development of visually guided behavior. J Comp Physiol Psychol 56:872–876

Hirai T, Jones EG (1989a) A new parcellation of the human thalamus on the basis of histochemical staining. Brain Res Rev 14:1–34

Hirai T, Jones EG (1989b) Distribution of tachykinin- and enkephalin-immunoreactive fibers in the human thalamus. Brain Res Rev 14:35–52

Hyvärinen J (1982) The parietal cortex of monkey and man. Springer, Berlin Heidelberg New York

Jones EG, Powell TPS (1968) The ipsilateral cortical connexions of the somatic sensory areas in the cat. Brain Res 9:71–94

Jones EG, Wise SP, Coulter JD (1979) Differential thalamic relationships of sensory-motor and parietal cortical fields in monkeys. J Comp Neurol 183:833–882

Kakei S, Shinoda Y (1990) Parietal projection of thalamocortical fibers from the ventroanterior-ventrolateral complex in the cat thalamus. Neurosci Lett 117:280–284

Kakei S, Yagi J, Wannier T, Na J, Shinoda Y (1995) Cerebellar and cerebral inputs to corticocortical and corticofugal neurons in areas 5 and 7 in the cat. J Neurophysiol 74:400–412

Kasdon DL, Jacobson S (1978) The thalamic afferents to the inferior parietal lobule of the rhesus monkey. J Comp Neurol 177:685–706

Kawamura K (1973) Corticocortical fiber connections of the cat cerebrum. II. The parietal region. Brain Res 51:23–40

Keele SW, Ivry R (1990) Does the cerebellum provide a common computation for diverse tasks? Ann N Y Acad Sci 608:179–211

Kievit J, Kuypers HGJM (1977) Organization of thalamo-cortical connexions to the frontal lobe in the rhesus monkey. Exp Brain Res 29:299–322

Kim SG, Ugurbil K, Strick PL (1994) Activation of a cerebellar output nucleus during cognitive processing. Science 265:949–951

Kyuhou S, Kawaguchi S (1987) Cerebellocerebral projection from the fastigial nucleus onto the frontal eye field and anterior ectosylvian visual area in the cat. J Comp Neurol 259:571–590

Matelli M, Luppino G, Fogassi L, Rizzolatti G (1989) Thalamic input to inferior area 6 and area 4 in the macaque monkey. J Comp Neurol 280:468–488

Mesulam MM (1978) Tetramethylbenzidine for horseradish peroxidase neurohistochemistry. A non-carcinogenic blue reaction-product with superior sensitivity for visualizing neuronal afferents and efferents. J Histochem Cytochem 26:106–117

Miyata M, Sasaki K (1983) HRP studies on thalamocortical neurons related to the cerebellocerebral projection in the monkey. Brain Res 274:213–224

Mizuno N, Mochizuki K, Akimoto C, Matsushima R, Sasaki K (1973) Projections from the parietal cortex to the brain stem nuclei in the cat, with special reference to the parietal cerebro-cerebellar system. J Comp Neurol 147:511–522

Mountcastle VB, Lynch JC, Georgopoulos A, Sakata H, Acuna C (1975) Posterior parietal association cortex of the monkey: command function for operations within extrapersonal space. J Neurophysiol 38:871–908

Nakano K, Takimoto T, Kayahara T, Takeuchi Y, Kobayashi Y (1980) Distribution of cerebellothalamic neurons projecting to the ventral nuclei of the thalamus: an HRP study in the cat. J Comp Neurol 194:427–439

Nakano K, Tokushige A, Kohno M, Hasegawa Y, Kayahara T, Sasaki K (1992) An autoradiographic study of cortical projections from motor thalamic nuclei in the macaque monkey. Neurosci Res 13:119–137

Oka H, Sasaki K, Matsuda Y, Yasuda T, Mizuno N (1975) Responses of pontocerebellar neurones to stimulation of the parietal association and the frontal motor cortices. Brain Res 93:399–407

Petersen SE, Fox PT, Posner MI, Mintun M, Raichle ME (1989) Positron emission tomographic studies of the processing of single words. J Cogn Neurosci 1:153–170

Reinoso-Suárez F (1984) Connectional patterns in parietotemporooccipital association cortex of the feline cerebral cortex. In: Reinoso-Suárez F, Ajmone-Marsan C (eds) Cortical integration. Raven, New York, pp 255–278

Robertson RT, Rinvik E (1973) The corticothalamic projections from parietal regions of the cerebral cortex. Experimental degeneration studies in the cat. Brain Res 51:61–79

Sasaki K, Kawaguchi S, Matsuda Y, Mizuno N (1972a) Electrophysiological studies on cerebello-cerebral projections in the cat. Exp Brain Res 16:75–88

Sasaki K, Matsuda Y, Kawaguchi S, Mizuno N (1972b) On the cerebello-thalamo-cerebral pathway for the parietal cortex. Exp Brain Res 16:89–103

Sasaki K, Oka H, Matsuda Y, Shimono T, Mizuno N (1975) Electrophysiological studies of the projections from the parietal association area to the cerebellar cortex. Exp Brain Res 23:91–102

Sasaki K, Kawaguchi S, Oka H, Sakai M, Mizuno N (1976) Electrophysiological studies on the cerebellocerebral projections in monkeys. Exp Brain Res 24:495–507

Schell GR, Strick PL (1984) The origin of thalamic inputs to the arcuate premotor and supplementary motor areas. J Neurosci 4:539–560

Schmahmann JD (1991) An emerging concept. Arch Neurol 48:1178–1187

Schmahmann JD, Pandya DN (1989) Anatomical investigation of projections to the basis pontis from posterior parietal association cortices in rhesus monkey. J Comp Neurol 289:53–73

Schmahmann JD, Pandya DN (1990) Anatomical investigation of projections from thalamus to posterior parietal cortex in the rhesus monkey: a WGA-HRP and fluorescent tracer study. J Comp Neurol 295:299–326

Seltzer B, Pandya DN (1980) Converging visual and somatic sensory cortical input to the intraparietal sulcus of the rhesus monkey. Brain Res 192:339–351

Shinoda Y, Kakei S (1989) Distribution of terminals of thalamocortical fibers originating from the ventrolateral nucleus of the cat thalamus. Neurosci Lett 96:163–167

Shinoda Y, Kano M, Futami T (1985a) Synaptic organization of the cerebello-thalamo-cerebral pathway in the cat. I. Projection of individual cerebellar nuclei to single pyramidal tract neurons in areas 4 and 6. Neurosci Res 2:133–156

Shinoda Y, Futami T, Kano M (1985b) Synaptic organization of the cerebello-thalamo-cerebral pathway in the cat. II. Input-output organization of single thalamocortical neurons in the ventrolateral thalamus. Neurosci Res 2:157–180

Shinoda Y, Ohgaki T, Sugiuchi Y, Futami T (1992) Morphology of single medial vestibulospinal tract axons in the upper cervical spinal cord of the cat. J Comp Neurol 316:151–172

Shinoda Y, Kakei S, Futami T, Wannier T (1993) Thalamocortical organization in the cerebello-thalamo-cortical system. Cereb Cortex 3:421–429

Stanton GB (1980) Topographical organization of ascending cerebellar projections from the dentate and interposed nuclei in *Macaca mulatta:* an anterograde degeneration study. J Comp Neurol 190:699–731

Stanton GB, Orr A (1985) [^{3}H] choline labeling of cerebellothalamic neurons with observations on the cerebello-thalamo-parietal pathway in cats. Brain Res 335:237–243

Stoney SD, Thompson WD, Asanuma H (1968) Excitation of pyramidal tract cells by intracortical microstimulation: effective extent of stimulating current. J Neurophysiol 31:659–669

Strick PL, Kim CC (1978) Input to primate motor cortex from posterior parietal cortex (area 5). I. Demonstration by retrograde transport. Brain Res 157:325–330

Updyke BV (1983) A reevaluation of the functional organization and cytoarchitecture of the feline lateral posterior complex, with observations on adjoining cell groups. J Comp Neurol 219:143–181

Wannier T, Kakei S, Shinoda Y (1992) Two modes of cerebellar input to the parietal cortex in the cat. Exp Brain Res 90:241–252

Waters RS, Favorov O, Mori A, Asanuma H (1982) Patterns of projection and physiological properties of cortico-cortical connections from the posterior bank of the ansate sulcus to the motor cortex, area 4γ, in the cat. Exp Brain Res 48:335–344

Wiesendanger R, Wiesendanger M (1985) The thalamic connections with medial area 6 (supplementary motor cortex) in the monkey (*Macaca fascicularis*). Exp Brain Res 59:91–104

Yamamoto T, Oka H (1993) The mode of cerebellar activation of pyramidal neurons in the cat parietal cortex (areas 5 and 7): an intracellular HRP study. Neurosci Res 18:129–142

Yamamoto T, Samejima A, Oka H (1987) Morphological features of layer V pyramidal neurons in the cat parietal cortex: an intracellular HRP study. J Comp Neurol 265:380–390

Yamamoto T, Yoshida K, Yoshikawa H, Kishimoto Y, Oka H (1992) The mediodorsal nucleus is one of the thalamic relays of the cerebellocerebral responses to the frontal association cortex in the monkey: horseradish peroxidase and fluorescent dye double staining study. Brain Res 579:315–320

Yeterian EH, Pandya DN (1985) Corticothalamic connections of the posterior parietal cortex in the rhesus monkey. J Comp Neurol 237:408–426

Direct Bilateral Cortical Projections to the Vestibular Complex in Macaque Monkey

S. FAUGIER-GRIMAUD[1], C. BALEYDIER[1], M. MAGNIN[2], and M. JEANNEROD[1]

[1]INSERM U 94, Vision et Motricité, Bron, France
[2]INSERM U 371, Cerveau et Vision, Bron, France

Introduction

Defining he role of the cerebral cortex in vestibular function is a long standing physiological problem. It has often been postulated that integration of vestibular input with other sensory inputs (e.g. visual, somatosensory) is a prerequisite for spatial orientation. Neurons fulfilling this criterion of plurimodality were recorded in several cortical areas, and particularly in monkey posterior parietal cortex. With regard to the visual modality, neurons responding to both natural vestibular stimulation and moving visual (optokinetic) stimuli were found at the lower end of the intraparietal sulcus (Büttner and Buettner 1978) in a small area next to the mouth region in area 2 (hence designated as area 2v by Schwartz and Fredrickson 1971). Visual vestibular neurons were also found in the posterolateral part of the inferior parietal lobule (area 7a, Kawano et al. 1980, 1984). Finally, another focus with similar properties was identified at the parietoinsular junction (parietoinsular vestibular cortex of Grüsser et al. 1982 and Akbarian et al. 1988). This area might correspond in man to the cortical zone located in the superior part of the temporal lobe near the temporoparietal border, where regional cerebral blood flow was found to increase during caloric vestibular stimulation (Friberg et al. 1985, Bottini et al. 1994). Convergence of vestibular and somatosensory (mostly proprioceptive) inputs was also observed at the single neuron level, not only in area 2v (Schwartz and Fredrickson 1971, in squirrel monkeys), but also in area 3a at the depth of the central sulcus (Odkvist et al. 1974). Area 3a had been classified early as a "primary vestibular field" by Hassler (1964).

Lesion experiments suggest that at least some of these areas influenced by vestibular input might in turn control vestibular functions. Ventre and Faugier-Grimaud (1986) found that lesions restricted to the surface of the inferior parietal lobule and to the upper part of the anterior bank of the superior temporal sulcus produced a strong vestibular imbalance. Following the lesion, spontaneous nystagmus was present for more than one week; the vestibuloocular response remained asymmetrical for about one month (see also Takemori et al. 1979, for a similar effect of parietal lesions in man). In an attempt to describe the descending

In: Parietal Lobe Contributions to Orientation in 3D Space (1997). P. Thier and H.-O. Karnath (eds). Springer-Verlag, Heidelberg.

posterior parietal projections responsible for these effects, Faugier-Grimaud and Ventre (1989) injected anterograde tracers (horseradish peroxidase and tritiated aminoacids) in the inferior parietal lobule in four monkeys. These authors observed descending projections to vestibular complex, originating from the retroinsular cortex and the medial superior temporal area (area MST).

More recently, Akbarian et al. (1994), using a retrograde tracer injected into the vestibular nuclei, found that, in addition to the parietal cortex, projections to the vestibular complex also originate from parts of areas 3a and 6. Our study, carried out with larger injections of retrograde tracer into the vestibular complex, not only confirms but also expand the Akbarian's findings, showing that the prefrontal cortex also projects to the vestibular complex.

Material and Methods

Experiments were conducted on three adult cynomolgus monkeys (Macaca fascicularis). Injections into the vestibular complex were performed according to the following procedure. Anesthesia was induced by ketamine hydrochloride (20mg/kg IM) and maintained during surgery by IV injections of Sodium pentobarbital. After lateral reflection of neck muscles, a major craniotomy was performed caudally to the occipital lobe. Once the dura was opened, the cerebellum was lifted up from back to front by progressive insertion of cotton pellets carefully placed on both sides of the floor of the fourth ventricle. Rostrally and laterally, the cerebellar peduncles became visible and provided a landmark for the location of the rostral part of the vestibular complex.

The tracer injected was either free horseradish peroxidase (HRP, 20%) or wheat-germ agglutinin HRP (WGA-HRP) plus unconjugated HRP in equal volumes of 10% solutions. Although WGA-HRP appears to be 20 to 30 times as potent as free HRP, we used free HRP in one animal. This was done to control our results concerning transcellular transport, which is unlikely to occur with free HRP (Mesulam 1982).

Injections were performed with a glass micropipette (tip diameter 60 um) sealed to the needle of a Hamilton syringe. The tracer was delivered in six injections of 0.05 microliter each.

Following a survival period of 48 h, the animals were given a lethal dose of Sodium pentobarbital and perfused transcardiacally with 1l of saline followed by 2l of fixative containing 1.25% glutaraldehyde plus 1% paraformaldehyde in 0.1 M phosphate buffer and a subsequent post-rinse consisting of 0.1 M buffer plus 10% sucrose. Brains were stored in sucrose buffer overnight and frozen sectioned at 60 um. A series of free floating sections were processed according to the low artifact tetramethylbenzidine (TMB) method recommended by Gibson et al. (1984). These sections were then mounted, and some were counterstained with neutral red. Alternate sections were not processed with TMB but stained either for myelin with the method of Gallyas (1979) or for cytochrome oxidase according to Wong-Riley (1979).

Spatial distribution of labeled cells within each section was transferred to paper with the aid of an XY plotter electronically coupled to the microscope stage. Architectonic boundaries were identified by projecting adjacent sections stained for Nissl, myelin or cytochrome oxidase onto the drawings representing the injection sites and the distribution of the label. For nomenclature of vestibular nuclei, we used the atlas of the brainstem by Smith et al. (1972).

Result

Injection sites

In accordance with Mesulam (1982), we considered the opaque and purple area immediately surrounding the point of tracer deposit to be the effective injection site, whereas the paler halo surrounding this area should represent the outcome of local transport.

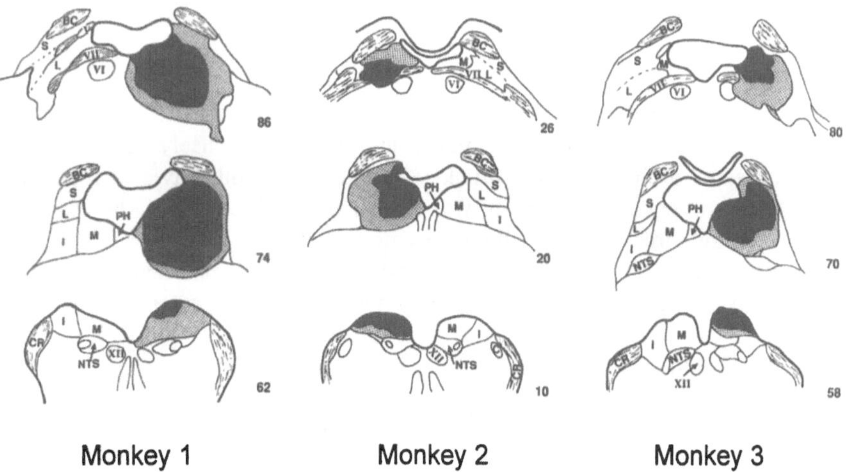

Fig. 1. Injection sites in the vestibular complex of monkeys 1, 2 and 3, represented on drawings of frontal sections. *Dark area*, effective injected site; *gray area*, outcome of local transport of the tracer (see text). *Numbers* indicate the position of the section in the series. *BC*, brachium conjonctivum; *CR*, corpus restiformis; *I*, inferior vestibular nucleus; *L*, lateral vestibular nucleus; *M*, medial vestibular nucleus; *NTS*, nucleus tractus solitari; *PH*, nucleus prepositus hypoglossi; *S*, superior vestibular nucleus

In the three animals, the injections were found to involve most of the anteroposterior extent of the vestibular complex (Fig.1). Rostrally, all the injection sites reached the level of the genu of the seventh nerve. In monkeys 1 and 3, however, they abutted onto the mesencephalic root of the fifth nerve, which was completely spared in monkey 2. Caudally, injections extended to the very end of the medial vestibular nucleus.

Dorsally, the areas of injection were bordered by the brachium conjonctivum which was never included in the injection site. Ventrally, all the injections involved the lower border of the vestibular complex. In monkeys 2 and 3, they slightly encroached on the nucleus abducens and the contiguous reticular formation. In monkey 1, the injection site included the abducens nucleus, a large part of the adjacent reticular formation and the nuclei of the trigeminal nerve (main, mesencephalic, motor and spinal).

In these three cases, all vestibular nuclei were injected but to different extents. The larger injection was found in monkey 1 where the whole vestibular complex was entirely injected except for a small area in the dorsal part of the superior vestibular nucleus. In the other two animals, the injections were about the same size and spared the lateral part of the inferior vestibular nucleus. In monkey 3, large portions of the superior vestibular nucleus and of the ventral part of the lateral vestibular nucleus were also spared.

Cortical labeling

The present study deals exclusively with retrograde labeling of cortical neurons. Although marked cells were less numerous and less intensely labeled in the animal injected with free HRP (monkey 3) than in the other two injected with WGA-HRP, the pattern of labeling was similar in the three cases. Therefore, the results from the three animals will be presented together.

Three main cortical territories were found to contain labeled neurons (Fig. 2). One was located at the parieto-temporal junction and more anteriorly along the lateral sulcus, another one along the central sulcus and in the cingulate sulcus, and the third one in the frontal cortex.

It is worth noting that, in these three territories, labeled neurons were found in both hemispheres, ipsilateral and contralateral to the injection sites, and at nearly equivalent amounts.

Parietotemporal Region and Lateral Sulcus. The main zones of labeling consistently present in the three monkeys were found in the depth of the caudal part of the superior temporal sulcus (STS) and at the depth of the lateral sulcus (Fig. 2).

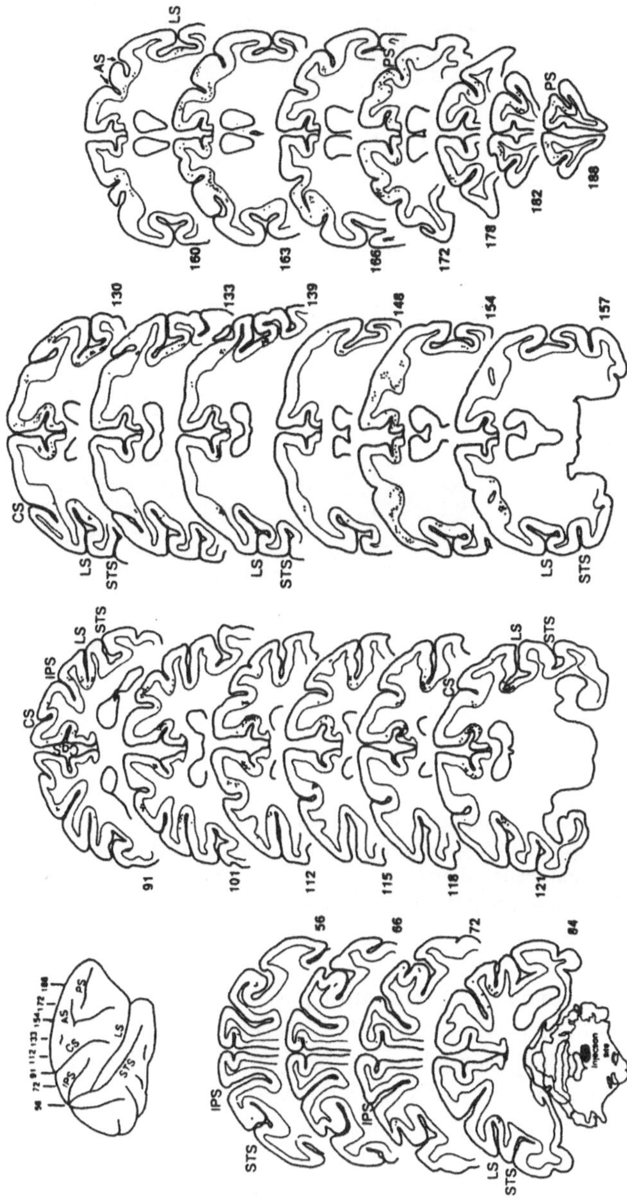

Fig. 2. Distribution of retrogradely labeled neurons (*black dots*) in the cerebral cortex after injection of tracer (*in black*) in the vestibular complex of monkey 1. As shown in the lateral view of the brain, drawings of frontal sections are presented from back to front, the *numbers* indicating the position of the section in the series. *AS*, arcuate sulcus; *CS*, central sulcus; *CgS*, cingulate sulcus; *IPS*, intraparietal sulcus; *LS*, lateral sulcus; *PS*, principal sulcus; *STS*, superior temporal sulcus

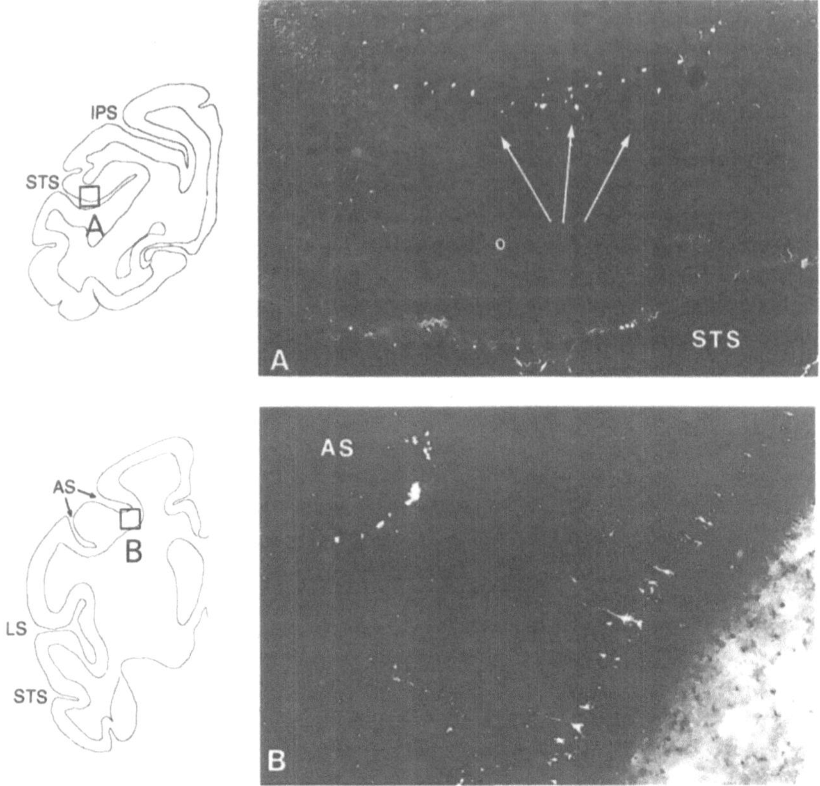

Fig. 3. Dark field microphotographs showing labeled neurons in the medial superior temporal area (area MST) located in the dorsal bank of the posterior part of *STS* (**A** corresponding to insert *A* of the frontal section drawing), and in the deep layers of cortex of the arcuate sulcus (**B** corresponding to insert *B* of the frontal section drawing). Abbreviations as in Fig. 2

Area MST in caudal STS. The most posterior labeling was observed in the dorsal bank of the posterior part of the STS (Fig. 2, levels 56-72; Fig. 3A). This area extended caudally along 2 mm to the level of the STS lateral sulcus junction. This part of the cortex corresponds to the MST area described by Desimone and Ungerleider (1986). In monkey 2, labeled neurons, which were more numerous in this area than they were in the other two animals, were distributed almost across the entire extent of area MST. In the other two monkeys, marked cells were restricted to that part of the MST area that, on myelin stained sections, exhibited a dense network of myelinated fibers throughout the whole of the cortex, and was thus identified as the dense myelinated zone by Ungerleider and Desimone (1986a). The labeling consisted of a group of medium-sized cells lined up in the deep cortical layers of area MST.

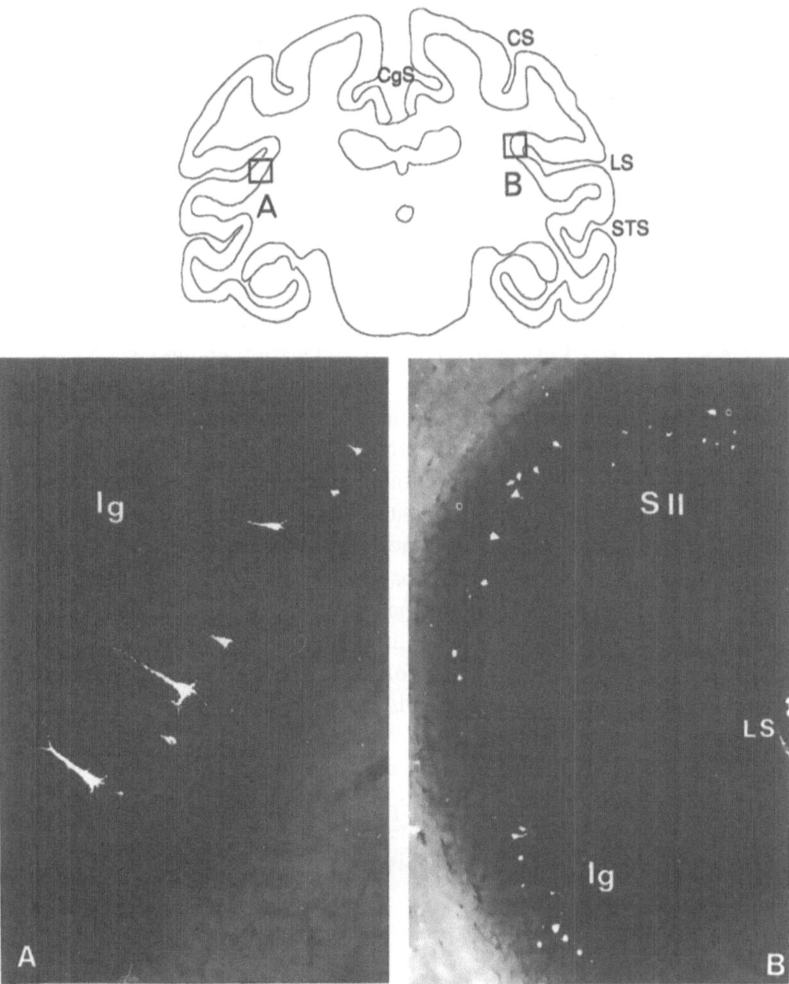

Fig. 4. Dark field microphotographs of labeled pyramidal cells in the granular insular cortex *Ig* (**A** corresponding to *insert A* of the frontal section drawing), and in *area SII* and *Ig* (**B** corresponding to *insert B* of the frontal section drawing). Abbreviations as in Fig. 2 and *Ig*, granular insular cortex

Retroinsular Cortex (Ri) in Caudal Lateral Sulcus. Further labeling was observed more rostrally, in the depth of the very caudal end of the lateral sulcus (Fig. 2 levels 84-91). It extended along 2 mm. Labeled cells were medium-sized neurons of the pyramidal type and were regularly spaced in layer 5. On the basis of cytoarchitectural criteria, this zone of the cortex was identified by Pandya and Sanides (1973), Roberts and Akert (1963) and Jones and Burton (1976) as the retroinsular cortex (Ri). However, in the folding of the sulcus, the cytoarchitectural features of this cortex become so subtle that Burton and Jones (1976) instead identified Ri on the basis of thalamocortical connections. In fact, it is mainly because of its location at the bottom of the caudal end of the lateral sulcus that we identified this area as the Ri cortex.

Area SII in Caudal Lateral Sulcus. Labeling was also detected more rostrally in the lateral sulcus, about 2.5 mm in front of the Ri labeling described above (Fig. 2 levels 115-118; Fig. 4B). This labeling involved the bottom of the lateral sulcus just caudal to the insula proper; dorsally, it overlapped slightly on the very medial part of the inner parietal operculum, and ventrally, it abutted the dorsal insular cortex. The labeled zone corresponds to the most medial part of area SII. Marked cells were medium-sized pyramidal neurons and were found in layer 5.

Granular Insular Cortex. Proceeding forward, a fourth zone of labeling was observed in the fundus of the lateral sulcus, extending along 6 mm length in the dorsal part and the floor of the insula proper (Fig. 2 levels 121-154; Fig. 4 A, B). The majority of marked neurons were found in the part of the insular cortex where layers 2 and 4 are both clearly granulated and easily demarcated from adjacent layers, i.e., in the granular insular cortex (Ig). More anteriorly in the insula, in monkey 1 only, sparse neurons were observed in the dysgranular cortex which, in alternate sections treated for histochemistry, had a higher density of acetylcholinesterase in layer 4 than in the granular cortex (see Jones and Burton 1976 and Mesulam and Mufson 1982 for identification of the insular cortices). In the granular insular cortex, labeled cells were large pyramidal neurons regularly spaced along layer 5.

Finally, sparse labeled neurons were observed in the banks of the intraparietal sulcus and on the gyral surface of the superior parietal lobule in monkey 1 only. A small patch of labeling was found at the tip of the intraparietal sulcus in the two monkeys injected with WGA-HRP. This zone corresponds to the area defined as area 2v.

Central Sulcus and Cingulate Sulcus. In the three animals, a stripe of labeled cells was found to underline the bottom and medial bank of the central sulcus and to occupy the middle third of its mediolateral length (Fig. 2 levels 112-121; Fig. 5 A and C). Another labeled zone was found in the mesial wall of the hemisphere and depth of the cingulate sulcus (Fig. 2 levels 101-121; Fig. 5, B). In both areas, labeled neurons were of the pyramidal type and located in layer 5, in continuity with the row of giant pyramidal cells of adjacent area 4 and clearly identifiable in Nissl sections (Fig. 5) and sections stained with cytochromoxidase. These two labeled regions were identified as belonging to area 3a. Area 3a is described as a thin band of transitional cortex that borders the agranular motor cortex of area 4 in the medial bank of the central sulcus and extends into the internal aspect of the hemisphere down to the cingulate sulcus (Burton and Jones 1976).

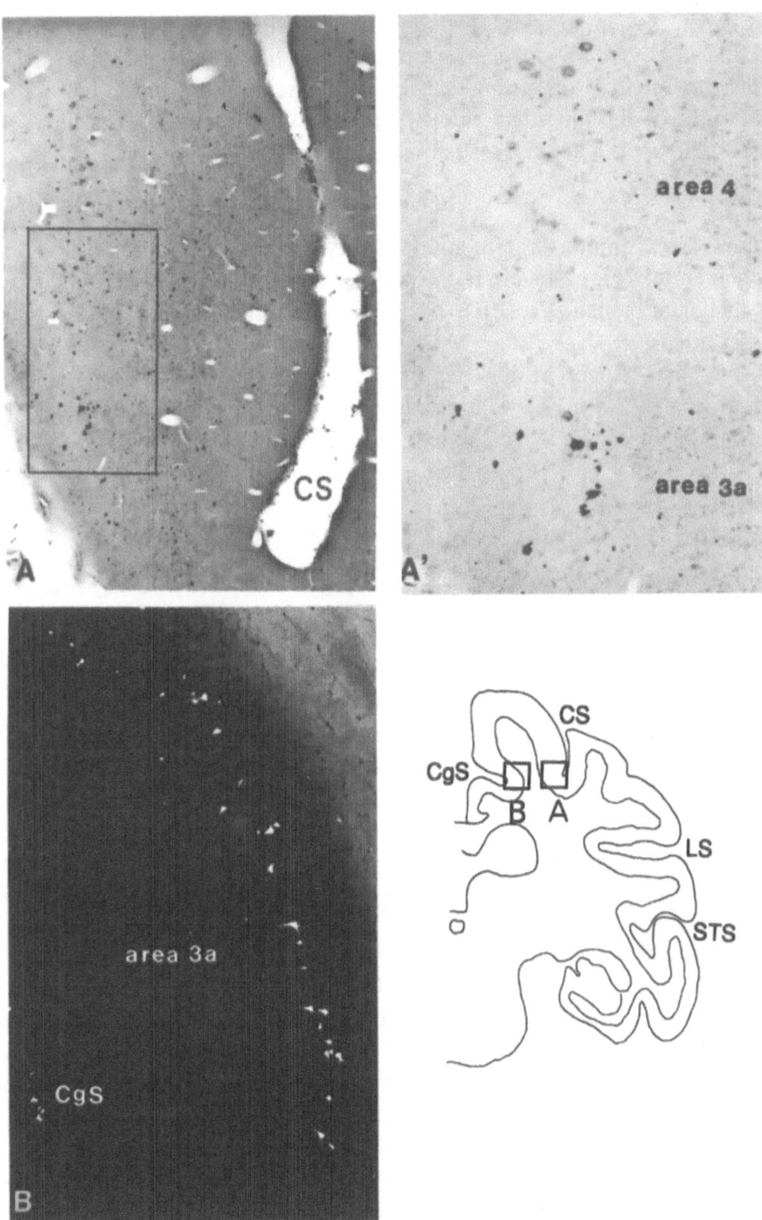

Fig. 5. Light field microphotographs (corresponding to *insert A* of the frontal section drawing) of *area 3a* in the depth of *CS*. The row of horseradish peroxidase (HRP)- labeled neurons in *area 3a* is in continuity with the Nissl stained giant pyramidal cells of *area 4*. *A'* is an enlarged view of the region outlined on microphotograph **A**. Darkfield microphotograph (corresponding to *insert B* of the frontal section drawing) of *area 3a* in the dorsal bank of *CgS* showing aligned HRP positive cells. Abbreviations as in Fig.2

More rostrally, in monkey 1 only, at the level where area 4 is no longer detectable in the interhemispheric aspect of the hemisphere, a patch of labeled neurons was found in the mesial surface of the hemisphere and in the dorsal and ventral bank of the cingulate sulcus (Fig. 2 level 130-148). These loci should correspond to areas 6 and 23, respectively.

Frontal Cortex. In the three monkeys, many labeled cells were observed in the prefrontal cortex (Fig. 2 levels 154-188), in the curve of the arcuate sulcus and all along its dorsal branch (Fig. 3B). The labeling extended along both banks of the principal sulcus.

On the inner aspect of the hemisphere, marked neurons were consistently observed in area 24, in the ventral bank of the anterior cingulate sulcus (Fig. 2 level 154-172).

Discussion

In the present study, the use of a retrograde tracer made it possible to precisely identify the cortical regions sending direct input to the vestibular complex. These areas, which for the most part are hidden in the banks of the sulci, are depicted in Fig.6 in a semischematic diagram representing an unfolded cortex of monkey brain. Some of them are located at the temporo-parietal junction : area MST in the dorsal bank of the superior temporal sulcus, area Ri in the caudal part of the lateral sulcus, and areas SII and Ig in the dorsal bank and floor of the lateral sulcus, respectively; others are located in the depths of the central sulcus (area 3a), and in the concavity of the arcuate sulcus and the cortex of both banks of the principal sulcus.

In the literature, only a few studies have described direct projections from cortex to the vestibular complex. After a brief mention of the existence of a direct projection from parietal cortex to vestibular nuclei by Kuypers and Lawrence (1967), Faugier-Grimaud and Ventre (1989) gave a description of this connection, based on an anterograde tracing technique. Their injection was performed in the posterior parietal cortex but more precisely involved areas MST and Ri and resulted in the labeling of the vestibular complex on both sides.

Recently, Grüsser's group (Akbarian et al. 1994) by using the same method (HRP injections into the vestibular nuclei), described projections from the cerebral cortex to the brainstem vestibular nuclei in macaque. Our results are basically similar to those reported by these authors. The projections are bilateral. They originate from areas SII, Ri and Ig, these three regions being mentioned as PIVC (parietoinsular vestibular cortex) by the authors, and from part of area 3a along

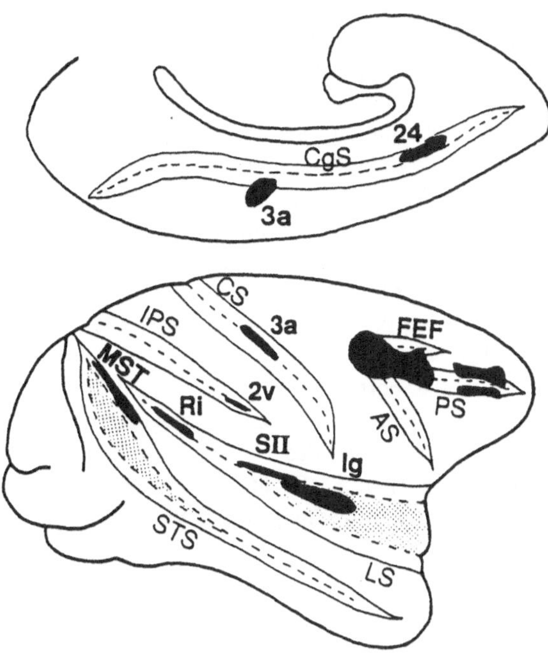

Fig. 6. Regions of cortex sending direct input to the vestibular complex are shown (*in black*) on a semischematic diagram of an unfolded cortex of monkey brain. These are (1) areas involved in somatosensory processing (areas *SII*, *Ri*, *Ig* and *3a*), and (2) areas involved in the control of eye movement and in visuomotor processing (medial superior temporal area, *MST* and areas surrounding the arcuate and principal sulci). Abbreviations as in Fig. 2. FEF, frontal eye field; *Ig*, granular insular cortex; *Ri*, retroinsular cortex

the central sulcus and from area MST (Akbarian et al. named this area T3, in accordance with Jones and Burton 1976). In the two studies, the projections observed from area 2v are quantitatively very poor.

However, some differences are to be mentioned. Apart from the same label in a part of area 3a hidden in the middle of the curve of the central sulcus, on the convexity, we found a very clear additional projection originatingfrom a small part of area 3a located in the mesial surface of the hemisphere. This is interesting if we consider the somatotopy of area 3a (see below). We found the projection originating from area 6 (mesial wall) in one animal only.

The main difference between the results of Akbarian et al. (1994) and ours is the significant labeling we observed along the prefrontal cortex. The difference in the injection sites could explain such a discrepancy. Our injections always involved the nucleus prepositus hypoglossi. Those of Akbarian et al. are mostly situated lateral to the midline except in their case M4 which received HRP in the nucleus prepositus hypoglossi and in which they observed labeling in the frontal eye field. The fact that Leichnetz and Goldberg as well as Stanton et al. described in 1988 such direct projections from the frontal eye field (FEF) to the nucleus prepositus hypoglossi made us confident in this result. Moreover, Akbarian et al. (1994) also suggested that the labeling they observed in the prefrontal cortex could be due to a HRP outcome into the reticular formation. We are skeptic about this statement : in a study by Cowie et al. (1994) in which HRP was injected into the reticular formation ventral to the vestibular complex, the authors found cortical labeling in areas 4 and 6 of the precentral gyrus, but never in the prefrontal cortex. Although indirect, another argument suggested by an experiment in man is worth noting : Bottini et al. (1994), using positron emission tomography (PET), reported that caloric vestibular stimulation induced activation of the anterior cingulate cortex.

We can notice that the cortical regions described in the present paper as projecting to the vestibular complex may be categorized into two main groups according to their functional properties : 1) areas SII, Ri, Ig, and 3a, involved in somatosensory processing; 2) area MST and areas surrounding the arcuate sulcus, involved in the control of eye movements and in visuomotor processing.

Somatosensory Areas

Grüsser et al. (1990 a,b) described the presence of vestibular neurons in areas Ri and Ig (their PIVC; 50% of the recorded neurons), and in area 3a (10%-20%). They also noticed that "with very few exceptions, all vestibular neurones tested responded to visual and somatosensory stimulation, therefore being classified as polymodal vestibular units". According to Robinson and Burton (1980a), the vast majority of neurons in areas SII (93%), Ig (76%) and Ri (74%) are activated by somatosensory stimuli. It is noteworthy that the parts of cortical areas that we found to project to the vestibular complex are involved in the axial representation of the body. From the somatotopic maps established by these authors (Robinson and Burton 1980b), the caudal part of area Ri, in which we found labeled cells after HRP injections into the vestibular nuclei, contains the representation of the tail, foot and trunk. The labeled part of SII, which we found to be located not on the inner bank of the parietal operculum but in the fold of the cortex adjacent to area Ig, corresponds to the foot and upper and lower trunk areas. Few data are available concerning the somatotopy in area Ig, except that some of the neurons recorded by Robinson and Burton (1980b) had receptive fields corresponding to the trunk or to the entire body. The neurons described by Grüsser's group (Grüsser et al. 1982, Akbarian et al. 1988) at the depth of the caudal half of the sylvian fissure responded not only to dynamic vestibular stimulation, but also to proprioceptive stimulation of deep muscle receptors of the neck and of the shoulder girdle.

It is still a matter of debate as to whether area 3a should be classified within the motor, sensory or transitional cortex. Hassler (1964) suggested that area 3a was a vestibular area, and it was found early on to be connected with the vestibular nuclei. Odkvist et al. (1974), studying the squirrel monkey, recorded responses to electrical stimulation of the vestibular nerve in a restricted part of area 3a corresponding to neck representation. This area is also known to be a preferential receiving zone for impulses conveyed by group Ia fibers from muscle spindles and tendon organs. In the present study, two zones of area 3a were found to contain labeled cells after HRP injections into the vestibular nuclei; one is situated halfway along the length of the central sulcus on the convexity, and the other is on the inner side of the hemisphere. If area 3a is organized along the same somatotopic pattern as the adjacent area 3b, which seems to be the case according to Nelson et al. (1980) and Pons et al. (1985), these two zones should correspond to the trunk-neck-shoulder area and to the tail area, respectively. The fact that the labeled cortical zones were mostly linked to representations of axial and proximal muscles and joints seems to be confirmed by the complete lack of labeled cells in the large parts of areas SII and 3a devoted to the representation of digits.

These sets of somatosensory areas, concerning axial parts of the body and showing direct projections to the vestibular complex, may participate in regulation of posture.

Visual and Oculomotor Areas

A consistent number of labeled neurons were observed in a specific area of the parietal lobe, area MST, located in the medial bank of the superior temporal sulcus, and in the arcuate cortex or FEF. Although these areas are all involved in visual-oculomotor activity, their contribution to the generation of eye movements is not equivalent. Area MST is primarily involved in visual motion processing and is related to smooth-pursuit eye movements (SPEM). MST neurons discharge during SPEM (Wurtz and Newsome 1985) and many of them are directionally selective (Desimone and Ungerleider 1986). Dürsteler and Wurtz (1988) showed that lesions of area MST impair SPEM. Area MST has a key position in the succession of the different steps leading to SPEM: it is interconnected with area MT that encodes the direction and velocity of targets moving within the contralateral visual field (Ungerleider and Desimone 1986b); it sends descending projections to the dorsal lateral pontine nuclei which, through the vestibular cerebellum and the vestibular nuclei, have access to the oculomotor nuclei (see Tusa and Zee 1989 for a review). In addition, area MST is the origin of direct afferent fibers to the vestibular nuclei (this study; Faugier-Grimaud and Ventre 1989).

The cortex located in the concavity of the arcuate sulcus (or FEF), has been known for many years (since Ferrier 1875) to be involved in the generation of saccades. Most of the prearcuate neurons have saccade-related activity, and many of them have visual receptive fields (Mohler et al. 1973). Neuronal discharge often precedes visually guided and purposive saccadic eye movements (Goldberg and Bruce 1986). According to several studies (e.g., Leichnetz and Goldberg 1988 and Stanton et al. 1988) the FEF has direct projections to the prepositus hypoglossi nucleus which is considered by Canon and Robinson (1986) as the "neural integrator of the oculomotor system". One may thus speculate that this cortical area is involved in orientation behavior and participates in generating purposive saccades toward visual targets.

Oculomotor commands are generated for SPEM in area MST and for saccades in the FEF. Direct transmission of these command signals to the vestibular nuclei could have important implications in visual vestibular interactions. It could, for example, represent the anatomical basis of vestibuloocular reflex suppression during fixation of a visual target.

A Functional Hypothesis

A possible interpretation of the role of these direct cortico-vestibular projections seems to emerge from the pattern of afferent connections and the functional specialization of the labeled areas. These areas are able to integrate signals concerning trunk position, head position (of both vestibular and neck-proprioceptive origin) and eye position. The general idea is that the influence they exert on the vestibular nuclei could contribute to the determination of an egocentric reference used for directional coding of movements and spatial orientation (Ventre et al 1984; Jeannerod and Biguer 1987).

Under normal conditions, in which the sensory inputs (vestibular, somatosensory, visual) contributing to the activity of the brain areas involved are distributed symmetrically, the egocentric reference is aligned with body midline and roughly splits personal and extrapersonal space into two equal halves. This may not be the case when this distribution becomes asymmetrical, for example, during prolonged unilateral stimulation of one of the contributing sensory inputs. Indeed, unilateral vestibular stimulation (caloric or rotational), prolonged head or gaze deviation, unidirectional optokinetic stimulation and visual exposure to laterally displacing prisms are situations which create systematic deviation of reaching movements or locomotion or deviation of the perceived direction of the egocenter (tested in man by pointing "straight ahead"; for a review see Howard, 1982). Similar effects can be obtained by vibratory stimulation of neck muscles (Biguer et al. 1988; Taylor and McCloskey 1991).

Following the same line of reasoning, unilateral lesion of neural structures where these inputs are processed should deviate egocentric reference in one direction and produce a directional bias in spatially oriented behavior. This prediction is confirmed by a large body of experimental data. With regard to cortical structures in monkeys, unilateral lesion of the FEF produces ipsilesional forced circling (Kennard and Ectors 1938); lesion of posterior parietal areas produces a reaching bias toward the lesion side (Faugier-Grimaud et al. 1978), as well as spontaneous nystagmus and asymmetrical vestibuloocular responses (Ventre and Faugier-Grimaud 1986). In man, unilateral parietal lesion also often produces an ipsilesional reaching bias (Ratcliff and Davies-Jones 1972; Allison et al. 1969; Jeannerod 1986; Tropper et al. 1991).

Cortical lesions are not the only ones to produce a directional bias: the syndrome resulting from vestibular lesions offers another well known example. This suggests that the position of the egocentric reference might be controlled by a broader system including cortical and subcortical structures. The specific contribution of the cortical step within this system is illustrated by recent work on the phenomenon of unilateral spatial neglect in man. Patients with posterior parietal lesions, especially those with lesions located in the right hemisphere, may present inattention (neglect) to stimuli arising from contralesional personal and extrapersonal space. These patients simultaneously present deviation of the perceived direction of their egocenter toward the non neglected side (Heilman et al. 1983). This observation indicates that the same posterior parietal lesion can simultaneously produce an attentional bias (due to alteration of cortical mechanisms related to attention) and a directional bias (due to an asymmetry of cortical influence on the vestibular nuclei). Other data, however, indicate that the relationship between deviation of the egocentric reference and neglect is more complex. It has been shown that caloric vestibular stimulation in patients presenting unilateral neglect, when applied on the appropriate side (e.g., cold water in the left ear if the neglected side is to the left), not only (transiently) restores the position of egocentric reference, but also simultaneously reduces neglect (Rubens 1985; Cappa et al. 1987; Bisiach et al. 1991; Rode et al. 1992; Vallar et al. 1993). Similar effects have been obtained by vibratory stimulation of the neck muscles on the appropriate side (Karnath et al. 1993), and optokinetic stimulation in the appropriate direction (Pizzamiglio et al. 1990).

Directional biases and neglect produced by cortical lesions are usually rapidly overcome. This may be explained by the distribution of cortical control among the vestibular nuclei. Also, the fact that corticovestibular projections are bilateral (as shown by bilateral labeling of cortical neurons after HRP injections into the vestibular nuclei) is a direct explanation of recovery following unilateral lesion.

Acknowledgements. We are grateful to Noèlle Boyer and Sandrine Richard for histology, to Pascale Giroud for the graphic illustrations and to Jean-Louis Borach for the microphotographs.

References

Akbarian S, Grüsser OJ, Guldin WO (1994) Corticofugal connections between the cerebral cortex and brainstem vestibular nuclei in the macaque monkey. J Comp Neurol 339:421–437

Akbarian S, Berndl K, Grüsser OJ , Guldin W, Pause M, Schreiter U (1988) Responses of single neurons in the parietoinsular vestibular cortex of primates. Ann New York Acad Sci 545:187–202

Allison RS, Hurwitz LJ, Graham White J, Wilmot TJ (1969) A follow-up study of a patient with Balint's syndrome. Neuropsychologia 7:319–333

Barbas H , Mesulam MM (1985) Cortical afferent input to the principalis region of the rhesus monkey. Neuroscience 3:619–637

Biguer B, Donaldson IML, Hein A, Jeannerod M (1988) Neck muscle vibration modifies the representation of visual motion and direction in man. Brain 111:1405–1424

Bisiach E, Rusconi ML, Vallar G (1991) Remission of somatoparaphenic delusion through vestibular stimulation. Neuropsychologia 29:1029–1031

Bottini G, Sterzi R, Paulesu E, Vallar G, Cappa SF, Erminio F, Passingham RE, Frith CD, Frackowiak G (1994) Identification of the central vestibular projections in man : a positron emission tomography activation study. Exp Brain Res 99:164–169

Burton H, Jones EG (1976) The posterior thalamic region and its cortical projection in new world and old world monkeys. J Comp Neurol 168:249–302

Büttner U, Buettner UW (1978) Parietal cortex (2v) neuronal activity in the alert monkey during natural vestibular and optokinetic stimulation. Brain Res 153:392–397

Canon SC, Robinson DA (1986) The final common integrator is in the prepositus and vestibular nuclei. In : Keller EL, Zee DS (Eds) Adaptive processes in visual and oculomotor systems. Advances in the Biosciences, vol 57. Pergamon, Oxford, pp 307–311

Cappa S, Sterzi R, Vallar G, Bisiach E (1987) Remission of hemineglect and anosognosia during vestibular stimulation. Neuropsychologia 25:775–782

Cowie RJ, Smith MK, Robinson DL (1994) Subcortical contributions to head movements in Macaques. II. Connections of a medial pontomedullary head-movement region. J Neurophysiol 72:2665–2682

Desimone R, Ungerleider L (1986) Multiple visual areas in the caudal superior temporal sulcus of the macaque. J Comp Neurol 248:164–189

Dürsteler MR , Wurtz RH (1988) Pursuit and optokinetic deficits following chemical lesions of cortical areas MT and MST. J Neurophysiol 60:940–963

Faugier-Grimaud S, Ventre J (1989) Anatomic connections of inferior parietal cortex (area 7) with subcortical structures related to vestibulo-ocular function in a monkey (Macaca fascicularis). J Comp Neurol 280:1–14

Faugier-Grimaud S, Frenois C, Stein DG (1978) Effect of posterior parietal lesions on visually guided behavior in monkeys. Neuropsychologia 16:151–168

Ferrier D (1875) Experiments on the brains of monkeys. Philos Trans R Soc Lond, B Biol Sci 165:433–488

Friberg L, Olsen TS, Roland PE, Paulson OB, Lassen NA (1985) Focal increase of blood flow in the cerebral cortex of man during vestibular stimulation. Brain 108:609–623.

Gallyas F (1979) Silver staining of myelin by means of physical development. Neurology Res 1:203–209

Gibson AR, Hansma DI, Houk JC, Robinson FR (1984) A sensitive low artefact TMB procedure for the demonstration of WGA-HRP in the CNS. Brain Res 298:235–241

Goldberg ME, and Bruce CJ (1986) The role of the arcuate frontal eye fields in the generation of saccadic eye movements. Progr Brain Res 64:143–154

Grüsser OJ, Pause M, Schreiter U (1982) Neuronal responses in the parietoinsular vestibular cortex of alert Java monkeys (Macaca fascicularis). In: Roucoux A, Crommelink M (eds) Physiological and pathological aspects of eye movements. W Junk, pp 251–270

Grüsser OJ, Pause M, Schreiter U (1990a) Localization and responses of neurons in the parieto-insular vetibular cortex of awake monkeys. J Physiol (Lond) 430:537–557

Grüsser OJ, Pause M, Shreiter U (1990b) Vestibular neurones in the parieto-insular cortex of monkey : visual and neck receptor responses. J Physiol (Lond) 430:559–583

Hassler R (1964) Spezifische und unspezifische Systeme des menschlichen Zwischenhirns. Progr Brain Res 5:1–32

Heilman KM, Bowers D, Watson RT (1983) Performance on hemispatial pointing task by patients with neglect syndrome. Neurology 33:661–664

Howard I (1982) Human visual orientation. Wiley, Toronto

Jeannerod M (1986) Mechanisms of visuo-motor coordination. A study in normal and brain-damaged subjects. Neuropsychologia. 24:41–78

Jeannerod M, Biguer B (1987) The directional coding of reaching movements. A visuomotor conception of spatial neglect. In: Jeannerod M (ed) Neurophysiological and neuropsychological aspects of spatial neglect. Elsevier, Amsterdam pp 87–113

Jones EG, Burton H (1976) Areal differences in the laminar distribution of thalamic afferents in cortical fields of the insular, parietal and temporal regions of primates. J Comp Neurol 168:197–248

Karnath HO, Christ K, Hartje W (1993) Decrease of contralateral neglect by neck muscle vibration and spatial orientation and trunk midline. Brain 116:383–396

Kawano K, Sasaki M, Yamashita M (1980) Vestibular input to visual tracking neurons in the posterior parietal association cortex of the monkey. Neurosci Lett 17:55–60

Kawano K, Sasaki M, Yamashita M (1984) Response properties of neurons in posterior parietal cortex of monkey during visual-vestibular stimulation. I. Visual tracking neurons. J Neurophysiol 51:340–351

Kennard MA, Ectors L (1938) Forced circling in monkeys following lesion of the frontal lobes. J Neurophysiol 1:45–54

Kuypers HGJM, Lawrence DG (1967) Cortical projections to the red nucleus and the brain stem in the rhesus monkey. Brain Res 4:151–188

Leichnetz GR, Goldberg ME (1988) Higher centers concerned with eye movement and visual attention: cerebral cortex and thalamus. In: Büttner-Ennever JA (ed) Neuroanatomy of the oculomotor system. Elsevier, Amsterdam, pp 365–429

Mesulam MM (1982) Principles of horseradish peroxidase neurohistochemistry and their applications for tracing neural pathways. Axonal transport, enzyme histochemistry and light microscope analysis. In:Mesulam M M (ed) Tracing neural connections with horseradish peroxidase. Wiley Chichester, pp 1–151.

Mesulam MM, Mufson EJ (1982) Insula of the old world monkey. I Architectonics in the insulo-orbito-temporal component of the paralimbic brain. J Comp Neurol 212:1–22

Mohler CW, Goldberg ME, Wurtz RH (1973) Visual receptive fields of frontal eye field neurons. Brain Res 61:385–389

Nelson RJ, Sur M, Felleman DJ, Kaas JH (1980) Representations of the body surface in postcentral parietal cortex of Macaca mulatta. J Comp Neurol 192:611–643

Odkvist LM, Schwartz DWF, Fredrickson JM, Hassler R (1974) Projection of the vestibular nerve to the area 3a arm field in the squirrel monkey (Saimiri sciureus). Exp Brain Res 21:97–105

Pandya DN, Sanides F (1973) Architectonic parcellation of the temporal operculum in the rhesus monkey and its projection pattern. Z Anat Enturickl Gesch 139:127–161

Pizzamiglio L, Frasca R, Guariglia C, Incoccia C, Antonucci G (1990) Effect of optokinetic stimulation in patients with visual neglect. Cortex 26:535–540

Pons TP, Garraghty PE, Cusik CG, Kaas JH (1985) The somatotopic organization of area 2 in macaque monkeys. J Comp Neurol 241:445–466

Ratcliff G, Davies-Jones GAG (1972) Defective visual localization in focal brain wounds. Brain 95:49–60

Roberts T, Akert K (1963) Insular and opercular cortex and its thalamic projection in Macaca mulatta. Schweiz Arch Neurol Neurochir Psychiat 92:1–43

Robinson CJ, Burton H (1980a) Somatic submodality distribution within the second somatosensory (SII), 7b, retroinsular, postauditory, and granular insular cortical areas of Macaca fascicularis. J Comp Neurol 192:93–108

Robinson CJ, Burton H (1980b) Organization of somatosensory receptive fields in cortical areas 7b, retroinsula, postauditory and granular insula of Macaca fascicularis. J Comp Neurol 192:69–92

Rode G, Charles N, Perenin MT, Vighetto A, Trillet M, Aimard G (1992) Partial remission of hemiplegia and somatoparaphrenia through vestibular stimulation in a case of unilateral neglect. Cortex 28:203–208

Rubens AB (1985) Caloric stimulation and unilateral visual neglect. Neurology 35:1019–1024

Schwartz DWF, Fredrickson JM (1971) Rhesus monkey vestibular cortex: a bimodal primary projection field. Science 172:280–281

Smith OA, Kastella KG, Randall DC (1972) A stereotaxic atlas of the brainstem of Macaca mulatta in a sitting position. J Comp Neurol 145:1–24

Stanton GB, Goldberg ME, Bruce CJ (1988) Frontal eye-field efferents in the Macaque Monkey: II. Topography of terminal fields in midbrain and pons. J Comp Neurol 271:493–506

Takemori S, Uchigata M, Ishikawa M (1979) Eye movements in cerebral lesions. Neurosci Lett 2:530–531

Taylor JL, McCloskey DI (1991) Illusions of head and visual target displacement induced by vibration of neck muscles. Brain 114:755–759

Tropper J, Melvill Jones G, Bloomberg J, Fadlallah H (1991) Vestibular perceptual deficits in patients with parietal lobes lesions. Acta Otolaryngol Suppl 481:528–533

Tusa RJ, Zee DS (1989) Cerebral control of smooth pursuit and optokinetic nystagmus. In: Lessell S, Van Dalen JTW (eds) Current neuro-ophtalmology. Year Book, Chicago, pp 115–146

Ungerleider L, Desimone R (1986a) Projections to the superior temporal sulcus from the central and peripheral field representation of V1 and V2. J Comp Neurol 248:147–163

Ungerleider LG, Desimone R (1986b) Cortical connections of visual area MT in the Macaque. J comp Neurol 248:190–222

Vallar G, Bottini G, Rusconi ML, Sterzi R (1993) Exploring somatosensory hemineglect by vestibular stimulation. Brain 116:71–86

Ventre J, Faugier-Grimaud S (1986) Effects of posterior parietal lesions (area 7) on VOR in monkeys. Exp Brain Res 62:654–658

Ventre J, Flandrin JM, Jeannerod M (1984) In search for egocentric reference. A neurophysiological hypothesis. Neuropsychologia 22:797–806

Wong-Riley M (1979) Changes in the visual system of monocularly sutured or enucleated cats demonstratable with cytochrome oxidase histochemistry. Brain Res 171:11–28

Wurtz RH, Newsome WT (1985) Divergent signals encoded by neurons in extrastriate areas MT and MST during smooth pursuit eye movements. Soc Neurosci Abstr 11:1246

Superior Parietal Lobule Involvement in the Representation of Visual Space: a PET Review

L. PETIT[1,2], C. ORSSAUD[1,2], N. TZOURIO[1], B. MAZOYER[1], and A. BERTHOZ[2]

[1] Groupe d'Imagerie Neurofonctionnelle, Service Hospitalier Frédéric Joliot, Orsay, France, and EA 1555, Université Paris 7, Paris, France

[2] Laboratoire de Physiologie de la Perception et de l'Action, CNRS-Collège de France, Paris, France

Introduction

A fundamental capability in vision is to process and act on objects at different spatial locations. This requires that a scene be segmented into objects, for interesting objects to be identified, for the spatial locations of the objects to be computed, for attention to be shifted to the appropriate location and engaged on the object, and an appropriate response to be planned. The analysis of space has long been considered to be an important role of the parietal cortex based on clinical (Critchley 1953), behavioral (Mishkin et al. 1983) and physiological (Mountcastle et al. 1975; Lynch et al. 1977; Robinson et al. 1978; Duhamel et al. 1992) studies. The exact nature of the processing in this region has been the subject of discussion in the literature, because single neurons in parietal cortex of the monkey discharge under a number of different circumtances and tend to yield the answer that any given experiment is designed to produce. Thus, different studies have come to the conclusion that parietal cortex is important in the generation of motor command for operations in immediate extrapersonal space (Mountcastle et al. 1975), for visual attention (Robinson et al. 1978), and for motor planning (Gnadt and Andersen 1988).

In such schemes, one of the main problems of perception of visual space is how the spatial locations of objects can be coded in the brain. The retinotopic representation of space can be transformed in several ways: relative to some head, body, or egocentric location (Sakata and Kusunoki 1992; Pouget and Sejnowski, this volume) or relative to the current or anticipated center of gaze, providing a continuous remapping of the representaion of visual space (Duhamel et al. 1992; Colby et al. 1993). As suggested by numerous physiological studies in the monkey, the posterior parietal cortex is the most likely area of the brain involved in such transformations.

In humans, the role of the superior parietal cortex also appears to be the organization of the action directed towards one or several objects. For instance, that the superior parietal lobule is involved in the processing for eye movements

In: Parietal Lobe Contributions to Orientation in 3D Space (1997). P. Thier and H.-O. Karnath (eds). Springer-Verlag, Heidelberg.

has been appreciated for some time. Balint was the first to describe bilateral lesions to the superior parietal cortex in human patients, which resulted in the inability to will saccades, although spontaneous saccades were unaffected (Balint 1909; Pierrot-Deseilligny and Müri, this volume; Heide and Kömpf, this volume). Recently, it has been shown that the gripping and the handling of objects are also impaired as a result of superior parietal lesions (Goodale 1993, Goodale and Murphy, this volume).

Interestingly, by measuring local hemodynamic changes associated with specific visuospatial processes, the functional brain imaging technique of positron emission tomography (PET) makes it possible to map the organization of the superior parietal cortex with far greater precision than is possible with human lesion studies. Up to now, the exploratory approach which was developed using PET tended to identify the cerebral regions involved in the control of a particular cognitive function in their entirety. The PET data accumulated on the superior parietal lobule through the use of this technique in the last 5 years allow us to present current knowledge of the functional anatomy of the human superior parietal lobule. Furthermore, these PET data allow us to apply the data accumulated in the monkey of how the parietal cortex processes space to healthy humans.

Anatomical Definition

In nonhuman primates, the part of the parietal lobe involved in the representation of visuospatial information is called the posterior parietal cortex, which is located in the caudal region of the parietal lobe. This cortical area contains the superior and inferior parietal lobules. Currently, in monkey studies, the posterior parietal cortex designates mainly the inferior parietal lobule, i.e., Brodmann's area 7 and the intraparietal sulcus (for discussion, see Milner, this volume).

The homologies between areas in the posterior parietal cortex in monkeys and humans are still unclear. Eidelberg and Galaburda (1984) reappraised the cytoarchitectonic map of the posterior parietal cortex in humans. They claimed that the human posterior parietal cortex comprises both the superior and inferior parietal lobules. The superior parietal lobule, in humans, consists generally of both the superior parietal gyrus and the intraparietal sulcus, both structures corresponding to the posterior parietal cortex in the monkey. The inferior parietal lobule, on the other hand, can be divided into the supramarginal and angular gyri. In this review, we use the anatomical term “superior parietal lobule” to describe activations located in the superior parietal gyrus and/or in the intraparietal sulcus rather than the “posterior parietal cortex”, which has a cytoarchitectonic connotation.

General PET Methods and Rationale

The past 15 years have seen the development of functional imaging techniques such as PET. Using tomographic mathematical reconstruction methods, the distribution in the brain of positron-emitting molecules can be measured at several horizontal levels (Raichle 1989). [^{15}O]water is used to directly measure local blood flow, which has been shown to vary precisely with local changes in neuronal activity. For most [^{15}O]water studies, a single scan lasts 40–90 s. As a scan can be acquired every 15 min or less, due to the short measurement time and physical half-life of ^{15}O (123 s), multiple scans can be completed in a single recording session (Raichle 1989). The spatial resolution of PET images depends upon a number of factors. One limiting factor is the intrinsic resolution of the PET scanner used for data collection. The newest generation of scanners has intrinsic resolutions as high as 3–5 mm in the plane; the theoretical limit is thought to be near 2 mm. Nowadays, images are usually reconstructed and/or smoothed to in-plane resolutions between 10–20 mm. As the signal-to-noise ratio of PET images is relatively low, methods to increase the ratio are usually applied, such as averaging of images within and across subjects (Fox et al. 1988; Friston et al. 1991).

Differences in brain activity are measured for different tasks. By mathematically computing the differences in the scans across the different activation conditions, changes in brain responses can be correlated with changes in task conditions; the basic rationale is that during the performance of a task, information-processing demands produce changes in neuronal activity in various regions in the brain and this neuronal activity induces changes in regional cerebral blood flow (rCBF). A variety of approaches have been used to identify and assess the significance of regional changes (Fox et al. 1988; Friston et al. 1994; Worsley et al. 1992). Once created, images are usually searched with computerized routines to localize areas of change in standard stereotactic coordinates within Talairach's atlas of the human brain (Herscovitch et al. 1983) (Fig. 1). Others have developed a method aimed at detecting increases in cerebral blood flow in cerebral structures with anatomical boundaries based on sulcal and gyral parcelling of the brain (Mazoyer et al. 1993). In this review, we focus on the different PET studies describing superior parietal activation expressed in Talairach's coordinates.

Activation of the Superior Parietal Lobule in Different Visuospatial Tasks

Table 1 summarizes for both hemispheres the location of the local maxima (in Talairach's coordinates) within the superior parietal lobule, demonstrating significant rCBF increase during the performance of different visuospatial tasks.

Table 1. Talairach coordinates (x, y, z) of local maxima detected within the superior parietal activations related to the performance of various visuospatial tasks

	Spatial tasks	Left	Right	References
Visually cued saccades				
1.	Visually guided saccades		+27, -54, +45	Paus et al.1993
2.	Visually guided saccades	-18, -68, +36	+2, -74, +36	Anderson et al.1994
3.	Visually guided saccades	-14, -68, +32	+20, -76, +32	Sweeney et al. 1996
4.	Remembered single saccades	-18, -56, +48		Anderson et al.1994
5.	Remembered single saccades	-2, -66, +44	+10, -68, +44	O'Sullivan et al.1995
6.	Prelearned sequence of saccades	-16, -72, +44	+34, -44, +40	Petit et al.1996a
7.	Antisaccades	-20, -59, +48	+27, -59, +47	Paus et al.1993
8.	Antisaccades		+15, -62, +52	O'Driscoll et al.1995
Visually cued arm movements				
9.	Visually guided hand movement	-16, -64, +52	+22, -68, +52	Deiber et al.1991
10.	Visually guided finger movement	-44, -32, +40	+42, -62, +40	Deiber et al. 1996
11.	Hands, mental rotation	-46, -30, +42		Bonda et al. 1995
Visuospatial dorsal pathway				
12.	Spatial location matching	-16, -62, +44	+26, -58, +44	Haxby et al.1991
13.	Spatial location matching	-16, -64, +48	+10, -58, +44	Haxby et al.1994
14.	Perceptual maze test	-25, -58, +50	+45, -55, +52	Ghatan et al.1995
15.	Spatial mental imagery	-20, -68, +40	+14, -70, +36	Mellet et al.1995
Visuospatial attention				
16.	Exogenous shift of attention	-33, -35, +46	+25, -45, +50	Corbetta et al.1993
17.	Endogenous shift of attention	-23, -39, +44	+23, -43, +48	Corbetta et al.1993
18.	Visual serial search	-31, -53, +44	+17, -59, +44	Corbetta et al.1995
Visuospatial working memory				
19.	Spatial working memory		+42, -40, +36	Jonides et al.1993
20.	Spatial working memory		+37, -42, +38	Smith et al.1995
21.	Spatial working memory	-16, -62, +44	+14, -62, +44	Courtney et al.1996

Distances are in millimeters to the right (+) and left (–) of the midline for x coordinates, anterior (+) and posterior (–) to the VAC line for y coordinates, and above (+) and below (–) the AC-PC line for the z coordinates. VAC, vertical plane passing through the anterior commissure; AC, anterior commissure; PC, posterior commissure

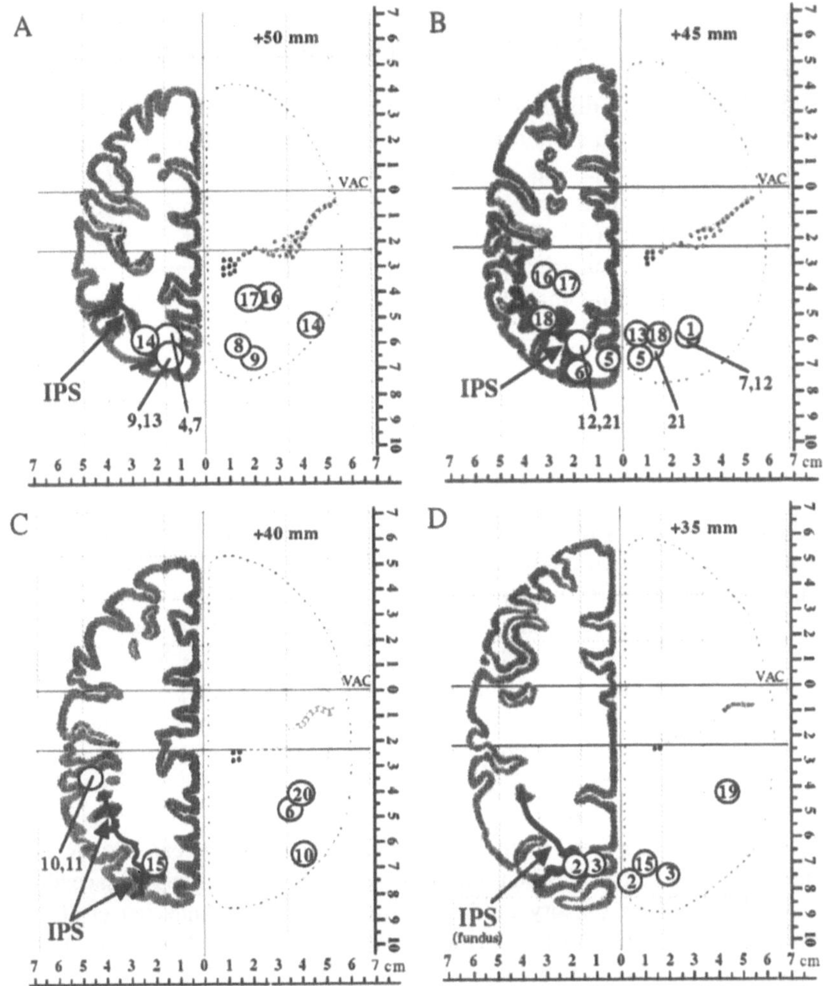

Fig. 1 A-D. Superior parietal activations observed during performance of various visuospatial tasks, as demonstrated by increased regional cerebral blood flow measured with PET. *Numbers* in *symbols* indicate the study reporting each activated focus listed in Table 1. These results are displayed on four axial stereotactic grids of Talairach and Tournoux atlas (1988), +50, +45, +40 and +35 mm above the anterior commissure-posterior commissure line. Distances are in millimeters anterior (+) and posterior (–) to the vertical plane passing through the anterior commissure line, and to the right (+) and left (–) of the midline. *IPS*, intraparietal sulcus

These parietal activations do not correspond just to a single point in the Talairach's stereotactic space. In each of the studies mentioned, the superior parietal activations extended between +32 mm and +55 mm above the bicommissural plane. This area corresponds to the superior parietal lobule in Talairach's atlas of the human brain, comprising both the superior parietal gyrus and the intraparietal sulcus.

Figure 1 shows four different Talairach's axial slices where these parietal local maxima are located. This figure clearly demonstrates that all the superior parietal activations described in this review overlap the intraparietal sulcus in both hemispheres. Axial slices at +50 and +45 mm above the anterior commissure-posterior commissure (AC-PC) plane accommodate most of the local maxima observed within the superior parietal lobule during performance of different visuo-spatial tasks. Anatomically, this corresponds precisely to the intraparietal sulcus location.

Visually Guided Saccadic Eye Movements

Saccades are used to catch an image of interest on the fovea rapidly. Reflexive saccades are externally triggered by the sudden appearance of a visual target in the peripheral visual field. Voluntary saccades are internally triggered to catch a target on the fovea, either one that has been visible on the peripheral visual field for a period of time (visually guided saccades) or a position which was previously perceived and remembered (memory guided saccades). Self-paced saccades are internally triggered, but without a target (i.e., without sensory input). Animal studies, and observations following cerebral ablations in man, suggest that saccades performed under different behavioral circumstances are controlled by different cortical and subcortical oculomotor regions (for review, see Pierrot-Deseilligny and Müri, this volume). In particular, it has long been known that the posterior parietal cortex plays an important role in the processing of saccadic eye movements based on sensory guidance (Andersen 1989; Andersen et al. 1992, Pierrot-Deseilligny and Müri, this volume; Heide and Kömpf, this volume).

Paus et al. (1993) studied the cortical regions involved in the execution of visually guided saccades toward a visual stimulus presented within the right or the left hemifield. By comparison with the rCBF variations measured when subjects were blindfolded, the execution of symmetrical visually guided saccades showed a rCBF increase in the right superior parietal lobule (Table 1), centered in the intraparietal sulcus (Fig. 1B, 1). Anderson et al. (1994) described a bilateral activation of the superior parietal lobule, medial to the intraparietal sulcus, resulting from the execution of visually guided saccades (Table 1; Fig. 1D, 2). The same authors also observed a left superior parietal activation during the execution of remembered saccades toward locations of previous target appearance (Fig. 1A, 4). Such results were confirmed recently by O'Sullivan et al. (1995), showing bilateral activation of the superior parietal lobule during remembered saccades (Fig. 1B, 5). Moreover, Sweeney et al. (1996) described activation in the inferior part of the superior parietal lobule (Table 1, Fig. 1D, 3) during visually guided saccades as well as during a delayed saccadic task.

Interestingly, we have also observed that the repetition of a prelearned sequence of horizontal saccades in total darkness led to specific activations in the superior parietal lobule in the depth of the intraparietal sulcus (Fig. 1B,C, 6, Fig. 2) (Berthoz et al. 1992; Petit et al. 1993a, 1996b). In such a task, the subject had to produce successive saccadic eye movement based strictly on an internal representation of spatial information.

Antisaccade tasks were also studied using the PET technique (Table 1; Paus et al. 1993; O'Driscoll et al. 1995; Sweeney et al. 1996). During such a task subjects had to inhibit a saccade toward a briefly appearing peripheral target and immediately generate a saccade to an equivalent point in the opposite hemifield. In particular, Paus et al. (1993) showed that the execution of antisaccades led to a bilateral superior parietal activation centered in the intraparietal sulcus, at the same location that activation was observed during the execution of visually guided saccades (Fig. 1A,B, 7).

Finally, we have previously shown that the control of executed, imagined, and suppressed self-paced horizontal saccades in total darkness did not lead to superior parietal activation when the saccadic tasks were not sensory guided (Lang et al. 1994; Petit et al. 1993b, 1995, 1996a).

To summarize, the fact that the superior parietal activation centered in the intraparietal sulcus was observed only during the execution of saccades set on the basis of either internally or externally cued visual targets confirms the crucial role of this structure in the coding of visually guided saccades. More specifically, these findings may suggest that the intraparietal sulcus in human plays a role in higher level processes related to planning of saccades rather than to the control of the execution of such movements.

Visually Guided Limb Movements

In everyday life, we interact continuously with objects. We reach for them, we grasp them, and we manipulate them. In all these actions, the role of vision is crucial in order to determine the relative positions of the hand and the object to be used. In such a scheme, there is evidence that the superior parietal lobule is involved in the visual control of limb movements (Sakata and Taira 1994; Jeannerod et al. 1995; Faugier-Grimaud et al., this volume).

Grafton et al. (1992) investigated the functional anatomy of visually guided movement in subjects performing visuomotor tracking with the index finger during PET measurements. The most interesting finding was the bilateral activation of the superior parietal lobule during cued visually guided movement. Parietal activations were present for all the visuomotor tasks, irrespective of the limb used or whether there was somatosensory feedback. In addition, the superior parietal activations were maximal when a directional cue was introduced. This latter task was designed so that the subject would have to select and integrate useful visual information into a preexisting, goal-directed motor plan. Previously, another PET study of the selection of movement in which these potentially confounding variables were carefully controlled for also demonstrated robust responses of rCBF in the superior parietal lobule (Table 1; Fig. 1A, 9; Deiber et al. 1991). Activations in this area were present whether the task was driven by internal or external (auditory) cues compared to fixed movements, suggesting a critical role of the superior parietal lobule for movement selection rather than movement execution.

Recently, Deiber et al. (1996) described an increase in rCBF in both the left anterior and right posterior part of the parietal cortex during preparation for finger movements based on visual information (Table1). The left activation was located in the most anterior part of the intraparietal sulcus, whereas the right one was found in the depth of the sulcus (Fig. 1C, 10). This finding supports a primary role of the parietal cortex in using visual information for specific motor preparation. Such a study also confirms the previous results of anterior intraparietal sulcus activation during a mental rotation task of the hand (Table 1; Fig. 1C, 11; Bonda et al. 1995).

In summary, these studies show that the superior parietal lobule is critical for hand movement based on the integration of visual information.

Visuospatial Perception

Space perception provides information about where objects are in the environment and how they move. It may be dissociated from pattern perception, which provides information about what the objects are and how they are characterized. The original evidence for separate processing streams for object vision and spatial vision was the contrasting effects of inferior temporal and posterior parietal lesions in monkeys (Mishkin et al. 1983). Only posterior parietal lesions caused a severe deficit in visuospatial performance while having no effect on visual discrimination performance.

In experiments designed to investigate the possible existence of separate object vision and spatial vision pathways in humans, changes in rCBF were measured using PET while subjects performed object identity and spatial location match-to-sample tasks (Haxby et al. 1991, 1994). The pattern of rCBF changes associated with performance of object matching and location matching demonstrated a clear dissociation of visual functions associated with extrastriate visual areas in ventral occipitotemporal and dorsal occipitoparietal cortex. In particular, the location-matching task selectively activated the superior parietal lobule near the fundus of the intraparietal sulcus (Table 1; Fig. 1B,C, 12-13).

More complex visuospatial cognitive tasks such as the Perceptual Maze Test (PMT; Ghatan et al. 1995) or a verbally guided construction of three-dimensional (3D) mental objects (Mellet et al. 1995) led to the activation of the dorsal occipitoparietal pathway, with robust activations in the superior parietal lobule (Table 1). The PMT consists of cognitive processing with components of visual perception, spatial decoding, selection of route, motor planning, and execution. Ghatan et al. (1995) showed that such a test causes strongest changes in rCBF bilaterally in the superior parietal lobule (Table 1; Fig. 1A, 14). The PET study of construction of 3D mental objects consisted of spatial mental imagery using verbal rather than visual material. Mellet et al. (1995) observed a bilateral superior parietal activation centered in the depth of the intraparietal sulcus (Table 1; Fig. 1C, D, 15).

Therefore, the various PET studies have accumulated evidence demonstrating the existence in humans, similar to that in monkeys, of an occipitoparietal visual pathway for processing spatial location information. Interestingly, as described for saccadic and hand movements, the intraparietal sulcus may represent the superior parietal component of such a dorsal pathway.

Visuospatial Shift of Attention

Corbetta et al. (1993) have studied the regions of the brain related to shifting of attention to different locations in the visual field. Two issues were addressed. The first was whether the shifts of visual attention are computed in terms of the direction of the shift or in terms of the location of the object in space. The second issue was to address whether the same mechanisms are used when shifts of attention are driven by external input (exogenous) or by internal knowledge (endogenous).

Three different tasks were used. The first task ("shifting attention") was designed to induce endogenous shifts of attention to peripheral locations of visual stimuli. The second task ("central detection") was intended to prevent orienting of attention to peripheral stimuli by engaging the subject's attention at the center of gaze. The third "passive" task was used to explore exogenous shifts of visual attention. Stimuli were flashed at random locations in either the left or right hemifield. Both central detection and passive tasks were designed to match with the shifting attention task in total peripheral visual stimulation and arousal.

A superior parietal region was more active in both the shifting tasks and the passive task, when attention was presumably shifted to peripheral stimuli. In the central detection task, when attention was maintained at the center of gaze and peripheral stimuli were effectively ignored, superior parietal activation was not seen. The superior parietal activation was primarily contralateral to the location of

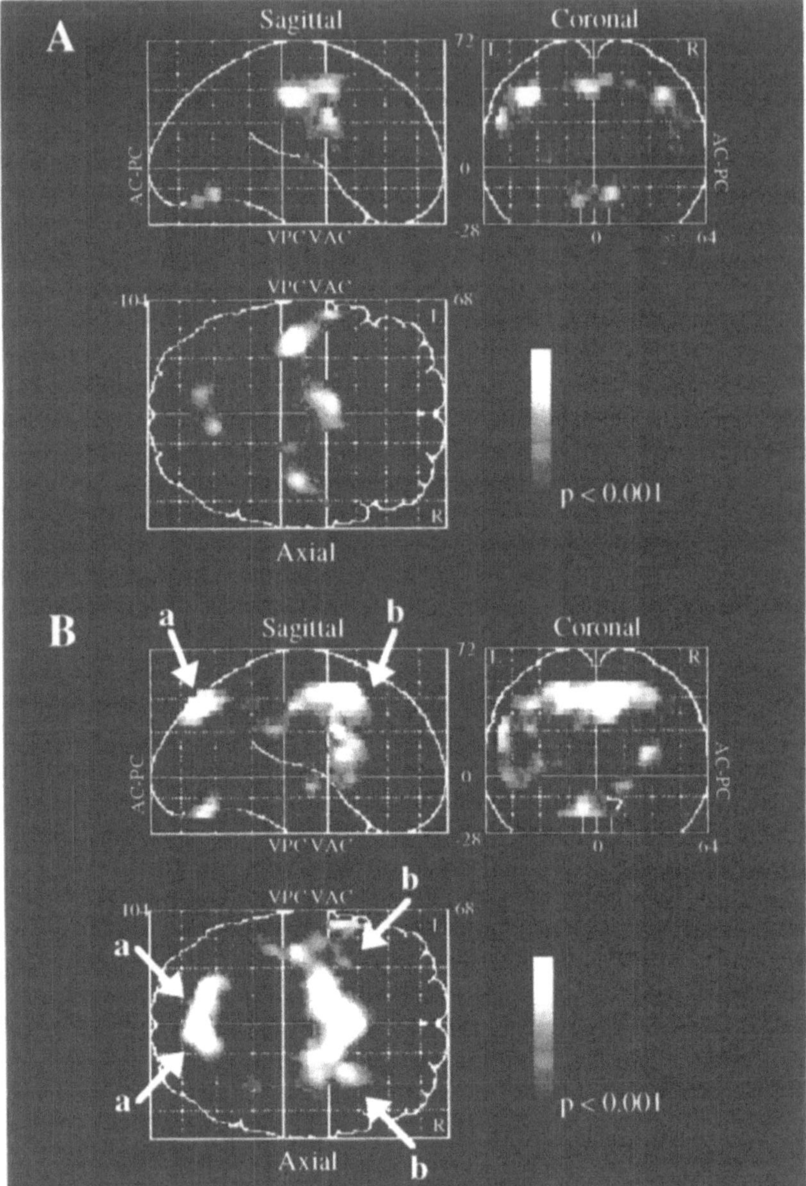

Fig. 2 A, B. Statistical parametric maps (SPM) for two comparisons. **A** The execution of voluntary self-paced saccades was compared to a resting condition (eyes open in total darkness). **B** The repetition of a prelearned sequence of six saccades was contrasted with the same resting condition. Significant pixels at the given threshold of $p<0.001$ not corrected for multiple comparisons are displayed on single sagittal, coronal, and axial projections of the brain (most significant in white). The spatial location of each activated area can be established

by comparing its position in three views. Note the additional activation related to the prelearned saccade sequences execution in both the superior parietal (*a*) and the superior frontal cortex (*b*). *AC-PC*, anterior commissure - posterior commissure plane; *VAC, VPC*, vertical plane passing through the anterior and posterior commissure, respectively

the shifts and not significantly influenced by their direction. These results show that a region in the superior parietal cortex is activated when attention is shifted endogenously or exogenously (Table 1; Fig. 1A, B, 16-17).

Recently, the same authors showed changes in the rCBF while subjects searched visual displays for targets defined by color, motion, or a combination of color and motion while they fixated a central cross (Corbetta et al. 1995). The behavioral data of this study indicated that the conjunction task required serial search and, therefore, shifts of visuospatial attention. A region in the superior parietal cortex was activated only in the combination task at a location that had previously been shown to be engaged by successive shifts of spatial attention (Table 1; Fig. 1B, 18).

Visuospatial Working Memory

The concept of working memory, as originally proposed by Baddeley (1992), has three distinct components: a phonological rehearsal loop for the storage and manipulation of verbal information; a visuospatial sketch pad for visual and spatial information; and a central executive for attentional control. Confirming the previous data from nonhuman primates and the dissociation of object and spatial information in extrastriate areas in the humans, it has been recently demonstrated by PET studies that the visuospatial sketch pad could be dissociable into two subsystems: one for object-based information and one for spatial information (Smith et al. 1995; Courtney et al. 1996). In the Courtney et al. study, subjects were instructed to remember either three faces in a memory set or the three locations in which the faces appeared. For the face memory task, the subjects indicated whether the test face was the same as one of the three faces seen in the memory set, regardless of the location in which the face appeared. For the location memory task, the subjects indicated whether the test location was the same as one of the three locations indicated in the memory set, regardless of which face appeared. Interestingly, only the location working memory task led to robust bilateral activation of superior parietal lobule (Table 1; Fig. 1B, 21). In addition, this bilateral activation was nearly identical to the right superior parietal area found in previous PET studies of location working memory (Fig. 1C, D, 19-20; Jonides et al. 1993; Smith et al. 1995).

Another Cortical Visuospatial Structure: the Superior Frontal Cortex

One cannot neglect to note that each PET study presented in this review described several regions activated outside the superior parietal lobule. In particular, there is now strong evidence that the superior frontal cortex is also involved in visuospatial information processing. The comparison of the various PET studies (Jonides et al. 1993; Corbetta et al. 1993; Haxby et al. 1994; Mellet et al. 1995; Petit et al. 1995a; Courtney et al. 1996) corroborates the presence of both intraparietal and superior frontal activations and further suggest an extension of the dorsal spatial visual pathway into the frontal lobe.

We have recently emphasized a bilateral activation in the depth of the superior frontal sulcus during the execution of sequence of prelearned saccades (Petit et al. 1996b). Such an activation was not detected during the execution of voluntary self-paced saccades in total darkness (Petit et al. 1993b, 1996a; Lang et al. 1994). The results obtained with the statistical parametric map (SPM) analysis are illustrated in Fig. 2 and enabled us to distinguish a set of cortical and subcortical activations during both the repetition of prelearned saccade sequences and the execution of self-paced saccades. At the cortical level, this network was composed of the supplementary eye fields (SEF), the median part of the cingulate gyrus, the frontal eye fields (FEF), and the insula. At the subcortical level, the lenticular nucleus (putamen and globus pallidus) and the thalamus were consistently activated as was the cerebellar vermis. In addition, the execution of prelearned saccade sequences led to specific rCBF increases deep in the superior frontal sulcus and intraparietal sulcus (Table 1; Fig, 1B, C, 6). These observations show that the determining factor is the existence of sensory information that has to be used for remembering and reproducing a saccadic movement and thus reveal a possible involvement of the superior frontal cortex.

Conclusion

Previous neurophysiological and neuropsychological studies have indicated that spatial representations are derived by integrating visual signals with information about eye, head, and limb position and movement. As recently claimed by Andersen et al. (1993), the lateral bank of the intraparietal sulcus in the monkey is fascinating in that it appears to be one of the most important regions of the brain where these signals are brought together. The PET data summarized in this review suggest that the neural substrates of space perception and spatial control of movements in healthy humans are inseparably connected within the superior parietal lobule and more specifically within the intraparietal sulcus. Thus, the PET technique offers us a means of checking and extending these data from animal to human, even when, because of the limited PET spatial resolution, it is not possible to state positively that the different superior parietal activations observed belong exactly in the lateral bank of the intraparietal sulcus. Thus, further studies are necessary and both functional magnetic resonance imaging and 3D-PET techniques provide now a better spatial and temporal resolution. In addition, in future studies individual functional neuroanatomy will offer a useful complement to intersubject averaging that may help to segregate functional areas within the intraparietal sulcus in healthy human.

References

Andersen RA (1989) Visual and eye movement functions of the posterior parietal cortex. Annu Rev Neurosci 12:377–403

Andersen RA, Brotchie PR, Mazzoni P (1992) Evidence for the lateral intraparietal area as the parietal eye field. Curr Opin Neurobiol 2:840–846

Andersen RA, Snyder LH, Li CS, Stricane B (1993) Coordinate transformations in the representation of spatial information. Curr Opin Neurobiol 3:171–176

Anderson TJ, Jenkins IH, Brooks DJ, Hawken MB, Frackowiak RSJ, Kennard C (1994) Cortical control of saccades and fixation in man: a PET study. Brain 117:1073–1084

Baddeley A (1992) Working memory. Science 255:556–559

Balint R (1909) Seelenlähmung des "Schauens", optische Ataxie, räumliche Störung der Aufmerksamkeit. Monatsschr Psychiatr Neurol 25:51–81

Berthoz A, Mazoyer B, Petit L, Orssaud C, Raynaud L, Tzourio N (1992) Bilateral parietal involvement in the execution of a sequence of memorized saccades in man. Soc Neurosci Abstr 18:214

Bonda E, Petrides M, Frey S, Evans AC (1995) Neural correlates of mental transformations of the body-in-space. Proc Natl Acad Sci USA 92:11180–11184

Colby CL, Duhamel JR, Goldberg ME (1993) The analysis of visual space by the lateral intraparietal area of the monkey: the role of extraretinal signals. Prog Brain Res 95:307–316

Corbetta M, Miezin FM, Shulman GL, Petersen SE (1993) A PET study of visuospatial attention. J Neurosci 13:1202–1226

Corbetta M, Shulman GL, Miezin FM, Petersen SE (1995) Superior parietal cortex activation during spatial attention shifts and visual feature conjunction. Science 270:802–805

Courtney SM, Ungerleider LG, Keil K, Haxby JV (1996) Object and spatial visual working memory activate separate neural systems in human cortex. Cereb Cortex 6:39–49

Critchley M (1953) The parietal lobes. Edward Arnold, London.

Deiber MP, Passingham RE, Colebatch JG, Friston KJ, Nixon PD, Frackowiak RSJ (1991) Cortical areas and the selection of movement: a study with positron emission tomography. Exp Brain Res 84:393–402

Deiber MP, Ibañez V, Sadato N, Hallett M (1996) Cerebral structures participating in motor preparation in humans: a positron emission tomography study. J Neurophysiol 75:233–247

Duhamel JR, Colby CL, Goldberg ME (1992) The updating of the representation of visual space in parietal cortex by intended eye movements. Science 255:90–92

Eidelberg D, Galaburda AM (1984) Inferior parietal lobule. Divergent architectonic asymmetries in the human brain. Arch Neurol 41:843–852

Fox PT, Mintun MA, Reiman EM, Raichle ME (1988) Enhanced detection of focal brain responses using intersubject averaging and change-distribution analysis of subtracted PET images. J Cereb Blood Flow Metab 8:642–653

Friston KJ, Frith CD, Liddle PF, Frackowiak RSJ (1991) Plastic transformation of PET images. J Comput Assist Tomogr 15:634–639

Friston KJ, Worsley KJ, Frackowiak RSJ, Mazziotta JC, Evans AC (1994) Assessing the significance of focal activations using their spatial extent. Hum Brain Mapping 1:210–220

Ghatan PH, Hsieh JC, Wirsen-Meurling A, Wredling R, Eriksson L, Stone-Elander S, Levander S, Ingvar M (1995) Brain activation induced by the perceptual maze test: a PET study of cognitive performance. Neuroimage 2:112–124

Gnadt JW, Andersen RA (1988) Memory related motor planning activity in posterior parietal cortex of macaque. Exp Brain Res 70:216–220

Goodale MA (1993) Visual routes to knowledge and action. Biomed Res 14:113–123

Grafton ST, Mazziotta JC, Woods RP, Phelps ME (1992) Human functional anatomy of visually guided finger movements. Brain 115:565–587

Haxby JV, Grady CL, Horwitz B, Ungerleider LG, Mishkin M, Carson RE, Herscovitch P, Schapiro MB, Rapoport SI (1991) Dissociation of object and spatial visual processing pathways in human extrastriate cortex. Proc Natl Acad Sci USA 88:1621–1625

Haxby JV, Horwitz B, Ungerleider LG, Maisog JM, Pietrini P, Grady CL (1994) The functional organization of human extrastriate cortex: a PET-rCBF study of selective attention to faces and locations. J Neurosci 14:6336–6353

Herscovitch P, Markham J, Raichle ME (1983) Brain blood flow with intraveinous $H_2{}^{15}O$. J Nucl Med 24:782–798

Jeannerod M, Arbib MA, Rizzolatti G, Sakata H (1995) Grasping objects: the cortical mechanisms of visuomotor transformation. Trends Neurosci 18:314–320

Jonides J, Smith EE, Koeppe RA, Awh E, Minoshima S, Mintun MA (1993) Spatial working memory in humans as revealed by PET. Nature 363:623–625

Lang W, Petit L, Höllinger P, Pietrzyk U, Tzourio N, Mazoyer B, Berthoz A (1994) A positron emission tomography study of oculomotor imagery. Neuroreport 5:921–924

Lynch JC, Mountcastle VB, Talbot WH, Yin TCT (1977) Parietal lobe mechanisms for directed visual attention. J Neurophysiol 40:362–389

Mazoyer BM, Tzourio N, Frak V, Syrota A, Murayama N, Levrier O, Salamon G, Dehaene S, Cohen L, Mehler J (1993) The cortical representation of speech. J Cogn Neurosci 5:467–479

Mellet E, Crivello F, Tzourio N, Joliot M, Petit L, Laurier L, Denis M, Mazoyer B (1995) Construction of mental images based on description : Functional neuroanatomy with PET. Hum Brain Mapping [Supp] 1:273

Mishkin M, Ungerleider LG, Macko KA (1983) Object vision and spatial vision: two cortical pathways. Trends Neurosci 6:414–417

Mountcastle VB, Lynch JC, Georgopoulos AP, Sakata H, Acuna C (1975) Posterior parietal association cortex of the monkey: command function for operations within extrapersonal space. J Neurophysiol 38:871–908

O'Driscoll GA, Alpert NM, Matthysse SW, Levy DL, Rauch SL, Holzman PS (1995) Functional neuroanatomy of antisaccade eye movements investigated with positron emission tomography. Proc Natl Acad Sci USA 92:925–929

O'Sullivan EP, Jenkins IH, Henderson L, Kennard C, Brooks DJ (1995) The functional anatomy of remembered saccades: a PET study. Neuroreport 6:2141–2144

Paus T, Petrides M, Evans AC, Meyer E (1993) Role of the human anterior cingulate cortex in the control of oculomotor, manual, and speech responses: a positron emission tomography study. J Neurophysiol 70:453–469

Petit L, Orssaud C, Lang W, Pietrzyk U, Höllinger P, Tzourio N, Raynaud L, Mazoyer BM, Berthoz A (1993a) PET activations of voluntary, memorized and imagined saccades. J Cereb Blood Flow Metab 13[Suppl1]:S535

Petit L, Orssaud C, Tzourio N, Salamon G, Mazoyer B, Berthoz A (1993b) PET study of voluntary saccadic eye movements in humans: basal ganglia-thalamocortical system and cingulate cortex involvement. J Neurophysiol 69:1009–1017

Petit L, Tzourio N, Orssaud C, Pietrzyk U, Berthoz A, Mazoyer B (1995) Functional neuroanatomy of the human visual fixation system. Eur J Neurosci 7:169–174

Petit L, Orssaud C, Tzourio N, Mazoyer B, Berthoz A (1996a) Do executed, imagined and suppressed saccadic eye movements share the same neuronal mechanisms in healthy human ? In: Viviani P, Lacquaniti F (eds) Multi-sensory control of movement. Kluwer Academic, Amsterdam

Petit L, Orssaud C, Tzourio N, Crivello F, Mazoyer B, Berthoz A (1996b) Functional anatomy of a prelearned sequence of horizontal saccades in humans. J Neurosci 16:3714–3726

Raichle ME (1989) Developing a functional anatomy of the human brain with positron emission tomography. Curr Opin Neurol 9:161–178

Robinson DL, Goldberg ME, Stanton GB (1978) Parietal association cortex in the primate: Sensory mechanisms and behavioral modulations. J Neurophysiol 41:910–932

Sakata H, Kusunoki M (1992) Organization of space perception: neural representation of three-dimensional space in the posterior parietal cortex. Curr Opin Neurobiol 2:170–174

Sakata H, Taira M (1994) Parietal control of hand action. Curr Opin Neurobiol 4:847–856

Smith EE, Jonides J, Koeppe RA, Awh E, Schumacher EH, Minoshima S (1995) Spatial versus object working memory: PET investigations. J Cogn Neurosci 7:337–356

Sweeney JA, Mintun MA, Kwee S, Wiseman MB, Brown DL, Rosenberg DR, Carl JR (1996) Positron emission tomography study of voluntary saccadic eye movements and spatial working memory. J Neurophysiol 75:454–468

Talairach J, Tournoux P (1988) Co-planar stereotaxic atlas of the human brain. Thieme Medical Publishers, New York

Worsley KJ, Evans AC, Marrett S, Neelin P (1992) A three-dimensional statistical analysis for CBF activation studies in human brain. J Cereb Blood Flow Metab 12:900–918

Eye Movements and Beyond

Multiple Parietal "Eye Fields": Insights from Electrical Microstimulation

P. THIER[1] **and R. A. ANDERSEN**[2]

[1] Sektion für Visuelle Sensomotorik, Neurologische Universitätsklinik, Tübingen, Germany

[2] California Institute of Technology, Pasadena, USA

Introduction

Research into the functional role of the parietal lobe of primates began with the pioneering experiments of David Ferrier, who studied the functional consequences of lesions and the effects of electrical stimulation of monkey cortex, including parts of the parietal lobe (Ferrier 1876). Lesions of the angular gyrus surrounding the dorsal end of the superior temporal sulcus disrupted visually guided behavior, a deficit Ferrier erroneously mistook for blindness, leading him to suggest that the angular gyrus was the visual area of the brain. This misinterpretation brought about a fierce dispute with Munk (see Glickstein 1985), who, based on his own lesion work, had correctly identified the visual area with the occipital lobe. Ferrier's bias for a primary visual role of the parietal lobe was also reflected in his narrow interpretation of the effects of electrical stimulation of this part of the brain. Fig. 1 shows a reproduction of figures taken from his work which summarize his observations based on cortical stimulation. Electrical stimulation of sites 13 and 13′on the anterior and the posterior limbs of the angular gyrus, respectively, caused eye movements towards the opposite side, occasionally accompanied by head movements in the same direction. Ferrier interpreted these movements as reflexive reactions to a stimulation-induced visual sensation, also being responsible for the observed contraction of the pupils and "a tendency to closure of the eyelids as if under the stimulus of a strong light." In spite of his narrow interpretation of the stimulation effects, the observation of these movements was instrumental for the development of the concept of a parietal eye field, contributing to visually guided eye movements. Ferrier had to rely on stimulation techniques which from our present vantage point necessarily must look crude and ill-defined. The use of comparatively large surface electrodes confined to the cortical opercula and the lack of objective and reliable means to control and quantify the currents[1] produced observations that were the result of

[1] The appropriate amount of current was detected by noting that the electrode caused sensory sensations when attached to the investigator's tongue.

In: Parietal Lobe Contributions to Orientation in 3D Space (1997). P. Thier and H.-O. Karnath (eds). Springer-Verlag, Heidelberg.

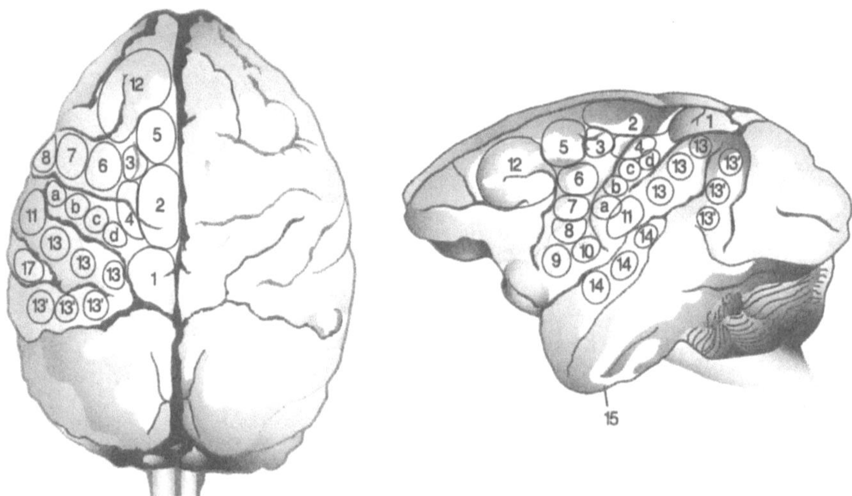

Fig. 1. Ferrier's figure from the first edition of his book *The Function of the Brain* summarizing the effects of electrical stimulation of the monkey cortex. Sites *13* and *13'* are on the anterior and posterior limbs of the angular gyrus. Their stimulation evoked eye movements to the contralateral side, occasionally accompanied by head movements in the same directions. Furthermore, pupillary contraction and eyelid closure were observed. These effects were interpreted as reflexive reactions to strong visual sensation, contributing to the view that the angular gyrus was the visual area of the brain. (From Ferrier 1876 with slight modifications)

simultaneous activation of widespread parts of the parietal lobe. The lack of any differences between the various stimulation sites along the angular gyrus directly reflects these limitations.

Areas LIP and MST

More specific stimulation effects both in anatomical and functional terms were only obtained almost 100 years later when surface electrodes were replaced by microelectrodes, allowing for a more localized application of electrical current to patches of parietal cortex. The first stimulation study of the posterior parietal cortex based on microelectrodes was presented by Shibutani et al. (1984), who were able to demonstrate that evoked eye movements were restricted to a part of the posterior parietal cortex they referred to as area 7. Actually, a look at the electrode penetrations reproduced in their original publication makes it clear that most of the effective sites were located on the posterior bank of the intraparietal sulcus, many of them within a cortical area we today refer to as the lateral intraparietal area (LIP) rather than area 7 (Fig. 2). The eye movements evoked

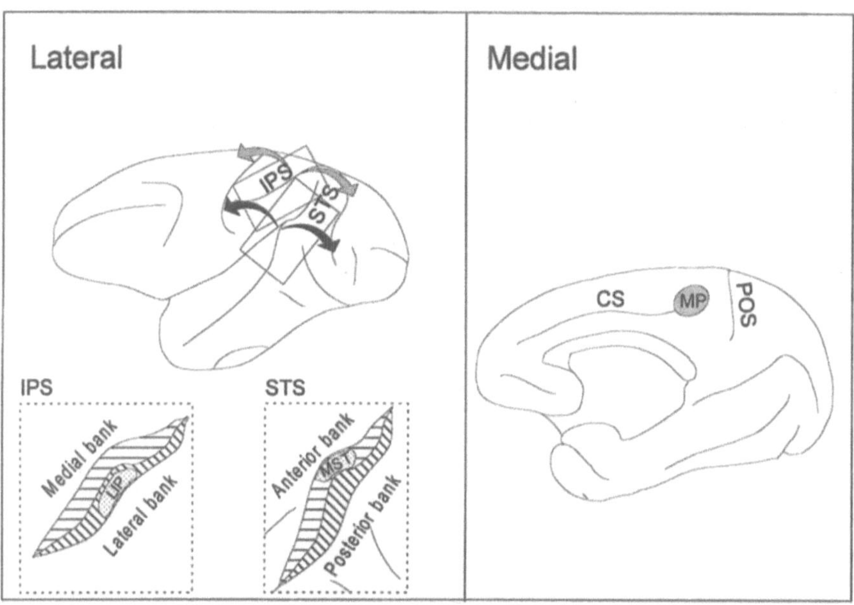

Fig. 2. Views of the lateral and medial aspects of the monkey cortex showing the locations of the lateral intraparietal area (*LIP*), the medial parietal area (*MP*), the *intercalated zone* and the medial superior temporal area (*MST*). *IPS*, intraparietal sulcus; *STS*, superior temporal sulcus; *POS*, parietooccipital sulcus; *CS*, cingulate sulcus

were always fast, resembling saccades, and eye movements as slow as smooth pursuit eye movements were not found although some of their penetrations may have hit pursuit-related medial superior temporal area (MST).

Our own experiments (Thier and Andersen 1996) with electrical microstimulation corroborate and extend the conclusions suggested by the Shibutani et al. study. With stimulation currents restricted to 200 μA or less, saccade-like eye movements could be evoked from the lateral aspect of the posterior parietal cortex (Fig. 3 A) only when the microelectrode hit a small patch of cortex on the posterior bank of the intraparietal sulcus (IPS). This stimulation-sensitive region is congruent with area LIP as defined on the basis of single-unit recordings and anatomical markers but smaller by roughly a factor of 2 (refer to Fig. 2 for the location of LIP and the other areas mentioned). In contrast, no eye movements could be evoked from the posterior parietal operculum, corresponding to area 7a, nor did stimulation at the few sites tested inside the superior temporal sulcus (STS), caudal to the IPS, evoke any behavioral response.

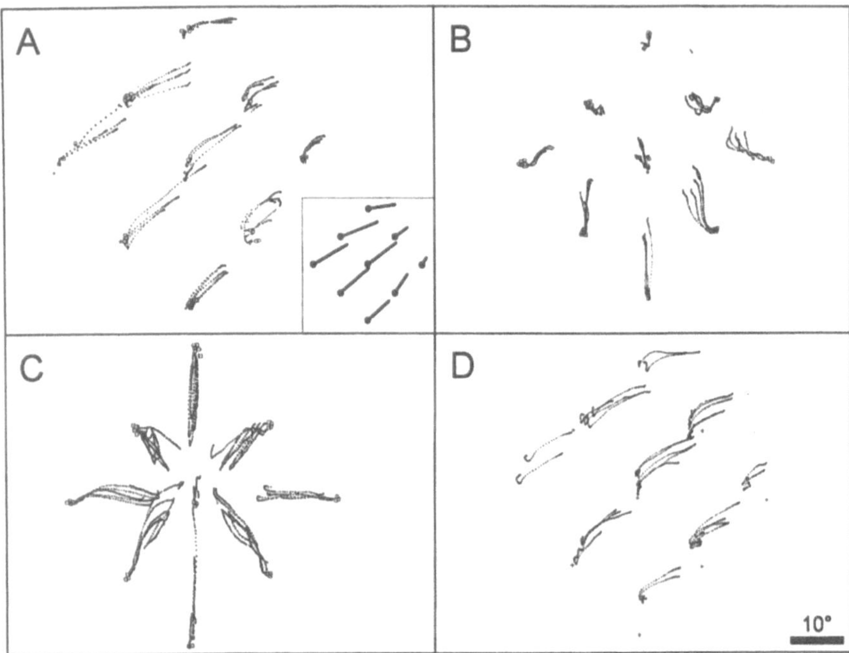

Fig. 3 A-D. *X*, *Y* plots of saccades evoked from three different parietal areas. The plots start at stimulation onset and end 150 ms later. *Small squares* indicate the initial eye position on each trial before electrical stimulation. **A** Saccades evoked from the *modified vector saccade* representation in the lateral intraparietal area (LIP). Note that varying the initial orbital position affected saccade amplitude, whereas the effect on direction was minor. The *inset* in **A** shows linear approximations of the mean evoked saccades used in some of the later figures. **B** Saccades evoked from the intercalated zone of the IPS, separating the representations of modified vector saccades with an upward and downward component. Stimulation drives the eyes into a head-centered location (goal zone) independent of their orbital position at the time of stimulation onset. **C** Goal-directed saccade evoked from a site in white matter underneath the intercalated zone. **D** Modified vector saccades evoked from the medial parietal area (MP) on the medial aspect of the parietal lobe. (Parts **A-C** from Thier and Andersen 1996). *Experimental paradigm*: monkeys were rewarded for keeping their line of sight within an eye position window of 5° centered on a memorized location in the frontoparallel plane cued by the presentation of a small fixation spot. If the monkey kept fixation of the spot for 500 ms, the spot was turned off for another 500 ms (gap) and 300 ms after onset of the gap period, electrical microstimulation was applied on half of the trials

The latter finding is in accordance with results presented earlier by Komatsu and Wurtz (1989) who were unable to evoke significant eye movements by stimulating sites in the STS if the eyes fixated a stationary location during stimulation. On the other hand, Komatsu and Wurtz showed that electrical microstimulation with currents as low as 25–50 μA was able to modify the velocity of ongoing smooth pursuit eye movements, provided the microelectrode stimulated sites within the lateral part of parietal area MST (= MSTl) on the anterior bank of the STS. This finding shows directly that fixation is not simply smooth pursuit at zero velocity. In other words, this result supports the idea of two at least partially independent systems for smooth pursuit eye movements and fixation, respectively. Area LIP contacts the superior colliculus, the key structure in the control of both fixation and saccades (Munoz and Wurtz 1993). It is tempting to speculate that evoked saccades following stimulation of LIP may be the combined result of an activation of collicular saccade representations going hand in hand with a shutdown of the collicular fixation center, vetoing eye movements. The absence of significant evoked smooth pursuit following stimulation of area MST may be the consequence of an inability of MST stimulation to shut the collicular fixation center down. Selective influence of electrical microstimulation on smooth pursuit is in full accordance with the results of single-unit recordings and the effects of lesions, which have suggested that this caudalmost part of the posterior parietal lobe is a key element in circuits organizing slow eye movements such as smooth pursuit (Thier and Erickson 1992; Komatsu and Wurtz 1988; see also Ilg and Thier, this volume) while not contributing to saccades or fixation. The effects of microstimulation therefore suggest the existence of two anatomically and functionally distinct representations for eye movements within the classical parietal eye field on the lateral aspect of the parietal lobe, area LIP, the more rostral one contributing to saccades and area MST, right at the end of the posterior parietal lobe involved in smooth pursuit eye movements. While the stimulation effects obtained for LIP are compatible with a contribution to saccades, they do not rule out an additional contribution to smooth pursuit. We did not test electrical microstimulation of sites in LIP during ongoing smooth pursuit eye movements. An additional contribution to smooth pursuit does not seem to be very likely though, considering the absence of convincing evidence for smooth pursuit-related single-unit activity in LIP.

Areas VIP and MP

Areas MST and LIP are not the only components of the classical parietal eye field of Ferrier electrical microstimulation is able to distinguish. Electrical microstimulation suggests a third eye movement area incorporating a strip of cortex within VIP as defined by Maunsell and Van Essen (1989). This *intercalated zone* has properties very different from those of neighboring cortex. Microstimulation of the *intercalated zone* evokes goal-directed saccades.

Electrical microstimulation suggests the existence of still another representation of saccadic eye movements which is completely confined to the medial aspect of the posterior parietal lobe. We have termed this medial representation of saccades located close to the end of the cingulate sulcus area MP (for *medial parietal*; Thier and Andersen 1993). MP is probably within area PGm of Pandya and Seltzer (1982) or area 7m of Cavada and Goldman-Rakic (1989, 1993), known to be richly interconnected with area LIP on the lateral aspect (Cavada and Goldman-Rakic 1989, 1993). Electrical microstimulation of sites in MP evokes saccadelike eye movements (Fig. 3D), which at first glance look very similar to those evoked from LIP. Stimulation of both sites in LIP, and MP evokes short-latency (about 30 ms) saccades whose amplitude and direction is largely determined by the location of the site stimulated. However, cortical location is not the only variable influencing the saccade trajectory. The other variable is orbital position at stimulation onset. While the direction of evoked saccades does not depend on orbital position, their amplitude is modified in a characteristic manner (Fig. 3 A, D), as discussed later in this chapter, reflecting a specific representation of space in these areas. MP is not just a redundant duplicate of LIP. A difference between MP and LIP, which sheds first light on the question of why there is a medial parietal eye field in addition to LIP, relates to the distribution of saccade amplitudes. Unlike LIP, which is characterized by a conspicuous emphasis on small amplitudes, the corresponding distribution for MP is much more widespread. Therefore, MP might make a more substantial contribution than LIP to the exploration of the visual periphery. This view is supported by the results of tracing studies, which show that mesial parietal cortex is preferrentially connected to the large amplitude saccade representation of the frontal eye field (FEF; Stanton et al. 1995; Schall et al. 1995).

Modified vector saccades, i.e., saccades showing a modification of their amplitude but not of their direction with orbital position, were the dominant type of response to microstimulation in the IPS. However, there was a second type of response found in the IPS characterized by an orbital position dependency not only of the amplitude of the evoked saccades but also of their direction. The modification of saccade amplitude and direction was such as to move the eyes into a particular region in head-centered space (the *goal zone*), largely independent of the starting position of the eyes (Fig. 3 B, C). Two very different interpretations of such *goal-directed* saccades have been suggested. The first one, put forward by Robinson (1972), is that electrical microstimulation mimicks the natural activation of a small group of cells, representing a location in head-centered space. The second one, based on experiments on the caudal superior colliculus of the cat (Roucoux et al. 1980), is that sites which seem to be encoding goal-directed saccades actually encode vector gaze shifts, i.e., shifts based on both eye and head movements. Such vector gaze shifts could be programmed by directly converting a retinal vector into a gaze vector of the same amplitude and direction. The important distinction is that the latter view does not require a nonretinal (i.e., head-centered) representation of the desired location. While stimulation of sites in the intercalated zone may evoke head movements accompanying the evoked saccades, the two do not add up to fixed gaze vectors (Thier and Andersen 1996).

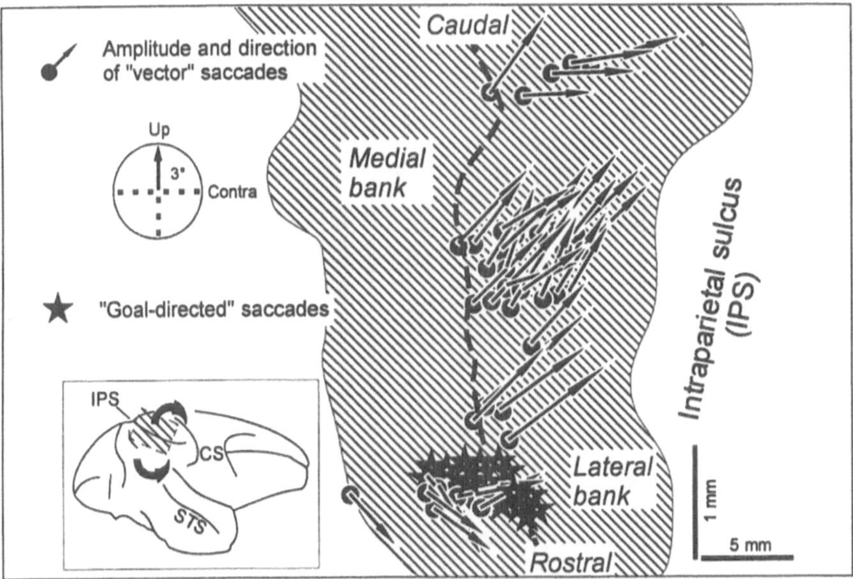

Fig. 4. Flattened reconstruction of the intraparietal sulcus (*IPS*) and its neighboring structures of one of the monkey hemispheres used by Thier and Andersen (1996). The *arrows* indicate the amplitude and the direction of the saccades evoked at a particular site using straight-ahead as a starting position. *Asterisks* indicate sites where goal-directed eye movements were evoked. *STS*, superior temporal sulcus; *CS*, cingulate sulcus. (From Thier and Andersen 1996)

We therefore interpret the occurrence of goal-directed saccades in the intercalated zone as an indication of a localized representation of head-centered space with neighboring groups of cells representing different locations relative to the head (Fig. 4).

Modified Vector Saccades

What is the explanation of the amplitude modification of saccades evoked fromarea LIP? Again, two qualitatively different interpretations have to be considered. The first one assumes that the group of cells stimulated represents a location on the retina relative to the fovea, defining a vector in retinal coordinates. Activation of this group of cells, either by the occurrence of a visual target at this location or by electrical microstimulation, would elicit the conversion of the retinal vector into a saccade vector of equal amplitude and direction. In the case of

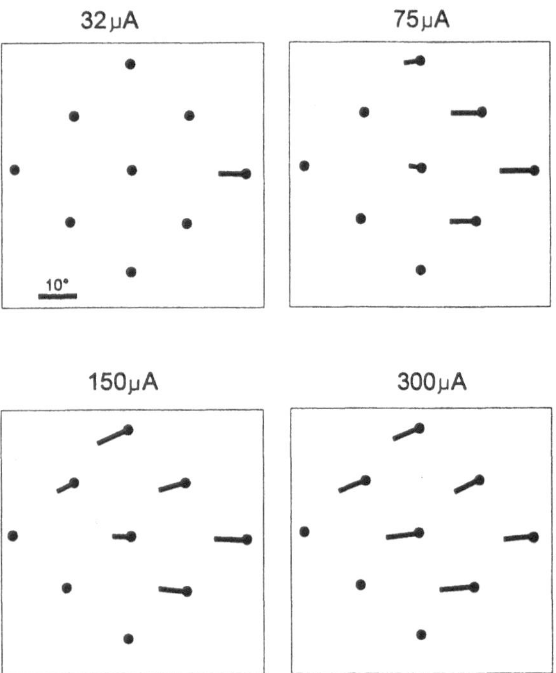

Fig. 5. Stimulation of site in the modified vector saccade representation of the lateral intraparietal area. Thresholds for evoked saccades depend on orbital position

a natural target this would move its image into the fovea. Modification of the amplitudes of evoked saccades could result from downstream mechanisms, such as constraints imposed by orbital mechanics, having the effect that centripetally directed saccades require less muscular effort than centrifugally directed saccades of the same amplitude and direction (Bruce 1990). Alternatively, the modification of the saccade vectors could be a consequence of neural signals related to eye position having a direct effect at the site of stimulation. The eye position-related neural signals required are indeed available. There is ample evidence for LIP cells having access to information on both eye position and retinal location, the two variables needed in order to encode locations relative to the head (Andersen et al. 1990). However, the two variables are usually not combined in such a way as to allow single LIP cells to unambiguously encode locations in head-centered space. Location relative to the head can only be recovered if the compound activity of large groups of cells is considered, suggesting that head-centered space is encoded in the activity of a population of neurons spread out over larger parts of LIP.

There are three reasons which led us prefer the view that the modification of vector saccades is indeed a consequence of an integration of eye position-related information at the level of the site being stimulated. The first one relates to the pattern of orbital position dependency. Although for most sites tested, the amplitudes of the saccades evoked were decreased when the eyes were shifted in the direction of the saccade, there were also exceptions to this rule. Such exceptions, even if rare, are not compatible with the view that the amplitude modification results from orbital mechanics, which should be the same independent of the site stimulated.

A second finding not compatible with a modification of fixed vector saccades by orbital mechanics is the orbital position dependency of saccade thresholds exemplified in Fig. 5, with smaller thresholds for centripetally directed saccades in most cases. This change of threshold may be expected if eye position-related information is available at the site of stimulation, modifying the state of the cells activated by the artificial stimulus in a gradual fashion. Conversely, it is hard to conceive of a position-dependent mechanical mechanism located downstream of the cortical trigger, modifying the thresholds of the properties of the cortical trigger.

A third argument comes from preliminary results of double stimulation experiments. The rationale of these experiments is outlined on the left side of Fig. 6. We assume that stimulation site 1 represents upward saccades and site 2 rightward saccades. If activation of these sites results in modified vector saccades as a consequence of a local integration of eye position information as suggested in Fig. 6A (= hypothesis 1), the result of simultaneous activation of the two sites will correspond to the vectorial summation of the individual patterns of modified vector saccades. The outcome will be a pattern whose overall direction will be in between the individual ones. Moreover, the resulting pattern, unlike the ones resulting from isolated activation of either site, will show a dependence of saccade direction on orbital position. This orbital position dependency of saccade direction will introduce convergence into the pattern. This is the necessary consequence of the eye position-dependent amplitude modification enhancing the contribution of modified vector sites representing centripetally directed saccades, while conversely reducing the contribution of centrifugally directed saccades. As illustrated in Fig. 6B (= hypothesis 2), no convergence would be introduced into the pattern of saccades evoked by double stimulation if the stimulated sites represented fixed vector saccades with downstream modification of saccade amplitude. Double stimulation will shift the overall direction of evoked saccades to a direction in between the two represented by the two sites, but the direction will remain independent of orbital position. Figure 6C shows the results of one of the few experiments on LIP in which we succeeded in placing one electrode into the representation of upward saccades, while the second electrode was placed into a site representing rightward saccades. Simultaneous stimulation of these two sites resulted in a pattern of evoked saccades whose general direction was to the upper right, and exhibited an increased amount of convergence or centering in full accordance with the idea of a direct influence of eye position-dependent information at the sites of stimulation.

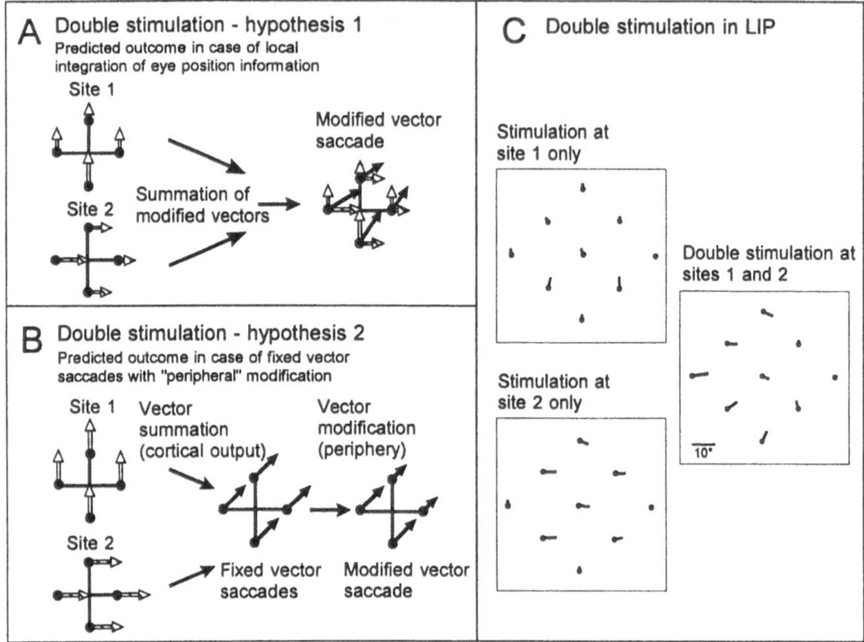

Fig. 6 A-C. Simultaneous electrical microstimulation at two IPS sites clarifies the nature of the orbital position dependency of the amplitude of "modifed vector saccades." **A** Modified vector saccades result from an integration of eye position information at the site of stimulation. Double stimulation shifts the direction and induces centering. **B** The cortical sites represent fixed vector saccades whose amplitude is modified by an eye position-dependent mechanism downstream of the site of stimulation. In this case, double stimulation shifts the direction but does not induce centering. **C** Double stimulation of sites in the lateral intraparietal area (*LIP*) results in a pattern of evoked saccades which is fully compatible with hypothesis 1

Studies of parietal single-units have resulted in a good understanding of how eye position information affects the firing of parietal neurons. Typically, saccade-related responses of parietal neurons are modulated by eye position in a monotonic fashion (Andersen et al. 1990). As shown by Zipser and Andersen (1988), the same pattern of modification characterizes the hidden units emerging at the intermediate level of 3-layered network models, whose output layer represents target location relative to the head and whose input layer consists of two sets of units, encoding retinal location and eye position respectively. While the output units encode head centered space in a localized manner, the hidden units (considered models of parietal neurons) encode head-centered space in a distributed fashion. Electrical microstimulation of parietal cortex can be simulated in such networks by fully activating these hidden layer units (Goodmann and Andersen 1989). This results in evoked saccades which show the amplitude modification by orbital position found in our experiments. In other words, the demonstration of modified vector saccades in LIP is in full accordance with the concept of a distributed representation of head-centered space in this part of the brain.

Possible Functional Relation Between LIP and the Intercalated Zone

Is there a relationship between the distributed representation of head-centered space in LIP and the localized representation in the intercalated zone? A possible answer is suggested by our double stimulation experiment, which has shown that one can increase centering by adding input from sites representing modified vector saccades with sufficiently different directions. As illustrated in Fig. 7, eye movements directed to a goal zone in head-centered coordinates would be expected to occur if the site being stimulated added inputs from sites representing modified vector saccades with several directions more or less equally distributed between 0° and 360°. As already mentioned, centering would be the necessary consequence of the eye position-dependent modification enhancing the contribution of modified vector sites representing centripetally directed saccades, while conversely reducing the contribution of centrifugally directed saccades. Goal zones other than straight ahead could be easily realized by adjusting the size of the contributions of the different directions converging on the same cell or group of cells. The intercalated zone is located right at the border between the representations of upward and downward saccades in the IPS. If local connections were indeed available, the neurons or groups of neurons in the intercalated zone could easily assimilate the inputs required to form a localized representation of head-centered space. However, if sites in the intercalated zone integrated only inputs from the adjoining representations of upward-contralateral and downward-contralateral modified vector saccades, eye movements should always converge in zones located in contralateral head-centered space. Goal zones congruent with the straight ahead location or goal zones in ipsilateral head-centered space, as found in the intercalated zone, would require input reflecting modified vector saccades with an ipsilateral component. Since saccade vectors with this directionality are found in the opposite hemisphere, a specific and testable prediction derived from these considerations is that the intercalated zone should integrate callosal inputs in addition to inputs from the adjoining parietal cortex of the same side.

A distributed representation would be fully sufficient to encode target location relative to the head. Actually, it would have several advantages over a more localized representation such as a much larger resistance of function to a loss of neurons. Why then does the brain take the trouble to implement a second, localized representation of head-centered space in the IPS? A speculative answer to this question is suggested by our observation that stimulation of the intercalated

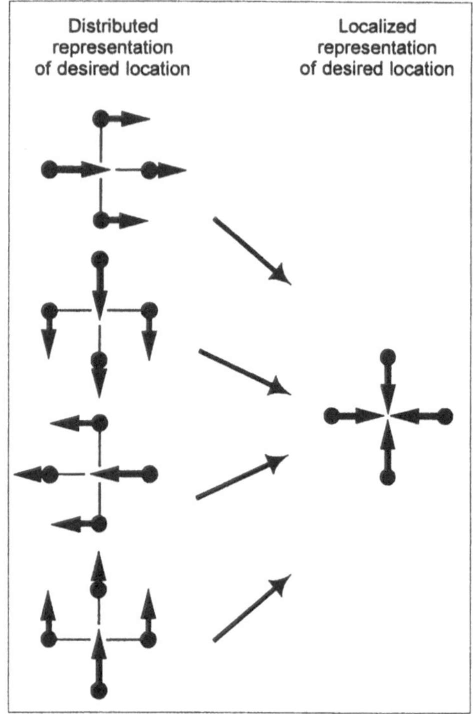

Fig. 7. Scheme suggesting how a localized representation of object location relative to the head might be derived from an earlier, distributed representation of head-centered space. (From Thier and Andersen 1996)

zone evokes combinations of non-eye movements such as movements of the head, shoulders, arms, parts of the face, or the pinnae accompanying the evoked saccades (Thier and Andersen 1996). The intercalated zone could therefore be a first stage contributing to the organization of coordinated movements of different body parts which may require access to multiple spatial representations. Integration of a signal derived from a localized head-centered representation, rather than independent inputs of eye position and retinal position, might help to reduce the complexity of such a system.

Conclusion

Experiments based on electrical microstimulation clearly question the concept of a parietal eye field as inaugurated by the studies of Ferrier. Rather than demonstrating a unitary parietal representation of eye movements, they suggest that the posterior parietal cortex of monkeys houses at least four distinct areas whose functional architecture suggests specific contributions to goal-directed eye movements. It is plausible from a phylogenetic point of view to find evidence for a parietal eye field specializing in smooth pursuit, namely, area MST, and, on the other hand, to find evidence for distinct areas specifically contributing to saccadic eye movements found in VIP, LIP or MP. Differences in the phylogeny of small and large amplitude saccades may also account for the development of two widely segregated representations for saccades such as LIP and MP. While their role in the organization of eye movements cannot be disputed, it would be a misleading simplification to understand these parietal areas as *eye fields* in the sense of areas making exclusive contributions to eye movements, similar to the well-established brainstem centers for saccades. The demonstration that stimulation of a saccade representation such as the tiny intercalated zone in VIP is able to evoke a complex motor synergy comprising saccades and non-eye movements as well is a good case for contributions of these parietal eye fields beyond eye movements.

The view then that electrical microstimulation is suggesting is that of a network of interdependent parietal modules offering different spatial representations shaped by behavioral context and phylogenetic history. The elements of this parietal network in monkey are small, with the intercalated zone representing an extreme case, having a size which is probably on the order of only 10 mm^2. This is far beyond the dimensions of the parietal lesions, the study of which is still the major basis of our view of the role of the function of human parietal cortex, and it is probably also still beyond the resolution of the functional imaging techniques available. This is a necessary qualification one should bear in mind before jumping to premature conclusions on similarities of parietal lobe functions in experimental animals and humans.

Acknowledgements. Supported by grants from the National Institute of Health and the Office for Naval Research to R.A. Andersen and by grants from the Deutsche Forschungsgemeinschaft to P.Thier.

References

Andersen RA, Bracewell RM, Barash S, Gnadt JW, Fogassi L (1990) Eye position effects on visual, memory, and saccade-related activity in areas LIP and 7a of macaque. J Neurosci 10:1176–1196

Bruce CJ (1990) Integration of sensory and motor signals in primate frontal eye fields. In: Edelman GM, Einar Gall W, Maxwell Cowan W (eds) Signal and sense. Wiley-Liss, New York, pp 261–314

Cavada C, Goldman-Rakic P.S. (1989) Posterior parietal cortex in rhesus monkey: I. parcellation of areas based on distinctive limbic and sensory corticocortical connections. J Comp Neurol 287:393–421

Cavada C, Goldman-Rakic PS (1993) Multiple visual areas in the posterior parietal cortex of primates. Prog Brain Res 95:123–137

Ferrier D (1876) Functions of the brain. Smith, Elder and Co, London

Glickstein M (1985) Ferrier's mistake. Trends Neurosci 8:341–344

Goodman SJ, Andersen RA (1989) Microstimulation of a neural network model for visually guided saccades. J Cogn Neurosci 1:317–326

Komatsu H, Wurtz RH (1988) Relation of cortical areas MT and MST to pursuit eye movements. I. Localization and visual properties of neurons. J Neurophysiol 60:580–603

Komatsu H, Wurtz RH (1989) Modulaton of pursuit eye movements by stimulation of cortical areas MT and MST. J Neurophysiol 62:31–47

Maunsell JRH, Van Essen DC (1983) The connections of the middle temporal visual area (MT) and their relationship to a cortical hierarchy in the macaque monkey. J Neurosci 3:2563–2586

Munoz DP, Wurtz H (1993) Fixation cells in monkey superior colliculus. I. Characteristics of cell discharge. J Neurophysiol 70:559–575

Pandya DN, Seltzer B (1982) Intrinsic connections and architectonics of posterior parietal cortex in the rhesus monkey. J Comp Neurol 204:196–210

Robinson DA (1972) Eye movements evoked by collicular stimulation in the alert monkey. Vision Res 12:1795–1808

Roucoux A, Guitton D, Crommelinck M (1980) Stimulation of the superior colliculus in the alert cat. II. Eye and head movements evoked when the head is unrestrained. Exp Brain Res 39:75–85

Schall JD, Morel A, King DJ, Bullier J (1995) Topography of visual cortex connections with frontal eye field in macaque: convergence and segregation of processing streams. J Neurosci 15:4464–4487

Shibutani H, Sakata H, Hyvärinen J (1984) Saccade and blinking evoked by microstimulation of the posterior parietal association cortex of the monkey. Exp Brain Res 55:1–8

Stanton GB, Bruce CJ, Goldberg ME (1995) Topography of projections to posterior cortical areas from the macaque frontal eye fields. J Comp Neurol 353:291–305

Thier P, Erickson RG (1992) Responses of visual-tracking neurons from cortical area MSTl to visual, eye and head motion. Eur J Neurosci 4:539–553

Thier P, Andersen RA (1993) Electrophysiological evidence for a second, medial "parietal eye field". Soc Neurosci Abstr 19:27

Thier P, Andersen RA (1996) Electrical microstimulation suggests two different forms of representation of head-centered space in the intraparietal sulcus. Proc Natl Acad Sci USA 93:4962–4967

Zipser D, Andersen RA (1988) A back-propagation programmed network that simulates response properties of a subset of posterior parietal neurons. Nature 331:679–684

Area LIP and the Population Vector: Single and Double Targets

S. BARASH

Department of Neurobiology, Weizmann Institute of Science, Rehovot, Israel

Ubiquity of Visual Scenes and of Sensorimotor Tasks Involving Multiple Locations

Our world is filled with numerous objects. Almost any visual scene contains several significant objects or object parts, inevitably, involving multiple locations.

Almost all natural sensorimotor tasks involve computations based on the locations of multiple objects. For example, our ability to avoid obstacles as we move around is based on comprehending these obstacles simultaneously, at least comprehending their locations. A football player passing the ball to a team mate may have to take into account not only the positions of several other players, but also their movements, even their ability to rapidly change their movement.

Maintaining information about multiple "pure" locations, independent of any object, is frequently also necessary, as when a predator might appear from two specific directions, or in the clinical line bisection task, and in any other situation where the relative configuration of several locations must be considered.

In this chapter we will discuss saccadic eye movements and, in particular, intentions of making saccades. Most of the visual fixations in typical visual scenes are concentrated in few locations, which might be the locations of the most informative details (Yarbus 1967). Presumably, prior to any given saccade, several of these locations are considered as multiple potential targets, the actual target only later being selected.

Synopsis

At issue is the neuronal activity in the lateral intraparietal area (area LIP) in the presence of multiple potential targets for saccades. The last section of this chapter will describe the neuronal activity evoked by a novel task, designed specifically to study the neuronal coding of simultaneous multiple targets. The activity in this

In: Parietal Lobe Contributions to Orientation in 3D Space (1997). P. Thier and H.-O. Karnath (eds). Springer-Verlag, Heidelberg.

task does not appear to follow predictions based on the standard, single-target tasks.

First we will survey evidence obtained with single targets. The close relationship of area LIP to saccades, suggested by its anatomical connectivity, is supported by the presence of strong saccadic activity. Moreover, strong neuronal activity marks also the intention to make the next saccade. Neuronal fields of activity in LIP are frequently broad. Similar broad neuronal fields in primary motor cortex (M1) led Georgopoulos to formulate the population vector theory. Georgopoulos's theory is briefly surveyed, as well as another well-known example of population coding, the work of Sparks and colleagues on the superior colliculus. Afterwards the adequacy of the population vector approach is considered for area LIP. The adequacy is not self-evident, as the general pattern of activity in LIP is multi-phased and more complex than in M1. Nevertheless, for single targets, the requirements for the applicability of the theory hold in LIP. However, the population vector theory cannot directly explain the results with multiple targets. In fact, this theory in principle cannot directly account for the simultaneous coding of multiple targets.

Finally, the neuronal activity recorded in the new task is consistent with extinction. This neuropsychological deficit, commonly found in parietal damage, specifically requires simultaneous comprehending of two objects.

Intended Saccade and Parietal Area LIP Neuronal Activity

Early Electrophysiological Studies of the Inferior Parietal Lobule and the Discovery of Area LIP

Initial single-unit recording studies of the inferior parietal lobule clearly indicated that this cortical region is not somatosensory, as originally suspected. Hyvarinen (1982) described how he discovered almost by chance the visual responsiveness of this region. Hyvarinen and Poranen (1974) and Mountcastle et al. (1975) described single-neuron activity related to active vision, mostly to smooth pursuit of visual stimuli of interest. Linking single-neuron activity to specific types of eye movements required simultaneous measurement of eye position, which was accomplished by Lynch et al. (1977), who reported on neurons related to saccadic eye movements. These neurons were active before and during visually evoked saccadic eye movements, but not before spontaneous saccades. This distinction, observed in the neuronal activity, nicely complements a distinction made earlier for eye movement deficits after cerebral, mainly parietal lesions. Cogan and Adams (1953) and Cogan (1965), coined the term "ocular motor apraxia" to refer to the state in which volitional movements of the eyes are impaired, but random movements are retained.

At this time doubts were raised by Robinson et al. (1978) whether what seemed like a saccadic discharge is indeed saccadic, or actually the discharge reflects covert shifts of attention. Bushnell et al. (1981) showed that many parietal neurons that show a saccadic discharge, do indeed discharge also in relation to covert shifts of attention.

The relationship between attention and saccades is both subtle and complex. The parietal cortex is clearly related to visual attention. Nevertheless, substantial evidence was later provided that neurons in the parietal cortex are also related to the intention of making saccades. Two major developments led to this evidence.

First, in the early 1980s the parcelation of the cerebral cortex into a hierarchical system of areas based on task connectional criteria became prevalent. In the parietal cortex, Andersen and collaborators (Andersen et al. 1985; Asanuma et al. 1985; Andersen et al. 1990) focused on the connections of subregions of the parietal cortex with a variety of other regions, including established saccadic centers such as the frontal eye fields and the superior colliculus. They showed that one parietal subregion, which they named LIP, has very strong connections to these saccadic centers. These anatomical connections suggested that area LIP might be functionally related to saccades. The second development was the introduction of memory-guided saccades.

Memory-Guided Saccade Task Reveals Separate Phases of Activity

The activity pattern of a typical saccadic neuron during a standard visual saccade, made in its preferred direction, is the following: The neuron begins to discharge after the target appears, sometime during the latent period of the movement; the discharge increases towards the onset of the movement, and continues throughout the saccade. Unfortunately, this pattern of discharge does not allow determining whether the discharge is visual - a response to the appearance of the target - or motor, a preparation for the upcoming saccade.

This problem was resolved with the introduction of memory-guided saccades (Hikosaka and Wurtz 1983). Andersen and his colleagues (Andersen et al. 1987; Gnadt and Andersen 1988; Barash et al. 1991a,b; Mazzoni et al. 1996) used memory-guided saccades to study the parietal cortex. They showed that LIP neurons have separate visual, saccadic, and also intention-related activity occurring during the memory period.

Fig. 1 shows a typical response of an LIP neuron recorded from a monkey while the monkey was performing memory-guided saccades. All the trials in Fig. 1 were made at one target location, positioned in the neuron's preferred direction. At the beginning of the trial, a fixation spot appears on the screen, in front of the animal. The monkey must fixate this spot as long as it is on the screen. Midway through this fixation, a target spot is flashed somewhere else on the screen. The monkey must remember the location of this target spot after its offset, until the offset of the

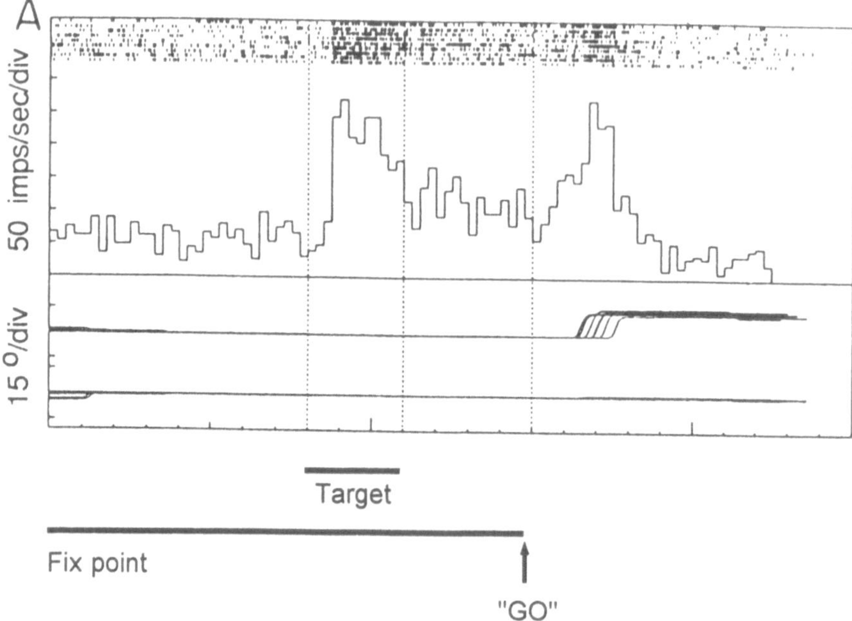

Fig. 1. The memory-guided saccade task: sequence of events and the response of a typical lateral intraparietal area (LIP) neuron. Onset and offset times of both fixation and target spots are indicated both in schematics at the bottom of Fig. 1, and by the *dotted vertical lines* above. Shown from the top are the spike rasters, the resulting histogram, and the superimposed traces of the horizontal and vertical components of eye position. Trials are aligned on sensory events (note variable saccadic latencies). From Barash et al. (1991a)

fixation spot which is, by definition, a memory period. After the offset of the fixation spot (marked as "GO" in Fig. 1), the monkey must make a saccade towards the remembered location of the previously shown target. This movement is made in the dark.

Figure 1 shows the superimposed traces of the vertical and horizontal components of the eye position. These traces demonstrate that the monkey does indeed accomplish the task previously described. The raster display (at the top) shows the times that spikes occurred in the single LIP neuron being recorded; each row in the raster represents one trial. Trials are aligned with respect to the sensory stimulation; the spikes are summed in the histogram below.

At the beginning of the trial, while the monkey looks at the fixation spot, there is a low base-level of activity. The appearance of the target stimulus triggers a strong discharge. Because eye movements are made neither during this visual stimulation, nor for some time after it, this discharge is a visual response to the target ("light sensitive").

A second prominent burst of activity can be observed around the time the saccadic movements are made. It begins some time before the movement, but after the fixation spot offset, and continues until after the saccade is completed. Because the saccade is made in the dark, long after the target stimulus is extinguished, this burst cannot be visual and is indeed related to the saccadic movement.

During the memory period, the neuron maintains a sustained level of discharge (compared to the base activity in Fig. 1). This activity is related to the monkey's intention to make the next saccade, as will be discussed later.

Thus, the memory-guided saccade task enables differentiation of three types of activity that cannot be distinguished on the basis of recordings in the visually-guided saccade task. The activity (of the typical neuron illustrated in Fig. 1) begins with a sensory event and is maintained throughout the duration of the motor intention, and until the completion of a motor act; this activity is indeed sensorimotor.

The neuron illustrated in Fig. 1 is excited in all three phases of the response. Other LIP neurons may respond in any one, two, or all three phases; also, neurons may not be excited, but inhibited. The overall pattern of activity may occasionally appear quite complex. This should be kept in mind in the discussion of the adequacy of the population vector approach.

Neuronal Activity in Area LIP Related to Saccades

The anatomical definition of area LIP, as the parietal region densely connected to saccadic centers, suggested that area LIP is strongly related to saccades - more strongly than, say, neighboring area 7a, which was previously considered part of the same cortical area. Are there differences in the neuronal activity between these areas consistent with this anatomical makeup?

One difference between areas LIP and 7a is in the distribution of the neuronal activity throughout the phases of the trial. Area LIP neurons show higher activity in the visual and intention (memory) phases. They are also more active in the beginning of the saccadic burst, throughout the saccadic movement. Afterwards, however, activity in area LIP declines. In contrast, area 7a shows weaker visual, intentional, and pre-saccadic activities, but intense post-saccadic activity (Barash et al. 1991a).

The latency of the saccadic burst is a particularly important parameter to study because a necessary condition for an area to participate in preparing a movement is that a significant fraction of its neurons has to be activated before the initiation of the movement. Fig. 2 illustrates analysis of the saccadic latency and compares areas LIP and 7a. The analysis of an individual LIP neuron is illustrated in panel A of Fig. 2. The trials are aligned on the beginning of the saccade. The activity at

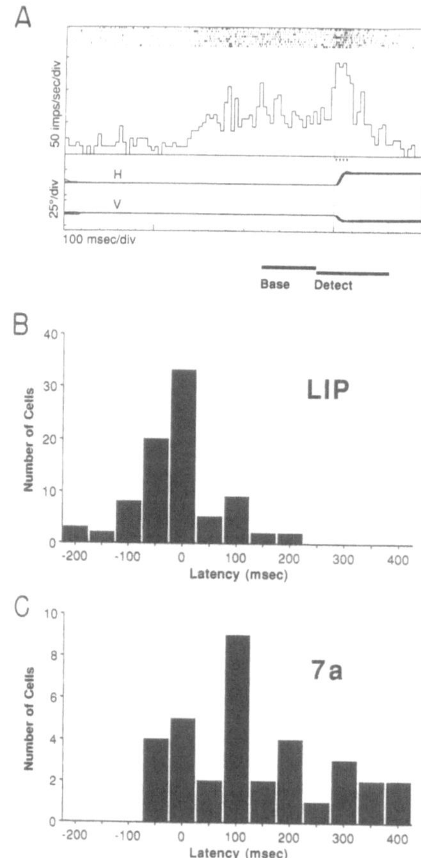

Fig. 2 A-C. Latency of saccadic burst in areas LIP and 7a. **A** determination of the latency for an individual LIP neuron. Same format as Fig. 1, but trials are aligned on the beginning of the saccadic movement. Histogram bins within the detect interval that showed significantly higher activity than the "base" period are marked with *arrows*. The latency was defined as the time from the beginning of the saccade to the midpoint of the first marked bin. (Hence negative latency implies pre-saccadic start of saccadic burst). **B, C** Distributions of latencies of the saccadic bursts in areas LIP and 7a, respectively. From Barash et al. (1991a)

each time-bin, after the beginning of the saccade, is compared to that in a base interval (as marked in Fig. 2). For the neuron illustrated in the top panel, the latency was 10 ms, that is, 10 ms after the movement had begun, its activity became significantly different from that of the baseline.

Panels B and C of Fig. 2 compare the distribution of saccadic latencies for the neurons recorded in areas LIP and 7a. In LIP, 72% of the neurons were significantly more active than baseline by the time the saccade had begun. In contrast, only 18% of the neurons in area 7a were significantly activated at that time.

The functional separation between LIP and 7a was corroborated in another study in which electrical microstimulation was applied to several parietal areas (Thier and Andersen, this volume). Stimulating sites in LIP yielded saccadic eye movements with low current thresholds. The saccades were generally in the same directions as the fields of single neurons in the immediate vicinity of the stimulation sites. In contrast, stimulation of area 7a did not generate eye movements, only blinks (at least if currents were less than 200 μA).

These results show that, within the inferior parietal lobule, area LIP is specifically closely related to saccades and to intentions to make saccades, which is consistent with its anatomical connections.

Neuronal Activity in Area LIP Related to the Intention to Make a Saccade

The sustained activity in the memory phase of the memory-guided saccade task could in principle be either sensory or motor. That is, the activity could either be related to the visual impression, or, alternatively, to a motor program to make the required saccade formed early in the trial, and maintained in memory throughout the memory interval, until the eye is allowed to move. (More briefly, the second alternative is that the activity is related to the motor intention to make the subsequent saccade).

Just as the memory-guided saccade task was needed to dissociate visual from saccadic activity in the visually-guided saccade task, here a new task is needed to dissociate the two possible explanations for the sustained memory activity. The back-saccade task illustrated in Fig. 3 accomplishes this dissociation. Panels A and B sketch the task. The initial fixation spot is replaced by a brief presentation of a peripheral target; the monkey must first saccade to the locations of this target, and then saccade back to the initial fixation position. Both movements are made in the dark, because the target is extinguished before the first saccade begins (during the latent period). The essential point is that no retinal stimulation occurs in the trial in the direction of the second saccade.

Figure 3 illustrates the activity of an LIP neuron in the back-saccade task. Mapping the activity of this neuron by memory-guided saccades showed that it is strongly excited in the visual, memory, and saccadic phases of saccades in the upwards direction. (Post-saccadic activity followed saccades in the downwards direction). Panels C-E of Fig. 3 illustrate back-saccade trials in which the target was in the upwards direction, that is, in the response field of the neuron. In this case the target onset is followed by a discharge which continues through the upward saccade. After the end of the first saccade the activity declines (panel D)

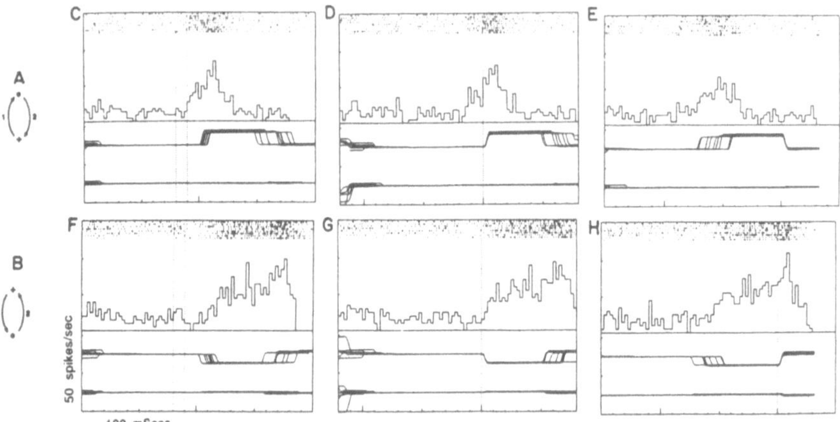

Fig. 3 A-H. Back-saccade task. **A, B** Sketch of the task. The first saccade is to the single target, the second saccade is made in the dark back to the location of the original fixation spot. **C- H** Activity of a single neuron in the back-saccade task. The preferred direction of this neuron is upwards. In **C** and **F** trials are aligned on the sensory stimuli. First *vertical dashed line* represents offset of the fixation spot and simultaneous onset of target; second *vertical dashed line* represents target offset. In **D** and **G** trials are aligned on start of first saccade, and in **E** and **H** on start of second saccade. Same format as Fig. 2. From Barash et al. (1991b)

and remains low throughout the second saccade (panel E), until the end of the trial. This pattern of activity is fully accountable on the basis of the results of the memory-guided saccades.

In contrast, the pattern of activity evoked by a target in the downward direction (panels F-H of Fig. 3) resolves the issue of the sustained activity in memory-guided saccades. After the target is presented (outside the neuron's receptive field) the activity is slightly inhibited (panel F), and the activity remains low until after the first saccade, made in the downward direction, outside the neuron's movement field (panel G). Immediately after the completion of the first saccade the neuron's activity is elevated (panel G) and it is sustained until after the second, upward saccade (panel H). Because no visual stimulus was presented in the upward direction in this trial, the sustained activity is not visual; it is related to the intention to make a movement into the saccadic field.

Mazzoni et al. (1996) reinforced the conclusion that the sustained activity is intentional by reporting that the activity can be evoked in total absence of visual targets, by auditory memory-guided saccades. The majority of the neurons with responses in the auditory memory-saccade task also responded in the visual memory-guided saccades. The spatial tuning for the two modalities was the same in 85% of the tested neurons.

An important property of intentions is that they can be cancelled in the absence of overt action. What would happen in these circumstances to the intentional activity? Bracewell et al. (in press) put this to the test, by introducing a novel task. Trials began like standard memory-guided saccade trials, but in half the trials an alternative target was flashed midway through the trial. The monkey was trained to change his motor plan and, after the offset of the fixation spot, saccade to the location in which the second target had been flashed. The neuronal activity in area LIP followed the monkey's motor intention. If the first target was in a neuron's preferred direction and the second target away from it, the sustained activity evoked by the first target was abolished by the onset of the second target. Alternatively, if the first target was positioned away from the preferred direction, and the neuron consequently inactive, a second target in the preferred direction generated sustained activity that continued until the saccade. These on and off switches of the activity could be repeated in the same trial by changing and re-changing the monkey's motor plan.

What would happen to the intentional activity if the monkey planned a sequence of saccades? Mazzoni et al. (in press) introduced a memory-guided version of the double-saccade task, which enabled the careful testing of this question without mixing the visual with the intentional and saccadic activities. The results were that sustained activity in most neurons (81%) encoded specifically the next intended saccade, but not the visual or the subsequent saccade. It is as yet unclear how (and where) the motor plan for the second saccade is maintained.

In sum, several studies explored the neuronal activity in area LIP using memory-guided saccades and related tasks that all use single targets and target sequences (but not simultaneous targets). These studies demonstrated strong sustained neuronal activity, reflecting the intention for the single next saccade.

The analysis presented in this section was based solely on the activity in the neuron preferred directions of the analyzed neurons. The next set of questions concerns the spatial characteristics of these neurons.

The Population Vector Approach Is Adequate for Single-Target-Related Neuronal Activity in Area LIP

The neuronal responses in area LIP are multi-phasic and complex. The fields of activity are broad. What is the relationship of the neuronal activity in LIP to the emergent eye movement? The population vector theory is a highly-influential theory, offering a specific, quantitative hypothesis of the relationship of broad neuronal fields to behavior. The ubiquity of broad fields in the mammalian brain, and the general recognition that cortical functions are mediated by neuronal populations, may lead to the assumption that the population vector approach is a priori valid. However, certain conditions must hold for this approach to be applicable. Also, even if the population vector approach is adequate, it is not clear if it is consistent with the activity evoked by double simultaneous stimulation. The present section examines whether the pre-conditions for the population vector approach are satisfied by the activity evoked in area LIP by single-target tasks.

The Population Vector Theory in Primary Motor Cortex

The population vector theory of Georgopoulos offers an algorithm, whose input is an array of parameters derived from standard single-unit recordings that characterize the response of single neurons; and whose output is a single vector which is a predictor of behavior. This section briefly describes the population vector theory.

The population vector theory is based on the spatial activity profiles of individual neurons recorded in the primary motor cortex, M1 (Georgopoulos et al. 1982). M1 neurons were recorded from monkeys making arm movements from a central position to eight equidistant peripheral targets. The discharge of most of the neurons was determined primarily by the direction of movement. For most neurons, there was some direction in which the discharge was most intense; this was called the neuron's "preferred direction". When movements were made in directions farther away from the preferred direction, the discharge was gradually reduced. (Figure 5 illustrates the response of a typical M1 neuron to movements in eight directions).

The direction tuning of most M1 neurons is generally similar: discharge is a sinusoidal function of the arm movement's direction. More specifically, three parameters – preferred direction $\overline{PD}$, baseline, and maximal levels of activity – are sufficient to predict the mean discharge related to *any* movement $\overline{M}$:

$$TotalDischarge(\overline{M}) = baseline + D \cdot \cos(\overline{M} - \overline{PD})$$

The net discharge D in the formula is defined as the difference between baseline and maximal levels of activity. (In more detail, the maximal level is found by sinusoidal regression, not by actual measurement). This tuning characterization was later extended to movements in three-dimensional space (reviewed, Georgopoulos 1995) and to a number of activities and cortical areas, even inferotemporal activity related to the perception of faces (Young and Yamane 1992).

The most prominent aspect of the sinusoidal character of direction tuning is that it is broad. A neuron's discharge changes with movements in a wide range of directions, rather than a single direction. This implies that movements in any particular direction are not subserved by motor cortical neurons uniquely related to that direction. Instead, neurons with overlapping tuning curves cooperate to generate the movement. What is the relationship of individual neurons to the emerging motor command? Can the relationship be described quantitatively? Georgopoulos et al. (1983) addressed this issue by defining the "population vector" $\overline{PV}$ as a vector sum. The summands are the preferred directions of individual neurons, weighted by the strength of their discharge; that is, the population vector $\overline{PV}$ is defined for any condition *cond* in which the discharge of each neuron, in the sample D_n is known.

$$\overline{PV}(cond) = \sum_{\substack{n \in sampled \\ neurons}} D_n(cond) \cdot \overline{PD_n}$$

The population direction that is defined in this way is, of course, a vector. Therefore, the summand characterizing a given neuron, that is, the contribution of a given neuron to the action of the population, is always made by exerting "influence" in the direction of its preferred direction (or opposite the preferred direction, for inhibition).

The basic condition in which the discharge of M1 neurons has been characterized is the condition during reaching movements, as described previously. Then, the sinusoidal tuning formula can be substituted in the population vector formula, yielding a population vector $\overline{PV}$ for a movement in direction $\overline{M}$:

$$\overline{PV}(\overline{M}) = \sum_{\substack{n \in sampled \\ neurons}} D_n \cdot \cos(\overline{M}, \overline{PD_n}) \cdot \overline{PD_n}$$

Clearly, the closer the preferred direction of a given neuron is to $\overline{M}$, the more this neuron will contribute to the population vector. The preferred directions are distributed in space roughly equally. Therefore, the population vector will be directed close to $\overline{M}$. Calculating directly with their recorded data, Georgopoulos et al. (1988) found that about 100 neurons have to be summed up for the variation of the resultant population vector to be reduced to the variation of the movement end-point itself.

Population Coding in the Superior Colliculus

Another prominent and well-explored case of population coding is the discharge of superior colliculus neurons related to saccadic eye movements (reviewed, Sparks and Groh 1995). The main distinctive feature of the superior colliculus as a model for population coding, in comparison to both M1 and LIP, is that the superior colliculus is arranged as a precise polar map of target positions (with respect to current eye position). Also, the collicular discharge relates to saccadic eye movements, which lack the complexities specific to arm or body movements.

In the superior colliculus, as in M1, the movement fields of individual neurons typically occupy a rather broad region in space. The maximal activation is yielded by a direction roughly at the center of the movement field, called the neuron's preferred direction (Wurtz and Goldberg 1972; Sparks et al. 1976; Sparks and Mays 1980). Hence, movement fields of neighboring neurons generally overlap. The question has been raised about which collicular neurons determine a saccade's direction? Is it the whole population of activated neurons, or only the relatively few neurons that are maximally activated (McIlwin 1976; Sparks et al. 1976). Sparks et al. (1976) suggested that: “precise saccadic movements are not produced by the discharge of a small population of finely tuned neurons, but result from the weighted sum of the simultaneous movement tendencies produced by the activity of a large population of less finely tuned neurons”.

Lee et al. (1988) elegantly used the special topographical ordering of the colliculus to examine the hypothesis of Sparks et al. (1976). Because of this topographical ordering, the whole population of neurons activated by any given saccade is concentrated in some region of the collicular map. This region, which spans a finite extent of the collicular map, must contain neurons with a range of preferred directions. Preferred directions at the borders of the activated region are particularly prone to vary from the neuronal preferred directions at the center of the activated region. Lee et al. (1988) explored why these differences in preferred direction within the activated neuronal population do not seem to affect the movement direction. They reasoned that, normally, these deviations in preferred directions would cancel out each other if summed over the whole border because the region is nearly symmetrical relative to its center. They then tested this idea by injecting lidocaine and thus reversibly deactivating known locations of the collicular map. Saccades made very close to the location that was inactivated in the collicular map were quite accurate. Conversely, saccades made slightly off the inactivated location were deformed so that the movement end-points were shifted away from the focus of deactivation. These results show that the whole population of activated neurons collaborate in determining the movement, not only the few neurons whose preferred directions are exactly identical to the planned movement.

The Population Vector and Double Simultaneous Stimulation: Presentation of the Problem

For any given set of neuronal preferred directions and activations, the population vector is always defined, and it is always a single vector. Therefore, by definition, the population vector cannot simultaneously represent locations, or directions, of multiple targets.

There are two levels for which the population vector scheme might have to be changed to accommodate multiple targets. The first level is that of the single neurons. What happens to the spatial tuning of single neurons in the presence of two targets? Is the response to a target in the preferred direction modulated by the existence of a second target? In particular, is the sinusoidal direction tuning still valid for M1 neurons?

The second level concerns the joint action of the neuronal population. The vector summation, an operation that always yields a single resultant vector, will have to be replaced by another operation that does not have this restriction.

Glimcher and Sparks (1992) studied saccadic superior colliculus neurons in the presence of two stimuli. This is perhaps the only single-neuron study with multiple targets. However, the results were that the neurons were activated only after the required (single) saccadic movement was made known to the animal. Thus the the superior colliculus may be not involved in maintaining multiple potential targets.

Adequacy of the Population Vector Approach for the Saccadic and the Intentional Neuronal Activity in Area LIP

Most LIP neurons are spatially selective. They are not active during all memory-guided saccades, only during saccades made to certain parts of space. Fig. 4 shows the response of an individual LIP neuron that was activated when a saccade was made in either up-and-right, up, up-and-left, left, or left-and-down directions, but remained inactive during saccades made in other directions. Spatial selectivity is required for the neurons of any region that may be related to spatial processing. Clearly, area LIP satisfies this condition, but the very broad direction tuning of the neuron illustrated in Fig. 4 echoes the questions raised regarding the relationship of neuronal activity to movement in the studies of Georgopoulos.

Fig. 5 shows a typical spatial tuning of an M1 neuron studied by Georgopoulos et al. (1982). The overall similarity of the direction tuning in Figs. 4 and 5 strongly suggests that the population vector theory, formulated on the basis of responses such as in Fig. 5, will also be adequate to describe the activity in Fig. 4. However, some differences do exist in specific characteristics of the activity in M1 and LIP. Three points will be discussed in this context.

First, the activity in area LIP is of at least the three types described above - visual, intentional, and saccadic. A close examination of Fig. 4 shows that there are differences in direction tuning between these types of activity, even if only a subtle one. For example, the upwards-directed trials show strong saccadic activity but little intentional (memory period) activity. In the leftwards trials, the intentional activity is stronger than the saccadic. In many neurons, not all three types of activity are present. Furthermore, some neurons are excited in one phase, and inhibited in another. These variations might be more complex than the analogous pattern of activity in M1. Fig. 6, from Smyrnis et al. (1992), shows the activity of an M1 neuron tested while a monkey performed memory-guided reaching, analogous to memory-guided saccades. The neuron begins discharging

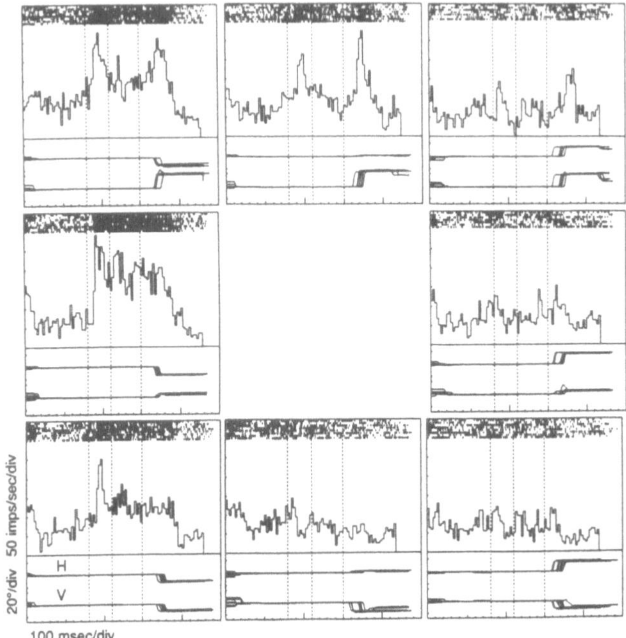

Fig. 4. Activity of an area LIP neuron made in eight equally spaced directions, positioned on a 15° grid. Each panel illustrates the trials made in the direction the panel occupies relative to the center of Fig. 4 (Trials appeared in the experiment in a randomly interleaved order, and reordered post-hoc.) Same format as Fig. 2. From Barash et al. (1991b)

after the flashing of the cue and continues up to the arm movement. This sustained activity, reflecting motor intention, is analogous to the intentional activity in LIP. Nevertheless, the activity in Fig. 6 is more uniform than that of Fig. 5; the tuning of the sustained intentional activity is quite similar to the "visual" activation and to the final motor discharge.

To clarify the extent to which the tuning of the different types of activity do vary, Barash et al. (1991b) compared the preferred directions of the different types of activity for each neuron. The preferred directions were nearly aligned. The visual, intentional, and also the pre-saccadic "time-slice" of the saccadic activity, were well aligned. The saccade-coincident activity is somewhat less well aligned with the earlier phases. The post-saccadic activity is not well-aligned. This difference in alignment might reflect additional sources of activity that join in during the saccadic phase, such as corollary discharge, feedback, or activity dependent on static eye position.

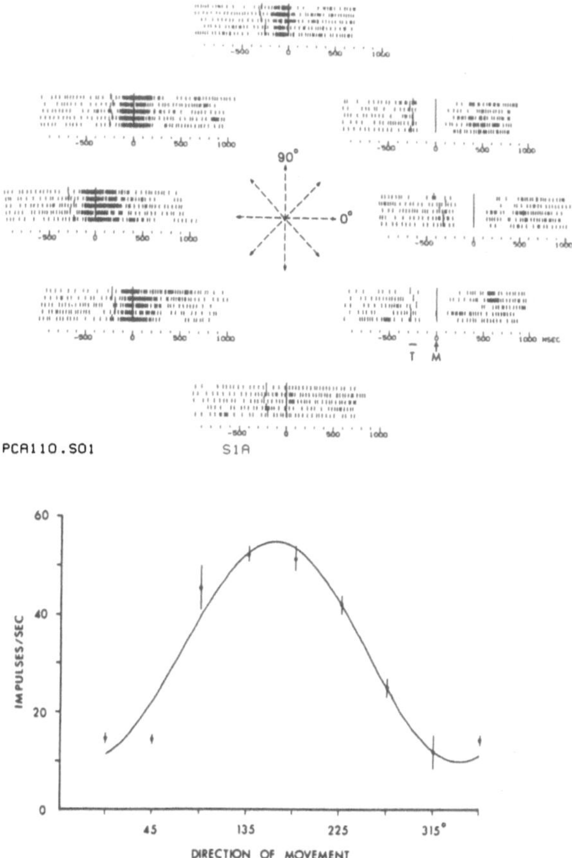

Fig. 5 A, B. Activity of a motor cortex neuron varying with the direction of arm movement. **A** Rasters are aligned on the movement onset, *M*, and show spike record during five repetitions of movements made in each of the eight directions indicated by the center diagram. **B** Directional tuning curve of the same neuron. A sinusoid is fitted to the neuron's responses by regression. From Georgopoulos et al. (1982), by permission

Translated to the population level, these results mean that, as a trial proceeds from the visual to the pre-saccadic stages, the population of active neurons will be quite stable, though the activity of some neurons will wax and wane. During the saccade, the set of bursting neurons will still be similar to the set of neurons active before. After completing the saccade, the set of active neurons will change.

Another apparent difference between areas LIP and M1 concerns tuning width. The cosine approximation for the tuning of M1 neurons implies that tuning width is similar for different neurons. In contrast, in area LIP tuning width varies considerably between neurons, as shown in Fig. 7. Some neurons are very sharply

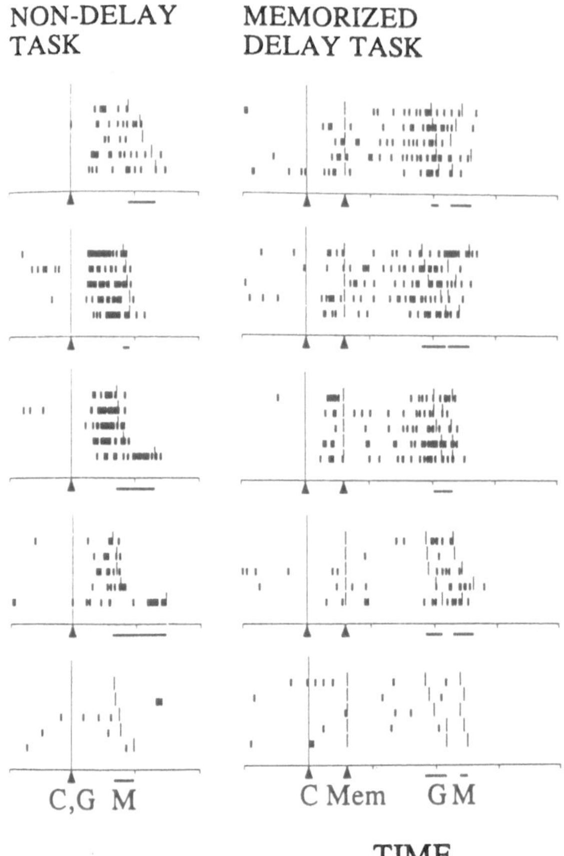

Fig. 6. Spike record of an M1 neuron during five trials in simple, non-delay reaching and in a memorized delay reaching task that is analogous to the memory-guided saccade task. The *arrows* on the left indicate the movement's direction. Longer *vertical bars* indicate the times of behavioral events. Rasters are aligned on the visual cue (*C*). In the non-delay task, this is also the go signal (*G*); *Mem* is cue offset; *M* is onset of movement. *Horizontal bars* indicate the range of times above; *filled triangles* indicate fixed events. From Smyrnis et al. (1992)

tuned, activated in virtually a single direction (Fig. 7a). Other neurons lack direction selectivity (Fig. 7c). The typical tuning is, however, intermediate, spanning roughly a quadrant (Fig. 7b): the median tuning width, at 50% maximum net activation, was 90° for both visual and saccadic types of activity.

Most LIP neurons are broadly tuned, even though some are narrowly tuned, as previously discussed. Consequently, the set of neurons activated before any given

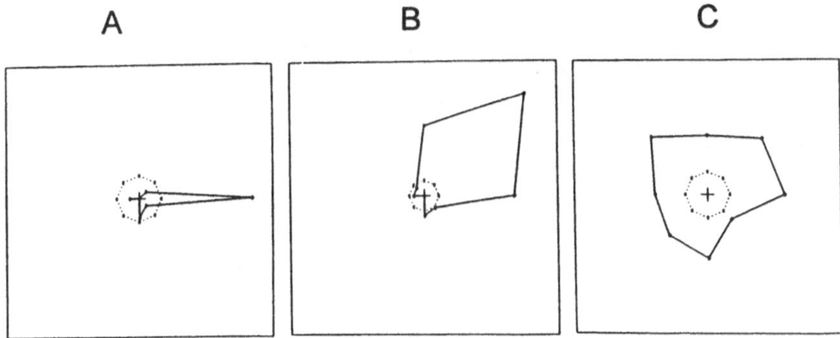

Fig. 7 A-C. Variation in tuning width: direction tuning curves recorded from three LIP neurons. Polar plot depicts, for each direction in which a saccade was made, the mean response of the neuron at the relevant part of the trial. *Bottom curve* is saccadic, the other visual. The *dotted line* represents mean baseline activity (recorded during the initial fixation, before target presentation). Calibration: the baseline activity was, **A-C**, 12, 5, and 7 spikes/s, respectively. From Barash (1990)

saccade contains not only the few narrowly tuned neurons whose preferred directions exactly match the planned saccade. Rather, the activation will encompass a much larger population of broadly tuned neurons, neurons whose fields cover the target of the planned saccade. Is the population vector a reasonable representative of the activity of the population in LIP as well?

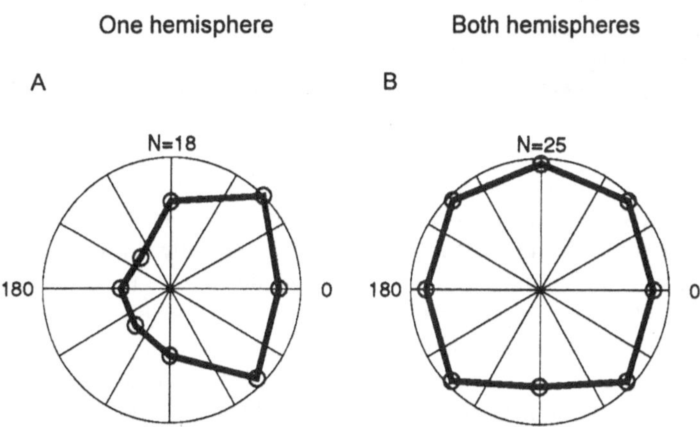

Fig. 8 A-B. Distributions of saccadic preferred directions of LIP neurons. **A** For one hemisphere. Here 0° is contralateral, 180° ipsilateral. Most of the neurons are contralateral. **B** Derivation from **A** of the distribution for the two areas LIP of both hemispheres of a monkey. The distribution is close to uniform. Based on Barash et al. (1991b)

At first sight, this condition does not hold for LIP because most of the neurons in any given hemisphere are contralateral (Fig. 8a). However, a distribution of preferred directions for both hemispheres can be derived by adding the single-hemisphere distribution to its mirror image (Fig. 8b). The resulting distribution, which reflects both LIP areas of a given monkey together, is remarkably uniform.

In sum, the three requirements that must hold for the population vector theory to be applicable do hold in areas LIP. They include: firstly, generally broad spatial selectivity, so that before any given movement a large population of neurons is activated, with varying preferred directions; secondly, unimodal, roughly symmetric direction tuning for the individual neurons; thirdly, the distribution of preferred directions is nearly uniform.

Nevertheless, the situation in LIP is apparently more complex than in M1 in at least two ways. Firstly, the visual, intentional, and saccadic components of the activity combine in a complex manner. Secondly, the tuning width is more variable. These complexities clearly do not undermine the basic validity of the population vector approach, they may yet be of significance in the coding of simultaneous stimuli.

The Neuropsychological Syndrome of Extinction Also Indicates That the Parietal Cortex Is Involved in Processing Double Simultaneous Stimulation

Up to this point, we noticed that visually-evoked, volitional saccades are in deficit following parietal lesions, and that area LIP is strongly related to the intention to make saccades. We further noted the basic adequacy of the population vector approach for the neuronal activity in area LIP, and that this approach is not helpful in understanding simultaneous coding of multiple targets. But should this apparent difficulty concern us? Is simultaneous coding of target location relevant at all to neuronal processing in LIP?

Extinction is one of the characteristic deficits resulting from parietal lesions in humans, particularly unilateral parietal lesions. Two identical stimuli are positioned in front of the patient, typically symmetrically with respect to his vertical midline. One stimulus is in the intact field, the other in the neglected field. The patient with pure extinction deficits is capable of detecting and recognizing stimuli not only in his intact field, but also in his neglected field - if the stimuli appear in isolation. However, in the configuration of double simultaneous stimulation the patient reports only the stimulus in the intact field.

Extinction was discovered very early in the history of the study of occipito-parietal lesions. Husain and Stein (1988) quote a 1905 report of a presentation by Balint of a dog with "occipital injury" in which, when presented with two pieces of meat, "first caught the meat placed more to the right instead of the closer one". Until very recently, extinction was considered a sign of neglect (e.g., Rafal and

Robertson 1995). The reason is that one characteristic of neglect, made particularly evident if compared to homonymous hemianopia, is its intermittent character. In hemispatial neglect, objects in the "neglected" hemifield may sometimes be missed and at other times be detected, for no apparent reason. Intermittency is likely to reflect contingency. Whatever neglect is contingent upon must be absent in the circumstances of testing for extinction.

Since extinction is an acquired incapacity to comprehend double simultaneous stimuli, that parietal damage frequently leads to extinction suggests that the intact parietal cortex is instrumental in the comprehension of double simultaneous stimuli. Translated to LIP, this may suggest that the activity of LIP neurons will be specifically modulated in the circumstances of double simultaneous stimuli in ways not obvious from single-targets.

Double Simultaneous Stimuli: Activity in LIP

In the final section of this chapter, I will briefly allude to some new data that directly probe the activity of parietal neurons when a monkey must relate to two locations. Armenuhy Melikyan and Alexey Sivakov participated in the experiments in which these data were collected.

We chose to focus on the representation of targets for saccades in area LIP. As discussed previously, the activity of single neurons in this cortical area is related to the intention to make saccades. The visual and intentional fields of these neurons are generally broad, and the population vector approach is appropriate for understanding how the overall pattern of activity in this area conveys the direction of the intended saccade. But what happens if there is more than one target for a saccade?

First, a behaviorally controlled task must be conceived in which the monkey must relate to two target locations. We have achieved this in the following manner: Instead of the single target followed by a "memory" interval of the memory-guided saccade task, in the newly conceived task two pairs of targets are flashed on the screen – a first pair of targets, then a first memory interval, then a second pair of targets followed by a second memory interval. All four targets are identical spots. They differ only in location (and time) of presentation. The targets are placed so that, in each trial, one (and only one) target location is common to both the first and second pair. The monkey has to make a saccadic eye movement towards this common target location. Of course, several locations are used and each one of them has an equal probability of holding the common target, and all target combinations are randomly ordered. Also, a small fraction of the trials of the new task (up to one third), are memory-guided saccades, so that the intention field of the neuron can be mapped at the same time that its response to target pairs is characterized.

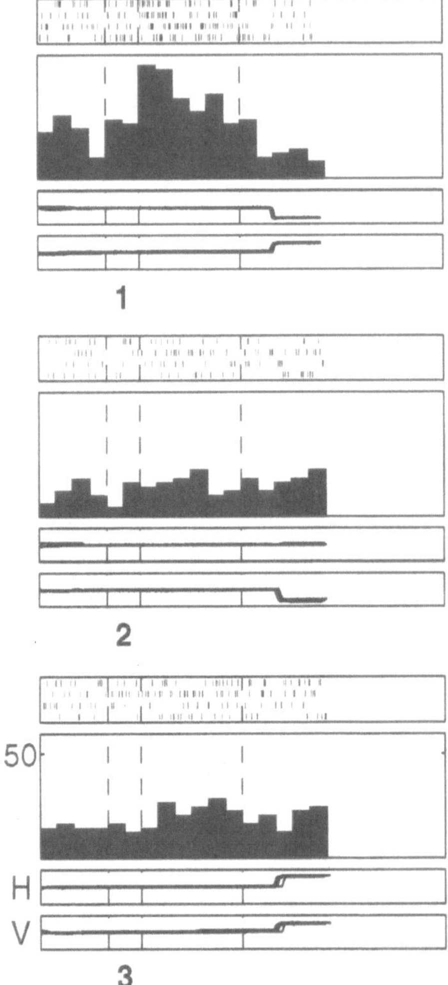

Fig. 9. Sequence of events and response of an LIP neuron in memory-guided saccades made in three equally spaced directions. Part of a block in a new task designed to explore simultaneous targets for saccades. Same format as Fig. 2. Time scale: stimulation lasts 250 ms, memory 750 ms. Response of an LIP neuron shows activation in direction "1" (150°, up and left), but not in directions "2" and "3". Same neuron and same block as Fig. 10

The close relationship of parietal cortex to extinction suggests that the response of parietal neurons to multiple targets may not be a simple combination of their responses to single targets. Indeed, this is so in area LIP, as indicated by the single-neuron recordings carried out in this area while the monkey performs the new task described previously. One rather puzzling result is that the activity of many neurons is suppressed while the monkey is performing the multiple-target trials. An example is illustrated in Figs. 9 and 10. The data shown in both Figs. 9 and 10 were collected in the same block of trials from one LIP neuron. (The block also contained other trials, not shown.) All conditions were randomly interleaved; they are rearranged into the two figures (Figs. 9 and 10) only for the convenience of the reader.

Fig. 9 shows the response of the neuron while the monkey is performing memory-guided saccade trials to three targets, 120° apart. The directions, marked in Fig. 9 by sequential numbers, are: 1, up and left (150°); 2, down (270°); 3, right and up (30°). The two bottom boxes in each panel show the superimposed vertical and horizontal eye position records. These records show that the monkey fixates the central fixation-spot at the center of the screen. Fixation is maintained throughout the flashing of the targets (pair of dashed lines at each panel) and throughout the memory period, until fixation offset (rightmost dashed line). Shortly afterwards, the monkey makes a saccade in the direction of the memorized target location.

Onset of the target in location 1 is followed by a sustained discharge. This is the intentional discharge that was described above, and that can be evoked even in the absence of visual stimulation of the receptive field - if the monkey is planning a saccade in the neuron's preferred direction. The intentional discharge continues throughout the memory period, until well after the fixation offset, and dies out just before the saccade gets under way. In other directions there is little, if any, activation. Therefore, this neuron conveys the intention of the monkey to make a saccade in (or close to) direction 1.

Fig. 10 shows the activity of this neuron in trials of the new task. The two bottom panels of Fig. 10 show two target-pair combinations in which location 1 was the common target location, and therefore this location is the target also for the illustrated compound-stimulation trials. The monkey performed his task adequately, as shown by the eye-position records. The monkey maintained fixation of the fixation-spot, at the center of the screen, throughout the two target-pairs (two pairs of dashed lines in each panel of Fig. 10) and the memory periods, until after the fixation-spot was turned off (rightmost dashed line). Then the monkey made a saccade in the direction of the target-location that was common to both pairs. (The top panel redisplays the responses in memory-guided saccade trials in direction 1 for ease of comparison.)

Even though target 1 appeared in all four stimulation phases shown (two phases per each class of trials), not even once did it result in a response similar to the one that occurred when target 1 appeared on its own. This was true not only for both the first stimulation and first memory intervals in these trials, but also for the second stimulation and memory intervals. This is significant because after the second pair of targets is flashed, the monkey already has all the information needed to decide where to look at the end of the trial. It could be expected that, at this time, the neuron will "convert" to the single-target situation and discharge unambiguously as in the top panel. Indeed, while some neurons do just that, others, as the one shown here, do not.

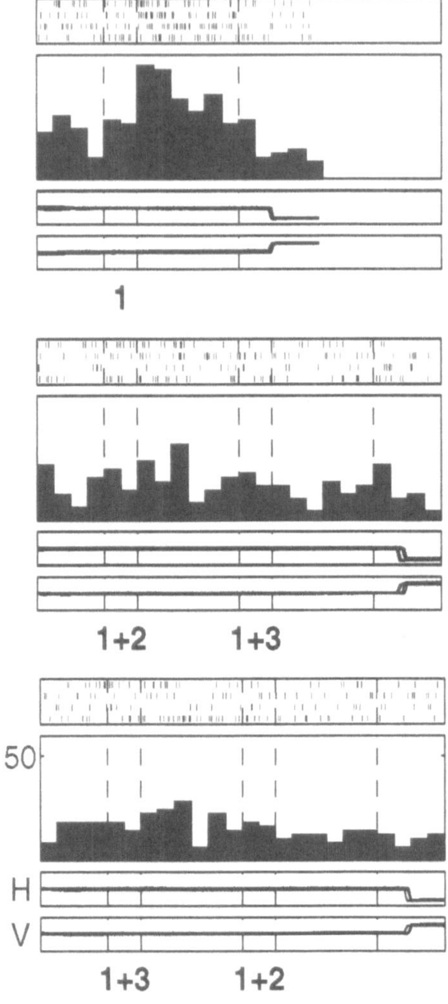

Fig. 10. Sequence of events and response of an LIP neuron in a new task designed to explore neuronal response to simultaneous targets for saccades. Same format as Fig. 2. Time scale: stimulation lasts 250 ms, memory 750 ms. Fig. 10 shows three types of trials in which the requested movement is the same, in direction "1". The *top panel* illustrates regular memory-guided saccades in direction "1". The *middle* and *bottom panels* illustrate trials in which two pairs of stimuli appeared, separated by memory intervals. Response of an LIP neuron shows strong activation in the memory-guided saccade trials, but not in the combined task trials. Same neuron and same block as Fig. 9

To fulfill the task, the monkey must maintain both locations of the first pair of targets throughout the first memory period. Indeed, at this time the monkey can begin to form motor intentions to look at these locations, because he will eventually have to make a saccade towards one of these two locations. Whether these two locations are reflected in the activity at this phase of the trials is as yet unclear. To clarify this issue, we investigate whether the directions of the two targets are reflected in the population of preferred directions.

This new task enables us to examine the activity of single neurons related to two locations, or two directions. Evidence suggests that LIP neurons are activated very differently in this situation (as compared to single targets).

Summary

In this chapter we saw, using tasks with single targets or a sequence of targets, that area LIP contains strong activity related to the intention to make the (single) next saccade. The broad neuronal activity fields are adequate for applying the population vector approach. The population vector approach cannot directly account for comprehension of simultaneous multiple targets. Neuropsychological evidence (extinction) suggests that the parietal cortex is particularly related to comprehension of multiple targets. LIP neurons, tested in a novel task that requires the monkeys to simultaneously maintain double targets, are suppressed before movement selection, and sometimes also after the selection, even though the precise quantitative proportion of the post-selection inhibition is as yet unclear.

It is as yet unclear why the activity in LIP is suppressed in this task. One possibility is that rate coding holds only part of the information in the multiple-target situation. Another possibility is that the lower rates are still high enough to determine the movement. Whatever the explanation, the surprising character of the activity in LIP, while maintaining double simultaneous targets, complements the neuropsychological deficit of extinction, that is typical of parietal lesions.

Acknowledgments. Supported by the Israel Science Foundation.

References

Andersen RA, Asanuma C, Cowan WM (1985) Callosal and prefrontal associational projecting cell populations in area 7a of the macaque monkey: A study using retrogradely transported fluorescent dyes. J Comp Neurol 232:443–455

Andersen RA, Essik GK, Siegel RM (1987) Neurons of area 7 activated by both visual stimuli and oculomotor behavior. Exp Brain Res 67:316–322

Andersen RA, Asanuma C, Essik G, Siegel RM (1990) Corticocortical connections of anatomically and physiologically defined subdivisions within the inferior parietal lobule. J Comp Neurol 296:65–113

Asanuma C, Andersen RA, Cowan WM (1985) The thalamic relation of the caudal inferior parietal lobule and the lateral prefrontal cortex in monkeys: Divergent cortical projections from cell clusters in the medial pulvinar nucleus. J Comp Neurol 241:357–381

Barash S (1990) Relatively local neurons in a distributed representation: A neurophysiological perspective. Behav Brain Sci 13:489–491

Barash S, Bracewell RM, Fogassi L, Gnadt JW, Andersen RA (1991a) Saccade-related activity in the lateral intra-parietal area (LIP) I. Temporal properties; comparison to area 7a. J Neurophysiol 66:1095–1108

Barash S, Bracewell RM, Fogassi L, Gnadt JW, Andersen RA (1991b) Saccade-related activity in the lateral intra-parietal area (LIP). II. Spatial properties. J Neurophysiol 66:1109–1124

Bracewell RM, Mazzoni P, Barash S, Andersen RA (in press) Motor intention activity in area LIP. II. Changes of motor plan. J Neurophysiol

Bushnell MC, Goldberg ME, Robinson DL (1981) Behavioral enhancement of visual responses in monkey cerebral cortex. I. Modulation in posterior parietal cortex related to selective visual attention. J Neurophysiol 46:755–772

Cogan DG (1965) Ophthalmic manifestations of bilateral non-occipital cerebral lesions. Br J Ophthalmol 49:281–297

Cogan DG, Adams RD (1953) A type of paralysis of conjugate gaze ("ocular motor apraxia"). Arch Ophthalmol 53:434–442

Georgopoulos AP (1995) Motor cortex and cognitive processing. In: Gazzaniga M (ed) The cognitive neurosciences. MIT Press, Cambridge

Georgopoulos AP, Kalaska JF, Caminiti R, Massey JT (1982) On the relations between the direction of two-dimensional arm movements and cell discharge in primate motor cortex. J Neurosci 2:1527–1537

Georgopoulos AP, Caminiti R, Kalaska JF, Massey JT (1983) Spatial coding of movement: A hypothesis concerning the coding of movement direction by motor cortical populations. Exp Brain Res [Suppl] 7:327–336

Georgopoulos AP, Kettner RE, Schwartz AB (1988) Primate motor cortex and free arm movements to visual targets in three-dimensional space: II. Coding of the direction of movement by a neuronal population. J Neurosci 8:2928–2937

Glimcher PW, Sparks DL (1992) Movement selection in advance of action in the superior colliculus. Nature 355:542–545

Gnadt JW, Andersen RA (1988) Memory related motor planning activity in posterior parietal cortex of macaque. Exp Brain Res 70:216–220

Hikosaka O, Wurtz RH (1983) Visual and oculomotor functions of the monkey substantia nigra pars reticulara III. Memory-contingent visual and saccade responses. J Neurophysiol 49:1268–1284

Husain M, Stein J (1988) Rezso Balint and his most celebrated case. Arch Neurol 45:89–93

Hyvarinen J (1982) The parietal cortex of monkey and man. Springer, Berlin Heidelberg New York

Hyvarinen J, Poranen A (1974) Function of the parietal associative area 7 as revealed from cellular discharges in alert monkeys. Brain 97:673–692

Lee C, Roher WH, Sparks DL (1988) Population coding of saccadic eye movements by neurons in the superior colliculus. Nature 332:357–360

Lynch JC, Mountcastle VB, Talbot WH, Yin TCT (1977) Parietal lobe mechanism for related attention. J Neurophysiol 40:362–389

Mazzoni P, Bracewell RM, Barash S, Andersen RA (1996) Spatially tuned auditory responses in area LIP of macaques performing delayed memory saccades to acoustic targets. J Neurophysiol 75:1233–1241

Mazzoni P, Bracewell RM, Barash S, Andersen RA (in press) Motor intention activity in area LIP. I. Most neurons encode the next planned saccadic eye movement and not the locations of sensory stimuli. J Neurophysiol

McIlwain JT (1976) Large receptive fields and spatial transformations in the visual system. Int Rev Physiol 10:223–248

Mountcastle VB, Lynch JC, Georgopoulos A, Sakata H, Acuna C (1975) Posterior parietal association cortex of the monkey: command functions for operations within extrapersonal space. J Neurophysiol 38:871–908

Rafal R, Robertson L (1995) The neurology of visual attention. In: Gazzaniga M (ed) The cognitive neurosciences. MIT Press, Cambridge, pp 625–648

Robinson DL, Goldberg ME, Stanton GB (1978) Parietal association cortex in the primate: Sensory mechanisms and behavioral modulations. J Neurophysiol 41:910–932

Smyrnis N, Taira M, Ashe J, Georgopoulos AP (1992) Motor cortical activity in a memorized delay task. Exp Brain Res 92:139–151

Sparks DL, Holland R, Guthrie BL (1976) Size and distribution of movement fields in the monkey superior colliculus. Brain Res 113:21–34

Sparks DL, Mays LE (1980) Movement fields of saccade-related burst neurons in the monkey superior colliculus. Brain Res 190:39–50

Sparks DL, Groh JM (1995) The superior colliculus: A window for viewing issues in integrative neuroscience. In: Gazzaniga M (ed) The cognitive neurosciences. MIT Press, Cambridge, pp 565–584

Wurtz RH, Goldberg ME (1972) Activity of superior colliculus in behaving monkey. III. Cells discharging before eye movements. J Neurophysiol 35:575–586

Yarbus AL (1967) Eye movements and vision. Plenum Press, New York

Young MP, Yamane S (1992) Sparse population coding of faces in the inferotemporal cortex. Science 256:1327–1331

Posterior Parietal Cortex Control of Saccades in Humans

C. PIERROT-DESEILLIGNY[1] and R. MÜRI[2]

[1]Hôpital de la Salpêtrière, Neurologie, Paris, France
[2]Neurologische Klinik, Inselspital, Bern, Switzerland

Electrophysiological studies in the monkey have shown that the inferior parietal lobule in the posterior parietal cortex (PPC) contains areas involved in saccade control (Andersen et al. 1990, 1992). The lateral intraparietal area, currently called the parietal eye field (PEF), is involved in both the visual and motor aspects of saccade paradigms, controlling the preparation and triggering of different types of saccades. Area 7a may be involved in the visual aspects of saccade paradigms and could also contribute to the preparation of saccade amplitude, but such a role has yet to be convincingly demonstrated. In humans, three complementary methods have been used to study the parietal ocular motor areas: functional imaging studies to locate these areas, lesion studies to demonstrate their involvement in the control of the different types of saccades, and, more recently, transcranial magnetic stimulation (TMS) studies to determine more precisely their functions during the preparation and execution of these saccades. This paper will cover the authors' personal studies and review the literature involving these different methods.

Functional Imaging Studies

In order to localize precisely the human cortical ocular motor areas, positron emission tomography (PET) scan studies and, more recently, functional magnetic resonance imaging (fMRI) studies have been used. Despite the existence of several studies that used PET scans during saccades, the location of the human PEF remains imprecise, either because these studies concerned voluntary saccades (Melamed and Larsen 1977; Petit et al. 1993), i.e., a type of saccade in which parietal control is not crucial (Pierrot-Deseilligny et al. 1995), or because they resulted in relatively widespread activation in the parietooccipital region (Fox et al. 1985; Anderson et al. 1994). However, in an abstract (Sweeney et al. 1994), it has been mentioned that the "lateral intraparietal area" is involved in saccades. Preliminary reports of studies using fMRI during saccades have not yet, to the best

In: Parietal Lobe Contributions to Orientation in 3D Space (1997). P. Thier and H.-O. Karnath (eds). Springer-Verlag, Heidelberg.

of our knowledge, yielded a more precise localization of the human PEF (Kleinschmidt et al. 1994; Lee et al. 1995).

Recent lesion studies of parietal lobe control of saccades have assumed, by cytoarchitectural analogy with the relevant areas in the monkey (von Economo, 1929), that the human PEF is located in the part of the PPC corresponding to the angular gyrus and the adjacent intraparietal sulcus. In order to specifically examine this question, we used fMRI to study activation of the cerebral cortex in normal subjects performing visually guided saccades (Müri et al. 1996). These subjects, who were placed in darkened MRI equipment, were instructed to follow a luminous target moving with small successive jumps from 30° on the left to 30° on the right and then back in the other direction. This cycle was repeated for 47s. The direction and timing of target displacement were predictable, but amplitude varied slightly (from 3° to 6°) at each jump. It has been suggested that such visually guided saccades are controlled both by the PEF and the frontal eye field (FEF; see Pierrot-Deseilligny 1994). Functional MRI was performed using a spoiled GRASS sequence. Six successive horizontal sections passing through the different parts of the parietooccipital and frontal lobes were acquired. In a first experiment performed in three subjects, activity was compared for two conditions: a "saccade" period versus a rest condition (with the eyes closed) of the same duration. Significant bilateral activity was observed on the lower sections, i.e., in the medial and lateral occipital visual areas. Besides this occipital activity, bilateral activity was observed on the upper sections in two other cortical areas: (1) the precentral gyrus and sulcus at the level of the middle frontal gyrus and (2) the intraparietal sulcus. The pooled activity of these two areas observed in the three subjects is shown in Fig. 1. The frontal activity corresponds to the location of the human FEF (Petit et al. 1993; Rivaud et al. 1994). The intraparietal activity was present and bilateral in each subject, but varied slightly from one subject to another along the anteroposterior axis, lying either in area 39 or area 40 of Brodmann, or even on the border between these two areas. Although no quantitative study comparing activity in both hemispheres could be performed here, there was no obvious qualitative difference between the two sides (the three subjects being right-handed).

In order to ensure that this parietal activity was saccade-related, control experiments were performed. In a second experiment concerning three other subjects, the cerebral activity existing during the fixation of a fixed target (identical in size to the "jumping" target) was compared to that of the rest condition. In this procedure, the occipital visual areas were again significantly active, but there was no significant activity in the FEF and intraparietal sulcus. Finallly, in a third experiment concerning three further subjects, the cerebral activity during saccades was compared to that during fixation of a fixed target. A significant level of activity, similar to that existing in the first experiment, existed both in the FEF and intraparietal sulcus, but no activity was observed in the occipital visual areas. This absence of activity in the occipital cortex in the third

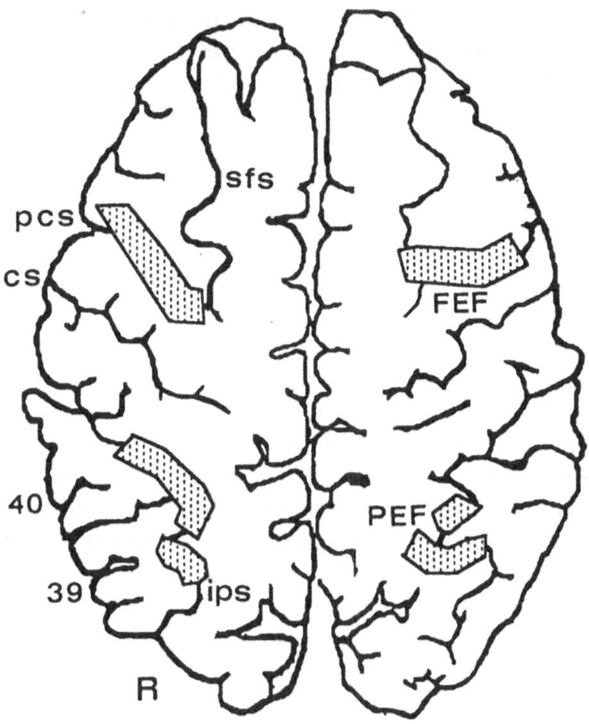

Fig 1. Functional magnetic resonance imaging during visually guided saccades. The pooled cerebral activity (*gray zones*) of three normal subjects performing a visually guided saccade task (compared to a rest condition with the eyes closed) is shown on a horizontal section passing above the lateral ventricles. *cs*, central sulcus; *FEF*, frontal eye field; *ips*, intraparietal sulcus; *pcs*, precentral sulcus; *PEF*, parietal eye field; *R*, right side; *sfs*, superior frontal sulcus; *39* and *40*, areas 39 and 40 of Brodmann

experiment could result from the nature of the comparison which involved two visual activities cancelling each other (jumping visual target versus fixed visual target).The activity observed both in the FEF and intraparietal sulcus in the first and third experiments (but not in the second experiment) did correspond to saccades and was not simply due to visual fixation or even to visual attention.

It should be noted that some other cortical areas were slightly active in each of the three procedures used here. However, the absence of significance for these areas suggests that the most important areas involved in the control of visually guided saccades are the occipital visual areas for the visual aspect of the paradigm and both the FEF and intraparietal sulcus for the motor aspect (i.e., saccades). These results confirm experimental findings and the results of human lesion studies, showing that both the FEF and an area in the PPC are involved in the control of visually guided saccades. Taken together, our results suggest that the human PEF is located in the intraparietal sulcus bordering areas 39 and 40 of Brodmann. Finally, the precise location of the human equivalent of area 7a in the monkey is not known. However, the absence of further significant parietal activity observed in our study during saccades suggests that such an area could also be located in the intraparietal sulcus near the PEF, as in the monkey.

Lesion Studies

Activity in a cortical area shown by functional imagery or even electrophysiological experimental studies does not necessarily imply that this area is crucial in the control of the ensuing or current movement. Lesion studies are necessary to ascertain the crucial nature of this role, by showing that the movement is impaired after damage to the cortical area under consideration.

After bilateral PPC lesions resulting in Balint's syndrome, some types of saccades are disturbed, in particular those made under the control of vision (Pierrot-Deseilligny et al. 1986). In a patient with bilateral cerebral hemispheric infarctions affecting both the FEF and the PPC (including the PEF), all types of

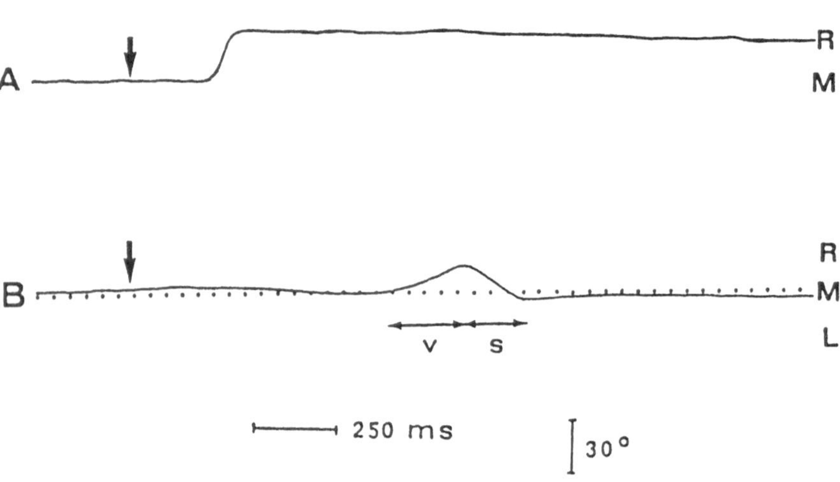

Fig 2 **A**, **B**. Visually guided saccade in a patient with bilateral frontoparietal lesions. Recordings were performed using direct-current electro-oculography. **A** Normal subject. **B** Patient. The *arrow* indicates the occurrence of the lateral visual target. In the patient, note the long latency and marked decreased velocity of the saccade (*s*) occurring after a head movement (not shown) and the associated vestibular eye movement (*v*). *L*, Left; *M*, Midline; *R*, Right

saccades were severely impaired (Pierrot-Deseilligny et al. 1988). The rare residual, visually guided saccades observed in this patient with "acquired ocular motor apraxia" had considerably increased latency (over 1s, with normal latency being around 200 ms) and markedly decreased velocity (Fig. 2). Furthermore, most of these saccades were triggered only after a vestibular ocular reflex movement accompanying a head movement. These residual saccades could be triggered by the supplementary eye field, i.e., an area (spared in this patient) that appears to control saccades made after a vestibular input (Pierrot-Deseilligny et al. 1993). Such clinical cases of Balint's syndrome or acquired ocular motor apraxia show the involvement of the PEF and the FEF in the control of saccades in humans and furthermore suggest that these two areas have complementary roles in the triggering of visually guided saccades. However, such cases are rare and those reported so far were studied with only a few saccade paradigms.

Unilateral focal PPC lesions are much more frequent but result in relatively subtle saccade impairment, requiring quantitative analysis and comparison with a control group. All these lesions usually affect the region of the posterior part of the intraparietal sulcus, i.e., probably the PEF and the equivalent of area 7a. After unilateral PPC lesions, reflexive visually guided saccades (Fig. 3) are impaired. Saccade amplitude is decreased and latency increased bilaterally, but with impairment more marked contralateral to the lesion (Pierrot-Deseilligny et al. 1987, 1991a; Braun et al. 1992; Heide and Kömpf 1994; Cochin et al. 1996). Latency is increased in the gap task (i.e., with extinguishing of the central fixation point 200 ms before the onset of the lateral target) as well as in the overlap task (i.e., with no extinguishing of the central point), in which the increase is significantly greater than in the gap task (Walker et al. 1991; Heide and Kömpf 1994). It should be noted that FEF lesions result in normal latency in the gap task but increased latency in the overlap task (Rivaud et al. 1994), and that a lesion affecting the superior colliculus (SC) results in a similar disturbance (Pierrot-Deseilligny et al. 1991c). Longer latency in the gap task after PPC lesions, but not after FEF lesions, suggests that the PPC is predominantly involved in the triggering of such saccades, probably through the direct projection of the PEF to the SC (for a review see Pierrot-Deseilligny et al. 1995). The parietal lobe could facilitate the excitation of the SC neurons via this pathway. Longer latency in the overlap task than in the gap task after PPC, FEF, and SC lesions also suggests that these three areas are involved in the control of disengagement from fixation. Anatomical connections existing in the monkey are compatible with a pathway successively involving the PEF, the FEF, and the SC for the control of this process (for a review see Rivaud et al. 1994).

Latency disturbances of reflexive, visually guided saccades are more marked after right than after left parietal lesions (Pierrot-Deseilligny et al. 1991a). This indicates that a certain degree of lateralization exists in the human parietal lobe for this control mechanism, as for many other visuospatial functions (Weintraub and Mesulam, 1987; Meador et al. 1989). There is not always a correlation between latency increase and visual neglect (Pierrot-Deseilligny et al. 1987, 1991a; Heide et al. 1995; Cochin et al. 1996). It should be noted that visual attention shifts appear to be controlled by the posterior part of the superior parietal lobule, i.e., an area located just medially to the intraparietal sulcus and the PEF (Corbetta et al. 1993, 1995). Therefore, after PPC damage, visual neglect may be associated with a saccade deficit, or each of the two disorders may occur independently, probably depending upon the location and extent of the lesion in the PPC. The frequency with which increased latency is not accompanied by an attentional disorder tends to confirm experimental results suggesting that the PEF is involved in motor planning rather than in attentional processes (Andersen 1989). Finally, although the latency of single, reflexive visually guided saccades of neglect patients may be normal, the general ocular motor behavior of such patients is usually disturbed: for example, in a free exploratory ocular motor task, these patients make successive intentional saccades that go in all directions but remain largely within the ipsilateral visual space (Chédru et al. 1973; Karnath and Fetter 1995). Such an abnormality of motor behavior resulting from a disturbance of visual attention should not be confused with disorders specifically affecting the preparation or triggering of a given type of single saccades.

Memory-guided saccades made after visual input (Fig. 3) are also impaired after PPC lesions (Pierrot-Deseilligny et al. 1991b; Israël et al. 1995). The latency of such saccades is increased, possibly due to PEF damage. The pathophysiological mechanism involved in this disorder is probably analogous to that resulting in the increase of reflexive visually guided saccade latency: it has been hypothesized that there is a pre-excitation of the SC by the PEF as soon as the lateral visual target appears (Pierrot-Deseilligny et al. 1991a; see Fig. 5). Furthermore, saccade amplitude is decreased. This impairment could result from a defect in visuomotor integration, i.e., the calculation of saccade amplitude corresponding to target location. In contrast, memory-guided saccades made after vestibular input (Fig. 3) are preserved after PPC lesions (Israël et al. 1995). This suggests that such saccades are predominantly controlled by the vestibular cortex, located in humans in the posterior part of the superior temporal gyrus (Friberg et al. 1985). Therefore, the cortical pathways controlling memory-guided saccades by visual input and those employing vestibular input could be quite different, with only the prefrontal cortex involved in both cases.

In the double-step saccade paradigm (Fig. 3), extraretinal signals are required to calculate the correct amplitude of the second saccade, since this amplitude depends upon the amplitude of the first saccade just performed. It has been shown that the FEF is not crucial for the preparation of the second saccade (Rivaud et al. 1994). In contrast, this second saccade is impaired after fronto-parietal (Duhamel et al. 1992) or PPC lesions (Heide et al. 1995). This suggests that the PPC controls the integration process of retinal and extraretinal signals (for more details, see Heide and Kömpf, this volume). Extraretinal information could originate in the brain stem after the first saccade and reach the parietal lobe via the central thalamus (Gaymard et al. 1994).

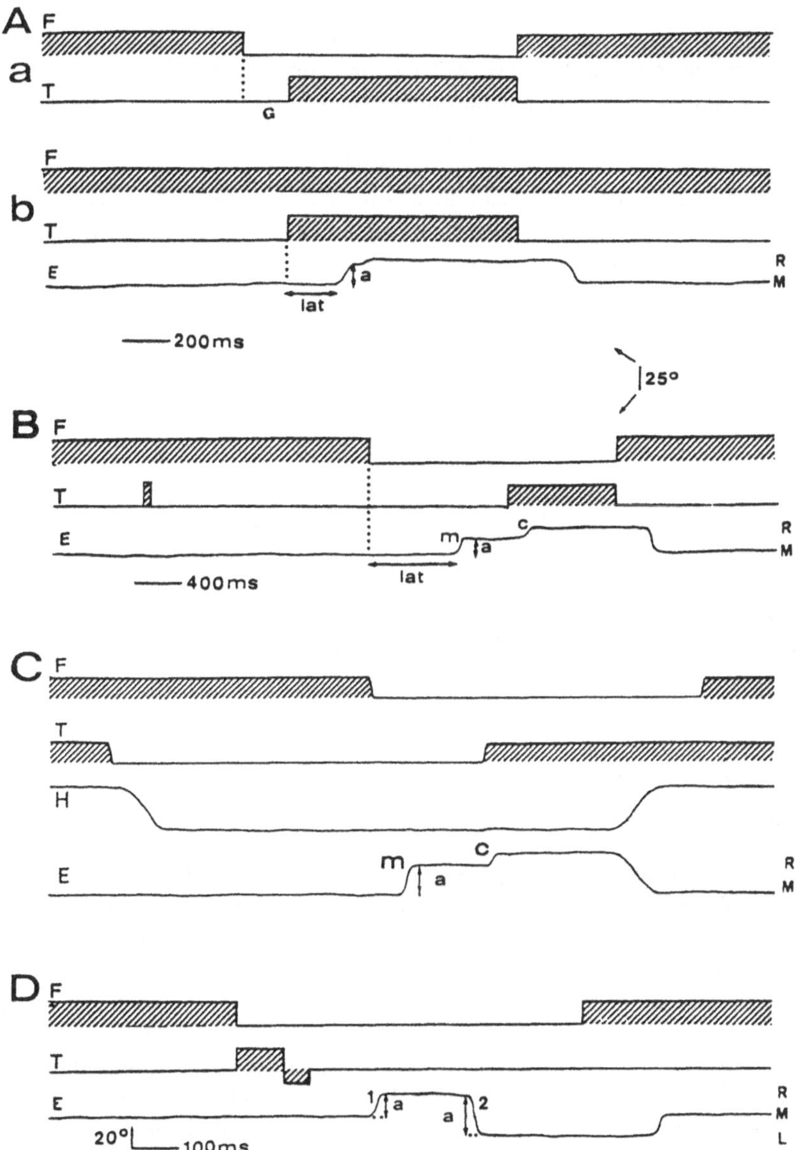

Fig 3 A-D. Saccade paradigms. **A** Visually guided saccades, in the gap task (a) and the overlap task (b). **B** Memory-guided saccades with visual input. **C** Memory-guided saccades with vestibular input, i.e., with rotation of the body and head (*H*). **D** Double-step saccades. *a*, amplitude; *c*, corrective saccade; *E*, eye movement; *F*, fixation point (rotating with the subject in **C**); *G*, gap; *H*, head movement; *L*, left; *lat*, latency; *M*, midline; *m*, memory-guided saccade; *R*, right; *T*, lateral target (central target at the beginning of the task, before body rotation, in **C**); *1*, first saccade; *2*, second saccade

Therefore, human lesion studies have shown that the PPC is involved (1) in disengagement from central fixation, (2) in triggering of certain types of saccades (reflexive visually guided saccades and memory-guided saccades with visual input), (3) in amplitude calculation of saccades made under the control of vision (visually guided and memory-guided sacccades), including saccades in which both retinal and extraretinal signals are used (second saccade of the double step paradigm), but (4) not in saccades made only with vestibular input. Human lesion studies cannot determine the respective roles of the PEF and the equivalent of area 7a, since the location of the latter is not precisely known and could well be close to the former. However, experimental studies suggest that the PEF controls saccade triggering (Andersen et al. 1989) and, therefore, that increased latency probably results from PEF damage. Furthermore, since the PEF and/or the equivalent of area 7a could be involved in visuomotor integration, amplitude disorders may be due to lesions affecting either structure.

Transcranial Magnetic Stimulation Studies

Transcranial magnetic stimulation (TMS) is a novel neurophysiological method useful for exploring the functioning of the human cerebral cortex. In the case of the motor cortex, TMS elicits motor responses in the limbs. In contrast, the effect of TMS on eye movements seems to be mainly a disturbance of cortical activity for a few milliseconds, since (1) it is impossible to elicit saccades by a single TMS, and (2) saccade programming and execution are affected by TMS (Wessel and Kömpf 1991; Müri et al. 1991). Therefore, applying TMS over a cortical area at different times during a paradigm allows us to determine the precise time(s) at which the execution of the paradigm is disturbed, i.e., the time(s) when the cortical area specifically controls the paradigm. Such findings supplement human lesion studies, which can establish the role of a cortical area in a paradigm but cannot determine the moment in time crucial for this control.

TMS has so far been used to examine the role of the PPC in saccade control in only a few studies. TMS applied over the PPC 80 ms after a lateral target had been flashed (for 100 ms) resulted in increased latency and decreased amplitude of such reflexive visually triggered saccades (Elkington et al. 1992). This study thus confirms the role of the PPC in the triggering and amplitude preparation of these saccades. Memory-guided saccades with visual input have also been studied using TMS over the PPC. In a study in which TMS was applied over the posterior part of the intraparietal sulcus at different times after extinguishment of the central fixation point (i.e., during the period of saccade latency), saccade accuracy was impaired when TMS occurred 100 ms after extinguishment of the central point (Oyachi and Ohtsuka 1995). Saccade latency was unaffected, but saccade accuracy was impaired after right parietal stimulation. Therefore, these results confirm that the PPC is also involved in the execution of memory-guided saccades and that a lateralization of the human brain exists for this control, as already reported in lesion studies (Pierrot-Deseilligny et al. 1991b). Furthermore, it can also be inferred that the triggering of memory-guided saccades depends upon another region, probably the FEF (see Pierrot-Deseilligny et al. 1995). Finally, in another study presented in abstract form (Brandt et al. 1995), repetitive stimulation over the PPC resulted in saccade impairment when applied shortly after target presentation. This study confirms that the PPC controls memory-guided saccades early after target presentation. However, the method used (repetitive pulses) means that the crucial time for PPC control within this period cannot be determined.

In order to study further the parietal role in saccade control, we studied ten normal subjects who performed a memory-guided saccade task with visual input (Fig. 3B) while undergoing TMS at different times during the paradigm (Müri et al. 1995b, in press). Stimulation (single pulse; for more technical details, see Müri

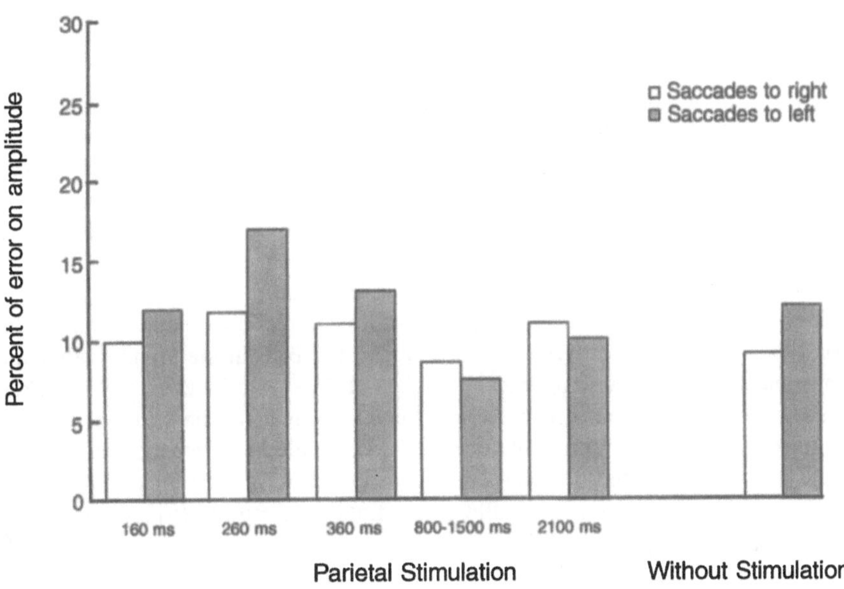

Fig 4. Amplitude error of memory-guided saccades after transcranial magnetic stimulation of the right posterior parietal cortex. The results concerning ten normal subjects, were compared to those of the same subjects without stimulation. The percentage of error in amplitude of memory-guided saccades was significantly increased (Wilcoxon signed-rank test for pairwise comparison) for contralateral (leftward) saccades when stimulation was applied 260 ms after flashing of the target. Note that the flashed target occurred at 0 ms and extinguishment of the central point (which was the signal for the subject to perform a memory-guided saccade) at 2000 ms (see also the text)

et al. 1995a) was applied over the right PPC 160, 260, and 360 ms after the flashed target (i.e., during the visuomotor integration phase) and also during a period of time between 800 and 1500 ms after the appearance of this target (i.e., during the memorization phase) and, finally, 2100 ms after the flashed target, namely, 100 ms after extinguishment of the central fixation point (i.e., during saccade latency). All stimulation times were randomly studied. An increase in the percentage of error in amplitude of memory-guided saccades was observed when stimulation was applied 260 ms after the onset of the flashed target, but this increase existed only for saccades performed contralaterally to stimulation (Fig. 4). Directly before (160 ms) and directly after (360 ms) this crucial time, amplitude was not significantly impaired. Another experiment with prefrontal cortex (PFC) stimulation (with the same stimulation times) resulted in amplitude impairment of contralateral saccades when stimulation was applied during the memorization period (between 800 and 1500 ms). Saccade amplitude was normal at other stimulation times. These results suggest that these two disturbances of saccade amplitude observed after TMS over the PPC and the PFC were specific to the stimulated area. The disturbance induced during the memorization phase by PFC stimulation, like that observed in the study previously mentioned (Brandt et al. 1995), confirms experimental results suggesting that a spatial working memory is organized in this area (Goldman-Rakic 1987). The time of 260 ms, corresponding to the saccade disturbance observed after PPC stimulation, is probably later than the beginning of the visuomotor integration phase. However, this period of time perhaps corresponds to the end of this phase, namely, when integrated information could be definitively transmitted to the PFC, where it is memorized for a longer period. The absence of disturbance induced by stimulation at 160 ms is perhaps the result of reafferences of sensory information in the PPC during the first 200 or 300 ms. This period of time could correspond to that of the visual excitation phase observed in area 7a and the PEF in the monkey after the appearance of a lateral visual target (Barash et al. 1991; Colby et al. 1995). Taken together, our results confirm that the different cortical mechanisms controlling memory-guided saccades probably operate successively in different areas (Fig. 5): visual information in the occipital cortex, visuomotor (spatial) integration in the PPC, spatial memorization in the PFC, and, finally, execution (triggering) in the FEF (and also perhaps in the PEF).

The spatial precision of TMS is not sufficient to determine the respective roles of two areas as close together as the PEF and the equivalent of area 7a. However, experimental results have shown that area 7a projects principally to the PFC and the PEF, whereas the PEF projects principally to the FEF and the SC (for a review see Pierrot-Deseilligny et al. 1995). Therefore, our results concerning memory-guided saccades are compatible with an early visuomotor integration performed in the equivalent of area 7a and a subsequent short spatial memorization in the PFC before a triggering which could be shared by the FEF and PEF (Fig. 5).

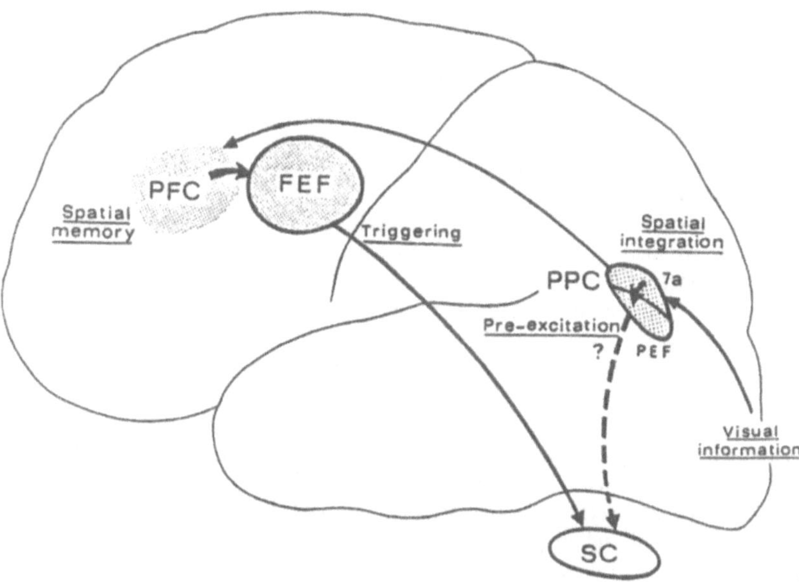

Fig 5. Hypothetical cortical circuitry involved in memory-guided saccades with visual input. *FEF*, frontal eye field; *PEF*, parietal eye field (i.e., the human equivalent of area LIP in the monkey); *PFC*, prefrontal cortex (area 46 of Brodmann); *PPC*, posterior parietal cortex; *SC*, superior colliculus; *7a*, area corresponding to area 7a in the monkey

Conclusion

Human studies using functional imagery, lesions, and TMS have confirmed experimental results showing that the PPC plays a crucial role in saccade control. The PEF and the equivalent of area 7a in the monkey could be located in the intraparietal sulcus bordering areas 39 and 40 of Brodmann, whereas the area controlling visual attention could be located more medially, in the superior parietal lobule. The PPC (probably via the PEF) controls the triggering of reflexive visually guided saccades and memory-guided saccades with visual input as well as disengagement from central fixation. This area (PEF and/or equivalent of area 7a) also controls the amplitude preparation of all saccades performed with retinal signals alone or with combined retinal and extraretinal signals, but probably not of saccades performed with vestibular input. This visuomotor integration preparing saccade amplitude could be completed in the PPC between 200 and 300 ms after visual stimulation, and if subsequent cortical control involving spatial memory is required, this control appears to be carried out by the PFC and not by the PPC.

References

Andersen RA (1989) Visual and eye movement functions of the posterior parietal cortex. Annu Rev Neurosci 12:377–403

Andersen RA, Bracewell RM, Barash S, Gnadt JW, Fagassi L (1990) Eye position effects on visual memory, and saccade-related activity in area LIP and 7a of macaque. J Neurosci 10 :1176–1196

Andersen RA, Brotchi PR, Mazzoni P (1992) Evidence for the lateral intraparietal area as the parietal eye field. Curr Opin Neurobiol 2:840–846

Anderson TJ, Jenkins IH, Brooks DJ, Hawken MB, Frackowiak RSJ, Kennard C (1994) Cortical control of saccades and fixation in man: a PET study. Brain 117:1073–1084

Barash S, Bracewell RM, Fogassi L, Gnadt JW, Andersen RA (1991) Saccade related activity in the lateral intraparietal area. I. Temporal properties. J Neurophysiol 66:1095–1108

Brandt SA, Ploner CJ, Meyer BU, Stoerig P, Villringer A (1995) Differential effects of repetitive transcranial magnetic stimulation over cortical visuomotor areas in prefrontal and parietal cortex in man. Soc Neurosci Abstr 21:1196

Braun D, Weber H, Mergner T, Schulte-Mönting J (1992) Saccadic reaction times in patients with frontal and parietal lesions. Brain 115:1359–1386

Chédru F, Leblanc M, Lhermitte F (1973) Visual searching in normal and brain-damaged subjects : contribution to the study of unilateral inattention. Cortex 9:94–111

Cochin JP, Hannequin D, Auzou P, Fodil D, Dreano E, Mihout B, Augustin P (1996) Latences saccadiques et négligence spatiale unilatérale par lésion pariétale. Rev Neurol (Paris) 152:32–37

Colby CL, Duhamel JR, Goldberg ME (1995) Oculocentric spatial representation in parietal cortex. Cereb Cortex 5:470–481

Corbetta M, Miezin FM, Shulman GL, Petersen SE (1993) A PET study of visuospatial attention. J Neurosci 13:1202–1226

Corbetta M, Shulman GL, Miezin FM, Petersen SE (1995) Superior parietal cortex activation during spatial attention shifts and visual feature conjunction. Science 270:802–805

Duhamel JR, Goldberg ME, Fitzgibbon EJ, Sirigu A, Grafman J (1992) Saccadic dysmetria in a patient with frontoparietal lesion. Brain 115:1387–1402

Elkington PTG, Kerr GR, Stein JS (1992) The effect of electromagnetic stimulation of the posterior parietal cortex on eye movements. Eye 6:510–514

Fox PT, Fox JM, Raiche ME, Burde RM (1985) The role of cerebral cortex in the generation of voluntary saccades: a positron emission tomographic study. J Neurophysiol 54:348–369

Friberg L, Olsen TS, Roland PE, Paulson DB, Lassen NA (1985) Focal increase of blood flow in the cerebral cortex of man during vestibular stimulation. Brain 108:609–623

Gaymard B, Rivaud S, Pierrot-Deseilligny C (1994) Impairment of extraretinal eye position signals after central thalamic lesions in humans. Exp Brain Res 102:1–9

Goldman-Rakic PS (1987) Circuitry of primate prefrontal cortex and regulation of behavior by representational memory. In: Plum F (ed) Handbook of American physiology, sect 1. The nervous system, vol 5. American Physiological Society, Bethesda, pp 373–414

Heide W, Kömpf D (1994) Saccades after frontal and parietal lesions. In: Fuchs AF, Brandt T, Büttner U, Zee D (eds) Contemporary ocular motor and vestibular research: a tribute to David A. Robinson. Thieme, New York, pp 225–227

Heide W, Blankenburg M, Zimmerman E, Kömpf D (1995) Cortical control of double-step saccades. Implications for spatial orientation. Ann Neurol 38:739–748

Israël I, Rivaud S, Gaymard B, Berthoz A, Pierrot-Deseilligny C (1995) Cortical control of vestibular-guided saccades in man. Brain 118:1169–1184

Karnath H-O, Fetter M (1995) Ocular space exploration in the dark and its relation to subjective body orientation in neglect patients with parietal lesions. Neuropsychologia 33:371–377

Kleinschmidt A, Merboldt KD, Requardt M, Hänicke W, Fralm J (1994) Functional MRI of cooperative cortical activation patterns during eye movements. Soc Neurosci Abstr 20:1402

Lee KM, Hirsch J, Kim K, De Lapaz RL, Relkin N (1995) Visually-guided saccadic eye movements elicit activation in frontal oculomotor areas in human brain using functional magnetic resonance imaging (f MRI). Soc Neurosci Abstr 21:1196

Meador KJ, Loring DW, Lee GP, Brooks BC, Nichols PT, Thompson WO, Heilman KM (1989) Hemisphere asymmetry for eye gaze mechanisms. Brain 112:103–111

Melamed E, Larsen B (1977) Cortical activation pattern during saccadic eye movements in humans: localization by focal cerebral blood flow increases. Ann Neurol 5:79–88

Müri RM, Hess CW, Meienberg O (1991) Transcranial magnetic stimulation of the human frontal eye field by magnetic pulses. Exp Brain Res 86:219–223

Müri RM, Rivaud S, Vermersch AI, Léger JM, Pierrot-Deseilligny C (1995a) Effects of transcranial magnetic stimulation on the supplementary area region during sequences of memory-guided saccades. Exp Brain Res 401:163–166

Müri RM, Vermersch AI, Rivaud S, Gaymard B, Pierrot-Deseilligny C (1995b) Transcranial magnetic stimulation of the posterior parietal cortex and prefrontal cortex during memory-guided saccades. Soc Neurosci Abstr 21:1197

Müri RM, Iba-Zizen MT, Derosier C, Cabanis EA, Pierrot-Deseilligny C (1996) Location of the human posterior eye field with functional magnetic resonance imaging. J Neurol Neurosurg Psychiatry, 60:445–448

Müri RM, Vermersch AI, Rivaud S, Gaymard B, Pierrot-Deseilligny C (in press) Effects of single pulse transcranial magnetic stimulation over the prefrontal and posterior parietal cortices during memory-guided saccades in humans. J Neurophysiol

Oyachi H, Ohtsuka K (1995) Transcranial magnetic stimulation of the posterior parietal cortex degrades accuracy of memory-guided saccades in humans. Invest Ophthalmol Vis Sci 36:1441–1448

Petit L, Orssaud C, Tzourio N, Salamon G, Mazoyer B, Berthoz A (1993) PET study of voluntary saccadic eye movements in humans: basal ganglia-thalamocortical system and cingulate cortex involvement. J Neurophysiol 69:1009–1017

Pierrot-Deseilligny C (1994) Saccade and smooth pursuit impairment after cerebral hemispheric lesions. Eur Neurol 34:121–134

Pierrot-Deseilligny C, Gray F, Brunet P (1986) Infarcts of both inferior parietal lobules with impairment of visually guided eye movements, peripheral visual inattention and optic ataxia. Brain 109:81–97

Pierrot-Deseilligny C, Rivaud S, Penet C, Rigolet MH (1987) Latencies of visually-guided saccades in unilateral hemispheric cerebral lesions. Ann Neurol 21:138–148

Pierrot-Deseilligny C, Gautier JC, Loron P (1988) Acquired ocular motor apraxia due to bilateral frontoparietal infarcts. Ann Neurol 23:199–202

Pierrot-Deseilligny C, Rivaud S, Gaymard B, Agid Y (1991a) Cortical control of reflexive visually-guided saccades in man. Brain 114:1473–1485

Pierrot-Deseilligny C, Rivaud S, Gaymard B, Agid Y (1991b) Cortical control of memory-guided saccades in man. Exp Brain Res 83:607–617

Pierrot-Deseilligny C, Rosa A, Masmoudi K, Rivaud S (1991c) Saccade deficits after a unilateral lesion affecting the superior colliculus. J Neurol Neurosurg Psychiatry 54:1106–1109

Pierrot-Deseilligny C, Israël I, Berthoz A, Rivaud S, Gaymard B (1993) Role of the different frontal lobe areas in the control of the horizontal component of memory-guided saccades in man. Exp Brain Res 95:166–171

Pierrot-Deseilligny C, Rivaud S, Gaymard B, Müri R, Vermersch AI (1995) Cortical control of saccades. Ann Neurol 37:557–567

Rivaud S, Müri RM, Gaymard B, Vermersch AI, Pierrot-Deseilligny C (1994) Eye movement disorders after frontal eye field lesions in humans. Exp Brain Res 102:110–120

Sweeney JA, Mintun MA, Carl JR, Kwee S, Steinkopf MB, Rosenberg DR (1994) A positron emission tomography study of voluntary saccadic eye movements. Soc Neurosci Abstr 20:234

von Economo C (1929) The cytoarchitectonics of the human cerebral cortex. Oxford University Press, London

Walker R, Findlay JM, Young AW, Welch J (1991) Disentangling neglect and hemianopia. Neuropsychologia 29:1019–1027

Weintraub S, Mesulam MM (1987) Right cerebral dominance in spatial attention. Arch Neurol 44:621–625

Wessel W, Kömpf D (1991) Transcranial magnetic stimulation: lack of oculomotor response. Exp Brain Res 86:216–218

Specific Parietal Lobe Contribution to Spatial Constancy Across Saccades

W. HEIDE and D. KÖMPF

Medizinische Universität zu Lübeck, Klinik für Neurologie, Lübeck, Germany

Disorders of Visuospatial Orientation Following Posterior Parietal Lesions

Since the historical cases reported by Balint (1909) and Holmes (1918, 1919), much clinical evidence (Critchley 1953; De Renzi 1982) has been collected from patients with focal lesions of the posterior cerebral hemispheres, confirming a key role of the posterior parietal lobe in visuospatial perception and orientation. Deficits of these functions were more severe with bilateral than with unilateral parietal lesions and more pronounced with right hemispheric than with left hemispheric lesions (Brain 1941; Paterson and Zangwill 1944; Meerwaldt and van Harskamp 1982; Heilman et al. 1993). In cases of unilateral lesions, such deficits might be discrete and require specific neuropsychological testing for their diagnosis. According to the level of which the processing of space information is defective, three classes of visuospatial disorders have been identified (De Renzi 1988):

1. Disorders of *space exploration*, including saccadic scanning (gaze paresis, ocular motor apraxia, deficits of exploratory saccades), shifting of visual attention (narrowing of attention, called "simultanagnosia", or unilateral neglect; Heilman et al. 1993), and visuomotor coordination (optic ataxia; Balint 1909; Jeannerod 1986)
2. Disorders of the *perceptual analysis of visual space*, including basic space-perceptual features like the coordinates of the egocentric spatial reference (subjective straight ahead, subjective visual vertical and horizontal), the localization of targets in personal or extrapersonal space and *spatial constancy* despite eye movement-induced shifts of the retinal image, furthermore the analysis of spatial relationships (estimation of distances, angles, orientations, length or depth) and spatial operations (such as mental rotation). As a consequence of these deficits, such patients are impaired at copying or drawing tasks (constructive apraxia, Fig. 1).

In: Parietal Lobe Contributions to Orientation in 3D Space (1997). P. Thier and H.-O. Karnath (eds). Springer-Verlag, Heidelberg.

Fig. 1. Copy of the house drawn on the *left*, performed by a patient with bilateral posterior parietal lobe lesions centered around the intraparietal sulcus

3. Disorders of *spatial memory*: These patients get lost even in a familiar environment, despite clear consciousness and preservation of general memory. Either they have completely lost their mental maps, i.e., the inner representation of extrapersonal space (topographical amnesia), or they are no longer able to match their visuospatial percept with these maps (topographical agnosia), even though they can reconstruct them from memory.

In the latter group of disorders, lesions are located in the mesial occipitotemporal cortex involving the parahippocampal gyrus, whereas disorders of space exploration are caused by damage to the posterior parietal lobe, the temporoparietal junction or the frontal eye fields. For the perceptual analysis of visual space, however, the posterior parietal cortex (PPC) is the only critical structure. This view has been confirmed by behavioral, anatomical, and neurophysiological evidence in monkey and man during the past 20 years: The main stream of cortical visual projections involved in spatial vision ("where"-pathway) leads to the PPC (Mishkin et al. 1983; Ungerleider and Desimone 1986; Haxby et al. 1991) which plays a central role in directing visual attention and actions in extrapersonal space (Mountcastle et al. 1975; Lynch et al. 1977; Bushnell et al. 1981; Hyvärinen 1982). Because of its multimodal sensory input and its involvement in visuomotor control of eyes and hands, the PPC is the appropriate cortical region for performing the supramodal sensorimotor integration that is necessary for the control of spatial behavior (Goldberg et al. 1990; Stein 1992; Andersen 1989, 1995).

In cases of unilateral posterior parietal lesions, disorders of the perceptual analysis of visual space do not lead to complete spatial disorientation, and deficits may be subtle and difficult to recognize for the clinician, although they are a handicap for these patients. Their assessment requires specific tests; copying and drawing tasks (Figs. 1, 8) are quite sensitive in this respect, but unspecific. Rather, tests should be aimed at basic space-perceptual features, as mentioned earlier; this is also necessary for a specific therapy in rehabilitation. One of these features is spatial constancy across saccades, the topic of this chapter. Until recently, there were no specific tests of this function available to the clinician.

Spatial Constancy Across Saccadic Eye Movements

When we move our eyes, the image of the visual world shifts on our retina, yet we are nevertheless able to maintain a stable percept of visual space. This spatial constancy across saccadic eye movements is a basic function of visuospatial perception and a prerequisite for an accurate localization of visual targets in space. As information about a target's location on the retina varies with each eye movement, it cannot be sufficient for accurate spatial localization, but it must be combined with information about current eye position or at least about the preceding eye displacement. Three sources of information subserve this purpose (for a review see Bridgeman et al. 1994):

1. *Retinal* information about the structure of the visual world surrounding the target is a cue for reorientation after a saccadic eye movement by reevaluation of the visual scene. Due to the processing time of the retinal signal to the cortex, however, this information is not available for a postsaccadic delay of at least 100 ms, and this would transiently disrupt the stability of visuospatial perception and localization. Another mechanism contributing to visual stability is a reduction of visual sensitivity during and directly prior to a saccade: It is called *saccadic suppression* and prevents the perception of the saccade-associated jump (motion) of the visual scene across the retina, but per se this does not help to localize the target more accurately because it bears no information about eye position.
2. *Proprioceptive input* from the eye muscles is an important source of *extraretinal information* about eye position (Steinbach 1987; Gauthier et al. 1990), but not sufficient to maintain spatial constancy. First, analogous to retinal input, there is an afferent processing delay of at least 20 ms after the start of the saccade. Second, proprioceptive input does not explain the perceptual findings in the historical eye-press experiment (Purkinye 1825; von Helmholtz 1962): A passive eye movement induced by tapping lightly on the canthus of one eye evokes the illusion that the world has moved, in contrast to a comparable active saccadic eye movement that is associated with the same *inflow* of proprioceptive and retinal information. Conversely, a retinal afterimage in darkness (Grüsser et al. 1987) jumps with each saccade (active situation), but remains stationary during the eye-press experiment (passive situation).

3. As an explanation of these findings, von Helmholtz (1962) had already (in 1866) postulated a signal in the brain that is specifically linked to an *active* eye displacement, monitoring "the effort of will" that is, the saccade-associated neural *outflow*. Since 1950, when von Holst and Mittelstaedt published their model of the "reafference principle," this signal is called "*efference copy*" because it is considered to be a copy of the central motor command that is fed back into the afferent limb. There it interacts with the inflow of reafferent signals that result from the performed movement. Thus a saccade-associated neural discharge reflecting the efference copy signal (a so-called corollary discharge, according to Sperry 1950, Guthrie et al. 1983) would cancel the neural excitation caused by reafference of retinal image displacement, thereby signaling visual stability across the saccade without delay and providing the neural basis for an accurate egocentric localization of objects in extrapersonal space.

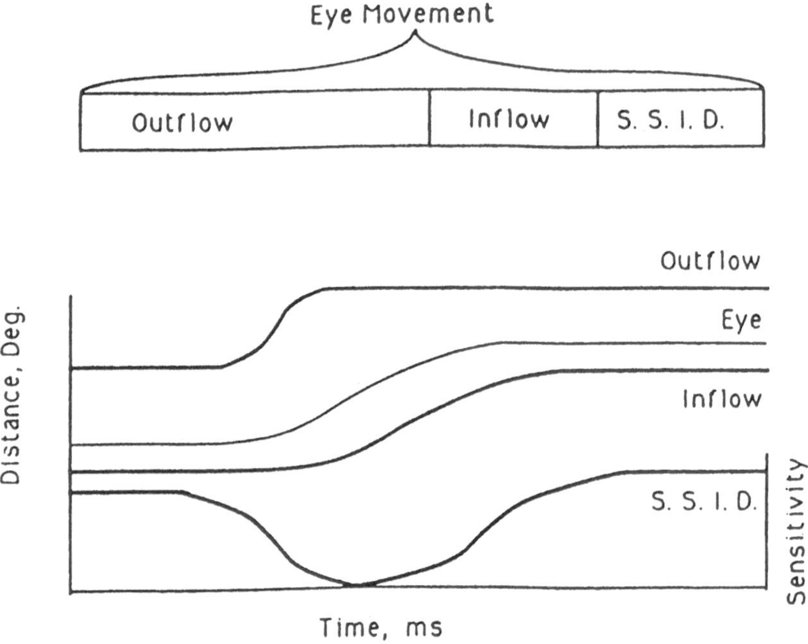

Fig 2. Three components contributing to perceptual stability across a saccade, namely, outflow (efference copy), inflow (proprioception), and saccadic suppression of image displacement (*S.S.I.D.*). The *bottom panel* shows the time course of these components' contribution around the time of a saccade. The duration along the *horizontal axis* is 100 ms. (From Bridgeman et al. 1994)

None of these mechanisms can fully compensate for the retinal image shift across saccades, but obviously they all work together, each of them having a different time course (Fig. 2). The outflow signal (efference copy) provides the most important contribution during saccades in an unstructured visual environment, but its gain is not higher than 0.6, as can be calculated from pointing accuracy to a foveal afterimage in darkness (Grüsser et al. 1987; Bridgeman 1995). It is even lower if saccades are of large amplitude or high frequency, resulting in the perception of oscillopsia during their performance.

It is not yet clear how these different signals are represented and computed in the brain to achieve spatial constancy. There are several alternatives: The brain could use the extraretinal signal of saccadic eye displacement to eliminate or cancel retinal image displacement ("elimination solution"); yet perceptual stability would then be incomplete, as the extraretinal signal with a gain of about 0.7 (comprising both outflow and proprioceptive inflow) is not fully compensatory. Alternatively, retinal and eye position signals could be translated into an invariant stored map representing space in head-, body-, or world-centered coordinates ("translation solution"). There is, however, no convincing psychophysical evidence for the existence of such a fixed spatiotopic visual memory in the brain (Irwin et al. 1990; Bridgeman et al. 1994). Conversely, this solution is in accordance with Robinson's (1973) classical model of saccade generation, in which the relevant input to the pontine saccade generator is a "spatial error" signal (in head-centered coordinates), calculated by combining retinal error with eye position feedback. Another model suggests that information is assigned to the correct spatial location by evaluation of its content rather than by extraretinal signal corrections alone ("evaluation solution," MacKay 1973).

Alternatively the brain might recalibrate and update retinocentric maps of visual space in association with each saccade according to the current center of gaze ("calibration solution," Bridgeman et al. 1994). In the respective "eye displacement" models of saccade generation, target location is not coded in spatial, but in retinal coordinates, and it is repeatedly updated by extraretinal feedback on saccadic eye displacement (Jürgens et al. 1981; Goldberg et al.1990; Moschovakis and Highstein 1994). If there is a stable spatiotopic map as proposed by the translation solution, spatial localization of visual targets should be perfectly accurate even around the time of a saccade, but this is actually not the case: If a visual target is flashed (for 2 ms) shortly before or during a saccade, there is considerable perceptual and saccadic mislocalization of this target in the direction of the saccade (Honda 1989, 1990; Dassonville et al. 1992; Fig. 3). This implies a sluggish and dampened representation of the eye position signal around the time of a saccade (see also Gellman and Fletcher 1992): It seems to start moving into the direction of the impending saccade about 80 ms prior to its onset, and this is attributed to the afferent visual processing delay of the flashed target. At the end of the saccade, the eye position signal again coincides with the actual eye position, thus having moved much slower than the trajectory of the saccade. This slowness

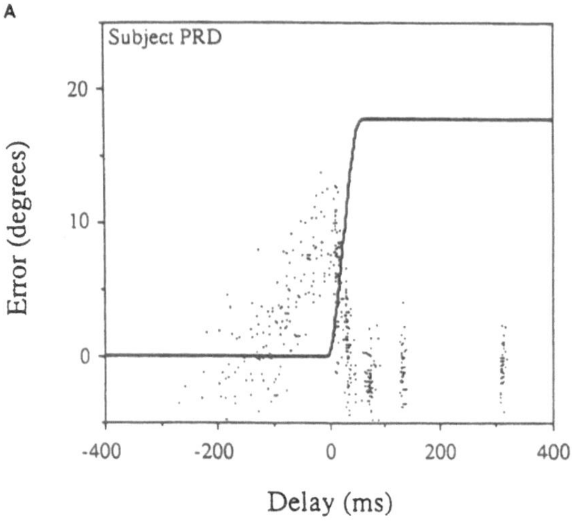

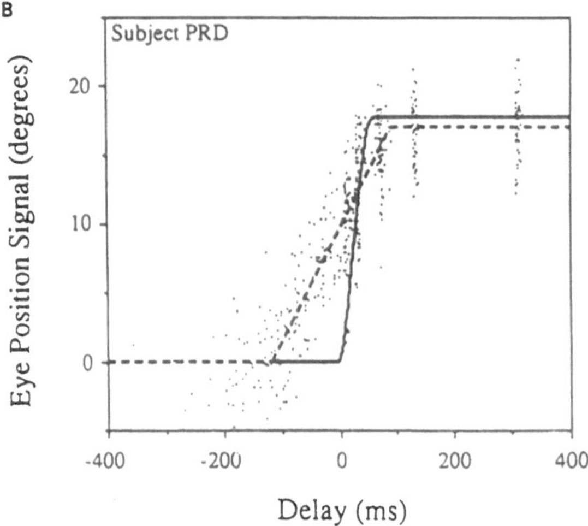

Fig 3 A, B. A Saccadic mislocalization of a visual target flashed (for 2 ms) around the time of a horizontal saccade (*solid curve*). The magnitude of position error depends on the delay between the flash and saccade onset. **B** For the same data, the diagram shows the representation of the extraretinal eye displacement signal used to localize the flash; it was derived by subtracting the retinal error vector from the endpoint of the targeting saccade. The *dashed curve* is the best-fit three-segment curve describing the time course of the extraretinal eye position signal in this subject. It is slower than the saccade and seems to start earlier. (From Dassonville et al. 1992)

might be due to the time needed for the "recalibration" process. During the perisaccadic instability of the eye position signal, spatial localization of visual targets could possibly rely on allocentric cues of the visual environment more than on extraretinal signals (Schlag et al. 1994; Dassonville et al. 1995).

In conclusion, efference copy is the most important factor for maintaining spatial constancy across saccades under natural conditions, although it does not fully compensate saccade-associated retinal image displacement. There is no agreement on how this information is computed in the brain, and some psychophysical evidence might contradict the view of constantly stable spatiotopic coding of target location by reliance on the extraretinal eye position signal. The questions are how this signal is represented on the neuronal level and what is the role of the parietal lobe in this respect.

Representation of Saccade-Associated Extraretinal Signals in the Brain

With respect to saccadic eye movements in monkeys, evidence for the presence of extraretinal signals has been found in several areas of the brain. One line of evidence is derived from electrical stimulation which not only evoked *fixed-vector saccades* (characterized by the same change of position in amplitude and direction, irrespective of primary eye position), but also *goal-directed saccades* which terminate at the same point in space so that their direction depends on their starting point, with eye position being taken into account. The latter saccades were found in the superior colliculus (SC; Roucoux and Crommelinck 1976) and the central thalamus (Maldonado et al. 1980) of cats as well as in the posterior parietal lobe (Shibutani et al. 1984) and the supplementary motor area (SMA; Schlag and Schlag-Rey 1987) of monkeys. But even if fixed-vector saccades were elicited, the presence of the extraretinal eye position signal could be demonstrated by combining electrical microstimulation with visual stimulation: If the visual target was flashed in the latency period of an electrically elicited fixed-vector saccade, the vector of the second (visually-elicited) saccade was corrected according to the electrically induced saccadic eye displacement. This was the case for neurons in the SC (Sparks and Mays 1983) and the frontal eye fields (FEF; Schiller and Sandell 1983); analogous findings were obtained with the reverse order of stimulation, i.e., the "colliding saccade paradigm" (Schlag and Schlag-Rey 1990), in neurons of the superficial SC, the internal medullary lamina of the central thalamus, the FEF and the supplementary eye field (SEF) in the SMA (Schlag et al. 1994). These findings imply that the site of stimulation is necessarily upstream of the source of the extraretinal signal used to compensate initial saccadic eye displacement. In contrast to this, such a compensation did not occur with stimulation of some sites in the pons (Sparks et al. 1987), the cerebellar vermis (Gochin and McElligott 1987), and the fastigial nucleus (Noda et al. 1991). These structures must be situated downstream of the source of the extraretinal eye position or eye displacement signal, which consequently seems to arise somewhere in the pons.

Furthermore, the presence of both retinal and extraretinal information in association with saccades was demonstrated by single-neuron recordings in the superior colliculus (Mays and Sparks 1980), the central thalamus (Schlag-Rey and Schlag 1989), the FEF (Goldberg and Bruce 1990), the SEF (Schlag et al. 1990), and the posterior parietal cortex (PPC) of monkeys (Andersen et al. 1985; Goldberg et al. 1990). Even though neurons in the PPC (in the lateral intraparietal area [LIP], in area 7a, and area V3a) have retinotopically organized receptive fields, their visual and saccade-related responses are modulated by eye and head position (Andersen et al. 1985; Andersen 1995). Thus, the entire population of these neurons contains all of the information necessary for computing spatiotopic coding of target location (Zipser and Andersen 1988), but the "hidden units" that represent this computation have not yet been found. There is only one report in which parietal neurons seemed to respond to a fixed location in craniotopic space, independent of eye position (Galletti et al. 1993).

In contrast, an important correlate of the efference copy signal has been found in LIP neurons with retinotopic receptive fields (Duhamel et al. 1992a): These neurons anticipate the retinal consequences of an intended saccade by shifting their receptive fields accordingly, thus updating any remapping the retinal representation of the visual world for a continuous maintenance of spatial accuracy. As they do this up to 80 ms prior to the saccade, they must use an efference copy signal providing information about the intended eye displacement. This could explain why flashed visual targets are mislocalized in the perisaccadic period, starting about 80 ms prior to saccade onset: Remapping has already occurred during this period, and there is no more compensation for the saccade-associated retinal image shift of these flashed targets. These results can be explained by the "calibration solution" and by the eye displacement model of saccade generation better than by Robinson's spatiotopic model. An even more critical test between these two models would be to assess the type of deficits in cases of lesions: Is the result of lesions a deficit in the extraretinal eye displacement signal or is it a deficit in coding spatiotopic locations of saccade targets in the contralateral hemispace? This can be investigated by applying saccadic "double-step" stimuli.

The Double-Step Saccadic Paradigm

Since Westheimer's first report in 1954, visual double-step stimuli have frequently been used to investigate the spatiotemporal programming of saccades. It has been shown that the two saccades are programmed in parallel and can interfere with

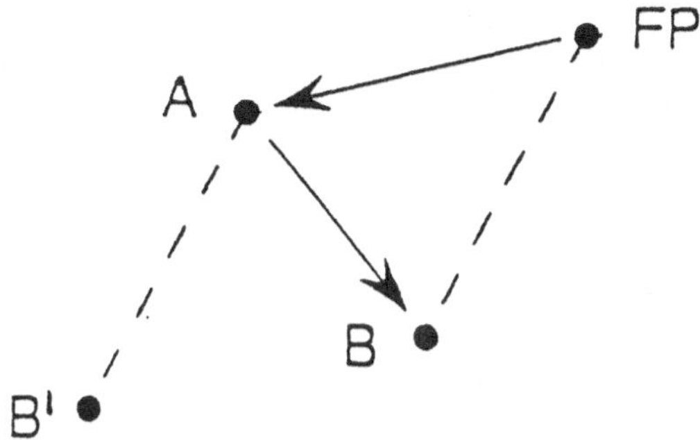

Fig 4. Example of a double-step stimulus with the two targets *A* and *B* being flashed successively while the gaze is directed to a central fixation point (*FP*). When both saccades are performed after all of the targets have disappeared, the motor vector of the second saccade (*A-B*) is different from the retinal vector of the second target (*FP-B* or *A-B'*)

each other because the saccadic system, although reacting in a discontinuous manner, almost continuously processes afferent visual information (Becker and Jürgens 1979; Gellman and Carl 1991). For example, the amplitude of the first saccade is influenced more by the location of the second target, the later it starts after its appearance (amplitude transition function). Furthermore, spatial programming of the second saccade is particularly interesting in Hallett and Lightstone's (1976) flashed version of the double-step saccadic paradigm. It allows separation of a target's retinal vector from its saccadic motor vector: Two sequentially flashed visual targets have to be fixated by two consecutive saccades (Fig. 4). If both targets have disappeared before the first saccade, the second saccade (from A to B) will not start at the location (FP) from which the second target was seen, and thus a spatial dissonance is created between the retinal coordinates of the second target (FP-B or A-B') and the motor coordinates (A-B) of the necessary second saccade ("retinospatial dissonance"). Nevertheless, normal human subjects direct this saccade accurately to the spatial location of the second target (B) and not to its retinal location (B'), which would be grossly dysmetric. This implies the use of extraretinal information about eye displacement associated with the first saccade (from FP to A) for updating the spatial (retinocentric) representation of the second target (eye displacement model). Alternatively, according to the spatiotopic model, the location of the second target could be coded in stable craniotopic coordinates by using the extraretinal eye position signal, with the need for continuous coordinate transformations from the retinotopic to the craniotopic mode.

The results from single-neuron studies with double-step saccades do not permit differentiation among these two antagonistic models. It turned out that some visually responsive neurons in the SC (Mays and Sparks 1980), the FEF

(Goldberg and Bruce 1990), and parietal area LIP (Barash et al. 1991) fire whenever the vector of the second-step saccade terminates within their receptive field, even though no stimulus ever appears in that field during the double-step task. Such a spatially accurate neuronal response in connection with the second saccade implies the presence of the extraretinal signal in these neurons, but can be explained by either of the two models. As mentioned earlier, however, a lesion study could be the critical test, as the two models make different predictions concerning the type of resulting deficits. Furthermore, such a study could reveal which of the three mentioned cerebral structures (SC, FEF, or the parietal eye field) in the human brain is really essential for the control of the extraretinal signal that is needed to achieve spatial accuracy of double-step saccades.

Double-Step Saccades in Patients with Cerebral Lesions

Patients and Methods

In our present study, the problem of cerebral lesions was addressed by recording horizontal double-step saccades in 35 human patients with focal unilateral hemispheric lesions involving cortical areas that participate in the control of saccades (Pierrot-Deseilligny et al. 1995; see also Pierrot-Deseilligny and Müri, this volume), namely, the PPC (19 cases, 14 of them right- and five left-hemispheric), the right FEF (four cases), the left SMA (four cases) and the dorsolateral prefrontal cortex (PFC; eight cases, three of them right- and five left-hemispheric). Lesions were due to ischemic infarction or tumor surgery several weeks prior to examination. In each of these groups, the location of the common lesioned area concerned only one of the saccade-related cortical areas, thus being selective for the group. In the PPC the common lesioned area was located in the inferior parietal lobule along the border between the angular and supramarginal gyrus (Brodmann's areas 39 and 40), extending cranially toward the intraparietal sulcus, caudally to the temporo-parietal junction, and posteriorly into the angular gyrus. In some of the frontal cases, it involved the assumed location of the right FEF (n=4) in the premotor region around the middle portion of the precentral sulcus and the adjacent precentral gyrus. Other lesions (five left, three right) overlapped in the dorsolateral PFC (area 46), anterior to the FEF, or (n=4) in the anterior portion of the SMA (the assumed location of the supplementary eye field) in the dorsomedial part of the superior frontal gyrus. Thirty-two healthy human adults served as a control group.

The severity of visual hemineglect in these patients was assessed using simple neuropsychological tests like bisection, cancellation, and copying and drawing tasks (Halligan et al. 1991). Eye movements were recorded using infrared reflection oculography. In double-step trials, a central fixation point disappeared

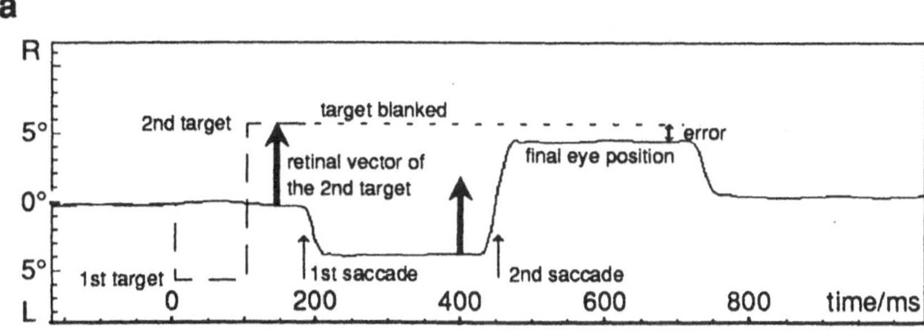

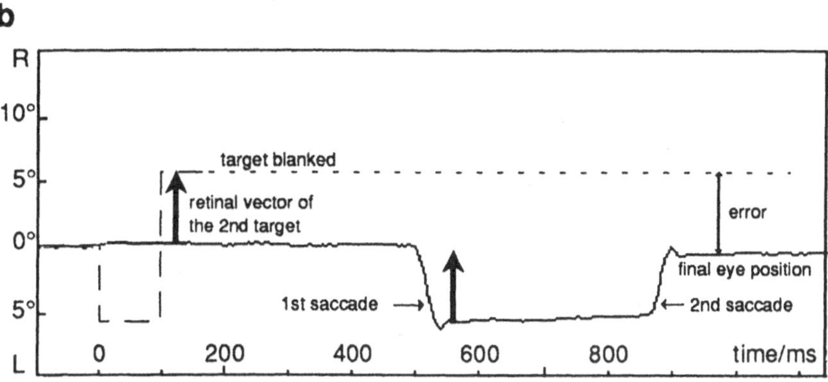

Fig 5 a, b. Examples of double-step saccades in a normal subject (**a**) and a patient with a right parietal lesion (**b**), with the two targets presented at 6° in opposite hemifields. The *traces* in each panel show the horizontal position (*R*, rightward; *L*, leftward) of the right eye (*solid line*) and the target (*broken line*). Retinospatial dissonance occurred after the first saccade, as the second target was not visible any more at that time. The location of the blanked second target is marked by the *fine dashed line*. Ideally it should be met by the final eye position, the error of which is indicated by *thin biheaded arrows*. The error is rather small in the normal subject, but large in the patient who erroneously performs the second saccade according to the retinal vector of the second target, indicated by *thick vertical arrows*

after an unpredictable delay from stimulus onset, and the two targets were flashed successively for 140 ms (first step) and 100 ms (second step), respectively. Alternative presentation times were 100/80 ms and 180/100 ms. Each pair of targets was located either in the same hemifield at horizontal eccentricities of 10° and 5°, respectively, or in different hemifields at 6°/6° (Figs. 5, 6A). After presentation of the second target, the stimulus was extinguished and reappeared at the fixation point 1 s later so that at any rate the second saccade was performed while the screen was blank, thus never being visually guided. Subjects were instructed to fixate the two target locations in the order of their appearance as

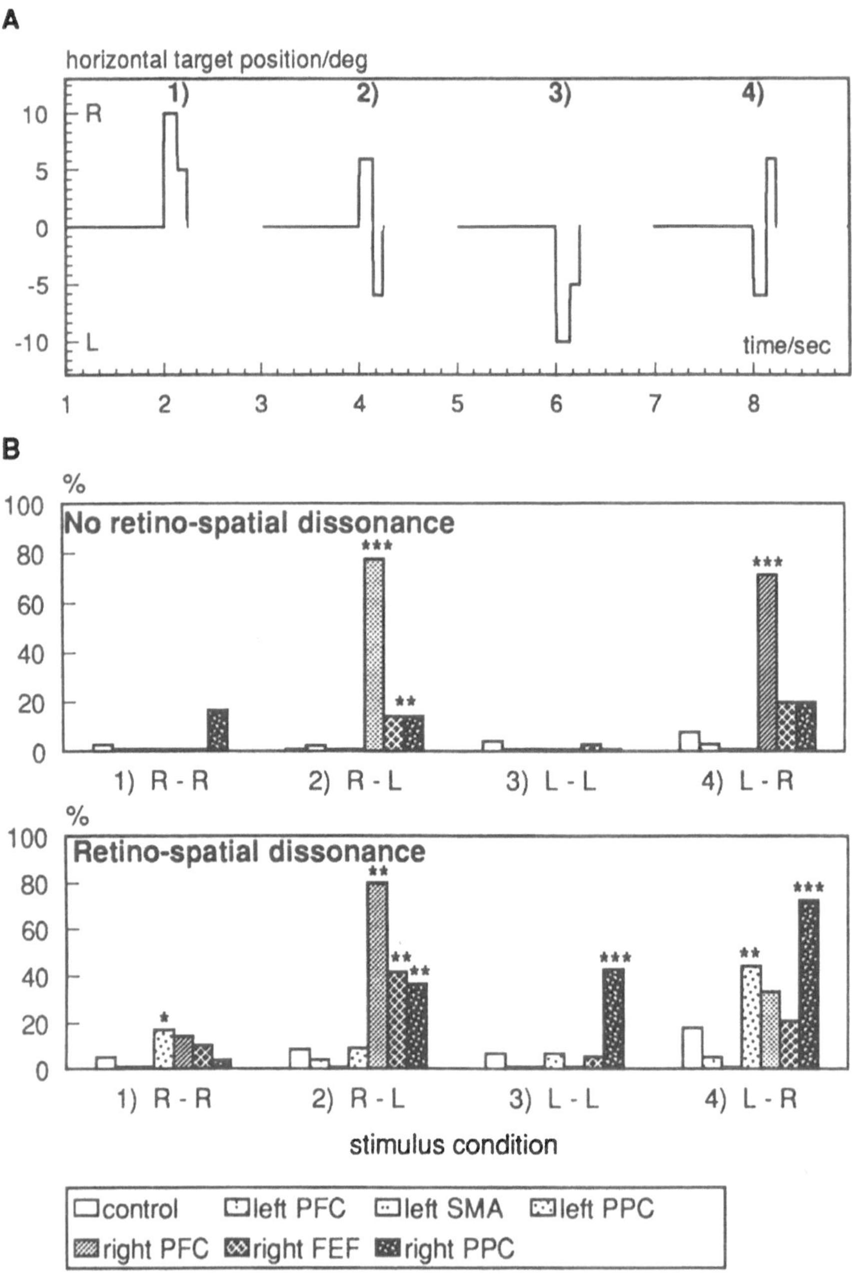

Fig 6 A, B. **A** The four different types of double-step stimuli in the right (**R**) or left (**L**) hemifield. **B** Percentage of missing or dysmetric second saccades with (*central panel*) and without (*bottom panel*) retinospatial dissonance in each subgroup of subjects, plotted

separately according to the four stimulus conditions of **A**. In all figures, *, **, and *** indicate a significant difference from the control group, correponding to $p<0.05$, $p<0.01$, and $p<0.005$, respectively. Although right prefrontal patients showed most errors with crossed stimuli (conditions 2 and 4), this was independent of whether dissonance occurred or not, whereas parietal patients had elevated error rates specifically in the case of dissonance when the first target was located in the contralesional hemifield (condition 1 for left and conditions 3 and 4 for right parietal lesions). *PFC*, prefrontal cortex; *SMA*, supplementary motor area; *PPC*, posterior parietal cortex; *FEF*, frontal eye fields

exactly as possible. Data were analyzed separately for those trials in which the condition of retinospatial dissonance was supplied with respect to the second saccade (i.e., all trials in which the second target had disappeared before the end of the first saccade) and for those trials in which it was not supplied. In the following section, we present an outline of the relevant results. Further details of this study have been published elsewhere (Heide et al. 1995).

Dysmetria of Double-Step Saccades After Parietal and Central Thalamic Lesions: Failure of the Extraretinal Signal

In all groups of subjects, the spatial accuracy of the *first saccade* was comparable to that of visually-guided single-step saccades, with a mean amplitude gain of 0.89 ± 0.15 in the control group, similar results in the frontal subgroups, and hypometria of contralateral saccades in parietal patients (mean gain of 0.79; $p<0.05$). This retinotopic deficit of first saccades into the contralateral hemifield in cases of PPC lesions was also reflected in increased latencies of these saccades (mean latency of 337 ms versus 229 ± 65 ms in the control group; $p<0.01$). Analogous results have been reported for simple visually guided saccades as a specific parietal deficit (Pierrot-Deseilligny et al. 1991; Heide and Kömpf 1994).

In contrast to the first saccade in the double-step paradigm, spatial accuracy of the *second saccade* in the case of retinospatial dissonance was lower than that of visually guided single-step saccades even in *normal subjects*. Although their mean final eye position was close to target location, standard deviation was rather high. There are two reasons for this lower accuracy: First, this saccade is directed to the remembered location of the second target, and such saccades have been demonstrated to be less accurate than those with visual feedback; second, this saccade has to rely on extraretinal information about eye position, and this information has been shown to be inaccurate around the time of a saccade, as mentioned earlier (Dassonville et al. 1992).

Considering the *patients*' performance of the *second saccade*, our most important finding was failure or gross dysmetria of this saccade specifically in parietal patients in the case of retinospatial dissonance whenever the first saccade had been directed into the hemifield contralateral to the side of the lesion. This happened even when the second target had appeared in the healthy ipsilesional hemifield, as illustrated in Fig. 5 for a patient with a right parietal lesion: In comparison with the normal subject (a), the patient's (b) first saccade was delayed, but not dysmetric, whereas his second saccade was markedly dysmetric, with a large error of final eye position. It did not even reach the right hemifield where the second target had been presented, but it was directed back towards the invisible fixation point. Obviously, it was performed according to the initial retinal vector of the second target (black vertical arrow in Fig. 5), thus not taking into account extraretinal information about the contralateral displacement of the eyes associated with the first saccade.

The specificity of this disorder as a parietal deficit of the spatial programming of saccades is evident from Fig. 6B, in which the percentage of failure or gross dysmetria of second saccades is plotted for each group of subjects. Exclusively in parietal patients, the percentage was above the level of the control group only in trials with retinospatial dissonance (upper panel), but not without dissonance (bottom panel), whenever the first saccade had been directed into the contralateral hemifield. This was the case in conditions 3 and 4 (of the stimuli plotted in Fig 6A) for right PPC lesions ($p<0.001$) and, to a lesser extent, in condition 1 for left PPC lesions ($p<0.05$). In the frontal groups, however, the percentage of erroneous second saccades was either within normal limits (left SMA, left PFC) or elevated (right PFC, right FEF), irrespective of whether there was retinospatial dissonance or not.

Figure 7 shows the mean absolute error of final eye positions after the second saccade in each trial, reflecting saccadic dysmetria. It is obvious that the error was markedly elevated after PFC or FEF lesions in conditions 2 and 4 (crossed stimulus), in which these patients tended to abort the second saccade, but this happened both with and without retinospatial dissonance. Only parietal patients, however, selectively showed significant dysmetria of second saccades in the case of retinospatial dissonance following contralateral first saccades (conditions 3 and 4 for right and 2 for left PPC lesions). This dysmetria did not correlate with the severity of visual hemineglect, but it did correlate with right parietal patients' impairment in a global neuropsychological test of visuospatial orientation, namely, in copying Rey's complex figure (examples in Fig. 8); performance in this test had been quantified using a score, as suggested in the literature (Lezak 1983); the correlation coefficient amounted to 0.72 ($p<0.05$).

In conclusion, the dissociation of a visual target's retinal and spatial coordinates during the double-step task made it possible to identify a deficit of saccade metrics that was specific for parietal lesions and more marked in right than left hemispheric lesions: Following a first saccade into the contralesional hemifield, the second saccade was often dysmetric or missing when extraretinal information was needed for spatial accuracy. This occurred even when the second target was located in the healthy ipsilateral hemifield or when the second saccade had to be performed into the ipsiversive direction, indicating that it is neither a retinotopic

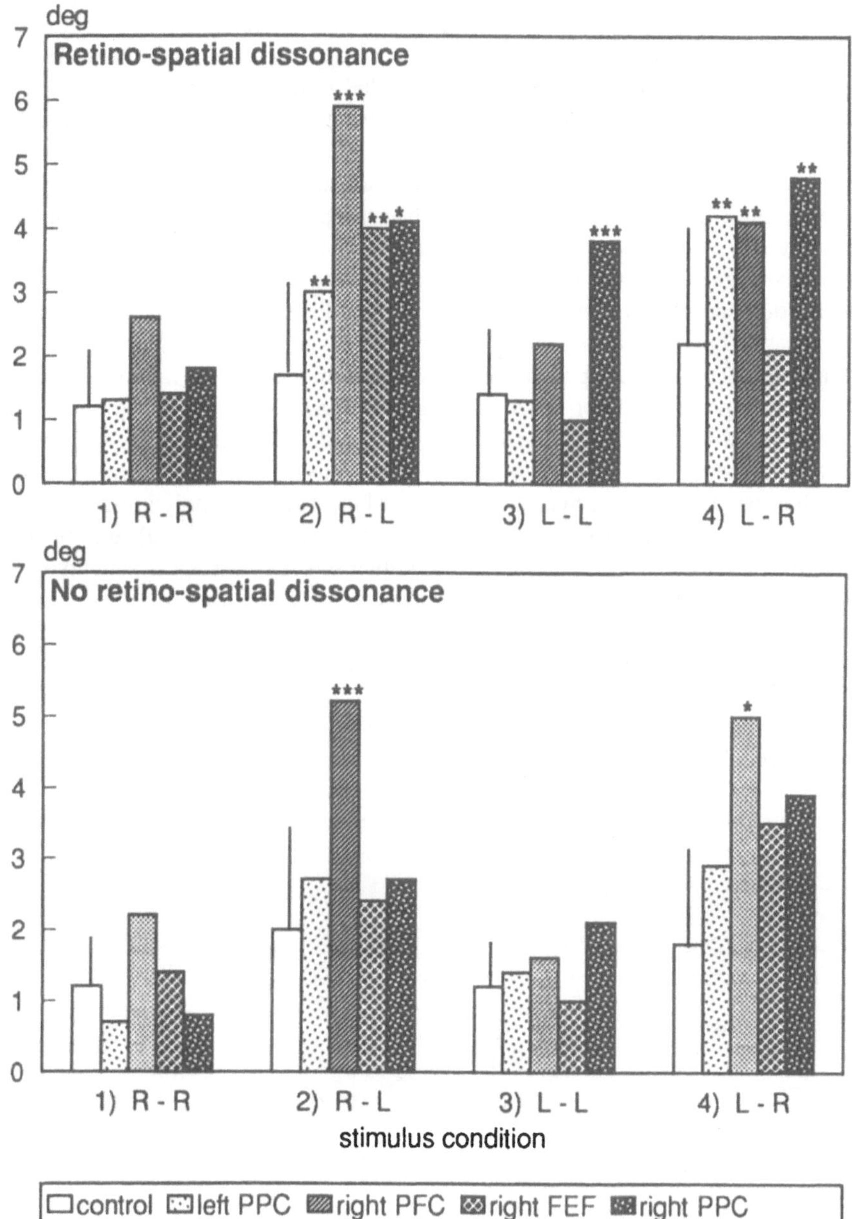

Fig 7. Mean absolute error of final eye positions after double-step trials, plotted separately for the four different stimulus conditions of Fig 6A. Standard deviations of the control group are indicated by *vertical lines*. Results of lesions in the left supplementary motor area (*SMA*) and in the left prefrontal cortex (*PFC*) are not shown in this diagram, because they were not significantly different from the control group. *R*, right; *L*, left; *PPC*, posterior parietal cortex; *FEF*, frontal eye fields

nor a directional deficit nor one of spatiotopic coding. Rather, it is a failure of efference copy (corollary discharge) because PPC patients are impaired in registering extraretinal information about the motor vector (amplitude and direction) of a contralateral saccade and using it for updating the spatial representation of the next target. It is not yet clear if this deficit of the extraretinal signal concerns only the programming of saccades or perceptual stability in general. The latter might be speculated on because parietal patients often complain of unspecific visual disturbances such as blurred vision or dizziness, but this has to be investigated by specific psychophysical tests for visual stability and spatial localization across saccades. At any rate, we could show that the saccadic deficit in our parietal patients is of some relevance for visuospatial orientation in general, which is reflected in these patients' impaired performance in copying Rey's figure. The same type of disorder (even more severe) has already been reported in a patient with a large right frontoparietal lesion (Duhamel et al. 1992b). Our study has shown that it has to be attributed predominantly to the parietal lesion. Even though a spatially accurate neuronal signal has been found not only in parietal, but also in FEF neurons of monkeys (Goldberg and Bruce 1990) during the performance of double-step saccades, the deficit did not turn up in patients with FEF lesions. There is only one other site in the human brain where a similar deficit was reported in cases of lesions (Gaymard et al. 1994): the central thalamus, particularly the internal medullary lamina (IML). Two patients with unilateral lesions in this region had markedly impaired saccadic accuracy in conditions in which extraretinal information about a previous eye displacement was needed for the spatial programming of a memory-guided saccade. In this study, eye displacement took place during the memorization period of a memory-guided saccade, like in the classica double-step paradigm, and the deficit

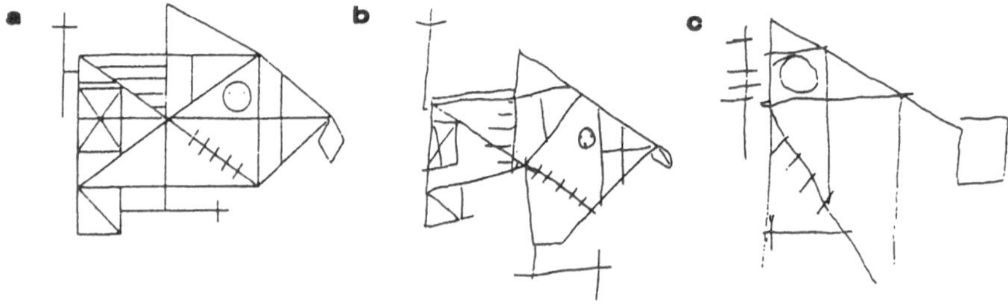

Fig 8 a-c. Performance of two patients with right posterior parietal lesions in copying Rey's complex figure (**a**), one of them with mild deficits (**b**) and the other with severe impairment (**c**)

occurred whether the eye displacement was a saccade, a smooth pursuit eye movement or a vestibularly induced eye movement. Obviously, it was a general deficit of the eye movement-associated extraretinal signal. One of these two patients was able to take the eye displacement into account, but overestimated it; thus these authors conclude it might be a deficit of calibrating the extraretinal signal. These findings and evidence from monkey experiments lead to the conclusion that the extraretinal signal is conveyed from the pons (Büttner-Ennever and Henn 1976; McCrea and Baker 1985), its probable origin, via the central thalamus, where it might be calibrated, to the posterior parietal cortex, where it is integrated with retinotopic information. The IML is also connected with frontal saccade areas, namely the FEF and the SEF (Schlag-Rey and Schlag 1989). In both of these areas, the presence of the extraretinal signal has been demonstrated, as mentioned earlier.

Contralateral Neglect of Double-Step Saccades

When the first target appeared in the ipsilateral and the second target in the contralateral hemifield (condition 2 in right- and condition 4 in left-hemispheric lesions), the percentage of aborted second saccades (Fig. 6) was elevated in parietal patients ($p<0.01$) as well as in right FEF and in right prefrontal (PFC) cases ($p<0.01$), with *and* without retinospatial dissonance. In parietal and FEF patients, the percentage correlated with the neglect score ($r=0.61$; $p<0.05$), obviously reflecting hemineglect (or extinction) of a contralateral visual target. Correspondingly, when the two targets were presented in reverse temporal order, the contralateral first target was neglected by parietal and FEF patients (in about 45% patients versus 10% normal subjects) so that only one saccade was performed directly to the second target.

Neglect of the contralateral target occurred only with crossed double-step stimuli when there was an ipsilateral target to attract attention, analogous to Posner's cueing task (Posner et al. 1984). This might be due to an impaired disengagement of saccade-relevant visual attention from ipsilateral targets. These findings confirm the close relationship between the orienting of attention and the initiation of saccadic eye movements that specifically characterizes neurons in the PPC coding "attentional vectors" (Lynch and McLaren 1989; Goldberg et al. 1990; Sheliga et al. 1994).

The Contribution of the Frontal Lobes

In our study, frontal lesions did not significantly affect spatial accuracy of saccades, but rather the initiation of the second saccade in the double-step task, *independent of retinospatial dissonance*. Patients with prefrontal (PFC) lesions who had no visual hemineglect often aborted the second saccade in conditions 2 and 4 (Fig. 6) when it had to cross the vertical meridian, independent of the hemifield where the second target had appeared. Instead of performing the second saccade, they tended to repeat the first saccade (Fig. 9a). This reflects a motor impairment of crossed saccadic sequences and can be explained with respect to the role of the PFC in spatial working memory (Funahashi et al. 1991, 1993), as the memory trace of the second target (located in the opposite visual hemifield) could have been lost during the processing time of the first saccade.

Furthermore, patients with PFC or right FEF lesions and to a lesser extent patients with SMA lesions generally tended to execute the saccade to the location of the first target following the saccade to the second target, i.e., in the reverse temporal order, irrespective of the targets' retinotopic locations (Fig. 9b). Obviously, frontal lesions generally impair the temporal control and triggering of voluntary double-step saccadic sequences, independent of visual hemineglect or retinospatial dissonance.

Patients with lesions of the left SMA showed a prolonged intersaccadic interval (527 ms versus 310 ± 215 ms in normal subjects; $p<0.05$) concerning only contraversive second saccades that had to cross the vertical meridian (condition 4), whether retinospatial dissonance had occurred or not. This delay might reflect a specific deficit in *timing* voluntary saccadic sequences in correspondence with the role of the SMA for the execution of motor sequences, particularly if they are internally triggered or memorized, which has been shown for skeletal movements (Deecke et al. 1985; Mushiake et al. 1990) and for sequences of memory-guided saccades (Gaymard et al. 1990).

Conclusion

In conclusion, the frontal and parietal cortices complement each other in the control of saccadic sequences: *Frontal* areas are more involved in *temporal* aspects, and the *posterior parietal cortex* in *spatial* programming. Thus, our data confirm the key role of the PPC in the analysis of visual space with a dominance of the right hemisphere. The double-step paradigm is helpful in identifying a specific deficit of the saccade-associated efference copy signal (corollary discharge) in patients with posterior parietal or central thalamic lesions. Thus, these brain structures, and not the frontal lobes, turned out to be critical for the processing of extraretinal information about eye displacement, making them essential for the maintenance of spatial constancy across saccades.

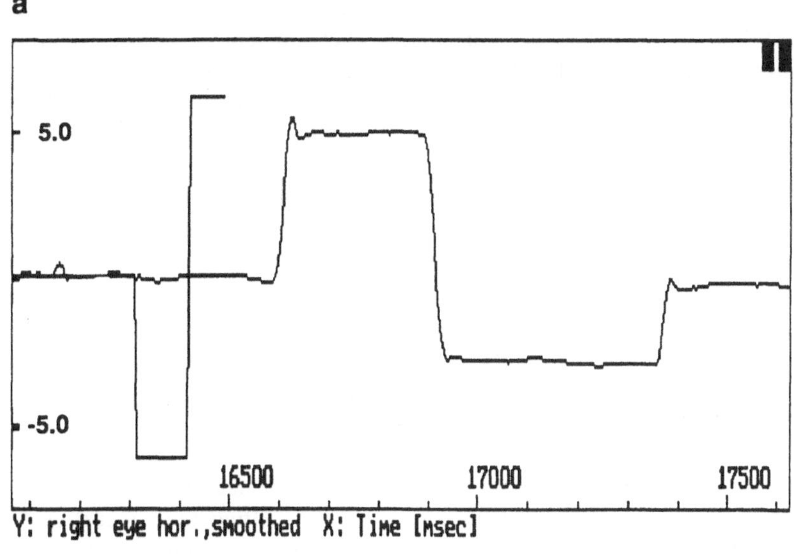

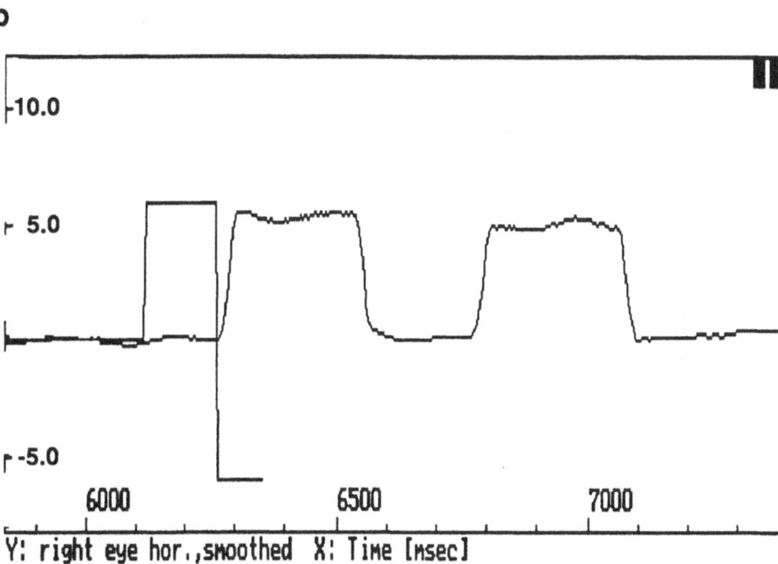

Fig 9 a, b. Examples of impaired performance of horizontal double-step saccades in patients with lesions of the right frontal eye field (**a**) and the right dorsolateral prefrontal cortex (**b**). The two targets are presented at 6° in opposite hemifields (crossed stimulus, condition 2 of Fig. 6A). The *traces* in each panel show the horizontal position (positive Y-axis = rightward, negative Y-axis = leftward) of the right eye, superimposed with target position. *FEF*, frontal eye field; *PFC*, prefrontal cortex

However, our findings do not support the view that the PPC contains a spatially invariant target representation in stable head- or body-centered coordinates because unilateral parietal lesions did not cause spatiotopic deficits restricted to saccade targets in the contralateral hemispace. Rather, these patients also failed to aim at ipsilateral targets that should be accessible in the case of a stable spatial map. Yet this does not exclude the spatiotopic model, as dysmetria of the second saccade could alternatively be due to a lack of eye position information *after* the first saccade when the eyes are directed towards the contralateral hemispace. This possibility could be tested by recording double-step saccades from different initial eye positions in the orbita so that a contraversive first saccade is not necessarily directed towards the contralateral craniotopic hemispace. Nevertheless, our results are in better agreement with the eye displacement model in which target position is represented in retinocentric coordinates and updated according to the current direction of gaze. The parietal patients in our study were impaired in updating their internal representation of target location for a *contralateral eye displacement*, and this could just be a deficit of those neurons that anticipate the retinal consequences of an intended saccade by shifting their receptive fields accordingly, thus remapping the retinal representation of the visual world (as found in area LIP of monkeys; Duhamel et al. 1992a). Tus, parietal lesions in our patients seem to have predominantly damaged the homologue of area LIP which is involved particularly in the control of saccades and saccade-related spatial computations (Andersen 1995).

References

Andersen RA (1989) Visual and eye movement functions of the posterior parietal cortex. Ann Rev Neurosci 12:377–403

Andersen RA (1995) Coordinate transformations and motor planning in posterior parietal cortex. In: Gazzaniga MS (ed) The cognitive neurosciences. MIT Press, Cambridge, pp 519–532

Andersen RA, Essick GK, Siegel RM (1985) Encoding of spatial location by posterior parietal neurons. Science 230:456–458

Balint R (1909) Seelenlähmung des "Schauens", optische Ataxie, räumliche Störung der Aufmerksamkeit. Monatsschr Psychiatr Neurol 25:51–181

Barash B, Bracewell RM, Fogassi L, Gnadt JA, Andersen RA (1991) Saccade-related activity in the lateral intraparietal area II: spatial properties. J Neurophysiol 66:1109–1124

Becker W, Jürgens R (9179) An analysis of the saccadic system by means of double step stimuli. Vision Res 19:967–983

Brain WR (1941) Visual disorientation with special reference to lesions of the right cerebral hemisphere. Brain 64:224–272

Bridgeman B (1995) A review of the role of efference copy in sensory and oculomotor control systems. Ann Biomed Eng 23:409–422

Bridgeman B, van der Heijden AHC, Velichkovsky BM (1994) A theory of visual stability across saccadic eye movements. Behav Brain Sci 17:247–292

Bushnell MC, Goldberg ME, Robinson DL (1981) Behavioral enhancement of visual responses in monkey cerebral cortex. II. Modulation in posterior parietal cortex related to selective visual attention. J Neurophysiol 46:755–772

Büttner-Ennever JA, Henn V (1976) An autoradiographic study of the pathways from the pontine reticular formation involved in horizontal eye movements. Brain Res 108:155–164

Critchley M (1953) The parietal lobes. Arnold, London

Dassonville P, Schlag J, Schlag-Rey M (1992) Oculomotor localization relies on a damped representation of saccadic eye displacement in human and nonhuman primates. Visual Neurosci 9:261–269

Dassonville P, Schlag J, Schlag-Rey M (1995) The use of egocentric and exocentric location cues in saccadic programming. Vision Res 35:2191–2199

Deecke L, Kornhuber HH, Lang W, Schreiber H (1985) Timing function of the frontal cortex in sequential motor and learning tasks. Human Neurobiol 4:143–154

De Renzi E (1982) Disorders of space exploration and cognition. Wiley, New York

De Renzi E (1988) Visuo-spatial disorders. In: Kennard C, Rose CF (eds) Physiological aspects of clinical neuro-ophthalmology. Chapman and Hall, London, chap 9

Duhamel J-R, Colby CL, Goldberg ME (1992a) The updating of the representation of visual space in parietal cortex by intended eye movements. Science 255:90–92

Duhamel J-R, Goldberg ME, Fitzgibbon EJ, Sirigu A, Grafman J (1992b) Saccadic dysmetria in a patient with a right frontoparietal lesion. Brain 115:1387–1402

Funahashi S, Bruce CJ, Goldman-Rakic PS (1991) Neuronal activity related to saccadic eye movements in the monkey's dorsolateral prefrontal cortex. J Neurophysiol 65:1464–1483

Funahashi S, Bruce CJ, Goldman-Rakic PS (1993) Dorsolateral prefrontal lesions and oculomotor delayed-response performance: evidence for "mnemonic" scotomas. J Neurosci 13:1479–1497

Galletti C, Battaglini PP, Fattori P (1993) Parietal neurons encoding spatial locations in craniotopic coordinates. Exp Brain Res 96:221–229

Gauthier G, Nommay D, Vercher J-L (1990) The role of ocular muscle proprioception in visual localization of far targets. Science 249:58–61

Gaymard B, Pierrot-Deseilligny C, Rivaud S (1990) Impairment of sequences of memory-guided saccades after supplementary motor area lesions. Ann Neurol 28:622–626.

Gaymard B, Rivaud S, Pierrot-Deseilligny C (1994) Impairment of extraretinal eye position signals after central thalamic lesions. Exp Brain Res 102:1–9

Gellman RS, Carl JR (1991) Early responses to double-step targets are independent of step amplitude. Exp Brain Res 87:433–437

Gellman RS, Fletcher WA (1992) Eye position signals in human saccadic processing. Exp Brain Res 89:425–434

Gochin PM, McElligott JG (1987) Saccades to visual targets are uncompensated after cerebellar stimulation. Exp Neurol 97:219–224

Goldberg ME, Bruce CJ (1990) Primate frontal eye fields. III. Maintainance of a spatially accurate saccade signal. J Neurophysiol 64:489–508

Goldberg ME, Colby CL, Duhamel J-R (1990) Representation of visuomotor space in the parietal lobe of the monkey. Cold Spring Harb Symp Quant Biol 55:729–739.

Grüsser O-J, Krizic A, Weiss L-R (1987) Afterimage movement during saccades in the dark. Vision Res 27:215–226

Guthrie BL, Porter JD, Sparks DL (1983) Corollary discharge provides accurate eye position information to the oculomotor system. Science 221:1193–1195

Hallett PE, Lightstone AD (1976) Saccadic eye movements to flashed targets. Vision Res 16:107–114

Halligan PW, Cockburn J, Wilson BA (1991) The behavioural assessment of visual neglect. Neuropsychol Rehabil 1:5–32

Haxby JV, Grady CL, Horwitz B, Salerno J, Ungerleider LG, Mishkin M, Carson RE, Herscovitch P, Schapiro MB, Rapoport SI (1991) Dissociation of spatial and object visual processing pathways in human extrastriate cortex. Proc Natl Acad Sci USA 88:1621–1625

Heide W, Kömpf D (1994) Saccades after frontal and parietal lesions. In: Fuchs AF, Brandt T, Büttner U, Zee DS (eds) Contemporary ocular motor and vestibular research: a tribute to David A. Robinson. Thieme, Stuttgart, pp 225–227

Heide W, Blankenburg M, Zimmermann E, Kömpf D (1995) Cortical control of double-step saccades: implications for visuo-spatial orientation. Ann Neurol 38:739–748

Heilman KM, Watson RT, Valenstein E (1993) Neglect and related disorders. In: Heilman KM, Valenstein E (eds) Clinical neuropsychology, 3rd edn. Oxford University Press, New York, pp 279–336

Holmes G (1918) Disturbances of visual orientation. Br J Ophthalmol 2:449–486, 506–516

Holmes G (1919) Disturbances of visual space perception. Br Med J 2:230–233

Honda H (1989) Perceptual localization of visual stimuli flashed during saccades. Percept Psychophys 45:162–174

Honda H (1990) Eye movements to a visual stimulus flashed before, during or after a saccade. In: Jeannerod M (ed) Attention and performance, vol 13. Erlbaum, Hillsdale, pp 567–582

Hyvärinen J (1982) The parietal cortex of monkey and man. Springer, Berlin Heidelberg New York

Irwin DE, Zacks JL, Brown JS (1990) Visual memory and the perception of a stable visual environment. Percept Psychphys 47:35–46

Jeannerod M (1986) Mechanisms of visuomotor coordination: a study in normal and brain-damaged subjects. Neuropsychologia 24:41–78

Jürgens R, Becker W, Kornhuber HH (1981) Natural and drug-induced variations of velocity and duration of human saccadic eye movements: evidence for a control of the neuronal pulse generator by local feedback. Biol Cybern 39:87–96

Lezak MD (1983) Neuropsychological assessment, 2nd edn. Oxford University Press, New York

Lynch JC, McLaren JW (1989) Deficits of visual attention and saccadic eye movements after lesions of parieto-occipital cortex in monkeys. J Neurophysiol 61:74–90

Lynch JC, Mountcastle VB, Talbot WH, Yin TCT (1977) Parietal lobe mechanisms for directed visual attention. J Neurophysiol 40:362–389

MacKay DM (1973) Visual stability and voluntary eye movements. In: Jung R (ed) Handbook of sensory physiology, vol 7/3. Springer, Berlin Heidelberg New York, pp 307–331

Maldonado H, Joseph J-P, Schlag J (1980) Types of movement evoked by thalamic microstimulation in the alert cat. Exp Neurol 70:613–625

Mays LE, Sparks DL (1980) Dissociation of visual and saccade-related responses in superior colliculus neurons. J Neurophysiol 43:207–232

McCrea RA, Baker R (1985) Anatomical connections of the nucleus prepositus of the cat. J Comp Neurol 237:377–407

Meerwaldt JD, van Harskamp F (1982) Spatial disorientation in right-hemisphere infarction. J Neurol Neurosurg Psychiatr 45:586–590

Mishkin M, Ungerleider LG, Macko KA (1983) Object vision and spatial vision: two cortical pathways. Trends Neurosci 6:414–417

Moschovakis AK, Highstein SM (1994) The anatomy and physiology of primate neurons that control rapid eye movements. Annu Rev Neurosci 1:465–488

Mountcastle VB, Lynch JC, Georgopoulos A, Sakata H, Acuna C (1975) Posterior parietal association cortex of the monkey: command functions for operations within extrapersonal space. J Neurophysiol 38:871–908

Mushiake H, Inase M, Tanji J (1990) Selective coding of motor sequence in the supplementary motor area of monkey cerebral cortex. Exp Brain Res 82:208–210

Noda H, Murakami S, Warabi T (1991) Effect of fastigial stimulation upon visually directed saccades in macaque monkey. Neurosci Res 10:188–199

Paterson A, Zangwill OL (1944) Disorders of visual space perception associated with lesions of the right hemisphere. Brain 67: 331–358

Petit L, Orssaud C, Tzourio N,Salamon G, Mazoyer B, Berthoz A (1993) PET study of voluntary saccadic eye movements in humans: basal ganglia-thalamocortical system and cingulate cortex involvement. J Neurophysiol 69:1009–1017

Pierrot-Deseilligny C, Rivaud S, Gaymard B, Agid Y (1991) Cortical control of reflexive visually guided saccades. Brain 114:1473–1485

Pierrot-Deseilligny C, Rivaud S, Gaymard B, Müri R, Vermersch A-I (1995) Cortical control of saccades. Ann Neurol 37:557–567

Posner MI, Walker JA, Friedrich FJ, Rafal RD (1984) Effects of parietal injury on covert orienting of attention. J Neurosci 4:1863–1874

Purkinye J (1825) Über die Scheinbewegungen, welche im subjectiven Umfang des Gesichtssinnes vorkommen. Bulletin der naturwissenschaftlichen Section der Schlesischen Gesellschaft 4:9–10

Robinson DA (1973) Models of the saccadic eye movement control system. Kybernetik 14:71–83

Roucoux A, Crommelinck M (1976) Eye movements evoked by superior colliculus stimulation in the alert cat. Brain Res 106:349–363

Schiller PH, Sandell JH (1983) Interactions between visually and electrically elicited saccades before and after superior colliculus and frontal eye field ablations in the rhesus monkey. Exp Brain Res 49:381–392

Schlag J, Schlag-Rey M (1987) Evidence for a supplementary eye field. J Neurophysiol 57:179–200

Schlag J, Schlag-Rey M (1990) Colliding saccades reveal the secret of their marching orders. Trends Neurosci 13:410–415

Schlag J, Schlag-Rey M, Pigarev I (1990) Supplementary eye field: influence of eye position on neural signals of fixation. Exp Brain Res 90:302–306

Schlag J, Schlag-Rey M, Dassonville P (1994) For and against spatial coding of saccades. In: d'Ydewalle G, van Rensbergen J (eds) Visual and oculomotor functions. Elsevier, Amsterdam, pp 3–17

Schlag-Rey M, Schlag J (1989) The central thalamus. In: Wurtz RH, Goldberg ME (eds) The neurobiology of saccadic eye movements. Rev Oculomotor Res, vol 3. Elsevier, Amsterdam, pp 361–390

Sheliga BM, Riggio L, Rizzolatti G (1994) Orienting of attention and eye movements. Exp Brain Res 98:507–522

Shibutani H, Sakata H, Hyvärinen J (1984) Saccades and blinking evoked by microstimulation of the posterior association cortex of the monkey. Exp Brain Res 55:1–8

Sparks DL, Mays LE (1983) Spatial localization of saccade targets. I. Compensation for stimulus induced perturbations in eye position. J Neurophysiol 49:45–74

Sparks DL, Mays EL, Porter JD (1987) Eye movements induced by pontine stimulation: interaction with visually triggered saccades. J Neurophysiol 58:300–318

Sperry RW (1950) Neural basis of the spontaneous optokinetic response produced by visual inversion. J Comp Physiol Psychol 43:482–489

Stein JF (1992) The representation of egocentric space in the posterior parietal cortex. Behav Brain Sci 15:691–700

Steinbach MJ (1987) Proprioceptive knowledge of eye position. Vision Res 27:1737–1744

Ungerleider LG, Desimone R (1986) Cortical connections of visual area MT in the macaque. J Comp Neurol 248:190–222.

von Helmholtz H (1962) Treatise on physiological optics, vol III. The perceptions of vision. Dover, New York. Translation of: Handbuch der physiologischen Optik, vol III. Voss, Hamburg, 1866

von Holst E, Mittelstaedt H (1950) Das Reafferenzprinzip (Wechselwirkungen zwischen Zentralnervensystem und Peripherie). Naturwissenschaften 37:464–476

Westheimer G (1954) Eye movement resonses to a horizontally moving stimulus. AMA Arch Ophthal 52:932–941

Zipser D, Andersen RA (1988) A back propagation programmed network that simulates response properties of a subset of posterior parietal neurons. Nature 331:679–684

MST Neurons Are Activated by Smooth Pursuit of Imaginary Targets

U. J. ILG and P. THIER

Sektion für Visuelle Sensomotorik, Neurologische Universitätsklinik, Tübingen, Germany

Research on the function of the primate posterior parietal cortex has been dominated for many years by a dispute between conflicting interpretations of the response properties of posterior parietal cortex neurons, leading to very different and even exclusive interpretations of parietal lobe functions. In the 1970s, neuroscientists began to study the properties of neurons recorded from area 7 of primate posterior parietal cortex using paradigms which required monkeys to direct either their eyes or their hands to visual targets in extrapersonal space (Hyvärinen and Poranen 1974; Mountcastle et al. 1975). Since many of the neurons studied were active in conjunction with distinct oculomotor behaviors such as fixation, saccades, or smooth pursuit, or alternatively in conjunction with visually directed hand movements, Mountcastle et al. (1975) suggested that the observed neuronal activation reflected commands for the execution of movements to objects in extrapersonal space. Not much later, Robinson and coworkers arrived at a very different interpretation of eye or hand movement related single unit responses (Robinson et al. 1978). Instead of interpreting posterior parietal cortex as a command device programming movements directed to targets in extrapersonal space, they suggested, alternatively, that the posterior parietal cortex of monkeys should be understood as a sensory structure extracting relevant information by focusing spatial attention to relevant sensory stimuli. Today, these exclusive views, localizing the functional role of the posterior parietal cortex on opposite sides of the sensory vs. motor dividing line, are no longer tenable. Rather than being either sensory or motor in the strict sense, recent work on visually guided saccades has suggested that parietal neurons involved in these types of visually guided behavior are better described as elements of networks underlying intermediate stages of the sensory-to-motor transformations leading to these behaviors (see Thier and Andersen, this volume; Barash, this volume). While this view is gaining ground, attempts to extend it to smooth pursuit eye movements, the other type of visually guided eye movements, whose parietal implementation was suggested by the early studies of Mountcastle and coworkers (1975), have faced especially strong resistence. This is a direct consequence of the influence of models of smooth pursuit eye movements, which from a computational point of view have successfully reduced cortical contributions to smooth pursuit to the

In: Parietal Lobe Contributions to Orientation in 3D Space (1997). P. Thier and H.-O. Karnath (eds). Springer-Verlag, Heidelberg.

extraction of the relevant sensory information such as retinal image slip. In contrast, the actual sensory-to-motor transformation was proposed to be mediated by subcortical structures such as the cerebellum (Lisberger et al. 1987). The alternative possibility that parietal smooth pursuit-related single units are actually part of a network underlying the sensory-to-motor transformation, rather than being mere sensory error detectors, has proven to be hard to test. The reason is that smooth pursuit eye movements are much more sensory driven than saccades or reaching hand movements. This makes it impossible to disentangle the influence of retinal and nonretinal factors on the discharge of smooth pursuit-related parietal neurons by simply separating the presentation of the visual target and the motor response required in time, a standard strategy in the study of saccade- and reaching-related single-unit activity. Nevertheless, recent work on smooth pursuit-related single-unit activity has come up with ways to demonstrate the involvement of nonretinal smooth pursuit-related signals. This chapter discusses these techniques and their limitations, which have prompted us to use a new approach based on the presentation of *imaginary targets* guiding smooth pursuit eye movements. Our results with this new technique support the view that parietal cortex contributions to smooth pursuit cannot be reduced to the extraction of retinal error signals.

Evidence for Nonretinal Signals Affecting the Discharge of Smooth Pursuit-Related Parietooccipital Neurons

In the 1980s, both electrophysiological observations and the study of the effects of experimental lesions suggested that two parietooccipital areas are involved in the generation of smooth pursuit eye movements (Dürsteler and Wurtz 1988; Komatsu and Wurtz 1988; Erickson and Dow 1989). Both areas are located within the superior temporal sulcus (STS) of rhesus monkeys, the middle temporal area (area MT) on the posterior bank of the sulcus and the middle superior temporal area (area MST) confined to the anterior bank and the fundus, the latter marking the caudal end of the posterior parietal cortex. In both areas, single units were found which were activated by smooth pursuit eye movements. Newsome and colleagues (1988) set out to reveal the nature of the pursuit-related activity found in the STS of rhesus monkeys by trying to separate the retinal and eye movement-related factors involved. The first step taken towards this end was to rule out the possible impact of the visual background, whose image is shifted across the retina by the pursuit eye movement. This was achieved by presenting the target in an otherwise dark environment and by avoiding dark adaptation, possibly rendering remaining low contrast contours visible by switching background lights on in the intertrial intervals. While these choices were able to eliminate the influence of self-induced background image slip on the discharge of a single unit under study, they, of course, did not interfere with the other retinal stimulus involved, the target itself. Even during steady state pursuit, well after the onset of target movement, eye velocity does not necessarily match the velocity of the visual target (see Fig. 4 for an example). The resulting residual retinal slip of the target image, however, might activate a motion-sensitive mechanism, giving rise to the discharge of pursuit-related STS neurons. In order to test whether this was the case, Newsome et al. (1988) adopted two paradigms, which had in common that they tried to temporarily remove target image slip while smooth pursuit continued. The first paradigm was based on the assumption that steady state smooth pursuit is carried on even in the absence of a visible target, provided the target is taken away only briefly. This was achieved in their study by simply switching the target off for 150 ms (*blink paradigm*). The second approach (*stabilization paradigm*) was less radical. Rather than turning the target off completely, they tried to eliminate residual target image slip during steady state pursuit by temporarily clamping the target image to the retina by electronic means, thus compensating for the insufficiencies of the pursuit system. The rationale underlying these two paradigms was as follows: If the activation of a pursuit-related STS neuron during on-going pursuit was not affected by the removal of target image slip *(=target gap)* by either technique it obviously could not have a retinal cause and the persisting activation necessarily had to reflect the impact of a nonretinal pursuit-related signal on the neuron. When Newsome et al. (1988) applied the blink and the stabilization paradigm to pursuit-related neurons in the representation of the foveal parts of the visual field in MT, they found that none of them continued firing during the target gap. In other words, the activation of these neurons by smooth pursuit resulted from retinal image slip. In contrast, when these authors explored area MST on the anterior bank of the STS, they found many neurons in both the dorsal and lateral parts of this area whose activation was not affected by the target gap, a finding corroborated by later work (Thier and Erickson 1992). Under different circumstances, the same neurons could be shown to be sensitive to retinal image slip. Their ability to respond to smooth pursuit even in the absence of retinal image slip suggested that these neurons were driven by two pursuit-related signals, namely, target image slip and a nonretinal signal related to eye velocity. While the sensitivity to target image slip would enable these neurons to contribute to the initiation of smooth pursuit eye movement, their sensitivity to eye velocity would allow them to maintain smooth pursuit even in the absence of significant image slip, a property with profound ecological significance. Absence of retinal image slip is by no means a laboratory artifact, resulting from the usage of the blink paradigm. Rather, under natural conditions, the object of interest will often be hidden temporarily by nonattended objects such as trees or bushes falling in the line of sight while keeping on moving.

Even a Brief Disappearance of the Target Affects Smooth Pursuit Eye Movements

The interpretation of the experiments with target gaps presented in the preceding paragraph was based on the assumption that smooth pursuit is not affected by the disappearance of the target or the stabilization of its image on the retina. If, however, the target gap affected smooth pursuit, any concomitant change in neuronal activitation could either result from the elimination of retinal image slip or, alternatively, from the change in eye velocity. Consequently, the conclusion that a drop in discharge rate during the gap reflected removal of retinal stimulation and the absence of nonretinal input would no longer be legitimate. The complementary conclusion would also become less straightforward. If the firing rate of a neuron depended on a nonretinal signal related to the smooth pursuit eye movement in the absence of retinal stimulation, why should it remain unaffected if eye velocity changed. Actually, we had convinced ourselves earlier (unpublished observations) that the smooth pursuit eye velocity of monkeys almost always drops if the target image is stabilized by electronic means. Several years ago Becker and Fuchs (1985) reported that human smooth pursuit eye velocity drops if the target is turned off for periods of time similar to the ones prevailing in the experiments by Newsome et al. (1988) and Thier and Erickson (1992). Our own experiments demonstrate that monkeys, similarly to humans, show a drop of their smooth pursuit eye velocity if the target is turned off. One might have expected that monkeys, usually highly overtrained, might be less susceptible to target gaps than humans. However, our results show that even overtrained monkeys are not able to bridge the target gap without significant loss in smooth pursuit eye velocity, thereby complicating the interpretation of the responses of pursuit-related single units for the reasons outlined before. When the target disappeared unpredictably for 300 ms during steady state pursuit of a target moving at 10°/s on a homogeneous background, mean eye velocity of our human subjects (n=11) dropped to 55% of the eye velocity prevailing before the disappearance of the target, while mean eye velocity of the monkey subjects (n=2; see Fig. 1 for an example) dropped somewhat less to only 63% of the pregap velocity. This already small difference between monkeys and humans was further reduced when the demands on the pursuit system were increased by introducing a structured, stationary background on which the target moved. In this case, both monkeys and humans were even less able to bridge target gaps and the performance of monkeys and humans became almost idential (mean eye velocity during the gap of humans 40% and of monkeys 44% of pregap velocity). Sixty-two pursuit-related single units recorded from the STS of one of the two monkeys showed a persistent discharge rate during the gap, although, as mentioned before, the drop in eye velocity was in the order of 1/3 (homogeneous background during single-unit recordings). Obviously, the discharge of such neurons cannot be the neuronal

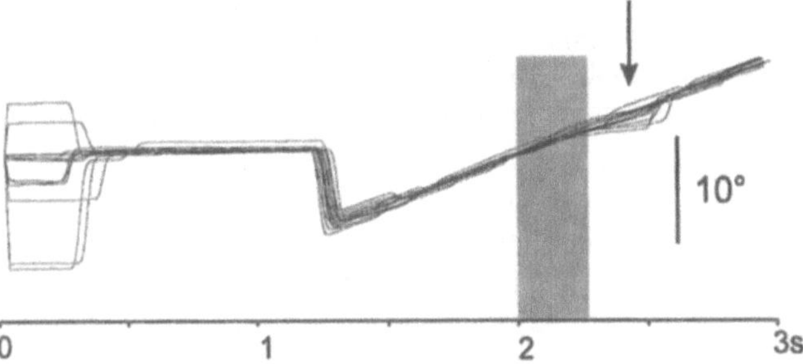

Fig. 1. Single trials (n=15) of horizontal eye position of a rhesus monkey pursuing a target moving at 10°/s towards the right after an initial target step 10° to the left. The *shaded vertical bar* marks the period of time (=300 ms) the target was blinked out. Note the pronounced decline in eye velocity after disappearance of the target (*arrow*)

replica of eye velocity assumed by previous work. However, although not offering a true replica of eye velocity, such neurons might still offer a coarse representation of eye velocity. For instance, they might already exhibit response saturation at the velocities prevailing in the gap period (which were in the order of 5°/s). In this case, we would not expect to see much of an effect of the the change of velocity induced by the target gap. Actually, there is evidence that pursuit-related MST neurons are indeed characterized by response saturation at target velocities well below 10°/s (unpublished observation). However, these considerations, plausible as they may be, cannot rule out a radically different explanation for the persistence of discharge despite changes in eye velocity, namely, a neuronal memory of the (para-)foveal image slip prevailing before the introduction of the target gap. In an attempt to critically test this possiblity we sought a way to evoke smooth pursuit eye movements without presenting foveal image slip in the first place, thus avoiding the charging of a putative image slip memory element. If the activation of pursuit-related MST neurons depended on foveal image slip, either presented in real-time or memorized, these neurons should be shut down in the case of pursuit without foveal stimulation. Our experiments with *imaginary* targets sparing the fovea show that this is not the case, lending new support to the idea that pursuit-related MST neurons depend on nonretinal information.

Smooth Pursuit of Imaginary Figures

Human subjects are able to track a wide variety of moving objects whose location and velocity cannot be directly derived from perifoveal retinal information. Examples are the lower corner of a moving diamond whose lower part is hidden by a stationary foreground or the invisible hub of a spinning wheel (Steinbach 1976). The common denominator of this class of pursuit targets is the fact that they are hidden parts of objects readily completed by our visual system using top-down knowledge of the shape of the object. However, even if we do not complete a meaningful object from sparse retinal cues we are able to evoke and maintain high-gain smooth pursuit eye movements. An example investigated in detail by Wyatt et al. (1994) is the imagined midpoint between two dots, presented extrafoveally and moving in synchrony. Such figures have in common that they lack any foveal information, thereby allowing us to critically test the hypothesis of a neuronal memory of foveal image slip in area MST. In order to use this approach, we first had to clarify whether monkeys are able to pursue such *imaginary*[1] target as well. Initially, we trained rhesus monkeys to track the center of an hourglass-like figure (target *R*; Fig. 2A) given by the intersection of the two diagonals defining the hourglass. After the monkeys had learned to track the center, we removed the central 12°x12° of the hourglass centered on the intersection, while continuously asking the monkey to keep his line of sight on the now invisible center (target *I1*; Fig. 2B). Finally, the monkeys learned to use *I1* not only to maintain smooth pursuit eye movements but also to initiate smooth pursuit or to saccade to the invisible center if presented in the visual periphery. One might object at this point that our monkeys simply learned to stay away evenly from the four line ends delimiting the void center of the hourglass rather than perceiving and fixating an imagined intersection of the two diagonals. The important difference between these two interpretations obviously is that the first one assumes that smooth pursuit would be generated by the direct conversion of information derived from extrafoveal retinal cues without any need for a filling in of hidden object parts by top-down processes. This objection is irrelevant for the single-unit experiment we had in mind, whose only requirement was the absence of foveal retinal information. However, we sought a way to clarify which of the two possible interpretations was correct. A possible way was suggested by the following consideration: If the monkeys′ pursuit indeed depended solely on the use of extrafoveal retinal information without any help from imagination, a reduction of the amount of extrafoveal information available should necessarily result in deterioration of smooth pursuit. A triangle whose lower, hidden corner (target *I2*) serves as a target for pursuit (Fig. 2D) is an example of an object offering less extrafoveal cues than the hourglass with a void center. Since the monkey would have to stay away evenly from only two line ends rather than from four as in the case of the hourglass with a void center, the amount of useful retinal information would be exactly halved. We trained two monkeys on pursuit of the imaginary targets and an additional control target, a moving dot presented at an

---

[1] We use this term for the sake of brevity, although, as discussed later, the view that monkeys actually look at an imagined target is debatable.

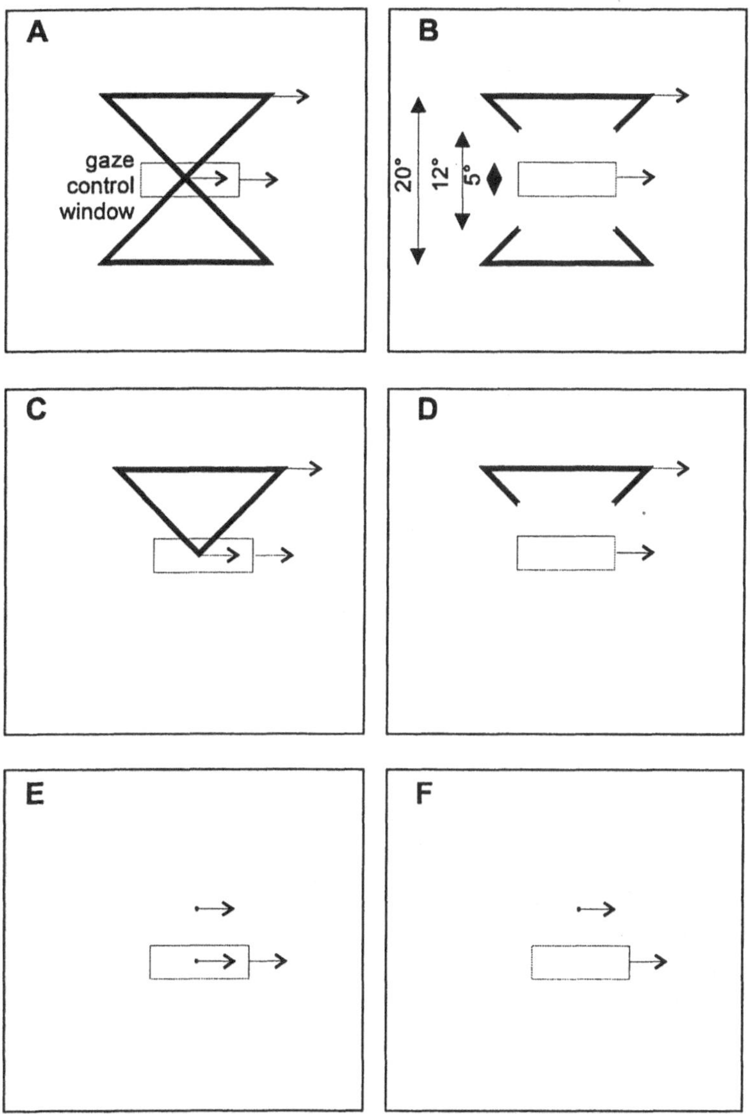

Fig. 2 A-F. Figures used in experiments to clarify whether monkeys are able to direct their gaze to imarginary targets. **A**, **B** hourglass-like figure. In **A** the visible intersection of the two diagonals serves as target (target R); in **B** the central parts of the hourglass are void and the monkey is asked to look at the now invisible intersection (target I1). **C**, **D** Triangle, whose lower corner serves as target. In **C** this corner is visible, in **D** the corner is hidden (target I2). **E**, **F** Dot target presented at 6° eccentricity. The monkey was trained to pursue this target without trying to fovealize it by first presenting the extrafoveal cue in combination with a foveal target (**E**), which was later left out (**F**). The *hair line rectangles* in **A** through **F** indicate the location and the extent of the eye position windows used

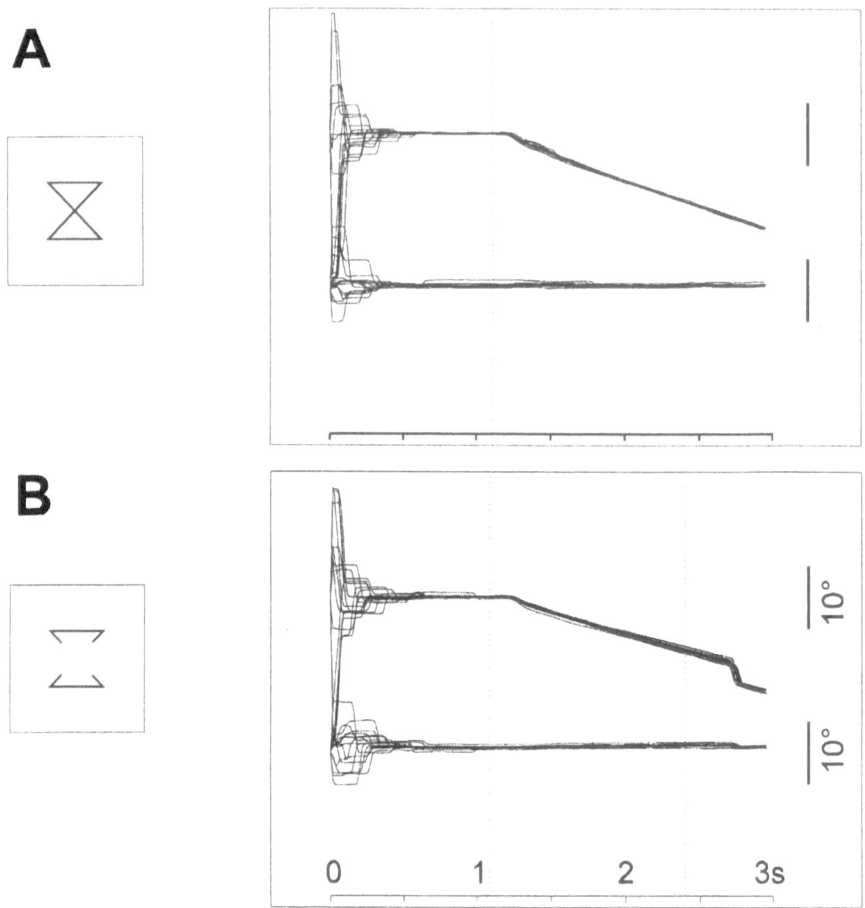

Fig. 3 A, B. Single trials (n=20 each) of horizontal and vertical eye position of a rhesus monkey pursuing target R (**A**) and target I1 (**B**), moving at 10°/s towards the left. The onset of target motion is indicated by the *first vertical dashed line*. The *second vertical dashed* line in **B** indicates the time target I1 transforms into target R

eccentricity of 6° (Fig. 2F), the monkeys were required to pursue without foveation. Figure 3 compares representative trials of smooth pursuit of targets *R* (A) and *I1* (B). The examples shown do not suggest major differences in smooth pursuit performance for the real and the *imaginary* target (*R* and *I1*, respectively), an impression which was fully supported by a quantitative analysis of smooth pursuit performance of this monkey. The gain of steady state pursuit was 0.99 for target *R*, 0.94 for target *I1* and 0.97 for target *I2* (no statistically significant difference, p>0.1, Wilcoxon-Rank-test). Only when this monkey was asked to pursue a single dot, presented at 6° eccentricity, did the gain decrease to 0.72

(statistically significant difference when compared with the gain of the real target, $p<0.01$, Wilcoxon-Rank-test). In other words, pursuit of the two *imaginary* figures was clearly better then pursuit based solely on extrafoveal information, suggesting that this monkey actually pursued a mental image of the invisible intersection in the case of the hourglass and the hidden corner in the case of the partially covered triangle. In contrast, the results obtained for the second monkey indicated that he preferred the alternative strategy which was based on the use of extrafoveal retinal information, making the completion of a full mental image of the object dispensable. This conclusion is suggested by the fact that we found a gradual decline in steady state pursuit gain paralleling the decline in the amount of retinal information (target *R*: 0.92, target *I1*: 0.83, target *I2*: 0.76, control with isolated dot at 6° eccentricity: 0.72; all values are significantly different from the gain of the real target, $p<0.01$, Wilcoxon-Rank-test). Irrespective of the strategy chosen to determine the location and velocity of the target, the processing of *imaginary* targets takes more time. This is indicated by the fact that saccades directed towards the *imaginary* target *I1* took significantly longer in both monkeys than saccades directed towards the real target *R* presented at 5°, 10°, and 20° eccentricity (saccade latencies monkey I: 181 *R* vs. 160 ms *I1*; monkey II: 160 *R* vs. 146 ms *I1*; statistically significant difference $p<0.01$, t-test).

Responses of STS Neurons to Smooth Pursuit of Imaginary Targets

Monkey area MT is characterized by a high percentage of neurons responding in a directionally selective manner to visual motion (Dubner and Zeki 1971; Zeki 1974). MT neurons with (para-)foveal receptive fields are typically activated by smooth pursuit eye movements. *Imaginary* targets like the ones discussed in the preceding section allow us to demonstrate that the pursuit-related responses indeed result from visual stimulation of the receptive field center.

Figure 4A shows the direction tuning of a typical MT neuron taken from the representation of the central parts of the visual field which preferred upward motion. When the monkey was asked to pursue the *real* target moving upward, this neuron showed a conspicuous oscillatory activation (Fig. 4B). A look at the eye velocity record suggests that bursts of activation occurred whenever the eye velocity fell below target velocity, resulting in image slip in the preferred (= upward) direction of this neuron. The purely retinal nature of this pursuit-related reponse is readily demonstrated by the fact that no significant response was obtained when the monkey pursued the *imaginary* target *I1* whose central void region was much larger than the receptive field center of this cell. The same cell also stopped discharging when the conventional dot target was blinked out temporarily (not shown). Similar observations on a large number of MT neurons fully support the view that the absence of activation during the target gap first

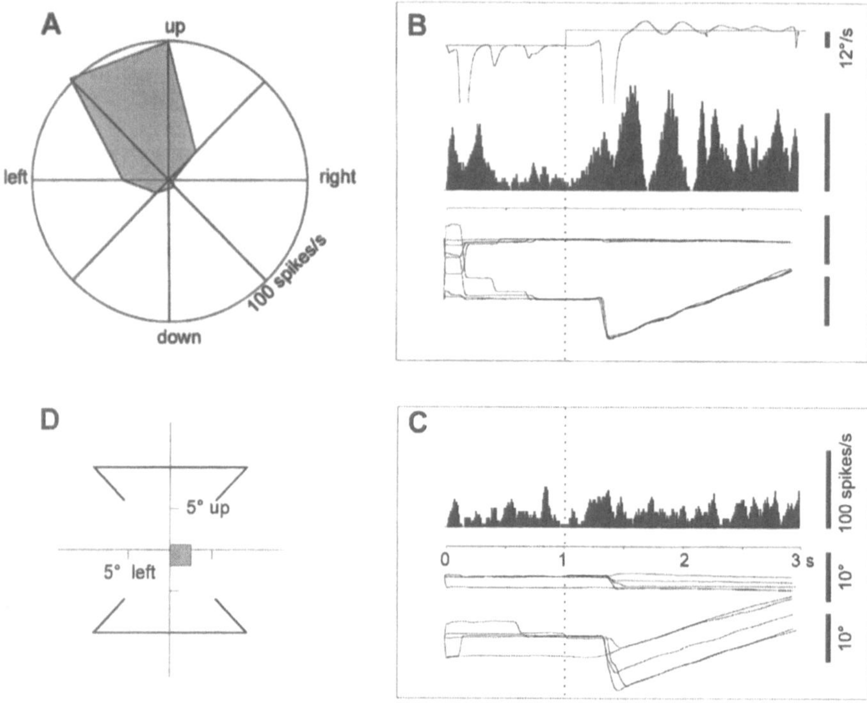

Fig. 4 A-D. Directionally selective neuron from the representation of the central parts of the visual field in the middle temporal area. **A** Direction tuning curve. **B** The monkey tracks target R, moving upward. The horizontal and vertical eye position records (five trials) are displayed together with the average eye velocity and a peristimulus time histogram (bin width 10 ms) characterizing the discharge of this neuron. The neuron discharged whenever the eye speed fell short of target speed, resulting in an upward retinal image slip. **C** The monkey tracked target I1. Note the absence of any significant neuronal response. **D** Size and location of the receptive field relative to the hourglass figure serving as target (I1)

noted by Newsome and colleagues (1988) is indeed a consequence of the exclusively retinal nature of pursuit-related responses of MT neurons and not a result of the inevitable drop in eye velocity during the gap. Our experiments with imaginary pursuit targets on one monkey so far also demonstrate clearly the integration of nonretinal information by neurons in neighboring area MST. In this monkey, we recorded from 20 pursuit related neurons located in area MST which were activated by pursuit of both imaginary and real targets. The same neurons showed resistence to the blinking of the target during steady state pursuit although

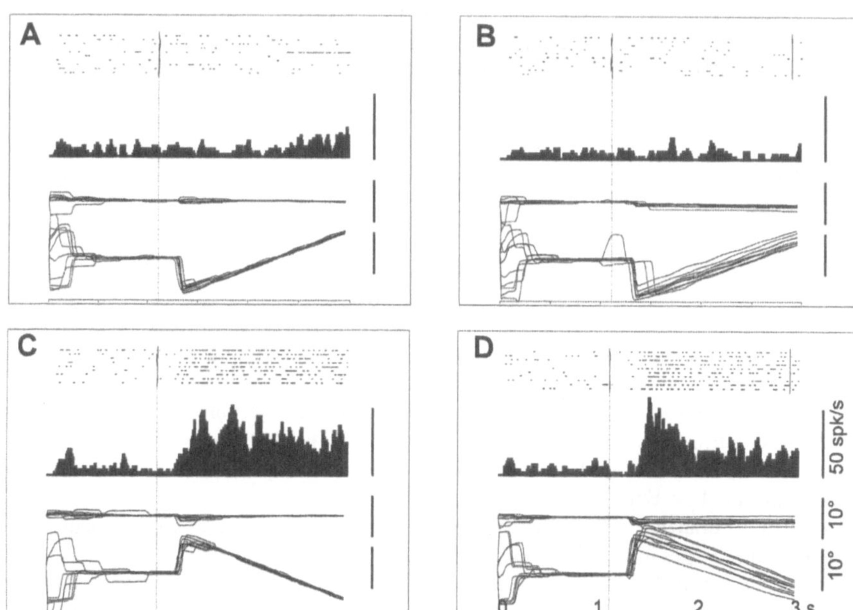

Fig. 5 A-D. Directionally selective neuron from the middle superior temporal area. **A, C** The monkey tracks target R, moving upward in **A** and downward in **C**. The horizontal and vertical eye position records (ten trials each) are displayed together with the peristimulus time histogram (bin width 10 ms) and a raster plot of the discharge of this neuron. **B, D** The monkey tracks target I1, moving upward in **B** and downward in **D**. Format of plots as in **A, C**. Target speed 12°/s in **A-D**. Note that this neuron was activated by downward pursuit independent of whether the target was visible (as in **A, C**) or not (as in **B, D**)

eye velocity dropped. Figure 5 shows a representative of this group of cells. Receptive fields of MST neurons are typically larger than those of MT neurons at comparable eccentricities and in the case of the cell shown in Fig. 5 we were actually unable to identify a receptive field by probing the visual field with moving patterns while the eyes were stationary.

In summary, our findings with imaginary targets fully support the view that area MST, unlike area MT, is a cortical area which integrates nonretinal information related to smooth pursuit eye movements useful for maintaining smooth pursuit even in the absence of foveal retinal information. This extraretinal signal is available only to a subset of pursuit-related neurons, being absent from MT and also from a substantial number of pursuit-related neurons in MST. If the target disappears as a consequence of occluding objects, the population response based on all pursuit-related cells should show a clear decline, possibly underlying the drop in eye velocity observed in the experiments in which we simulated occlusion by temporarily blinking out the target.

Acknowledgements. We wish to thank Martina Keller for her contributions to some of the experiments presented in this chapter. Supported by grants from the European Community (HCM CHRX-CT93-0267) and the Deutsche Forschungsgemeinschaft (SFB 307 A1).

References

Becker W, Fuchs AF (1985) Prediction in the oculomotor system: smooth pursuit during transient disappearance of a visual target. Exp Brain Res 57:562–575

Dubner R, Zeki SM (1971) Response properties and receptive fields of cells in an anatomically defined region in the superior temporal sulcus in the monkey. Brain Res 35:528–532

Dürsteler MR, Wurtz RH (1988) Pursuit and optokinetic deficits following chemical lesions of cortical areas MT and MST. J Neurophysiol 60:940–965

Erickson RG, Dow B (1989) Foveal tracking cells in the superior temporal sulcus of the macaque monkey. Exp Brain Res 78:113–131

Hyvärinen J, Poranen A (1974) Function of the parietal association area 7 as revealed from cellular discharges in alert monkeys. Brain 97:673–692

Komatsu H, Wurtz RH (1988) Relation of cortical areas MT and MST to pursuit eye movements. I. Localization and visual properties of neurons. J Neurophysiol 60:580–603

Lisberger SG, Morris EJ, Tychsen L (1987) Visual motion processing and sensory-motor integration for smooth pursuit eye movements. Annu Rev Neurosci 10:97–129

Mountcastle VB, Lynch JC, Georgopoulos A, Sakata H, Acuna C (1975) Posterior parietal association cortex of the monkey: command functions for operations within extrapersonal space. J Neurophysiol 38:871–908

Newsome WT, Wurtz RH, Komatsu H (1988) Relation of cortical area MT and MST to pursuit eye movements. II. Differentiation of retinal from extraretinal inputs. J Neurophysiol 60:604–620

Robinson DL, Goldberg ME, Stanton GB (1978) Parietal association cortex in the primate: sensory mechanisms and behavioral modulations. J Neurophysiol 41:910–932

Steinbach MJ (1976) Pursuing the perceptual rather than the retinal stimulus. Vision Res 16:1371–1376

Thier P, Erickson RG (1992) Responses of visual-tracking neurons from cortical area MST-l to visual, eye, and head motion. Eur J Neurosci 4:539–553

Wyatt HJ, Pola J, Fortune B, Posner M (1994) Smooth pursuit eye movements with imaginary targets defined by extrafoveal cues. Vision Res 34:803–820

Zeki SM (1974) Functional organization of a visual area in the posterior bank of the superior temporal sulcus of the rhesus monkey. J Physiol (Lond) 236:549–573

A Cortically Mediated Visual Stabilization Mechanism with Ultrashort Latency in Primates

K. KAWANO[1], Y. INOUE[1], A. TAKEMURA[1], T. KITAMA[1], and F. A. MILES[1,2]

[1] Neuroscience Section, Electrotechnical Laboratory, Tsukubashi, Ibaraki, Japan
[2] Laboratory of Sensorimotor Research, National Eye Institute, National Institutes of Health, Bethesda, USA

Short-Latency Ocular Following: A Complex Visual Stabilization Mechanism

Eye movements exist to aid vision and do this in part by preventing excessive motion of images on the retina. A major potential source of such retinal image motion comes from the observer's own movements, and there are a number of visual and vestibular mechanisms that help to offset these by generating compensatory eye movements. This article will concentrate on one of the visual tracking mechanisms which responds to image motion by negative feedback and thus helps to stabilize gaze with respect to stationary surroundings. A hallmark of this system is that it responds to the motion of large textured images with machine-like consistency at ultrashort latencies – less than 60 ms in monkeys (Miles et al. 1986) and less than 85 ms in humans (Gellman et al. 1990). It has been known for some time that these reflex-like ocular following responses – as they are known – have some of the properties generally attributed to low-level motion detectors. For example, when the textured images consist of sine wave grating patterns, the latency of ocular following is solely a function of contrast and temporal frequency (Miles et al. 1986). Yet there is increasing evidence that this ultrarapid tracking system also has a number of complex properties that appear to have evolved specifically for dealing with the optic flow associated with translational movements of the observer. By way of illustration, consider the retinal events associated with a lateral (sideways) translation of the eyes, as for instance when one looks out from a moving train: distant objects appear relatively stationary whereas nearby ones seem to rush by (Fig. 1a). It can be shown in this case by simple optical geometry that image motion on the retina is in inverse proportion to viewing distance. In this example the observer does not compensate for the motion of the train, and the only stable images on his/her retina are those at a great distance while his/her visual world pivots about infinity – see the mountains in Fig. 1a. If the observer wishes to stabilize the retinal image of an object in the middleground – the tree in Fig. 1 for example – then he/she must

In: Parietal Lobe Contributions to Orientation in 3D Space (1997). P. Thier and H.-O. Karnath (eds). Springer-Verlag, Heidelberg.

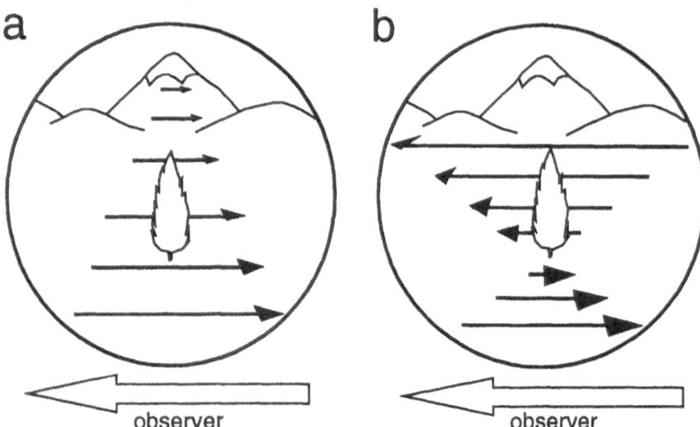

Fig. 1 a, b. Retinal image motion associated with leftward translation of the observer. **a** The observer makes no compensatory eye movements so that retinal image motion pivots about infinity: retinal image motion is inversely proportional to viewing distance. (The length of the arrows indicates the speed of retinal image motion.) **b** The observer makes compensatory eye movements so that the optical pivot shifts forward to the tree in the middleground, thereby reversing the apparent motion of the more distant objects and creating a swirling pattern of optic flow. (After Miles et al. 1992b)

generate compensatory eye movements that are inversely proportional to the tree's viewing distance and must, in effect, shift the optical pivot forwards to coincide with the tree (see Fig. 1b). How does the ocular following system deal with such complex yet commonplace situations and single out the tree? The motion signals in the foreground and background would surely tend to confuse any simple tracking system. Clearly, some kind of target selection mechanism is required. However, if the system is to respond with ultrashort latencies it is clear that selection must be rather machine-like and cannot involve time-consuming cognitive decisions. Recent experiments indicate that one way the ocular following system achieves this selectivity is by stereoscopic depth mechanisms that exploit the fact that we have two eyes with slightly differing viewpoints. A defining characteristic of objects in the plane of fixation – the ones presumably of primary interest to the observer in Fig. 1 – is that their retinal images occupy corresponding positions in the two eyes: images of objects that are nearer or farther occupy noncorresponding positions (and are said to have binocular disparity). Figure 2 is a binocular version of Fig. 1, and indicates how the binocular disparity of images provides a robust cue to delineate the objects residing in the plane of fixation – such as the mountain in Figs. 1a and 2a and the tree in Figs. 1b and 2b. Busettini et al. (1996) have recently shown that early ocular following is selectively responsive to moving images that have disparities close to zero and ignores images that have disparities of more than a few degrees.

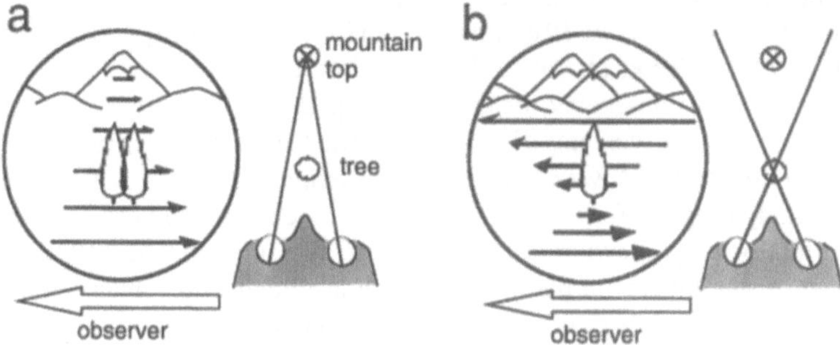

Fig. 2 a, b. Retinal image motion associated with lateral translation of the observer (binocular version). The cartoons at the left mirror those in Fig. 1, except that they indicate the cyclopean view of the binocular observer. The diagrams on the right show a plan view of the optical situation. **a** The distant mountains are in the plane of stabilization which is also the plane of fixation. Because both eyes are aligned on the distant peak, it is imaged in the same position on the two retinas (shown in this cyclopean view as single), whereas the nearer tree is imaged in different positions (shown in this cyclopean view as double). The tree is said to have "crossed disparity": from the right eye the tree is seen to the left of the mountain peak whereas from the left eye the tree is seen to the right of the peak. **b** The tree in the middleground is in the plane of stabilization which is also the plane of fixation. Because both eyes are aligned on the tree, it is imaged in the same position on the two retinas (shown in this cyclopean view as single), whereas the more distant mountains are imaged in different positions (shown in this cyclopean view as double). The mountains are said to have "uncrossed disparity": from the right eye the mountain peak is seen to the right of the tree whereas from the left eye the peak is seen to the left of the tree. Note that the dimensions of the eyes and their separation have been exaggerated to illustrate the disparity more clearly. In fact, disparity is much more evident with near viewing, which is also associated with the most vigorous optic flow and requires the most vigorous tracking from the observer to compensate for his/her own motion. For this reason, near viewing is used in most experiments. (Busettini et al. 1996)

The clear suggestion is that the motion detectors mediating initial ocular following have binocular receptive fields that occupy roughly matching positions on the two retinas.

It is important to remember that visual mechanisms, such as ocular following, represent only one aspect of gaze stabilization by the moving observer. In many situations the primary mechanism is vestibular: the rotational vestibuloocular reflex (RVOR) compensates for angular accelerations of the head, which are sensed through semicircular canals, and the translational vestibuloocular reflex (TVOR) compensates for linear accelerations of the head, which are sensed through the otolith organs. The output of the TVOR is inversely proportional to viewing distance (as required by optical geometry), though it is far from perfect, tending to overcompensate at far and undercompensate at near distances (Busettini et al. 1994; Bush and Miles 1996; Schwarz and Miles 1991; Schwarz et al. 1989).

It has been suggested that one of the major factors in the evolution of the short-latency ocular following system was the need to make up for such shortcomings in the operation of the TVOR (Busettini et al. 1991; Miles 1993, 1995; Miles and Busettini 1992; Miles et al. 1992a, b). One piece of evidence in support of this suggestion is that, remarkably, ocular following shares the TVOR's dependence on the inverse of the viewing distance even when care is taken to keep retinal stimulation constant (Schwarz et al. 1989; Busettini et al. 1991, 1994). This is a rather unlikely property for a visual stabilization mechanism and has been taken to indicate that ocular following operates in close synergy with the TVOR even to the extent of sharing central pathways – which just happen to contain the variable gain element that modulates the TVOR with viewing distance.

Such complexities in a visual tracking system that responds with such ultrashort latencies might seem rather surprising, raising the issue of its neural mediation. The system's disparity selectivity points to cortical mechanisms, the extraordinarily short latency and machine-like quality notwithstanding. Current estimates of minimum latencies for the visual initiation of cell discharge in various regions of the visual cortex seem to suggest that the very earliest ocular following responses occur too soon for them to be cortically mediated (see, for example, Raiguel et al. 1989 and Nowak et al. 1995). However, these estimates used small targets, either flashed or moved at various speeds, whereas ocular following is initiated by movements of large textured patterns involving abrupt onsets. The remainder of this paper will summarize the recent single unit recordings and local chemical lesion studies in our laboratory (Kawano et al. 1990, 1992, 1994; Shidara et al. 1991; Shidara and Kawano 1993) which suggest that the earliest ocular following is mediated by a pathway that includes the medial superior temporal (MST) area of the cortex, the dorsolateral pontine nucleus, and the ventral parafloccular lobes of the cerebellum.

Neural Mediation of Short-Latency Ocular Following: MST Area of the Cerebral Cortex

Chemical lesions in the MST area produced by local injections of ibotenic acid reduce the amplitude of initial ocular following towards the side of the lesion (Shidara et al. 1991). Even the earliest responses are affected, consistent with the idea that they are cortically mediated. However, one cannot rule out the possibility that the lesion removes a cortical facilitatory influence on a subcortical pathway, in which event the cortex would play only a modulatory role and would not be a necessary link in the pathway. For this reason, we undertook an examination of the discharge properties of neurons in the MST area to see if there were cells with the requisite properties to mediate ocular following: activation at short latencies by sudden movements of large textured patterns, presumably in a directionally selective manner.

Figure 3a shows the activity of an MST neuron in the form of a raster display and peristimulus histogram together with the ocular following responses elicited by downward-velocity steps (160°/s) applied to a large random dot pattern (Kawano et al. 1994). It is evident that the firing rate of the neuron increased abruptly approximately 40 ms after the onset of stimulus motion and the eyes began moving about 16 ms later. The directional preferences of this neuron were analyzed for eight directions of motion (right, left, up, down, and four diagonals), and it is clear from the polar plot in Fig. 3b that the neuron was directionally selective with a preference for downward movements. The speed dependence of the neuron's responses was studied using downward ramps and, over the range examined (10, 20, 40, 80, 160°/s), its discharge rate increased monotonically with increases in speed, showing only minor saturation (Fig. 3c). These are exactly the

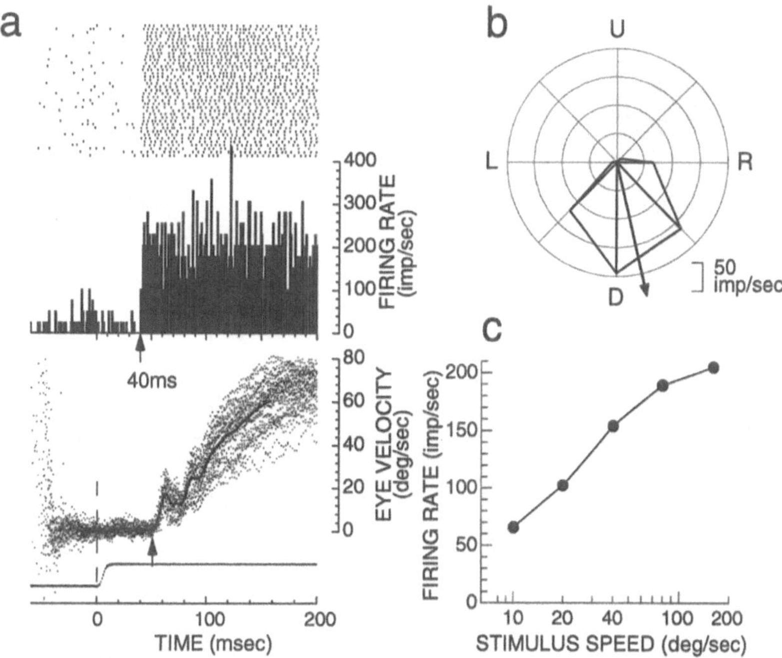

Fig. 3. a Discharges of an medial superior temporal (MST) neuron in association with the ocular following responses to multiple presentation of a 160°/s downward test ramp (Kawano et al. 1994). The responses are aligned at stimulus onset. From *top to bottom*: impulse rasters, peristimulus histogram, superimposed vertical eye velocity, and stimulus velocity. *Arrows* indicate latencies. **b** Directional selectivity of the same MST neuron shown in a polar diagram. *Arrow* indicates the "preferred direction" calculated by summing the vectors of the firing frequencies for eight directions. *D*, down; *L*, left; *R*, right; *U*, up. **c** Speed tuning of the neuron

properties expected of neurons mediating the very earliest ocular following responses: vigorous activation by movements of large patterns at latencies that precede eye movements by 10 ms or more and strong directional preferences together with a preference for high speeds (Kawano et al. 1994).

Such neurons were not unusual. Most MST neurons (~80%) showed strong directional preferences, and when their preferred directions (directions of movement associated with the most vigorous discharges) are plotted together in polar form it is clear that all directions of motion are represented about equally, with only minor anisotropies (Kawano et al. 1994, see Fig. 4 left column, top row). Direction-selective MST neurons generally responded best to high stimulus speeds (Fig. 4 left column, bottom row). Response latencies were measured when the large-field random dot pattern was moved in the preferred direction and at the

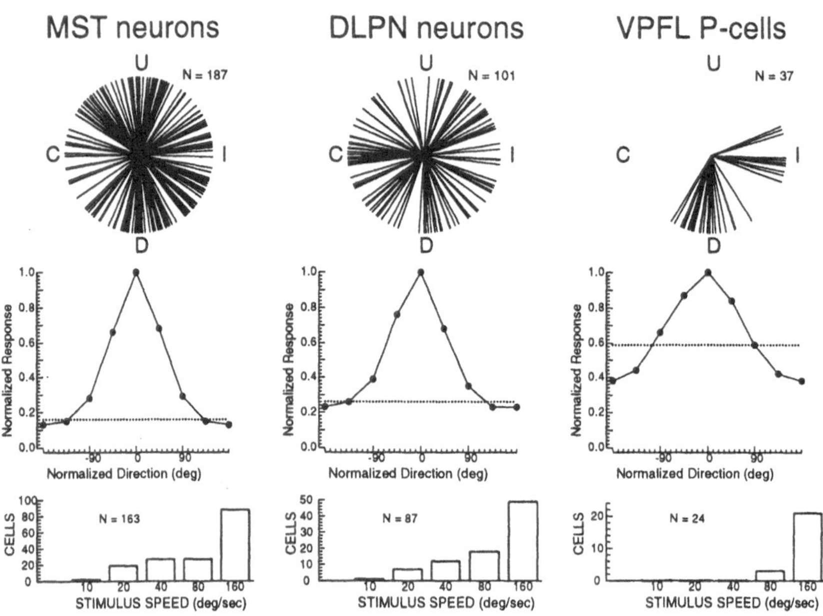

Fig. 4. Comparison of response properties of neurons in three regions. (Kawano et al. 1996) From *left to right*: medial superior temporal (MST) neurons (Kawano et al. 1994), dorsolateral pontine (DLPN) neurons (Kawano et al. 1992), and simple spike activity of ventral parafloccular Purkinje cells (VPFL P-cells, Shidara and Kawano 1993). *Top row*: estimated preferred directions of the neurons. *C*, contralateral; *D*, down; *I*, ipsilateral; *U*, up. *Middle row*: normalized, direction-tuning profile for MST neurons, DLPN neurons, and VPFL Purkinje cells. Direction-tuning profile for each neuron was normalized to its best response, and the best response directions were shifted and aligned to 0°. *Bottom row*: distribution of the preferred speeds of the neurons. (Stimulus speeds used were 10°, 20°, 40°, 80°, and 160°/s)

preferred speed for each neuron, and the distribution of latencies for the entire population is shown in the histogram in Fig. 5. In most cases (~90%), increases in firing rate commenced before ocular following, with approximately 60% of cells leading the very earliest tracking eye movements by more than 10 ms, early enough for them to be causally involved in the genesis of the response.

The recent discovery that the earliest ocular following responses are sensitive to binocular disparity – consistent with their mediation by direction-selective motion detectors that are also selectively sensitive to images moving in immediate vicinity of the plane of fixation (Busettini et al. 1996) – would lead one to expect that putative ocular-following-neurons would also be sensitive to the disparity of textured patterns. Disparity-selective neurons have been described in various cortical areas of the monkey and have been grouped into two broad categories: "tuned" neurons, which are selectively sensitive to images moving in a very restricted range of depths in immediate vicinity of the plane of fixation, and "reciprocal" neurons activated by binocular images that are merely nearer (or farther) than the plane of fixation (for a review see Poggio 1995). Clearly, ocular following's preference for images in the plane of fixation would suggest that its neural mediation most likely involves disparity-selective neurons of the tuned type. We have yet to examine this question experimentally, but there is evidence that most neurons in the MST area are sensitive to the disparity of large random dot patterns moving in the frontoparallel plane – the plane of motion in our studies of ocular following (Roy et al. 1992). However, very few of the neurons in that study were of the tuned type (5%), though some (17%) were "mixed" in that they were of the reciprocal type, but their disparity-tuning curves showed a subsidiary

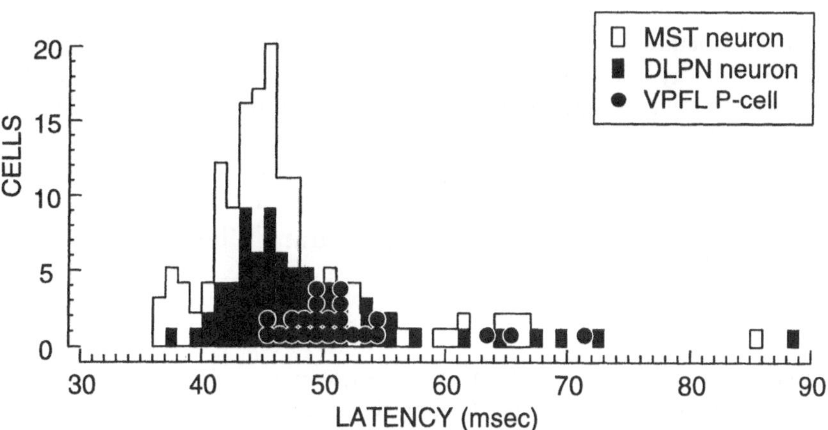

Fig. 5. Comparison of latencies of onset for neurons in medial superior temporal (*MST*) area, dorsolateral pontine nucleus (*DLPN*), and ventral parafloccular Purkinje cells (*VPFL P-cells*) (after Kawano et al. 1996). Latencies were estimated from the onset of the visual stimulus

peak in the vicinity of the plane of fixation. Forty percent of the disparity-selective neurons in the MST area were most unusual in showing the opposite preference for motion of images in front of and beyond the plane of fixation, consistent with "a role in signaling the direction of self-motion of the observer through the environment" (Roy et al. 1992). However, the patterns in that study were rather impoverished compared with the ones we used: even while those patterns were moving, animals were able to maintain fixation of a stationary target spot, which was not true of our patterns – ocular following was quite compelling with our densely textured stimuli. Also, dichoptic viewing in the study of Roy et al. was achieved using red and green images viewed through matching filters, whereas Busettini et al. (1996) used crossed polarized images in combination with matching polarizing spectacles. This raises the possibility that some of the selectivity for depth and direction of motion in the study of Roy et al. resulted from differential sensitivity of the neurons to red and green. Some years ago, Cynader et al. (1978) and Gardner et al. (1985) presented evidence that latency differences between inputs from the two eyes could account for the disparity-selectivity of some neuronal responses to motion in cat visual cortex. Consider a "tuned" neuron that is not normally directionally selective (that is, for black/white images, for example) but responds at shorter latency to the motion of green images than to the motion of red ones. In the experimental situation used by Roy et al. such neurons might be expected to show optimal binocular summation when the green pattern leads the red – that is, respond in a directionally selective manner – for moving images presented outside the plane of fixation. Furthermore, such a neuron might be expected to show a preference for one direction of motion inside the plane of fixation and the reverse direction outside the plane of fixation because the green image will lead for one direction of motion inside the plane of fixation and for the reverse direction of motion outside. Clearly, in regard to the question of whether or not the putative ocular-following-neurons in the MST area show the disparity-selectivity characteristic of ocular following, it would be best to examine the question using the exact same stimuli as in the original study of Busettini et al. (1996). Such an approach has yet to be tried.

Neural Mediation of Short-Latency Ocular Following: Parietopontocerebellar Pathway

Anatomical studies have suggested that the parietopontocerebellar pathway has an important role in visuomotor coordination (Glickstein, this volume). Neurons in the MST area are known to project to the dorsolateral pontine nuclei (Glickstein et al. 1985), which in turn projects to the ventral paraflocculus (Langer et al. 1985). In fact, there is a preliminary report mentioning that lesions of the ventral paraflocculus impair early ocular following (Miles et al. 1986). We used local injections of lidocaine to reversibly inactivate the dorsolateral pons and observed reductions in the amplitude of the earliest ocular following responses (Kawano et al. 1990). In addition, we found that the response properties of dorsolateral pontine neurons during ocular following were very similar to those of MST neurons (Kawano et al. 1992). Thus, most of these pontine neurons showed strong directional preferences – the population extending over a wide range of directions with only minor anisotropies – and a clear preference for higher speeds. It is evident from Fig. 4 (middle column) that the population data for neurons in the dorsolateral pons closely match those for MST neurons. However, the simple spike responses of Purkinje cells in the ventral paraflocculus showed much narrower response ranges in terms of both directional preferences – restricted to ipsiversive and downward motion – and speed preferences – responses were always best at the very highest speeds we tested (see right column in Fig. 4; Shidara and Kawano 1993). It is also evident from the middle row in Fig. 4 that modulation of the Purkinje cells' simple spike activity differed from that of neurons in the MST area and the dorsolateral pons by showing decreases during antipreferred motion in addition to the usual increases during preferred motion. Thus, neurons in the ventral paraflocculus are much more selective in terms of their directional and speed preferences than are neurons in the MST area and the dorsolateral pontine nucleus. The latency distributions for the MST area, dorsolateral pontine neurons, and ventral parafloccular Purkinje cells (simple spikes) are shown in Fig. 5 and are consistent with the idea that they constitute a serial linkage in a pathway culminating in ocular following. Significantly, electrical microstimulation at the recording sites of Purkinje cells in the ventral paraflocculus evoked short-latency (8–10 ms) eye movements in the directions preferred by the cells (Shidara and Kawano 1993). A comparison of the response properties of the neurons in the three regions suggests that MST neurons abstract sensory information about the motion of elements in the visual scene and relay this via the dorsolateral pons to the ventral paraflocculus which computes the motor information needed to drive the eyes (Kawano et al. 1996).

The Site at Which Ocular Following Acquires Its Dependence on Viewing Distance

As mentioned earlier, it has been shown that the magnitude of the ocular following responses elicited by a given visual stimulus is inversely proportional to the viewing distance (Schwarz et al. 1989; Busettini et al. 1991, 1994). In a first step towards determining the site of changes underlying this dependence on distance in the central neuronal pathways mediating ocular following, we have recently examined the effect of varying the vergence angle of the two eyes on ocular following and on the associated activity of MST neurons in monkeys. We chose to manipulate the vergence angle rather than the viewing distance because it was much easier to do so while recording single neurons. However, the use of this

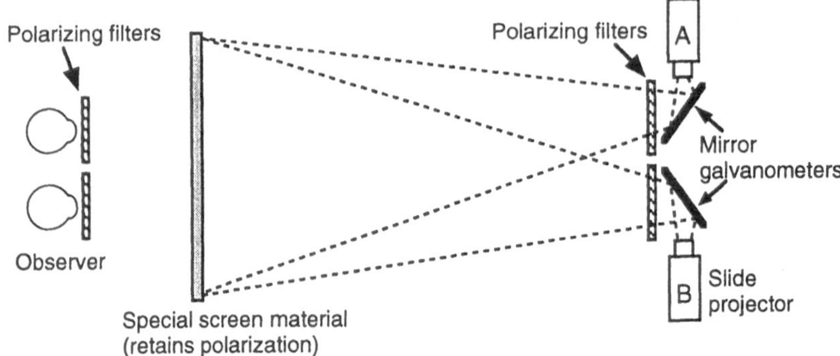

Fig. 6. Diagram of the optical arrangements. The two identical visual images were produced by two identical projectors (*A, B*) with orthogonal polarizing filters. The animal viewed the scene through matching polarizing filters so that the left eye saw only the image produced by projector A and the right eye saw only the image produced by projector B

approach necessitated our first demonstrating a clear dependence of ocular following on the vergence angle. This made it necessary to be able to move the images seen by each of the two eyes independently, and our method is illustrated in Fig. 6. Two identical overlapping images (random dots) were back-projected directly onto the screen facing the animal, and orthogonal polarizing filters were placed in the two projection paths so that when the animal viewed the screen through matching filters, each of the two eyes saw only one of the two projected patterns. Mirror galvanometers in an X/Y configuration were used to control the horizontal and vertical positions of the two images. Each trial commenced with all images stationary and in the plane of the screen, and then the image seen by the left eye was slowly moved to a new position to induce the animal to adopt a new vergence angle. Once this new vergence angle was established, the two images were moved together at constant speed to elicit ocular following in the usual way.

Initial ocular following showed clear dependence on the desired vergence angle, and this is apparent from the superimposed mean eye velocity profiles shown in Fig. 7a. These profiles were obtained from one monkey when the desired vergence angle (specifying the distance at which the two eyes were required to converge in order to maintain binocular alignment on the two stimulus patterns) was infinity, 1/2, and 1/4 m, and it is evident that the responses were strongest for 1/4 m (equivalent to near viewing) and weakest for infinity (equivalent to distant viewing). Note that the screen was at 1/2 m. Quantitative estimates of the magnitude of this effect (based on the change in eye position over a period of 50 ms to 100 ms, measured from ramp onset) indicate that initial ocular following increased monotonically with increases in the desired vergence angle (see Fig. 7b). The mean percentage modulation with vergence ([(max - min)/min] x 100%) averaged 176% for the three animals.

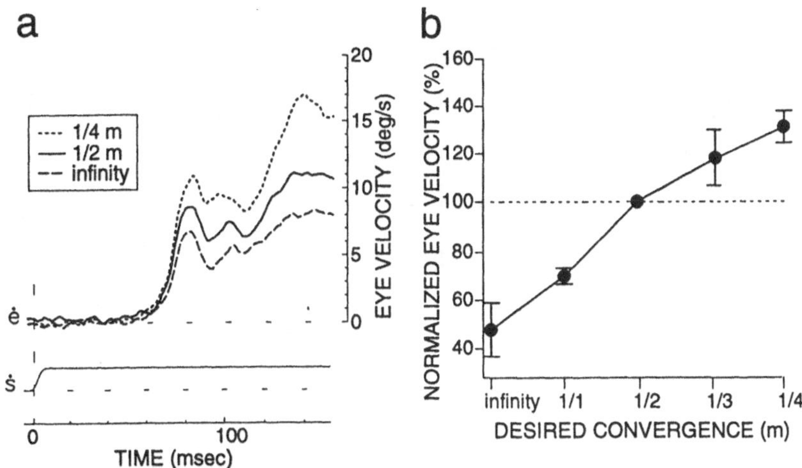

Fig. 7 a, b. Dependence of ocular following on the desired convergence. **a** Mean eye-velocity response profiles. The profiles associated with three vergence states (see key) are shown together with the stimulus velocity profile. **b** Average ocular following responses for three monkeys (normalized with respect to the responses in the case of overlapping binocular images and thus seen in the plane of the screen, i.e., 1/2 m) are plotted against the desired convergence

Using visual stimuli moving in the preferred direction and at the preferred speed for each neuron, it was clear that the discharge modulations of many MST neurons (associated with ocular following) showed clear dependence on convergence. Figure 8 shows the superimposed mean discharge rate profiles for two such neurons when the desired vergence angle was infinity, 1/2, and 1/4 m: the activity of the neuron in Fig. 8a showed increased modulation (in association with ocular following) with increased convergence, and the neuron in Fig. 8b showed the reverse tendency. To quantify such effects for each neuron, we averaged the firing rate of a number of trials over a time period extending from 40 ms to 100 ms (measured from the onset of the test ramp) for each of the three vergence conditions. The percentage modulation with vergence ([(near – far)/far] x 100%) was calculated for each cell, and the distribution for the entire population of cells is shown in Fig. 9. Thirty-six percent of the MST neurons resembled the neuron in Fig. 8a and responded significantly more vigorously when the animal was converged ("near viewing" neurons, black bars in Fig. 9, $p<0.05$, Student's t-test), while 23% resembled the neuron in Fig. 8b and responded significantly more vigorously when the animal was not converged ("far viewing" neurons, hatched bars in Fig. 9, $p<0.05$). The remaining cells showed no significant modulation with vergence state ("insensitive" neurons, white bars in Fig. 9). The mean percentage modulation with vergence was +43% for "near viewing" neurons and -26% for "far viewing" neurons.

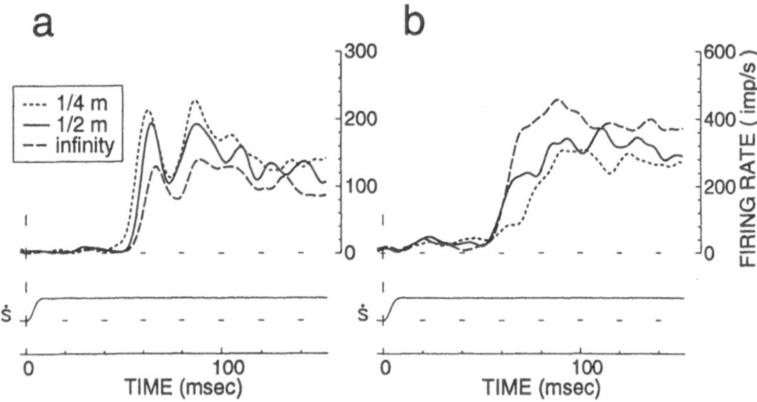

Fig. 8 a, b. Sample medial superior temporal discharge profiles associated with ocular following: dependence on the desired convergence. The superimposed mean spike density functions for a "near viewing" neuron (**a**) and for a "far viewing" neuron (**b**) are shown for three vergence states (see key)

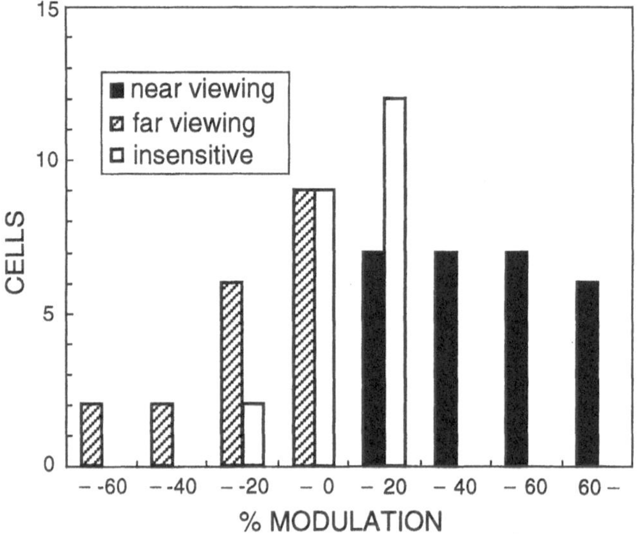

Fig. 9. Distribution of percentage modulation with vergence for the 69 medial superior temporal neurons. Statistical significance (Student's t -test) was assèssed from the averaged firing rates for "near viewing" (target vergence, 1/4 m) and for "far viewing" (infinity). "Near viewing" neurons responded significantly more vigorously when the animal was converged (*black bars*). "Far viewing" neurons responded significantly more vigorously when the animal was unconverged (*hatched bars*). "Insensitive" neurons showed no significant modulation (*white bars*)

There are two clear conclusions from these last experiments. Firstly, initial ocular following was directly related to the vergence angle of the two eyes, suggesting that vergence can be a major cue in the modulation of ocular following with viewing distance described by Schwarz et al. (1989) and Busettini et al. (1991, 1994). Secondly, although modulation of the activity of many MST neurons in association with ocular following also showed clear dependence on the vergence angle, none of the neurons were as sensitive to vergence as ocular following was. In fact, on average, the "near viewing" neurons showed a dependence on vergence that was only 24% (43%/176%) of that shown by ocular following under the same stimulus conditions, suggesting that either MST neurons mediate only a proportion of the ocular following response, and/or most of the modulation of ocular following with vergence occurs downstream of the MST area, i.e., in the dorsolateral pons, the ventral paraflocculus or the relay between the latter and ocular motoneurons. It has been suggested that modulation of ocular following with distance occurs in neuronal pathways that are shared by the TVOR, whose responses show similar modulation with viewing distance and vergence (Busettini et al. 1994; Paige and Tomko 1991; Schwarz et al. 1989; Schwarz and Miles 1991). A relevant factor here is that, in monkeys, the component of the TVOR that is sensitive to viewing distance has recently been shown to have a latency of <20 ms (Bush and Miles 1996), clearly indicating subcortical mediation. Interestingly, the maintained simple spike activity of Purkinje cells in the ventral paraflocculus has been shown to modulate with viewing distance (Miles et al. 1980), suggesting that this structure might be the site at which ocular following (and the TVOR?) acquires its sensitivity to viewing distance.

References

Busettini C, Miles FA, Schwarz U (1991) Ocular responses to translation and their dependence on viewing distance. II. Motion of the scene. J Neurophysiol 66:865–878

Busettini C, Miles FA, Schwarz U, Carl JR (1994) Human ocular responses to translation of the observer and of the scene: dependence on viewing distance. Exp Brain Res 100:484–494

Busettini C, Masson GS, Miles FA (1996) A role for stereoscopic depth cues in the rapid visual stabilization of the eyes. Nature 380:342–345

Bush GA, Miles FA (1996) Short-latency compensatory eye movements associated with a brief period of free fall. Exp Brain Res 108:337–340

Cynader MS, Gardner JC, Douglas RM (1978) Neural mechanisms underlying stereoscopic depth perception in cat visual cortex. In: Cool S, Smith EL (eds) Frontiers of visual science. Springer, Berlin, Heidelberg, New York, pp 373–386

Gardner JC, Douglas RM, Cynader MS (1985) A time-based stereoscopic depth mechanism in the visual cortex. Brain Res 328:154–157

Gellman RS, Carl JR, Miles FA (1990) Short-latency ocular following responses in man. Visual Neurosci 5:107–122

Glickstein M, May J, Mercer BE (1985) Corticopontine projection in the macaque: the distribution of labeled cortical cells after large injections of horse-radish peroxidase in the pontine nuclei. J Comp Neurol 235:343–359

Kawano K, Shidara M, Yamane S (1990) Relation of the dorsolateral pontine nucleus of the monkey to ocular following responses. Neurosci Res Suppl 16:902

Kawano K, Shidara M, Yamane S (1992) Neural activity in dorsolateral pontine nucleus of alert monkey during ocular following responses. J Neurophysiol 67:680–703

Kawano K, Shidara M, Watanabe Y, Yamane S (1994) Neural activity in cortical area MST of alert monkey during ocular following responses. J Neurophysiol 71:2305–2324

Kawano K, Shidara M, Takemura A, Inoue Y, Gomi H, Kawato M (1996) Inverse-dynamics representation of eye movement by cerebellar Purkinje cell activity during short-latency ocular following responses. Ann NY Acad Sci 781:314–321

Langer T, Fuchs AF, Scudder CA, Chubb MC (1985) Afferents to the flocculus of the cerebellum in the rhesus macaque as revealed by retrograde transport of horseradish peroxidase. J Comp Neurol 235:1–25

Miles FA (1993) The sensing of rotational and translational optic flow by the primate optokinetic system. Rev Oculomot Res 5:393–403

Miles FA (1995) The sensing of optic flow by the primate optokinetic system. In: Findlay JM, Walker R, Kentridge R (eds) Eye movement research: mechanisms, processes and applications. Elsevier, Amsterdam, pp 47–62

Miles FA, Busettini C (1992) Ocular compensation for self motion: visual mechanisms. In: Cohen B, Tomko DL, Guedry F (eds), Sensing and controlling motion: vestibular and sensorimotor function. Ann NY Acad Sci 656:220–232

Miles FA, Fuller JH, Braitman DJ, Dow BM (1980) Long-term adaptive changes in primate vestibulo-ocular reflex. III. Electrophysiological observations in flocculus of normal monkeys. J Neurophysiol 43:1437–1476

Miles FA, Kawano K, Optican LM (1986) Short-latency ocular following responses of monkey. I. Dependence on temporospatial properties of visual input. J Neurophysiol 56:1321–1354

Miles FA, Busettini C, Schwarz U (1992a) Ocular responses to linear motion. In: Shimazu H, Shinoda Y (eds) Vestibular and brain stem control of eye, head and body movements. Japanese Scientific Societies, Tokyo, pp 379–395

Miles FA, Schwarz U, Busettini C (1992b) The decoding of optic flow by the primate optokinetic system. In: Berthoz A, Graf W, Vidal PP (eds) The head-neck sensory-motor system. Oxford University Press, New York, pp 471–478

Nowak LG, Munk MHJ, Girard P, Bullier J (1995) Visual latencies in areas V1 and V2 of the macaque monkey. Visual Neurosci 12:371–384

Paige GD, Tomko DL (1991) Eye movement responses to linear head motion in the squirrel monkey. II. Visual-vestibular interactions and kinematic considerations. J Neurophysiol 65:1183–1196

Poggio GF (1995) Mechanisms of stereopsis in monkey visual cortex. Cereb Cortex 3:193–204

Raiguel SE, Lagae L, Gulyas B, Orban GA (1989) Response latencies of visual cells in macaque areas V1, V2, and V5. Brain Res 493:155–159

Roy J-P, Komatsu H, Wurtz RH (1992) Disparity sensitivity of neurons in monkey extrastriate area MST. J Neurosci 12:2478–2492

Schwarz U, Miles FA (1991) Ocular responses to translation and their dependence on viewing distance. I. Motion of the observer. J Neurophysiol 66:851–864

Schwarz U, Busettini C, Miles FA (1989) Ocular responses to linear motion are inversely proportional to viewing distance. Science 245:1394–1396

Shidara M, Kawano K (1993) Role of Purkinje cells in the ventral paraflocculus in short-latency ocular following responses. Exp Brain Res 93:185–195

Shidara M, Kawano K, Yamane S (1991) Ocular following response deficits with chemical lesions in the medial superior temporal area of the monkey. Neurosci Res 14:S69

Eye Movement Control in Spatial and Object-Based Neglect

R. WALKER[1] and J. M. FINDLAY[2]

[1]Department of Clinical Neuroscience, Charing Cross Hospital Medical School, London, UK
[2]Department of Psychology, University of Durham, Durham, UK

Introduction

The parietal lobe is an area of association cortex involved in the integration of sensory and motor functions for the control of complex behavioral responses (see Hyvärinen 1982 for a review). The sensory functions are known to include visual, somaesthetic and auditory processing, and the motor functions include the control of limb and eye movements. Examples of the behavioral functions thought to be under the control of the parietal lobe include reaching for an object under visual guidance, and orienting the eyes to locate a novel stimulus. Therefore, the functional role of the parietal lobe is a complex one, and it is not surprising that damage to this region can lead to a wide range of sensory and motor deficits in man. One disorder commonly encountered following damage to the parietal lobe in man is 'unilateral neglect'. Patients with neglect fail to respond to, and appear to be unaware of, stimuli located in the side of space contralateral to the side of their lesion.

Neglect is most commonly encountered following a lesion in the region of the right parietal lobe which results in left neglect. Examples of neglect are seen in the patient's everyday behavior. A patient may fail to eat the food from the left side of a plate, shave only the right side of his own face and ignore people who approach him from the left side. Patients with neglect may deviate to the right when walking (Robertson et al. 1994), and their arm movements may deviate to the right when reaching for an object (Goodale et al. 1990; Chieffi et al. 1993).

The presence of neglect can be demonstrated in several simple tests. One such test is the cancellation task (Albert 1973) in which a patient is given a page onto which a large number of short lines are randomly distributed and is asked to cross out all of the lines. Patients with neglect often fail to cross out the lines and letters that are located on the contralesional side of the page. In line bisection the patient is asked to mark the centre of a horizontal line; the presence of neglect is demonstrated when the bisection is placed to the right of true centre. When drawing from memory, a patient with neglect may omit to draw the features on the left side. When asked to copy a simple line drawing containing a horizontal array

In: Parietal Lobe Contributions to Orientation in 3D Space (1997). P. Thier and H.-O. Karnath (eds). Springer-Verlag, Heidelberg.

of objects, patients may omit to draw whole objects from the left side of the model, and may also omit features from the left sides of the individual objects (Gainotti et al. 1972). Neglect patients are also found to make errors when reading and may fail to read the first (left most) letters from a word (neglect dyslexia). A second form of reading disorder is shown when patients omit to read whole words located on the left side of a page of text. Single word neglect dyslexia errors and whole word omissions have been shown to be doubly dissociable, which is taken as indicating that separate processes underlie these two reading disorders (Ellis et al. 1987; Riddoch and Humphreys 1991; Young et al. 1991).

Similar dissociations may be found between other tasks used in the assessment of neglect. For example, a patient may show neglect on line cancellation, but not when drawing or reading. It has come to be realised that the dissociation on different tasks implies that there are a number of different forms of neglect and that neglect cannot be regarded as being a unitary "syndrome" (Halligan and Marshall 1992; Milner and Harvey 1994; Walker 1995). One current distinction is neglect for *near* (personal) space and neglect for *far* (extra personal) space. Halligan and Marshall (1991) reported details of a case of neglect that was more severe for tasks performed in near space than for tasks performed in far space. The opposite dissociation of neglect being more severe in far space than in near space has been reported by Cowey et al. (1994). The double dissociation between these two reports supports the view of separate forms of personal and extra personal neglect. A second distinction is that of separate *perceptual* and *motor* forms of neglect. In many tasks used to assess the presence of neglect perceptual and motor factors are often confounded. For example, in line cancellation, a failure to cross out the lines on the left side of the page may reflect either contralesional perceptual neglect or contralesional motor neglect. Two studies in which the perceptual and motor components involved in a task have been separated have revealed impairments consistent with the view that there are separate perceptual and motor forms of neglect (Bisiach et al. 1990; Tegnér and Levander 1991).

Another important distinction which forms the main focus of interest in this chapter is the separation of neglect into *spatial* and *object-based* forms (Gainotti et al. 1972; Humphreys and Riddoch 1994). Spatial neglect is indicated by a tendency to omit stimuli located on the contralesional side of space, space being defined in egocentric co-ordinates with respect to the patient's retinotopic frame of reference (Ládavas 1987) and also with respect to the patient's own head and body (Ládavas 1987; Farah et al. 1990; Karnath et al. 1991). Object-based neglect is shown when the patient neglects the contralesional side of an individual object, regardless of the spatial position of the object (Gainotti et al. 1972; Young et al. 1992). Neglect for the contralesional side of an object involves an allocentric co-ordinate system in which left and right sides of an object midline are assigned with respect to the viewer's position. Reports of neglect being modulated by the frame of reference involved in the task have provided results consistent with the view that neglect may be fractionated into spatial or object-based forms.

The following section reviews some of the evidence which provides support for the distinction of separate spatial and object-based forms of neglect. Particular emphasis will be placed on detailed studies of the eye movements made by neglect patients when viewing arrays containing more than one object, and when viewing single objects. Two published reports have examined the eye movements made by neglect patients when they viewed simple scenes. Both of these studies described patients whose eye movements were largely restricted to the right side of the midline of these images. These patients therefore show problems orienting their eyes in a spatial (or between-object) frame of reference, but no problems making eye movements in an object-based frame of reference. These findings are contrasted with our own detailed study of the eye movements made by a patient with object-based neglect. The implication of this double dissociation in eye movement deficits for theories of parietal lobe functioning are discussed in the final section.

Spatial and Object-Based Neglect

Patients with unilateral neglect typically ignore stimuli located in the left (contralesional) side of space. It has frequently been observed that there is more than one definition of 'left' depending on the exact frame of reference involved (see Ládavas 1993). In terms of an egocentric frame of reference, *left* is defined with respect to the viewer's own body co-ordinates. Thus, left can be defined in retinal co-ordinates (the patient's left visual field), with respect to the patient's own body trunk (left of the body midline), or with respect to the patient's head (left of the head midline). Furthermore, it has been demonstrated that neglect can operate in terms of more than one of these frames of reference, and that the degree of neglect can be influenced by changes in the patient's viewing position (Ládavas 1987; Farah et al. 1990; Karnath et al. 1991).

Although neglect has often been regarded as being a spatial deficit, it has been known for a number of years that neglect can also arise in 'object-based' co-ordinates (Gainotti et al. 1972). Gainotti et al. performed a detailed analysis of the drawing performance of patients with right brain damage, and reported deficits consistent with neglect occurring in an object-based frame of reference. In their study patients were asked to copy a picture of a simple scene which contained a number of objects located along a horizontal axis. Some patients failed to draw the left sides of the individual objects, but continued to draw objects located further to the left of their incomplete copies. This is interpreted as showing that these patients could locate each of the objects, but neglected the left sides of the individual objects. Thus, the patients showed little spatial neglect, but had more severe object-based neglect. The observation that patients may fail to copy the contralesional side of an individual object has been made in many other cases of neglect (Halligan and Marshall 1993, 1995; Hornak 1995; Walker et al. 1996).

The existence of separate between-object and object-based forms of neglect has been further supported by reports of both forms co-existing, but in opposite directions in a single patient. Costello and Warrington (1987) reported a case of a patient with bilateral lesions who made left-sided errors when reading single words, but who omitted stimuli from the right side of a page on letter cancellation. A similar case has been reported for the patient J.R. by Humphreys and Riddoch (1994). When J.R. was asked to read single words placed randomly on a page, he omitted letters from the *left* sides of words regardless of the spatial position of the words on the page. J.R. was also found to omit whole words located on the *right* side of the page. Cueing J.R. to the left sides of the individual words reduced the amount of left sided letter omissions, but had no effect on the amount of right sided whole word omissions. Conversely cueing him to the right side of the page reduced the amount of right sided word omissions, but had no effect on the number of left sided letter omissions. The selective cueing effects suggest that attention can operate separately in either 'within-object' or 'between-object' representations. Thus, the performance of J.R. is consistent with the distinction of two separate forms of spatial representation, one for the relations between parts of an object and one for the relations between separate objects. Humphreys and Riddoch termed these two forms of neglect *within-object* and *between-object* neglect.

The presence of object-based neglect has also been revealed in detailed studies which have used 'chimaeric' objects as stimuli. Chimaerics are made by joining the left side of one object with the right side of another object at the vertical midline, and have been found to be sensitive indicators of the presence of neglect. Young et al. (1992) performed an examination of visual recognition in a single case, B.Q., using chimaeric objects and chimaeric faces. When B.Q. was shown chimaerics comprising of the left half of one famous face joined at the vertical midline to the right half of a different famous face, she consistently named the face on the right side of the image only. In a second chimaeric 'gap' condition, B.Q. was shown the left and right half faces which had been separated by a spatial gap and named both the left and right half faces correctly (see also Buxbaum and Coslett 1994). It is important to realise that in the gap condition the face on the left is actually positioned further into the patientís contralesional ("bad") hemifield than when it forms part of a whole chimaeric. Thus, B.Q.'s neglect of the left side of a chimaeric object and her ability to recognise the left half in the gap condition are consistent with the view that her problems were enhanced when the material to be reported formed a single object. Young et al. also examined the effects of presenting chimaerics entirely in B.Q.'s right visual field under tachistoscopic view conditions. The use of brief tachistoscopic presentation prevents eye movements from being made, a possible confounding factor in many other studies of object-based neglect (Driver and Halligan 1991; Driver et al. 1994). Under stringent tachistoscopic viewing conditions with chimaerics presented entirely in her right visual field, B.Q. again named the object on the right side of the chimaeric only. Neglect for the left side of a chimaeric when presented entirely in the right visual field is consistent with the view that B.Q.'s neglect operates primarily in object-based co-ordinates.

The dissociations described above support the view that neglect can operate in terms of a spatial or an object-based frame of reference. Many models of neglect attribute the patients' deficits to a disturbance of an attentional process (e.g., Kinsbourne 1977; Posner et al. 1984, 1987; Rizzolatti and Berti 1993), but few have taken into account the different manifestations of neglect. Thus, there is a need to relate the deficits shown by neglect patients to current models of attentional orienting. Saccadic eye movements form one means of redirecting visual attention and it has been proposed that other, covert forms of orienting visual attention could also depend on the neural system involved in programming saccadic eye movements (Rizzolatti et al. 1987, 1994). Deficits in orienting attention in neglect should be manifested by similar deficits in eye movement control. The distinction between spatial and object-based neglect should therefore be apparent in dissociations in the eye movements made by neglect patients. In the following section studies of neglect that have revealed eye movement deficits in either a spatial or object-based frame of reference are described.

Eye Movements in Neglect

The observation that patients with neglect also have deficits in eye movement control has been known for many years (Silberpfennig 1941; Hécaen 1962; Girotti et al. 1983). These deficits are also consistent with animal lesion studies that have shown that the parietal lobes are involved in saccade generation (Lynch and McLaren 1989). Many of the studies of eye movements in neglect have been concerned only with the patients' ability to make simple target elicited saccades. There are a number of reports that neglect patients fail to make contralesional saccades on many trials and that any contralesional saccades made are usually of small amplitude and long latency (Meienberg et al. 1981, 1986; Girotti et al. 1983; Ishiai et al. 1987; Karnath et al. 1991; Walker et al. 1991). However, three recent studies of eye movements in patients with neglect have examined scanning eye movements using more complex material such as scenes. In the next section we show how these studies support the dissociation of between-object and object-based neglect.

Spatial (Between-Object) Neglect

Rizzo and Hurtig (1992) examined the eye movements in a number of cases of neglect and revealed patterns of deficits consistent with the view that neglect can impair eye movements involving a spatial (or between-object) frame of reference.

Rizzo and Hurtig recorded the eye movements made by neglect patients while they viewed a simple scene (cookie theft scene from Boston Diagnostic Aphasia Exam) and when viewing a famous face (President Nixon). When viewing the scene, the neglect patients rarely made saccades into the neglected hemifield, and restricted their fixations to the objects located on the right side of the image midline. However, the same patients did make saccades into the left side of the face of President Nixon, and made many transitions across the midline to fixate on both eyes. The patients spent much more time fixating on the left side of the face than on the left side of the scene. Rizzo and Hurtig suggested that the improvement shown in scanning the contralesional side of the face may be due to top-down processes, whereby the patients use their knowledge about faces to trigger search routines into both sides of the image (Walker-Smith et al. 1977). An equally plausible alternative, to which they allude, is the distinction between orienting attention to scenes, and orienting attention within an object. The performance of their patients is consistent with the idea that they had a more severe impairment when orienting their attention between objects, but were less impaired at orienting their attention within an object (such as a face).

A similar report of impaired scene scanning with unimpaired object scanning in a single case study of eye movements in neglect has been reported by Karnath (1994). Karnath recorded the eye movements made by a neglect patient H.S., while he viewed simple line drawings of scenes which contained more than one object. When viewing the scene, H.S.'s eye movements were found to be restricted to scanning the objects located on the right side of the image. However, H.S. scanned the left and right sides of the individual objects in the scene. Karnath also examined the effects of altering the importance of the information of the objects located in the left visual field on scanning behavior. In some scenes the object on the left side of the image had little relevance to the objects on the right; for example, a man on the left side and a crane lifting a box on the right side. In other drawings there was a strong connective element between the objects located in the left and right sides; for example, a car crash where the point of impact was located at the vertical midline. The theory suggested by Karnath was that, in the case of drawings with high connective elements (such as the car crash), the information in the right side of the image may guide scanning of the object in the left side. It was found that the strength of the connections between objects had no effect on the amount of time spent scanning the left side of the image. Therefore, H.S. did not appear to use top-down knowledge to guide his contralesional scanning. H.S.'s failure to scan the left sides of scenes, while being able to scan both sides of the individual objects located in the scenes, is consistent with the interpretation that he is impaired at shifting his attention 'between objects', but not at shifting his attention 'within' an object.

The deficits in eye movement control observed in the neglect patients, reported by Rizzo and Hurtig (1992) and Karnath (1994), are similar in that they show an impairment for orienting between-objects, but not for orienting within-objects. The distinction between the two separate forms of neglect would be strengthened by a case of neglect which reveals the opposite dissociation, namely an impairment in orienting the eyes within a single object but not in orienting between objects. In the following section we report the findings of a detailed study into the eye movements made by patient R.R., which are consistent with those of a patient showing object-based neglect without between-object neglect.

R.R.: A Case of Object-Based Neglect

We (Walker et al. 1996) have reported a detailed examination of the eye movements made by a neglect patient R.R. when he viewed simple scenes and objects. R.R. was of particular interest as he was a rare case of a non-hemianopic In our eye movement study R.R. was asked to view a number of images presented on a computer visual display screen. The images appeared following the presentation of a central fixation square. R.R. was instructed to look at the central fixation square and to scan each of the images upon presentation. Eye movements were recorded while each image was being viewed. To ensure that R.R. would be

Fig. 1. Copy of scene made by patient R.R. showing object-based neglect. (From Walker et al. 1996)

motivated to scan each image he was asked to verbally report what he had seen after each image was removed. Figure 2 shows one of the scan paths obtained during a trial in which R.R. viewed a simple scene (for 10 s) superimposed onto the original image. It can be seen from Fig. 2 that R.R. made saccades and fixations on the objects located to the left of the image midline. In fact, he was found to have spent more time fixating on the area to the left of the image midline (7 s), than on the area to the right of the image midline (3 s). Furthermore, R.R. can be seen to have made a large left saccade which crossed well into the left side of the scene, following a fixation on an object located on the far right side of the image. R.R.'s performance when viewing the scene shows that he is able to make contralesional saccades and fixations when scanning between-objects.

In contrast to his relatively unimpaired contralesional scanning when viewing the scene, R.R. did show defective scanning when viewing whole and chimaeric objects. The whole and chimaeric stimuli were presented centrally onto a VDU screen, for a viewing duration of five seconds each. Prior to the main testing session R.R. was shown an example of a chimaeric image and informed that similar images would appear on some trials. During the main testing session eye movements were again recorded and he was asked to verbally report what he had

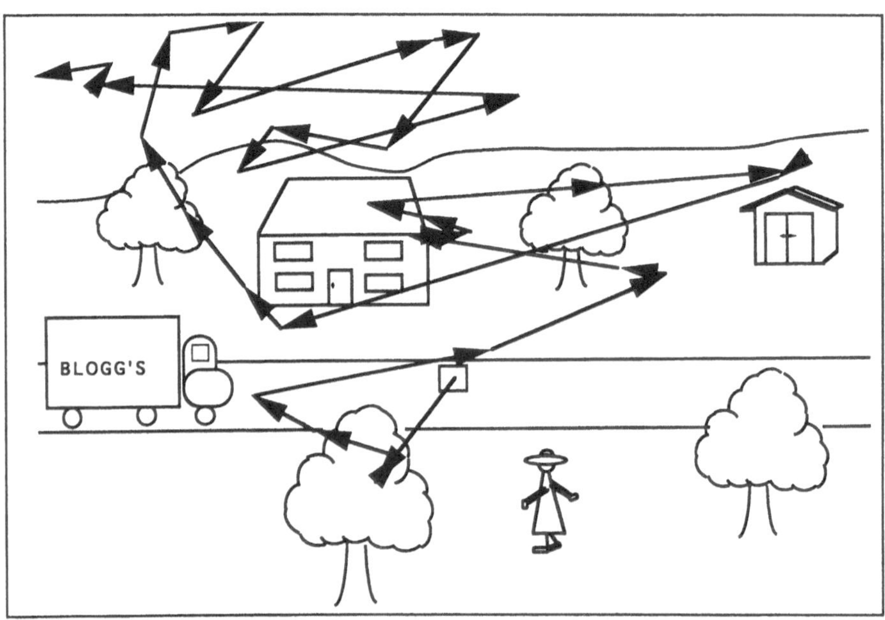

Fig. 2. R.R.'s scan path superimposed onto the original scene (10-s viewing time). The *square* denotes the location of the initial fixation point, and the *arrows* denote the location of fixations. (From Walker et al. 1996)

seen. Examples of scan paths obtained when R.R. viewed whole and chimaeric faces are shown in Fig. 3. It can be seen that R.R. made a number of saccades and fixations during the 5 s viewing period. However, these saccades and fixations were restricted to the right side of the image midline and the majority of R.R.'s fixations were made in the region of the right eye.

R.R.'s verbal reports were consistent with the observation that he scanned the right of each image only. Thus, he named all of the whole objects correctly (12/12), and all of the right sides of chimaerics (12/12), but failed to report any of the left sides (0/12) of chimaerics. This shows that R.R. is impaired at orienting his eyes into the contralesional side of an individual object, which is in contrast to his preserved ability to scan the contralesional side of the scene. The impaired scanning is associated with impaired awareness of material on the left side of chimaeric objects.

To confirm that R.R.'s inability to scan the left side of an object was not simply due to an impairment in making contralesional saccades, his eye movements were recorded while he viewed left-half objects presented to the left of fixation, and right half objects presented to the right of fixation. Examples of the scan paths obtained to left and right half faces are shown in Fig. 4. When viewing right-half faces, R.R. made saccades and fixations to the right side of the image. When viewing left-half faces, R.R. made contralesional saccades and fixated on the left

Fig. 3. R.R.'s scan paths made while viewing whole and chimaeric faces (following 5-s viewing time). R.R. correctly named the whole face but named the face on the right side of the chimaeric only

side. Furthermore, R.R. named almost all (10/11) of the left-half faces correctly. The important observation is that a viewing time of 5 s is sufficiently long to enable R.R. to make a number of contralesional saccades and fixations onto left-half images located to the left of fixation. Thus, R.R.'s failure to scan the left side of a chimaeric face cannot simply be attributed to an inability to make contralesional saccades.

It should be noted that R.R.'s scanning performance with regard to left-half images was not entirely normal. In some cases he occasionally made vertical saccades up and down the right hand edge of the left-half images. Furthermore, R.R. tended to make fixations on the region of empty space located to the right of left-half images. Although R.R.'s performance with left-half faces should not be regarded as being entirely normal, it does show that he can make contralesional saccades and that he is able to name left-half images correctly. As R.R. could make contralesional saccades to left-half images and also when viewing the scene, his failure to scan and report the left side of chimaeric images cannot simply be attributed to an inability to make contralesional saccades.

Previous studies of object-based neglect which have used chimaeric objects as stimuli (Young et al. 1992) have shown that neglect for the left side of a chimaeric occurs even when images are presented entirely in the patient's ipsilesional visual

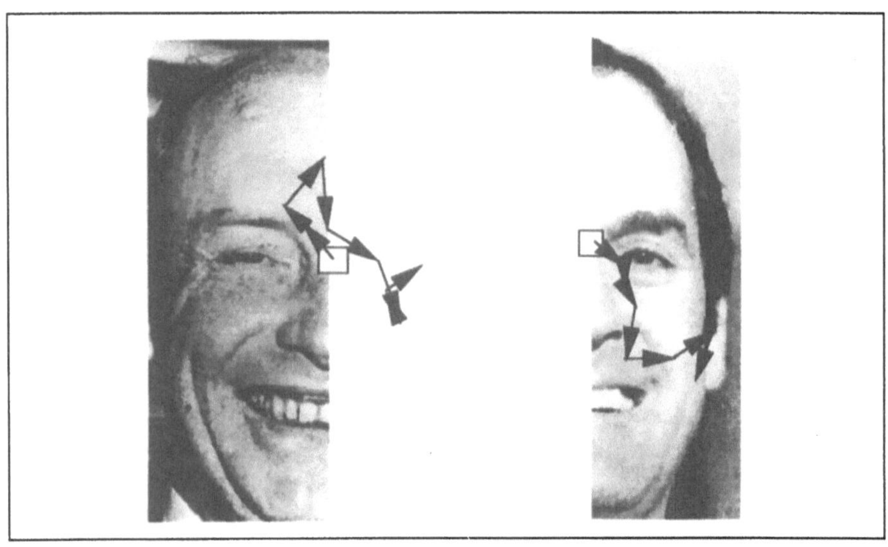

Fig. 4. R.R.'s scan paths made while viewing left and right half faces. Both half faces were correctly reported. (From Walker et al. 1996)

field. The failure to report the left side of a chimaeric object when it is presented to the right of fixation provides strong support for the view that neglect can operate in an object-based frame of reference. We therefore recorded R.R.'s eye movements in situations where the whole and chimaeric faces were presented entirely in his right visual field. Thus, the left sides of the chimaerics presented to the right of fixation corresponded to the same spatial location as the right side of centrally presented chimaerics. To rule out practice effects we also recorded R.R.'s eye movements with whole and chimaeric faces presented with the vertical midline aligned with the central fixation location. The eye movement sessions in which R.R. viewed chimaerics presented centrally or to the right of fixation were performed some months after the first study.

The first observation to note from these later studies was that R.R. had shown some degree of improvement in his ability to report the left sides of chimaeric faces compared to when first tested. With centrally presented chimaerics, R.R. correctly reported 27/31 faces on the right side, and 12/31 faces on the left side. This is in contrast to his failure to report any of the left sides of chimaeric faces in the first study. When chimaerics were presented to the right of fixation, R.R. correctly reported 30/35 faces located to the right of the image midline, and 21/35 faces located to the left of the image midline. This improvement in reporting the left sides of chimaeric faces in the second study may be due to practice effects, or

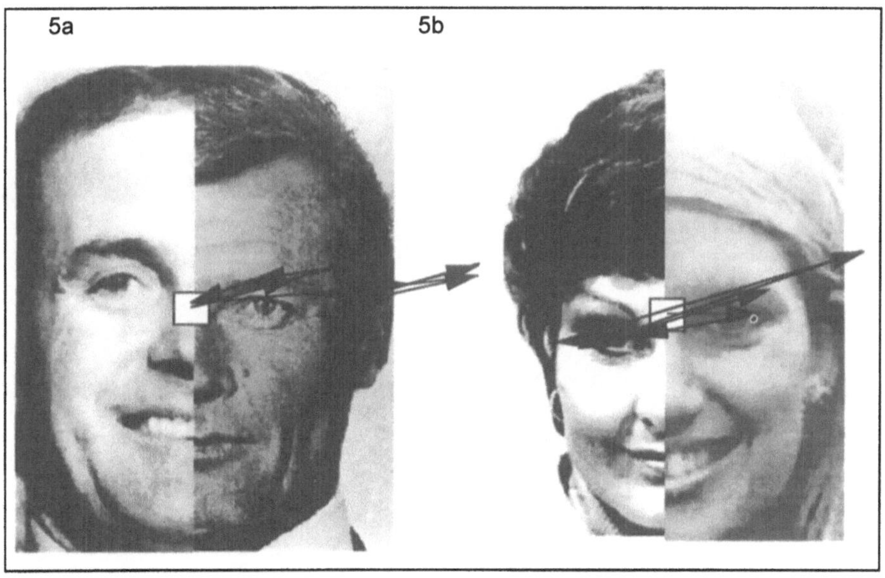

Fig. 5a, b. R.R.'s scan paths made with centrally presented chimaerics. **a** Face on right side only reported. **b** Both faces reported

possibly to fluctuations of the severity of his neglect over time. However, the dramatic observation is that, even with right visual field presentation, R.R. failed to report 14/35 of the left sides of chimaerics which were presented entirely in his ipsilesional visual field. The improvement in R.R.'s performance provided a fortuitous benefit, as it allowed us to compare his scanning performance on trials in which the left sides were neglected, to his scanning on trials in which the left and right sides were reported.

Fig. 5a shows an example of a scan path obtained for a centrally presented chimaeric obtained on a trial in which R.R. failed to report the face on the left side. This pattern of scanning was shown on ten occasions and it can be seen that R.R. restricted his fixations to the right side of the image midline. Fig. 5b shows an example of a scan path obtained for a centrally presented chimaeric where both the left and right sides were reported. There were nine similar instances, and it can be seen that R.R. scanned the right side of the chimaeric and only then made saccades back into the left side. A somewhat similar relationship between scanning and reporting was observed when the chimaerics were presented in R.R.'s right visual field. Fig. 6a shows an example of a scan path obtained in a trial in which R.R. did not report the face on the left side of a chimaeric. There were eight similar trials in which R.R. scanned the right side of the image and reported the face on the right side only. Fig. 6b shows the scan path obtained from

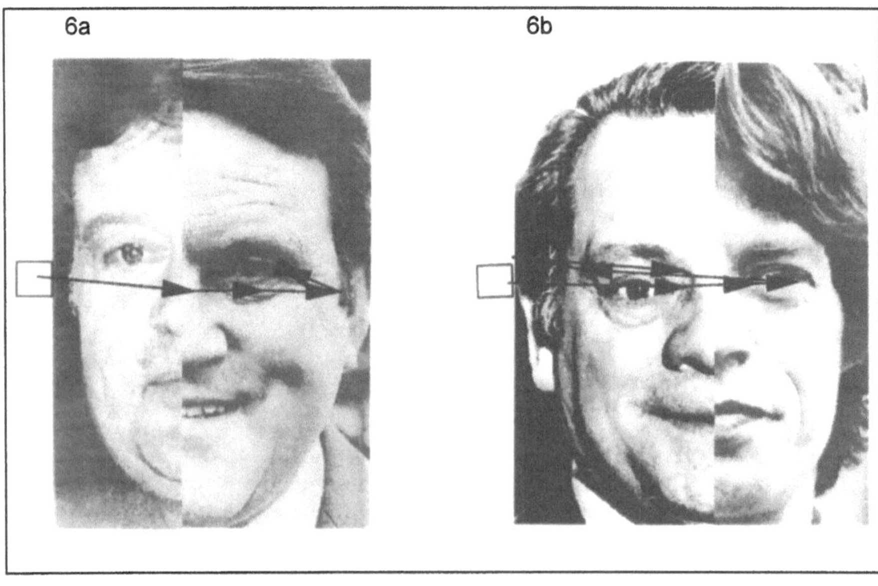

Fig. 6a, b. R.R.'s scan paths made while viewing chimaerics presented in the right visual field. **a** Face on right side only reported. **b** Both faces reported

a trial in which R.R. reported both the left and right sides of a chimaeric. There were 15 similar trials and once again he made saccades and fixations back into the left side of the image only after he had scanned the right side.

It was noted that on all occasions R.R.'s first fixation was made in the region of the eye located on the *left* side of the image when chimaerics were presented in his right visual field. The fixation made following the first one was then directed to the eye on the right side of the chimaeric. In 15 trials (15/29) in which the left side of the chimaerics were reported R.R. made saccades back into the left side of the image after scanning and fixating the right side. However, there were eight occasions (8/29) on which R.R. made a fixation of some 250 ms in the region of the left eye and still failed to report the face on the left side. We therefore performed an experiment in a separate study to examine whether or not a fixation of 200 ms was sufficient to enable R.R. to report the left side of a chimaeric. In this study left-half faces were presented tachistoscopically for 200 ms, and R.R. was asked to name each one. It was found that R.R. could name the majority (16/22) of left-half faces when they were presented for only 200 ms. Thus, R.R.'s failure to report, or to be aware of, the left side of the chimaerics presented to the right side of fixation cannot simply be attributed to a failure to fixate on the left side of the image. Similarly, Fig. 7 shows a scan path from a trial in which R.R.

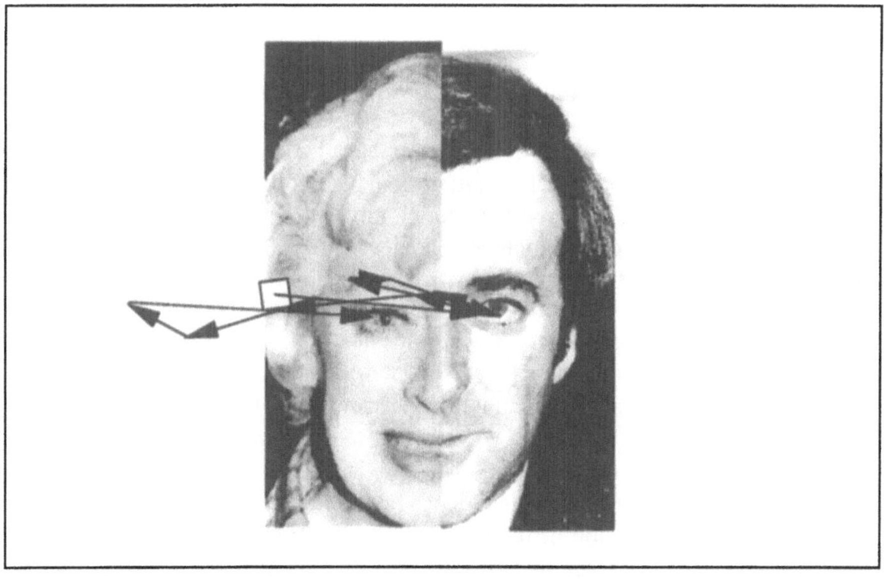

Fig. 7. R.R.'s scan path with a chimaeric presented to the right of fixation. R.R. scanned both sides of the image but reported the face on the right side only. (From Walker et al. 1996)

scanned both the left and right sides of a chimaeric, but reported the object on the right side only. In this case R.R. actually spent more time fixating the left side of the image (over 2.2 s) than on the right side (1.9 s) and yet he reported the face on the right side only.

There are a number of important points to note from the findings of our eye movement study. The first observation is that R.R. scanned and fixated both the left and right sides of the scene and was able to make saccades into the contralesional side of this image. However, when he viewed single objects presented centrally, R.R. restricted his fixations to the right side of the image midline. Since R.R. was able to make left saccades when left-half objects were presented in his left visual field, his failure to scan the left side of a centrally presented object cannot be attributed to a problem in making contralesional saccades. The pattern of impaired scanning of individual objects, with relatively unimpaired scanning of a scene, is consistent with the performance of a patient with object-based neglect without between-object neglect. The dissociation with the cases reported by Rizzo and Hurtig (1992) and Karnath (1994) provides additional support for the view that neglect may be fractionated into within-object and between-object based forms.

Conclusions

This chapter has described some of the evidence supporting the view that neglect may be a spatial or an object-based phenomenon (see Walker 1995 for a review). In contrast to the cases reported by Rizzo and Hurtig (1992) and Karnath (1994), our case R.R. was able to scan the contralesional side of a scene containing more than one object, but was impaired at scanning the contralesional side of individual objects. R.R. is a case of a patient with object-based neglect without between-object neglect. The cases reported by Rizzo and Hurtig and by Karnath are consistent with patients who show between-object neglect, but not within-object neglect. This double dissociation in eye movement deficits is consistent with the view that neglect can operate in either a spatial or object-based frame of reference. Although the double dissociation supports the view of separate spatial and object-based forms of neglect, it should be noted that these forms may not be mutually exclusive categories. For example, R.R. might manifest severe object-based neglect as well as mild spatial neglect which he is able to compensate for.

One important implication of the study is that R.R.'s abnormal eye movements are not easily accounted for in terms of the 'attentional disengagement' deficit model of neglect (Posner et al. 1984, 1987). According to this model, neglect results from an inability to 'disengage' attention from one location when it has to be shifted in the contralesional direction. However, in our study R.R. readily made contralesional eye movements when viewing the scene, but did not do so when viewing a single object. Furthermore, it was found that R.R. failed to report the face on the contralesional side of a chimaeric in some trials, even though he had fixated and scanned the left side. R.R.'s apparent unawareness of the left sides of chimaeric images cannot be accounted for in terms of a failure to scan the contralesional side of the object or by a deficit of attentional disengagement.

The finding that neglect can be manifest in object-based forms is consistent with models which have proposed that attention can be allocated to an object as well as to a spatial location. Baylis and Driver (1993) proposed a hierarchical model of position coding to account for this result which is an alternative to the purely spatial-based accounts of attentional allocation. Their model is based on the idea that the spatial locations of object parts are coded explicitly in relation to their parent object. They further proposed that spatial information is represented in two different ways. One representation codes the spatial location of objects in a scene, and a second representation codes the relative spatial position of the parts that make up an individual object. An important point to note about this model is that both forms of representation are spatial in nature; one codes the spatial location of objects in a scene, and the other codes the spatial relationship between object parts. Baylis and Driver further suggested that this hierarchical representation may parallel the processes thought to be performed by the primate dorsal and ventral visual pathways.

The spatial and object-based distinction observed in neglect would also appear to be consistent with Ungerleider and Mishkin's (1982) functional distinction of two separate visual pathways. According to Ungerleider and Mishkin, the ventral pathway is thought to be responsible for object-recognition ('what'), whilst the dorsal stream pathway is involved in mapping spatial location ('where'). One of the puzzling aspects of object-based neglect is that it appears to be associated with damage to the dorsal visual pathway, which, according to the functional dichotomy, is involved in spatial processing, and not object recognition. Milner and Goodale (1995) have addressed this issue and provide one possible solution to this apparent paradox. They observe that neglect in man occurs most frequently after damage to the inferior parietal or parietal temporal region. However, they discuss evidence which suggests that the human dorsal stream actually terminates in the superior parietal lobe and not in the inferior part. Milner and Goodale suggest that in cases of object-based neglect there may be damage to the representational system of the inferior parietal lobe which extends into the ventral pathway itself. They conclude that "the ventral stream must be crucially involved in providing visual representations in these patients (those with object-based neglect) and that their symptoms reflect a failure at a higher level where the brain operates upon such representations".

Finally, it should be stated that, although the distinction of spatial and object-based forms of neglect is an important one, it does not in itself provide an explanation of the phenomena of neglect. It therefore remains for future work to make progress in increasing our understanding of the nature of this disorder. However, the realisation that neglect is not a unitary syndrome and the fractionation of the disorder into separate forms should greatly aid the development of future models.

Acknowledgements. This work was supported by an MRC Grant awarded to J.M. Findlay and A.W. Young. We are grateful to Nadina Lincoln for providing access to the patient R.R.. Figures 1, 2, 4, and 7 are reprinted by permission of Erlbaum (UK) Taylor and Francis, Hove UK.

References

Albert ML (1973) A simple test of visual neglect. Neurology 23:658–664

Baylis B, Driver J (1993) Visual attention and objects: evidence for hierarchical coding of location. J Exp Psychol: Hum Percept Perform, 19:451–470

Bisiach E, Geminiani G, Berti A, Rusconi ML (1990) Perceptual and premotor factors of unilateral neglect. Neurology 40:1278–1281

Buxbaum LJ, Coslett BH (1994) Neglect of chimeric figures: two halves are better than a whole. Neuropsychologia 32(3):275–288

Chieffi S, Gentilucci M, Allport A, Sasso E, Rizzolatti G (1993) Study of selective reaching and grasping in a patient with unilateral parietal lesions. Brain 116:1119–1137

Costello A de L, Warrington E (1987) Word comprehension and word retrieval in patients with localised cerebral lesions. Brain 101:163–185

Cowey A, Small M, Ellis S (1994) Left visuo-spatial neglect can be worse in far than in near space. Neuropsychologia 32:1059–1066

Driver J, Halligan PW (1991) Can visual neglect operate in object-centred co-ordinates? An affirmative single-case study. Cogn Neuropsychol 8:475–496

Driver J, Baylis GC, Goodrich SJ, Rafal RD (1994) Axis-based neglect of visual shapes. Neuropsychologia 32:1353–1365

Ellis AW, Flude BM, Young AW (1987) "Neglect dyslexia" and the early visual processing of letters in words and nonwords. Cogn Neuropsychol 4:439–464

Farah MJ, Brunn JL, Wong AB, Wallace MA, Carpenter PA (1990) Frames of reference for allocating attention to space: evidence from the neglect syndrome. Neuropsychologia 28:335–347

Farah MJ, Wallace MA, Vecera SP (1993) "What" and "Where" in visual attention: evidence from the neglect syndrome. In: Robertson IH, Marshall JC (eds) Unilateral neglect: clinical and experimental studies. Erlbaum, Hillsdale, pp 123–136

Gainotti G, Messerli P, Tissot R (1972) Qualitative analysis of unilateral spatial neglect in relation to laterality of cerebral lesion. J Neurol Neurosurg Psychiatry 35:545–550

Girotti F, Casazza M, Musicco M, Avanzini G (1983) Oculomotor disorders in cortical lesions in man: the role of unilateral neglect. Neuropsychologia 5: 543–553

Goodale MA, Milner AD, Jakobson LS, Carey DP (1990) Kinematic analysis of limb movements in neuropsychological research: subtle deficits and recovery of function. Can J Psychol 44:180–195

Halligan PW, Marshall JC (1991) Left neglect for near but not far space in man. Nature 350:498–500

Halligan PW, Marshall JC (1992) Left visuo-spatial neglect: a meaningless entity? Cortex 28:525–535

Halligan PW, Marshall JC (1993) When two is one: a case study of spatial parsing in visual neglect. Perception 22:309–312

Halligan PW, Marshall JC (1995) Completion in visuo-spatial neglect: a case study. Cortex 30:685–694

Hécaen H (1962) Clinical symptomatology in right and left hemisphere lesions. In: Mountcastle VB (ed) Interhemispheric relations and cerebral dominance. Johns Hopkins University Press, Baltimore

Hornak J (1995) Perceptual completion in patients with drawing neglect: eye-movement and tachistoscopic investigation. Neuropsychologia 33:305–325

Humphreys GW, Riddoch MJ (1994) Attention to within-object and between-object spatial representations: multiple sites for visual selection. Cogn Neuropsychol 11:207–241

Hyvärinen J (1982) Posterior parietal lobe of the primate brain. Physiol Rev 62:1060–1129

Ishiai S, Furukawa T, Tsukagoshi H (1987) Eye-fixation patterns in homonymous hemianopia and unilateral spatial neglect. Neuropsychologia 25:675–679

Karnath H-O (1994) Spatial limitation of eye movements during ocular exploration of simple line drawings in neglect syndrome. Cortex 30:319–330

Karnath H-O, Schenkel P, Fischer B (1991) Trunk orientation as the determining factor of the 'contralateral' deficit in the neglect syndrome and as the physical anchor of the internal representation of body orientation in space. Brain 114:1997–2014

Kinsbourne M (1977) Hemi-neglect and hemispheric rivalry. In: Weinstein EA, Friendland RP (eds) Hemi-inattention and hemispheric specialization. Raven, New York, pp 41–49

Ládavas E (1987) Is the hemispatial deficit produced by right parietal lobe damage associated with retinal or gravitational coordinates? Brain 110:167–180

Ládavas E (1993) Spatial dimensions of automatic and voluntary orienting components of attention. In: Robertson IH Marshall JC (eds) Unilateral neglect: clinical and experimental studies. Erlbaum, Hillsdale, pp 193–209

Lynch JC, McLaren JW (1989) Deficits of visual attention and saccadic eye movements after lesions of parietooccipital cortex in monkeys. J Neurophysiol 61:74–90

Meienberg O, Zangemeister WH, Rosenberg M, Hoyt WF, Stark L (1981) Saccadic eye movement strategies in patients with homonymous hemianopia. Ann Neurol 9:537–544

Meienberg O, Harrer M, Wehren C (1986) Oculographic diagnosis of hemineglect in patients with homonymous hemianopia. J Neurol 233:97–101

Milner AD, Harvey M (1994) Towards a taxonomy of spatial neglect. In: Halligan PW, Marshall JC (eds) Spatial neglect: position papers on theory and practice. Erlbaum, Hove, pp 177–181

Milner AD, Goodale MA (1995) The visual brain in action. Oxford University Press, Oxford

Posner MI, Walker JA, Friedrich FJ, Rafal RD (1984) Effects of parietal injury on covert orienting of attention. J Neurosci 4:1863–1874

Posner MI, Walker JA, Friedrich FA, Rafal RD (1987) How do the parietal lobes direct covert attention? Neuropsychologia 25:135–145

Riddoch MJ, Humphreys GW (1991) Visual aspects of neglect dyslexia. In: Willows DM, Kruk RS, Corcos E (eds) Visual processes in reading and text disabilities. Erlbaum, New York, pp 111–136

Rizzo M, Hurtig R (1992) Visual search in hemineglect: what stirs idle eyes? Clin Vis Sci 7:39–52

Rizzolatti G, Riggio L, Dascola I, Umiltà C (1987) Reorienting attention across the horizontal and vertical meridians: evidence in favour of a premotor theory of attention. Neuropsychologia 25:31–40

Rizzolatti G, Berti A (1993) Neural mechanisms of spatial neglect. In: Robertson IH, Marshall JC (eds) Unilateral neglect: clinical and experimental studies. Erlbaum, Hillsdale, pp 87–105

Rizzolatti G, Riggio L, Sheliga BM (1994) Space and selective attention. In: Umiltà C, Moscovitch M (eds) Attention and performance XV. MIT Press, Cambridge MA, pp 231–265

Robertson IH, Tegnér R, Goodrich SJ, Wilson C (1994) Walking trajectory and hand movements in unilateral left neglect: a vestibular hypothesis. Neuropsychologia 32:1495–1502

Silberpfennig J (1941) Contributions to the problem of eye movements. III. Disturbances of ocular movements with pseudohemianopsia in frontal lobe tumors. Confin Neurol 4:1–13

Tegnér R, Levander M (1991) Through a looking glass. a new technique to demonstrate directional hypokinesia in unilateral neglect. Brain 114:1943–1951

Ungerleider LM, Mishkin M (1982) Two cortical visual systems. In: Ingle DJ, Goodale MA, Mansfield RJW (eds) Analysis of visual behavior. MIT Press, Cambridge MA, pp 549–587

Walker R (1995) Spatial and object-based neglect. Neurocase 1:189–207

Walker R, Findlay JM, Young AW, Welch J (1991) Disentangling neglect and hemianopia. Neuropsychologia 29:1019–1027

Walker R, Findlay JM, Young AW, Lincoln NB (1996) Saccadic eye movements in object-based neglect. Cogn Neuropsychol 13:569–615

Walker-Smith GJ, Gale AG, Findlay JM (1977) Eye movement strategies involved in face perception. Perception 6:313–326

Young AW, Newcombe F, Ellis AW (1991) Different impairments contribute to neglect dyslexia. Cogn Neuropsychol 8:177–193

Young AW, Hellawell DJ, Welch J (1992) Neglect and visual recognition. Brain 115:51–71

Reaching and Grasping

From Vision to Movement: Cortico-Cortical Connections and Combinatorial Properties of Reaching-Related Neurons in Parietal Areas V6 and V6A

P. B. JOHNSON[1], S. FERRAINA[1], M. R. GARASTO[1], A. BATTAGLIA-MAYER[1], L. ERCOLANI[1], Y. BURNOD[2], and R. CAMINITI[1]

[1]Istituto di Fisiologia umana, Università di Roma "La Sapienza", Rome, Italy
[2]INSERM CREARE, Université Pierre et Marie Curie, Paris, France

Introduction

Hand-reaching movements to visual targets located in extrapersonal space require the combination of different informational domains within a distributed cortical network which includes, at the least, occipital, parietal and frontal cortices, and their reciprocal association connections. Network models of visual reaching impose constraints on both the anatomical layout of this network and on its functional architecture. While the neural bases of reaching have been described in detail at the level of those cortical areas which are considered to be the output layer of the network, little is known of the early cortical operations underlying the composition of visuomanual commands and on the role of the re-entrant signals influencing, through cortico-cortical connections, the different nodes of the network.

In this work we will critically evaluate the most recent developments on the cortico-cortical connectivity of the different areas involved in the coding of reaching. At the same time, we will offer a preliminary description of the early operations in the occipito-parietal cortex which underlie eye-hand coordination required for reaching.

Network Models and Combinatorial Units

In recent years, experimental data from both behavioral and neurophysiological studies have provided a framework for the development of network models designed specifically to address certain spatial aspects of reaching movements. One noteworthy feature of these models is the emphasis placed on the integration of different types of spatial information; in particular, the combination of

In: Parietal Lobe Contributions to Orientation in 3D Space (1997). P. Thier and H.-O. Karnath (eds). Springer-Verlag, Heidelberg.

information about the location of objects in extrapersonal space with information about the current locations of the body segments which will be involved in the upcoming movement.

A distributed network architecture has been shown to be capable of learning to use retinal target position information, angle of gaze, vergence angles, and head angles in order to derive a representation of target position in three dimensional space (Grossberg et al. 1993; Guenther et al. 1994). This model first computes a head-centered representation in terms of distance (computed directly from eye vergence angles) and horizontal and vertical spherical angles from a cyclopean origin. By using head angle information, this head-centered target representation is then, in a second stage, converted into a similar body-centered spherical coordinate system which is independent of head and eye orientations. Subsequent network computations can combine such an internal spatial representation with visual and proprioceptive information regarding the position of the arm in order to first obtain a movement direction vector and then a motor command (Bullock et al. 1993). Such a body-centered internal representation of external space simplifies many aspects of motor programming (see Bullock et al. 1993) and is consistent with existing psychophysical (see Soechting and Flanders 1992) and neurophysiological (Lacquaniti et al. 1995) data. Another modeling study (Burnod et al. 1992) assumes the existence of a movement direction vector derived from visual information and is thus compatible with, but does not explicitly require an internal representation of target position in extrapersonal space. In this model, visual information on movement direction is encoded relative to the fixation point. Visual tuning properties of posterior parietal neurons in area 7 which are activated by visual stimuli moving toward the fovea (Motter and Mountcastle 1981) are consistent with such a model. It is likely that both schemes for using visual input are employed profitably by the nervous system, as in the model of Mel (1991).

A common feature of these models is that the combination of visual information on target location with proprioceptive information on arm configuration is performed in a single "combinatorial" layer of units ("matching layer" of Burnod et al. 1992; "Position-Direction map" of Bullock et al. 1993). In this layer, the units are tuned independently to specific visually-defined movement directions and specific arm configurations. As such neuronal activity is not known to exist in the cerebral cortex, these combinatorial neurons remain a prediction of the network models. Confirmation of the existence of such units would further our understanding of how the brain controls reaching and would guide the development of further generations of neural models.

The above network models place anatomical constraints on the units of the combinatorial layer which may help to guide our search for such units. Firstly, they must have connections to those motor output regions of the frontal lobe involved in the control of the arm musculature used for reaching. Secondly, they must have access to information regarding target location and arm configuration arriving from both visual and proprioceptive centers.

Parietal Projections to Frontal Lobe Reach Centers

The lateral aspect of the frontal lobe contains numerous motor fields (Matelli et al. 1985; Barbas and Pandya 1987). Among these, primary motor cortex (M1) and the dorsal premotor cortex (PMd; Weinrich and Wise 1982) are of particular interest to us, since they are involved in coding direction of arm reaching (Schwartz et al. 1988; Caminiti et al. 1990, 1991). It has been known for many years that there are no direct cortico-cortical projections between primary visual cortices and either M1 or PMd (Pandya and Kuypers 1969; Jones and Powell, 1970). The striate cortex (V1) was shown to project to the surrounding peristriate belt and this, in turn, to the caudal part of the inferior parietal lobule (IPL). As a result of these studies, the IPL was long regarded as the probable relay between visual and motor cortical areas. However, these same anatomical studies (Pandya and Kuypers 1969; Jones and Powell 1970) revealed that the IPL projections to the frontal lobe were addressed not to M1 or PMd, but to prefrontal cortex. This connectivity was later confirmed in more detail by many other studies (see Petrides and Pandya 1984; Schwartz and Goldman-Rakic 1984; Andersen et al. 1985; Cavada and Goldman-Rakic 1989b). Heavy parietal projections arising from area 5 in the superior parietal lobule (SPL), a cortical region believed devoid of any visual input, were shown to terminate in M1 and PMd (Pandya and Kuypers 1969; Jones and Powell 1970; Strick and Kim 1978).

In more recent studies, to further investigate this apparent paradox, we employed a combined neurophysiological and neuroanatomical approach (Johnson et al. 1993, 1996). An instructed-delay directional reaching task was used to characterize the arm-related zones of both frontal (M1 and PMd) and parietal [dorsal area 5 and medial intraparietal (MIP)] cortices. Then, injections of different retrograde tracers were made within different physiologically characterized regions of M1 and PMd to identify their cortical sources of afferent information. When neuronal activity was correlated with the tangential position of individual cells, a rostral to caudal gradient of functional properties emerged in the frontal lobe arm zones, with the dominant activity types changing gradually from those related to the input of visual information concerning target localization (signal- and set-related activities) to those mainly linked to movement generation (position- and movement-related activities). While the tracer injections confirmed that parietal input to the frontal lobe regions involved in reaching originates almost exclusively from the SPL, it also demonstrated a detailed pattern of parietal projections. The SPL projects in an orderly and partially overlapping fashion to these frontal regions (Johnson et al. 1993, 1996). Dorsal area 5 projects mainly to M1 and to a lesser extent to the PMd/M1 border; MIP area and area 7m project to the PMd/M1 border and to PMd, with the heaviest projection coming from MIP.

Visual and Proprioceptive Inputs to the Superior Parietal Lobule

Meanwhile, anatomical studies from other laboratories have been challenging the established dogma that the only parietal regions receiving visual information are those located in the IPL. Recent tracing experiments have shown that parietooccipital (PO) area (Covey et al. 1982; Gattass et al. 1985; Colby et al. 1988), a cortical region located deep in the medial wall of the hemisphere, in the rostral bank of the parietooccipital sulcus, receives direct projections from V1, V2, V3, V4 and the medial temporal visual area MT (Colby et al. 1988). Injections of retrograde tracers in physiologically defined regions of PO labeled neurons in MIP and medial dorsal parietal (MDP) area (Colby et al. 1988), while injections in area MIP labeled neurons in both areas PO and MDP (Blatt et al. 1990). Other studies (Cavada and Goldman-Rakic 1989a) have shown that tracer injections in area 7m labeled a reciprocal pathway to and from area PO. Furthermore, 7m projects to PMd and to the supplementary motor area (SMA; Cavada and Goldman-Rakic 1989b), both of which project to M1 (Johnson et al. 1993, 1996). A direct projection from PO to the rostral part of PMd has recently been reported (Tanné et al. 1995). This projection is addressed to a cortical region which is just rostral to the reach-related zone of PMd (Johnson et al. 1996). Therefore, area PO provides visual information to MIP, 7m, MDP, and also to the rostralmost part of PMd. In addition, area 7m receives a dense projection from area 7a and the lateral intraparietal (LIP) area (Cavada and Goldman-Rakic 1989a) in the inferior parietal lobule. Recent anatomo-physiological studies (Matelli et al. 1995; Galletti et al. 1996) have added further details by dividing the cortex of the rostral bank of the parietooccipital sulcus into a ventral region, area V6 (corresponding to PO), and a dorsal one, area V6A, but have not substantially changed the picture of the cortico-cortical connectivity of these superior parietal regions.

Within the SPL, area 7m receives projections from dorsal area 5 and MIP (Pandya and Seltzer 1982; Cavada and Goldman-Rakic 1989a). Projections to MIP arise from somatosensory area 2 and dorsal area 5 (Pandya and Seltzer 1982; Pons and Kaas 1986) and from ventral intraparietal (VIP) area (Lewis and Van Essen 1994).

Potential Sites of Combinatorial Units

From the studies mentioned above, a parieto-frontal network emerges as a possible substratum within which the different types of information relevant to reaching are combined. The main feature of this network is the parallel nature of the connections between its "layers" (parietal and frontal cortices) and the gradient architecture of these layers, by which information originally segregated can be progressively combined. This has led to a reconsideration of the role of the SPL, traditionally believed to be a somatosensory and motor center, and suggests a visuo-motor involvement of this structure (Johnson et al. 1993, 1996; Caminiti et al. 1996). Interestingly, studies using metabolic mapping with 2-deoxy-D-glucose (Savaki et al. 1993) have pointed to the same conclusion.

In particular, areas MIP, 7m, V6 and V6A of the SPL emerge as potential sites for the location of the hypothetical combinatorial units. Physiological studies have found neurons in area MIP to be responsive to both visual and somatosensory stimuli (Colby and Duhamel 1991). Quantitative studies (Johnson et al. 1996) have shown that MIP neurons can combine visual information on target location and somatic information on arm position for the generation of motor output signals for reaching. Activity believed to be an expression of the neural events involved in target localization predominates in the ventral part of area MIP. Activity types which may reflect the combination of information about target location with those about the position of the arm in space (Ferraina and Bianchi 1994), are more frequent at intermediate dorso-ventral locations of area MIP, while arm movement- and position-related activity are more frequently encountered in the more dorsal part of MIP and area 5. Other previous physiological analyses of SPL have not distinguished dorsal area 5 from MIP, although differences in functional properties of neurons in the tangential extent of area 5 have been long reported (Mountcastle et al. 1975; Seal et al. 1983; Crammond and Kalaska 1989; Burbaud et al. 1991; for reviews see Mountcastle 1995; Caminiti et al. 1996). Parietal and frontal regions displaying similar activity-types tend to be preferentially linked through parallel sets of association connections (Johnson et al. 1996).

We have recently studied the dynamic properties of neurons in area 7m during reaching to foveated targets and to targets located in the periphery of the visual field, to dissociate oculomotor from visuomanual neuronal activity (Ferraina et al., submitted). We have found that a large proportion of neurons in this area participate in eye-hand coordination during reaching. Furthermore, these neurons seem to combine eye- and hand-position information and to transform a dynamic retinal error signal into a motor command to move the hand toward the desired target location in space.

There is no previously existing data on the involvement of V6 and V6A in arm movement. These areas are interposed in the parieto-frontal pathway underlying visual reaching between visual cortices and movement-related parietal areas from which they receive re-entrant signals. Therefore, their functional properties during manual visuomotor behavior is of special interest for understanding the nature of the cortical operations at an early stage of the composition of motor commands for reaching.

Neural Activity in V6 and V6A during Reaching

During visually guided hand reaching, the target is first perceived, then the eyes saccade to the target and finally the hand moves to it. Therefore, in studying the substrata of reaching, it is important to distinguish the various factors potentially influencing neuronal activity. This becomes even more important when studying the coding of reaching in cortical areas such as V6 and V6A, where a relationship between cell activity and both oculomotor behavior (Galletti et al. 1996) and presentation of visual stimuli (Colby et al. 1988; Galletti et al. 1996) are well established.

We have approached this complex issue by using a directional instructed-delay reaching task (Fig. 1) which dissociates, in time, oculomotor from visuomanual behavior and the visually-derived activities linked to target presentation from those which are expression of hand movement planning and execution.

This task begins with the presentation of a center-light which the monkey fixates and on which she places her hand, for a variable center-holding time (CHT; Fig. 1A_1). Then, an instruction signal (IS; Fig. 1A_2) is randomly presented at one of eight possible locations located on a circle, at 45° angular intervals, and the center-light is extinguished. The IS instructs the animal on the direction of the next reaching movement. When the IS is presented, typically the eyes saccade to the target (Fig. 1A_3) and stay there for the rest of the behavioral trial (Fig. 1A_{4-7}) while the hand stays immobile at the origin of the movement for the remainder of this instructed delay-time (IDT; Fig. 1A_2-A_5). The IDT comprises the time from onset of IS to when the signal becomes the "go" signal (GS; Fig. 1A_5) by changing color. Following the GS, within pre-specified reaction and movement times (RT, MT; Fig. 1A_{5-6}) the hand moves to the target and remains there for a variable target holding time (THT; Fig. 1A_7). In the initial part of the IDT (IDT_1; Fig 1B, C) eye movement and position are dissociated from hand position, while in the second part of the IDT (IDT_2; Fig 1B-C), eye position is dissociated from the hand position. The time of transition from IDT_1 to IDT_2 is determined by using thresholds on eye velocity data collected from a scleral search coil.

As a consequence of this task's structure, the neural activities linked to planning and execution of hand reaching during IDT_2, RT and MT, can be separated in time by those linked to oculomotor behavior during IDT_1. Furthermore, the presentation of the visuo-spatial signal concerning target location is dissociated in time from the planning and execution of the reaching movement.

During reaching, when the hand is in the field of view, its position and movement can be controlled mainly by using visual information. When the hand is out of the field of view, its control must be proprioceptive in nature. The task just described was performed both under normal light conditions and in the dark. In the latter case, the animal had no visual control of either static position or movement of the hand, which could therefore be achieved only by using kinesthetic information.

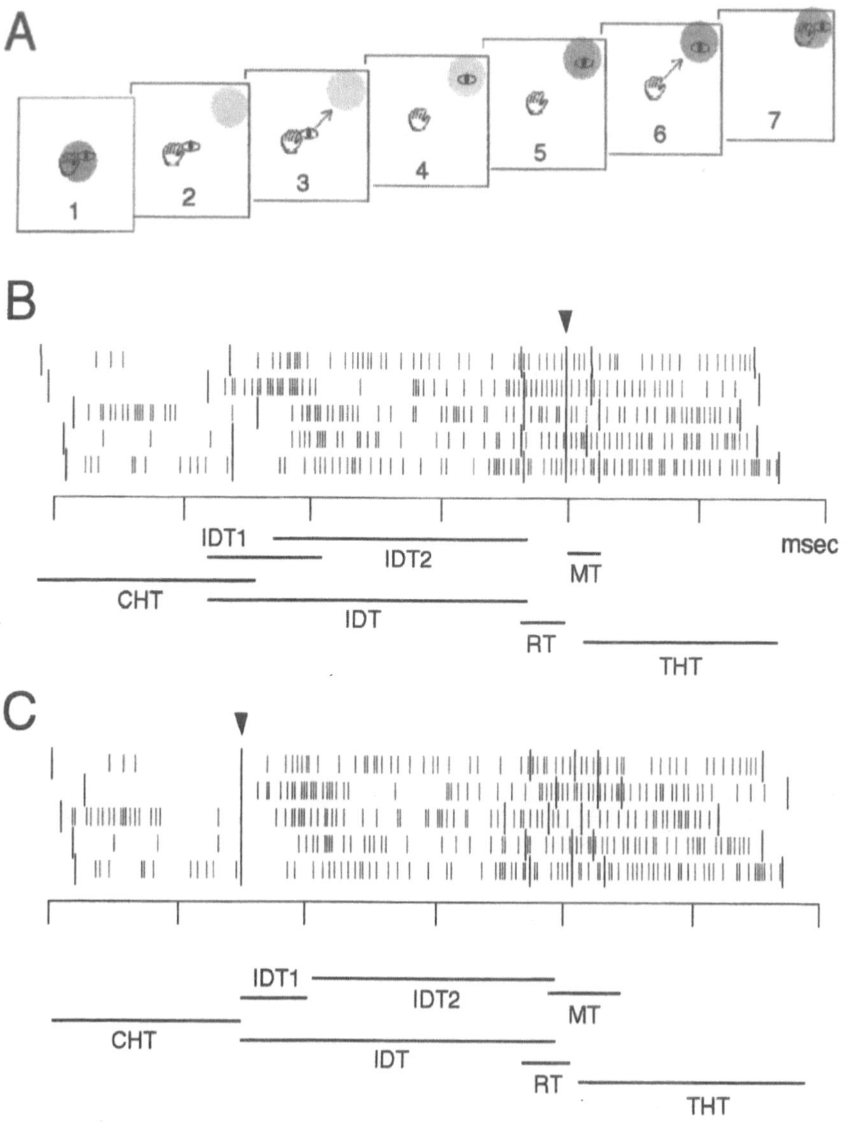

Fig. 1 A-C. A Temporal sequence of events in the instructed-delay reaching task. Rasters of five replications of neural activity are aligned (arrow) to the movement onset in **B** and to the presentation of the instruction signal (IS) in **C**, to better display the temporal structure of the different epochs of the task. *CHT*, center-holding time; *IDT1-2*, instructed delay times 1 and 2; *MT*, movement time; *RT*, reaction time; *THT*, target holding time

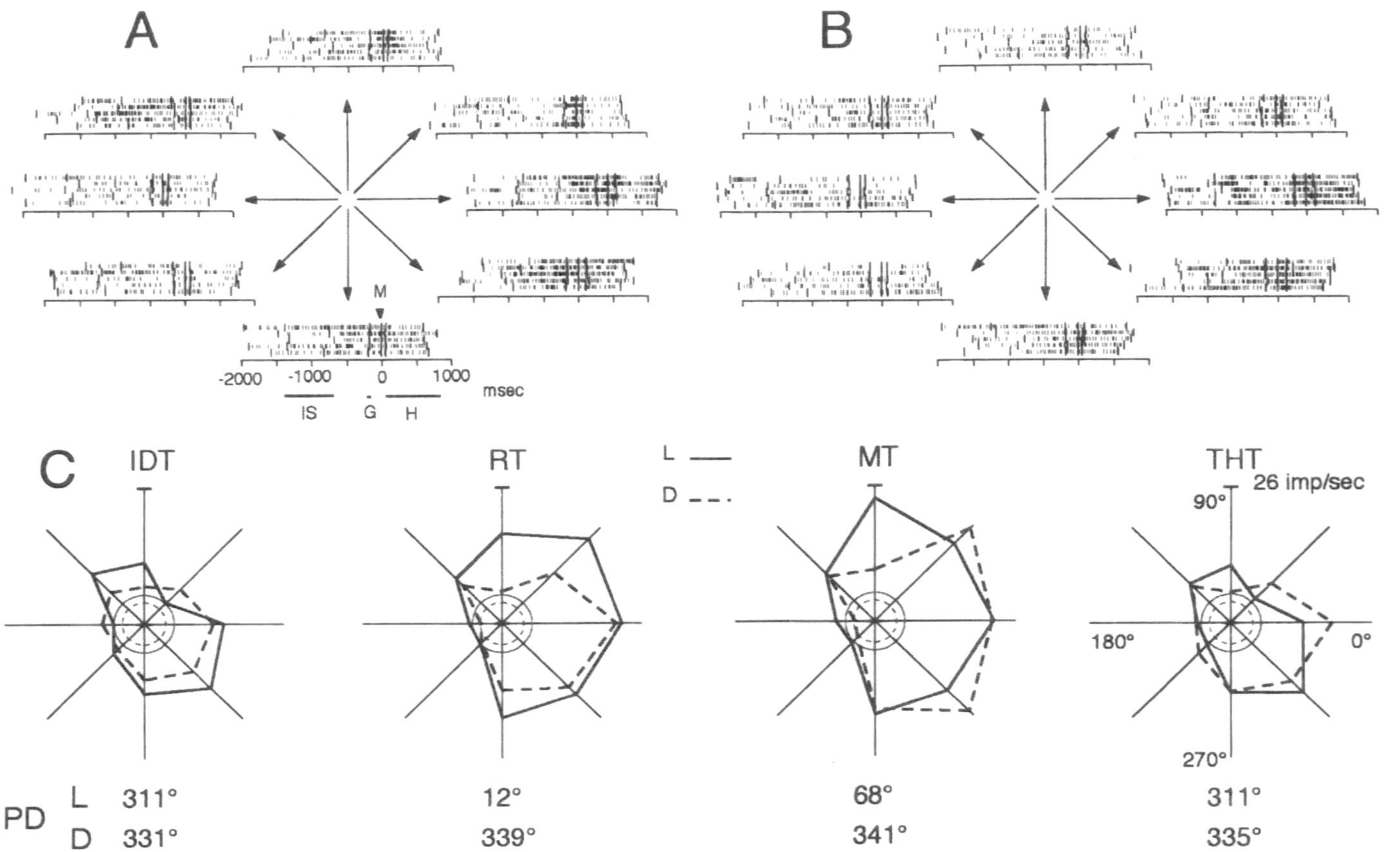
A
B
M
-2000
-1000
0
1000
msec
IS
G
H
C
IDT
RT
L
D
MT
THT
26 imp/sec
90°
180°
0°
270°
PD
L 311°
D 331°
12°
339°
68°
341°
311°
335°

Fig. 2 A-C. Impulse activity of a neuron in area V6 recorded during the instructed-delay reaching task under both light (**A**) and dark (**B**) conditions. Rasters of five replications for every movement directions were aligned to movement onset (M). On each replication, longer vertical bars indicate, from left to right, beginning of the trial, presentation of instruction signal (*IS*), presentation of go-signal (*G*), movement onset (*M*), beginning and end of target-holding time (*H*). The eight arrows indicate the directions of hand reaching movement. The positions of the rasters indicate the different target locations in space. **C** Polar plots of impulse activity during different epochs for both task conditions (*L*, light, continuous line; *D*, dark, interrupted lines). For both task conditions, circles indicate frequency of discharge during center holding time (*CHT*), taken as control time. The preferred direction (*PD*) of the cell during different epochs was calculated by using a two-harmonic Fourier expansion of the discrete firing frequency during each epoch, under both light (*L*), and dark (*D*) conditions

The repeated measures data concerning the average firing frequency of cell activity for each trial during these different epochs and across task conditions can be analyzed and compared by using standard statistical techniques. We employed the Wald tests of significance ($p<0.05$).

The activity of a prototypical neuron studied in area V6 is shown in Fig. 2. This cell displays a significant directional modulation during all epochs of the task, under both light and dark conditions. When its activity during IDT_1 and IDT_2 is compared, a significant difference emerges under both light and dark conditions. Neural activity during IDT_1 can be related to the visuo-spatial signal about target location (signal-related activity) and/or to the saccadic eye movement to the target. During IDT_2 activity is probably linked to both static eye position and/or to the preparation for the upcoming reaching movement. During visuomanual behavior (RT-MT), when the eyes fixate the target, the cell's activity changes significantly from that observed during IDT_2, under both light and dark conditions. This change indicates a relation to planning and execution of hand reaching movement.

Interestingly, during THT, when the hand attains the target upon which the eyes have been fixed since the beginning of IDT_2, cell activity is not significantly different from that observed during IDT_2. This implies that a directional signal concerning eye-position influences this cell's firing rate on which, instead, hand-position signals exert no effect. We can conclude that the activity of this neuron, when coding visual reaching, combines both eye-position and eye-movement signals with hand-movement information.

This cell was not modulated by conventional visual stimuli during a visual fixation task. However, neuronal activity was significantly higher in the light than in the dark. This occurred only during RT, suggesting that a "retinal error" signal related to the difference between the actual and desired location of the hand (in other words the difference between hand and target position) also influences the activity of this cell just before movement onset.

This property is also a characteristic of neurons in area V6A (Fig 3). The firing rate of the neuron shown here is directionally modulated during all epochs of the

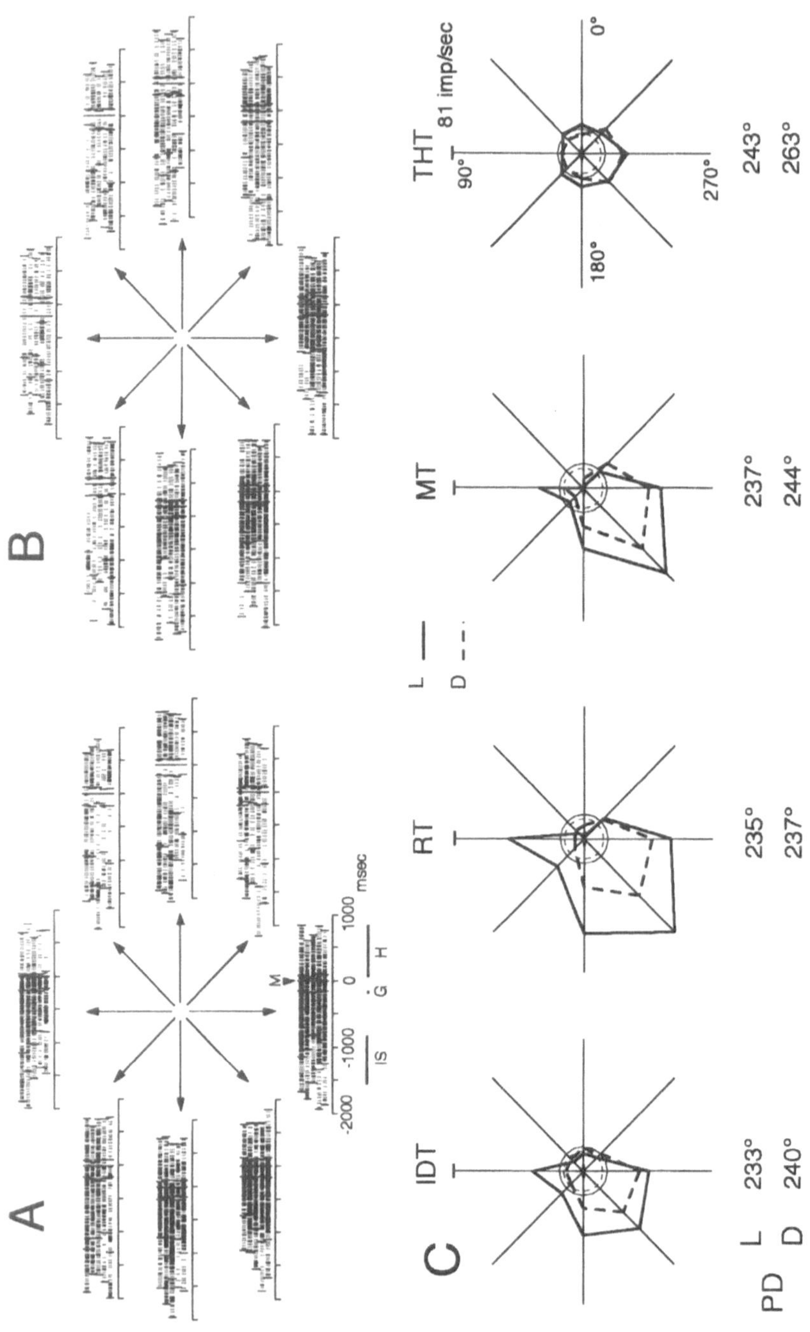
A
B
C
M
-2000
-1000
0
1000
msec
IS
G
H
IDT
RT
MT
THT
81 imp/sec
90°
0°
180°
270°
L
D
PD
L 233°
D 240°
235°
237°
237°
244°
243°
263°

Fig. 3 A-C. Impulse activity of a neuron in area V6A recorded during the instructed-delay reaching task, under both light **(A)** and dark **(B)** conditions. Rasters of five replications for every movement directions were aligned to movement onset (*M*). **C** Polar plots of impulse activity during different epochs for both task conditions. PD, preferred direction of the cell during different epochs. Conventions and symbols as in Fig. 2

task. The comparison of the firing frequency during IDT_1 and IDT_2 reveals a significant difference both in the light and in the dark. Therefore, the neural activity during IDT_1 can be signal-related and/or may depend on the saccade to the target, while during IDT_2 it can be linked to the planning of the hand reaching movement. This relation to hand reaching is also suggested by the fact that during RT and MT of the hand, in both light and dark conditions, the firing rate of this cell is significantly different from that observed while the eyes are immobile on the target and the hand is immobile at the origin of the movement. This indicates that a dynamic signal about planning and movement of the hand influences cell activity. At variance with that observed in the previous cell, during THT, when both eyes and hand are on the target, cell activity changes significantly from that seen during IDT_2, when their spatial positions are dissociated. This indicates that a directional signal concerning the actively-held static position of the hand in space is encoded in the neuronal activity.

The activity of this cell is significantly higher in the light than in the dark during all epochs of the task, except for the THT. This suggests that a "retinal error" signal is probably encoded during IDT, RT and MT, when the hand is at the origin of the movement, but not during THT, when the spatial position of the hand and that of the target coincide. Activity during MT could also be due to the "retinal image" of the hand which is created when this moves inward toward the fovea which fixates the target. Neurons displaying similar directional properties to moving visual stimuli have been described in area 7 by Motter and Mountcastle (1981). They can participate in the visual monitoring of the hand trajectory in space. It is worth stressing that this cell of area V6A was not responsive to conventional visual stimuli during a visual fixation task. It may well be that the activity of these populations of neurons is highly context-dependent and that the movement of the hand in the visual field results in a modulation of cell activity which is not seen when using conventional laboratory visual stimuli. In this respect it is interesting that a recent study (Galletti et al. 1996) has stressed the complex nature of the visual stimuli necessary to modulate cell activity in V6A.

The similarity of the preferred direction of this cell across epochs and under both light and dark conditions implies that this neuron participates in the determination of the direction of hand movement during both planning and execution, by using different sources of information which are spatially congruent.

The re-entrant signals traveling the abundant reciprocal association connections linking parietal and frontal cortex are the likely source of the hand-position and movement information influencing neuronal activity at such an early processing stage of reaching in V6 and V6A.

Conclusion

It appears that the SPL, long held to be a site related primarily to somatosensory and motor integration, is a locus of some of the early computations in the processes controlling reaching to visual targets.

A current view of V6 and V6A has been to consider these areas as performing a preliminary step in the processes underlying spatially-guided behavior: the mapping of visual stimuli from retinotopic to body-centered coordinates. At least at the population level, neurons in these areas are capable of combining retinal, eye-position and oculomotor signals in order to encode the position of the stimulus with respect to the body (see Galletti et al. 1996, for a discussion). We have found that V6/V6A neurons can participate in the signaling of target position as derived from both retinal and oculomotor information.

However, our results suggest that neurons in V6 and V6A are not only involved in the process of target localization in space, but are also directly involved in the early computation of motor commands from sensory information. We found that many neurons in these regions appear to participate in the computation of movement direction using information from different sources (eye position with respect to the head or stimulus position on the retina, together with arm position), since they show similar modulations with movement direction as the directionally tuned neurons of the frontal cortex. This modulation is already present before movement onset, during the reaction time, even in the dark task condition, and cannot be due solely to a reafferent visual feedback. Furthermore, these neurons can participate in monitoring the appropriate movement direction (during MT and THT) using information from different sources: either proprioceptive (eye and/or hand relative positions in dark) or retinal (image of the hand with respect to the image of the target).

Our results do not imply that the visual-to-motor transformation in reaching is achieved in discrete steps in different cortical areas by a hierarchical network progressing from a retinal-frame (V1), through a body-frame (V6), an arm-centered frame and a motor command (M1) step. The current results, together with our previous data on cortical function and connectivity, suggest that this transformation is accomplished in a distributed fashion along a visual-to-somatic gradient in the parallel parietal and frontal pathway (caudal to rostral in the parietal lobe, and symmetrically rostral to caudal in the frontal lobe). An appropriate visual-to-motor transformation can be computed by this network from two operations learned and encoded in two sets of synaptic weights (Burnod et al.

1992). Distributed along this gradient, in the visual-to-somatic direction, is a *matching operation* that combines retinal, eye-position, oculomotor, and somatic hand position information. In the opposite, somatic-to-visual direction, a *synergy operation* that links together the results of sensory combinations that correspond to the same motor command is computed in a distributed manner. Thus, the neural populations along the visual-to-somatic gradient effect a *progressive match* between the two sensory modalities and the appropriate motor command. A simple interpretation is that the population activity in the whole parieto-frontal network is modified through an iterative process due to intra- and interareal cortico-cortical connections, resulting in a progressive computation of the movement direction consistent with the different available sources of information along the visual-to-somatic gradient.

An essential feature of this putative network is in the specific role of the frontal cortex. The distribution of functional properties across the arm-related regions of the frontal cortex and their pattern of association and intrinsic connectivity support a model where information throughout the entire visual to somatosensory continuum is combined (see Caminiti et al. 1996, for discussion), with further local computations occurring within the frontal cortical areas, once the information concerning target location becomes available to them.

Acknowledgments. This study was supported by the European Science Foundation, the Human Capital and Mobility Project, the Human Frontier Science Program, the Ministry of Scientific and Technological Research of Italy.

References

Andersen RA, Asanuma C, Cowan WM (1985) Callosal and prefrontal associational projecting cell populations in area 7A of the macaque monkey: A study using retrogradely transported fluorescent dyes. J Comp Neurol 232:443–455

Barbas H, Pandya DN (1987) Architecture and frontal cortical connections of the premotor cortex (area 6) in the rhesus monkey. J Comp Neurol 256:211–228

Blatt GJ, Andersen RA, Stoner GR (1990) Visual receptive field organization and cortico-cortical connections of the lateral intraparietal area (area LIP) in the macaque. J Comp Neurol 299:421–445

Burbaud P, Doegle C, Gross C, Bioulac B (1991) A quantitative study of neuronal discharge in area 5, 2, and 4 of the monkey during fast arm movements. J Neurophysiol 66:429–443

Bullock D, Grossberg S, Guenther FH (1993) A self-organizing neural model of motor equivalent reaching and tool use by a multijoint arm. J Cognitive Neurosci 5:408–435

Burnod Y, Grandguillaume P, Otto I, Ferraina S, Johnson PB, Caminiti R (1992) Visuomotor transformations underlying arm movements toward visual targets: A neural network model of cerebral cortical operations. J Neurosci 12:1435–1453

Caminiti R, Johnson PM, Urbano A (1990) Making arm movements in different parts of space: Dynamic aspects in the primate motor cortex. J Neurosci 10:2039–2058

Caminiti R, Johnson PB, Galli C, Ferraina S, Burnod Y (1991) Making arm movements within different parts of space: The premotor and motor cortical representation of a coordinate system for reaching to visual targets. J Neurosci 11:1182–1197

Caminiti R, Ferraina S, Johnson PB (1996) The sources of visual information to the primate frontal lobe: A novel role for the superior parietal lobule. Cerebral Cortex 6:319–328

Cavada C, Goldman-Rakic PS (1989a) Posterior parietal cortex in rhesus monkey: I. Parcellation of areas based on distinctive limbic and sensory cortico-cortical connections. J Comp Neurol 287:393–421

Cavada C, Goldman-Rakic PS (1989b) Posterior parietal cortex in rhesus monkey: II. Evidence for segregated corticocortical networks linking sensory and limbic areas with the frontal lobe. J Comp Neurol 287:422–485

Colby CL, Gattass R, Olson CR, Gross CG (1988) Topographic organization of cortical afferents to extrastriate visual area PO in the macaque: A dual tracer study. J Comp Neurol 269:392–413

Colby CL, Duhamel J-R (1991) Heterogeneity of extrastriate visual areas and multiple parietal areas in the macaque monkey. Neuropsychologia 29:517–537

Colby CL, Duhamel J-R, Goldberg ME (1993) Ventral intraparietal area of the macaque: Anatomic location and visual response properties. J Neurophysiol 69:902–914

Covey ER, Gattass R, Gross CG (1982) A new visual area in the parieto-occipital sulcus of the macaque. Neurosci (Abstr) 8:681

Crammond DJ, Kalaska JF (1989) Neuronal activity in primate parietal cortex area 5 varies with intended movement direction during an instructed-delay period. Exp Brain Res 76:458–462

Ferraina S, Bianchi L (1994) Posterior parietal cortex: Functional properties of neurons in area 5 during an instructed-delay reaching task within different parts of space. Exp Brain Res 99:175–178

Ferraina S, Johnson PB, Garasto M, Battaglia Mayer A, Ercolani L, Bianchi L, Lacquaniti F, Caminiti R (submitted) Neural correlates of eye-hand coordination during reaching: activity in area 7m in the monkey

Galletti C, Fattori P, Battaglini PP, Shipp S, Zeki S (1996) Functional demarcation of a border between areas V6 and V6A in the superior parietal gyrus of the macaque monkey. Eur J Neurosci 8:30–52

Gattass R, Sousa APB, Covey E (1985) Cortical visual areas of the macaque: Possible substrates for pattern recognition mechanisms. In: Chagas R, Gattass R, Gross CG (eds) Pattern recognition mechanisms. Pontificiae Acad Sci Scripta Varia: 54:1–20

Grossberg S, Guenther FH, Bullock D, Greve D (1993) Neural representation for sensorimotor control. 2. Learning a head-centered visuomotor representation of 3-D target positions. Neural Networks 6:43–67

Guenther FH, Bullock D, Greve D, Grossberg S (1994) Neural representations for sensorimotor control. 3. Learning a body-centered representation of a three-dimensional target position. J Cognitive Neurosci 6:341–358

Johnson PB, Ferraina S, Caminiti R (1993) Cortical networks for visual reaching. Exp Brain Res 17 (2):361–365

Johnson PB, Ferraina S, Bianchi L, Caminiti R (1996) Cortical networks for visual reaching. Physiological and anatomical organization of frontal and parietal lobe arm regions. Cereb Cortex 6:102–119

Jones EG, Powell TPS (1970) An anatomical study of converging pathways within the cerebral cortex of the monkey. Brain 93:793–820

Lacquaniti F, Guigon E, Bianchi L, Ferraina S, Caminiti R (1995) Representing spatial information for limb movement: The role of area 5 in the monkey. Cereb Cortex 5:391–409

Lewis JW, Van Essen DC (1994) Connections of area VIP with MIP and other architectonically identified areas of the intraparietal sulcus in the macaque monkeys. Soc Neurosci (Abstr) 20 (1): 774

Matelli M, Luppino G, Rizzolatti G (1985) Patterns of cytochrome oxidase activity in the frontal agranular cortex of the macaque monkey. Behav Brain Res 18:125–137

Matelli M, Luppino G, D'Amelio M, Fattori P, Galletti C (1995) Frontal projections of a visual area (V6A) of the superior parietal lobule in the macaque monkey. Soc Neurosci (Abstr) 21 (1): 410

Mel BW (1991) A connectionist model may shed light on neural mechanisms for visually guided reaching. J Cogn Neurosci 3:273–292

Motter BC, Mountcastle VB (1981) The functional properties of light-sensitive neurons of the posterior parietal cortex studied in waking monkeys: Foveal sparing and opponent vector organization. J Neurosci 1:3–26

Mountcastle VB (1995) The parietal system and some higher brain functions. Cereb Cortex 5:377–390

Mountcastle VB, Lynch JC, Georgopoulos AP, Sakata H, Acuña C (1975) Posterior parietal association cortex of the monkey: Command functions for operations within extrapersonal space. J Neurophysiol 38:871–908

Pandya DN, Kuypers HGJM (1969) Cortico-cortical connections in the rhesus monkeys. Brain Res 13:13–36

Pandya DN, Seltzer B (1982) Intrinsic connections and architectonics of posterior parietal cortex in the rhesus monkey. J Comp Neurol 204:196–210

Petrides M, Pandya DN (1984) Projections to the frontal cortex from the posterior parietal region in the rhesus monkey. J Comp Neurol 228:105–116

Pons TP, Kaas JH (1986) Corticocortical connections of area 2 of somatosensory cortex in monkeys: A correlative anatomical and physiological study. J Comp Neurol 248:313–335

Savaki HE, Kennedy C, Sokoloff L, Mishkin M (1993) Visually guided reaching with the forelimb contralateral to a "blind" hemisphere: A metabolic mapping study in monkeys. J Neurosci 13:2772–2789

Schwartz AB, Ketner RE, Georgopoulos AP (1988) Primate motor cortex and free-arm movements to visual targets in three-dimensional space. 1. Relations between single cell discharge and direction of movement. J Neurosci 8:2913–2927

Schwartz ML, Goldman-Rakic PS (1984) Callosal and intrahemispheric connectivity of the prefrontal association cortex in rhesus monkeys: relation between intraparietal and principal sulcal cortex. J Comp Neurol 226:403–420

Seal J, Gross C, Bioulac B (1983) Different neuronal populations within area 5 of the monkey. Exp Brain Res (Suppl) 7:156–163

Soechting JF, Flanders M (1992) Moving in three-dimensional space: Frames of reference, vectors, and coordinate systems. Ann Rev Neurosci 15:167–191

Strick PM, Kim R (1978) Input to primate motor cortex from posterior parietal cortex (area 5). I. Demonstration by retrograde transport. Brain Res 157:325–330

Tanné J, Boussaoud D, Boyer-Zeller N, Rouiller EM (1995) Direct visual pathways for reaching movements in the macaque monkey. Neuroreport 7:267–272

Weinrich M, Wise SP (1982) The premotor cortex of the monkey. J Neurosci 2:1329–1345

Parietal Visual Neurons Coding Three-Dimensional Characteristics of Objects and Their Relation to Hand Action

H. SAKATA[1], M. TAIRA[1], A. MURATA[1], V. GALLESE[2], Y. TANAKA[1], E. SHIKATA[1], and M. KUSUNOKI[1]

[1]Department of Physiology, School of Medicine, Nihon University, Tokyo, Japan
[2]Istituto di Fisiologia Umana, Università degli Studi di Parma, Parma, Italy

Introduction

One of the most remarkable symptoms resulting from parietal lobe lesion in humans is optic ataxia, first described by Bálint in 1909. Patients with optic ataxia cannot grasp objects presented visually within the reach of their arms, suggesting that the parietal cortex plays an important role in visually guided hand and arm movement. The disturbances due to parietal lobe lesion are not limited to misreaching but include disturbances in preshaping of the hand according to the shape, size and orientation of objects in space (Jeannerod 1986; Perenin and Vighetto 1988).

Neurophysiological studies of alert behaving monkeys have revealed that many neurons in the posterior parietal cortex are activated during visually guided reaching and grasping (Mountcastle et al. 1975; Hyvärinen and Poranen 1974; Kalaska et al. 1983; Taira et al. 1990). Neurons of the posterior parietal cortex are also known to be involved in the neural coding of the spatial position of the target (Sakata et al. 1980; Andersen and Mountcastle 1983; Galletti et al. 1993). However, little is known about the visual processing required for discriminating three-dimensional (3D) features of objects to enable matching of the shape and orientation of the hand to the object to be grasped. Moreover, according to the concept of two visual cortical pathways (Ungerleider and Mishkin 1982; Mishkin et al. 1983), "object vision" is attributed to the ventral pathway and the inferotemporal cortex, whereas "space vision" is attributed to the dorsal pathway and the parietooccipital cortex.

In our previous investigation of hand manipulation task-related neurons we found visually responsive neurons in the anterior part of the posterolateral bank of the intraparietal sulcus (anterior intraparietal area, area AIP) that were activated by the sight of objects for manipulation. Many of these neurons were selective for the shape, orientation and/or size of objects. Where are these visual signals of the spatial features of objects processed in the cerebral cortex? One possibility is that the visual signals concerning the shape of objects processed in the inferotemporal

In: Parietal Lobe Contributions to Orientation in 3D Space (1997). P. Thier and H.-O. Karnath (eds). Springer-Verlag, Heidelberg.

cortex are sent to area AIP by way of long association fibers (Seltzer and Pandya 1984). Another possibility is that the visual signals concerning 3D features are processed in the dorsal pathway including the parietal cortex itself.

We recently found binocular visual neurons in the lateral bank of the caudal-intraparietal-sulcus neighboring area V3a that responded preferentially to a bar, plate or solid object in a particular orientation in space (Kusunoki et al. 1993, Sakata et al. 1995b; Shikata et al. 1995). We studied the visual properties of such neurons using a stereoscopic display of 3D computer graphics. The results, although still preliminary, suggest that there is a higher area of stereopsis in the posterior parietal cortex which integrates binocular disparity signals with monocular cue signals of shape to encode 3D features of objects (Sakata et al. 1995b). In this article we will review briefly the visual properties of hand manipulation task-related neurons in area AIP and describe the functional properties of visual neurons in the area around the caudal intraparietal sulcus. We will also address the issue of whether or not the neural representation of the 3D structure of objects is processed in the dorsal cortical visual pathway separately from that of the 2D image of the objects processed in the ventral cortical visual pathway.

Visual Responses of the Hand Manipulation Task-Related Neurons

The cortical neurons that are involved in visually guided reaching and grasping were first recorded in the inferior parietal lobule by Mountcastle et al. (1975) and Hyvärinen and Poranen (1974). They classified these neurons into two groups: "arm-projection" and "hand-manipulation" neurons. More recently, neurons that are involved in hand movement were found to be concentrated in a small zone within the rostral part of the posterolateral bank of the intraparietal sulcus (IPS), designated area AIP (Sakata et al. 1995a; Jeannerod et al. 1995). This area is strongly interconnected with area F5 of the inferior premotor cortex (Matelli et al. 1986), in which Rizzolatti et al. (1988) recorded "grasping-with-the-hand" neurons. We recorded the activity of the neurons of this area in monkeys that had been trained to manipulate various types of switches (some are shown in Fig. 1; Taira et al. 1990; Sakata et al. 1995a). Many of these hand manipulation task-related neurons were highly selective and were preferentially activated during the manipulation of one of four routinely used objects.

Separation of Visual and Motor Components

In order to determine the contribution of visual factors to the activation of these neurons, we let the monkey perform the same task in the dark, guided only by a small spot of light on the object (Taira et al. 1990; Sakata et al. 1992b, 1995a).Thus, the task-related neurons were classified into three groups according to the difference between the level of activity during manipulation of objects in the light and that during manipulation of objects in the dark: "motor-dominant"

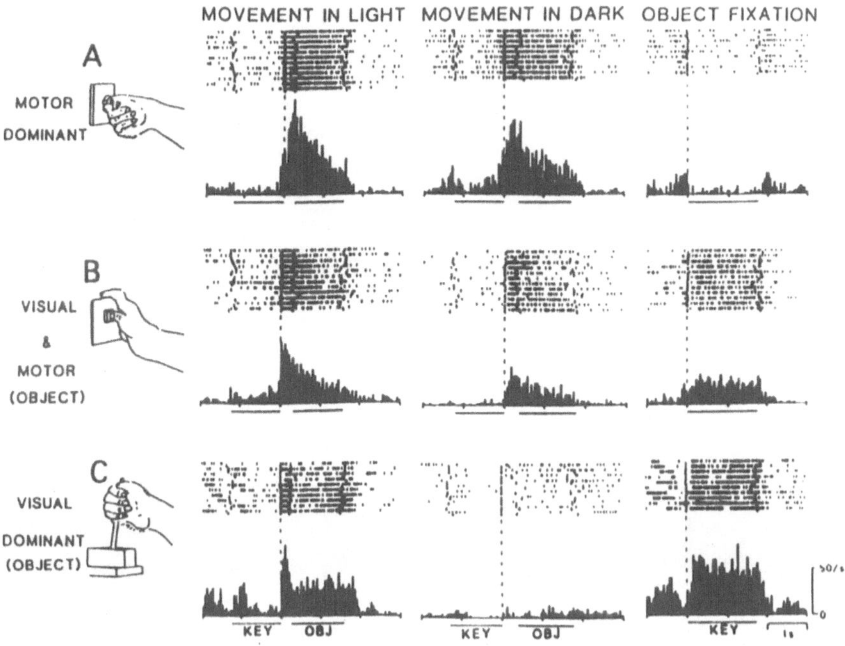

Fig. 1 A–C. Types of neurons in monkey anterior intraparietal area (area AIP) that are involved in hand manipulation. Activity of cells during hand manipulation in the light and in the dark, as well as during visual fixation of objects, is shown with *rasters* and *histograms*. **A** "Motor-dominant" neuron that preferred an open "pull knob". The cell showed almost equal levels of activity during manipulation in the light and in the dark but was not activated during object fixation. **B** An object-type "visual-and-motor" neuron that preferred a "push button". The cell was less active during manipulation in the dark than in the light, and was weakly activated during object fixation. **C** An object-type "visual-dominant" neuron that preferred an upright "pull lever". The cell was not activated during hand movement in the dark but was strongly activated during fixation of the object in the light. *Key* indicates the duration of pressing of the anchor key before movement toward the object. *Obj* indicates the duration of holding of the object to keep the switch on. (From Taira et al. 1990; Sakata et al. 1992b)

neurons (Fig. 1A) did not show any significant difference in the level of activity between manipulation in the light and that in the dark; "visual-and-motor" neurons (Fig. 1B) were less active during manipulation in the dark than during that in the light; and "visual-dominant" neurons (Fig. 1C) were active exclusively during manipulation in the light. Many of these visually responsive neurons were activated by the sight of objects during fixation without grasping (object-type, Fig. 1B, C). Other neurons were not activated during object fixation (nonobject type) but seemed to require other visual stimuli such as the sight of the moving hand to be activated.

Object Selectivity of Visual-Dominant Neurons

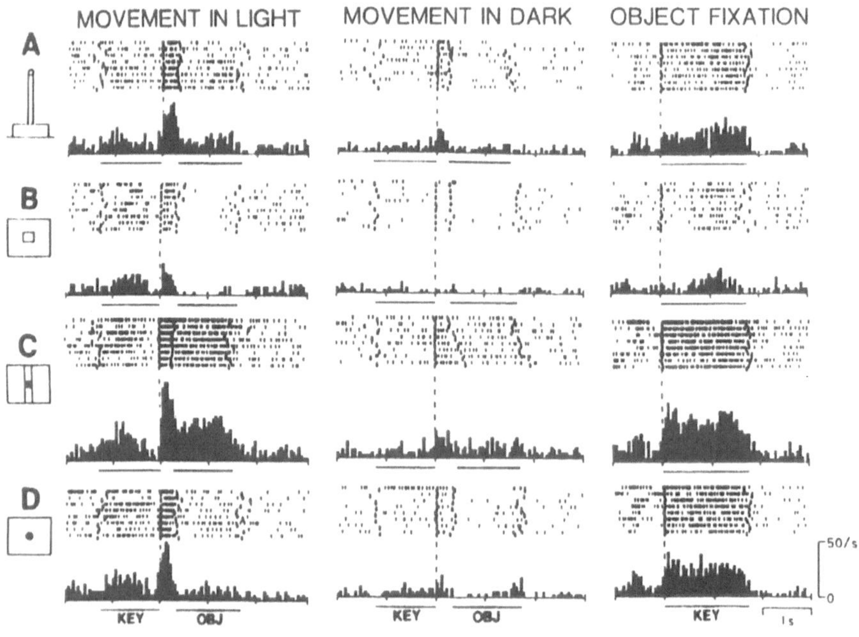

Fig. 2. Activity profile of an object-type visual-dominant neuron under three task conditions (*columns*) for four objects (*rows*) shown with *rasters* and *histograms*. *Left column*: object manipulation in the light; *middle column*: object manipulation in the dark; *right column*: object fixation in the light. All records are from the same neuron. The objects used for the tasks are shown on the *left*; from the *top*: pull-lever, push button, knob in groove, pull knob. *Key* and *Obj* as in Fig. 1

We examined activity profiles of these neurons across different objects under three task conditions: movement in light, movement in dark and object fixation in light. About one quarter of the cells were found to be highly selective; the activity of these cells for a particular object was significantly stronger than that for any other objects. The activity profile of one highly selective "visual-dominant" neuron is shown in Fig. 2. The cell was activated most strongly during manipulation of the knob in groove in the light (Fig. 2, left column). The activity profile of the cell across different objects during object fixation in the light (Fig. 2, right column) was similar to that during manipulation in the light. However, no significant activation occurred during manipulation in the dark (Fig. 2, center column). Thus, the cell was weakly active during fixation of the lever or the knob, but a maximum response was obtained during the fixation of the knob in groove. The activity during manipulation of the pull-lever or pull-knob was transient, probaly due to the occlusion of the object by the grasping hand. This example suggests that some visual-dominant neurons prefer complicated structures with two or more components. We also recorded highly selective visual-dominant neurons that preferred the pull-lever or push-button. Most of the cells that preferred the pull-lever showed selectivity in the orientation of the axis of the lever. However, it was not clear in this series of experiments whether or not the visual-dominant neurons were discriminating the 3D shape of the object as a whole, because these objects were designed primarily for eliciting different types of grips such as a precision grip, whole-hand grip, or side grip, and differed from each other only in the part to be pulled or pushed.

Selectivity in the 3D Shape of Objects

In more recent experiments on hand-manipulation task-related neurons (Sakata et al. 1992a; Murata et al. 1993) we used six different objects of simple geometric shape: sphere, cone, cylinder, cube, ring and square plate (Fig. 3). We used three sets of six shapes of different sizes: small, medium and large. These objects connected with microswitches were set in six sectors of a turntable and presented in random order one at a time. Since the objects were painted white they stood out against the black background. The animal was required to grasp and pull the object to turn the microswitch on or fixate the spot reflected on a half mirror and superimposed on the object. The following description is based on the observation of 112 hand manipulation task-related neurons recorded during this task (Murata et al. 1993). More than one quarter of the hand manipulation task-related neurons (32/112) were highly selective for one particular object. The activity profile of one highly selective object-type visual-dominant neuron is shown in Fig. 3. The cell responded preferentially to the view of the square plate among the six objects. We found highly selective visual-dominant neurons for every one of the six objects. The square plate and circular ring were the most commonly preferred

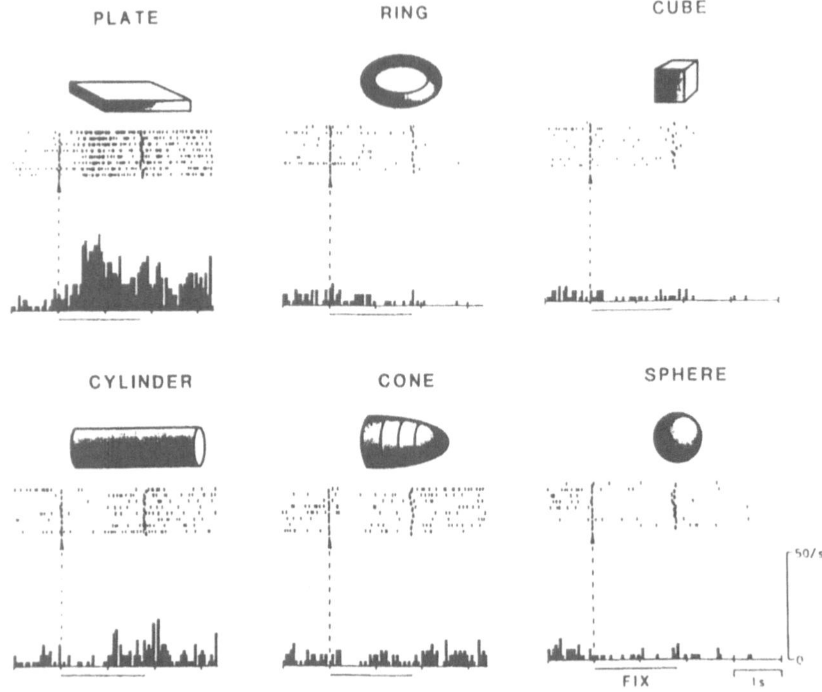

Fig. 3. Activity profile of an object-type visual-dominant neuron during object fixation (*Fix*) in the light for six objects of simple geometric shape: square plate, ring, cube, cylinder, cone and sphere

objects, and many of the cells that preferred these two objects showed selectivity in the orientation of the plate or ring. We also found moderately selective neurons that responded to two or more objects equally well (55/112). Some of these moderately selective neurons showed preference for a certain category of geometric shapes such as round objects (sphere, cone and cylinder), angular objects (cube and square plate), flat objects (plate and ring), or short small objects (small cube, sphere and cone). These results suggested that the visually responsive neurons in area AIP may represent spatial characteristics of objects for manipulation, and at least some of these neurons may represent the 3D shape of the object, categorized into a limited number of simple geometric shapes. This is consistent with the hypothesis of recognition-by-component (RBC) proposed by Biederman (1987). Biederman speculated that the perceptual recognition of objects is conceptualized as a process in which the image of the input is segmented into an arrangement of simple geometric components, such as (rectangular) blocks, cylinders, wedges and cones which are called geons.

However, it is not clear where and how the visual signals are processed in the cortical visual pathways to encode 3D shapes.

Axis-Orientation Selective Neurons

Discrimination of axis orientation in egocentric space or a viewer-centered coordinate system is important for manipulation of objects with the hand and may be dissociated from perception of the visual axis (Perenin and Vighetto 1988; Goodale et al. 1991; Milner et al. 1991). Indeed, we found that most of the cells

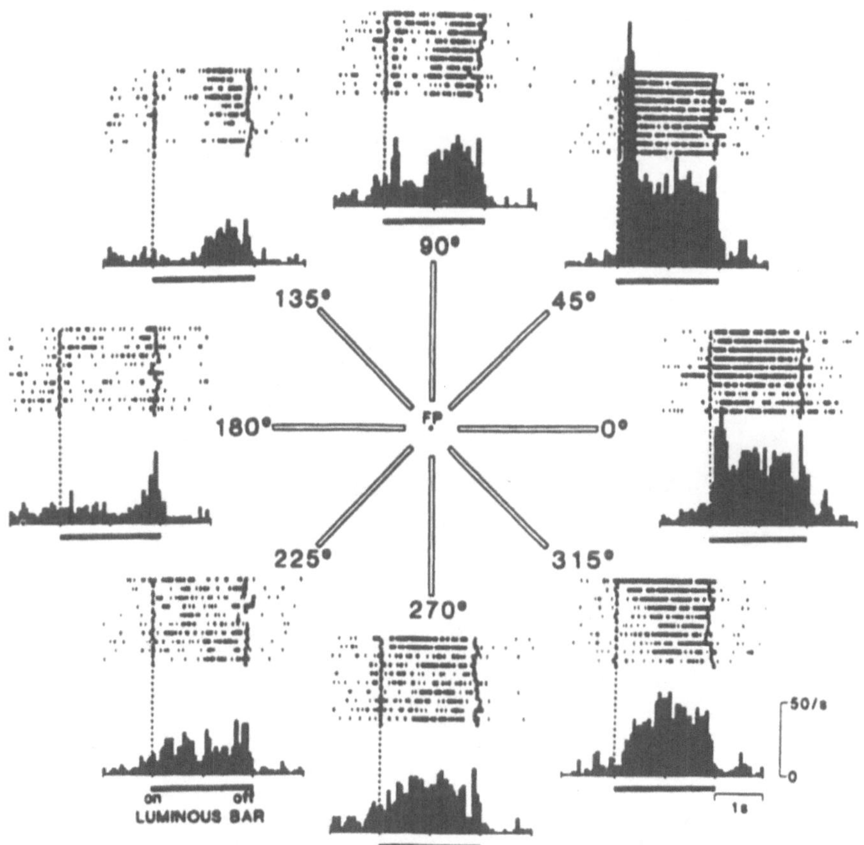

Fig. 4. Response of an object-type visual-dominant neuron, prefering the pull-lever, to the visual stimulus of a luminous bar presented in eight different orientations around the fixation point (*FP*)

that preferred the pull-lever were selective in the axis-orientation of the metal rod. Some of the visual-dominant neurons that were selective in the orientation of the lever were activated by the visual stimulus of a luminous acrylic bar and showed remarkable orientation preference as shown for one cell in Fig. 4. This cell responded most strongly to the bar tilted 45° upward and to the right of the fixation point, but much less to the bar tilted 45° down and to the left (225°). Later we changed the orientation of the bar at the same location in the visual field and in the sagittal or horizontal plane as well as in the frontoparallel plane. During a further investigation, we found a group of neurons that showed selectivity in the

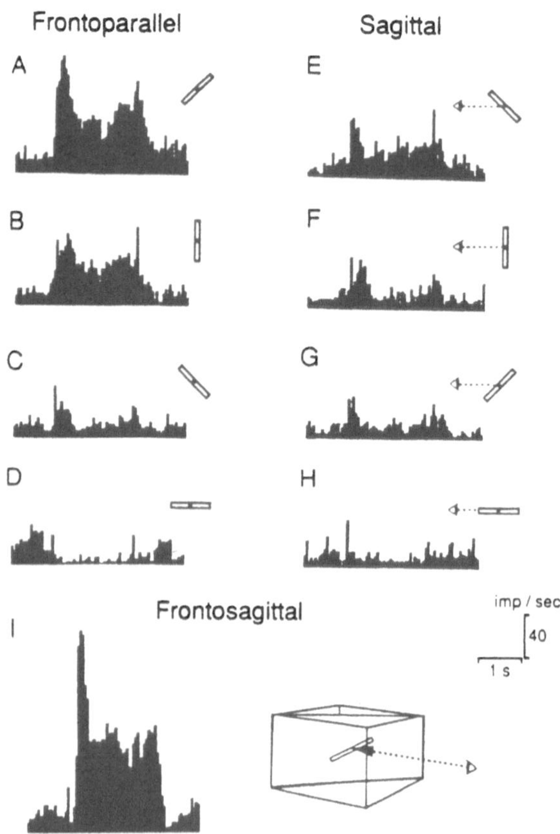

Fig. 5 A–I. An example of the responses of an axis-orientation selective (AOS) neuron of the inferior parietal lobule of the monkey. **A–D** Responses to the luminous bar in various orientations in the frontal plane (*left*) at 45° steps. **E–H** Responses to the luminous bar in various orientations in the sagittal plane at 45° steps. **I** The strongest responses, to the bar tilted 45° to the right and forward

orientation of the luminous bar in the lateral bank of the caudal part of the intraparietal sulcus. The following description is based on the observation of 23 axis-orientation selective (AOS) neurons recorded in this area (Kusunoki et al. 1993). Figure 5 shows an example of the responses of an AOS neuron that responded most strongly to a diagonal bar tilted 45° forward and 45° to the right (Sakata and Taira 1994). The cell preferred the bar tilted 45° to the right in the frontoparallel plane (Fig. 5a), but also responded to the bar tilted 45° forward in the sagittal plane (Fig. 5e). It showed the strongest response, however, to the bar tilted 45° forward in the frontosagittal plane (Fig. 5i). We also recorded AOS neurons that preferred the bar tilted either forward or backward in the sagittal plane, and one that responded most strongly to the bar along the sagittal axis. It was clear that these neurons had orientation selectivity in 3D space in a viewer-centered coordinate system. Most of these AOS neurons were binocular visual neurons, and they responded much less strongly under monocular viewing conditions. The discharge rate of most of the cells increased monotonically with increasing length of the stimulus, and when we changed the width of the slit or the thickness of the bar we found that most of the cells preferred a narrower (or thinner) stimulus. However, most of the AOS neurons showed the same orientation preference across a wide range of change in thickness and length. Most of them also had wide receptive fields, sometimes more than 50° x 50°, and their responses were position invariant.

These results suggest that the axis-orientation selective neurons represent the orientation of the longitudinal axes of objects in 3D space. In human parietal patients considerable shift of vertical and horizontal axes toward the contralesional side was reported by Bender and Jung (1948). McFie et al. (1950) described a similar symptom in several patients with a right occipitoparietal lesion. Similar deficits in the perception of line orientation due to parietal lobe lesion were reported recently by von Cramon and Kerkhoff (1993). Therefore, the discrimination of the axis orientation is one of the prominent functions of the parietal cortex in the domain of space perception. Perception of axis orientation may also be important for recognition of objects; Warrington and Taylor (1973) reported that their patients with right parietal lesion had difficulty recognizing common objects in unconventional views in which the main axis of the objects was foreshortened (see also Humphreys and Riddoch 1984). David Marr (1982) suggested that this difficulty may be due to the inability to perceive the main axis, and emphasized that axis-based description is important for the representation of the 3D structure of objects.

Surface-Orientation Selective Neurons

According to Marr's theory of vision (Marr 1982), the main purpose of vision is object-centered representation of the 3D shape and spatial arrangement of an object. The main stepping stone toward this goal is representing the geometry of the visible surface. Therefore, there should be some area in the visual cortical pathways to represent surface orientation and curvature, if 3D shape is to be represented somewhere in the cerebral cortex.

In our recent experiments on hand manipulation task-related neurons, we found that some visual-dominant neurons that preferred the square plate showed selectivity in the orientation of the plate. This suggested that some of the parietal visual neurons may discriminate surface orientation. Since a disparity gradient is the most important cue for perceiving the orientation of the surface in depth, we used a stereoscopic display of 3D computer graphics and performed an extensive study of binocular neurons in the caudal part of the lateral bank of the IPS that were sensitive to changes in surface orientation. The following description is based on a study of 36 parietal visual neurons recorded in three hemispheres of two Japanese monkeys (Macaca fuscata; Shikata et al. 1995). The monkey, wearing a pair of polarizing glasses, fixated on a small spot at the center of a stereoscopic display located 100 cm from the eyes (Fig. 6). We presented a square plate or checkerboard in various orientations and at various distances and changed stimulus parameters such as width and thickness.

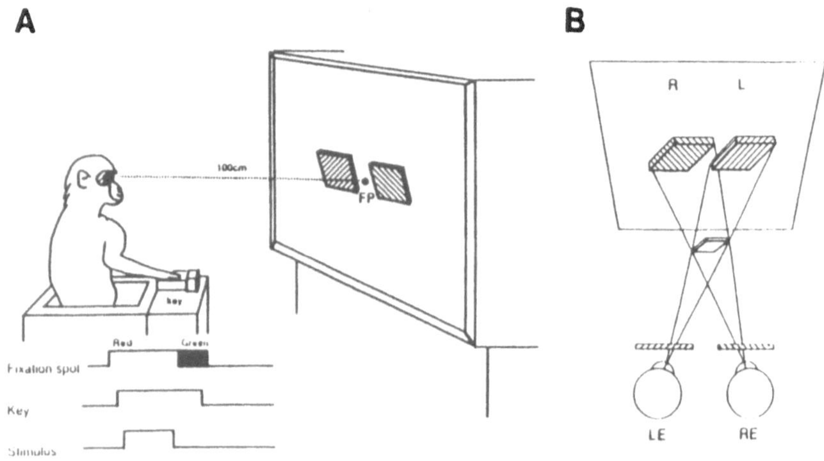

Fig. 6. A Experimental setup for stereoscopic stimulation using 3D computer graphics. The monkey wearing polarizing glasses fixated on a spot at the center (*FP*) of a stereoscopic display on which a pair of polarized images for the left eye and the right eye, respectively, were presented. **B** A pair of polarized images with disparity are fused at a depth closer to the eye than the screen. *R* and *L* are polarized stimulus for right eye (*RE*) and left eye (*LE*) respectively

Since we found the parietal visual neurons that responded to a square plate or luminous checkerboard in the same area as the axis-orientation selective neurons, we first compared the response of the cells to a flat stimulus with that to an elongated stimulus. Seventeen cells that preferred the flat to the elongated stimulus (17/32) showed selectivity in surface orientation of the stimulus presented on the screen of the stereoscopic display. Thirteen of these cells (13/17) were defined as surface-orientation selective (SOS) neurons with a criterion of high orientation index (O.I. 2.0), with O.I.=response to optimal surface/response to orthogonal surface. Figure 7 shows orientation tuning of an SOS neuron that responded most strongly to a square plate tilted 45° to the left (Fig. 7A). No response was obtained when a stimulus was presented either to the left (Fig. 7B) or to the right eye. Almost all SOS neurons responded more strongly to the binocular than to the monocular stimulus. The responses of all of the SOS neurons tested (N=6) changed with the depth on the stereoscopic display (we changed the

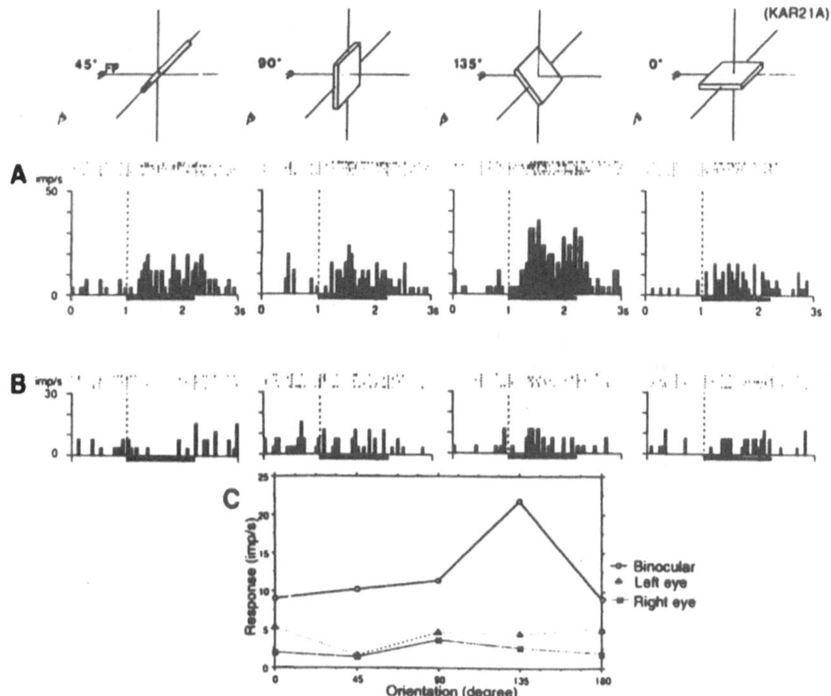

Fig. 7 A–C. Orientation tuning of a surface-orientation selective (SOS) neuron. **A** Responses to the square plate in various orientations around the sagittal axis at 45° steps, under binocular viewing conditions. **B** Responses to the same set of stimuli under monocular viewing conditions (left eye). **C** Orientation response curve for binocular and monocular viewing conditions

disparity while keeping the size of the stimulus on the screen constant). Half of the SOS neurons responded to the near stimulus with crossed disparity, and the other half responded to the far stimulus with uncrossed disparity. None of these neurons showed sharp tuning in depth, but their depth-tuning curves showed a plateau, similar to a near cell or far cell defined by Poggio and Fischer (1977), except that the range of disparity for excitatory response was larger than that in the case of the disparity-selective neurons in V1 and V3a (Poggio et al. 1988).

In order to verify that the SOS neurons responded to an extended surface rather than the edge or corner, we studied the effect of the change in the width of the stimulus. In all three cells tested, the magnitude of the response of the cell to the stimulus decreased gradually when the width was reduced and the stimulus changed from a square to a narrow rectangular plate or bar. Thus, the width-response curve of the SOS neurons was a monotonically increasing function in contrast to that of the AOS neurons which was a monotonically decreasing function. We also studied the effect of changes in the thickness of the stimulus in order to verify that SOS neurons preferred a flat stimulus to a thick voluminous one. When the thickness was increased, the magnitude of the response of two of the four cells tested decreased, whereas that of the other two cells remained unchanged. Therefore, the thickness was not as critical as the width and orientation for the activation of the SOS neurons.

There is not much clinical evidence that the parietal cortex is involved in the perception of surface orientation, except for the confusion of planes in the drawings of patients with right parietooccipital lesions (Paterson and Zangwill 1944; McFie et al. 1950; Ettlinger et al. 1957), e.g., the frame of a bicycle was drawn in the frontal plane while the wheels were drawn in the horizontal plane. However, disturbances in the ability to draw 3D shapes such as houses (Piercy and de Ajuriaguerra 1960) or copy block designs (Critchley 1953) may be partly due to a disturbance in the ability to perceive surface orientation.

Selectivity in 3D Shape

During the study of axis-orientation selective neurons and surface-orientation selective neurons we found in the same area a few neurons that were different in type from both of these types of neurons. These were the cells that preferred a solid stimulus to a flat or thin stimulus. Figure 8 shows the responses of one such cell which was considered a possible AOS neuron. When we examined the effect of changes in thickness, we found that the cell responded most strongly to a cylinder of intermediate thickness (10 cm in diameter; Fig. 8A). The length-response curve had a peak at an intermediate length (20 cm). A dramatic change was observed when we changed the shape of the stimulus (Fig. 8B). The cell responded most strongly to a cylinder (10 cm x 20 cm) tilted backward and to the right, but a square column of the same size and orientation elicited only a weak

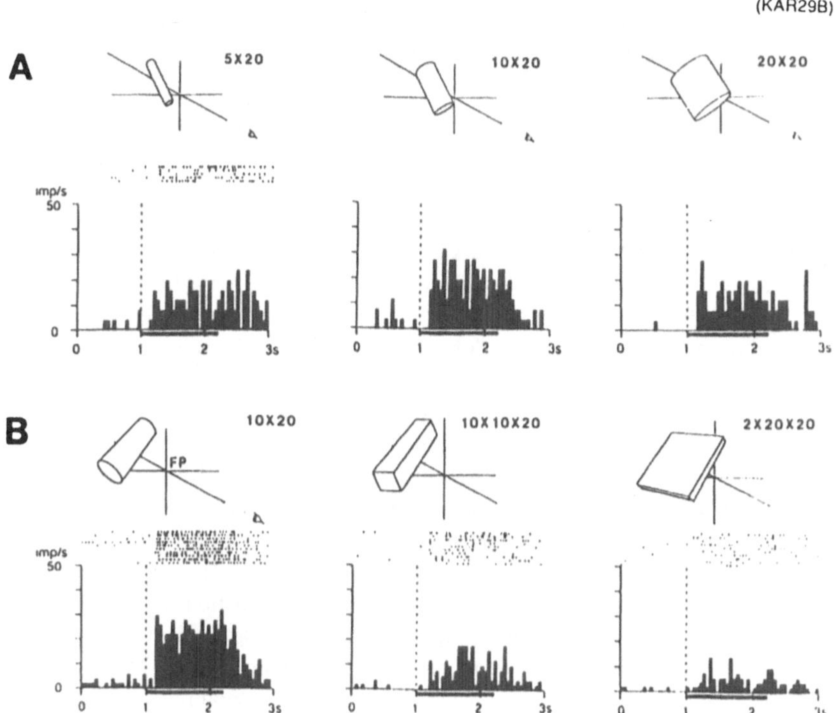

Fig. 8 A, B. Responses of an axis-orientation selective neuron that preferred a cylinder of intermediate thickness and length. **A** Responses to the 3D stimuli with various diameters (5–20 cm) at 100 cm. **B** Responses to various shapes: a cylinder, square column and plate. *P*, fixation point

response, and a square plate (20 cm x 20 cm) in the same orientation was even less effective than the square column. The cell was strongly binocular, showing almost no response to monocular stimulation, and was activated only by stimuli nearer than the fixation point. Thus, the cell was likely to discriminate a cylinder from a square column on the basis of surface curvature. This supports the hypothesis that there is a class of binocular stereoscopic neurons in the lateral bank of the caudal IPS that discriminates 3D shapes of objects by stereopsis.

There is no direct clinical neuropsychological evidence that the parietal cortex is involved in the perception of 3D shape. However, if one examines the drawings of a patient with a right parietal lobe lesion, disturbances in the ability to draw 3D objects such as a cube or a house may be clearly seen (for example Hécaen et al. 1956). The results of a number of studies on constructional apraxia due to parietooccipital lesion suggest that the right hemisphere supplies a perceptual component and the left hemisphere an executive component to the visuoconstructional task (Warrington 1969). Therefore, the symptom of constructional apraxia due to the parietal lesion may be indirect evidence that the parietal cortex is involved in the perception of 3D shape.

Summary and Conclusion

Responses of parietal neurons to the sight of a 3D object were first recorded in studies of hand-manipulation-task-related neurons in area AIP (Taira et al. 1990; Sakata et al. 1992b; Sakata et al. 1995a). We classified those cells which were activated during object fixation as "object-type", "visual-dominant" and "visual-and-motor" neurons. The selectivity of their responses among several different objects suggested that they may discriminate spatial characteristics of objects such as shape and/or orientation. In the second series of experiments (Sakata et al. 1992a; Murata et al. 1993) on hand-manipulation-task-related neurons, we used six different objects of simple geometric shape. More than a quarter of the object-type, visual-dominant and visual-and-motor neurons were highly selective in the shape of the object, preferring one of the six different shapes: cylinder, cone, sphere, cube, square plate and ring. The results suggested that at least some of these neurons may discriminate simple geometric shapes, as proposed by Biederman (1987) in his hypothesis of recognition by components. More recently we recorded a number of binocular visual neurons in the lateral bank of the caudal IPS that responded to static visual stimuli (Kusunoki et al. 1993; Sakata et al. 1995b; Shikata et al. 1995). One type of neurons were the AOS neurons that showed clear tuning in a particular orientation in 3D space in their response to elongated stimuli (luminous acrylic bar; Kusunoki et al. 1993). Most of the AOS neurons were binocular, responding more strongly under binocular than under monocular viewing conditions, and they preferred long and thin stimuli to short and thick ones.

The second type of neurons were the SOS neurons that preferred flat to elongated objects. We studied the functional properties of SOS neurons with the stimulus presented on a stereoscopic display. The SOS neurons were also binocular and responded to the stereoscopic stimulus with binocular disparity much more strongly than to the monocular stimulus. They responded to an extended surface, and the magnitude of their response increased with the width of the stimulus. The third type of neurons were the cells that preferred a solid stimulus to a flat or thin stimulus and showed selectivity in the 3D shape of the stimulus. We have recorded only a few of these neurons so far, but there may be many in the parietal cortex since many of the object-type visual-dominant neurons of area AIP showed high selectivity in simple 3D shape. These results suggest that the posterior parietal cortex is more important for the perception of 3D structure and orientation, mediated by stereopsis, than the inferotemporal cortex. Neurons

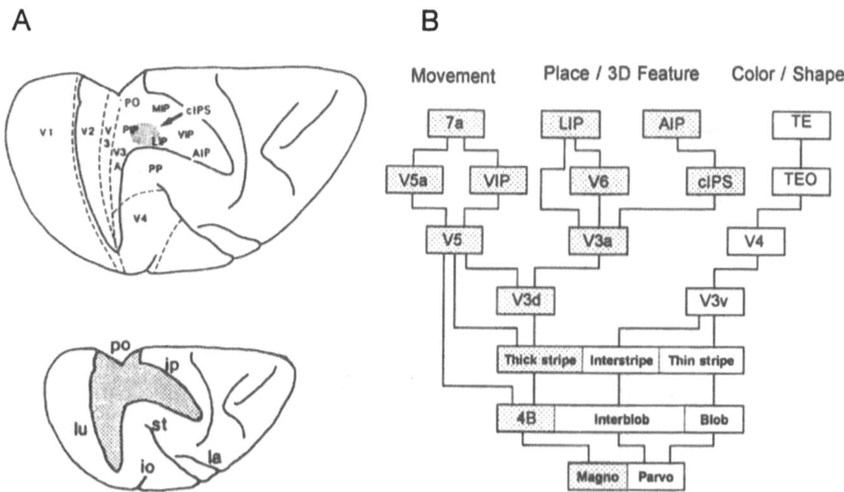

Fig. 9. A Site of recording of axis-orientation selective neurons and surface-orientation selective neurons. Lateral bank of the caudal intraparietal sulcus (*cIPS*) in a diagram showing the inside of *IPS* and lunate sulcus. Modified from Colby et al. 1988. **B** Hierarchy of visual areas, modified from the diagrams of Felleman and Van Essen 1991; and Zeki and Shipp 1988 *PO*: parieto occipital area. *PIP*, posterior intraparietal area; *LIP*, lateral intraparietal area; *MIP*, medial intraparietal area; *VIP*, ventral intraparietal area; *PP*, posterior parietal area; *V1-5*; primary, secondary, third, fourth and fifth visual areas; *lu*, lunate sulcus; *po*, parietooccipital sulcus; *ip*, intraparietal sulcus; *st*, superior temporal sulcus; *io*, inferior occipital sulcus; *la*: lateral fissure

of the inferotemporal cortex (areas TE and TEO) in anesthetized monkeys responded to simplified 2D images of objects almost as strongly as to the sight of real objects (Tanaka et al. 1993). Most of the inferotemporal neurons were view-selective, and the view-invariant recognition of objects might be achieved by interpolation of a set of 2D templates or prototypes (Logothetis and Pauls 1995). Although this may facilitate identification of objects, it is inconsistent with Marr's theory of vision postulating object-centered representation of 3D shape. According to Marr (1982), perception of surface geometry is an essential step in perceiving 3D structure, and perception of the orientation of the natural axis is important for the representation of 3D objects. Thus, the SOS and AOS neurons of the parietal cortex may help achieve the final goal of 3D representation.

Therefore, it may be concluded that the function of the parietal cortex in the dorsal cortical visual pathway is not limited to the perception of position and movement but also includes the perception of 3D shape and the orientation of objects in space mediated by stereopsis, and that the major purpose of the 3D representation of objects in the parietal cortex is the visual guidance of hand action as suggested by Goodale and Milner (1992; see also Goodale and Murphy, this volume) rather than the identification of objects. Binocular disparity signaling for stereopsis is mediated by the magnocellular system and the dorsal pathway through thick stripes of the V2 and V3-V3a complex in which many disparity-sensitive neurons were recorded (Hubel and Livingstone 1987; Poggio et al. 1988; Zeki 1978; Felleman and Van Essen 1987).

Thus, the dorsal cortical visual pathway projecting to the parietal cortex may be subdivided into two systems: one subsystem is relayed at area V5 (MT) and V5a (MST), subserving motion vision, and the other is relayed at the V3-V3a complex subserving the perception of spatial position and the 3D structure of objects (Fig. 9, modified from the diagrams of Zeki and Shipp 1988, De Yoe and Van Essen 1988 and Felleman and Van Essen 1991). The area in the lateral bank of the caudal IPS (cIPS) is especially important for the higher-order processing of stereopsis for the perception of 3D shape and orientation.

References

Andersen RA, Mountcastle VB (1983) The influence of the angle of gaze upon the excitability of the light-sensitive neurons of the posterior parietal cortex. J Neurosci 3:532–548

Bálint R (1909) Seelenlähmung des "Schauens", optische Ataxie, räumliche Störung der Aufmerksamkeit. Monatsschr Psychiat Neurol 25:51–81

Bender MB, Jung R (1948) Abweichungen der subjektiven optischen Vertikalen und Horizontalen bei Gesunden und Hirnverletzten. Arch Psychiat Nervenkr 181:193–212

Biederman I (1987) Recognition-by-component: a theory of human image understanding. Psychol Rev 94:115–147

Colby CL, Gattass R, Olson CR, Gross CG (1988) Topographical organization of cortical afferents to extrastriate visual area PO in the macaque: a dual tracer study. J Comp Neurol 269:392–413

Critchley M (1953) The parietal lobes. Arnold, London

De Yoe EA, Van Essen DC (1988) Concurrent processing streams in monkey visual cortex. Trends Neurosci 11:219–226

Ettlinger G, Warrington E, Zangwill OL (1957) A further study of visual-spatial agnosia. Brain 80:335–361

Felleman DJ, Van Essen DC (1987) Receptive field properties of neurons in area V3 of macaque monkey extrastriate cortex. J Neurophysiol 57:889–920

Felleman DJ, Van Essen DC (1991) Distributed hierarchical processing in the primate cerebral cortex. Cereb Cortex 1:1–47

Galletti C, Battaglini PP, Fattori P (1993) Parietal neurons encoding spatial locations in craniotopic coordinates. Exp Brain Res 96:221–229

Goodale MA, Milner AD (1992) Separate visual pathway for perception and action. Trends Neurosci 15:20–25

Goodale MA, Milner AD, Jakobson LS, Carey DP (1991) A neurological dissociation between perceiving objects and grasping them. Nature 349:154–156

Hécaen H, Penfield W, Bertrand C, Malmo R (1956) The syndrome of apractognosia due to lesions of the minor cerebral hemisphere. Arch Neurol Psychiatry 75:400–434

Hubel DH, Livingstone MS (1987) Segregation of form, color and stereopsis in primate area 18. J Neurosci 7:3378–3415

Humphreys GW, Riddoch MJ (1984) Routes to object constancy: implications from neurological impairment of object constancy. Q J Exp Psychol [A]36:385–415

Hyvärinen J, Poranen A (1974) Function of the parietal associative area 7 as revealed from cellular discharge in alert monkeys. Brain 97:673–692

Jeannerod M (1986) The formation of finger grip during prehension. A cortically mediated visuomotor pattern. Behav Brain Res 19:99–116

Jeannerod M, Arbib M, Rizzolatti G, Sakata H (1995) Grasping objects: the cortical mechanisms of visuomotor transformation. Trends Neurosci 18:314–320

Kalaska JF, Caminiti R, Georgopoulos AP (1983) Cortical mechanisms related to the direction of two dimensional arm movements: relations in parietal area 5 and comparison with motor cortex. Exp Brain Res 51:247–260

Kusunoki M, Tanaka Y, Ohtsuka H, Ishiyama K, Sakata H (1993) Selectivity of the parietal visual neurons in the axis orientation of objects in space. Soc Neurosci Abstr 19:770

Logothetis NK, Pauls J (1995) Psychophysical and physiological evidence for viewer-centered object representations in the primate. Cereb Cortex 5:270–288

Matelli M, Camarda R, Glickstein M, Rizzolatti G (1986) Afferent and efferent projections of the inferior area 6 in the macaque monkey. J Comp Neurol 251:281–298

Marr D (1982) Vision. Freeman, San Francisco

McFie J, Piercy MF, Zangwill OL (1950) Visual-spatial agnosia associated with lesions of the right cerebral hemisphere. Brain 73:167–190

Milner AD, Perrett DI, Johnston RS, Benson PJ, Jordan TR, Heeley DW, Bettucci D, Mortara F, Mutani R, Terazzi E, Davidson DLW (1991) Perception and action in 'visual form agnosia'. Brain 114:405–428

Mishkin M, Ungerleider LG, Macko KA (1983) Object vision and spatial vision: two cortical pathways. Trends Neurosci 6:414–417

Mountcastle VB, Lynch JC, Georgopoulos A, Sakata H, Acuna C (1975) Posterior parietal association cortex of the monkey: command functions for operations within extrapersonal space. J Neurophysiol 38:871–908

Murata A, Gallese V, Kaseda M, Kunimoto S, Sakata H (1993) Hand-manipulation-related neurons of the parietal cortex of the monkey: further analysis of selectivity in shape, size and orientation of objects for manipulation. Jpn J Physiol 43 [Suppl 2] S251

Paterson A, Zangwill OL (1944) A case of topographical disorientation associated with a unilateral cerebral lesion. Brain 68:188–212

Perenin MT, Vighetto A (1988) Optic ataxia: a specific disruption in visuo-motor mechanism. I. Different aspects of the deficit in reaching for objects. Brain 111:643–674

Piercy MH, de Ajuriaguerra J (1960) Constructional apraxia associated with unilateral cerebral lesions left and right sided cases compared. Brain 83:225–242

Poggio GF, Fischer B (1977) Binocular interaction and depth sensitivity in striate and prestriate cortex of behaving rhesus monkey. J Neurophysiol 40:1392–1405

Poggio GF, Gonzalaz F, Krause F (1988) Stereoscopic mechanisms in monkey visual cortex: binocular correlation and disparity selectivity. J Neurosci 8:4531–4550

Rizzolatti G, Camarda R, Fogassi L, Gentilucci M, Luppino G, Matelli M (1988) Functional organization of inferior area 6 in the macaque monkey. II. Area F5 and the control of distal movement. Exp Brain Res 71:491–507

Sakata H, Murata A, Luppino G, Kaseda M, Kusunoki M (1992a) Selectivity of hand-movement-related neurons of the parietal cortex for shape, size and orientation of objects and hand grips. Soc Neurosci Abstr 18:504

Sakata H, Shibutani H, Kawano K (1980) Spatial properties of visual fixation neurons in posterior parietal association cortex of the monkey. J Neurophysiol 43:1654–1672

Sakata H, Taira M (1994) Parietal control of hand action. Curr Opin Neurobiol 4:847–856

Sakata H, Taira M, Mine S, Murata A (1992b) Hand-movement-related neurons of the posterior parietal cortex of the monkey: their role in the visual guidance of hand movements. In: Caminiti R, Johnson PB, Burnod Y (eds) Control of arm movement in space: neurophysiological and computational approaches. Springer, Berlin Heidelberg New York, pp 185–198 (Experimental Brain Research Series 22)

Sakata H, Taira M, Murata A, Mine S (1995a) Neural mechanisms of visual guidance of hand action in the parietal cortex of the monkey. Cereb Cortex 5:429–438

Sakata H, Taira M, Tanaka Y, Shikata E (1995b) Neural representation of 3D features of visual objects in the parietal association cortex of the monkey. 4th IBRO World Congress in Neuroscience, p 45, S29.2

Seltzer B, Pandya DN (1984) Further observations on parieto-temporal connections in the rhesus monkey. Exp Brain Res 55:301–312

Shikata E, Tanaka Y, Nakamura H, Taira M, Sakata H (1995) Selectivity of the parietal visual neurons in 3D orientation of surface of stereoscopic stimuli. Soc Neurosci Abstr 21:665

Taira M, Mine S, Georgopoulos AP, Murata A, Sakata H (1990) Parietal cortex neurons of the monkey related to the visual guidance of hand movement. Exp Brain Res 83:29–36

Tanaka K, Saito H, Fukada Y, Moriya M (1993) Coding visual images of objects in the inferotemporal cortex of the macaque monkey. J Neurophysiol 66:170–189

Ungerleider LG and Mishkin M (1982) Two cortical visual systems. In: Ingle DJ, Goodale MA, Mansfield RJW (eds) Analysis of visual behavior. MIT Press, Cambridge, pp 549–586

von Cramon D, Kerkhoff G (1993) On the cerebral organization of elementary visuo-spatial perception. In: Gulyas B, Ottoson D, Roland PE (eds) Functional organization of the human visual cortex. Pergamon, Oxford, pp 211–231

Warrington EK (1969) Constructional apraxia. In: Vinken PJ, Bruyn GW (eds) Handbook of clinical neurology, vol 4. North-Holland, Amsterdam

Warrington EK, Taylor AM (1973) The contribution of the right parietal lobe to object recognition. Cortex 9:152–164

Zeki SM (1978) The third visual complex of rhesus monkey prestriate cortex. J Physiol (Lond) 277:245–272

Zeki S, Shipp S (1988) The functional logic of cortical connections. Nature 335:311–317

A Parietal-Frontal Circuit for Hand Grasping Movements in the Monkey: Evidence from Reversible Inactivation Experiments

V. GALLESE[1], L. FADIGA[1], L. FOGASSI[1], G. LUPPINO[1], and A. MURATA[2]

[1]Istituto di Fisiologia Umana, Università degli studi di Parma, Parma, Italy
[2]1st Department of Physiology, Nihon University, School of Medicine, Tokyo, Japan

Introduction

Grasping objects is an action that we perform many times every day. Its frequent occurrence and the ease with which we usually perform it overshadow the complex problems that grasping entails and that the brain has to solve for its implementation.

Leaving aside the problems of grasping execution (hand-arm coordination, postural adjustments, force deployment, recruitment of ordered muscles), the initial process of object grasping requires that three-dimensional (3D) features of an object are matched by the appropriate shaping and orientation of the hand. In other words, visual information about the intrinsic features of an object must be translated into the corresponding distal movements patterns (see Arbib 1981; Jeannerod 1988). Where does this process of visuo-motor integration occur? The data presented in this article are an attempt at answering this question.

In primates, including humans, correct execution of hand grasping movements requires integrity of the primary motor cortex. Lesions of this area, as well as lesions of the pyramidal tract, produce a severe deficit in the control of individual finger movements, with disruption of grasping movements, especially when grip has to be accomplished by opposing the thumb and index finger (Tower 1940; Fulton 1949; Lawrence and Kuypers 1968; Passingham et al. 1983). Since the primary motor cortex receives little, if any, visual information, visuomotor transformations required for the execution of grasping movements have to occur upstream in motor control.

The two a priori requisites that a cortical area should fulfill in order to control grip formation are: (1) It must receive a suitable visual input; and (2) it must be endowed with a representation of distal movements. Recent neurophysiological data showed that in the macaque monkey there are two areas that fulfill these requirements. They are the anterior intraparietal area (area AIP) and the rostral sector of inferior area 6 (F5). Their location is shown in Figure 1.

In: Parietal Lobe Contributions to Orientation in 3D Space (1997). P. Thier and H.-O. Karnath (eds). Springer-Verlag, Heidelberg.

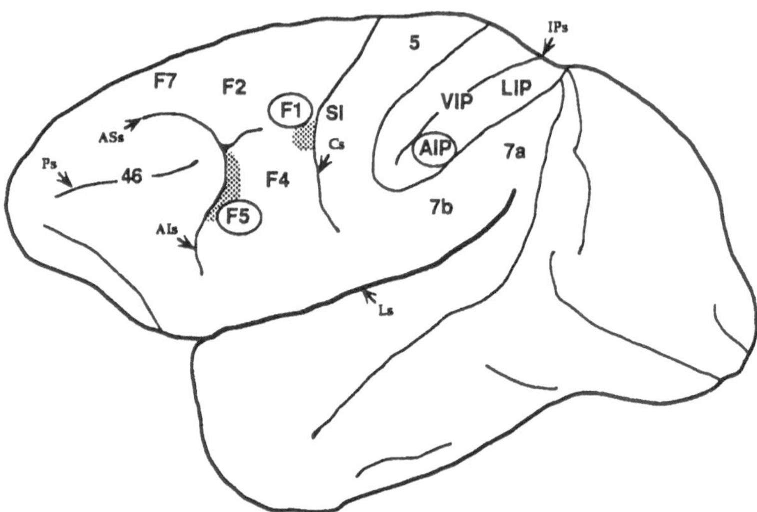

Fig. 1. Lateral view of the monkey brain. *Shaded areas* indicate the location of muscimol microinjections in areas *F5* and *F1* (primary motor cortex). Abbreviations: *AIP*, anterior intraparietal area; *AIs*, inferior arcuate sulcus; *ASs*, superior arcuate sulcus; *Cs*, central sulcus; *IPs*, intraparietal sulcus; *LIP*, lateral intraparietal area; *Ls*, lateral sulcus; *Ps*, principal sulcus; *VIP*, ventral intraparietal area. *IPs* has been opened to show hidden areas

In the present article we will review the functional properties of these two areas. In particular, the effect of their lesions will be discussed. The results of inactivation of F5, which have been published only in abstract form (Gallese et al. 1995), will be described in detail. Finally, after showing that the two areas are strictly connected and form a circuit involved in visuomotor transformations for grasping, we will present some hypotheses on how the AIP-F5 circuit might perform these transformations.

Posterior Parietal Cortex: Area AIP

Functional Properties

Earlier single unit studies of the parietal lobe in monkeys showed that many neurons in the inferior parietal lobule, along with other types of neurons, were related to hand manipulation (Hyvärinen and Poranen 1974; Mountcastle et al. 1975). The properties of these neurons have recently been extensively studied by

Sakata and his coworkers (Taira et al. 1990; Sakata et al. 1992; Sakata et al., this volume).

Monkeys were trained to manipulate, under visual guidance and in the dark, different types of objects, each requiring a peculiar type of grasping. By comparing neuron activity during task performance in the light and in the dark, task-related neurons were classified into three major categories: "motor dominant" neurons, whose activity was present and virtually equal in both conditions; "visual and motor" neurons that were less active during movement executed in the dark; "visual dominant" neurons active only during movement executed under visual guidance. Many of the neurons belonging to the last two categories also responded during object visual presentation. Particularly important was the observation that some neurons discharged selectively during a particular type of prehension (such as opposition of thumb and index finger) *and* during visual presentation of 3D

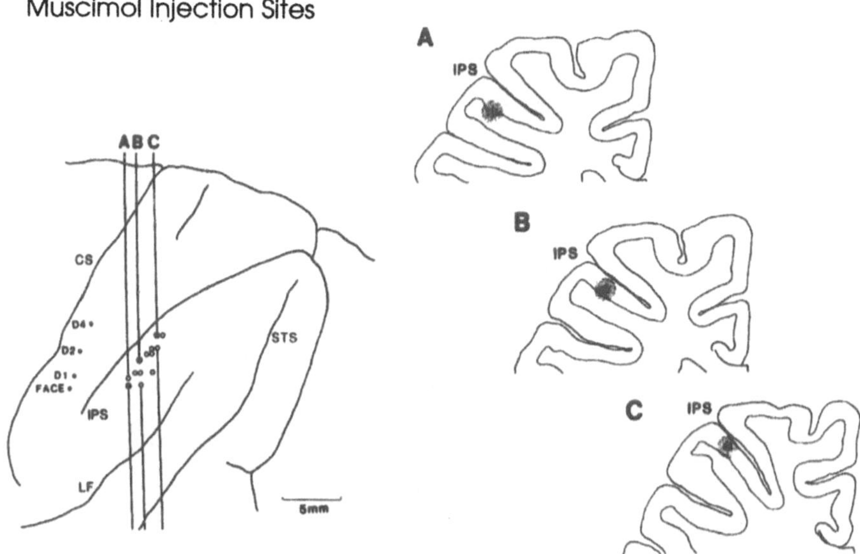

Fig. 2. Muscimol injection sites. *Left:* Lateral view of the parietal cortex of the monkey showing the location of single unit recordings and muscimol injection sites corresponding to levels *A*, *B*, and *C*. *Empty circles* indicate the loci where hand manipulation-related neurons were recorded. *Asterisks* indicate muscimol microinjection sites. Typically, a total amount of 15 μg muscimol (5 μg/μl) was injected at each site (for a more detailed description see Gallese et al. 1994). *Right*: Selected coronal sections *A*, *B*, and *C*, taken at the levels indicated by *vertical bars*, showing the microinjection sites in AIP. *Shaded areas* represent the local diffusion of Muscimol. Abbreviations: *CS*, central sulcus; *IPS*, intraparietal sulcus; *LF*, lateral fissure; *STS*, superior temporal sulcus. (Modified from Gallese et al. 1994)

objects whose size and/or shape was congruent with the coded type of grip (see Sakata and Taira 1994). As will be discussed later, these neurons might represent the neural correlate of the process of visuomotor transformation required for the execution of grasping movements.

Manipulation neurons are particularly concentrated within the rostral part of the lateral bank of the intraparietal sulcus (area AIP).

Inactivation Data

The effect on grasping behavior of the reversible inactivation of small sectors of AIP by microinjections of muscimol (a γ-aminobutyric acid, GABA, agonist) was recently studied by Gallese et al. (1994). They trained a monkey to reach for and grasp geometric solids of different sizes and shapes, each of which required a specific pattern of finger movements to be grasped smoothly. Hand movements during the task were recorded on videotape. Before starting the inactivation experiments, area AIP was functionally identified in single-unit recording experiments. Figure 2 shows the anatomical location of injection sites.

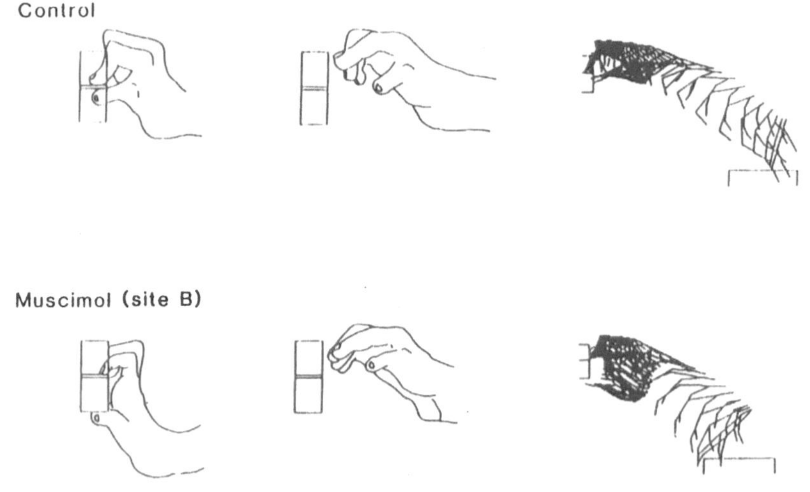

Fig. 3. Preshaping and actual grasping of a small plate in a groove. *Above*: Single frame images redrawn from video and stick diagram of a typical control trial. During preshaping the monkey extends its index finger and simultaneously flexes the last three fingers. Actual grasping is achieved by opposing the thumb and index finger (precision grip). *Below*: Single frame images redrawn from video and stick diagram of a single trial performed after muscimol injection into site B. During preshaping, flexion of the last three fingers did not occur and thumb-index opposition could not be executed. (Modified from Gallese et al. 1994)

Following AIP inactivation, grasping behavior of the hand contralateral to the injected hemisphere was impaired. Severe disruption of preshaping of the hand was constantly observed. This deficit produced a *mismatch* between the 3D features of some of the objects used in the task and the posturing of finger movements, leading either to complete failure of the task or to awkward grasping. When the monkey was successful in grasping the object, the grip was very often achieved only after several correction movements which relied on tactile exploration of the target. A deficit in the accuracy of reaching was not observed.

Although severity of the deficit varied following various AIP injections, impairment of precision grip was a constant feature. Figure 3 (upper part) shows hand preshaping and grip of an object (a small plate in a groove) that the monkey normally grasped using a precision grip. After muscimol injection (Fig. 3, lower part) preshaping of the hand during the phase of approaching the object was completely disrupted: the monkey did not flex its last three fingers and therefore very often failed to insert the index finger into the groove. When, occasionally, the monkey succeeded in inserting it, it was nevertheless not able to oppose the index finger to the pulpar surface of the thumb (Fig. 3, lower part). In one inactivation experiment (see Fig. 2, site A) the deficit observed was particularly severe. In this case side grip, which the monkey normally used to grasp small objects such as spheres or cubes, was also affected.

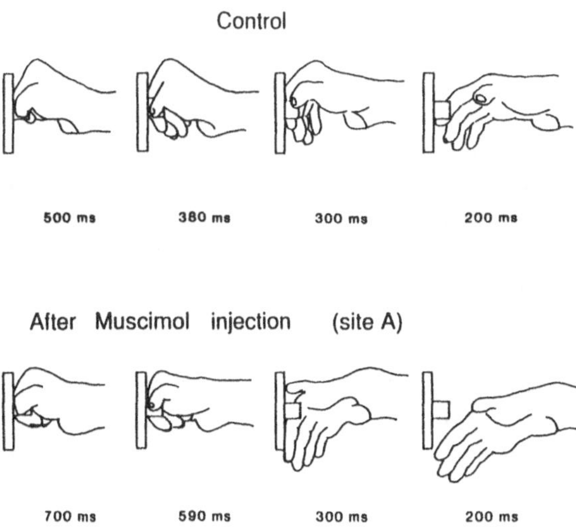

Fig. 4. Preshaping and actual grasping of a small cube performed by using a side grip. *Above*: Control trial. During the approaching phase, the thumb and other fingers flex around the object. *Below*: Single trial performed after muscimol injection into site A. During preshaping hand aperture was increased and prolonged until the backplate was touched by the index finger. Frames in both trials are synchronized at movement onset. (Modified from Gallese et al. 1994)

Figure 4 shows that after muscimol injection grasp time and maximal finger aperture increased considerably compared with the control situation. In many trials the monkey failed to grasp small objects. When grasping was achieved this was only after prolonged tactile exploration of the object. It is worth stressing that this abnormal behavior sharply differs from that observed by Hikosaka et al. (1985) after reversible inactivation of the hand region in area 2. In that case, finger movements were impaired only after contact with objects, when visual information was not available anymore and movement guidance had to rely only on tactile information. The inadequate shaping of the hand after inactivation of area AIP and the behavior often shown by the monkey to explore objects with its fingers after contact strongly suggested that a *visuomotor deficit* had been observed.

In conclusion, inactivation of small sectors of AIP produced an impairment of contralateral hand preshaping and actual grasping without any deficit in arm reaching.

Inferior Premotor Cortex: Area F5

Functional Properties

Area F5 occupies the posterior bank of the inferior limb of the arcuate sulcus and the cortical convexity adjacent to it (Matelli et al. 1985). Microstimulation and recording experiments showed that area F5 is functionally related to goal-directed hand movements (Gentilucci et al. 1988; Rizzolatti et al. 1988; Gentilucci and Rizzolatti 1990). Most neurons in this area are selectively activated during actions such as grasping, holding and manipulating. Many F5 neurons are specific for particular types of hand prehension (e.g., precision grip). It was therefore proposed that there is a "vocabulary" of elementary motor acts in F5, coded at the single neuron level, specifying the type of hand movement necessary to interact with the objects (Rizzolatti 1987; Rizzolatti and Gentilucci 1988). Many F5 neurons also respond to visual presentation of 3D objects whose size is congruent with the type of grip coded by the same neurons. Given the similarities between the functional properties of F5 and AIP, it was interesting to compare the effects of their reversible inactivation on grasping behavior.

Inactivation Data

The effect of reversible inactivation of sectors of F5 was studied by Gallese et al. (1995). They trained a monkey to reach for and grasp objects of different sizes,

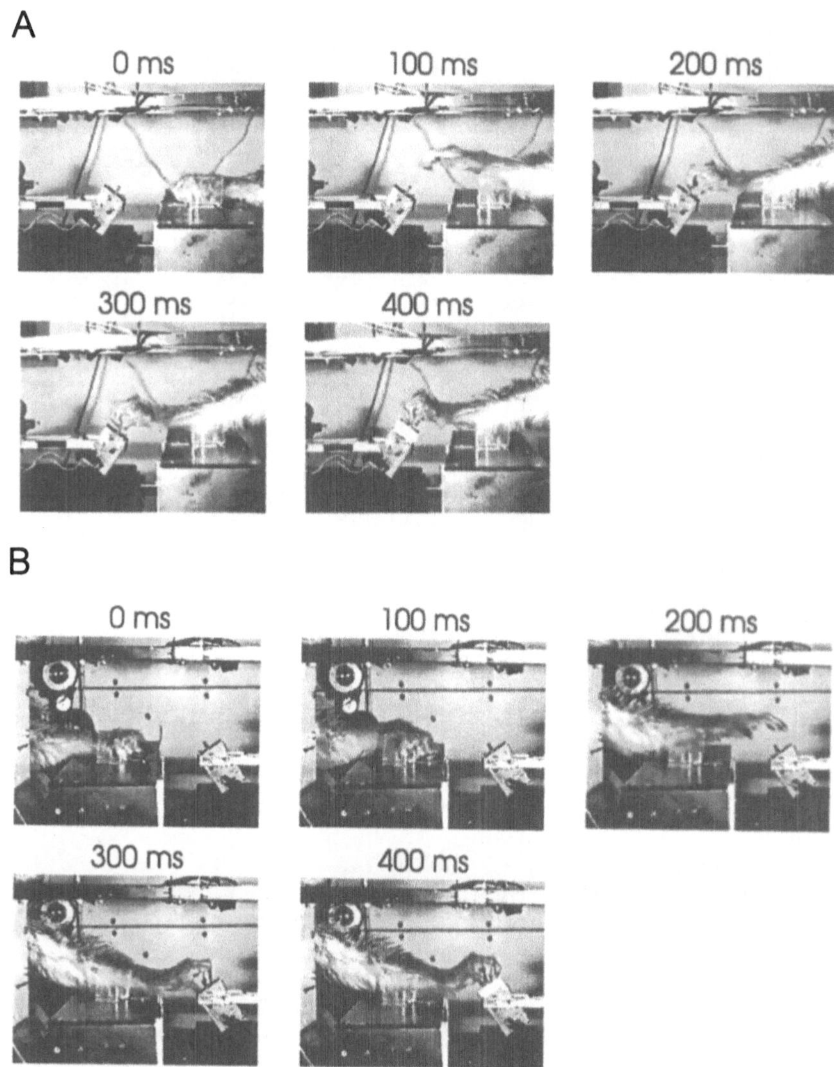

Fig. 5 A, B. Control trials in which the monkey grasped a thin plate using precision grip with its right (**A**) and left (**B**) hand. Behavioral paradigm and apparatus: single objects were randomly presented by a computer driven turntable. Each trial started when the monkey closed a switch with its hand. After 200 ms a liquid crystal screen, located between the monkey's head and the object, became transparent. After another variable period of time (1.2-1.8 s) a barrier interposed between the monkey's hand and the object was lowered, allowing the animal to reach for and grasp the object, and pull it. A liquid reward was delivered upon the correct performance of each trial. Timing above each photogram was calculated from the onset of hand movement

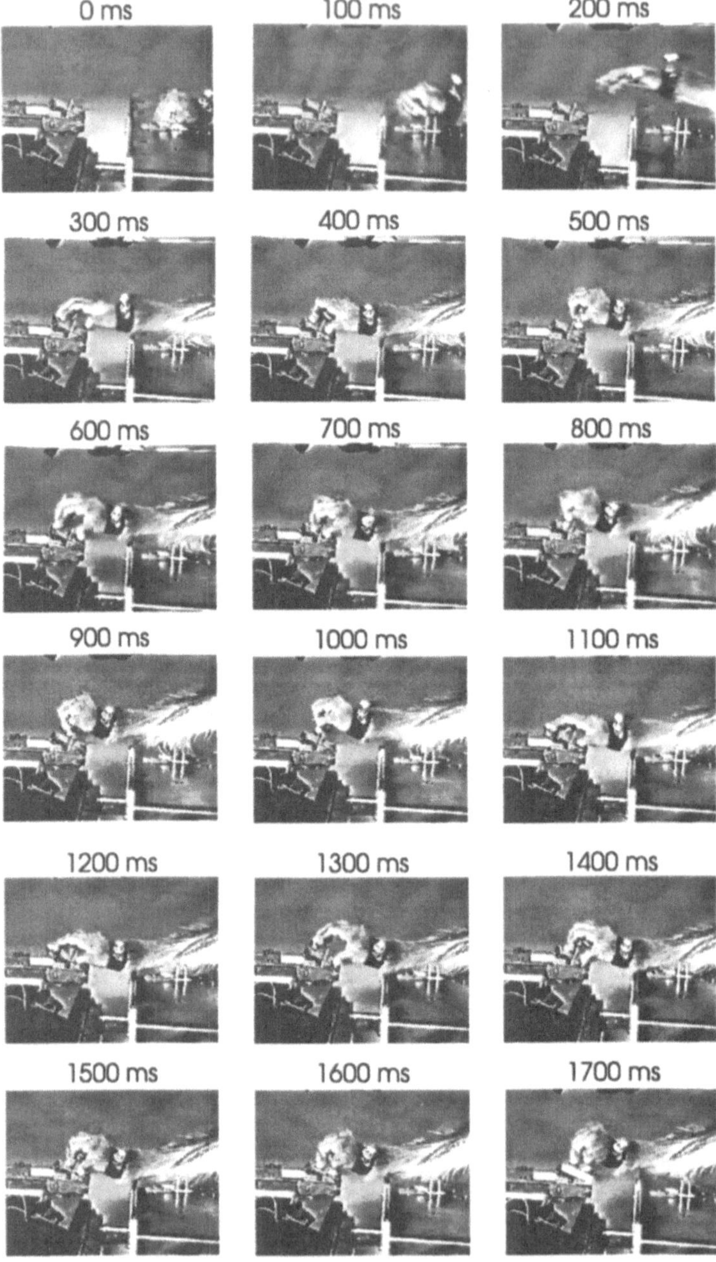
0 ms
100 ms
200 ms
300 ms
400 ms
500 ms
600 ms
700 ms
800 ms
900 ms
1000 ms
1100 ms
1200 ms
1300 ms
1400 ms
1500 ms
1600 ms
1700 ms

Fig. 6. Muscimol single injection in area F5. Preshaping and actual grasping of a thin plate were performed with the hand contralateral to the injection site. Movement time was extremely prolonged compared with the control situation. The monkey approached the object with its fingers exaggeratedly open and did not close them appropriately around the object. Only after contact and tactile exploration of the object could the latter be grasped by means of several hand-opening and closure phases

shapes and orientation, each requiring different types of hand prehension (i.e., precision grip, finger prehension, whole hand prehension, etc.). Selection of injection sites was determined after experimental sessions during which the hand representation region of the primary motor cortex (area F1) and area F5 were identified and functionally characterized, using single unit recording and intracortical microstimulation (for details see Rizzolatti et al. 1988).

Muscimol was injected during three distinct experimental sessions: (1) single injection of 3 µl (5 µg/µl) into the posterior bank of the inferior limb of the arcuate sulcus (area F5); (2) multiple injection (3 x 3 µl) into three distinct sites, each 2 mm away from the others, in the posterior bank of the inferior limb of the arcuate sulcus (area F5); (3) single injection of 3 µl (5 µg/µl) into the hand representation region of the primary motor cortex (area F1). Control sessions were performed before and after each injection experiment. During control and experimental sessions the reaching-grasping movements executed by the monkey in the behavioral task were recorded using a videocamera. Muscimol-induced effects developed in all experiments about 30 min after the injection. A complete recovery was always observed within 24 h.

Figure 5 shows selected single frames of two control trials illustrating the monkey's normal grasping movements. After switch release, the monkey opened its fingers and then started the closure phase while approaching the object. Some 300 ms after movement onset, the fingers were already closed around the thin plate.

After single injection into the posterior bank of the arcuate sulcus the main result was a severe deficit in preshaping and execution of *precision grip* (see Fig. 6). There was also some clumsiness during the execution of *side grip* of a small sphere with the hand contralateral to the injection site. Prehension of larger stimuli such as a large sphere, and vertical and horizontal cylinders was preserved. Performance using the ipsilateral hand appeared to be normal.

In order to assess the effects of a more global inactivation of F5, the grasping behavior of the monkey was tested in a second experiment following multiple microinjections of muscimol into the posterior bank of the inferior limb of the arcuate sulcus. The deficit observed was much more severe than in the previous experiment. The hand *ipsilateral* to the injected hemisphere was strongly impaired during grasping of most of the objects used in the task. Figure 7 shows how the monkey executed precision grip with the ipsilateral hand. Also in this case the final correct grip of the object could be accomplished only after tactile exploration. At variance with the single injection experiment, a deficit in wrist

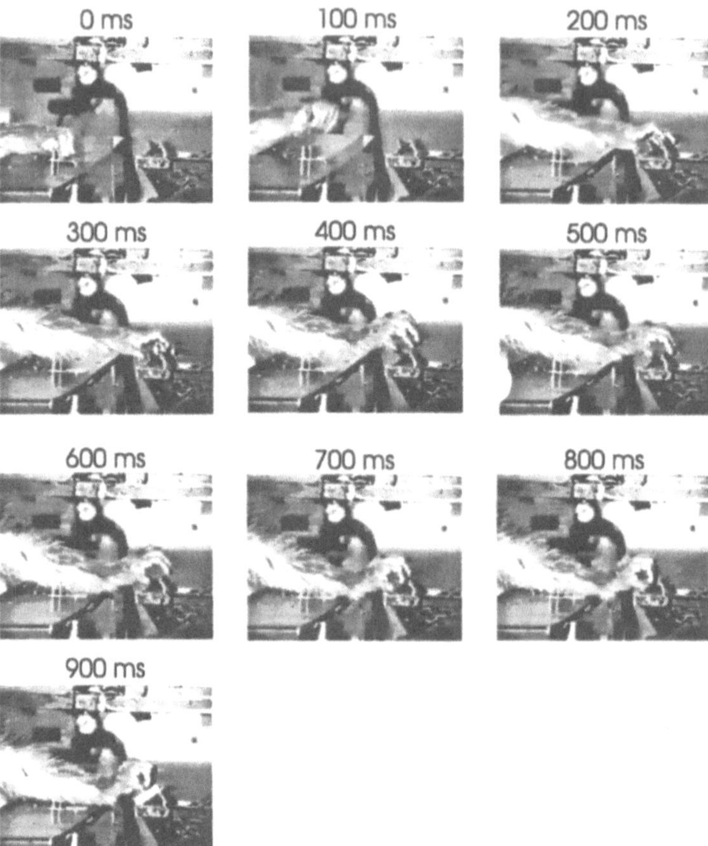

Fig. 7. Muscimol multiple injection in area F5. Preshaping and actual grasping of a thin plate were performed with the hand ipsilateral to the injection site. The time required to achieve the grip of the object was more than twice that of the control situation (see Fig. 5B). Furthermore, shaping of the hand was incorrect, with a prolonged and exaggerated fingers opening phase

orientation was also observed. An example was the prehension of the vertical cylinder. This object required wrist rotation in order to be grasped correctly. After inactivation, the monkey failed to orient its hand properly and grasped the upper part of the cylinder with the wrist pronated. The *contralateral* hand was even more severely impaired. The monkey grasped, very often incorrectly, only the largest objects, but did not even try to reach for small objects that required precision grip or side grip to be grasped.

In conclusion, inactivation of a large part of F5 produced a *bilateral deficit* in preshaping and grasping. The presence of a grasping deficit of the ipsilateral hand,

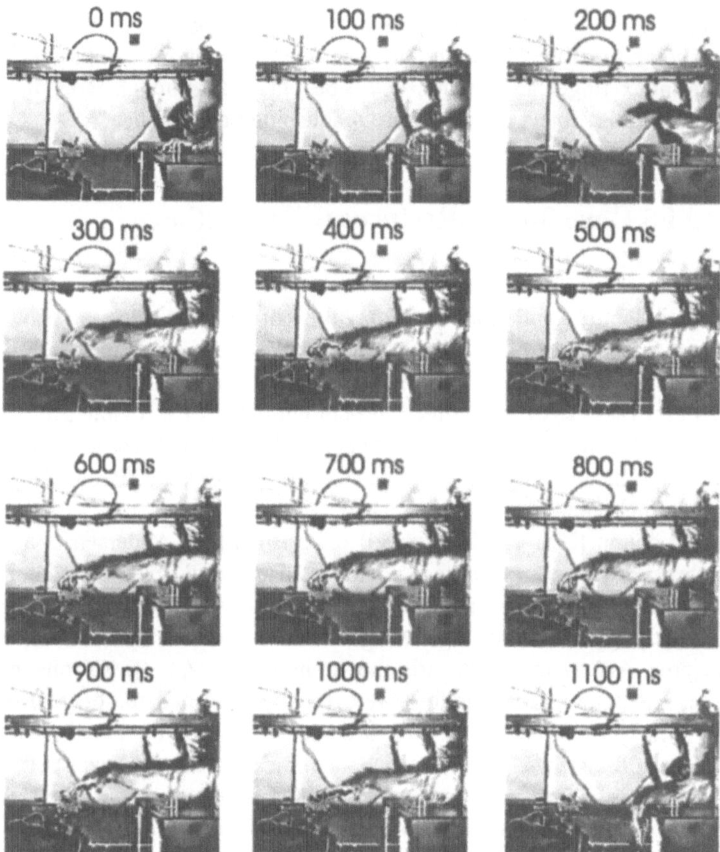

Fig. 8. Muscimol injection in the hand region of area F1 (primary motor cortex). Preshaping and actual grasping of a thin plate were performed with the hand contralateral to the injection site. Movement time was increased compared to the control situation. The fingers were constantly kept in a semi-flexed posture both during reaching and after contact with the object. The object could not be grasped

whose finger movements were nevertheless fairly normal when the monkey relied on tactile information in order to correct errors of hand shaping, renders motor interpretation of the deficit very unlikely.

In a third experiment grasping behavior of the monkey was studied following reversible inactivation of the hand region of the primary motor cortex (area F1). The main result was a severe deficit in the use of discrete finger movements of the hand contralateral to the injected hemisphere. The ipsilateral hand did not show any kind of deficit. After muscimol injection the monkey approached all the objects with a stereotypic hand configuration. Figure 8 shows an example of how the monkey tried to grasp the thin plate, that is, the same object shown in

Figs. 5, 6, and 7. Grasping of small objects could not be executed. Among large objects only the horizontal cylinder, the grasping of which did not require discrete use of the index finger and thumb, could be grasped, although with diminished force. Touching the object did not improve grasping performance.

AIP and F5 Have Strong Reciprocal Connections

According to Pandya and Seltzer (1982) the cortex of the lateral bank of the intraparietal sulcus constitutes a distinct cytoarchitectonic area (area POa). From a functional point of view, however, this area does not appear to be homogeneous. Its caudal part (area LIP) is involved in oculomotion (for a review see Andersen 1989; Barash, this volume; Thier and Andersen, this volume), while its rostral part (area AIP) is related to hand movements (for a review see Sakata and Taira 1994; Sakata et al., this volume).

While the connections of LIP are well established (see Andersen et al. 1990), rather little is known about those of area AIP. Recently, however, Matelli et al. (1994) addressed this problem. After a functional characterization of neuron properties they injected wheat germ agglutinin-horseradish peroxidase (WGA-HRP) in area AIP. The results are shown in Fig. 9 A1, 2. As one can see, retrograde and anterograde labeling were mostly confined to the sector of F5 located in the posterior bank of the arcuate sulcus. Some labeling was also observed along the crown. The selective anatomical linkage between AIP and F5 was confirmed in a further experiment in which a fluorescent tracer (DY) was injected in F5. Following this injection, labeling was concentrated in the anterior part of the lateral bank of the intraparietal sulcus, i.e., in a location that appears to coincide with the location of AIP (Fig. 9 B2). Taken together, these results demonstrate that F5 and AIP form an anatomical circuit. The way this circuit may mediate the visuomotor transformations underlying grasping will be discussed in the next section.

AIP and F5: Two Models for One Circuit

Areas AIP and F5 share many common features: they both control hand movements; they both are activated during visual presentation of objects; and the effects of their reversible inactivation on grasping behavior are strikingly similar. These two areas, however, also have different functional properties. Firstly, the percentage of neurons activated by visual presentation of 3D objects is higher in AIP than in F5 (see Jeannerod et al. 1995). Secondly, neurons in AIP are devoid of somatosensory properties (Gallese et al. 1994) while in F5 42% of grasping

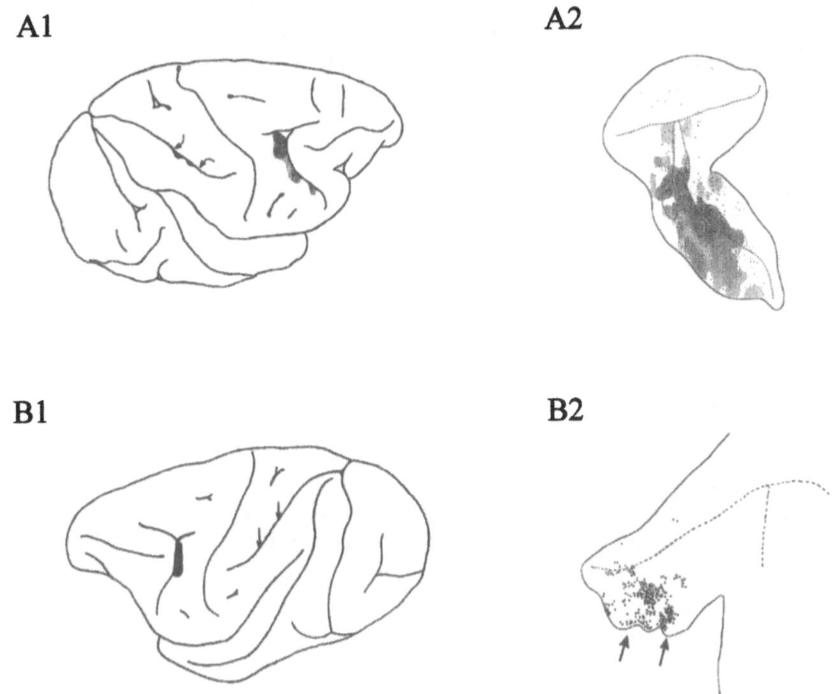

Fig. 9 A, B. Distribution of efferent projections and retrogradely labelled neurons following an HRP-WGA injection in area AIP (**A**) and a Diamidino Yellow (DY) injection in area F5 (**B**). **A1** Lateral view of the monkey's right hemisphere showing the injection site previously functionally characterized by single-unit recording experiments. Note that the core of the injection (delimited by *curved arrows*) is not visible since it is buried deep in the lateral bank of the intraparietal sulcus. Retro-anterograde labelling was present in the convexity of F5. Density of the *dots* represents the density of retrogradely labelled neurons. Intensity of *shading* represents the density of anterogradely labelled fibers. **A2** Reconstruction of the distribution of anterogradely labelled fibers and retrogradely labelled neurons in the inferior limb of the arcuate sulcus, which is shown opened, after injection of WGA-HRP in area AIP (conventions as in A1). **B1** Lateral view of the monkey's left hemisphere showing the DY injection site in area F5. *Arrows* delimit the location of the highest concentration of retrogradely labelled neurons in the lateral bank of the intraparietal sulcus. **B2** Reconstruction of the distribution of retrogradely labelled neurons in the anterior part of the lateral bank of the intraparietal sulcus, which is shown opened. *Arrows* delimit the region recipient of the heaviest projection from F5

neurons are activated by tactile stimulation or by passive joint displacements (Rizzolatti et al. 1988). Thirdly, despite the fact that there are motor neurons in both areas discharging prior to and during grasping execution, the timing of the discharge of these motor neurons varies between the two areas. In AIP most neurons are active during the entire grasping phase and often continue to discharge as long as the monkey is holding the object in its hand. Conversely, in F5 many neurons are activated only during some of the fractions into which grasping action can be segmented, such as fingers-opening phase, closure phase, or the time period around maximal grip aperture.

The major involvement of AIP in visual analysis of the intrinsic properties of the stimulus, on the one hand, and the direct link of F5 to final motor output, on the other, led Sakata et al. (1992) to suggest a conceptual model for the cortical programming and control of visually guided hand grasping movements. According to this model visual information about the intrinsic properties of objects to be manipulated is processed by "visual dominant" neurons of AIP. The result of this analysis is sent to F5, where it triggers the command for the appropriate pattern of hand movements. A corollary discharge, sent from F5 back to AIP, is fed into the "visual and motor" cells, by way of the "motor dominant" cells, allowing matching between the selected movement pattern and the visual signal. Such an interplay between the two areas would control and, if necessary, correct ongoing movement. The grasping deficit observed following AIP inactivation would be caused by the fact that in this condition F5 no longer receives the visual information about the object intrinsic properties necessary to trigger the appropriate hand movement pattern. Conversely, a lesion of F5 would have two effects: it would prevent the occurrence of correct visuomotor transformations, because the corollary discharge from F5 to AIP would be interrupted; it would impair the access of visual information to the final motor output.

This model posits that the visuomotor transformations required in order to accomplish grasping movements are the final product of the interplay between AIP and F5, assigning to AIP a basically visual role. An alternative model, however, can be suggested. According to it, the intrinsic visual object properties encoded by AIP "visual dominant" neurons would be translated into various possible types of grip within AIP itself. This process would occur in the "visual and motor" neurons, which, according to this hypothesis, analyze the "graspability" of an object in terms of affordances. This information would then be sent to F5. Thus, the main role of AIP would be that of "proposing" various possible ways the same object could be grasped. F5 would select the most appropriate type of grip on the basis of contextual information (quality of the object, its spatial relationships with the environment, purpose of the grasping action, internal drive, etc). Finally, F5 would fragment the selected grip into different phases (aperture, closure) and simultaneously keep active that set of parietal "visual and motor" neurons that, previously activated by the visual object, also encode the appropriate grip.

According to this model the lesion of AIP would determine the observed grasping deficit by preventing the visuomotor transformations necessary for this action. The deficits seen after lesion of F5, though, would result from the loss of the motor vocabulary for grasping.

Both of the models described postulate the integrity of the cortical circuit formed by F5 and AIP in order to accomplish grasping tasks on the basis of visual information, yet they assign different roles to the two cortical areas. The inactivation results presented here do not enable us to judge which of the two models is correct.

References

Andersen RA (1989) Visual and eye movement functions of the posterior parietal cortex. Ann Rev Neurosci 12:377–403

Andersen RA, Asanuma C, Essick G, Siegel RM (1990) Corticocortical connections of anatomically and physiologically defined subdivision within the inferior parietal lobule. J Comp Neurol 296:65–113

Arbib MA (1981) Perceptual structures and distributed motor control. In: Brooks VB (ed) Handbook of physiology, section 1, vol 2, part 2, vol 2. American Physiological Society, Bethesda, Maryland, pp 1449–1480

Fulton JF (1949) Physiology of the nervous system. Oxford University Press, Oxford

Gallese V, Murata A, Kaseda M, Niki N, Sakata H (1994) Deficit of hand preshaping after muscimol injection in monkey parietal cortex. Neuroreport 5:1525–1529

Gallese V, Buccino G, Fadiga L, Fogassi L, Rizzolatti G, Tedeschi P (1995) Reversible inactivation of inferior premotor areas (F4 and F5) and primary motor cortex (F1) after muscimol injection in the macaque monkey. Eur J Neurosci [Suppl 8]:18.27

Gentilucci M, Rizzolatti G (1990) Cortical motor control of arm and hand movements. In: Goodale MA (ed) Vision and action: the control of grasping. Norwood, Ablex, pp 147–162

Gentilucci M, Fogassi L, Luppino G, Matelli M, Camarda R, Rizzolatti G (1988) Functional organization of inferior area 6 in the macaque monkey: I. Somatotopy and the control of proximal movements. Exp Brain Res 71:475–490

Hikosaka O, Tanaka M, Sakamoto M, Iwamura Y (1985) Deficits in manipulative bahaviors induced by local injections of muscimol in the first somatosensory cortex of the conscious monkey. Brain Res 325:375–380

Hyvärinen J, Poranen A (1974) Function of the parietal associative area 7 as revealed from cellular discharges in alert monkeys. Brain 97:673–692

Jeannerod M (1988) The neural and behavioral organization of goal-directed movements. Clarendon, Oxford

Jeannerod M, Arbib MA, Rizzolatti G, Sakata H (1995) Grasping objects: the cortical mechanisms of visuomotor transformation. Trends Neurosci 18:314–320

Lawrence DG, Kuypers HGJM (1968) The functional organization of the motor system in the monkey. I. The effects of bilateral pyramidal lesions. Brain 91:1–14

Matelli M, Luppino G, Rizzolatti G (1985) Patterns of cytochrome oxidase activity in ther frontal agranular cortex of the macaque monkey. Behav Brain Res 18:125–136

Matelli M, Luppino G, Murata A, Sakata H (1994) Independent anatomical circuits for reaching and grasping linking inferior parietal sulcus and inferior area 6 in macaque monkey. Soc Neurosci Abstr 20:404.4

Mountcastle VB, Lynch JCGA, Sakata H, Acuna C (1975) Posterior parietal association cortex of the monkey: command functions for operations within extrapersonal space. J Neurophysiol 38:871–908

Pandya DN, Seltzer B (1982) Intrinsic connections and architectonics of posterior parietal cortex in the rhesus monkey. J Comp Neurol 204:204–210

Passingham RE, Perry H, Wilkinson F (1983) The long-term effects of removal of sensorimotor cortex in infant and adult rhesus monkeys. Brain 106:675–705

Rizzolatti G (1987) Functional organization of inferior area 6. In: Bock G, O'Connor M, Marsh J (eds) Motor areas of the cerebral cortex, Ciba Foundation Symposium 132. Wiley, Chichester, pp 171–186

Rizzolatti G, Gentilucci M (1988) Motor and visual-motor functions of the premotor cortex. In: Rakic P, Singer W (eds) Neurobiology of neocortex. Wiley, Chichester, pp 269–284

Rizzolatti G, Camarda R, Fogassi M, Gentilucci M, Luppino G, Matelli M (1988) Functional organization of inferior area 6 in the macaque monkey: II. Area F5 and the control of distal movements. Exp Brain Res 71:491–507

Sakata H, Taira M (1994) Parietal control of hand action. Curr Op in Neurobiol 4:847–856

Sakata H, Taira M, Mine S, Murata A (1992) Hand-movement related neurons of the posterior parietal cortex of the monkey: their role in visual guidance of hand movements. In: Caminiti R, Johnson PB, Burnod Y (eds) Control of arm movement in space. Exp Brain Res Suppl 22. Springer-Verlag, Berlin, pp 185–198

Taira M, Mine S, Georgopulos AP, Murata A, Sakata H (1990) Parietal cortex neurons of the monkey related to the visual guidance of hand movement. Exp Brain Res 83:29–36

Tower SS (1940) Pyramidal lesions in the monkey. Brain 63:36–90

Reaching, Grasping, and Bimanual Coordination with Special Reference to the Posterior Parietal Cortex

M. WIESENDANGER[1], O. KAZENNIKOV[1], S. PERRIG[2], E. ROUILLER[3], and I. KERMADI[3]

[1] Department of Neurology, University of Berne, Laboratory of Motor Systems, Inselspital, Bern, Switzerland

[2] Department of Neurology, University of Geneva, Hôpital cantonal, Geneva, Switzerland

[3] Institute of Physiology, University of Fribourg, Fribourg, Switzerland

The Control System for Manual Skills Has Large and Distributed Representations in Frontal and Parietal Cortical Areas

Our interest in the posterior parietal cortex (PPC) was prompted by our attempt to understand the neuronal organization of bimanual coordination (Kazennikov et al. 1994) and of reaching and grasping. Lesion studies have shown that the hand area of the primary motor cortex is a crucial control structure for skillful manipulations (e.g., Glees and Cole 1950; Denny-Brown 1966). Transection of bulbar pyramids leads to an even more severe deficit in skillful hand movements, as shown by Tower (1940) and Lawrence and Kuypers (1968). Since the pyramidal tract originates from a large cortical region including a number of frontal and parietal areas, it is not surprising that the deficit seen with pyramidal tract lesions is more severe, and it may be explained by additional interruption of pyramidal tract fibers from these secondary sensorimotor areas. Recent anatomical and physiological research has indeed revealed a number of hand-related, nonprimary motor areas, or at least neuronal clusters within premotor areas (Rizzolatti et al. 1988; Hepp-Reymond et al. 1994), the supplementary motor area (Brinkman and Porter 1979; Tanji et al. 1988; Wiesendanger 1986; unpublished observations), and the anterior cingulate area (Shima et al. 1991; He et al. 1995; Cadoret and Smith 1995).

In monkeys, there is a considerable restitutional capacity of hand function after near-total or even total unilateral pyramidotomy (Hepp-Reymond and Wiesendanger 1972; Hepp-Reymond et al. 1974; Schwartzman 1978; Chapman and Wiesendanger 1982). A remarkable and rather unusual recovery of hand function has been reported in a patient with a pericentral lesion who was observed by Foerster (1936) over a prolonged period. Several years after the cortical lesion, the patient died of an unrelated cause; at autopsy, the bulbar pyramid turned out to be completely degenerated. These lesion data of cases exhibiting functional

In: Parietal Lobe Contributions to Orientation in 3D Space (1997). P. Thier and H.-O. Karnath (eds). Springer-Verlag, Heidelberg.

recovery are often interpreted in terms of vicarious functioning of parallel systems and are in line with the concept of control systems having powerful and widely distributed representations. It thus appears that the highly complex hand function with its many degrees of freedom requires similar powerful and distributed representations as vision. The multiple cortical hand areas are to a large extent interconnected. However, it is still unknown how cortical elaboration in multiple areas leads to "...a smooth and neatly ordered sequence of skilled movements" (Mountcastle 1995). It is even uncertain whether all hand areas converge to the primary motor cortex as final common path addressing the spinal cord. Alternatively, these cortical hand areas that all have access to the spinal cord via direct and indirect descending fiber systems may also operate in parallel (as suggested by lesion studies).

The past 20 years of basic research have shown that hand function is also well represented in the PPC (Mountcastle 1995; Sakata et al. and Gallese et al., this volume). Clinical studies indicate furthermore that the PPC, especially of the dominant hemisphere, is implicated in "higher-order" deficits of intended goal-oriented movements (synergistic dyscoordination, optic ataxia, apraxia, neglect). A brief outline of the potential role of the PPC in interlimb coordination and in reaching and grasping is given below.

Does the PPC have Bilateral Coupling Functions?

Whether the PPC in monkeys has bilateral functions is an unresolved issue. Some authors found that, even with extensive lesions of the PPC, deficits were confined to the contralateral limb (e.g., LaMotte and Acuña 1978); others, however, also described ipsilateral effects (Hartje and Ettlinger 1973; Stein 1978). Recordings from single neurons showed mostly contralateral relationships, but, clear evidence for some bilateral associations was found in the lower parietal lobule (Hyvärinen and Poranen 1974; Mountcastle et al. 1975; MacKay 1992). Relevant to this observation is the presence of a strong transcallosal linkage between areas 7a of both hemispheres (Cavada and Goldman-Rakic 1989). Thus, previous observations in monkeys partly point to bilateral relationships that might play a role in interlimb coordination. In patients with PPC lesions of the dominant hemisphere, apraxic syndromes may concern bimanual actions (Hécaen 1978). Since the neural mechanism(s) subserving bimanual coordination in primates is a central issue of our research, we have undertaken a lesion study in monkeys to assess the possible role of the PPC in bimanual control, as will be reported below.

We have devised a manipulandum that allows us to test bimanual coordination quantitatively in a drawer pull-and-grasp task (Kazennikov et al. 1994). We have already subjected two well-trained monkeys (M1 and M2) to bilateral lesions of the supplementary motor area (SMA), as will be reported separately. Contrary to our expectation, these lesions had only minimal and transient effects on bimanual coordination (for preliminary data see also Wiesendanger et al. 1994), whereas there was a substantial delay in the movements of the two arms. This was the case irrespective of whether the workspace was visible or not. Both animals recovered from the SMA lesions and no persistent deficits were observed 6 months postoperatively. We therefore concluded that, inspite of parametric changes in the individual arm movements, bimanual coordination for goal achievement remained largley unaffected and that the SMA is not an obligatory prerequisite for bimanual coordination. However, the possibility has to be considered that the SMA might contribute to bimanual coordination in cooperation with other cortical and subcortical networks. Therefore, the unresolved question concerning possible involvement of the PPC in bimanual coordination in monkeys will be addressed below.

The Hypothesis of Separate Representations for Reaching and Grasping in the PPC

It is well established that lesions of the PPC in monkeys often lead to dysfunction in visually-guided reaching and prehension (Peele 1944; Cole and Glees 1955; Hartje and Ettlinger 1973; Faugier-Grimaud et al. 1978; LaMotte and Acuña 1978; Deuel and Regan 1985). The major deficits include misreaching, abnormal shaping of the hand with misorientation of fingers while grasping objects, and neglect of contralateral space. As first reported by Faugier-Grimaud et al. (1978), monkeys initially refused to use the affected hand for grasping when the other arm was restrained. However, several authors noted the remarkable recovery within only a few weeks. From the initial single unit studies in monkeys performing various behavioral tasks, it became clear that a class of neurons in the PPC is specifically activated during visually-guided reaching and manipulative movements (e.g. Mountcastle et al. 1975; Leinonen et al. 1979). Together with the results of lesion experiments, these single unit data strongly suggested that the PPC *functions as a spatial control and coordinate- transformation system for sensory driven synergies of the eyes, arm, and hand* (e.g., Sakata et al. 1995, this volume).

It is also well established that the rather large PPC can be subdivided into subareas differing in their structural relationships and/or in their functional properties. The question then arose whether the proximal transport component and the distal grasp component are differentially represented in the PPC (for a discussion of the "two-channels hypothesis" see Jeannerod and Biguer 1982). Although there is still some uncertainty regarding the hypothesis (Perenin and Vighetto 1988), it is now clear that posterior parietal lesions may severely compromise object-related preshaping and manipulation without affecting the reaching component (Jeannerod et al. 1994). Deficits of complex explorative and manipulative finger movements as a consequence of PPC lesions may indeed be a dominant feature (Pause et al. 1989). Yet we are not aware of human or monkey cases with lesions in the inferior parietal lobule exhibiting pure misreaching without a deficit in grasping and other manipulations. The question whether distinct neuronal networks for the two components are spatially separated in the inferior parietal lobule therefore remains to be clarified.

Following the early discovery of manipulation-related neurons by Mountcastle et al. (1975), Sakata and his associates (1995, this volume) have contributed immensely in further characterizing such neurons in the intraparietal sulcus (IPS), classically considered as the parietal eye field (Thier and Andersen 1996, this volume). Within the posterior bank of the IPS, Sakata and coworkers detected many neurons that were activated during grasping. The activation pattern was typically dependent on the form, size, and three-dimensional orientation of the grasped object. The ensemble of these neurons was considered to "...play an important role in matching the pattern of hand movement to the visuospatial characteristics of the object manipulated". "Visual-dominant neurons" were activated only if grasping was performed under visual guidance; other neurons, however, were equally active when grasping was performed in darkness ("motor-dominant"), and still others exhibited reduced activation in the dark ("visual and motor"). The two latter types of neurons were concentrated in a lateral and rostral portion of the IPS which was termed "anterior intraparietal area" or area AIP, whereas "visual dominant" neurons were concentrated in a more medial and posterior region of the IPS, termed area LIP (lateral intraparietal area). However, it was clear from the plot of the recording sites in the IPS that segregation is far from complete (as also emphasized by Sakata et al. 1995). Nevertheless, the team of Sakata (Gallese et al. 1994) recently reported that local reversible inactivation of area AIP by means of the γ–aminobutyric-acid (GABA) agonist muscimol resulted in a specific dysfunction of hand shaping, without loss of precision in reaching for objects. The severity of the deficit critically depended on the precise site of the cortical microinfusion in area AIP. One control microinfusion in area LIP (which has a larger proportion of "visual dominant" neurons) resulted in misreaching whereas hand shaping was not affected. It was therefore provisionally concluded that the global reach-and-grasp function (which breaks down with large lesions in the PPC) is spatially distributed in distinctive compartments of the IPS cortex: the distal component is located in area AIP and the proximal component in area LIP. Interestingly, area AIP has been found to be interconnected with the ventral premotor area F5 which also represents the distal grasp component (Rizzolatti et al. 1988). Thus, the two areas AIP and F5 have been considered as a parieto-frontal circuit involved in the visuomotor transformation process necessary for matching hand configuration to object configuration (Jeannerod et al. 1995; Gallese et al., this volume). However, it remains to be examined in more detail whether the functional localization of reaching in area LIP, as tentatively proposed by Gallese et al. (1994), is as specific as the distal grasping representation in area AIP. It should be kept in mind that "reach-neurons" were found in a number of areas of the PPC, including area 5 (Mountcastle et al. 1975;

Kalaska et al. 1983; MacKay 1992; Caminiti et al. 1996), and that lesions of area 5 may also result in a reaching deficit (Nixon et al. 1992). In the present chapter, we will further address the problem of topology of reaching and grasping and report our own results obtained with irreversible parietal lesions as well as with reversible and restricted inactivation of the PPC.

Thus, the aim of the experiments described below is twofold: (1) to test the hypothesis that the PPC in monkeys contributes to bimanual coordination, and (2) to examine further the idea of a differential spatial distribution of arm and hand functions in the parietal association cortex.

The PPC Contributes to Bimanual Coordination

As stated above, two previously used monkeys (M1, M2) had a complete recovery from SMA lesions inflicted more than half a year ago. The same monkeys have recently been subjected to an additional unilateral PPC lesion in order to test whether superposition of the new lesion would have greater consequences on bimanual coordination. In the first monkey M1, a large lesion was placed in the right inferior parietal lobule also including areas AIP and LIP (and also superficially encroaching on lateral area 5). In monkey M2, the lesion was relatively superficial and occupied a more caudal-medial portion of the inferior parietal lobule, as well as part of the superior parietal lobule. In the IPS, the lesion damaged parts of area LIP, but left area AIP intact. For better comparison, both lesions in Fig. 1A, B are drawn in the right hemisphere, although in M2 the lesion was on the left side.

The striking deficits in contralateral reaching and grasping observed after renewed surgical intervention in the PPC will be described below. First, the focus will be on the effects on bimanual performance. With intensive postoperative training, the formal bimanual task was again performed 3 weeks and 10 days postoperatively (monkeys M1 and M2, respectively), and the quantitative assessment of bimanual coordination was resumed. The task was to reach through a window with one hand, to grasp the handle of a drawer and to pull it open while the other hand picks up a small food morsel from a recess of the drawer (Fig. 2C). Since the drawer was spring-loaded, the pulling hand had to keep the drawer open against the load, allowing the other hand to pick up the food with the precision grip. In daily sessions, movement parameters and the time structure of the bimanual synergy were measured and compared with control data obtained 3 weeks before the parietal lesion. As described in detail for intact animals (Kazennikov et al.1994), criteria for bimanual coordination were the degree of synchronization and its variability (mean, SD) in goal-reaching, i.e., the interval between instances of picking up food and drawer opening. The other behavioral measure was the degree of temporal correlation between the goal-related events of

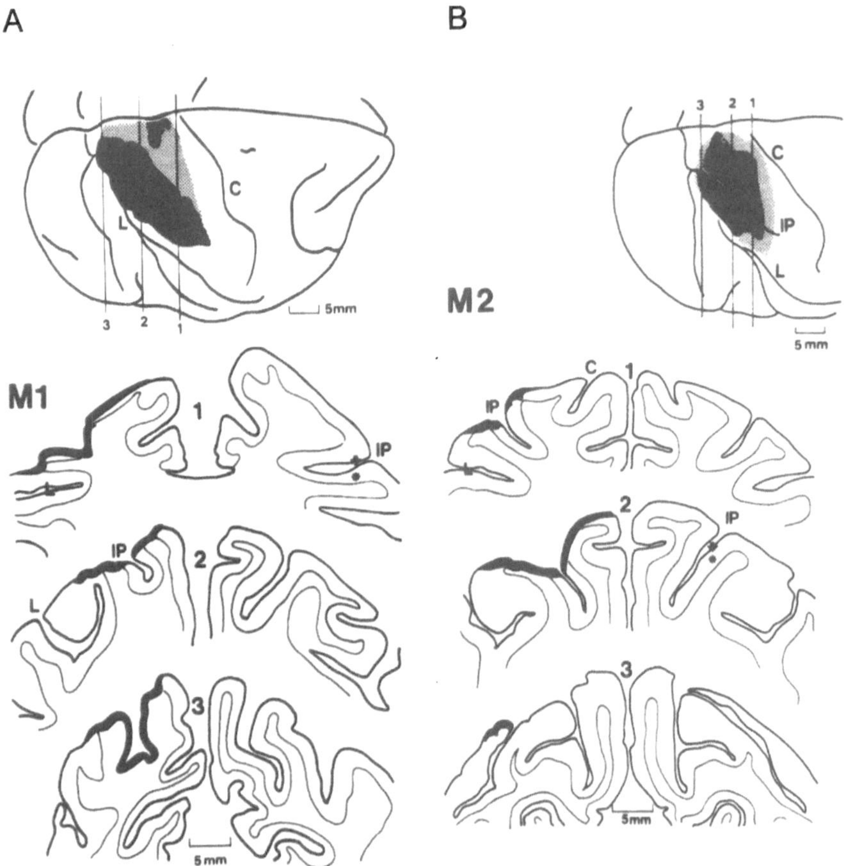

Fig. 1. A Lesion of posterior parietal cortex in monkey 1 (*M1*). **B** Lesion of monkey 2 (*M2*). For better comparison, both lesions are drawn on the right hemisphere, but the lesion of *M1* was on the left. *Numbered vertical lines* in the hemisphere correspond to the *numbered coronal sections* shown below. Note total removal of the intraparietal sulcus(*IP*) rostromedially in *M1*, whereas the IP was largely spared in *M2* at this rostral level. *C*, central sulcus; *IP*, intraparietal sulcus; *L*, lateral sulcus

the two hands, i.e., the linear correlation coefficients r. These goal-related criteria were relatively invariant compared with the large variance in timing of the individual hand components. The two goal-related quantitative measures were taken as the *critical goal invariance for rating bimanual coordination.*

Figure 3 shows the result of the right-sided PPC lesion on goal coordination in M1, based on cumulative data from the pre- and postoperative periods. In Fig. 3A, B the plots display the change in synchronization of the two hands as

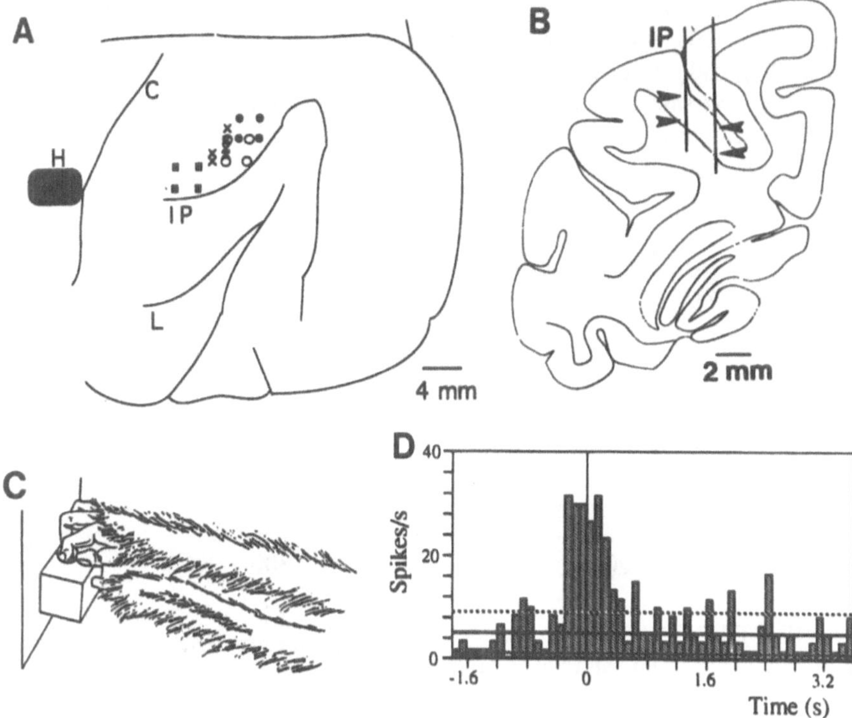

Fig. 2. A Lateral view of the posterior half of the left hemisphere showing the entry points of vertical tracks of muscimol microinfusions (see text). *Crosses, solid squares, solid circles,* and *open circles* indicate entry points of four microinfusion sessions. *C,* central sulcus; *IP,* intraparietal sulcus; *L,* lateral sulcus; *H,* hand area of primary motor cortex, as assessed from intracortical microstimulation. **B** Frontal section showing the location of two tracks, reconstructed from their entry point (putatively in area AIP). *Arrow-heads* delimit the portion of the tracks along which unitary activity related to the bimanual pull-and-grasp task, illustrated in **C**, was obtained. The perievent histogram of neuronal activity shown in **D** was aligned with the signal generated when the index finger, ready for the precision grip, entered the drawer recess

they reach the goal, i.e., the right-left time intervals as a time series in A and as interval histograms in B. The former illustrates particularly well the increase in variability after the lesion and the latter the shift to longer right-left intervals. The regression plot of the arrival times of the two hands at the goal (Fig. 3C) shows a marked increase in scattering and a decrease of the linear correlation coefficient (r). These data were obtained while the monkey performed the task in darkness; a similar deficit was observed when the monkey performed with full vision. This thus indicates that the parietal cortex lesion, which was superimposed on a bilateral SMA lesion from which the monkey had recovered, *resulted in a significant decay of bimanual coupling*. In M2, whose left-sided PPC-lesion was smaller (Fig. 1B), synchronization changed from 13 ± 53 ms to 69 ± 162 ms (mean, SD), and the linear correlation coefficient decreased from r_{pre} = 0.951 to r_{post} = 0.848. It should be noted that postlesion data were collected starting only from the third postoperative week when the monkeys had once again learned how to execute the bimanual task.

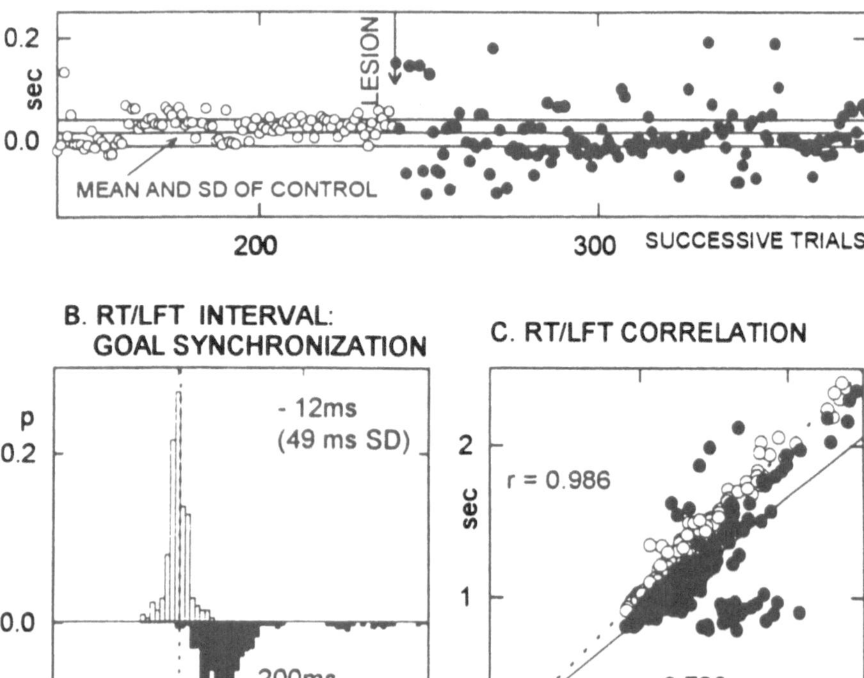

Fig. 3 A-C. Bimanual coordination in monkey M1 before (*open symbols* and *histograms*) irreversible lesion of the posterior parietal cortex (PPC) and after lesion (*filled_symbols* and *histograms*). **A**. The right-left hand (*RT/LFT*) differences in arrival times at the goal are plotted *serially* and **B** as *interval histograms* (with means and standard deviations). **C** *linear correlogram* of arrival times of both hands at the goal. Note increase of goal-related left-right intervals with substantial increase in variance and drop in correlation coefficients (*r*)

Keeping in mind the problematic rationale of deriving normal functions from two-stage lesion effects, *the conclusions* one can draw about the role of the PPC in bimanual coordination necessarily have to be cautious. In particular, the significant deficit observed may not be specific for the PPC. Furthermore, it was rather surprising for us that the reduced score in bimanual coordination was still relatively high, with many well-synchronized trials (Fig. 3C). Nevertheless, it was clear that the lesion effects from the previous large, double-sided lesion of the SMA were more transient and smaller than the deficit obtained by the combined two-stage lesion. Taken together, the results thus indicate that the posterior parietal cortex participates, possibly together with the SMA, in bimanual endpoint coordination. They also suggest that this coordination is a distributed process that may include a number of cortical and subcortical structures, such as the cerebellum, the basal ganglia, and perhaps even spinal networks.

To What Extent Are the Functions of Reaching and Shaping Separately Represented in the PPC?

Evidence from Irreversible Lesions. Data from the same monkeys M1 and M2 as above were obtained during the immediate postoperative periods of 3 weeks and 10 days, respectively. Monkey M1 with a right-sided lesion (Fig. 1A) exhibited a dramatic shaping deficit as well as misreaching of the contralateral left arm. The animal refused to use the affected hand for grasping during the first few postoperative days (the typical "parietal hand" described by Faugier-Grimaud et al. 1978). Gradually, bizarre and unsuccessful attempts were made to reach and grasp for food, but the left hand attempted to grasp with widely fanned and extended fingers, and the morsels were often missed. In the formal bimanual test situation, the monkey initially failed to pass the hand through the narrow window that provides access to the baited drawer because the hand was not properly shaped and oriented for passing. When the panel with the window was removed, the affected pull-hand reached out, yet too far to the left of the drawer handle. When food morsels were presented near the affected left hand, grasping was attempted, often without success, with a clumsy whole-hand finger closure. This was in sharp contrast to the precise index-thumb pincer of the unaffected right hand. When the animal, the right arm restrained, was exposed to a platform with rather large banana pieces, it consistently misreached with the left arm. When the hand contacted the food, it once again closed in a clumsy mass grip with extended fingers, often losing the morsel (Fig.4). Neglect to the left (i.e., contralateral to the lesion) was evident. Thus, M1 presented the full-fledged syndrome of large PPC lesions as reported in the literature (see above). With intensive daily training sessions, recovery was impressive, as also reported by previous authors. Within 3 weeks, M1 was again able to reach correctly for a piece of food and to handle it

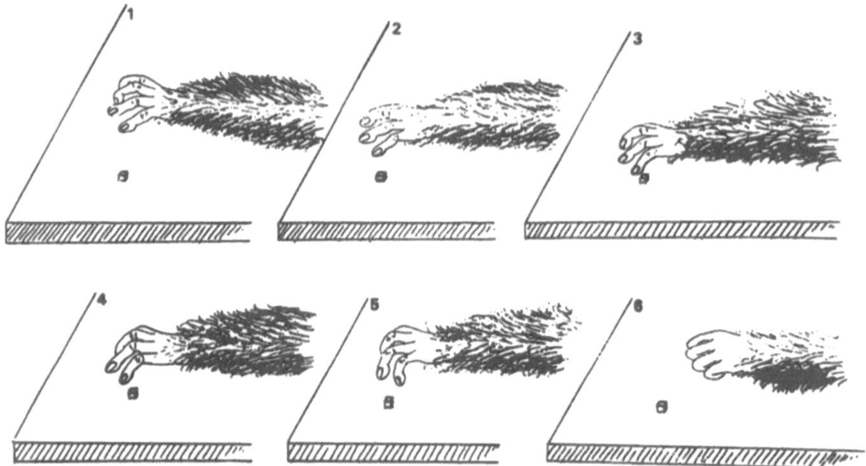

Fig. 4. Performance of monkey M1, two weeks after right posterior parietal cortex (PPC) lesion: note misreaching and missing of the banana piece as well as fanning of fingers and lack of shaping of the left hand. Redrawn from a series of digitized video frames (*1–6*)

with the precision grip. During this period, the monkey also gradually relearned to perform the bimanual task, including proper orientation of the arm and hand for reaching through the window and correct grasping of the drawer handle.

In M2, the effect of the PPC lesion was less dramatic. This was not surprising since the lesion was smaller and more superficial. For example, the monkey used the affected right arm more freely and successfully for grasping food morsels. Nevertheless, during the first 3-4 postoperative days, misreaching was present (Fig. 5), even so, it was less pronounced than in M1. Most importantly, shaping and grasping with the precision grip were performed flawlessly. The evidence from M2 therefore lends some support to the preliminary conclusion of Gallese et al. (1994) that precise spatial orientation in reaching depends on a more medial and caudal sector of the IPS. Another interpretation of our results might be, however, that goal-oriented reaching is simply more widely represented, also including areas outside the IPS region. The parietal lesion of monkey M1 with reaching difficulties was indeed considerably larger, also including some of area 5.

Evidence from Reversible Inactivation Experiments with Muscimol Microinfusions. One unoperated monkey has been conditioned to sit in a primate chair and to perform reaching and grasping movements in the drawer pulling task (Fig. 2C) while single unit activity was recorded from the parietal cortex. The electrode tracks were aimed at area AIP of the intraparietal sulcus (Fig. 2A, B) in

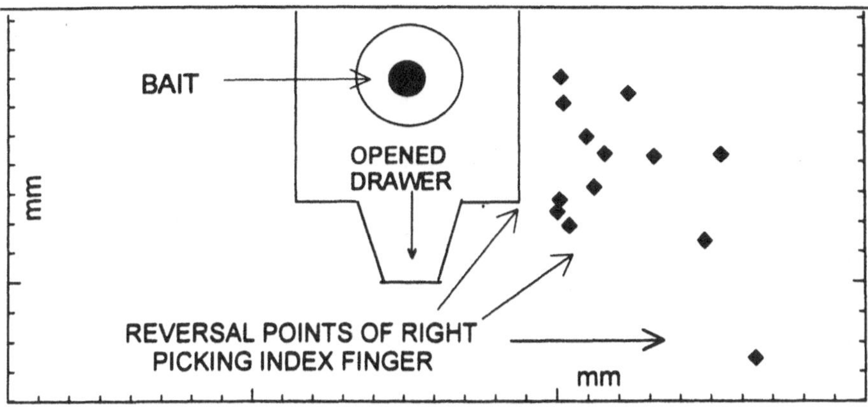

Fig. 5. Monkey M2, first week after left posterior parietal (PPC) lesion. *Top view* on opened drawer with *arrows* pointing to the misreaches to the right of the baited recess in the drawer. The *filled quadrangles* are the reversal points of the reaching trajectories (reconstructed from video frames; 1 division = 2 mm)

order to identify neurons related to grasping and reaching as a guide for the inactivation experiments. In Fig. 2D, a neuron in the presumed area AIP (microelectrode track marked by arrow in Fig. 2A) discharged in relation to the act of grasping small pieces of a cookie with the precision grip from the recess of the opened drawer. In the first experiment, five injections with a 5 μl Hamilton syringe were carried out to deliberately microinfuse a large total volume of 15 μl of muscimol solution (1 μg/μl). Entry points of these 5 injections are represented by circles in Fig. 2A. In three subsequent experiments, further microinfusions were aimed at entry points shown by filled squares, filled circles, and open circles, respectively, in Fig. 2A. The total dose for each of these three applications was reduced to 8 μl of muscimol. After initial testing of muscimol effects in the drawer pull-and-grasp paradigm, it turned out that video recordings were not optimal and that the effect of misreaching could not be assessed adequately. For optimal illumination, the animal was therefore placed outside the experimental cubicle. The monkey was required to reach for food placed at a reaching distance on a semi-circle platform. Shaping and delicate grasping were examined by means of a food board, similar to that used by Brinkman (1984) which was placed in front of the animal and required retrieval of raisins from small vertical and horizontal slots with appropriate hand orientation and use of the precision grip.

Results from Inactivation Experiment 1. The large infusion of muscimol, placed between the presumed areas LIP and AIP (crosses in Fig. 2A), had pronounced effects on the contralateral right arm-hand functions: the right hand was not used when both arms were free ("the parietal hand", Fig. 6A); misreaching of the right

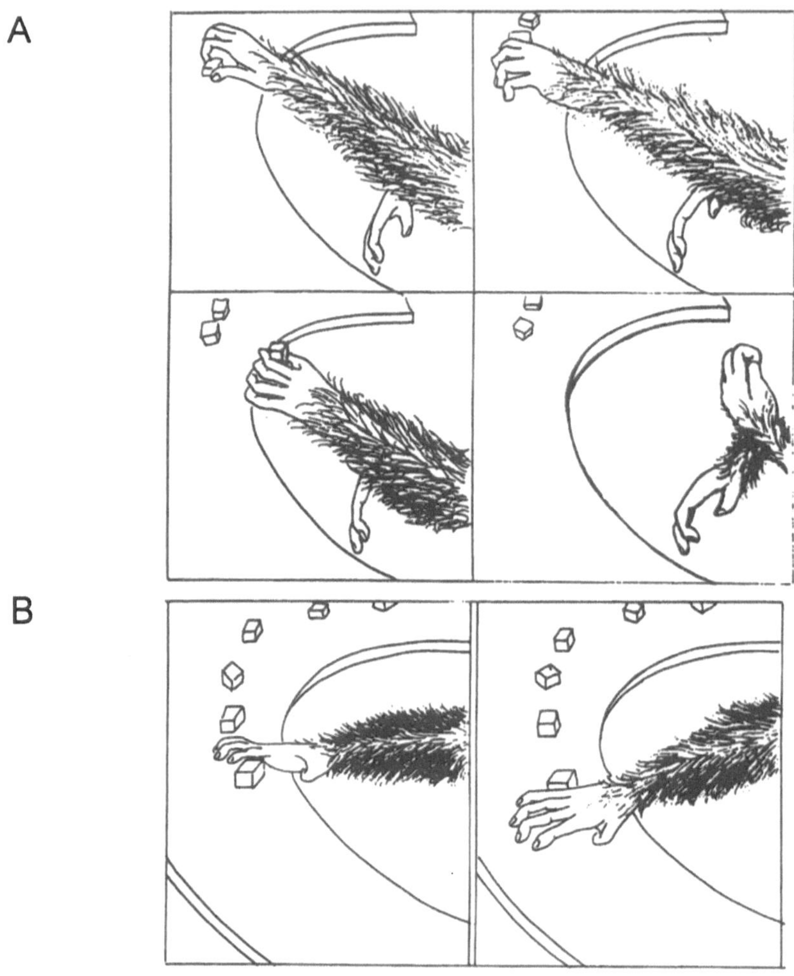

Fig. 6 A, B. Effect of a large left-sided microinfusion in an unoperated monkey at sites indicated by crosses in Fig. 2A. **A** Both hands are free to reach for food arranged on a semicircular platform mounted in front of the monkey. Note immobility of right arm with the typical posture of the "parietal hand" while the free hand, ipsilateral to the muscimol infusion, correctly reaches for and grasps the food morsels. **B** Attempts of right arm to reach for food when the ipsilateral unaffected arm is restrained: note misreaching and abnormal shaping with global finger closure; the banana pieces to the right are neglected. Redrawn from a series of digitized video frames

arm was prominent when the left arm was restrained; food morsels on the right side of the platform were neglected; the fine control of shaping and grasping was completely lost, with fingers spread apart and closing together without forming the pinch (Fig. 6B). Many attempts to pick up raisins were unsuccessful, and they were often lost if grasped. It was also obvious that the monkey tried to compensate with tactual explorations, often in holes that were already empty. Such behavior was never seen in control sessions without infusions or in sessions with infusions when the monkey was allowed to work with the ipsilateral hand. We conclude that this first large muscimol infusion mimicked the deficit observed with the large irreversible parietal lesion in monkey M1 described above. This massive effect is particularly remarkable since even the relatively large infusion must have inactivated a much smaller tissue volume than the permanent large parietal lesion produced in monkey M1 who had a similar deficit. The deficit in unimanual reaching and grasping observed during inactivation in this unoperated monkey was strikingly similar to the deficit in monkey M1 produced by an irreversible parietal lesion and furthermore suggests that the symptoms in the latter monkey M1 were not compromised by the SMA lesion that preceded the parietal lesion.

Results from Inactivation Experiment 2. This smaller infusion was aimed at the center of area AIP, i.e., somewhat more laterally and rostrally than in Experiment 1. It produced similar effects as Experiment 1 with respect to hand shaping and grasping raisins from the horizontal and vertical slots. Figure 7A illustrates the elegant shaping of the control hand (ipsilateral to the microinfusion) that was also properly oriented to the slots. This contrasts with the gross disturbance in shaping and lack of wrist orientation of the hand contralateral to the muscimol infusion (Fig. 7B). This deficit was as severe as that observed following the infusion in Experiment 1. In contrast to the first experiment, however, *reaching was always performed in the correct direction of the target.*

Results from Inactivation Experiments 3 and 4. These infusions were aimed at the area LIP. No effects on either reaching or grasping were observed (up to 1 h after injection).

The following *conclusions* can be drawn from the present muscimol inactivation experiments: infusions in our most lateral target in the IPS (Experiment 2), which probably corresponded to area AIP of Sakata, resulted in a selective disruption of hand shaping and abnormal configuration of the fingers for picking up the raisins from the grooves. This is in agreement with the observations of Gallese et al. (1994). Conversely, we failed to exclusively produce misreaching by muscimol inactivation. With infusions only a few millimeters more mediocaudally, neither hand shaping nor reaching were affected. It was only with the massive infusions of Experiment 1 (crosses in Fig. 2A) that misreaching and contralateral neglect were clearly present together with gross disturbances of shaping and object manipulation.These findings obtained with the larger infusion are compatible with

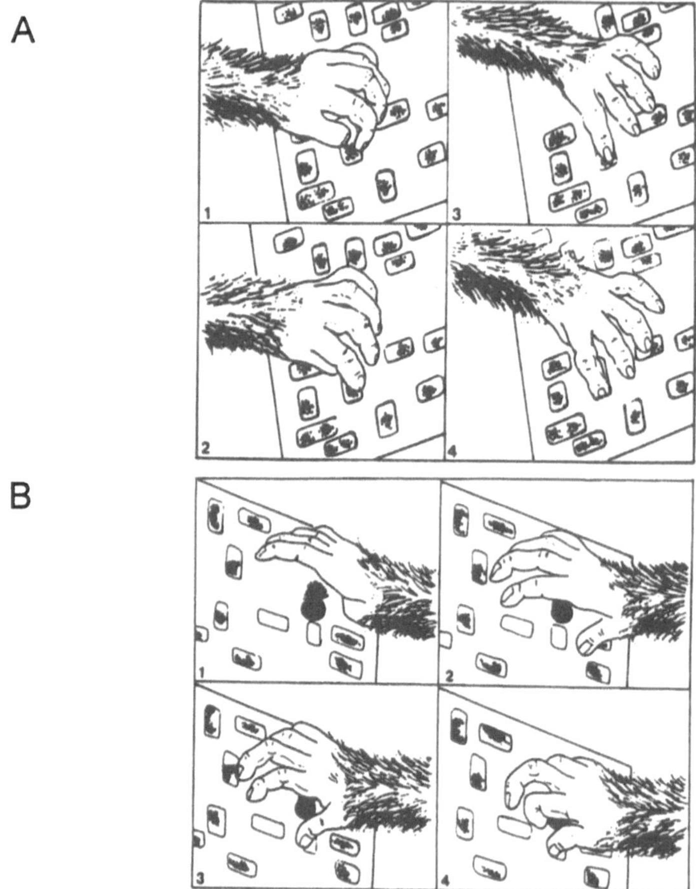

Fig. 7 A, B. Microinjection with smaller dose at sites indicated by *solid squares* in Fig. 2A, aimed at the anterior parietal area (AIP). **A** Hand ipsilateral to the infusion retrieves raisins with the precision grip from grooves of a board placed in front of the monkey. Note accurate shaping and wrist orientation for picking up the raisin from a horizontal slot with the hand ipsilateral to the microinfusion. **B** Performance of hand contralateral to the muscimol infusion with ipsilateral arm restrained. Highly abnormal posturing of fingers as the hand approaches the target: fanning of fingers with bizarre grasp configurations and lack of wrist adjustments to orientation of baited slots. Numbers 1–4 indicate the movement sequence. Redrawn from a series of digitized video frames

the idea of a wider distribution of the population of "reach neurons" outside area AIP, possibly involving area LIP or other foci near the IPS which required large infusions for their inactivation.

Summary and General Conclusion

The problem of how multiarticulate, purposeful arm-hand synergies are coordinated has been a focus of much interest in recent years. We addressed the question of the role of the posterior parietal cortex in the coordination of uni- and bimanual reaching and grasping. Although reaching and grasping, or right- and left-hand movements, can be executed independently, the reach-grasp synergies and bimanual goal-directed actions show impressive well-coordinated unitary performance. It is well known that brain lesions, for example in the cerebellum (Dichgans and Diener 1984; Inhoff et al.1989) can lead to a decomposition of synergies. With respect to the *unimanual reach-grasp synergy*, decomposition was clearly seen in a patient with a PPC lesion (Jeannerod et al. 1994), and similar decomposition was recently achieved experimentally with reversible inactivation techniques (Gallese et al. 1994; see also this volume).

In the present study, we confirmed the strong functional representation of the hand in the IPS of the PPC. A large lesion in this area resulted in a severe dysfunction in preshaping, i.e., in proper orientation of the hand and fingers when approaching an object; in addition, actual grasping was clumsy with whole-hand closure, utterly lacking the finesse of the precision grip. In the same monkey, the classical sign of misreaching and neglect towards the contralateral hemifield was also present. In contrast, a lesion that was more superficial and left the IPS intact in its anterior-lateral end did not interfere with hand function, but merely produced a short-lived deficit of directional errors in reaching. Whether this was due to a mass effect (small versus large lesion), rather than to distinct representations of the two components, was not clear. We therefore pursued the question further with the more elegant method of local and reversible inactivation. To our surprise, local infusion of muscimol in or near area AIP interfered with both the distal and the proximal component. The amount of injected muscimol was, however, large, and the dysfunction may well have involved neuronal populations outside area AIP in the posterior bank of the IPS. Infusions of lower doses placed slightly further medially and caudally (about 2 mm) had no effect on hand function, but also did not interfere with reaching. Finally, the most rostral-lateral infusion affected hand function only, without any sign of misreaching. Taken together, these findings are in good agreement with observations of the Sakata group (Sakata et al. 1995) that the hand area responsible for shaping and finger orientation in a grasping task is confined to a relatively small territory in the lateral-anterior portion of the IPS, i.e., in area AIP. We failed, however, to dissociate a clear-cut and specific area responsible for correct directional reaching. We submit that its representation is more widely distributed and that a relatively extended lesion is required for producing a reaching deficit. In monkey M2, the minor, short-lasting misreaching observed may have been caused by more superficial damage to the IPS region and/or by dorsal extension of the irreversible lesion encroaching upon area 5. This area in the superior parietal lobule has recently been considered as an important parietofrontal link executing visual-motor transformations for visually guided reaching (Caminiti et al. 1996).

As to the *role of the PPC in bimanual coordination*, the outcome of the lesion study suggests that the PPC contributes to interlimb coupling. Although monkey M1 with the large PPC lesion showed a significant reduction in the score of bimanual goal-related coupling not seen before with the SMA lesions, many trials still remained relatively well coordinated. Thus, since the two monkeys had previously also been subjected to bilateral lesions of the supplementary motor area, the lack of total breakdown of bimanual coordination with the additional PPC lesions is indicative of a very robust control system that may not depend exclusively on the investigated cortical areas, but also on coordinating mechanisms distributed in a number of cortical and perhaps also subcortical structures. The robustness and precision of bimanual coordination may be explained by its necessity for many skillful actions that are of great survival value. Every small deficits in bimanual coordination may dangerously impair the life of free-living primates.

Acknowledgements. We thank Aldo Tempini for his skillful training of monkeys and assistance during experimental sessions; Véronique Moret for histology and photography; Rita Wiesendanger for drawings in Figs. 1, 2, 4, 6, 7. Financial support was received from the Swiss National Science Foundation (grants # 31-36 183.92 and 4038-044053 to M.W. and grants # 31-28572.90, 31-43422.95, and 31-25128.88 to E.R.).

References

Brinkman C (1984) Supplementary motor area of the monkey's cerebral cortex: short-and long-term deficits after unilateral ablation and the effects of subsequent callosal section. J Neurosci 4:918–929

Brinkman C, Porter R (1979) Supplementary motor area in the monkey: activity of neurons during performance of a learned motor task. J Neurophysiol 42:681–709

Cadoret G, Smith AM (1995) Input-output properties of hand-related cells in the ventral cingulate cortex in the monkey. J Neurophysiol 73:2584–2590

Caminiti R, Ferraina S, Johnson PB (1996) The sources of visual information to the primate frontal lobe: a novel role for the superior parietal lobule (feature article). Cereb Cortex 6:319–328

Cavada C, Goldman-Rakic PS (1989) Posterior parietal cortex in rhesus monkey. I. Parcellation of areas based on distinctive limbic and sensory corticocortical connections. J Comp Neurol 287:393–421

Chapman CE, Wiesendanger M (1982) Recovery of function following unilateral lesions of bulbar pyramid in the monkey. Electroencephalogr Clin Neurophysiol 53:374–387

Cole J, Glees P (1955) Effects of lesions in posterior parietal lobe in trained monkeys. J Physiol 129:49–50P

Denny-Brown D (1966) The cerebral control of movements. Liverpool University Press, Liverpool, pp 110–142

Deuel RK, Regan DJ (1985) Parietal hemineglect and motor deficits in the monkey. Neuropsychologia 23:305–314

Dichgans J, Diener HC (1984) Clinical evidence for functional compartmentalization of the cerebellum. In: Bloedel JR, Dichgans JD, and Precht W (eds) Cerebellar functions. Springer, Berlin Heidelberg New York, pp 126–147

Faugier-Grimaud S, Frenois C, Stein DG (1978) Effects of posterior parietal lesions on visually guided behavior in monkeys. Neuropsychologia 16:151–168

Foerster O (1936) Motorische Felder und Bahnen. In: Bumke O, Foerster O (eds) Handbuch der Neurologie VI. Springer, Berlin, pp 1–357

Gallese V, Murata A, Kaseda M, Niki N, Sakata H (1994) Deficit of hand preshaping after muscimol injection in monkey parietal cortex. NeuroReport 5:1525–1529

Glees P, Cole J (1950) Recovery of skilled motor functions after small repeated lesions of motor cortex in macaque. J Neurophysiol 13:137–148

Hartje W, Ettlinger G (1973) Reaching in light and dark after unilateral posterior parietal ablations in the monkey. Cortex 9:346–354

He SQ, Dum RP, and Strick PL (1995) Topographic organization of corticospinal projections from the frontal lobe: motor areas on the medial surface of the hemisphere. J Neuroscience 15:3284–3306

Hepp-Reymond MC, Wiesendanger M (1972) Pyramidotomy in monkeys: effect on force and speed of a conditioned precision grip. Brain Res 36:117–131

Hepp-Reymond MC, Trouche E, Wiesendanger M (1974) Effects of unilateral and bilateral pyramidotomy on a conditioned rapid precision grip in monkeys (Macaca fascicularis). Exp Brain Res 21:519–527

Hepp-Reymond M-C, Hüsler EJ, Maier MA, Qi H-X (1994) Force-related neuronal activity in two regions of the primate ventral premotor cortex. Can J Physiol Pharmacol 72:571–579

Hécaen H (1978) Les apraxies idéomotrices, essai de dissociation. In: Hécaen H, Jeannerod M (eds) Du contrôle moteur à l'organisation du geste. Masson, Paris, pp 343–358

Hyvärinen J, Poranen A (1974) Function of the parietal association area 7 as revealed from cellular discharges in alert monkeys. Brain 97:673–692

Inhoff AW, Diener HC, Rafal RD, Ivry R (1989) The role of cerebellar structures in the execution of serial movements. Brain 112:565–581

Jeannerod M, Biguer B (1982) Visuomotor mechanisms in reaching within extrapersonal space. In: Ingle MA, Goodale MA, Mansfield RJW (eds) Advances in analysis of visual behavior. MIT Press, Cambridge, pp 387–409

Jeannerod M, Decety J, Michel F (1994) Impairment of grasping movements following a bilateral posterior parietal lesion. Neuropsychologia 32:369–380

Jeannerod M, Arbib MA, Rizzolatti G, Sakata H (1995) Grasping objects: the cortical mechanisms of visuomotor transformation. Trends Neurosci 18:314–320

Kalaska JF, Caminiti R, Georgopoulos AP (1983) Cortical mechanisms related to the direction of two-dimensional arm movements: relations in parietal area 5 and comparison with motor cortex. Exp Brain Res 51:247–260

Kazennikov O, Wicki U, Corboz M, Hyland B, Palmeri A, Rouiller EM, Wiesendanger M (1994) Temporal structure of a bimanual goal-directed movement sequence in monkeys. Eur J Neurosci 6:203–210

LaMotte RH, Acuña C (1978) Defects in accuracy of reaching after removal of posterior parietal cortex in monkeys. Brain Res 139:309–326

Lawrence DG, Kuypers HGJM (1968) The functional organization of the motor system. I. The effects of bilateral pyramidal lesions. Brain 91:1–14

Leinonen L, Hyvärinen J, Nyman G, Linnankoski I (1979) I. Functional properties of neurones in lateral part of associative area 7 in awake monkeys. Exp Brain Res 34:299–320

MacKay WA (1992) Properties of reach-related neuronal activity in cortical area 7A. J Neurophysiol 67:1335–1345

Mountcastle VB (1995) The parietal system and some higher brain functions. Cereb Cortex 5:377–390

Mountcastle VB, Lynch JC, Georgopoulos AP, Sakata H, Acuña C (1975) Posterior parietal association cortex of the monkey: command functions for operations within extrapersonal space. J Neurophysiol 38:871–908

Nixon PD, Burbaud P, Passingham RE (1992) Control of arm movement after bilateral lesions of area 5 in the monkey (*Macaca mulatta*). Exp Brain Res 90:229–232

Pause M, Kunesch E, Binkofski F, Freund H-J (1989) Sensorimotor disturbances in patients with lesions of the parietal cortex. Brain 112:1599–1625

Peele TL (1944) Acute and chronic parietal lobe ablations in monkeys. J Neurophysiol 7:269–286

Perenin MT, Vighetto A (1988) Optic ataxia: a specific disruption in visuomotor mechanisms. I. Different aspects of the deficit in reaching for objects. Brain 111:643–674

Rizzolatti G, Camarda R, Fogassi L, Gentilucci M, Luppino G, Matelli M (1988) Functional organization of inferior area 6 in the macaque monkey: II. Area F5 and the control of distal movements. Exp Brain Res 71:491–507

Sakata H, Taira M, Murata A, Mine S (1995) Neural mechanisms of visual guidance of hand action in the parietal cortex of the monkey. Cereb Cortex 5:429–438

Schwartzman RJ (1978) A behavioral analysis of complete unilateral section of the pyramidal tract at the medullary level in *Macaca mulatta*. Ann Neurol 4:234–244

Shima K, Aya K, Mushiake H, Inase M, Aizawa H, Tanji J (1991) Two movement-related foci in the primate cingulate cortex observed in signal-triggered and self-paced forelimb movements. J Neurophysiol 65:188–202

Stein J (1978) Effect of parietal lobe cooling on manipulative behaviour in the conscious monkey. In: Gordon G (ed) Active touch. Pergamon, Oxford, pp 79–90

Tanji J, Okano K, Sato KC (1988) Neuronal activity in cortical motor areas related to ipsilateral, contralateral, and bilateral digit movements of the monkey. J Neurophysiol 60:325–343

Thier P, Andersen RA (1996) Electrical microstimulation suggests two different forms of representation of head-centered space in the intraparietal sulcus of rhesus monkeys. Proc Natl Acad Sci USA 93:4962–4967

Tower SS (1940) Pyramidal lesion in the monkey. Brain 63:36–90

Wiesendanger M (1986) Recent developments in studies of the supplementary motor area of primates. Rev Physiol Biochem Pharmacol 103:1–59

Wiesendanger M, Wicki U, Rouiller E (1994) Are there unifying structures in the brain responsible for interlimb coordination? In: Swinnen SP, Heuer H, Massion J, Casaer P (eds) Interlimb coordinaton: neural, dynamical and cognitive constraints. Academic, San Diego, pp 179–207

Optic Ataxia and Unilateral Neglect: Clinical Evidence for Dissociable Spatial Functions in Posterior Parietal Cortex

M.-T. PERENIN

Vision et Motricité, INSERM U 94, Bron, France

The parietal lobe syndrome in humans typically appears as a multiplicity of disorders, mostly affecting spatially oriented behavior (for reviews see, e.g., De Renzi 1982; Milner and Goodale 1995). Although often associated, these disorders may sometimes appear in isolation. As will be seen in this chapter, this is the case in particular for misreaching and unilateral neglect. Such neuropsychological dissociations as well as neurophysiological findings obtained in the monkey during the past 20 years have largely improved our understanding of the functional organization of the posterior parietal lobe. This cortical region appears to subserve two main types of spatial functions respectively related to sensorimotor transformations and building abstract internal representations of space.

We will review below the clinical evidence for dissociable spatial functions in the human posterior parietal cortex and critically compare it with related neurophysiological findings in the monkey.

Optic Ataxia and Unilateral Neglect: Two Dissociable Clinical Entities

Optic ataxia and unilateral neglect were first observed as parts of more complex syndromes including a variety of other spatial disorders and often resulting from bilateral parietooccipital lesions. However, optic ataxia and neglect were then reported separately in patients with unilateral lesions. We will provide arguments showing that, although they both include directional biases, these two disorders disrupt different spatial functions.

Optic Ataxia: A Specific Visuomotor Disorder

Since the original observation of Balint (1909), "optic ataxia" has designated a disorder of coordination and accuracy of visually elicited hand movement not

In: Parietal Lobe Contributions to Orientation in 3D Space (1997). P. Thier and H.-O. Karnath (eds). Springer-Verlag, Heidelberg.

related to motor, somatosensory, visual acuity, or visual field deficits. In Balint's patient, who suffered from a bilateral infarction of the parietooccipital junction, optic ataxia was associated with major impairments in visual attention. These included "psychic paralysis of gaze" (the patient could not shift his gaze fixation) and neglect of the left hemispace. However, these disorders could not account for optic ataxia, which only affected the right hand of the patient. Even in this early case, optic ataxia appeared as a specific deficit of the visual control of movement, since the patient could perform movements directed to his own body quite accurately.

In most further reports of bilateral parietooccipital syndromes, epitomized as "Balint's syndromes", reaching disorders did not appear as specific: they usually affected both hands and could hardly be dissociated from oculomotor and visual-space perceptive impairments. In fact, these bilateral syndromes are most often closer to the "visual disorientation" of Holmes (1918). In this situation, inaccurate visuomotor behavior is observed not only in reaching but also in avoiding obstacles during walking. The first reported cases of misreaching following unilateral damage to the posterior parietal region were interpreted in this latter framework. Visual disorientation was observed in one hemifield in those cases (Riddoch 1935, Brain 1941, Cole et al. 1962).

Optic ataxia has later been recognized as a specific entity that may occur without perceptual, oculomotor, or visual attention disorders (Garcin et al. 1967). In most reported cases, optic ataxia results from a unilateral lesion centered on the posterior parietal region. Typically, patients are impaired in reaching with either hand for objects located in the contralesional visual field. However, other patterns of deficit can be seen; in particular, misreaching may affect only one hand within one hemifield (Castaigne et al. 1975; Tzavaras and Masure 1976; Rondot et al. 1977). In fact, as shown in a quantitative study of a series of ten patients (Perenin and Vighetto 1988), the distribution of errors (of both the reaching and grasping components of prehension) depends on the side of the lesion. While right-damaged patients show a deficit mostly related to a field effect (both hands affected in the left field), left-damaged patients show a deficit mostly related to a hand effect (right hand affected in both fields). This difference in the distribution of errors according to the side of the lesion most likely results from an asymmetry of functional organization between right and left parietal areas subserving visually-directed movements.

Unilateral Neglect: A Cognitive Disorder of Mental Representation of Space

Unilateral neglect is a complex and multiform syndrome whose manifestations may largely vary from one case to another. Patients with this disorder usually behave as if the left side of space does not exist anymore: they ignore, forget, or turn away from objects and people located in that part of space. They lack awareness of stimuli on their left even when able to see and shift their gaze to this side. In fact, although primary sensory or motor deficits are often associated, they cannot account for neglect.

In early reports, neglect manifestations were variously combined with other disorders of spatial cognition such as loss of topographical memory, dressing apraxia, and constructive apraxia following extensive right-hemisphere lesions, or combined with visuomotor disorders in cases of bilateral Balint-Holmes syndromes. All these different symptoms were sometimes regarded as part of a supposed syndrome of "visual-spatial agnosia" as opposed to visual object agnosia (reviewed in Weinstein and Friedland 1977; De Renzi 1982). Apart from a few exceptions [e.g., Zingerle's (1913) interesting but largely ignored case of "dyschiria"; for an appraisal see Bisiach and Berti (1987)], spatial neglect was not seen or recognized as a neurological entity until the middle of this century. A first attempt was made by Brain (1941) to subdivide "visual disorientation" from unilateral lesions into several kinds of disorders; in particular, this author claimed that "defective localization of objects" and "agnosia for the left half of space" could occur independently.

During the last two decades, an increasing number of studies have been devoted to unilateral neglect, which has proved a fascinating field for exploring spatial cognition and awareness (Bisiach and Vallar 1988, Jeannerod 1987, Robertson and Marshall 1993). From all of these studies, neglect, mostly observed following right hemisphere lesions, appears as a heterogeneous syndrome, in much the same way as aphasia following left hemisphere lesions. It may affect to various degrees extrapersonal and personal space but also mental space (e.g., when patients are asked to evoke mentally a familiar place: see Bisiach and Luzzatti 1978). It may be variously related to the different sensory modalities (visual, auditory, tactile) and to the different spatial frames of reference (head-, body-, or object-centered etc.). Motor components of neglect may predominate over perceptual components or vice versa. As already mentioned, primary sensory or motor disorders (hemianopia, hemianesthesia, hemiplegia) and contralesional gaze palsy are frequently associated with neglect. Mostly in the acute stage, other cognitive disorders concerning the contralesional space are typically observed. Anosognosia for hemiplegia and hemianopia, i.e. the denial or at least reluctance to admit the motor or visual deficit, is likely to be encountered most frequently. Patients may also fail to recognize their left hemibody as their own and they may sometimes show delusions and confabulations concerning the affected side (asomatognosia and somatoparaphrenia). They may perceive tactile or auditory stimuli produced on the contralesional side in a symmetrical position in the ipsilateral half of the body or surrounding space ("allochiria").

Despite the many dissociations observed between all these manifestations, there have been continuous attempts to ascribe them to a common basic disturbance. Two main interpretations of neglect have been proposed, the one relating it to spatial attention (Kinsbourne 1987, 1993; Heilman et al. 1987), the other to mental representation of space (Bisiach and Berti 1987; Bisiach and Vallar 1988;

Bisiach 1993). On the other hand, it was suggested that no single mechanism could account for neglect that was in fact seen as a heterogeneous set of disorders rather than a genuine clinical entity or syndrome (Halligan and Marshall 1992).

Kinsbourne's model assumes that two opposite and competitive vectors direct attention and orientation behavior to the two opposite sides of egocentric space. In addition, the righward vector, depending on left-hemispheric structures, would be functionally steeper than the leftward, right-hemispheric-dependent vector. Several empirical facts can be explained by this model, among them the higher incidence of right-hemispheric lesions responsible for neglect. Damage to this hemisphere not only impairs the leftward vector but also releases from inhibition the steeper rightward vector.

However, only part of the neglect phenomena and related symptoms can be explained by an attentional and/or orientational bias. A more systematic account is provided by Bisiach's representational model. This was originally prompted by the famous finding that Milanese patients also displayed neglect in their descriptions of the Piazza del Duomo when asked to visualize the square from a given vantage point (Bisiach et al. 1981). Since then, "imaginal" neglect has been found several times in descriptions of interiors or geographic maps (Halsband et al. 1985; Meador et al. 1987; Barbut and Gazzaniga 1987; Rode and Perenin 1994). In our own series of neglect patients, asked to evoke mentally the map of France and to name as many towns as possible in a limited period of time (see Fig. 1), imaginal neglect was found in a large majority of cases (27 out of 34 patients) (Perenin et al., in preparation; Rode and Perenin 1994). These data have pointed out space representation (closely related to the classical notions of "body scheme" and "scheme for external space") as the level at which disrupted mechanisms responsible for neglect should be found. In this model, unilateral neglect and related phenomena are viewed as a disorder of the conscious representation of one side of the body and of extracorporal space, accounting for both defective (neglect) and productive (e.g., somatoparaphrenia) manifestations. They were grouped many years ago under the name of "dyschiria" by Zingerle (1913), who at that time proposed an interpretation in a similar vein. In recent years, further insight into the neurobiological bases of the present model has been provided by the remission of virtually all components of dyschiria under vestibular or other sensory manipulations (see Vallar et al., this volume). The spectacular effect of vestibular stimulation on somatoparaphrenic delusions (Bisiach et al. 1991; Rode et al. 1992) and on imaginal neglect (Geminiani and Bottini 1992; Rode and Perenin 1990, 1994, see Fig. 1) suggests that the neural substrate of the representational scheme is a distributed and dynamic network: following a lesion in a given part of this network, mental representation of space can be restored by increasing activation in another part.

In the context of the attentional theory of neglect, misreaching following posterior parietal lesions (optic ataxia), which appears as an effector-specific orientation bias (see below), may be regarded as a minor form of unilateral neglect. We will provide arguments against this view by showing evidence for

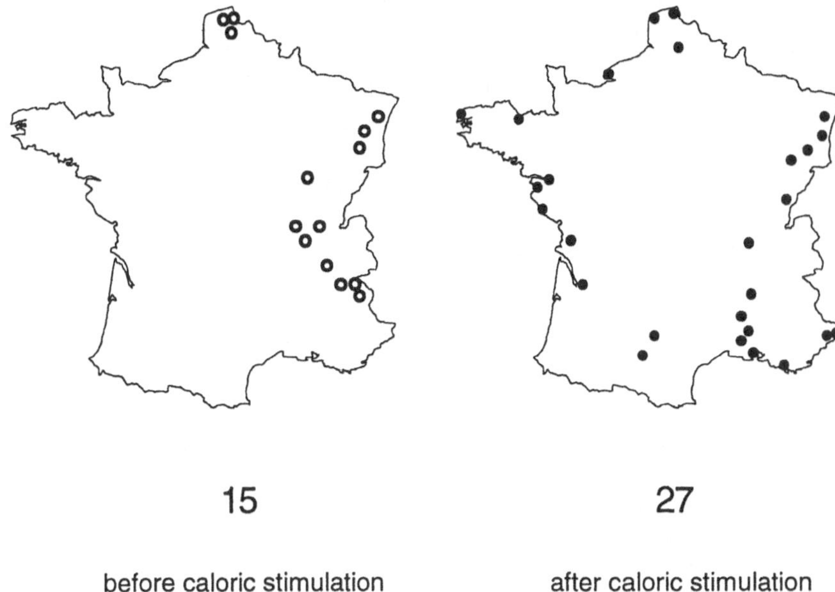

Fig 1. Example of "imaginal" neglect in a patient asked to visualize mentally the map of France and to name as many towns as possible during a limited period of time (2 min). Temporary remission through vestibular stimulation is seen in the *right part* of the figure

both clinical and anatomical dissociations between these two disorders. In addition, we will report psychophysical results on pointing at visual targets and on straight ahead judgment in the dark, leading to further insight on the functional organization of the posterior parietal cortex in humans.

Clinical Dissociations

As mentioned previously, early reports on "visuo-spatial agnosia" or "visual disorientation" already showed possible dissociations between misreaching and unilateral neglect (Brain 1941, Ettlinger et al. 1957).

In the first cases of pure optic ataxia reported in the French literature, no neglect symptom was ever observed. However, sensory and/or visual extinction were present in five out of seven cases (Garcin et al. 1967; Rondot et al. 1977; Tzavaras and Mazure 1976). Neither of these disorders were mentioned in the other four reported cases (Levine et al. 1978; Auerbach and Alexander 1981; Ferro et al. 1983; Ferro 1984). In our series of ten optic ataxia patients, only two of them with a left parietal lesion showed minimal signs of neglect that disappeared in 1 month after onset (in drawing or writing, they only used the leftmost part of the sheet of paper, although they performed correctly in other tasks such as crossing-out line segments and line bisection); only one patient displayed transient visual extinction (Perenin and Vighetto 1988). In eight subsequent cases, six did not show any neglect symptoms or extinction (four left, two right lesions), while the other two (right posterior parietal lesions) had trimodal extinction and mild and marked extrapersonal neglect respectively. However, probably due to the nature of the lesions (vascular infarct and intracerebral haematoma), unilateral neglect disappeared completely while optic ataxia remained; extinction was found less consistently.

Occurrence of unilateral neglect without optic ataxia is much more difficult to demonstrate. Most often, visual field, sensory, and/or motor deficits prevent examination of a specific visuomotor incoordination. In our series of neglect patients, only two showed virtually no sensorimotor deficit, but they had a complete left hemianopia; they could reach for objects in their right visual field accurately. Among the few patients without hemianopia or with a partial quadrantanopic deficit, three could be tested for reaching with their right hand only and three with both hands. Only one showed a slight misreaching with the right hand in the left visual field; this patient was in fact much more impaired when reaching with the left hand in the left field. This case can be considered as an association of optic ataxia and neglect. The other two patients showed a transitory misreaching with their left hand, while signs of neglect remained (e.g., in drawing and line bisection). Thus, at least in these two patients unilateral neglect could be seen in the absence of optic ataxia.

Anatomical Dissociations

It has been well established that unilateral neglect is much more frequent and severe following right hemisphere lesions. Optic ataxia, although first reported in right brain-damaged patients (Rondot et al. 1977), can in fact be seen after lesion of either hemisphere (Perenin and Vighetto 1988). There are other major differences: although in the majority of cases neglect results from lesions of the parieto-temporo-occipital junction, it may be observed following either more anterior or deeper lesions invading, e.g,. the fronto-capsular region, the caudate nucleus, or the thalamus (Vallar and Perani 1986); in contrast, optic ataxia has always been related to damage of the posterior parietal region. There is only one recent report of a case of misreaching resembling optic ataxia related to a thalamic lesion (Classen et al. 1995). In fact, even more interesting regarding the functional organization of the posterior parietal cortex, studies using reconstructions and superimpositions of lesions from CT scan images have shown that damage of

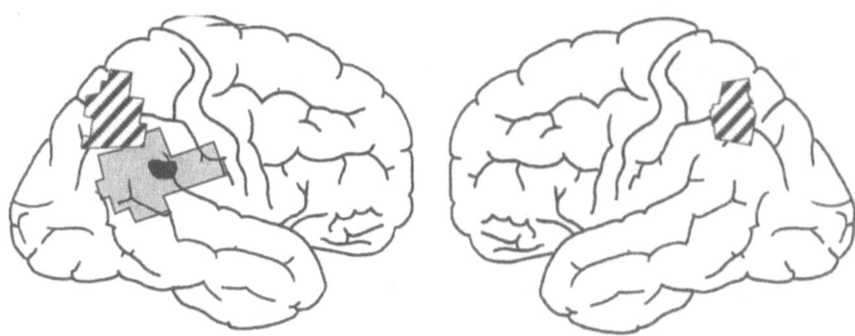

Fig 2. Regions of overlap of lesions responsible for optic ataxia (*hatched*) or unilateral neglect (*grey and black*). Optic ataxia lesions are drawn from CT reconstructions according to Perenin and Vighetto (1988); six left hemisphere and four right hemisphere lesions are superimposed. Neglect lesions are all located in the right hemisphere. Areas of overlap corresponds to lesions of the four patients studied in the present paper (*grey*) and the seven patients with severe neglect (*black*) from the study of Vallar and Perani (1986)

optic ataxia tends to lie in the upper part of the parietal lobe around the intraparietal sulcus (Perenin and Vighetto 1988). In contrast, as illustrated in Fig. 2, similar studies have shown that the focus for unilateral neglect lies more ventrally, in the region of the supramarginal and angular gyri (Heilman et al. 1983b; Vallar and Perani 1986).

Directional Biases in Optic Ataxia and Neglect: Disruption of Different Spatial Functions

Directional biases of orientation behavior can be observed in both optic ataxia and unilateral neglect. We will examine the hypothesis that, partly due to the different lesion sites, the bias would be mostly effector-specific in optic ataxia and more widely distributed in unilateral neglect. An additional question is whether or not the spatial cognition deficit underlying neglect affect the "on-line" visuomotor transformation processes taking place in the dorsal parietal cortex and subserving goal-directed movements.

Spatially oriented behavior and spatial cognition require sensory information from the periphery (visual, proprioceptive, vestibular) to be integrated and transformed in nonsensory (e.g., body-centered, world-centered) coordinates. These are used by the brain to build multiple neural representations of space that serve as interfaces between sensory inputs and motor outputs, mostly in a specific way, i.e., for a given action and which also underlie high-order spatial functions (e.g., Graziano and Gross 1995; Rizzolatti et al. 1983; Stein 1992; Vallar et al.

1993b). An important role of these nonsensory spatial coordinate systems is to provide cues on the position of the body (or of a body part) with respect to surrounding space, which are needed for accurate motor behavior in space. The fact that we can still reach a visual target even in the absence of vision of our hand and despite eye and/or head movements is a clear demonstration of these central mechanisms of coordinate transformations.

It has been argued recently that the main cause of neglect lies in a disturbance of these mechanisms: coordinate transformation supposedly works with a systematic error resulting in a deviation of the spatial reference frame to the side of the lesion (Karnath et al. 1991,1993; Karnath 1994). As a consequence of the shift of the reference frame underlying the internal representation of space, the subjective localization of body orientation (as part of this representation) would be separated from its objective position and displaced to the ipsilesional side (Ventre et al. 1984). This has been found indeed in several studies (Heilman et al. 1983a; Karnath 1994; Mark and Heilman 1990; Perenin 1992; complementary results shown below).

On the basis of current ideas on dissociable mechanisms subserving perception and action (for a review on normal subjects and brain-damaged patients see Jeannerod 1994; Milner and Goodale 1995), it could be expected that a disturbed mental representation of space would not affect the rapid input-output specific sensorimotor transformations such as those involved in reaching movements and vice versa. In order to answer this question, we have compared the performance of patients with either optic ataxia or unilateral neglect in two different tasks: pointing at visual targets in open loop condition (in the absence of vision of the hand) and subjective localization of the body orientation in total darkness.

Pointing experiments were performed by applying a paradigm already used in normal subjects (Prablanc et al. 1979) and in three optic ataxia patients (Perenin and Vighetto 1983). The patients were required to point at small target lights that appeared randomly along a frontoparallel plane. The present study was performed in complete darkness, with the head immobilized and the eyes free to move in the direction of the target (open loop, foveal vision condition). Instruction was given to point as fast and accurately as possible. Four optic ataxia patients, two with left and two with right parietal lesions [(JR, AV, and JP from Perenin and Vighetto (1983, 1988), and another right damaged patient BT)], and four neglect patients with lesions including the right parietotemporal junction participated in this experiment. In the latter group, two patients showed a left visual field defect (incomplete hemianopia and inferior quadrantanopia in RA and PW respectively), and all of them had motor and somatosensory deficits on the left hemibody; thus, they could perform the pointing task only with the right hand. It should be noticed that none of the neglect patients exhibited misreaching with the right hand on clinical examination.

In the usual clinical testing conditions of optic ataxia (peripheral vision of both target and hand) no true systematic error was ever noticed in the (about) twenty cases published until today nor in our eight last patients (but see Ratcliff and

Optic Ataxia

a

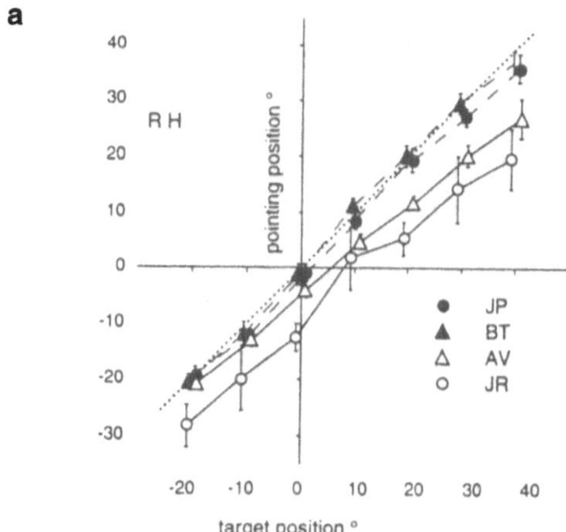

b

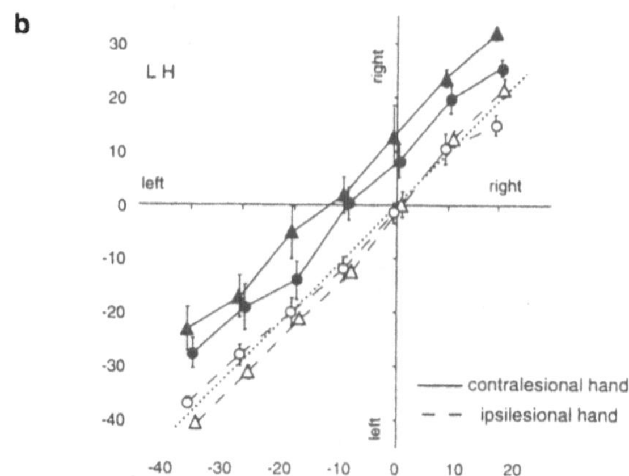

Fig 3 a, b. Pointing responses at visual targets in open loop condition in four patients with optic ataxia. *Open symbols*: patients with left hemisphere lesions (*AV*, *JR*); *filled symbols*: patients with right hemisphere lesions (*JP*, *BT*). Patients had to direct eyes and hand at targets that appeared randomly in darkness along a frontoparallel plane. **A** Pointing responses of the right hand (*RH*). **B** Pointing responses of the left hand (*LH*). All four optic ataxia patients show a directional error of more than 10° toward the side of the lesion with their contralesional hand, while performing normally with the ipsilesional hand

Davies-Jones 1972, who mentionned a trend to undershoot targets in the contralesional visual field). However, in the open loop condition we used, pointing responses with the contralesional hand showed a very clear constant error of more than 10° to the side of the lesion in both half-spaces of all four patients. Responses of the ipsilesional hand remained in the normal range (Fig. 3).

In the neglect patients, we did not find any evidence for a systematic ipsilesional bias on the right hand. Two patients (GS and RA) showed a pattern of responses similar to that of normal subjects in the same open loop condition (Prablanc et al. 1979): they tended to overshoot near targets and undershoot far targets. Yet the other two patients (PW and JO) showed a leftward error at nearly all eccentricities (Fig. 4).

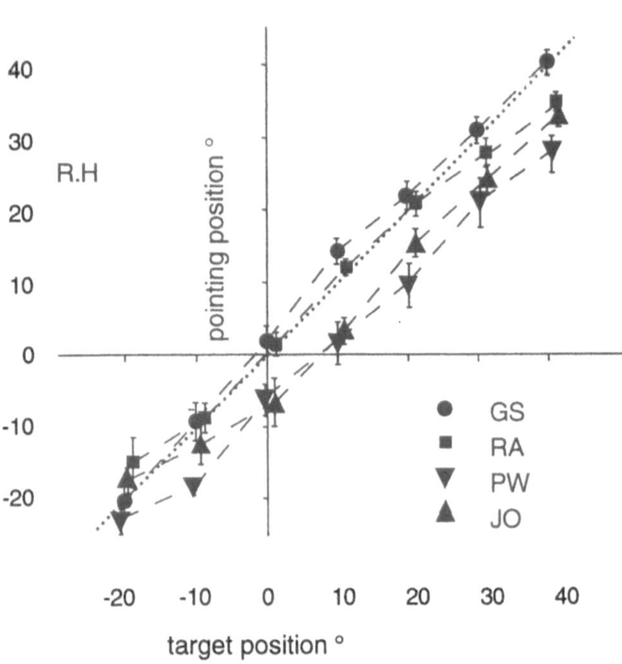

Fig 4. Pointing responses at visual targets in open loop condition in four patients with unilateral neglect (*GS*, *RA*, *PW*, *JO*). Patients could only be tested with their ipsilesional right hand. None of them show a systematic error toward the side of the lesion. Two patients (*GS* and *RA*) display normal responses, while the other two err to the left of targets at nearly all eccentricities

The subjective position of body orientation was assessed by means of two different tasks. Patients were seated in complete darkness, with their trunk and head restrained. In the first task, they had to point at an imaginary target located in front of the middle of their trunk, by displacing their ipsilesional hand on a recording table. In the second task, a small laser spot was moving horizontally on a hemicylindrical screen at a speed of 1°/s, starting randomly from 20° to the right or left. Patients had to stop the moving beam, by pressing a button as soon as the beam appeared located in front of the middle of their trunk. In a later version of this task, the laser spot was displaced by the patient himself by means of a cursor until he was satisfiedthat it was straight ahead.

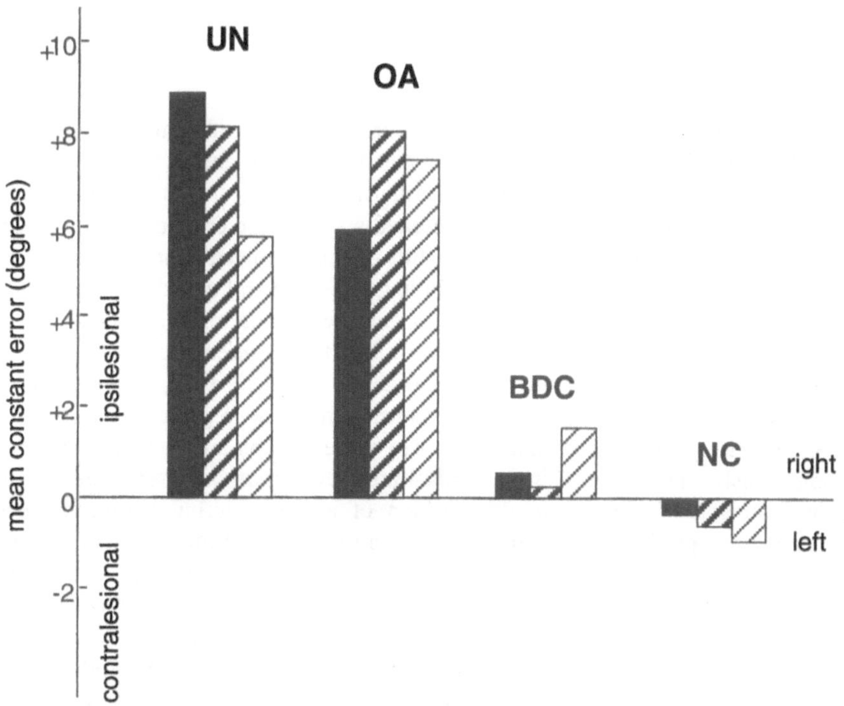

Fig 5. Subjective "straight ahead" judgements performed in darkness by patients with unilateral neglect (*UN*, *n*=25), optic ataxia (*OA*, *n*=8), brain-damaged controls (*BDC*, *n*=10) and normal controls (*NC*, *n*=10). Subjects were required to point in front of the middle of their trunk (*black bars*) or to perform a visual "straight ahead" judgement (*hatched bars*) using a laser spot either passively moved by the experimenter (*thick stripes*) or actively displaced by the subjects themselves (*thin stripes*)

Eight optic ataxia patients (six with a left and two with a right parietal lesion; among them JR and BT tested for pointing in open loop condition) and 25 neglect patients (among them the four patients tested in the previous condition) went through this experiment. Their performance was compared to that of ten age-matched brain-damaged controls and ten normal controls. The results are shown in Fig. 5.

Apart from three who did not differ from the brain-damaged controls despite moderate to severe signs of neglect, all neglect patients displayed a significant shift of their subjective midsagittal plane toward the side of the lesion (ranging from 4° to 26°, mean 6° to 9°, depending on the task); this was observed in both types of tasks in 15 cases and in only one of them in ten cases.

Less expected were the results of the optic ataxia patients. They all showed a shift of their subjective straight ahead (mean 6° to 8°), although less pronounced in the pointing task as compared to the neglect patients. A significant shift was observed in both types of tasks in five patients and in only one task in three patients.

These results partly fit with our expectation: patients with unilateral neglect, who showed a rightward shift of their subjective straight ahead, did not show a similar bias to the right in pointing at visual targets with the ipsilesional hand. However, optic ataxia patients, who exhibited a systematic error of pointing with the contralesional hand, also showed a shift of their subjective straight ahead in the same ipsilesional direction. These findings raise several questions concerning in particular the relationship betwen space perception and action and the functional significance of the subjective position of body orientation.

Although they usually collaborate in order to produce a single phenomenal percept of the object to be reached, mechanisms involved in perception and goal-directed actions are quite distinct. Many of these actions are in fact performed unconsciously. This aspect is well illustrated by perturbation experiments in normal subjects. A change in the position of a target during a saccadic eye movement can produce an "on-line" adjustment of the hand motor response even though that change is not perceived by the subject (e.g., Goodale et al. 1986). Yet even when the change in the position of a target object is perceived, time for awareness of the change is delayed by at least 300 ms with respect to adjustments of the trajectory of the motor response directed to that object (Castiello et al. 1991). In fact, object-oriented actions rely on fast, "on-line" processing of spatial information; when the motor response is artificially delayed, the performance is altered. Conversely, perception and identification of objects and places are time-consuming processes relying on more enduring characteristics (see review in Milner and Goodale 1995). In the optic ataxia patients, the directional errors of pointing with the contralesional hand can best be explained by a disruption of a specific sensorimotor transformation working with a systematic error toward the side of the lesion. The arm-specific character of the deficit may further support the idea of a shoulder-centered coordinate system underlying reaching movements, as previously argued (Soechting et al. 1990, Caminiti et al. 1992). The accuracy of pointing with the ipsilesional hand, despite a shift of the subjective body orientation, suggests that sensorimotor transformations rely on more ephemeral space representations probably incorporating other ongoing spatial cues, such as those related to saccadic eye movements. In the neglect patients, pointing responses were less consistent. Although this will need further investigation, a possible explanation lies in different movement times. In fact, the two patients who showed a leftward bias were tested only a few weeks postonset, while the other two were tested more than a year postonset. It might be that the former performed slower movements and relied on a perceptive mode for producing pointing responses, while the latter performed faster responses relying on mostly unconscious spatial representations. The performance of these two patients is reminiscent of the accurate object-oriented movements observed in a patient with visual agnosia due to a bilateral lesion of the ventral visual pathway partly including the inferior parietal lobe (Goodale et al. 1991) .

The subjective position of body orientation has been regarded as a perceptual parameter directly connected to an egocentric frame of reference required for accurate visually oriented actions and exploration of space (Jeannerod and Biguer 1987). This is indeed supported by the effects of sensory manipulations in normal subjects. Vibration of the left posterior neck muscles, for instance, produces a rightward illusory displacement of a visual target fixated in the dark and a systematic error of pointing with the unseen hand toward the perceived position of the target. In correlation, the subjective visual straight ahead is shifted by the same amount to the left side (Biguer et al. 1986). Using either vestibular or optokinetic stimulation we have obtained quite similar effects (unpublished results). Clearly, the subjective position of body orientation does not have the same predictive value on the visuomotor behavior of optic ataxia and neglect patients (only two of our patients erred to the left of visual targets, i.e., in a direction opposite to that of subjective straight ahead) or on the perceived localization of sounds or limb segments in neglect patients, as this was found systematically displaced to the right (Bisiach et al. 1984; Vallar et al. 1993a). As suggested above, sensorimotor transformations probably rely on transient, unconscious space representations, which do not necessarily coincide with subjective body orientation.

Displacement of the subjective body midline in neglect may be considered in fact as a consequence of an asymmetry of internal space representation that appears as an unawareness of the left side of space and as a correlated trend to orient to the right side. Two main (not mutually exclusive) mechanisms may account for this asymmetry. The unbalanced activity of the bilateral set of structures involved in building spatial coordinates (Kinsbourne 1987; Rizzolatti and Berti 1993) might be responsible for a left-right gradient of representation, with relatively degraded left-side representation and functionally predominant right-side representation. On the other hand, due to the right-hemisphere lesions, the mechanism by which various sensory coordinates (visual, vestibular, propioceptive) are integrated and transformed into spatial, body-centered coordinates might be disrupted, primarily at the expense of the left half-space representation. The hypothesis that this transformation process works with a systematic error, as argued by Karnath et al. (1993, Karnath 1994) may account for the ipsilesional shift of different types of orientation behavior as well as subjective localization of the midsagittal plane. However, misrepresentation of space in unilateral neglect does not only consist of a translation of the egocentric frame of reference toward the lesion side (Vallar et al., this volume). In fact, displacement of subjective body orientation can be observed outside the pathological condition of neglect. This is the case in normal subjects each time a strong imbalance is produced in a sensory system using vestibular, optokinetic or neck proprioceptive stimulation, as seen previously. Similarly, in the absence of sensory input on one side in patients with acute peripheral vestibular disorder (Hörnsten 1979), hemianopia of occipital origin or somatosensory deafferentation due to thalamocortical lesions (Perenin et al., in preparation) also induces a displacement of the subjective straight ahead. As shown in the present study, this is the case in optic ataxia, too. In all these conditions, the neural coordinate-transformation process is only partially impaired, which can be compensated for centrally without giving rise to the misrepresentation of space and the spatial cognition disorders observed in neglect (for further discussion of this topic see Karnath, this volume).

As shown by our results, the mechanisms of spatial coding required for action on the one hand and for perception and conscious representation of space on the other hand may be dissociated in pathological cases, suggesting different neural substrates. The upper part of the posterior parietal cortex including the intraparietal sulcus would be involved in short-living, unconscious spatial representations required for specific "on-line" sensorimotor transformations. The lower part as well as the adjacent occipitotemporal regions would be responsible for more enduring and conscious representations underlying spatial cognition and complex spatially oriented behavior.

We will provide below evidence for a similar dichotomy of spatial representations and spatial functions in the monkey, although a change of topography appears to have occurred between monkey and man.

Comparisons with Parietal Lobe Functions in the Monkey

Neurophysiological studies of the posterior parietal cortex (reviewed by several authors in this book) have shown that most neurons in this area have both sensory- and motor-related activities; many of them also respond to more than one sensory modality. This has suggested that posterior parietal cortex plays an essential role in sensorimotor transformations (Andersen et al. 1993). The latter would convert sensory coordinates into several effector-specific motor coordinate systems dealing with different components of actions such as eye and head movements, reaching and grasping at objects. These input-output specific transformations appear to rely on partially segregated subregions of the posterior parietal cortex (e.g., Andersen 1995; Sakata et al. 1995).

Higher cognitive functions such as attention, intention, and abstract representation of space have also been ascribed to the inferior parietal lobule of the monkey (Andersen 1995; Bushnell et al. 1981; Rizzolatti et al. 1994). It has been assumed in particular that abstract representation of space is constructed from the integration of visual, auditory, vestibular, eye-position, and proprioceptive head-position signals that reach the parietal cortex of the monkey. These functional properties of the monkey parietal cortex and in particular its role in spatial attention, mainly inferred from electrophysiological studies, have often been related to the occurrence of unilateral neglect following parietal damage in man. However, the effects of parietal lobe lesions in the monkey do not produce any homologue of the human neglect syndrome.

In fact, parietal lobe damage in the monkey primarily affects visually directed movements. Misreaching is certainly one of the most obvious disorders. Like optic ataxia in man, it appears as a visuomotor deficit rather than a deficit in spatial perception. Following unilateral lesions, the reaching disorder only affects the contralesional arm. The distal grasping component of prehension is also affected: monkeys fail to shape and orient their hand and fingers properly in anticipation of the grasp (Faugier-Grimaud et al. 1978; Haaxma and Kuypers 1975; Lamotte and Acuna 1978). Finally, posterior parietal lesions in monkey may also produce oculomotor impairments, such as increase in saccade latencies and asymmetry of optokinetic nystagmus (Lynch and McLaren 1983, 1989).

Although very similar visuomotor effects are observed in man and monkey following posterior parietal lesions, it has been very difficult to demonstrate any equivalent of hemispatial neglect after parietal damage in the monkey. Only milder disorders such as visual, auditory, and somatosensory extinction (bilateral stimuli elicit orienting only to the ipsilesional side) and a decreased blink at threat in the contralesional visual field have been found (Heilman et al. 1970). In the few studies in which neglect symptoms were reported, lesions also involved the superior temporal sulcus (STS; Deuel and Regan 1985). In fact, damage of this structure alone may be responsible for neglect in monkey (Luh et al. 1986, Watson et al. 1994). Watson et al. have contradicted the effects of the inferior parietal lobule (IPL) and STS lesions, consisting in misreaching and neglect respectively. Following STS lesions, monkeys are impaired in orienting to contralesional lights as well as to contralesional somatosensory and auditory stimuli; they show more or less marked limb akinesia on the contralesional side and sometimes tend to turn to the side of the lesion when walking. According to the authors, lesions of the STS induce neglect because this area is not only the recipient of multiple sensory inputs heavily connected with other polymodal areas (prefrontal cortex, limbic system) but also the convergence site of the two cortical systems involved in localization and identification. The failure of the STS to activate those areas that determine either the location of objects in space or their identity would prevent the animal from being aware of, or behaving within, contralateral hemispace.

Even though the intraparietal sulcus (IPS) is the surface landmark separating the superior and inferior lobules (SPL and IPL respectively) in both humans and monkeys, Brodmann's area 7 in humans is above the IPS, whereas in monkeys area 7 including areas PG and PF of Bonin and Bailey (1947) is below the IPS and thus part of the IPL. The cortex below the IPS in humans consists of areas 39 and 40. Whether there is a monkey homologue of these areas has been debated. Architectonic studies of human areas 39 and 40 have sometimes described these as homologues of areas PG (area 7a) and PF (area 7b) of the monkey (Eidelberg and Galaburda 1984), although other investigators have considered that the monkey homologue of human areas 39 and 40 is, at least in part, in the STS (Mesulam 1985). The fact that STS but not IPL lesions produce an equivalent of spatial neglect in the monkey, as mentioned previously, agrees better with the latter conception.

Conclusions

Although there is a change of topography between monkey and man, converging evidence in both species shows that separate regions of the posterior parietal lobe and/or the adjacent association cortex subserve two main types of spatial representations and functions .

In humans, the upper part of the posterior parietal cortex, which is damaged in optic ataxia patients, is mainly involved in coding space for action. This is performed by means of several effector-specific representations. These are short-living, unconscious processes underlying different "on-line" sensorimotor transformations each dedicated to a given goal-directed action.

The lower part of the posterior parietal cortex, as well as adjacent occipitoparietal regions, which are most often damaged in neglect patients, would be responsible for more enduring and conscious representations underlying spatial cognition and awareness. The various patterns of impairment observed in neglect suggest the existence of several representations dealing in particular with different reference frames.

Acknowledgements. I would like to thank M. Jeannerod for helpful discussions during the preparation of this paper. I am also gratefull to K. Knoblauch and Y. Rossetti who kindly revised the manuscript. The experiments were performed with the valuable collaboration of A. Vighetto, G. Rode, and P. Revol.

References

Andersen RA (1995) Encoding of intention and spatial location in the posterior parietal cortex. Cereb Cortex 5:457–469

Andersen RA, Snyder LH, Li CS, Stricanne BS (1993) Coordinate transformations in the representation of spatial information. Curr Opin Neurobiol 3:171–176

Auerbach SH, Alexander MP (1981) Pure agraphia and unilateral optic ataxia associated with a left superior parietal lobule lesion. J Neurol Neurosurg Psychiatry 44:430–432

Balint R (1909) Seelenlähmung des "Schauens", optische Ataxie, räumliche Störung der Aufmerksamkeit. Monatsschr Psychiatr Neurol 25:57–81

Barbut D, Gazzaniga MS (1987) Disturbances in conceptual space involving language and speech. Brain 110:1487–1496

Biguer B, Donaldson IML, Hein A, Jeannerod M (1986) Neck muscle vibration modifies the representation of visual motion and direction in man. Brain 111:1405–1424

Bisiach E (1993) Mental representation in unilateral neglect and related disorders: the twentieth Barlett Memorial Lecture. Q J Exp Psychol (A) 46:435–461

Bisiach E, Luzzatti C (1978) Unilateral neglect of representational space. Cortex 14:129–133

Bisiach E, Berti A (1987). Dyschiria: An attempt at its systemic explanation. In: Jeannerod M (ed.) Neurophysiological and neuropsychological aspects of spatial neglect. North-Holland, Amsterdam, pp 183–201

Bisiach E, Vallar G (1988) Hemineglect in humans. In Boller F, Grafman J (eds) Handbook of neuropsychology, vol. 1. Elsevier, Amsterdam, pp 195–222

Bisiach E, Capitani E, Luzzatti C, Perani D (1981) Brain and conscious representation of outside reality. Neuropsychologia 19:543–551

Bisiach E, Cornacchia L, Sterzi R, Vallar G (1984) Disorders of perceived auditory lateralization after lesions of the right hemisphere. Brain 107:37–52

Bisiach E, Rusconi ML, Vallar G (1991) Remission of somatoparaphrenic delusion through vestibular stimulation. Neuropsychologia 29:1029–1031

Bonin G von, Bailey P (1947) The neocortex of Macaca mulata. Illinois monographs in the medical sciences, vol 5, no 4. University of Illinois Press, Urbana

Brain WR (1941) Visual disorientation with special reference to lesions of the right cerebral hemisphere. Brain 64:244–272

Bushnell MC, Goldberg ME, Robinson DL (1981) Behavioral enhancement of visual responses in monkey cerebral cortex. I. Modulation in posterior parietal cortex related to selective visual attention. J Neurophysiol 46:755–772

Caminiti R, Johnson PB, Ferraina S, Burnod Y (1992) Reaching toward visual targets. 1 Neurophysiological studies. In: Caminiti R, Johnson PB, Burnod Y (eds) Control of arm movement in space. Springer, Berlin Heidelberg New York, pp 147–158

Castaigne P, Rondot P, Ribadeau Dumas JL, Tempier P (1975) Ataxie optique localisée au côté gauche dans les deux hémichamps visuels homonymes gauches. Rev Neurol (Paris) 131:23–28

Castiello U, Paulignan Y, Jeannerod M (1991) Temporal dissociation of motor responses and subjective awareness. Brain 114:2639–2655

Classen J, Kunesh E, Binkofski F, Hilperath F, Schlaug G, Seitz RJ, Glickstein M, Freund HJ (1995) Subcortical origin of visuomotor apraxia. Brain 118:1365–1374

Cole M, Schutta HS , Warrington EK (1962) Visual disorientation in homogenous half-fields. Neurology 12:257–263

De Renzi E (1982) Disorders of space exploration and cognition. Wiley, Chichester
Deuel RK, Regan DJ (1985) Parietal hemineglect and motor deficits in the monkey. Neuropsychologia 23:305–314
Eidelberg D, Galaburda AM (1984) Inferior parietal lobule. Divergent architectonic asymetries in the human brain. Arch Neurol 41:843–852
Ettlinger G, Warrington E, Zangwill DL (1957) A further study of visual-spatial agnosia. Brain 80:335–361
Faugier-Grimaud S, Frenois C, Stein DG (1978) Effects of posterior parietal lesions on visually guided behavior in monkeys. Neuropsychologia 16:151–168
Ferro JM (1984) Transient inaccuracy in reaching caused by a posterior parietal lobe lesion. J Neurol Neurosurg Psychiatry 47:1016–1019
Ferro J, Bravo-Marques JM, Castro-Caldas A, Antunes L (1983) Crossed optic ataxia: possible role of the dorsal splenium. J Neurol Neurosurg Psychiatry 46:533–539
Garcin R, Rondot P, Recondo J De (1967) Ataxie optique localisée aux deux hémichamps homonymes gauches (étude clinique avec présentation d'un film). Rev Neurol (Paris) 116:707–714
Geminiani G, Bottini G (1992) Mental representation and tempory recovery from unilateral neglect after vestibular stimulation. J Neurol Neurosurg Psychiatry 55:332–333
Goodale MA, Pélisson D, Prablanc C (1986) Large adjustments in visually guided reaching do not depend on vision of the hand or perception of target displacement. Nature 320:748–750
Goodale MA, Milner AD, Jackobson LS, Carey DP (1991) A neurological dissociation between perceiving objects and grasping them. Nature 349:154–156
Graziano MS, Gross CG (1995) The representation of extracorporal space: a possible role for bimodal, visual-tactile neurons. In: Gazzaniga MS (ed) The cognitive neurosciences. MIT Press, Cambridge, pp 1021–1034
Haaxma R, Kuypers HGJM (1975) Intrahemispheric cortical connexions and visual guidance of hand and finger movements in the rhesus monkey. Brain 98:239–260
Halligan PW, Marshall JC (1992) Left visuo-spatial neglect: a meaningless entity? Cortex 98:525–535
Halsband U, Gruhn S, Ettlinger G (1985) Unilateral spatial neglect and defective performance in one half of space. Int J Neurosci 28:173–195
Heilman KM, Pandya DN, Geschwind N (1970) Trimodal inattention following parietal lobe ablations. Trans Am Neurol Assoc 95:259–268
Heilman KM, Bowers D, Watson RT (1983a) Performance of hemispatial pointing task by patients with neglect syndrome. Neurology 33:661–663
Heilman KM, Watson RT, Valenstein E, Damasio AR (1983b) Localization of lesions in neglect In: Kertesz A (ed) Localization in neuropsychology. Academic, New York, pp 471–492
Heilman KM, Bowers D, Valenstein E, Watson RT (1987) Hemispace and hemispatial neglect. In: Jeannerod M (ed) Neurophysiological and neuropsychological aspects of spatial neglect. North Holland, Amsterdam, pp 115–150
Holmes G (1918) Disturbances of visual orientation. Br J Ophthalmol 2:449–168, 506–516
Hörsten G (1979) Constant error of visual egocentric orientation in patients with acute vestibular disorder. Brain 102:685–700
Jeannerod M (1987) Neurophysiological and neuropsychological aspects of spatial neglect. North-Holland, Amsterdam
Jeannerod M (1994) The representing brain: neural correlates of motor intention and imagery. Behav Brain Sci 17:187–245

Jeannerod M, Biguer B (1987) The directional coding of reaching movements. A visuomotor conception of spatial neglect. In: Jeannerod M (ed) Neurophysiological and neuropsychological aspects of spatial neglect. North-Holland, Amsterdam, pp 87–114

Karnath H-O (1994) Subjective body orientation in neglect and the interactive contribution of neck muscle proprioception and vestibular stimulation. Brain 117:1001–1012

Karnath H-O, Schenkel P, Fischer B (1991) Trunk orientation as the determining factor of the "contralateral" deficit in the neglect syndrome and as the physical anchor of the internal representation of body orientation in space. Brain 114:1997–2014

Karnath H-O, Christ K, Hartje W (1993) Decrease of contralateral neglect by neck muscle vibration and spatial orientation of trunk midline. Brain 116:383–396

Kinsbourne M (1987) Mechanisms of unilateral neglect. In: Jeannerod M (ed) Neurophysiological and neuropsychological aspects of spatial neglect. North-Holland, Amsterdam, pp 69–86

Kinsbourne M (1993) Orientation bias model of unilateral neglect: evidence from attentional gradients within hemispace. In: Robertson IH, Marshall JC (eds) Unilateral neglect: clinical and experimental studies. Erlbaum, Hove, pp 63–86

Lamotte RH, Acuna C (1978) Deficits in accuracy of reaching after removal of posterior parietal cortex in monkey. Brain Res 139:309–326

Levine DN, Kaufman KJ, Mohr JP (1978) Inaccurate reaching associated with a superior lobe tumor. Neurology 28:556–561

Luh KE, Butter CM, Buchtel HA (1986) Impairments in orienting to visual stimuli in monkeys following unilateral lesions of the superior sulcal polysensory cortex. Neuropsychologia 24:461–470

Lynch JC, McLaren JW (1983) Optokinetic nystagmus deficits following parieto-occipital cortex lesions in monkey. Exp Brain Res 49:125–130

Lynch JC, McLaren JW (1989) Deficits of visual attention and saccadic eye movements after lesions of parieto-occipital lesions in monkeys. J Neurophysiol 61:74–90

Mark VW, Heilman KM (1990) Bodily neglect and orientation biases in unilateral neglect syndrome and normal subjects. Neurology 40:640–643

Meador KJ, Loring DW, Bowers D, Heilman KM (1987) Remote memory and neglect syndrome. Neurology 36 (Suppl 1):170

Mesulam M-M (1985) Patterns in behavioral neuroanatomy: association areas, the limbic system and hemispheric specialization. In: Mesulam M-M (ed) Principles of behavioral neurology. Davis, Philadelphia, pp 1–70

Milner AD, Goodale MA (1995) The visual brain in action. Oxford University series no 27. Oxford University Press, Oxford

Perenin MT (1992) Référence égocentrique et cortex pariétal postérieur. Rev Oto-Neuro-Ophtalmol 16:14–16

Perenin MT, Vighetto A (1983) Optic ataxia: a specific disorder in visuomotor coordination. In: Hein A, Jeannerod M (eds) Visually oriented behavior. Springer, Berlin Heidelberg New York, pp 305–326

Perenin MT, Vighetto A (1988) Optic ataxia: a specific disruption in visuomotor mechanisms. I. Different aspects of the deficit in reaching for objects. Brain 111:643–674

Prablanc C, Echallier JF, Komilis E, Jeannerod M (1979) Optimal response of eye and hand motor systems in pointing at a visual target. I. Spatio-temporal characteristics of eye and hand movements and their relationships when varying the amount of visual information. Biol Cybern 35:113–124

Ratcliff G, Davies-Jones GAB (1972) Defective localization in focal brain wounds. Brain 95:49–60

Riddoch G (1935) Visual disorientation in homonymous half-fields. Brain 58:376–382

Rizzolatti G, Berti A (1993) Neural mechanisms of spatial neglect. In: Robertson IH, Marshall JC (eds) Unilateral neglect: clinical and experimental studies. Erlbaum, Hove, pp 87–102

Rizzolatti G, Matelli M, Pavesi G (1983) Deficits in attention and movement following the removal of postarcuate (area 6) and prearcuate (area 8) cortex in macaque monkeys. Brain 106:655–673

Rizzolatti G, Camarda R, Fogassi L, Gentilucci M, Luppino G, Matelli M (1994) Space and selective attention. In: Umiltà C, Moscovitch M (eds) Attention and performance XV. Conscious and nonconscious information processing. MIT Press, Cambridge, pp 231–265

Robertson IH, Marshall JC (1993) Unilateral neglect: clinical and experimental studies. Erlbaum, Hove

Rode G, Perenin MT (1990). Caloric stimulation and representational hemineglect. Poster presented at the symposium "Consciousness and Cognition: Neuropsychological Perspectives", St. Andrews, 3–5 September 1990

Rode G, Perenin MT (1994) Temporary remission of representational hemineglect through vestibular stimulation. Neuroreport 5:869–872

Rode G, Charles N, Perenin MT, Vighetto A, Trillet M, Aimard G (1992) Partial remission of hemiplegia and somatoparaphrenia through vestibular stimulation in a case of unilateral neglect. Cortex 28:203–208

Rondot P, De Recondo J, Ribadeau Dumas JL (1977) Visuomotor ataxia. Brain 100:355–376

Sakata H, Taira M, Murata A, Mine S (1995) Neural mechanisms of hand action in the parietal cortex of the monkey. Cereb Cortex 5:429–438

Soechting JF, Helms Tillery SI, Flanders (1990) Transformation from head to shoulder-centered representation of target direction in arm movements. J Cogn Neurosci 2:32–43

Stein JF (1992) The representation of egocentric space in the posterior parietal cortex. Behav Brain Sci 15:691–700

Tzavaras A, Masure MC (1976) Aspects différents de l'ataxie optique selon la latéralisation hémisphérique de la lésion. Lyon Medical 236:673–683

Vallar G, Perani D (1986) The anatomy of unilateral neglect after right hemisphere stroke lesions: a clinical CT-scan correlation study in man. Neuropsychologia 24:609–622

Vallar G, Antonucci G, Gariglia C, Pizzamiglio L (1993a) Deficit of position sense, unilateral neglect, and optokinetic stimulation. Neuropsychologia 31:1191–1200

Vallar G, Bottini G, Rusconi ML, Sterzi R (1993b) Exploring somatosensory hemineglect by vestibular stimulation. Brain 116:71–86

Ventre J, Flandrin JM, Jeannerod M (1984) In search for the egocentric reference. A neurophysiological hypothesis. Neuropsychologia, 22:797–806

Watson RT, Valenstein E, Day A, Heilman KM (1994) Posterior neocortical systems subserving awareness and neglect. Neglect associated with superior temporal sulcus but not area 7 lesions. Arch Neurol 51:1014–1021

Weinstein EA, Friedland RP (1977) Behavioral disorders associated with hemi-inattention. In: Weinstein EA, Friedland RP (eds) Hemi-inattention and hemisphere specialization. Adv Neurol, vol 18. Raven, New York, pp 51–62

Zingerle H (1913) Über Störungen der Wahrnehmung des eigenen Körpers bei organischen and Gehirnerkrankungen. Monatsschr Psychiatr Neurol 34:13–36

Distinguishing Sensory and Motor Deficits After Parietal Damage: An Evaluation of Response Selection Biases in Unilateral Neglect

J. B. MATTINGLEY[1] and J. DRIVER[2]

[1]Department of Experimental Psychology, University of Cambridge, Cambridge, England
[2]Department of Psychology, Birkbeck College, University of London, London, England

Introduction

Flexible and coherent control of goal-directed behavior requires that just a subset of the incoming sensory data be selected to guide action (Allport 1989; Desimone and Duncan 1995; Tipper et al. 1992). In order to modify the disposition of our sense organs and effectors appropriately in relation to the environment, the central nervous system must select just those inputs that are currently relevant, while simultaneously suppressing irrelevant inputs (Desimone and Duncan 1995; Posner and Dehaene 1994). In patients with the disorder of *unilateral neglect*, which in humans has traditionally been associated with right parietal lobe damage (Critchley 1953; Vallar and Perani 1986), such facilitatory and inhibitory processes are pathologically unbalanced. In right-parietal patients with left unilateral neglect, selective behavior is biased toward stimulus events occurring toward the right (ipsilesional) side, while stimuli occurring further to the left (contralesional) side no longer exert a comparable influence on selective behavior. Sometimes the lateral imbalance is only apparent when two events compete simultaneously for selection; this is known as *extinction*, which has also been associated with parietal lobe damage (see Milner, this volume) although, as with unilateral neglect, related impairments can follow lesions to other cortical and subcortical structures (Driver et al., this volume).

The study of selective deficits following parietal damage is itself curiously unbalanced. A substantial majority of existing research is concerned with biased selection of ipsilesional over contralesional *inputs* (see, for example, Robertson and Marshall 1993), with relatively few studies examining any effects of unilateral parietal damage on the selection of appropriate motor *outputs*. Indeed, many experiments with neglect or extinction patients have used somewhat arbitrary response measures, with any lateralised anomalies typically interpreted as reflecting biases in input selection, considered as somehow existing independently of the responses this selection subsequently guides. However, there is now abundant evidence that perceptual processes in humans and many other species are tightly constrained by the kinds of responses they control (Milner and Goodale

In: Parietal Lobe Contributions to Orientation in 3D Space (1997). P. Thier and H.-O. Karnath (eds). Springer-Verlag, Heidelberg.

1995). Moreover, there is an increasing realisation that many of the lateral biases apparent in neglect patients might be due to unbalanced selection of motor outputs, rather than of sensory inputs. For instance, when neglect patients fail to cancel lines toward the contralesional side of a page, their lateral bias might conceivably arise in the selection of appropriate responses with the hand, rather than in the selection of visual input (Tegnér and Levander 1991).

In this chapter, we review the existing literature on possible motoric deficits in patients (and animals) with unilateral neglect and describe some of our own recent experiments. We begin by considering evidence that unilateral damage may impair mechanisms responsible for selecting the appropriate effector for a particular task (e.g. left versus right hand). We then consider possible impairments in selecting the direction or amplitude of goal-directed movements to contralesionally located targets after the effector has been chosen. We shall see that although many of the data based on traditional clinical tests for neglect (cancellation, bisection, etc.) are difficult to interpret unambiguously, more recent experiments using temporal measures in simpler tasks are beginning to provide a clearer picture of response deficits in neglect. Since many of the paradigms now being used to examine visuomotor anomalies in neglect have been borrowed from studies of motor control in normals, it should ultimately be possible to integrate findings from these two areas. We critically assess recent claims (e.g., Bisiach 1993; Bisiach et al. 1990; Mattingley et al. 1992; Mesulam 1990; Tegnér and Levander 1991) that sensory and motor components of neglect can be directly dissociated. We note that it is somewhat harder to distinguish these components than has previously been thought.

Impairments in Effector Selection

One of the earliest studies which explicitly recognised the possible role of response factors in unilateral neglect was conducted by Watson and co-workers (1978). They trained monkeys to respond to tactile stimuli delivered to one leg by making a response with the contralateral hand, i.e. with the hand on the opposite side of the body to that stimulated. As we shall see later, this crossed task finds many echoes in more recent work on neglect in humans. After surgically lesioning the frontal arcuate region of one hemisphere, it was found that monkeys made normal responses with the ipsilesional hand to stimuli on the contralesional leg. However, when stimuli were delivered to the ipsilesional leg, the monkeys often failed to respond using the contralesional hand or responded incorrectly by using the ipsilesional hand. On the basis of these observations, Watson et al. (1978) concluded that their monkeys had "nonsensory neglect" arising from a failure of "intentional" mechanisms to activate an appropriate response by the contralesional limb. Similar observations of impaired contralesional responses for stimuli coming from the ipsilesional side were subsequently made in monkeys with unilateral parietotemporal lesions (Valenstein et al. 1982). It is important to note that the monkeys' failure to move their contralesional hand was not due to peripheral motor weakness, since separate tests indicated that they could use the contralesional limb when compelled to do so.

An analogous phenomenon to this nonsensory neglect in monkeys can also be observed in humans with so-called *motor neglect*. This disorder, which usually occurs in the context of visual neglect, manifests as a disturbance of spontaneous movement with the contralesional limbs which cannot be attributed to hemiplegia (Laplane and Degos 1983). The patient may use only the ipsilesional hand to perform tasks which would normally require the use of both hands, such as opening a jar or tying a shoelace. Indeed, such individuals are often very persistent in unsuccessfully using just the ipsilesional hand for simple bimanual tasks, even though they can show relatively intact performance when encouraged to use both hands (Laplane and Degos 1983). Patients with motor neglect do not deny the existence of their contralesional limb, nor do they claim that the limb is weak. Thus, on the face of it, such cases seem to exhibit a deficit in selecting the contralesional limb for action.

It has also been reported that, in some cases, impaired selection of the contralesional limb for action may only become apparent in the context of concurrent activation of the ipsilesional limb. Thus, when asked to move both arms simultaneously, the patient may move just the ipsilesional arm, even though requests to move either arm alone can elicit normal responses (Valenstein and Heilman 1981). This phenomenon seems strikingly analogous to the perceptual extinction that can be found with bilateral simultaneous stimuli (see Driver et al., this volume). To date, reports of such "motor extinction" have been based primarily on anecdotal evidence only.

However, in collaboration with Ian Robertson in Cambridge and Nicoletta Beschin in Verona, we recently tested motor extinction more formally in two patients with right parietal damage, both of whom were unimpaired on standard clinical tests of perceptual neglect and extinction. In separate blocks of trials, patients were asked to tap with their hands or feet at a rate of two beats per second for 20 s. In unilateral trials just one hand or foot was used, whereas in bilateral trials two limbs (both hands, both feet, or one hand and one foot) were used simultaneously, either in-phase (together) or out-of-phase (alternating). Although both patients had some contralesional motor weakness, we were able to use their unilateral performance as a baseline for assessing any decrement in contralesional limb performance due to competition from the simultaneously active ipsilesional limb. Performance was videotaped and analysed in terms of the number of taps per condition, the presence of dysrhythmias (switching inappropriately between in-phase and out-of-phase movements), and any reductions in the amplitude of taps.

When making bilateral movements, both patients were impaired in using their contralesional limbs, even though movements made by the same limbs in isolation were relatively normal. One patient was completely unable to tap her contralesional hand or foot in-phase with either ipsilesional limb, although she was entirely normal when making bilateral movements that were out of phase (i.e. alternating). Moreover, even when we began each trial by physically entraining in-phase movements of her two hands or feet, she quickly lapsed back into out-of-phase movements and thereafter could not re-establish the appropriate pattern. In contrast, our second patient was able to perform both in-phase and out-of-phase bilateral movements. However, the amplitude of her contralesional limb movements, particularly those made with her hand, was markedly reduced relative to the amplitude of such movements in the unimanual condition. Indeed, on some bimanual trials the contralesional hand ceased to move altogether, something which never happened during trials involving the contralesional hand alone.

These findings provide preliminary evidence for deficits in maintaining normal contralesional limb movements during repetitive bilateral activity in patients with right parietal damage. Future studies could focus on whether such anomalies also occur after damage to non-parietal regions and on the parameters of movement that modulate the extent of pathological response competition. The fact that one of our patients had problems with the temporal control of contralesional limb movements in bilateral conditions, whereas the other had problems in maintaining adequate force production, suggests that response competition may arise at various levels of bilateral limb co-ordination.

We have so far considered motoric deficits only in selecting the contralesional limb for action, either by itself or in the context of simultaneous ipsilesional limb movements. Recent evidence from a study by Làdavas et al. (1994) suggests that such selection problems can also extend to separate effectors even within one side of the body. They found that patients with left neglect were slower to make choice responses to visual targets when using the more contralesionally located of two adjacent fingers (either index or middle) on the ipsilesional hand, irrespective of the side of the display on which the visual target was presented. Thus, when the hand was held palm down, the index finger responded more slowly than the middle finger, whereas when the hand was held palm-up, the middle finger then responded more slowly. Since this effect was found regardless of visual target position, the slowed reaction times (RTs) for the more contralesional finger must reflect a bias at the level of effector specification. This asymmetry in selecting between fingers of the ipsilesional hand mirrors that discussed earlier *between* hands for patients with motor neglect (Laplane and Degos 1983). The finding that which finger was slower reversed with arm posture emphasises the importance of proprioceptive inputs in coding the spatial dispositions of effectors. The importance of proprioception for neglect in this respect has also been recently shown in the sensory domain for tactile extinction (see Driver et al., this volume).

In a further study of biases in response selection within the ipsilesional hand, Behrmann and coworkers (1995) used a centrally displayed arrow to indicate with high probability whether the index or middle finger of that hand should be used for response to a subsequent target stimulus that was also presented centrally. On most trials, the cue was valid, but on a few invalid trials the subsequent target required the other response. Neglect patients were particularly impaired for invalid trials if the finger located more ipsilesionally was cued, but the other finger then had to respond. This result seems to be a motor analogue of the so-called disengage deficit found in unilateral parietal patients by Posner et al. (1984), who cued shifts of attention to lateral visual targets, rather than cueing the motor response for a central target as in the study by Behrmann et al.

Premotor Theory: Possible Links Between Response Selection and Visual Selection

One major influence behind the recent interest in possible motor deficits in neglect has been the 'premotor theory' of attention put forward by Rizzolatti and colleagues (Rizzolatti and Camarda 1987; Rizzolatti and Gallese 1988; Rizzolatti and Berti 1993). This argues that shifts of focal attention are produced by programming of spatial responses, such as eye or limb movements towards objects in corresponding spatial locations. It argues that circuits connecting frontal and parietal areas subserve both selective attention and goal-directed movements, with distinct circuits for peripersonal and extrapersonal space. Evidence from various sources has been taken to support this theory, ranging from single-cell and lesion results in monkeys (see Rizzolatti and Camarda 1987) to findings in normal humans of interactions between saccades and visual attention (Sheliga et al. 1994).

The recent findings of motoric aspects to neglect, which we review throughout this chapter, have often been taken as providing further support for Rizzolatti's view. In particular, reports that motoric aspects of neglect can apparently be dissociated from sensory aspects (e.g. Bisiach et al. 1990; Tegnér and Levander 1991) have often been interpreted as direct evidence in favor of the premotor theory. However, it seems to us that the most basic prediction from the premotor theory for neglect is that any pathological bias in motor programming should lead to a corresponding bias in sensory attention and vice versa. In other words, spatially specific sensory and motor impairments should tend to be *associated*, rather than dissociated. On our reading, the dissociations predicted by the premotor theory should arise between different response systems, and their associated regions of peripersonal or extrapersonal space, rather than between perception and action (see Halligan and Marshall 1991; Cowey et al. 1994).

Nevertheless, the premotor theory remains justly influential, and several findings suggest that motoric factors can indeed affect visual performance, as the theory would expect. In a study by Bisiach et al. (1985), responses by a patient with left neglect to targets flashed in the right hemifield were normal if a button press with the right hand was required, but severely impaired when a left button press with the contralesional hand was required. The errors in this latter case were due mainly to incorrect right-hand button presses, but the patient occasionally failed to respond altogether. Interestingly, awareness of the visual stimulus was sometimes denied when a left button press was required but not executed, raising the possibility that motoric factors may have directly affected visual experience (although it does remain possible that this denial was merely a post hoc justification for a purely motoric failure to respond).

Several subsequent studies have reported a reduced severity of visual selection deficits for *contralesional* targets when the contralesional limb is used to respond. For example Joanette et al. (1986) measured right-hemisphere patients' detection rates in pointing to visual targets presented in either the left or right hemifield. They found that more targets were detected in the contralesional hemifield when the response involved pointing with the contralesional rather than the ipsilesional hand. This result was taken to imply that efferent or "premotor" processes involved in co-ordinating an appropriate response could modulate perceptual processes involved in stimulus detection. However, a possible alternative might be to explain this outcome in terms of impaired interhemispheric communication from the damaged hemisphere, as compared with intrahemispheric processing on that side. In a more recent study of pointing to visual targets by a patient with left neglect, Duhamel and Brouchon (1990) found faster responses with the contralesional hand than with the ipsilesional hand, but only for targets in the left hemifield. Moreover, slower responses for these targets were found if the responding hand started on the left versus on the right side of the body midline, ruling out the possibility that the left hand itself merely provided some helpful visual cue when used to respond. This effect of start position is rather harder to attribute to impaired interhemispheric communication alone.

These reports that self-initiated contralesional limb movements may modulate visual attention have led to studies of their possible rehabilitative potential for patients with neglect. Robertson and North (1993) found that contralesional omissions on a letter cancellation task were reduced when neglect patients made self-initiated movements with the fingers of their contralesional hand. As a control for proprioceptive influences, passive movement of the contralesional hand was found to produce no such effects. More recently, the so-called limb activation technique has been shown to reduce the frequency of reading errors in patients with left-sided neglect dyslexia (Worthington 1996) and to reduce left-sided collisions as patients negotiate their way through doorways (Robertson et al. 1994). Having patients practice contralesional limb movements over a period of weeks has been reported to produce long-lasting therapeutic benefits in patients with visual neglect (Robertson et al. 1992), though this technique can only be used in patients without full hemiplegia. The reported enhancement of contralesional vision is eliminated when the ipsilesional limb is moved simultaneously with the contralesional limb (Robertson and North 1994), suggesting that bimanual competition can also influence the allocation of visual attention.

However, rather than enhancing covert visual attention toward the affected side, as often claimed, contralesional limb activation may produce its reported effects simply by inducing more contralesional eye movements, with the result that more contralesional visual stimuli are foveated. Though such an effect might still remain therapeutically useful, from a theoretical standpoint it would merely reflect interactions between motor programs within two different effector systems (limbs and eyes), rather than the specific link between motor programs and sensory attention that is suggested by the premotor theory. The reported benefits of limb activation in neglect patients have so far only been observed in cancellation and reading tasks, both of which require voluntary eye movements. More convincing evidence that motor programming can lead to shifts in covert visual attention might be obtained in situations in which visual stimuli are presented so briefly that eye movements cannot be made.

In a recent experiment (Mattingley et al., in preparation) we implemented this by using a computerised extinction paradigm in which visual stimuli were displayed very briefly while a right-parietal patient made contralesional limb movements in some conditions. The patient maintained central fixation on a small cross in the centre of the computer display. On separate trials visual targets could appear in either hemifield alone, simultaneously in both hemifields, or in neither hemifield on catch trials; the patient was asked to detect any targets by saying "left", "right", "both", or "none". Limb activation was linked to the presentation of stimuli in the following way. The patient initiated each trial by pressing a key on a computer keyboard that was itself located to the left or right of the display. Both the keyboard and the patient's hands were occluded from view throughout the experiment. The patient detected all isolated ipsilesional stimuli and correctly reported the absence of stimuli on catch trials. As can be seen in Fig. 1, in the baseline condition without any limb movement her detection of contralesional visual stimuli was selectively impaired in bilateral trials, indicating visual extinction. There was a significant reduction in this extinction when the patient used her contralesional hand on the left side to initiate visual presentations. A slight beneficial effect was also obtained with ipsilesional hand movements on the right, perhaps indicating an alerting effect for self-initiated trials, but this effect appears smaller than for contralesional hand movements.

These results suggest that the competitive bias in favor of ipsilesional events in visual extinction can be reduced by self-initiated limb movements, especially when these are made by the affected limb on the contralesional side. This provides evidence that motor programming can indeed produce changes in covert selective attention, as the premotor theory suggests, even in tasks where no overt attention shifts via eye movements are possible. Future studies could directly measure any potential interactions between response coding of limb and eye movements by examining the saccades made by patients when contralesional limb movement is manipulated. It would also be of interest to determine the exact time course of any motoric influences on visual attention by manipulating the delay between limb movement and stimulus presentation. We suspect that such influences may be highly constrained in time in accordance with the need to shift spatial attention as a function of ongoing changes in goal-directed action.

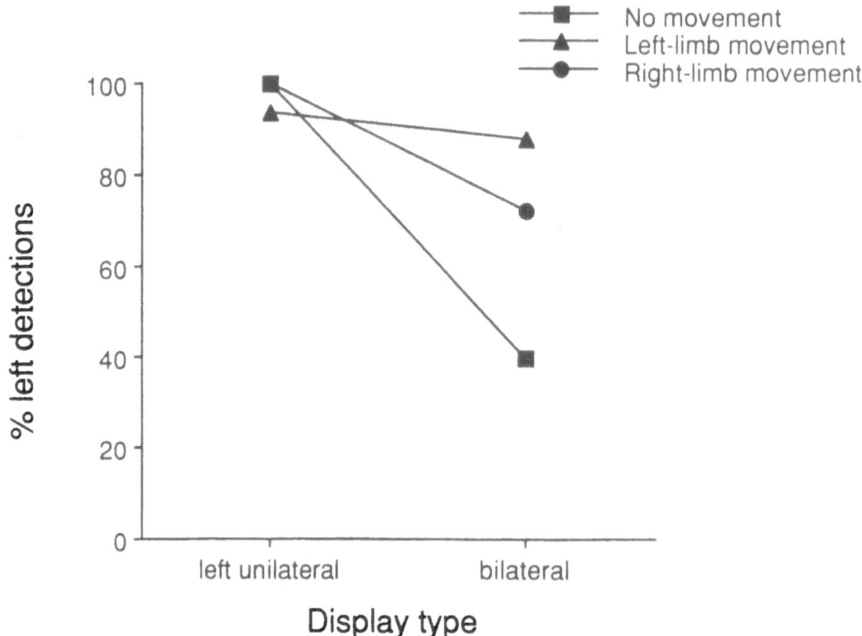

Fig. 1. The effects of left- and right-hand movement on detection of brief visual events in a right parietal-damaged patient with extinction. Data represent the percentage of correct contralesional stimulus detections for unilateral left-sided and bilateral target displays. Stimuli were grey circles (1.5° diameter) presented for 60 ms on a white background, 4.0° to the left and right of central fixation

Impairments of Visuomotor Exploration

In most clinical tests for unilateral neglect, patients are required to mobilise spatial attention and action towards a common target. In standard cancellation tests, patients must engage selective attention on each target in turn and then execute an appropriate limb movement to that target in order to cancel it. Similarly, in conventional line-bisection tests patients must first determine a point along the line that corresponds with the subjective midpoint and then move their hand to the desired location in order to place a transection mark. The potential advantage of such tests is that they mimic the demands of many everyday tasks involving visual exploration, such as finding a salt shaker on a cluttered dinner table or positioning the hand appropriately to pick up a pencil. However, the disadvantage of these measures is that, despite their apparent simplicity, they involve many separate components, so deficits in them can rarely be unambiguously attributed to just one component. Unfortunately, this problem arises even for some recent efforts to disentangle perceptual and motoric factors via modifications to the standard tasks.

Daffner and coworkers (1990) reported data on a woman who suffered sequential right hemisphere strokes, the first involving the frontal lobe and the second involving the parietal lobe. The initial frontal lesion resulted in left-sided omissions on a letter cancellation task and impaired tactile search in the contralesional hemispace. By contrast, her performance on visual, tactile, and auditory extinction tasks appeared normal. Following the parietal lesion, however, the patient showed contralesional extinction of visual and auditory stimuli, and her previously recovered impairment on letter cancellation re-emerged. On the basis of these findings it was suggested that the patient's frontal lesion induced an exploratory motor deficit, whereas her parietal lesion caused a "perceptual" (i.e. non-motoric) impairment. This was taken to support Mesulam's (1981, 1990) proposal that different brain regions subserve different aspects of spatial attention, with the dorsolateral frontal lobe mediating exploratory motor activity and the parietal lobe controlling shifts of sensory attention.

Further evidence in apparent support of Mesulam's model was claimed in a study by Liu et al. (1992), who examined two left neglect patients, one with frontal and the other with parietal damage. The frontal patient was selectively impaired on tests of visual scanning and blindfolded exploration of left hemispace, but showed no evidence of extinction on confrontation. In contrast, the parietal patient seemed normal on scanning and exploration, but showed left-sided sensory extinction. Finally, Bottini et al. (1992) found that a right frontal patient failed to cancel left-sided targets in a conventional cancellation task in which she had to make self-initiated exploratory movements, but was unimpaired in naming targets when they were pointed to by the examiner.

On the face of it, these studies provide direct evidence for a dissociation between exploratory and sensory deficits. However, the results seem less conclusive on closer inspection. In particular, the tests of "perceptual" and "motor" functions do not seem remotely equated in terms of task demands that are extrinsic to the sensory-motor distinction. Differences between performances on cancellation and extinction tests, for example, may reflect differences in patients' overall processing capacity and their attention-shifting ability, as well as any purported sensory-motor differences. Thus, cancellation requires continuous serial shifting of focal attention between many simultaneously present targets, in addition to precise localisation of individual stimuli. By contrast, standard tests of extinction require detection of only one or two brief target events, and attention must be divided between separate potential stimulus locations. Moreover, as we have noted elsewhere (Driver et al., this volume), the standard confrontation procedures used to measure sensory extinction in these studies and many others can suffer from floor and ceiling effects that render them insensitive. Such limitations can only be overcome by titrating exposure durations with computerised presentations. Finally, many of the putatively "motor" tests also demanded considerable processing of sensory information, such as proprioceptive inputs signalling hand position. Equally, the "perceptual" version of the cancellation test used by Bottini et al. (1992) would presumably still require some motoric components, such as contralesional eye movements.

Opposition Techniques: Setting Motoric Biases Against Sensory Biases

A number of recent papers have tried to separate sensory from motoric biases in neglect by setting them against each other, which we will refer to as the "opposition technique". The first such study examined line bisection performance under conditions in which the hemispace of bisection was dissociated from that in which visual feedback was provided, by occluding patients' direct vision of their responding hand and instead displaying a video image of it in the opposite hemifield (Coslett et al. 1990). Two of the four left neglect patients tested by Coslett et al. (1990) were more impaired in bisecting lines located in left hemispace, regardless of the hemifield from which visual feedback was provided. These patients, both of whom had frontal damage, were suggested to have a problem in directing motor responses within left hemispace. In contrast, the two patients with parietotemporal damage were more impaired when visual feedback came from left hemispace regardless of the side on which lines were located, suggesting a hemispatial perceptual bias. Unfortunately, the implications of these data are somewhat limited because of the authors' assumption that the body's midsagittal axis demarcates a boundary between normal and abnormal hemispatial attention. In fact, it is more likely that the two hemispheres have contralaterally opposing orienting tendencies that become pathologically unbalanced following unilateral damage, thereby leading to attentional biases on *both* sides of the midline (Kinsbourne 1993).

In an attempt to address the inherently directional nature of neglect, Tegnér and Levander (1991) placed a 90° angled mirror behind Albert's line cancellation test (Albert 1973), so that hand movements made in one direction would appear to be moving in the opposite direction in the mirror. Right-hemisphere patients failed to cancel left-sided targets in the standard (direct view) version of this task in which any movements of visual attention and the responding hand movement would be "congruent". In the mirror-reversed "incongruent" version of the task, most patients now cancelled targets reflected on the right half of the mirror by moving their ipsilesional responding hand into the left half of the page. A few patients, however, failed to move their hand into the contralesional hemispace even though such a manoeuvre would have been reflected on the "good" perceptual side by the mirror. The authors suggested that this latter group of individuals were specifically impaired in programming limb movements into the contralesional hemispace, i.e., that they suffered from motoric neglect rather than sensory neglect. These patterns of behavior were also observed by Bisiach et al. (1995), who used the same apparatus to test 36 right-hemisphere patients. They reported that patients with anterior cortical lesions, or subcortical damage, were more likely to show some neglect in the incongruent version of the task, whereas those with posterior damage (including parietal lesions) were more likely to show neglect in the congruent version. However, they also reported that the pattern shown by each patient could reverse, depending on factors such as where the responding hand started. This result calls into question any static diagnosis of a case as suffering from just motor neglect, or just sensory neglect, on the basis of the mirror-reversed task alone.

Although the results of these studies are intriguing, it is difficult to draw firm conclusions about their implications. In the mirror-reversed condition, patients were assumed to direct their sensory attention to just the reflected image of the stimulus page, and not to their own hand (which was occluded from view). However, it seems more likely that patients would switch attention between the visual information reflected in the mirror and kinaesthetic information from their responding hand as it moved beneath the occluding surface. Thus, different patterns of response might be expected, depending on whether any sensory neglect was more severe in vision or in the coding of hand position. Moreover, the mirror-reversed task will be highly sensitive to any difficulties in attentional switching (between the hand and the mirror) and likewise to any problems in adapting to incongruous or novel situations. Difficulties in both these abilities are well-known to exist in frontal patients (Duncan 1995). We would therefore suggest that the difficulties experienced by frontal patients in the mirror-reversed cancellation task may reflect more general problems of attentional modulation, rather than a specific impairment in producing contralesional limb movements. Any lateral bias in their finished cancellation might then simply be a function of where their hand starts. That is, they will simply fail to make it through the whole page, due to their general difficulty in performing novel and highly incompatible tasks.

Similar concerns arise for a closely related study of bisection in neglect by Bisiach et al. (1990), although they used a much simpler task in the sense that the possible movements were more constrained. Note that in the Tegnér and Levander (1991) study, as in standard cancellation tasks, patients were free to make exploratory hand and eye movements in any direction in order to cancel the individual targets. Any specific deficit in making movements in the contralesional direction may be easier to identify in simpler tasks in which possible movement direction is highly constrained. Bisiach et al. (1990) used a modified horizontal bisection test in which left neglect patients operated a pulley device to indicate the midpoint of line stimuli. In the congruent condition, patients had to move the pointer directly toward the midpoint of the line, and the result was that they bisected to the right of the true midpoint, as is normally found. In the novel incongruent condition, the patients had to operate the pulley such that the pointer moved in a direction opposite to the direction of the current hand movement. In this situation, most patients continued to bisect lines too far to the right, but some failed to move their ipsilesional hand sufficiently to the left, resulting in paradoxical leftward bisection errors. These patients tended to have lesions which included the frontal lobe, which were taken to produce a more motoric neglect.

As with the mirror-reversed cancellation task, however, this interpretation of the incongruent task can be queried. In particular, the incongruent condition does not merely oppose visual and motoric directions, but also visual and proprioceptive ones. Furthermore, the incongruent condition is both novel and highly incompatible, and it may be these aspects that primarily influence frontal patients, rather than just the direction of movement per se. Finally, it is unclear why a purely motoric bias would lead patients to misbisect lines in any case, given that the tasks used always provided them with visual feedback throughout. Such feedback should therefore have been available to correct any misreaching towards the midpoint if the only deficit did indeed lie in just this reaching movement, and not in the use of visual information.

Further evidence from an opposition technique comes from animal studies of neglect in rodents with unilateral lesions. Carli et al. (1985) studied the responses of rats to visual stimuli presented on the left or right of their heads. One group of rats was trained to move their heads to the same side as the stimulus (the congruent group), while the other group was trained to move their heads to the side opposite the stimulus (the incongruent group). Following unilateral striatal dopamine depletion, both groups showed an ipsilesional response bias, in addition to significant lengthening of initiation times when a correct response was made toward the contralesional side. Once initiated, however, the time required to complete the contralesional movement was not abnormally elevated by the lesion. The authors concluded that the neglect induced by unilateral dopamine depletion, traditionally attributed to sensory inattention, actually arises as a consequence of impairment in the initiation of contralesionally directed motor responses. A subsequent study ruled out the possibility that the rats were simply unable to localise response targets on the side opposite the lesion (Brown and Robbins 1989a).

Brown and Robbins (1989b) reported a dissociation between the patterns of response deficits exhibited by rats following unilateral dopamine depletion to different regions of the striatum. Lateral striatal depletions induced a strong ipsilesional response bias but did not elevate contralesional initiation times, whereas medial striatal depletions induced a relatively weak ipsilesional response bias but significantly elevated initiation times for contralesionally directed responses. This suggests that distinct components of contralesionally directed movements can be impaired. These animal data provide perhaps the best evidence from an opposition technique for the claimed dissociation between motoric neglect (for which the direction of response is most critical) and sensory neglect (for which the direction of the trigger stimulus is most critical). While the incongruent condition was again very incompatible spatially, as in the human studies, the rat work nevertheless has several methodological advantages which could be incorporated into subsequent patient work. First, the start position was always held constant at the centre for all conditions, so no lateral bias in the final response could emerge simply from a failure to complete the task leaving the effector too close to its original start position (cf. Bisiach et al. 1990; Tegnér and Levander 1991). Second, the initial component of the required movement was itself nondirectional (the rats had to pull their heads straight back from the hole inside which they were initially positioned before they could make their lateralised response). Hence, any impairments for movements in the contralesional direction that arise in the early stages of these movements cannot be attributed to subtle sensory impairments, such as impaired kinaesthetic feedback after effectors have been shifted contralesionally.

It would be interesting to adapt this aspect of the rat work for human research on neglect. For instance, patients could be required to make choice responses that each involve a sequence of submovements (e.g., successive key presses). The alternative sequences would both begin with a central submovement and only subsequently diverge into contralesional versus ipsilesional submovements. Any difference between initiation time for these sequences at the (nonlateralised) beginning could be unambiguously attributed to a deficit in *programming* movement sequences when these involve later components in the contralesional direction. By contrast, any deficit merely in the *execution* of contralesional components can always be attributed (in principle, at least) to a possible deficit in kinaesthetic feedback, arising only after the effectors have shifted contralesionally. Work in the motor control literature with normals already shows that later components of a motor sequence can influence earlier components, due to the existence of a motor program for the entire sequence (Rosenbaum 1991). It should be informative to pursue this issue further by testing any lateralised biases in the programming stages of motor control for neglect patients.

The results reviewed in this section demonstrate that the various opposition techniques have already proved informative. However, it should be noted that they all have several intrinsic limitations, even in the optimal case that was used by the rat experiments. First, when motor direction is set against visual direction, this also sets kinaesthetic direction against visual direction. This problem can be avoided by measuring just initiation time, at which point the movement has not yet led to any kinaesthetic inputs signalling a shift in the contralesional direction. Second, by their intrinsic nature, the congruent tasks are always more compatible than the incongruent tasks introduced by opposition methods. This means that any difficulty with contralesional movements in the incongruent task could be attributed to either of two problems. It might be due to a specific difficulty in initiating voluntary movements in the contralesional direction, as is usually claimed. Alternatively, it might be due to a particular difficulty in inhibiting the natural and compatible response for visual targets when they appear on the ipsilesional side. Moreover, this latter problem could even have a sensory basis. Targets on the ipsilesional side might simply be more salient perceptually (Milner and Harvey 1995), or they may have prior entry to response selection mechanisms (Mattingley et al., submitted; Rorden et al., in press). This should lead to a stronger tendency to make the compatible orienting response towards them, so this response will consequently be harder to inhibit, thus leading to the difficulty in making the contralesional movement *away* from them that is required by the incongruent task. This ambiguous aspect of the opposition technique can easily be avoided, simply by studying lateralised choice responses that depend on the identity of a *central* visual target, as in the work of Làdavas et al. (1994) and Behrmann et al. (1995) discussed earlier.

The final limitation of opposition techniques is that, by their very nature, they set any motoric bias *against* any sensory bias and, as such, can only measure which bias predominates. We suspect that most neglect patients will actually exhibit both sensory biases towards one side and corresponding motoric biases to that side. Moreover, as noted earlier, the premotor theory actually predicts just this, rather than the strict dissociations that have often been claimed. For these reasons, it would clearly be ideal to have some method for measuring any sensory biases and any motoric biases independently, rather than simply testing which bias is the stronger. We suggest one possible way to achieve this at the end of our chapter.

Direction-Specific Abnormalities of Movement in Non-opposition Situations

Patient studies examining motor impairments as a function of movement direction can be broadly divided into two categories: those employing temporal performance measures and those employing spatial measures. Using temporal measures such as RT and movement time, patients with left neglect have been found to be slow in initiating contralesionally directed movements. Heilman and coworkers (1985) used a task in which patients used their ipsilesional hand to move a handle along a fixed linear pathway in the horizontal plane. Patients with right hemisphere lesions showed significantly longer RTs for leftward versus rightward arm movements, even for movements carried out entirely within the right hemispace. Once a contralesional movement was initiated, however, there was no difference in the time required to *execute* leftward and rightward movements. However, it might be suggested that the initiation impairment was somehow due to a difficulty in perception at the beginning of the movement, perhaps for the target endpoint, especially since the task was conducted in free vision (Heilman, personal communication 1994).

In a more recent study, Mattingley et al. (1992) used a visually cued, sequential-movement task to measure initiation and execution times for contralesionally and ipsilesionally directed actions in right-hemisphere patients with left neglect. The apparatus used in this study is shown in Fig. 2. There were ten vertically aligned pairs of buttons with additional starting and finishing buttons at either end, each of

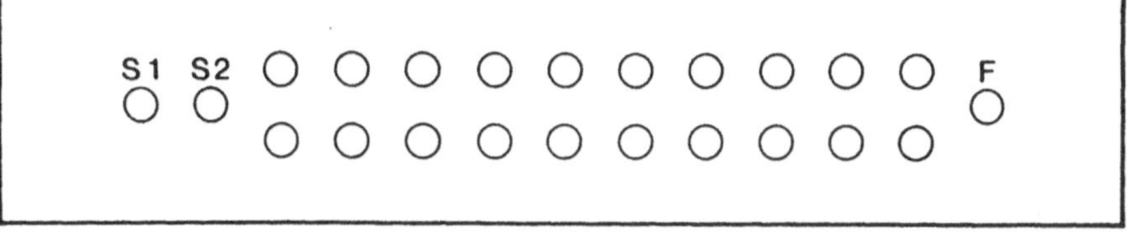

Fig. 2. Apparatus used to measure the timing of visually cued sequential movements. Subjects begin by pressing the buttons *S1* and *S2*, followed by a sequence of ten movements between successive pairs of buttons, finishing at *F*. Sequences proceed in just one direction, with the board being rotated 180° to yield either leftward or rightward movements

which could be illuminated independently. The subjects' task was to press successively a series of illuminated buttons as quickly as possible, using the index finger of the right hand. The exact sequence of cued buttons varied from trial to trial, so that each successive movement was somewhat unpredictable. However, the sequences always proceeded in just one direction, either contralesionally or ipsilesionally. Thus, a sequence of movements was elicited in which one member from each of the ten pairs of buttons was illuminated in orderly succession. Upon pressing each target button, the next cue in the sequence was activated, so that a single movement sequence consisted of ten cued movements in a consistent direction. All patients performed an equal number of leftward and rightward movement sequences, with the response board located across the body midline or entirely within the left or right hemispace.

When compared with age-matched healthy controls and non-neglecting patients who also had right-hemisphere lesions, patients with left neglect were impaired when making contralesional movements. As shown in Fig. 3A, left neglect patients were inordinately slow to initiate contralesional movements, indexed by the time for which they held each button down prior to movement initiation (the "down time"). In addition, whereas healthy controls and non-neglecting patients were faster to execute leftward versus rightward movements (as indicated by their movement times), neglect patients were slower for leftward movements (Fig. 3B). This direction-specific slowing of contralesional movements remained consistent across hemispatial locations, suggesting that the absolute location of the ipsilesional limb with respect to the body midline was relatively unimportant in determining the motoric bias. The performances of neglect patients did vary, however, with the site of anatomical damage. Patients with predominantly parietal damage were more impaired in initiating contralesional limb movements than those with frontal or subcortical (basal ganglia) lesions, though both groups were clearly slow to initiate leftward movements (Fig. 4A). The frontal patients, however, were also slow to execute contralesional movements, whereas those with parietal damage showed no such bias (Fig. 4B).

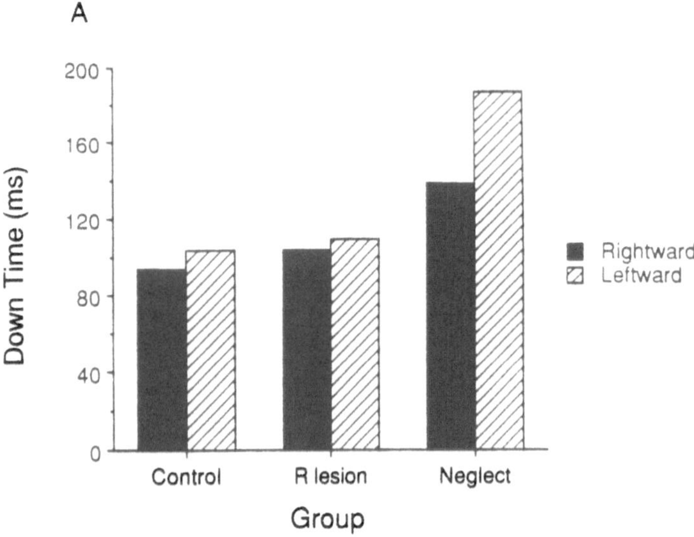

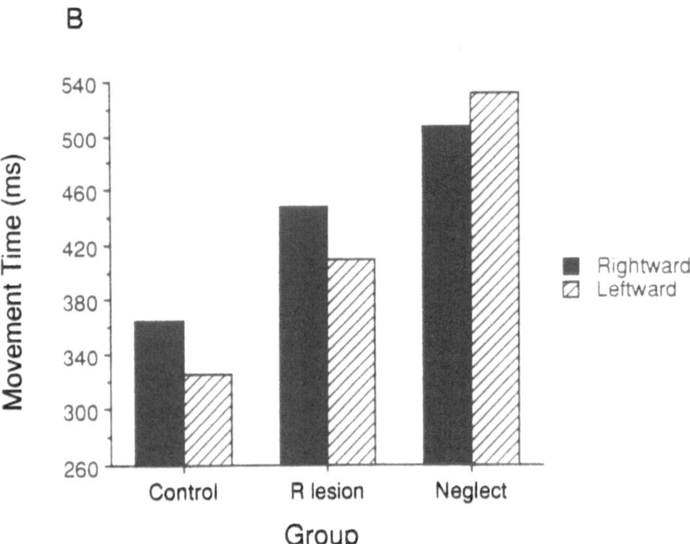

Fig. 3 A, B. Performances of healthy controls, right-hemisphere (*R*) patients *without* neglect, and right-hemisphere patients *with* neglect on the visually cued sequential-movement task shown in Fig. 2. Rightward movements are indicated by *solid bars*, leftward movements by *hatched bars*. Data have been collapsed across hemispatial position of the response board. **A** Mean down time. **B** Mean movement time. (Data from Mattingley et al. 1992)

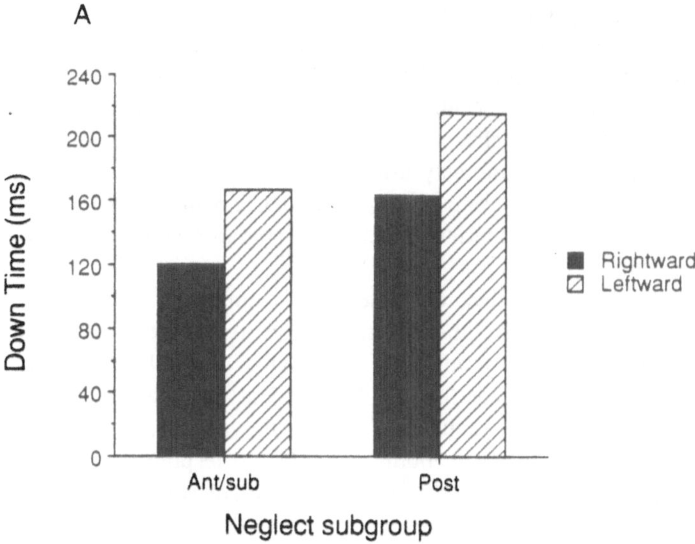

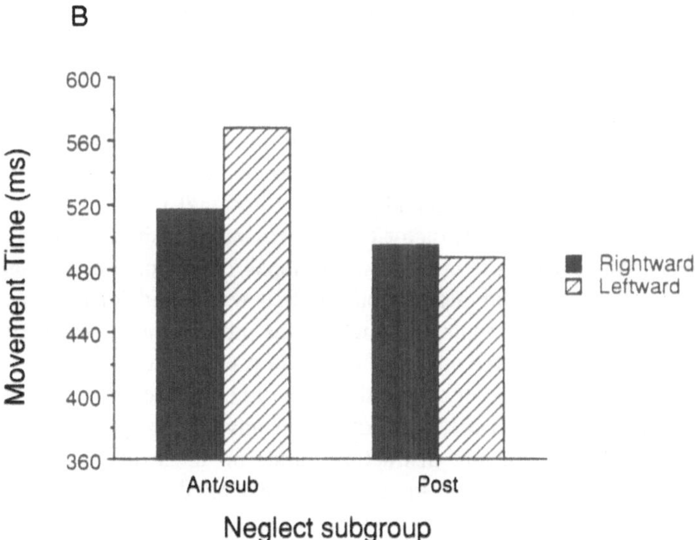

Fig. 4 A, B. Performances of left neglect patients on the visually cued sequential-movement task shown in Fig. 2. Rightward movements are indicated by *solid bars*, leftward movements by *hatched bars*. **A** Mean down time. **B** Mean movement time. *Ant/sub*, anterior and/or subcortical lesion subgroup; *Post*, posterior cortical lesion subgroup. (Data from Mattingley et al. 1992)

These results suggest that neglect patients have a visuomotor deficit in selecting responses to contralesional targets. It should be acknowledged that sensory impairments in the detection of visual targets or in the use of visual and kinaesthetic feedback to execute the movements may have contributed to the observed motoric abnormalities. Nevertheless, the anatomical distinction does suggest that different lesions may affect separate aspects of motor control, with parietal damage affecting primarily the planning and initiation of contralesional responses, while more anterior (frontal and basal ganglia) damage disrupts response execution.

More recently, we further examined the nature of this direction-specific impairment by modifying the sequential-movement task just described. We compared the effects of predictable and non-predictable sequences on the directional movements made by 12 left neglect patients (Mattingley et al., submitted). For predictable sequences, subjects moved to each adjacent button in succession along just the top or bottom row of the response board (see Fig. 2); while in the case of non-predictable sequences there was the usual two-choice decision with successive cued movements, as in the study of Mattingley et al. (1992). A further modification was the introduction of target-and-distractor trials in addition to the target-only trials used previously. Recall that in target-only trials just a single button at a time illuminated ahead in the sequence (Fig. 5A). In target-and-distractor trials, both the next button in the sequence and the previous button illuminated simultaneously (Fig. 5B). In these trials, successive responses had to be selected from a simultaneously competing distractor on the side opposite to the intended direction of movement.

As expected, patients were slower overall to execute movements to non-predictable versus predictable targets. More importantly, they were also slower to execute movements to contralesional versus ipsilesional targets, but only in the presence of a simultaneous distractor (which was more ipsilesional). In contrast, there was no significant effect of contralesional distractors on movements to ipsilesional targets. This direction-specific deficit was found for both predictable and non-predictable movements, although it was much stronger when successive movements had to be selected on-line during non-predictable sequences.

From these results we can conclude that even when the required movements in the contralesional direction are fully known in advance, response selection can still be disrupted by irrelevant distracting inputs on the ipsilesional side of the responding limb. Recent work in normals suggests that motor programs for responding to targets and separate programs for responding to distractors can both be activated in parallel and so compete for selection. Inhibitory mechanisms may be required to suppress the activity associated with distractors, so that just the target response emerges (Howard and Tipper, in press; Tipper et al. 1992; Tipper et al., in press). In patients with neglect, we suggest that these competitive interactions are unbalanced, such that ipsilesional stimuli have precedence over simultaneous contralesional stimuli in mediating response selection. In particular,

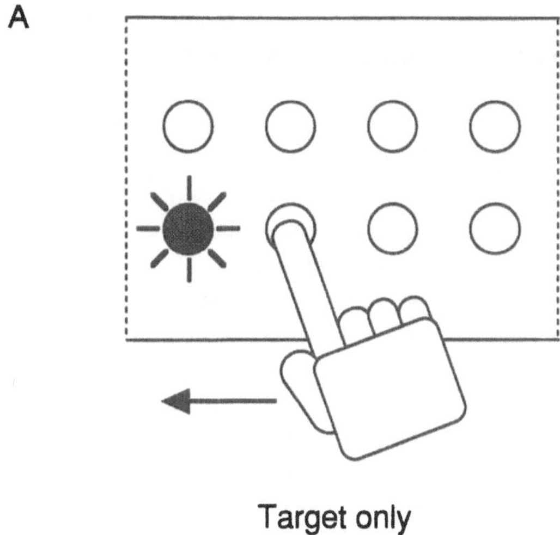

B

Target and distractor

Fig. 5 A, B. Task used to study the effects of distractor interference on visually cued movements. Subjects pressed a sequence of *illuminated buttons* one at a time, proceeding either leftward or rightward across the response board. The illustration shows part of a sample trial for a leftward (contralesionally directed) sequence. **A** Target-only condition. **B** Target-and-distractor condition

the increase in response times to contralesional targets in the presence of ipsilesional distractors seems most likely to reflect unbalanced competition between rival motor programs.

We turn now to studies that examined spatial parameters for reaching movements in neglect; these parameters cannot be assessed with the purely temporal measures described above. When reproducing horizontal displacements of a lever, patients with left neglect make systematic undershoot errors when directing their right arm toward the contralesional side (e.g. Meador et al. 1988). However, it remains unknown whether these errors in motor performance could be due to misperceptions of visual or kinaesthetic feedback, or of the required target position for the contralesional movements.

Recent technical advances in the analyses of movement kinematics now allow us to measure the spatial and temporal properties of natural reaching movements with exquisite precision (e.g., Massey et al. 1992; Wiegner and Wierzbicka 1992). One of the first studies to exploit such techniques in right-hemisphere patients was conducted by Goodale et al. (1990). They analysed the movement trajectories of recovered neglect patients as they pointed to a single visual target, or midway between two horizontally adjacent targets, with the ipsilesional hand. The most striking finding was that patients' hand trajectories were systematically distorted toward the ipsilesional side compared to normals (and left-hemisphere patients), even when targets were located contralesionally. These distortions were apparent in the initial phase of the reach but were corrected later in the movement, so that endpoint accuracy was generally within normal limits. This correction might be attributed to the availability of visual feedback signalling the error in the reach as the hand approached the target.

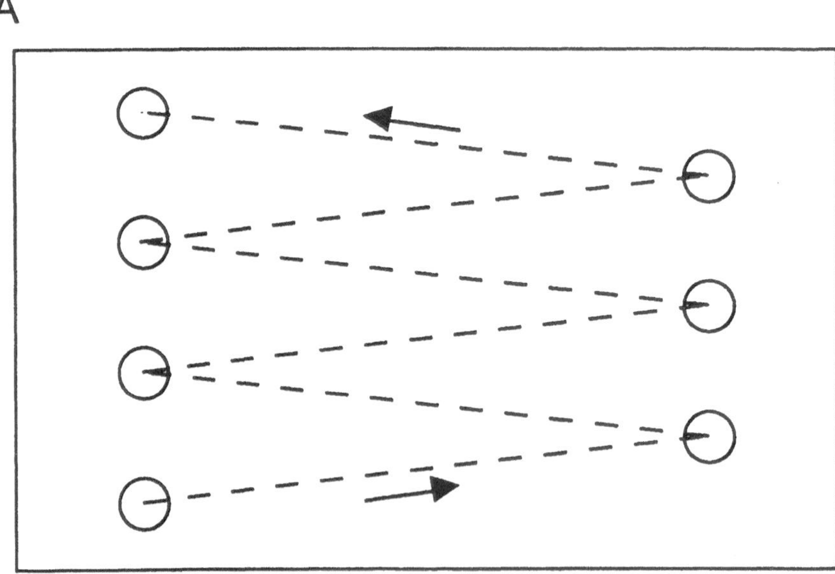

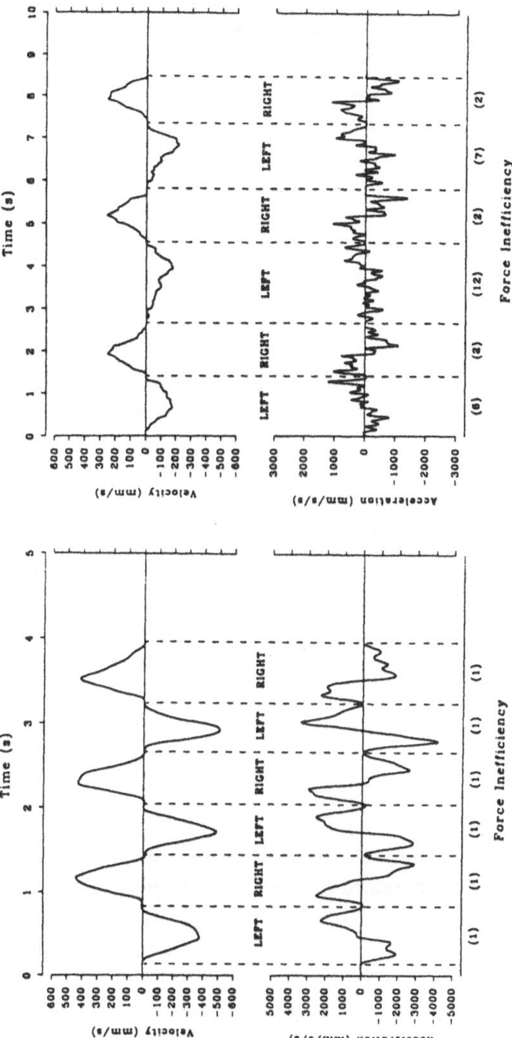

Fig. 6. A Task used to measure kinematics of visually guided hand movements. Subjects moved a non-inking electronic pen over the surface of a digitising pad, moving between left- and right-sided targets (indicated by *circles*) in a "zig-zag" pattern (*dotted line*). Side of start was counterbalanced across trials by rotating the target layout through 180°. *Arrows* indicate the direction of pen movement. **B** Sample velocity and acceleration profiles from a single trial of six zig-zag pen strokes. *LEFT*: healthy control subject. *RIGHT*: left neglect patient. Direction of movement (*left* or *right*) is indicated separately for each stroke. *Dashed lines* demarcate corresponding portions of velocity and acceleration functions. *Bracketed numbers* on the *lower axis* indicate the ratio of zero crossings in the acceleration and velocity functions (an index of force inefficiency). (Data from Mattingley et al. 1994)

Harvey et al. (1994) conducted a similar study, but manipulated whether visual feedback was available. The bias in trajectory towards the ipsilesional side was maximal under open-loop conditions, i.e. when no sight of the target or hand was available during the reach. Related findings were obtained in another kinematic study of a patient who had largely recovered from florid left neglect in which reaches were made to target objects in the presence of distractors positioned on the contralesional or ipsilesional side (Chieffi et al. 1993). In this case, an ipsilesional bias in the early phase of the reach was exacerbated by the presence of an ipsilesional distractor, whereas contralesional distractors had no effect (as similarly found by Mattingley et al., submitted, using a temporal measure).

A final kinematic study measured both spatial and temporal properties of neglect patients' movements as they made alternating ("zig-zag") leftward and rightward movements to circular targets that were fixed to the surface of a digitising tablet (Mattingley et al. 1994). Using their ipsilesional hand, patients moved a non-inking electronic pen over the surface of the digitiser as quickly as possible, with the constraint that each successive movement should terminate within the next target (Fig. 6A). Sample velocity and acceleration profiles from a single trial of six zig-zag movements (three leftward and three rightward) are shown in Fig. 6B. Each movement made by the healthy age-matched control is characterised by a symmetrical bell-shaped velocity curve, with zero-crossings indicating the end of each movement inside a target. The corresponding acceleration profile shows a single cycle of acceleration and deceleration for each movement in the sequence. Contrast this normal pattern with that obtained from a right-parietal patient with left neglect. For movements toward contralesional targets the peak velocity is lower and the time to reach peak velocity is longer than for movements to ipsilesional targets. Moreover, the corresponding acceleration function indicates poor force control, with many abnormal cycles of acceleration and deceleration for contralesionally directed movements.

To summarise these kinematic studies, fine-grained analyses of limb movements by neglect patients suggest a number of deficits in visuomotor control. Hand trajectories to discrete targets tend to be skewed ipsilesionally in the initial phase of movement, especially if a distractor object is positioned further toward the intact side. Distortions of ipsilesional limb trajectories are apparent for movements to targets located on either side of the body, but tend to be corrected later in the movement cycle if visual feedback concerning target and limb position is available. Lever movements may undershoot in the contralesional direction. Finally, movements to contralesional targets are characterised by prolonged movement time, lower peak velocity, and abnormal accelerative and decelerative phases which imply poor force control.

However, a possible difficulty in interpretation hangs over many of these studies. There has been a tendency to assume that since motoric parameters (e.g. the kinematics of a reach) are being measured, any abnormality that is found must reflect a disruption to strictly motoric processes. However, as we have noted at several points, one could explain many of the disturbances found for reaching in terms of perceptual impairments, either in the localisation of a visual target prior to initiating a reach or in the subsequent use of visual or proprioceptive feedback. On the one hand, reaching tasks have the advantage of being ecologically valid in the sense that they mimic real-life situations in which perception and action are mobilised toward a common goal. As we noted earlier, the incompatible tasks that opposition techniques have introduced with the aim of separating perceptual and motor factors can introduce unfortunate problems of their own. On the other hand, the very compatibility of natural reaching tasks makes it hard to tease apart visual and motoric contributions to visuomotor impairments. However, we now suggest one way that this might be achieved even within a natural reaching task.

Separating Sensory and Motor Biases Within the Same Natural Task

In collaboration with Masud Husain at Charing Cross Hospital, London, we are developing a potential method for distinguishing sensory and motoric biases in neglect *without* relying on spatially incompatible tasks as in standard opposition techniques. The approach we have adopted follows recent work in normals measuring interference effects in selective reaching (Howard and Tipper, in press; Tipper et al. 1992; Tipper et al., in press). Two bicolor light-emitting diodes are used, one in each visual field (see Fig. 7). The patient's basic task is to move the ipsilesional hand from a start-key positioned centrally, to press the button where a green light appears. The green target may appear alone in either visual field, or with a red distractor in the opposite field. On a few catch trials, only a red distractor appears, and the patient must not move from the start-key. Interference effects from the red distractor are measured by comparing target-alone initiation times to those for the same target with a distractor. For completeness we also measure movement time following initiation, but RT remains of primary interest. In normals, distractors can substantially delay RTs for target reaches (Howard and Tipper, in press; Tipper et al. 1992; Tipper et al., in press). We would expect patients with neglect to show dramatically more RT interference from an ipsilesional distractor upon a contralesional target than vice versa (as already found by Mattingley et al., submitted).

In order to examine whether any such asymmetry in distractor interference is due to motoric biases, to purely sensory biases, or to some combination of the two, we simply manipulate the start position of the ipsilesional responding hand (indicated by the peripheral grey boxes in Fig. 7) while the patient maintains central fixation. Consider the critical stimulus situation in which a contralesional (left-sided) target is presented together with an ipsilesional distractor. When the start-key is located on the left, the contralesional target and ipsilesional distractor

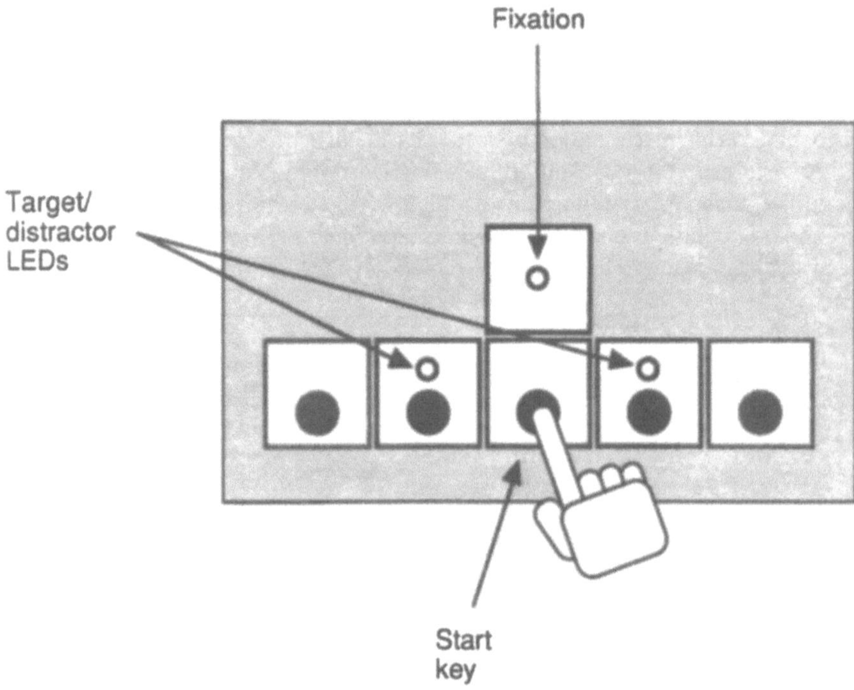

Fig. 7. Apparatus for distinguishing sensory and motor biases in neglect. In the reaching task patients move the ipsilesional hand from the *start key* to press a *button* beneath the target light emitting diode (LED). In the go/no-go task patients press the start-key when they detect the target LED

will both induce movements in the same ipsilesional direction (albeit with different amplitudes). With the start-key located centrally, however, the same ipsilesional distractor now induces a movement in the ipsilesional direction that will directly conflict with the required movement in the contralesional direction. Thus, a patient whose difficulty lies specifically in selecting movements in the contralesional direction should show drastically more interference from an ipsilesional distractor under the central start-key arrangement. By contrast, if the patient's difficulty lies only in perceiving contralesional visual events, the different start positions should have little or no effect, as the visual events remain the same irrespective of the starting hand position. When the start-key is moved to the right, the contralesional target and ipsilesional distractor will both induce movements in the contralesional direction, for which response selection may be impaired. Moreover, the distractor has to be passed over, which could make a motor response towards it compete especially strongly compared to a left-start position (see Tipper et al. 1992, for evidence that distractors that must be passed over produce particularly strong interference with target reaches in normals). It is important to note that even though a different amplitude of movement is required to reach the contralesional target in the various start positions, we only compare RT to a particular target against RT to that same target plus a distractor, so movement amplitudes remain constant across the critical comparisons.

It is possible that the different start positions could have some small effects on visual perception per se, in addition to their drastic alterations to the movements required to reach each position, by means of cueing visual attention in the direction where the hand starts (cf. Halligan et al. 1991). To control for any such purely visual effects from the start position, we also implement a second task comprising go/no-go responses to any green light, rather than reaching responses (no-go trials are those with only a red distractor displayed, as in the reaching task). The detection responses for green targets are made with the ipsilesional hand, and we vary start position as before. However, the go response is made simply by pressing the start-key where the hand is located, so no directional responses are ever required. Hence, this task should afford a pure measure of the ability to detect the green target visually, rather than to program a movement towards it. In this way we can measure any visual cueing effects from hand position, for comparison with the effects that hand position has on the directional reaching responses that must be selected in our reaching task. We have already tested six right-hemisphere patients on the reaching and go/no-go tasks just described. The results indicate that RTs are affected by hand position in patients with parietal lesions, but not in those with frontal lesions, thereby suggesting a motoric bias in parietal patients only on this task.

Conclusion

In addition to the more familiar attentional impairments associated with parietal neglect and extinction, there is an increasing realisation that motoric deficits may also be involved. In this chapter we have reviewed evidence suggesting that right parietal damage may lead to deficits in selecting the contralesional limb for action, either in isolation or in the context of simultaneous ipsilesional limb movement. Moreover, such problems in selecting an appropriate effector may even occur within a single hand, with proprioceptive inputs from the arm modulating the selection bias. There is also evidence for impaired selection of movements directed toward contralesional targets, though here the interpretation of existing data is far from straightforward. Studies claiming to find a dissociation between sensory and motor biases in neglect have typically used standard clinical tests of "perceptual" and "motor" functions. Yet, as we have seen, these tasks do not tap just one function or the other, nor are they equated in terms of their demands on overall processing capacity or selective attention. In general terms, clinical tasks such as line bisection and cancellation are appropriate for demonstrating neglect, but are often less suited for revealing the underlying cognitive impairments.

Attempts to circumvent these shortcomings by placing sensory and motoric biases in opposition within a single task are unfortunately prone to their own set of interpretative dilemmas. The incongruent tasks used in these experiments are likely to be sensitive to problems in inhibiting natural response tendencies and in adapting to unnatural visual-motor correspondences, which are extrinsic to any purported sensory-motor distinction. In any case, if Rizzolatti's premotor theory is true, lateralised sensory and motor biases should be *associated*, rather than dissociated as suggested by some studies of motoric deficits in neglect. We believe that progress in understanding response selection deficits in neglect is likely to come from using simple, more constrained tasks adapted from the normal motor-control literature. Recent studies characterising the temporal and kinematic properties of goal-directed movements in parietal patients have already made encouraging preliminary steps in this direction.

Acknowledgements. We wish to thank Nicoletta Beschin, Masud Husain, and Ian Robertson for their collaboration and encouragement. Jason Mattingley was supported by a National Health and Medical Research Council (Australia) Neil Hamilton Fairley Fellowship.

References

Albert ML (1973) A simple test of visual neglect. Neurology 23: 58–664

Allport DA (1989) Visual attention. In: Posner MI (ed) Foundations of cognitive science. MIT Press, Cambridge, pp. 658–664

Behrmann M, Black SE, Murji S (1995) Spatial attention in the mental architecture: evidence from neuropsychology. J Clin Exp Neuropsychol 17: 20–242

Bisiach E (1993) Mental representation in unilateral neglect and related disorders: the Twentieth Bartlett Memorial Lecture. Q J Exp Psychol [A] 46:435–461

Bisiach E, Berti A, Vallar G (1985) Analogical and logical disorders underlying unilateral neglect of space. In: Posner MI, Marin OSM (eds) Attention and performance XI. Erlbaum, Hillsdale, pp 239–246

Bisiach E, Geminiani G, Berti A, Rusconi ML (1990) Perceptual and premotor factors of unilateral neglect. Neurology 40:1278–1281

Bisiach E, Tegnér R, Làdavas E, Rusconi ML, Mijovic D, Hjaltason H (1995) Dissociation of ophthalmokinetic and melokinetic attention in unilateral neglect. Cerebral Cortex 5:439–447

Bottini G, Sterzi R, Vallar G (1992) Directional hypokinesia in spatial hemineglect: a case study. J Neurol Neurosurg Psychiatry 55:562–565

Brown VJ, Robbins TW (1989a) Deficits in response space following unilateral striatal dopamine depletion in the rat. J Neurosci 9:983–989

Brown VJ, Robbins TW (1989b) Elementary processes of response selection mediated by distinct regions of the striatum. J Neurosci 9:3760–3765

Carli M, Evenden JL, Robbins TW (1985) Depletion of unilateral striatal dopamine impairs initiation of contralateral actions and not sensory attention. Nature 313:679–682

Chieffi S, Gentilucci M, Allport A, Sasso E, Rizzolatti G (1993) Study of selective reaching and grasping in a patient with unilateral parietal lesion: dissociated effects of residual spatial neglect. Brain 116:1119–1137

Coslett HB, Bowers D, Fitzpatrick E, Haws B, Heilman KM (1990) Directional hypokinesia and hemispatial inattention in neglect. Brain 113:475–486

Cowey A, Small M, Ellis S (1994) Left visuo-spatial neglect can be worse in far than in near space. Neuropsychologia 32:1059–1066

Critchley M (1953) The parietal lobes. Edward Arnold, London

Daffner KR, Ahern GL, Weintraub S, and Mesulam MM (1990) Dissociated neglect behaviour following sequential strokes in the right hemisphere. Ann Neurol 28:97–101

Desimone R, Duncan J (1995) Neural mechanisms of selective visual attention. Ann Rev Neurosci 18:193–222

Duhamel JR, Brouchon M (1990) Sensorimotor aspects of unilateral neglect: a single case analysis. Cogn Neuropsychol 7:57–74

Duncan J (1995) Attention, intelligence, and the frontal lobes. In: Gazzaniga MS (ed) The cognitive neurosciences. MIT Press, Cambridge, pp 721–733

Goodale MA, Milner AD, Jakobson LS, Carey DP (1990) Kinematic analysis of limb movements in neuropsychological research: subtle deficits and recovery of function. Can J Psychol 44:180–195

Halligan PW, Marshall JC (1991) Left neglect for near but not far space in man. Nature 350:498–500

Halligan PW, Manning L, Marshall JC (1991) Hemispheric activation vs. spatio-motor cueing in visual neglect: a case study. Neuropsychologia 29:165–175

Harvey M, Milner AD, Roberts RC (1994) Spatial bias in visually guided reaching and bisection following right cerebral stroke. Cortex 30:343–50

Heilman KM, Bowers D, Coslett HB, Whelan H, and Watson RT (1985) Directional hypokinesia: prolonged reaction times for leftward movements in patients with right hemisphere lesions and neglect. Neurology 35:855–59

Howard LA, Tipper SP (in press) Hand deviations away from visual cues: indirect evidence of inhibition. Exp Brain Res

Joanette Y, Brouchon M, Gauthier L, Samson M (1986) Pointing with left vs. right hand in left visual field neglect. Neuropsychologia 24:391–396

Kinsbourne M (1993) Orientation bias model of unilateral neglect: evidence from attentional gradients within hemispace. In: Robertson IH and Marshall JC (eds), Unilateral neglect: clinical and experimental studies. Erlbaum, Hove, pp 63–86

Làdavas E, Farnè A, Carletti M, Zeloni G (1994) Neglect determined by the relative location of responses. Brain 117:705–714

Laplane D, Degos JD (1983). Motor neglect. J Neurology Neurosurg Psychiatry 46:152–158

Liu GT, Bolton AK, Price BH, Weintraub S (1992) Dissociated perceptual-sensory and exploratory-motor neglect. J Neurol Neurosurg Psychiatry 55:701–706

Massey JT, Lurito JT, Pellizzer G, Georgopoulos AP (1992) Three-dimensional drawings in isometric conditions: relation between geometry and kinematics. Exp Brain Res 88:685–690

Mattingley JB, Bradshaw JL, Phillips JG (1992) Impairments of movement initiation and execution in unilateral neglect: directional hypokinesia and bradykinesia. Brain 115:1849–1874

Mattingley JB, Phillips JG, Bradshaw JL (1994) Impairments of movement execution in unilateral neglect: a kinematic analysis of directional bradykinesia. Neuropsychologia 32:111–134

Mattingley JB, Corben LA, Bradshaw JL, Bradshaw JA, Phillips JG (submitted) The effects of competition and motor reprogramming on response selection in unilateral neglect.

Meador KJ, Loring DW, Baron D, Rogers OL, Kimpel TG (1988) Hemispatial-limb hypometria. Int J Neurosci 42:71–75

Mesulam MM (1981) A cortical network for directed attention and unilateral neglect. Ann Neurol 10:309–325

Mesulam MM (1990) Large-scale neurocognitive networks and distributed processing for attention, language, and memory. Ann Neurol 28:597–613

Milner AD, Goodale MA (1995) The visual brain in action. Oxford University Press, Oxford

Milner AD, Harvey M (1995) Distortion of size perception in visuospatial neglect. Curr Biol 5:85–89

Posner MI, Dehaene S (1994) Attentional networks. Trends Neurosci 17:75–79

Posner MI, Walker JA, Friedrich FJ, Rafal RD (1984) Effects of parietal injury on covert orienting of attention. J Neurosci 4:1863–1874

Rizzolatti G, Berti A (1993) Neural mechanisms of spatial neglect. In: Robertson IH, Marshall JC (eds) Unilateral neglect: clinical and experimental studies. Erlbaum, Hove

Rizzolatti G, Camarda R (1987) Neural circuits for spatial attention and unilateral neglect. In: Jeannerod M (ed) Neurophysiological and neuropsychological aspects of spatial neglect. North-Holland, Amsterdam, pp 289–313

Rizzolatti G, Gallese V (1988) Mechanisms and theories of spatial neglect. In: Boller F, Grafman J (eds) Handbook of neuropsychology, vol 1. Elsevier, Amsterdam, pp 223–246

Robertson IH, Marshall JC (1993). Unilateral neglect: clinical and experimental studies. Erlbaum, Hove

Robertson IH, North NT (1993) Active and passive activation of left limbs: influence on visual and sensory neglect. Neuropsychologia 31:293–300

Robertson IH, North NT (1994) One hand is better than two: motor extinction of left hand advantage in unilateral neglect. Neuropsychologia 32:1–11

Robertson IH, North NT, Geggie C (1992) Spatiomotor cueing in left unilateral neglect: three case studies of its therapeutic effects. J Neurol Neurosurg Psychiatry 55:799–805

Robertson IH, Tegnér R, Goodrich SJ, Wilson C (1994) Walking trajectory and hand movements in unilateral left neglect: a vestibular hypothesis. Neuropsychologia 32:1495–1502

Rorden C, Mattingley JB, Karnath H-O, Driver J (in press) Visual extinction and prior entry: impaired perception of temporal order with intact motion perception after unilateral parietal damage. Neuropsychologia

Rosenbaum DA (1991) Human motor control. Academic Press, San Diego

Sheliga BM, Riggio L, Rizzolatti G (1994) Orienting of attention and eye movements. Exp Brain Res 98:507–522

Tegnér R, Levander M (1991) Through a looking glass: a new technique to demonstrate directional hypokinesia in unilateral neglect. Brain 114:1943–1951

Tipper SP, Lortie C, Baylis GC (1992) Selective reaching: evidence for action-centred attention. J Exp Psychol: Hum Percept Perform 18:891–905

Tipper SP, Howard LA, Jackson SR (in press) Selective reaching to grasp: evidence for distractor interference effects. Vis Cogn

Valenstein E, Heilman KM (1981) Unilateral hypokinesia and motor extinction. Neurology 31:445–448

Valenstein E, Van Den Abell T, Watson RT, Heilman KM (1982) Nonsensory neglect from parietotemporal lesions in monkeys. Neurology 32:1198–1201

Vallar G, Perani D (1986) The anatomy of unilateral neglect after right-hemisphere stroke lesions. A clinical/CT-scan correlation study in man. Neuropsychologia 24:609–622

Watson RT, Miller BD, Heilman KM (1978) Nonsensory neglect. Ann Neurol 3:505–508

Wiegner AW, Wierzbicka MM (1992) Kinematic models and human elbow flexion movements: quantitative analysis. Exp Brain Res 88:665–673

Worthington AD (1996) Cueing strategies in neglect dyslexia. Neuropsychol Rehabil 6:1–17

On the Role of the Egocentric and the Allocentric Frame of Reference in the Control of Arm Movements

M. GENTILUCCI, E. DAPRATI, M. C. SAETTI, and I. TONI

Istituto di Fisiologia Umana, Università di Parma, Parma, Italy

Introduction

It is generally accepted that when a movement is directed towards a visual target, an egocentric frame of reference is used in order to encode its position in space. In contrast, an allocentric frame is more likely used in order to build a perceptual representation of the same object. We define as allocentric the frame of reference used to encode an object with respect to surrounding visual cues. It has been shown that when the two frames provide conflicting information about object position, information from the egocentric system is selected for motor planning. Bridgeman and coworkers (1981) have reported that even if a fixed visual target surrounded by a moving frame was judged as displaced in a direction opposite to that of the frame, subjects pointed to the true location of the target. Accordingly, Wong and Mack (1981) have shown that, when a target was represented after a 500-ms blank interval in the same location, but with the surrounding frame displaced a few degrees to the right or to the left, subjects had the strong illusion that the target had changed position. Despite this illusion, they directed their saccades to the true location of the target.

In order to control an arm movement towards a target, calculation of target position in space with respect to the body (egocentric frame of reference) is necessary. However, encoding the position of a visual target as well as planning the arm trajectory towards it can be influenced also by the spatial relationships that both the target and the hand have with surrounding visual cues (Foley 1977; Conti and Beaubaton 1980; Velay and Beaubaton 1986; Toni et al., in press). Object localization using multiple frames of reference may be useful in order to achieve a more stable representation of object position in space. Such a representation may be required in order to guide a movement when a decay of visual information is expected, i.e., during memory-driven movements. In fact, Elliot and Madalena (1987) observed that absolute errors increased when pointing to a memorized target was executed with increasing delay. Errors and kinematics of saccades differed depending on whether they were directed to a visual target or to a memorized target (Gnadt et al. 1991).

In: Parietal Lobe Contributions to Orientation in 3D Space (1997). P. Thier and H.-O. Karnath (eds). Springer-Verlag, Heidelberg.

In this paper we will attempt to discuss the respective roles of the egocentric and of the allocentric frames of reference in the control of arm reaching movements. We will provide evidence that visually guided movements rely more on an egocentric frame, whereas movements directed to internally represented objects (i.e., memory-driven movements) rely more on allocentric cues. The different involvement of the two frames will be discussed on the basis of behavioral data collected from healthy subjects.

It is possible that for both visually and internally driven movements, objects are represented by means of the balanced contribution of egocentric and allocentric information. Coding object position using different frames of reference can be performed by a single cortical system. It is generally accepted that the parietal cortex is involved in egocentric object representations, as neurophysiological (Hyvarinen and Poranen 1974; Mountcastle et al. 1975) and neuropsychological (Ratcliff and Davies-Jones 1972; Jeannerod 1986; Pierrot-Deseilligny et al. 1986; Perenin and Vighetto 1988; Rizzolatti and Berti 1990; Chieffi et al. 1993; Karnath et al. 1993) studies have shown. In addition, controversial data from neurophysiology (Haxby et al. 1991; Graziano et al. 1994) and neuropsychology (Calvanio et al. 1987; Farah et al. 1990; Driver and Halligan 1991; Mennemeier et al. 1994) support the notion that in the same area an object representation in allocentric coordinates is also present. In order to test this latter possibility, we studied the role of allocentric cues in the control of arm movements executed by an occipitoparietal patient and by control subjects.

Distance Reproduction in a Perceptual and Motor Task

Distance is an extrinsic property that is extracted from the object in order to calculate amplitude of arm movements and to plan forces and relative timings. Distance is also used for perceptual judgement in order to define the spatial relationships between objects. In a first control experiment we attempted to determine whether different representations of distance are used in motor and perceptual tasks (Gentilucci and Negrotti 1994). In a dark room subjects were presented with two visual stimuli whose distance was randomly varied. Figure 1 gives a schematic representation of the experimental apparatus.

Stimuli were a light emitting diode (LED) placed along the subject's midline and one of six LEDs placed to the right of the subject. They were presented for 700 ms. As soon as the stimuli were switched off, subjects were required to reproduce the interstimulus distances in two conditions. In one condition (reproduction by pointing) they pointed to a virtual position in space. In the other condition (visual reproduction) they matched the distance by means of two other visual stimuli. One of them was a fixed LED (L1, Fig. 1); the other was the beam of a laser (La, Fig. 1). Pointing kinematics and, in particular, errors were analyzed.

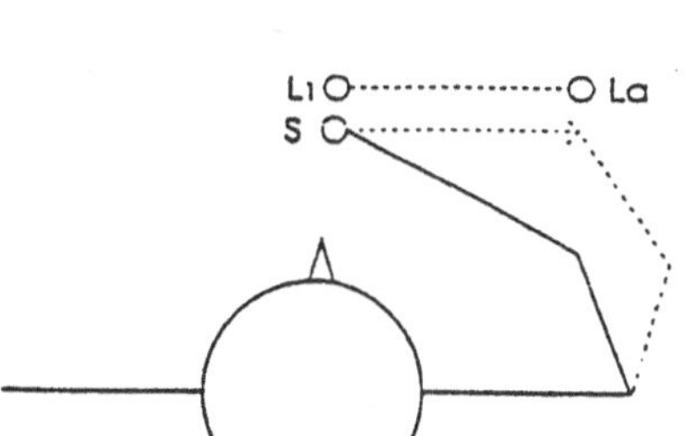

Fig. 1. Experimental setup during reproduction of a distance (L, raw of the visual stimuli). *L1* and *La* are the visual stimuli used during visual reproduction. *L1*, fixed stimulus; *La*, beam of a laser; *S*, starting position used during reproduction by pointing. (From Gentilucci and Negrotti 1994)

Subjects constantly overestimated distance during visual reproduction, whereas they underestimated distance during reproduction by pointing (see Fig. 2). Underestimation clearly increased with increasing distance. These results can be explained by assuming that different frames of reference were used to code distances during the two tasks. Their use produced different types of errors. The error of underestimation can be consequent to the use of an egocentric frame of reference. That is, it can depend upon linear approximations during transformation of visual information about target location in kinesthetic coordinates (Soechting and Flanders 1989). Overestimation during visual reproduction can be consequent to the fact that the two presented stimuli and/or the two reproducing visual stimuli were misplaced in depth. According to the task, indeed, they were referred each to another (allocentric frame of reference) rather than to the arm.

In a second experiment (doubling-distance experiment) a representation of distance common to both visual reproduction and reproduction by pointing was hypothesized (Gentilucci and Negrotti 1996). A common representation can be constructed when distance has to be deduced by mental elaboration. Subjects were required to reproduce the double of the presented distance. The distances between the two randomly presented stimuli were half of the distances used in the control experiment.

Subjects overestimated distance by the same amount during visual reproduction and reproduction by pointing (see Fig. 3). It is possible that in both tasks subjects, in order to localize end point, mentally rotated by 180° the medial stimulus with respect to the lateral one (see Fig. 1). During this process, the use of a reference fixed on the medial visual stimulus (allocentric frame of reference) induced distance overestimation, that is, an error typical of visual reproduction (see the

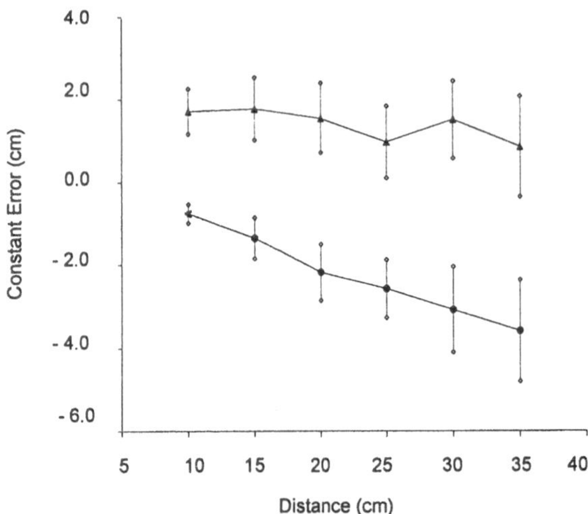

Fig. 2. Mean constant errors as a function of distance during the experiment of distance reproduction (control experiment). *Filled triangles* and *filled circles* show constant errors during visual reproduction and during reproduction by pointing, respectively. *Empty squares* show inter-subject standard errors (*SE*). (From Gentilucci and Negrotti 1996)

control experiment). However, during reproduction by pointing, the error decreased with increasing distance and inverted direction for the two longest distances (see Fig. 3). This was not the case during visual reproduction. It is commonly accepted that planning arm movements directed to a target requires two stages: target localization in space and transformation of visual object properties into movement parameters. Overestimation errors could be consequent to the process of movement end-point localization. Undershooting, deriving from the process of visuomotor transfor-mation (Soechting and Flanders 1989), could algebraically add to the localization error.

The second experiment shows that movement control can be affected by the same errors that can be observed during visual reproduction. This occurs when an object feature is the result of mental elaboration, requiring the use of an allocentric frame of reference. However, it is possible that a multiple coding of target position is used in simple visuomotor transformation only when the efficiency of the egocentric frame of reference is reduced. This was tested in a second study.

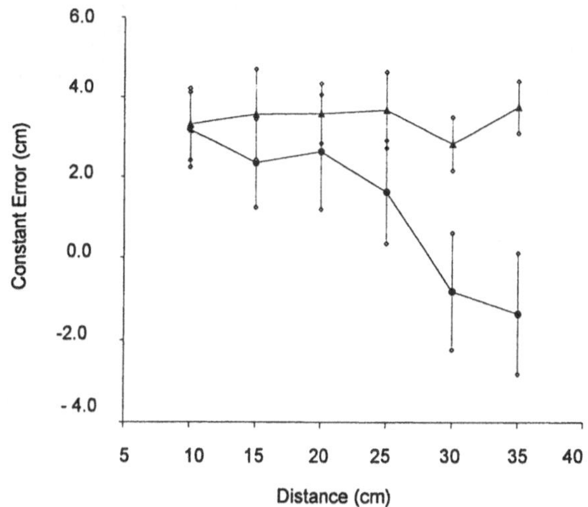

Fig. 3. Mean constant errors as a function of distance during the doubling-distance experiment. Conventions as in Fig. 2. (From Gentilucci and Negrotti 1996)

Pointing Movements and Visual Illusion in Healthy Subjects

We studied pointing movements directed to targets whose position in space could be perceived erroneously because of an illusion (Gentilucci et al. 1996). The Müller-Lyer illusion (Müller-Lyer 1889) was used. Figure 4 shows both the illusory and the control stimuli.

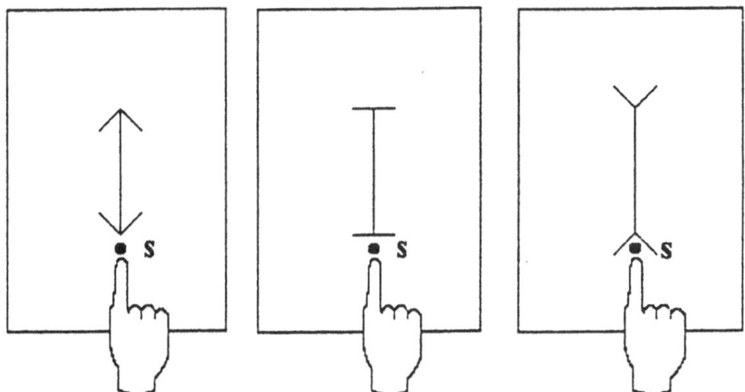

Fig. 4. Closed (*right*) and open (*left*) configurations of the Müller-Lyer illusion and control line (*middle*) presented during pointing to the distal vertex of the figure. *S*, starting position. (From Gentilucci et al. 1996)

As one can see, the shafts of the open and the closed configurations appear unequal in length although the two lines are identical. This illusion occurs mainly because experience has taught us to use shape as an indicator of size. The main axis of the stimulus was approximately aligned with the subjects' body midline. Subjects were required to execute a pointing movement towards the distal vertex of the configuration starting from a disk located near the proximal vertex of the same stimulus. Subjects were explicitly advised not to pay attention to the stimulus configuration.

Movements were executed in different experimental conditions during which the efficiency of the egocentric frame was expected to be gradually reduced. They were: full-vision, nonvisual feedback, and two no-vision conditions (i.e., from visual memory), namely, 0-s delay and 5-s delay conditions. In the nonvisual feedback condition, kinesthetic information, but not visual information about hand position was available to control movement execution towards an object that was

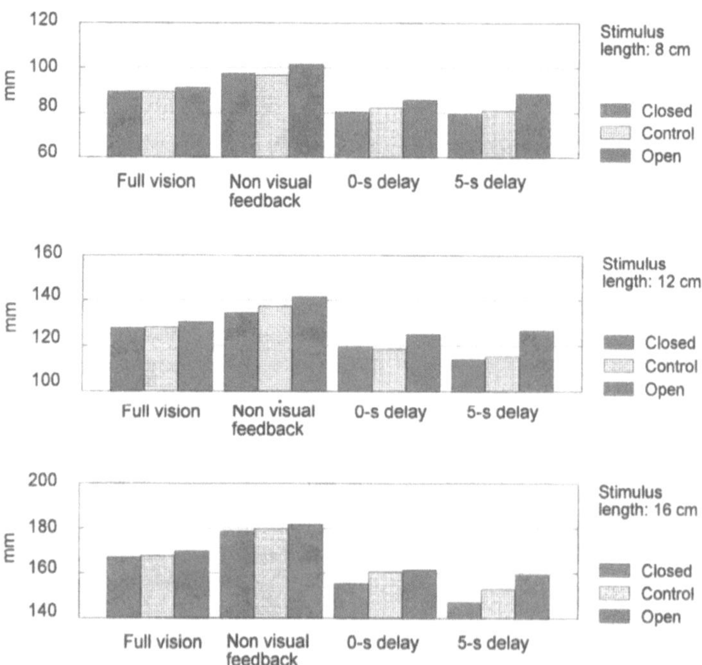

Fig. 5. Mean values of the movement amplitude during pointing to the distal vertex of the Müller-Lyer illusion. Values in the four experimental conditions measured for the three stimulus lengths and configurations are presented. Note that the starting position was displaced 1 cm from the proximal vertex of the stimulus. Since movement amplitude was calculated from this position, its value was longer than the reported stimulus length. (From Gentilucci et al. 1996)

still visible (for the role of proprioceptive information on motor control see Gentilucci et al. 1994). In the two memory conditions, stored visual information about both the object and the initial hand position was required for movement control. It can be assumed that in the first memory condition only execution, whereas in the second condition both planning and execution rely on this storage.

Figure 5 shows variations in movement amplitude as a function of the presented configuration (closed configuration, control lines, and open configuration) for each of the four experimental conditions. The illusion effect increased progressively moving from the visual condition towards the two memory conditions, being more evident for the open configuration. In other words, overestimation and underestimation were generally found for the open and the closed configuration with respect to the control lines. This effect was minimal in the full-vision condition, and slightly increased in the nonvisual feedback condition. In these conditions, a significant difference was found only between the open configuration and the two other figures. Conversely, in the two memory conditions both the open and the closed configuration had an effect on movement amplitude. In addition, overestimation during presentation of the open configuration in the 5-s delay condition was greater than in the 0-s delay condition.

The differential effect of the illusion observed across conditions can be explained by two alternative hypotheses on the visual analysis of the stimulus. According to the first hypothesis, a double coding of object position takes place. Information from the egocentric frame of reference is used to plan a visually guided movement. Both allocentric and egocentric references are used for coding object position in order to plan a memory-driven movement. The clear-cut difference of the illusion effect between vision- and memory-driven movements partially supports this view. The effect of the Müller-Lyer illusion should be absent when movement is executed in vision condition. The small effect observed in this condition can be due to the optical and retinal components of the illusion (Chaing 1968). These components could be mainly effective at the distal vertex of the illusion.

According to the second hypothesis, object position is always encoded by using both an egocentric and an allocentric frame of reference. In other words, relationships between object and body as well as those between object and surrounding visual cues are analyzed. In our experiment, localization of the vertex of the figure is thus affected by the Müller-Lyer illusion. We assume that the use of an egocentric frame of reference activates selection mechanisms allowing focus on the area around the distal vertex of the configuration. As a consequence of their activity, the influence of allocentric cues, and thus the effect of the illusion are reduced. Conversely, a progressive decay in the working memory of the efficiency of selection mechanisms induces an increasing effect of the illusion. In summary, distinct object representations are the consequence of the first hypothesis. On the contrary, a gradual penetrability between the two object representations is consequent to the second hypothesis.

Pointing Movements and Visual Illusion in an Occipito-Parietal Patient

If allocentric cues are analyzed for planning all movements, a deficit in their analysis should affect visually driven movements. This hypothesis can be tested by studying pointing movements directed to the vertex of the Müller-Lyer illusions in patients with visuospatial deficits. In this session preliminary results will be reported on pointing kinematics executed by an occipitoparietal patient and by two age-matched control subjects.

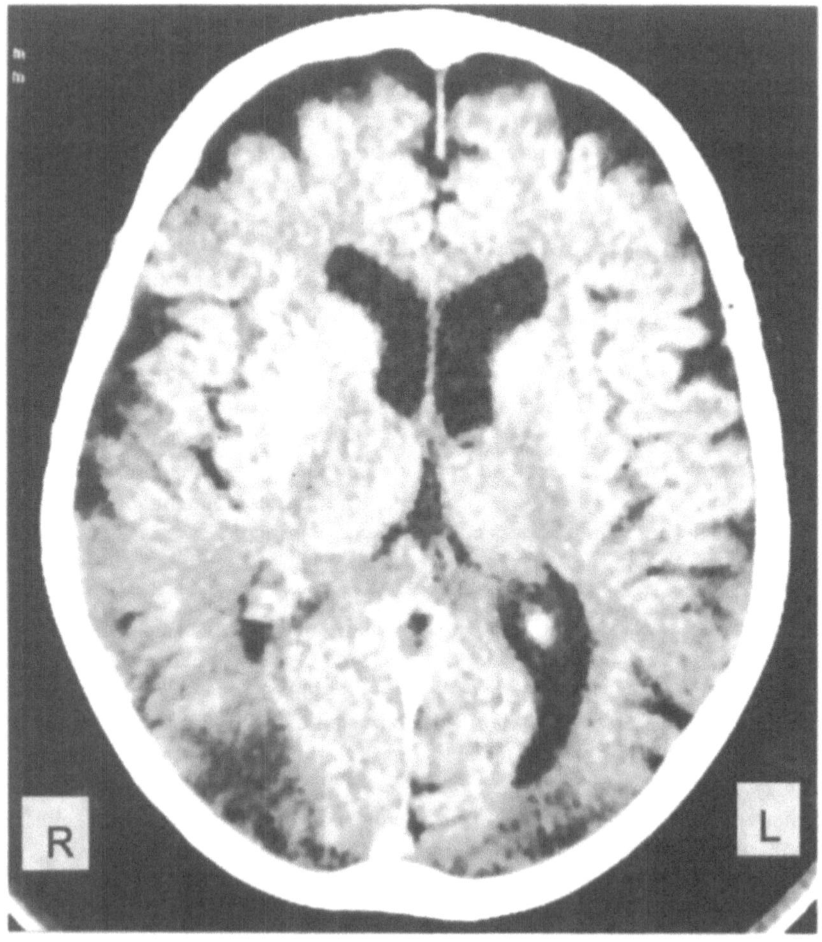

Fig. 6. CT scan of patient P.S.; *R*, right; *L*, left

The patient (P.S.), a 71-year-old right-handed woman, sustained a right occipito-parietal infarct in July 1995. She previously suffered a left occipito-parietal infarct (August 1992) from which she had completely recovered at the time of the second injury.

The CT scan (see Fig. 6) revealed the presence in both hemispheres of a clear-cut area of hypodensity which involved the occipital cortex and, to a lesser extent, the parietal cortex. The lesions in the two hemispheres were approximately symmetrical.

One week after admission to hospital, the patient presented neuropsychological deficits concerning evaluation of length (BORB: Length Match Task, Riddoch and Humphreys 1993), size (BORB: Size Match Task, Riddoch and Humphreys 1993), orientation (Judgement of Line Orientation, Benton et al. 1978; Hand Orientation Test, De Renzi and Lucchelli 1993; Unusual Views Test, Warrington and Taylor 1978) and background discrimination (X-O-N Discrimination Test, Warrington and Taylor 1973), whereas discrimination of figures requiring semantic representation (Talland Test, Talland 1958; Street Completion Test, De Renzi et al. 1969; Facial Recognition Test, Benton and Van Allen 1968) remained unaffected. Verbal memory was normal, whereas spatial memory (Corsi Test, De Renzi et al. 1977) was mildly impaired. Neither neglect (Albert Test, Albert 1973; Bells Test, Gauthier et al. 1989) nor extinction were found. Extinction was tested both by using standard clinical tests and by requiring the patient to respond to two visual stimuli simultaneously presented on a PC screen in both hemifields. Constructional apraxia (Geometrical Figures Copy Test, De Renzi and Faglioni 1967) was observed.

In a computerized perimetry test, visual field was normal on both sides, and visual discrimination of colors was preserved. Reading was not impaired, except that P.S. often lost the beginning of the next line in a paragraph. Elementary motor functions as well as tactile and kinesthetic sensations were intact. The patient showed visual disorientation and bilateral optic ataxia. Direction errors during reaching movements directed at targets located in both hemispaces were found. In particular, errors to the bottom and to the left were frequently observed in both hemispaces when the patient was required to point to circles of different sizes. In contrast, hand shaping while reaching to grasp an object was correct. In summary, deficits of visuomotor integration were associated with impaired perception of basic visuospatial features (Von Cramon and Kerkhoff 1993).

At the time of the experiment (October 1995) the patient had largely recovered from her perceptual deficits, in particular those concerning evaluation of length and size. The kinematics of reaching to grasp movements towards objects of different sizes presented in both hemispaces were studied. Both reaching and grasping movements were in normal range during movements executed in both full-vision and in the dark. However, arm trajectories were occasionally deviated to the left (ten trials out of 80). Errors were equally distributed in both vision and no-vision conditions. The patient was slightly impaired in the activities of her daily life. She reported to occasionally miss the target when she did not attentively fixate it.

Figure 7 gives a schematic representation of the experimental apparatus used in the present study. Both the patient and the two right handed controls were tested in three sessions. In two sessions they were required to point to a disk located on their midline (diameter 2 mm, see Fig. 7). Movement was initiated from a starting position located 1 cm from a reference disk either to the right or to the left of the body midline. The distance between the reference disk and the target disk could be at random either 8 or 12 cm. Both starting position and disks were projected onto the plane of a table in front of which the subjects were seated. Movements were executed in full-vision condition (first session) and in nonvision condition (second session). In both conditions the target and the reference stimulus were presented for 5 s. In full-vision condition a sound given by a PC was the signal to start the movement. In nonvision condition the sound was given after switching the light of the room off.

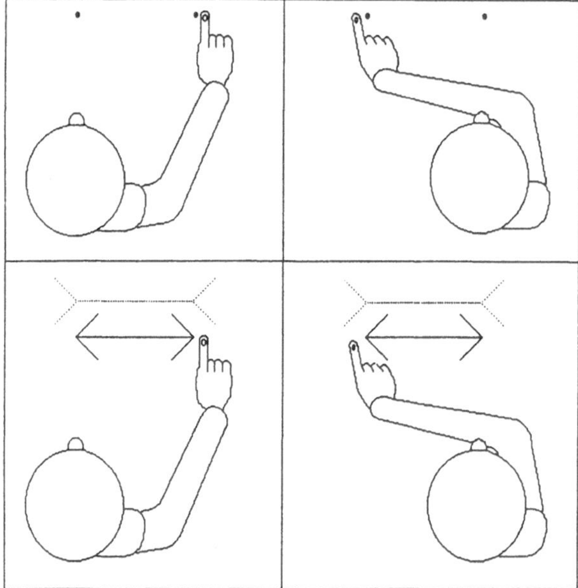

Fig. 7. Experimental setup used in the study of pointing movements executed by a parietal patient and control subjects. In the *upper row*, position of the hand and of the stimuli (midline target and peripheral reference disk) are presented. In the *lower row*, the Müller-Lyer configurations presented in the third session of the study are shown. In each trial either the closed (*solid lines*) or the open (*dashed lines*) configuration was presented. Movements were directed from the side vertex towards the midline vertex of each configuration

In the third session the presented stimuli were either the open or the closed configuration of the Müller-Lyer illusion projected onto the plane of the table. The main axis of the illusion was horizontally oriented. The vertices of the illusion lay in the same positions as the disks presented in the previous sessions. That is, the Müller-Lyer figures could be randomly presented either in the right or in the left hemispace (see Fig. 7). Their length could be at random either 8 or 12 cm. Subjects were required to point to the midline vertex of the configuration in full-vision condition. At the beginning of the trial the subjects' fingertips were placed on a starting position located 1 cm from the side vertex of the illusion. During the three sessions the patients and the controls were required to fixate the target and to execute the movement as fast and accurately as possible.

The closed and open configurations of the Müller-Lyer illusion were effective in the patient only when they were presented on the right side (see Fig. 8). In the control subjects the illusion effect was present on both sides, but to a larger extent on the left than on the right side. The illusion effect on the right side was greater in the patient than in the controls.

When pointing to the midline disk (see Fig. 9) as well as to the vertex of the illusions (see Fig. 8), the patient overestimated distance when movement was executed on the right side compared to when it was executed on the left side. In contrast, the controls overestimated distance when movement was executed on the left side. However, this effect was observed mainly in the second control subject (see Fig. 9).

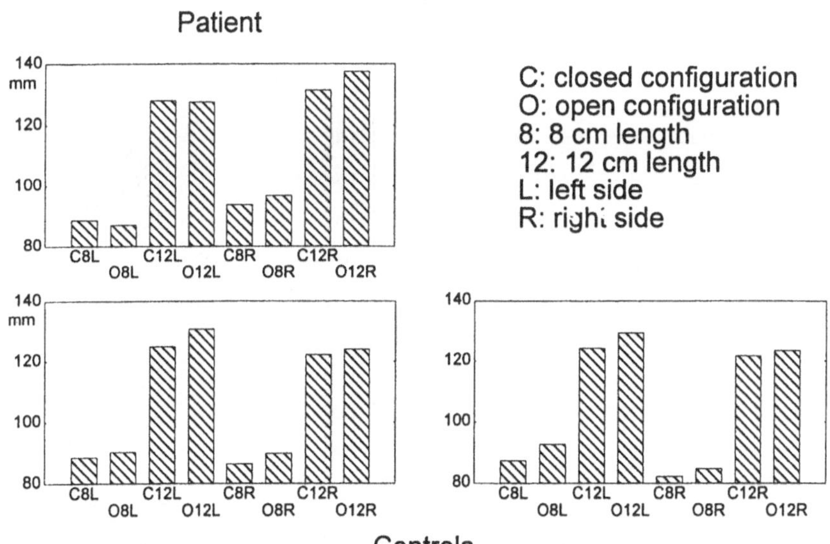

Fig. 8. Movement amplitude during pointing to the central vertex of the Müller-Lyer illusion. Values refer to the two illusory configurations, for each of the two lengths and sides. Each value is the median of five trials

Taken together, these results suggest that different visual analyses of the left and the right hemispace were performed by the patient and by the controls in order to plan a pointing movement. In the patient, the cues surrounding the target had no influence on pointing movements executed in the left hemispace, whereas they produced an effect in the right hemispace. This is supported by the data on distribution of the illusion effect. In the controls, the unbalance between the visual analyses of the two hemispaces was reversed. A more complete visual analysis of the scene was performed on the left side. Consequently, effects of surrounding visual cues on target localization were more evident on the left side. An alternative explanation of these results may be proposed. Movements in the right and in the left hemispace were executed with different directions, to the left and to the right, respectively. When the hand moves to the right, acceleration is greater than that of leftward movements (Gordon et al. 1994). Movements with greater accelerations can be less controlled and, in our experiment, the effects of surrounding visual cues on movement could be corrected less. That is, biomechanical constraints of the arm, rather than the accuracy of visual analysis, produced different movement amplitudes in the right and in the left hemispace. However, patient behavior does not support this hypothesis.

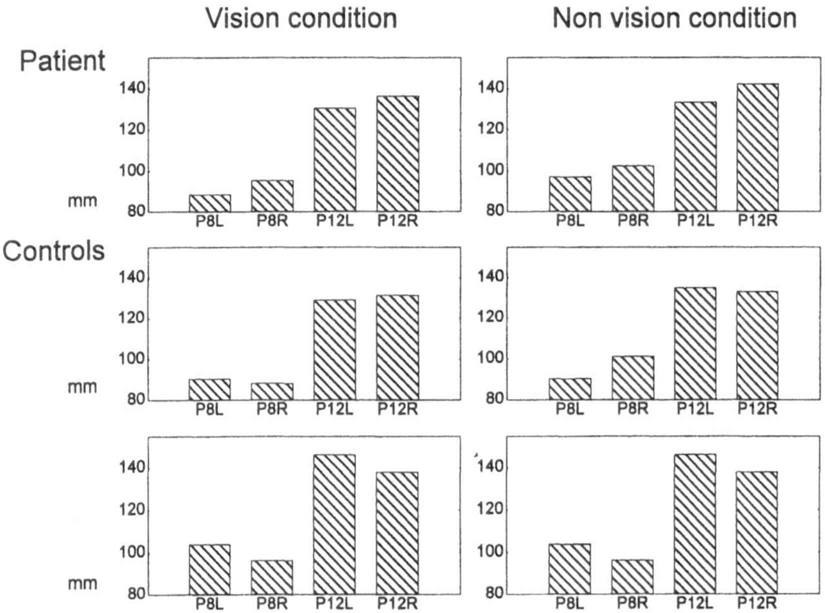

Fig. 9. Movement amplitude during pointing to the midline target. Median values of five trials measured for the two target distances and positions are presented. *Upper row* refers to movements recorded in the patient. *Middle and lower rows* refer to movements recorded in the control subjects. In the *left* and *right columns* movements executed in full-vision and non-vision condition are shown. *8* and *12* refer to distance of the target. *L* an *R* to the hemifields of stimulus presentation (left and right, respectively)

The results of the present study suggest that, for visuomotor transformation, prime information on object position is also coded in an allocentric frame of reference. In fact, if for visually driven movements target position were coded only in an egocentric frame of reference without taking into account allocentric cues, no differential effect should be found between pointing executed in the right and in the left hemispace. The target was centrally presented, and it was the same for both movements. Possible retinal effects of the midline vertex of the illusion, for example, should be effective for both types of movements. Thus, the results of the experiments on the healthy subjects and on the occipitoparietal patient indicate that a unique representation of space can be proposed for visually and internally driven movements of the arm. The different effects on pointing movements found between these two conditions can be due to a different efficiency of the system of target localization in egocentric coordinates. Further behavioral and neurophysiological experiments should be performed in order to study the possible selection mechanisms used to extract the target from surrounding visual cues. Foveal vision of the target could be responsible for the exclusion of allocentric information. The visuomotor integration for the control of eye movements can also be involved in these processes. It is interesting to note that in the parietal cortex neurones have been found whose receptive fields are in retinotopic coordinates, and, in addition, the magnitude of the visual response is modulated by eye position (Andersen et al. 1985). These neurones are very likely involved in the transformation of the retinotopic map into head (or body) centered map. However, their function might also be to either lower or enhance the analysis of the part of the visual field outside the fixation point, according to whether the eye is aligned with the target of an arm movement or not.

The data collected on the patient suggests that an allocentric representation of object position is present in the occipitoparietal visual stream. No clear deficit related to an impairment in the use of the egocentric frame of reference was observed in the patient. It is possible that the lesion affected the visual information necessary for constructing the visual scene, whereas the visual information and the mechanisms useful for localizing the object with respect to the body were spared. In other words, the dorsal visual stream of the cortex may transmit complex visual information such as induces visual illusions. In conclusion, visual information from both frames of reference can be present in the parietal cortex during visuomotor transformation of arm movements. In accordance with this hypothesis, multiple stimulus representations at the level of single unit have been observed in the parietal cortex for the visual control of eye movements. Parietal neurones that code stimulus in retinal coordinates in order to plan an eye movement have been found. Their discharge, in addition, is also modulated by head and eye position (Brotchie et al. 1995). These neurones are most likely involved in transforming target position from retinal to head-centered coordinates. One could speculate that a similar mechanism also exists for reaching neurones. That is, parietal neurones might exist whose discharge, in addition to being related to encoding target position with respect to allocentric cues, is also modulated by the spatial position of both target and allocentric cues with respect to the body.

Acknowledgements. We thank Dr. G. Pagnoni for the help in studying the occipitoparietal patient. The work was supported by Research Grant from Human Frontier Science Program, and by grants CNR (Centro Nazionale della Ricerca) and MURST (Ministero dell' Università e della Ricerca Scientifica e Tecnologica) to the Institute of Human Physiology of Parma and to M.G.

References

Albert ML (1973) A simple test of visual neglect. Neurology 23:658–664

Andersen RA, Essick GK, Siegel RM (1985) Encoding of spatial location by posterior parietal neurons. Science 230:456–458

Benton AL, Van Allen MW (1968) Impairment in facial recognition in patients with cerebral disease. Cortex 4:344–358

Benton AL, Varney NR, Hamsher KS (1978) Visuospatial judgement: a clinical test. Arch Neurol 35:364–367

Bridgeman B, Kirch M, Sperling A (1981) Segregation of cognitive and motor aspects of visual function using induced motion. Percept Psychophys 29:336–342

Brotchie PR, Andersen RA, Snyder LH, Goodman SJ (1995) Head position signals used by parietal neurons to encode locations of visual stimuli. Nature 375:232–235

Calvanio R, Petrone PN, Levine DN (1987) Left visual spatial neglect is both environment-centered and body-centered. Neurology 37:1179–1193

Chaing C (1968) A new theory to explain geometrical illusions produced by crossing lines. Percept Psychophys 3:174–176

Chieffi S, Gentilucci M, Allport A, Sasso E, Rizzolatti G (1993) Study of selective reaching and grasping in a patient with unilateral parietal lesion. Dissociated effects of residual spatial neglect. Brain 116:1119–1137

Conti P, Beaubaton D (1980) Role of structured visual field and visual reafference in accuracy of pointing movements. Percept Mot Skills 50:239–244

De Renzi E, Faglioni P (1967) The relationship between visuo-spatial impairment and constructional apraxia. Cortex 3:327–342

De Renzi E, Lucchelli F (1993) The fuzzy boundaries of apperceptive agnosia. Cortex 29:187–215

De Renzi E, Scotti G, Spinnler H (1969) Perceptual and associative disorders of visual recognition. Neurology 19:634–636

De Renzi E, Faglioni P, Previdi P (1977) Spatial memory and hemispheric locus of lesion. Cortex 13:424–433

Driver J, Halligan PW (1991) Can visual neglect operate in object-centered co-ordinates? An affirmative single-case study. Cogn Neuropsychol 8:475–496

Elliot D, Madalena J (1987) The influence of premovement visual information on manual aiming. Q J Exp Psychol [A] 39:541–559

Farah MJ, Brunn JL, Wong AB, Wallace MA, Carpenter PA (1990) Frames of reference for allocating attention to space: evidence from neglect syndrome. Neuropsychologia 28:335–347

Foley JM (1977) Error in visually directed manual pointing. Percept Psychophys 17:69–74

Gauthier L, Dehaut F, Joanette Y (1989) Bells test. A quantitative and qualitative test for neglect. Int J Clin Neuropsychol 11:49–54

Gentilucci M, Negrotti A (1994) Dissociation between perception and visuomotor transformation during reproduction of remembered distances. J Neurophysiol 72:2026–2030

Gentilucci M, Negrotti A (1996) Mechanisms for distance reproduction in perceptual and motor tasks. Exp Brain Res 108:140–146

Gentilucci M, Toni I, Chieffi S, Pavesi G (1994) The role of proprioception in the control of prehension movements: A kinematic study in a peripherally deafferented patient and in normal subjects. Exp Brain Res 99:483–500

Gentilucci M, Chieffi S, Daprati E, Saetti MC, Toni I (1996) Visual illusion and action. Neuropsychologia 34:369–376

Gnadt JW, Bracewell RM, Andersen RA (1991) Sensorimotor transformation during eye movements to remembered visual target. Vision Res 4:693–715

Gordon J, Ghilardi MF, Cooper SE, Ghez C (1994) Accuracy of planar reaching movements. II. Systematic extent errors resulting from inertial anisotropy. Exp Brain Res 99:112–130

Graziano MSA, Andersen RA, Snowden RJ (1994) Tuning of MST neurons to spiral motion. J Neurosci 14:54–67

Haxby JV, Grady CL, Horwitz B, Ungerleider LG, Mishkin M, Carson RE, Herscovitch P, Schapiro MB, Rapoport SI (1991) Dissociation of object and spatial visual processing pathways in human extrastriate cortex. Proc Natl Acad Sci USA 88:1621–1625

Hyvarinen J, Poranen A (1974) Function of the parietal associative area 7 as revealed from cellular discharge in alert monkey. Brain 97:673–692

Jeannerod M (1986) Mechanisms of visuomotor coordination: a study in normal and brain-damaged subjects. Neuropsychologia 24:41–78

Karnath H-O, Christ K, Hartje W (1993) Decrease of contralateral neglect by neck muscle vibration and spatial orientation of trunk midline. Brain 116:483–396

Mennemeier M, Chatterjee A, Heilman KM (1994) A comparison of the influence of body and environment centered reference frames on neglect. Brain 117:1013–1021

Mountcastle MB, Lynch JC, Georgopoulos A, Sakata H, Acuna C (1975) Posterior parietal association cortex of the monkey: command functions for operations within extra-personal space. J Neurophysiol 38:871–908

Müller-Lyer FC (1889) Dubloid-Reymonds Archive für Anatomie und Physiologie (suppl), pp 236–270

Perenin MT, Vighetto A (1988) Optic ataxia: a specific disruption in visuomotor mechanisms. I. Different aspects of the deficits in reaching for objects. Brain 111:643–764

Pierrot-Deseilligny CH, Gray F, Brunet P (1986) Infarcts of both inferior parietal lobules with impairment of visually guided eye movements, peripheral visual inattention and optic ataxia. Brain 109:81–97

Ratcliff G, Davies–Jones GAB (1972) Defective visual localization in focal brain wounds. Brain 95:49–60

Riddoch MJ, Humphreys GW (1993) BORB: Birmingham Object Recognition Battery. Erlbaum, Hove (UK)

Rizzolatti G, Berti A (1990) Neglect as a neural representation deficit. Rev Neurol (Paris) 10:626–634

Soechting JF, Flanders M (1989) Errors in pointing are due to approximations in sensorimotor transformation. J Neurophysiol 62:595–608

Talland GA (1958) Psychological studies of Korsakoff's psychosis: II Perceptual factors. J Nerv Ment Dis 127:197–219

Toni I, Gentilucci M, Jeannerod M, Decety J (in press) Differential influence of the visual framework on end point accuracy and trajectory specification of arm movement. Exp Brain Res

Velay JL, Beaubaton D (1986) Influence of visual context on pointing movement accuracy. Can Psychol Cogn 6:447–456

Von Cramon DY, Kerkhoff G (1993) On the cerebral organization of elementary visuo-spatial perception. In: Gulyas B, Ottoson D, Roland PE (eds) Functional organization of the human visual cortex. Pergamon, Oxford, pp 211–231

Warrington EK, Taylor AM (1973) The contribution of the right parietal lobe to object recognition. Cortex 9:152–164

Warrington EK, Taylor AM (1978) Two categorical stages of object recognition. Perception 7:695–705

Wong E, Mack A (1981) Saccadic programming and perceived location. Acta Psychol (Amst) 48:123–131

Attention and Perception

Attentional Modulation of Visual Signal Processing in the Parietal Cortex

S. TREUE[1] and J. H. R. MAUNSELL[2]

[1] Cognitive Neuroscience Laboratory, Sektion für Visuelle Sensomotorik, Department of Neurology, University of Tübingen, Tübingen, Germany
[2] S-603, Division of Neuroscience, Baylor College of Medicine, Houston, USA

Everyone knows what attention is. It is the taking possession by the mind, in clear and vivid form, of one out of what seems several simultaneously possible objects or trains of thought. Focalization, concentration, of consciousness are of its essence. It implies withdrawal from some things in order to deal effectively with others.
– William James: The Principles of Psychology

Attention is an important part of everyday life. We use it to concentrate on aspects of our sensory input that we deem worthy of further processing. Without such a selection process our visual system would be inundated with information. Instead, only a small fraction of the visual information received by our retinas reaches visual awareness. This important role of attention in visual processing is thoroughly documented in extensive *psychological* literature. Over the past years, there have also been a substantial number of *physiological* studies addressing the role of attention in visual information processing.

The representations that exist in the visual cerebral cortex reflect to a large extent a hierarchical processing that extracts increasingly complex information from the signals arising in the retinas. In addition to purely sensory processing, it has also been shown that representations in the visual cortex can be profoundly influenced by the organism's behavioral state, of which attention is an important component. The interplay between bottom-up sensory information and top-down effects of attention is likely to be an important aspect of cortical processing. A characterization of this interaction is essential for understanding cortical function.

Here we will give a brief overview of the physiological literature on attentional effects on single cell responses in the visual areas of the temporal and parietal cortex. Then we will present some results of our own work on attentional modulation of visual motion processing in the parietal cortex. In conclusion, we will outline some future questions and unresolved issues.

In: Parietal Lobe Contributions to Orientation in 3D Space (1997). P. Thier and H.-O. Karnath (eds). Springer-Verlag, Heidelberg.

Physiological Correlates of the Spotlight of Attention

The plethora of psychophysical studies of attention establishes (among many other things) two basic aspects of attention. First, they show that attention changes how sensory information is processed. Second, they demonstrate that this modulation is selective, i.e., that not all sensory signals are equally affected. The latter aspect distinguishes attention from arousal.

Many early physiological studies of attention using trained rhesus monkeys concentrated on establishing the neural correlates of these two features of attention. The experiment of Bushnell, Goldberg and Robinson (1981) exemplifies these efforts and points out some critical design elements of physiological studies of attention. These investigators trained a rhesus monkey to fixate a small spot while two stimuli were presented on a screen. The animal was trained at two tasks. In one, it had to pay attention to one of two stimuli and release a lever when it dimmed. In the other task, it had to direct its attention to the other stimulus and (again, without looking at it directly) respond to its dimming. By requiring identical fixation and presenting the same stimuli on all trials, sensory input was kept identical between the two conditions, an important requirement if one sets out to demonstrate that changes in neuronal firing rates are due to changes in the attentional state of the animal, and not merely to changes in sensory stimulation. One stimulus was positioned to fall on the receptive field of a neuron in the parietal cortex that had been isolated with a microelectrode. Testing eight cells in area 7 of the parietal cortex, the investigators found that when the animal was performing the task requiring attention to the stimulus *in* the receptive field, the neuron's response was stronger than in the other task. This shows that an increase in the neuronal firing rate is a physiological correlate of attentional *modulation* and that this enhanced response of only those neurons whose receptive fields overlap the location of the attended stimulus is a correlate of the *selective* aspect of attention.

Many other studies have described similar effects of attention (e.g., Robinson et al. 1978; Richmond et al. 1983; Mountcastle et al. 1981; Richmond and Sato 1987). These studies contributed to the notion of a "spotlight of attention" (Posner 1980; Broadbent 1982). Capturing both the spatially selective and the enhancing nature of attentional effects, this metaphor suggests that attention selects and enhances the processing of stimuli just like a flashlight in a darkened room. It has been proposed that this "spotlight" makes relevant signals more salient, thereby facilitating behavioral decisions (Crick 1984; Koch and Ullman 1985; Hulbert and Poggio 1985). To account for further psychophysical experiments on the spatial extent of attention, Eriksen and Yeh (1985) extended this metaphor into the "zoom lens" model, suggesting that attentional resources can be distributed across the visual field, but with low resolving power, or be restricted to small portions of the visual field with a concomitant increase in processing power.

The above metaphors imply a purely *spatial* organization of attention, but attention can be directed towards different *features* (such as color, motion, orientation, etc.) in the *same* part of the visual field (see, e.g., Corbetta et al. 1990 1991). This is only captured by the spotlight and zoom lens metaphors if one assumes that they operate on the feature-specific representations created in the various cortical areas specializing on selected attributes of the visual input, an approach reminiscent of the feature maps suggested by Treisman (1969).

Going Below the Scale of the Receptive Field Size

The spotlight and zoom-lens metaphors suggest a straightforward physiological implementation of attentional effects. Attention might simply be the enhanced response of neurons whose receptive field overlaps the current spotlight of attention. In such a purely spatial scheme, attention would be unable to selectively affect the processing of one of several stimuli if they all fell within a given receptive field and/or within the spotlight of attention, unless various cortical areas each had their own attentional spotlights and the attended and unattended stimuli belonged to different feature maps.

While such a scheme would account for results like Bushnell's, there is evidence that attention can affect neurons on spatial scales that are smaller than their receptive fields. Moran and Desimone (1985) recorded neuronal responses in areas V4 and the inferior-temporal area (IT) of macaques presented with two stimuli, both of which were placed within the receptive field of the neuron under study. The animal was cued to attend to one stimulus at a time. They found that the response of the neuron was largely determined by the attended stimulus, and that the unattended stimulus had little effect on the cell's response. Thus, if a stimulus with the neuron's preferred orientation fell on the attended location and a stimulus with a non-preferred orientation fell on the unattended location, the cell responded strongly, whereas the same stimulus configuration produced little response if the animal was attending to the location containing the non-preferred orientation. The simplistic attentional mechanism outlined above cannot account for Moran and Desimone's results since the spotlight of attention always overlapped the receptive field in their experiment. These researchers interpret their result as showing that attention could shrink the receptive field down to the spatial region of interest, thereby filtering out unattended stimuli in the vicinity. Closely related results from area IT reported by Richmond and his colleagues (1983) have also been interpreted as an influence of attention on receptive field size.

While these results suggest that attentional effects are not constrained by the scale of the large receptive fields found in many extrastriate cortical areas, they do not conclusively answer the question of whether the mechanism is purely spatial, such as shrinking the receptive field. Alternatively, the changes in neuronal responses described by Moran and Desimone might, for example, result from top-down influences acting to produce an internally generated representation of the animal's current target or from changes in the neurons' feature selectivity such that they become less responsive to the unattended stimulus.

Attentional Modulation in the Parietal Cortex

Despite the significance of such differential effects of attention within the spotlight of attention, they have so far only been examined in the *temporal ("what")* pathway concerned primarily with the analysis of color and form (Ungerleider and Mishkin 1982; Merigan and Maunsell 1993).

The objectives of our experiments in the parietal cortex were twofold. On the one hand, we wanted to establish how early the first robust attentional effects could be found along the *dorsal ("where")* pathway and if some of these effects go beyond merely enhancing or attenuating cellular responses. On the other hand, we wanted to know whether attention influences the processing of visual motion, a parameter that has so far received little interest from neurophysiologists studying the role of attention.

Most studies of attentional and other extraretinal effects in the dorsal visual pathway have concentrated on later stages of the dorsal pathway beyond the middle temporal visual area (MT). Studies using stationary spots of light in areas 7 and 7a have shown enhanced responses under some attentional conditions. The experiment of Bushnell and her colleagues (1981) described above is one such example. Mountcastle et al. (1981) also found enhanced responses in area 7a neurons compared to the responses to identical stimuli in inter trial intervals when they compared responses to stationary dots flashed within the receptive fields while the animal was performing a task. These studies also demonstrated that attentional responses are often hard to distinguish from presaccadic activity because many cells in these areas show responses to stimuli to which the animal intends to make an eye movement.

There is previous evidence of attentional enhancement in the medial superior temporal area (MST). Newsome et al. (1988) report that *stationary* dots that were to be used as saccade targets elicited the same response as identical dots when they were not saccade targets. However for six of 21 cells tested in area MST, the response to a *moving* stimulus was at least 50% greater when that dot was a pursuit target than when the animal was simply fixating under identical visual stimulation. Similarly, Andersen et al. (1990) report that many MST neurons show an enhanced response to a moving stimulus in the receptive field when the animal was required to detect a change in that stimulus.

Previous physiological studies failed to find evidence of appreciable systematic extraretinal effects in area MT (Newsome et al. 1988; Recanzone et al. 1993; Ferrera et al. 1994; Ilg and Thier, this volume).

Attentional Modulation of Motion Signals in Area MT and MST

We recorded responses from areas MT and MST in the superior temporal sulcus of two behaving macaque monkeys. Both areas contain a high proportion of direction-selective cells (Dubner and Zeki 1971; Van Essen et al. 1981; Maunsell and Van Essen 1983), and their sensory responses to moving stimuli have been extensively studied (Logothetis 1994). To prove attentional effects, we followed the criteria outlined above and designed a task that allowed comparison of the responses of individual neurons to identical visual stimuli under different attentional conditions.

The stimuli were small bright dots presented on an otherwise dark computer monitor in front of the animal. Each trial began with the appearance of a small fixation cross on the screen (see Fig. 1). After the monkey had foveated this cross, a stationary dot appeared somewhere on the screen, generally a few degrees to the

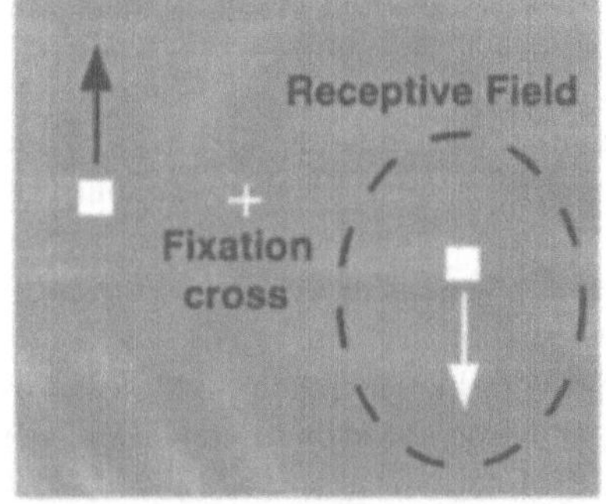

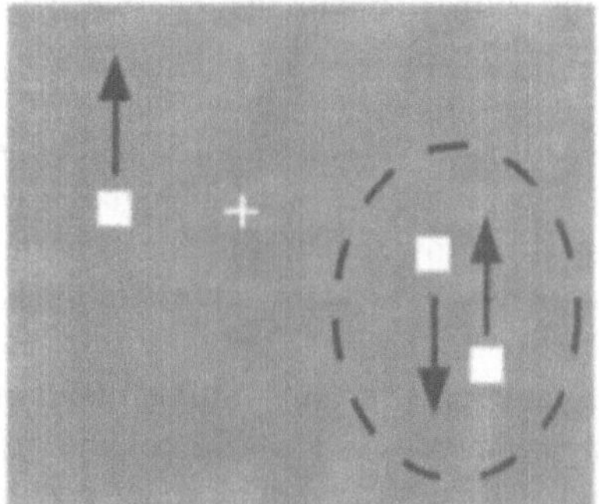

Fig. 1. A, B. Stimulus conditions employed in Experiment 1 (**A**) and 2 (**B**). The *dashed line* is the circumference of the receptive field, plotted by hand using a freely movable dot or light bar while the animal fixated a small spot. The *cross* marks the spot the animal had to fixate for the duration of each trial. In Experiment 1 one of the *dots* traveled back and forth through the receptive field along the cell's preferred and null directions while the other *dot* moved outside the receptive field. In Experiment 2 a further dot was added inside the receptive field, moving parallel to but in an opposite direction to the other dot. All dots (~0.5° x 0.5° in size) traveled back and forth along straight paths at a constant speed (roughly matched to a cell's preferred speed) and reversed their directions at the same time. On a given trial one of the dots was the target. The animal had to respond to a speed change of that dot by releasing a lever. Experiment 1 therefore consisted of two trial types that were identical as to the visual stimulus and only differed as to the dot the animal had been instructed to attend to. Experiment 2 employed one more dot and consisted of three such visually identical trial types. The animal was instructed on which dot to attend to by presenting the dot alone and stationary at the beginning of the trial. The animal had to depress the lever at this point, making the other dot(s) appear and all dots immediately started to move

left or right of the fixation point. The animal responded by depressing a lever which caused one (Experiment 1) or two (Experiment 2) additional dots to appear. All dots immediately started to move back and forth along straight, noncrossing paths at the same speed (but not necessarily in the same direction). The animal's task was to attentionally track the dot that had appeared first (the "target") and to release the lever quickly when this dot increased its speed. Sometimes one or both of the other dots ("distractors") changed speed first, but a response of the animal to a speed change of a distractor ended the trial without a reward. Throughout the trial, the animal had to maintain its gaze on the fixation cross. Only those portions of correctly completed trials before any dot had changed speed were analyzed. This insured that all trial periods analyzed were visually identical. The results presented in Fig. 2 are based on 65 MT cells, 21 MST cells, and three cells that could have been in either area MT or MST.

When a neuron was isolated, one (Experiment 1) or two (Experiment 2) dots were positioned to move back and forth within its receptive field, with the dot's axis of motion aligned to the cell's preferred direction. The other dot was placed outside the receptive field. Experiment 1 was designed to test the effect of directing attention either inside or outside the receptive field of the cell recorded from while maintaining identical visual stimulation. The upper panels in Fig. 2 show the response of a neuron in area MT to the back-and-forth motion of the dot in its receptive field under these two conditions. The left panel is a histogram of the cell's response during trials in which the animal was attending to the dot inside the receptive field, and the right panel shows the response when the animal was attending to the dot outside the receptive field. As mentioned above, visual stimulation was identical since both dots were present in both cases. Like most cells we encountered, this neuron showed a stronger response when the animal was attending to the stimulus inside its receptive field. The median value of this enhancement was 19% for cells in area MT and 40% for cells in area MST. For animal D, where we had about equal numbers of cells from areas MT and MST this difference was significant. The lower panel of Fig. 2 summarizes the strength of attentional modulation for all MT and MST cells.

We now asked whether the attentional effect we had observed operated at the scale of the receptive field or whether attention could differentially modulate the processing of several stimuli in the receptive field when they had different behavioral significance. Therefore, in the second experiment, an additional dot was presented inside the receptive field, moving parallel to the other dot, but always in the opposite direction. Any one of the three dots could be the target on given trial. The responses of most neurons depended greatly on which dot the animal attended to. The responses of one MT cell are shown in the upper panels in Fig. 3. The neuron responded strongly only while the attended dot moved in the cell's preferred direction (upward). The neuron's response was greatly reduced when the attended dot travelled in the antipreferred direction. Thus, the neuron encoded the movements of whichever dot the animal was attending to. When the animal attended to the third dot, which was outside the receptive field, the neuron

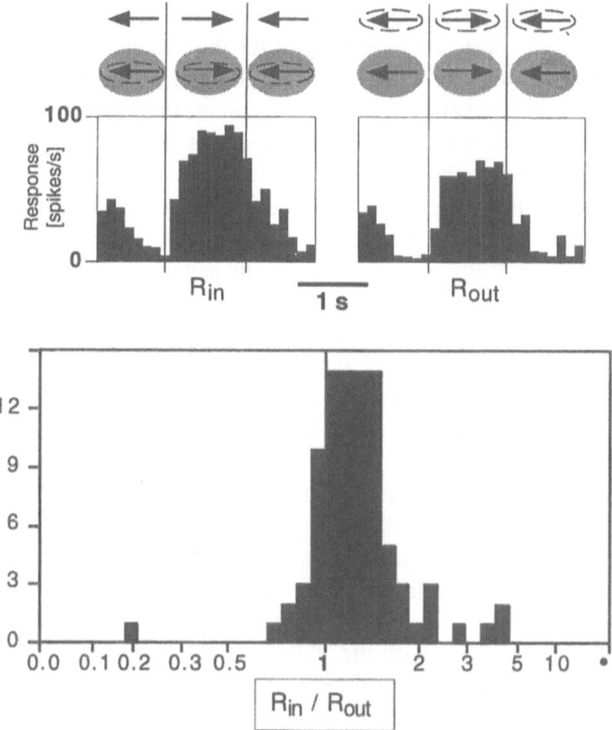

Fig. 2. Effects of attention on responses in Experiment 1 (see Fig. 1A). The *upper panels* show the responses of an isolated neuron in area MT when the animal attended either to a dot in the receptive field (*panel top left*) or to another that was outside the receptive field (*panel top right*). Drawings above each histogram schematize the stimulus motions, with the attended stimulus circled with the *dashed line* and the *shaded area* symbolizing the receptive field. *Vertical lines* in the histograms mark the times when the dots reversed direction. The response from the second period, during which the receptive field stimulus was moving in the neuron's preferred direction, was used for the analysis presented here. For this cell the response to the receptive field stimulus moving in the preferred direction was about 30% stronger when the animal was attending to that stimulus. The *lower panel* shows a stacked histogram of the strength of attentional modulation for all MT and MST neurons tested. An attentional index was computed: AI = R_{in} / R_{out}, with R_{in} as the response to the preferred motion inside the receptive field when the animal is attending to the stimulus inside the receptive field and R_{out} as the response to the same visual stimulation when the animal is attending to the stimulus outside the receptive field. The median modulation was 19% for MT cells and 40% for MST cells

maintained a relatively steady level of activity, between the level of responses to the preferred and antipreferred motions alone. This is to be expected from the previously observed response suppression in area MT using transparent stimuli (Snowden et al. 1991). It also demonstrates that the two dots in the receptive field were properly positioned to be about equally effective. Attending to the dot moving in the antipreferred direction inside the receptive field depressed the response of the neuron below the level evoked when the animal was directing its attention outside the receptive field, demonstrating that attentional modulation of sensory responses does not always enhance neuronal responses.

We quantified the strength of the attentional modulation for each neuron in Experiment 2 by comparing the response during the second phase of motion while the animal was attending to the dot travelling in the preferred direction through the receptive field with the response during the same phase while the animal was attending to the other receptive field dot, travelling in the antipreferred direction. The histogram in the lower panel of Fig. 3 shows that almost all MT and MST neurons responded most strongly when the attended dot was traveling in the preferred direction. The median enhancement was 86% for area MT and 113% for area MST, i.e., the neuronal response was about doubled when attention was on the stimulus moving in the preferred direction. While there was a trend toward stronger modulation in area MST than in area MT, this difference was not significant for our sample of cells.

These two experiments demonstrate a powerful effect of attention on the processing of visual motion information. When two stimuli are placed inside the receptive field of a neuron in area MT or MST, the response of the cells depends primarily on the movement of the attended dot, and directing attention to a stimulus outside the receptive field reduces the direction-selective modulation of these neurons. The influence of the ignored stimulus is greatly reduced, even if it is moving in the preferred direction and is thus a powerful sensory stimulus. Despite the presence of a dot moving in the preferred direction in the receptive field, many of our cells only responded strongly when the animal attended that dot and not when the animal attended the other dot in the receptive field.

This differential attentional effect we found in our Experiment 2 is much stronger than the modulation we found in our Experiment 1. This might be the reason why previous studies of attention in the parietal cortex, having employed designs more like our Experiment 1 have failed to find effects as strong, as common, and as early in the cortical hierarchy as ours. This finding also suggests that attention plays a more critical role under conditions in which signal (the "target") and noise (the "distractor") are close together, as is often the case under natural conditions.

Our findings are consistent with those of a positron emission tomography study of Corbetta et al. (1991). They found enhanced activity in the parietal cortex of humans during attentional tasks. Our results are also in agreement with a recent MRI study showing attentional modulation located in the human homologue of area MT/MST (O'Craven and Savoy 1995) in a motion attention task.

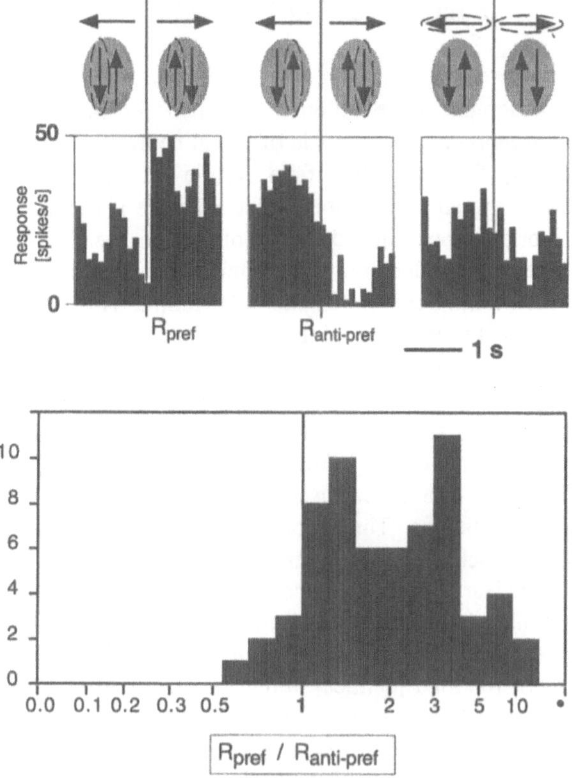

Fig. 3. Responses with two stimuli inside the receptive field. The *upper panels* show the responses of a neuron in area MT during Experiment 2 (see Fig. 1B) when three dots were presented. The two histograms to the *left* show responses while the animal attended to either of the two dots in the receptive field, and the *right* histogram plots responses when the animal attended to the dot outside the receptive field. The axis of motion of the dot outside the receptive field relative to the axis of motion of the dots inside varied from cell to cell. When one of the receptive field stimuli was the attended dot, the response of the neuron was strong whenever that dot (*circled*) moved in the preferred direction. The activity was relatively unmodulated at an intermediate level when the animal was attending to the dot outside the receptive field (this experimental condition is shown for reference only; it is not used for the analysis presented here). The *lower panel* shows a stack histogram of the attentional index for the subset of cells from Figure 2 from which we collected data in Experiment 2 (labels as in Fig. 2). Each index is computed using the average rate of firing when the animal attended to the dot moving in the preferred direction (marked R_{pref}) inside the receptive field, compared to the response when the animal was attending to the dot moving in the antipreferred direction (marked $R_{anti\text{-}pref}$) inside the receptive field. The median modulation was 86% for MT cells and 113% for MST cells

Does Attention Sharpen Tuning Curves?

The previous paragraphs have outlined evidence for the interaction between attention and an important characteristic of visual neurons: their spatial specificity as characterized by their receptive field. While receptive fields represent the tuning of a cell in spatial coordinates, most cells are also tuned to other visual features such as orientation, color, direction of motion, etc. Apart from enhancing neuronal responses to attended stimuli or filtering out unattended stimuli neurons might sharpen their tuning curves for attended stimuli. This is an important issue because of the link between neuronal tuning and psychophysical performance and the presumed role of attention in enhancing sensory performance.

There are no systematic studies of attentional effects on tuning properties in the parietal cortex, but two physiological studies have examined this issue in the temporal pathway. Spitzer et al. (1988) recorded from area V4 while varying the level of attention in an orientation discrimination task by running blocks of easy and difficult discriminations. They reported narrower orientation-tuning curves during the difficult blocks in which the monkey presumably paid more attention to the stimuli. It is unclear if other influnces, such as an increase in the background firing rate or the inclusion of cells with poor tuning fits in the analysis could have caused the appearance of sharpened tuning curves. Recently McAdams and Maunsell (in press) reported identical tuning widths in area V4 when comparing the tuning curves determined using attended versus unattended stimuli. Further research will be required to resolve this important issue.

Response Modulation in the Absence of Sensory Stimulation

The attentional modulation we have observed in areas MT and MST represents a shift away from a purely sensory neural representation of the visual input. There are other such extraretinal effects in the parietal cortex. Newsome et al. (1988) report that neurons in area MST (but not area MT) will fire when the animal is pursuing a moving target on the screen and that the neurons maintain a response even while the target is briefly blinked off (see also Ilg and Thier, this volume). Andersen and his colleagues and Goldberg and his coworkers report presaccadic activity of parietal cortical neurons during tasks in which the animal is instructed to delay an eye movement. These and other studies concerning eye-movement and eye-position related activity in the parietal cortex are the topic of other chapters in this volume.

While a strong neural response in our experiments is dependent on the presence of visual stimuli on the screen, it is nevertheless an actively constructed selective representation that accepts a loss of veracity in favor of a representation based on behavioral relevance. There is evidence that sometimes this process goes further.

Assad and Maunsell (1994) report neurons in the parietal cortex that respond to absent but imagined visual stimuli. They recorded from neurons from the superior temporal sulcus in a paradigm where a dot could be inferred to be moving behind an occluder. Using identical visual stimulation, they showed significantly higher activity during the occlusion phase than during conditions when the animal could assume that the dot was not moving behind the occluder. In a progression from the selective representation of sensory input demonstrated in our Experiment 2, the extraretinal signals of Assad and Maunsell seem to be able to drive visual neurons even in the absence of sensory stimulation, presumably creating a representation of the entirely imagined target.

Conclusion

Carefully designed physiological experiments have shown that the response of cortical neurons, both in the dorsal and temporal pathways, can be modulated by the attentional state of the animal. Early experiments demonstrated such modulation and suggested a spatial mechanism which increased the response of those neurons whose receptive field overlapped the "spotlight of attention." More recent work, including our own, has shown strong attentional effects that operate below the scale of the receptive field size and/or have non-spatial components. We have also extended the range of areas in the dorsal pathway with systematic attentional effects backwards towards the primary visual cortex by showing that the processing of motion information as early as in areas MT and MST is powerfully influenced by which of several stimuli the animal is attending to.

Our demonstration of robust attentional effects as early as in area MT – an area which receives direct input from the striate cortex (Maunsell and Van Essen 1983; Ungerleider and Desimone 1986) – suggests that responses of neurons throughout much of the extrastriate cortex may be profoundly influenced by behavioral state and that an understanding of visual information processing even in early extrastriate cortex requires an approach that does not concentrate solely on the sensory qualities of the visual input.

Acknowledgements. We wish to thank B. Noerager for excellent technical assistance in the experiments and J. Assad for helpful discussions and insightful comments.

References

Andersen RA, Graziano MSA, Snowden R (1990) Translational invariance and attentional modulation of MST cells. Soc Neurosci Abstr 16:7

Assad JA, Maunsell JHR (1994) Neuronal correlates of inferred motion in primate posterior parietal cortex. Nature 373:518–521

Broadbent DE (1982) Task combination and selective intake of information. Acta Psychologia (Amst) 50:253–290

Bushnell C, Goldberg ME, Robinson DL (1981) Behavioral enhancement of visual responses in monkey cerebral cortex. I. Modulation in posterior parietal cortex related to selective visual attention. J Neurophysiol 46:755–772

Corbetta M, Miezin FM, Dobmeyer S, Shulman GL, Petersen SE (1990) Attentional modulation of neural processing of shape, color, and velocity in humans. Science 248:1556–1559

Corbetta M, Miezin FM, Dobmeyer S, Shulman GL, Petersen SE (1991) Selective and divided attention during visual discriminations of shape, color, and speed: functional anatomy by positron emission tomography. J Neurosci 1:2383–2402

Crick F (1984) Function of the thalamic reticular complex: the searchlight hypothesis. Proc Natl Acad Sci USA 81:4586–4590

Dubner R, Zeki SM (1971) Response properties and receptive fields of cells in an anatomically defined region of the superior temporal sulcus in the monkey. Brain Res 35:528–532

Eriksen CW, Yeh Y (1985) Allocation of attention in the visual field. J Exp Psychol Hum Percept Perform 11:582–597

Ferrera VP, Rudolph KK, Maunsell JHR (1994) Responses of neurons in the parietal and temporal visual pathways during a motion task. J Neurosci 14:6171–6186

Hulbert A, Poggio T (1985) Spotlight of attention. Trends Neurosci 8:309–311

James W (1980) The principles of psychology. Henry Holt, New York

Koch C, Ullman S (1985) Shifts in selective visual attention: towards the underlying neural circuitry. Hum Neurobiol 4:219–227

Logothetis NK (1994) Physiological studies of motion inputs. In: Smith AT, Snowden RJ (eds) Visual detection of motion. Academic Press, New York, pp 177–216

Maunsell JHR, Van Essen DC (1983) The connections of the middle temporal visual area (MT) and their relationship to a cortical hierarchy in the macaque monkey. J Neurosci 3:2563–2586

McAdams CJ, Maunsell JHR (in press) Attention enhances neuronal responses without altering orientation selectivity in macaque area V4. Soc Neurosci Abstr

Merigan WH, Maunsell JHR (1993) How parallel are the primate visual pathways? Annu Rev Neurosci 16:369–402

Moran J, Desimone R (1985) Selective attention gates visual processing in the extrastriate cortex. Science 229:782–784

Mountcastle VB, Andersen RA, Motter BC (1981) The influence of attentive fixation upon the excitability of light-sensitive neurons of the posterior parietal cortex. J Neurosci 1:1218–1235

Newsome WT, Wurtz RH, Komatsu H (1988) Relation of cortical areas MT and MST to pursuit eye movements. II. Differentiation of retinal from extraretinal inputs. J Neurophysiol 60:604–620

O'Craven KM, Savoy RL (1995) Attentional modulation of activation in human MT shown with functional magnetic resonance imaging (fMRI). Invest Ophthalmol Vis Sci 36:S856

Posner MI (1980) Orienting of attention. Q J Exp Psychol 32:3–25

Recanzone GH, Wurtz RH, Schwarz U (1993) Attentional modulation of neuronal responses in MT and MST of a macaque monkey performing a visual discrimination task. Soc Neurosci Abstr 19:973

Richmond BJ, Sato T (1987) Enhancement of inferior temporal neurons during visual discrimination. J Neurophysiol 58:1292–1306

Richmond BJ, Wurtz RH, Sato T (1983) Visual responses of inferior temporal neurons in awake rhesus monkey. J Neurophysiol 50:1415–1432

Robinson DL, Goldberg ME, Stanton GB (1978) Parietal association cortex in the primate: Sensory mechanisms and behavioral modulations. J Neurophysiol 41:910–932

Snowden RJ, Treue S, Erickson RE, Andersen RA (1991) The response of area MT and V1 neurons to transparent motion. J Neurosci 11:2768–2785

Spitzer H, Desimone R, Moran J (1988) Increased attention enhances both behavioral and neuronal performance. Science 240:338–340

Treisman A (1969) Strategies and models of selective attention. Psychol Rev 76:282–299

Ungerleider LG, Desimone R (1986) Cortical connections of visual area MT in the macaque. J Comp Neurol 248:190–222

Ungerleider LG, Mishkin M (1982) Two cortical visual systems. In: Ingle DJ, Goodale MA, Mansfield RJW (eds) Analysis of visual behavior. MIT Press, Cambridge, pp 549–586

Van Essen DC, Maunsell JHR, Bixby JL (1981) The middle temporal visual area in the macaque: myeloarchitecture, connections, functional properties and topographic organization. J Comp Neurol 199:293–326

Attentional Modulation of Visual Receptive Fields in the Posterior Parietal Cortex of the Behaving Macaque

S. BEN HAMED, J.-R. DUHAMEL, F. BREMMER, and W. GRAF

C.N.R.S. Collège de France, Paris, France

Introduction

The neural mechanisms of attention, particularly with respect to visual and spatial information processing are a major focus of psychological and neurophysiological studies. Much of the ground work in this field of research comes from studies of deficits in patients with brain injuries. Attention can be directed to regions of space away from the central visual field, thus contributing to visual awareness and orientation toward peripheral objects. This process is impaired in patients with neglect or extinction on the side contralateral to their lesion. Attention can also move within the central visual field. During visual processing, consciousness is generally occupied by a single object at a time. However, if necessary, attention can rapidly be shifted to different parts of an object or to a neighboring object. Patients with simultanagnosia are unable to perceive two stimuli presented simultaneously at a given spatial location and are severely slowed in shifting attention to different parts of a foveal image. Neglect is frequently linked to damage in the right posterior parietal cortex. Simultanagnosia requires bilateral parietal damage, although it can also be a manifestation of temporal lobe lesions. The characteristics of these disorders and their possible underlying mechanisms are described in detail elsewhere in the present volume (see Karnath, this volume).

Electrophysiological experiments conducted in awake primates also emphasized the role of the posterior parietal cortex in spatial analysis and attentional processing (see also Treue and Maunsell, this volume). These studies generally employed tasks in which the direction of attention influences the detection of a sensory event in the visual periphery (Robinson et al. 1978; Colby et al. 1993; Steinmetz et al. 1994). This chapter will mostly be concerned with the representation of the central visual field in a single region of the parietal cortex, the lateral intraparietal area (LIP). The response properties of these neurons suggest the involvement of this area in visual analysis and attentional processing within both peripheral and central regions of visual space.

In: Parietal Lobe Contributions to Orientation in 3D Space (1997). P. Thier and H.-O. Karnath (eds). Springer-Verlag, Heidelberg.

Attentional Effects in Parietal Neurons.

The most frequently described neural manifestation of attention in single cell studies is expressed as changes in the intensity of responses to visual stimuli. Global enhancement of visual responsiveness in area 7a occurs when an animal is engaged in a fixation task, as opposed to a resting condition (Mountcastle et al. 1981). More selective enhancement effects have been described when a stimulus in the visual periphery is made relevant to behavior. For instance, in both area 7a (Robinson et al. 1978) and area LIP (Colby et al. 1993), neuronal activity increases during the task in which a foveal target has to be fixated, when the reward is made contingent upon releasing a lever at the dimming of a spot of light placed in a cell's receptive field as compared to the task when only foveal fixation is required. In this case, the enhancement can be assumed to reflect a heightened awareness of the peripheral spatial location.

Enhancement, however, may not be the only effect of attention at the single cell level. Selective suppression may also correlate with the locus of spatial attention under certain task conditions. Steinmetz et al. (1994) found that area 7a neurons may be inhibited in a task in which a spatial location is initially cued by a visual stimulus and the monkey has to respond to the reappearance of the stimulus at this location while refraining from responding to distracting stimuli at other locations.

Enhancement and suppression subsequent to spatial attention locking suggests that attention may play a role at the level of stimulus detection and analysis, but the central processing mechanisms are not yet understood. Extensive studies in different areas of the visual processing streams show that visual neurons are mainly characterized by their receptive fields (RFs) if the latter are considered as spatiotemporal dynamical structures. In this light, it seems reasonable to suggest that this is the entity that is modified by attentional processes, an idea supported by recent findings in ventral stream areas such as V4 and IT (inferiotemporal cortical area ; Desimone et al. 1991). We recently obtained evidence that it may also be true of parietal neurons. We suggest that the spatial and temporal dynamics of a receptive field – its variation in shape, size and spatial position – are a neural correlate of attention.

A New Approach to the Study of Receptive Fields

By definition, the receptive field of a neuron is the portion of the retina that elicits a physiological response when stimulated. RFs have been studied in the different areas of the visual pathway. RF properties have been most extensively studied in area V1, and the notion of a "classical receptive field" has emerged to characterize the interaction patterns between RF centers and surrounds in simple and complex cells (Allman et al. 1985). The organization of center/surround interactions that implies reciprocal connectivity between areas as well as lateral connectivity is being investigated in many extrastriate areas where effects of attended and unattended stimuli have also been noted. Furthermore, although for experimental convenience RFs have up to now been most often considered as two-dimensional structures, it is becoming more and more obvious that they are complex dynamic multidimensional structures that may be affected by parameters as diverse as eye position, vergence angle and head-on-trunk position (Brotchie et al. 1995; Gnadt and Andersen 1988; Gnadt and Mays 1995; Trotter et al. 1992).

By using a quantitative RF mapping technique, a clustering of RF types with clear temporal dynamics can be shown already in area V1 (DeAngelis et al. 1993a, b). A combination of physiological and modelling approaches suggests that RF organization might be crucial to the analysis and integration of all stimulus features. But are RF characteristics and attentional modulations two independent processes, or do they interact with each other? For instance, does attention influence visual responses globally, for example by modulating the overall gain of a cell's discharge, or does it interact in a more subtle manner with the spatial and temporal distribution of neuronal activity within the RF? If the latter is true, can these effects be investigated quantitatively at the neuronal level?

To investigate these questions, we used standard extracellular recording techniques to monitor the activity of single neurons in the macaque area LIP. This area is located on the lateral bank of the intraparietal sulcus (see also Thier and Andersen, this volume and Barash, this volume) and receives visual input from several extra-striate areas. It is reciprocally connected with area 7a, which is located on the convexity of the inferior parietal lobule, and also with the superior colliculus and the frontal eye fields, two areas that play a key role in saccadic eye movements (Andersen et al. 1990; Lynch et al. 1985).

We devised a technique to map receptive fields at high temporal and spatial resolution using computer-generated stimuli back-projected onto a tangent screen. The procedure involved defining a square array of regularly spaced positions (either 7 x 7 or 9 x 9 positions), which covered an area larger than the recorded cell's RF as defined by hand mapping. The distance between two adjacent positions of the array varied between 2° and 10°, depending on the RF size. Behaviorally irrelevant stimuli (i.e., which the monkey was free to ignore) consisted of white spots, 0.8° diameter, which were sequentially flashed at randomly selected positions of the array. Flashes usually lasted 100 ms and were presented 80–200 ms apart, allowing the presentation of 6–9 stimuli during a single trial. Each postion was sampled several times to obtain a reliable estimate of the cell's sensitivity. A map of the RF was subsequently constructed off-line by correlating stimulus position and spike trains using a shifted temporal window. The window was adapted to each neuron's typical response latency and discharge duration through preliminary examination of the cumulative peri-stimulus time histograms for a sample of effective array locations.

In order to assess the influence of the behavioral context on the structure of the RF, we carried out the mapping procedure in two different conditions: (1) during a foveal fixation and discrimination task, which consisted of fixating a small spot of light and releasing a handheld bar at the change of luminosity of the spot to obtain a liquid reward; (2) during free gaze, when the monkey was alert but not performing any task.

Continuous eye tracking with the magnetic search coil technique allowed to adjust the position of the visual stimuli on the screen to be adjusted in order to stimulate the same set of retinal locations as during the fixation task. Stimuli disappearing later than 50 ms before a saccade, or appearing earlier than 50 ms following a saccade were discarded from data analysis.

In vertebrates in general, and in primates in particular, precise visual analysis is associated with the analysis of the small portion of space that falls on the fovea. This analysis implies stabilizing stimuli on the retina, but also directing attention to that location. The periphery of the visual field must also be monitored in order to keep an updated representation of extrapersonal space. These two constraints might be met in parallel by dividing the spatial focus of attention between central and peripheral regions or serially by shifting attention over time to successive spatial positions. Our assumption is that, during the foveal fixation and discrimination task, attention is more focal and locked to the fovea, whereas during free gaze, attention is not locked to a particular location and its processing window might be wider and/or more mobile.

Receptive Fields Mapped During Foveal Attention and During Free Gaze

Of 119 cells studied in one monkey, we retained 88 RF maps obtained during the fixation period of the simple discrimination task on criteria of identifiable spatial and temporal organization. Under fixation and free gaze conditions, a total of 29 cells was studied.

Previous investigations of parietal areas had failed consistently to show any evidence of retinotopic organization, and the size of receptive fields had generally been described as quite large, typically from 10° wide up to a full quadrant or hemifield. In our study, RFs did not appear to be arranged retinotopically, but the presence of rather central and small RFs was unexpected. In a portion of area LIP, subtending approximately 2 mm by 5 mm, almost two thirds of the cells had a "fixation RF" within the central 5° of the visual field (Fig. 1A). Moreover, RFs are sometimes described as symmetrical circular structures. In our study, some RFs were of this kind but others were asymmetrical or elongated. In order to estimate RF size, the area where discharge rate rose above half of the peak firing was measured after subtraction of the background acitivty. To approximate a measure

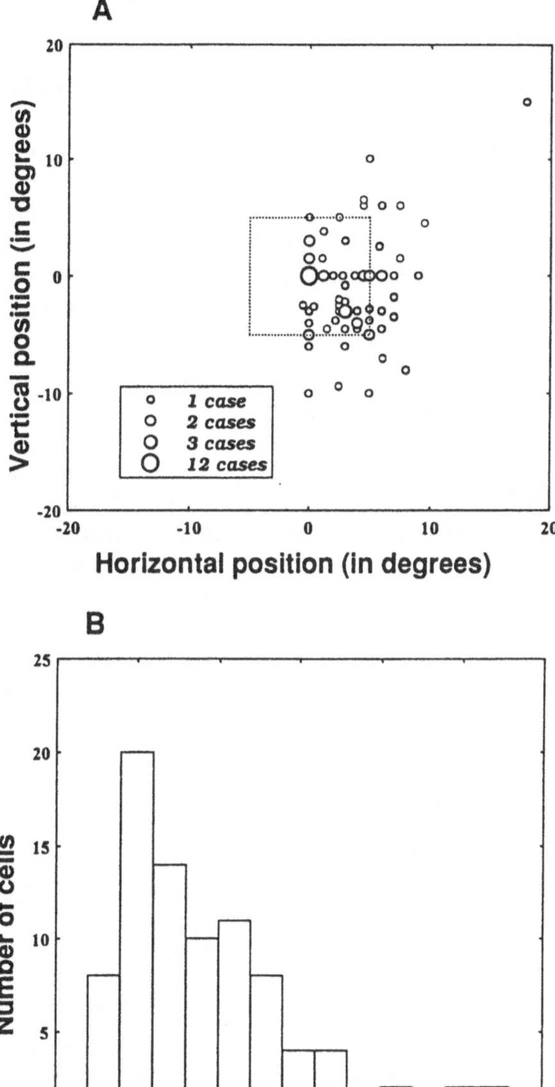

Fig. 1 A, B. Receptive field (*RF*) characteristics for a sample of 88 neurons, recorded from a 2 mm by 5 mm portion of area LIP and determined by electrode penetration records. **A** Spatial distribution of the peak in the computed receptive field profiles. Horizontal and vertical axes labels are in degrees. The position [0°, 0°] corresponds to the center of the screen and to primary eye position. 61.5% of the RF centers are located within the central 5° region. **B** Distribution of RF sizes, calculated as the average diameter of the RF at half of the height of the peak response

of receptive width, we made the assumption that this area was circular and computed its diameter. Consistent with previous studies, the median ''width'' of the RF was 8°, with a range of 2°-24°, suggesting that many LIP cells can perform a relatively fine-grained analysis of visual space (Fig. 1B).

How do RFs obtained during foveal fixation vary with respect to those obtained for the same cell during free gaze? We quantified these changes using three parameters: (1) maximum firing rate, (2) RF width, and (3) location of center of mass. When comparing RFs as mapped during fixation and during free gaze, we observed that three cells no longer had a RF during free gaze condition: the signal to noise ratio dropped to a very low level uniformly throughout the whole visual field, and the RFs' 3D representation looked like an eroded landscape from the primary era, with no outstanding peak. The remaining 26 cells had free gaze RFs that often differed markedly from the fixation RF.

Variation of Activity at Peak. The maximum level of discharge, i.e., the peak of the RF map, is a salient characteristic of most neurons. This value varies according to the ongoing task. Table 1 shows the proportion of cells for which the activity at peak during fixation is increased, decreased, or kept constant, with respect to that during free gaze: a majority of cells show activity that increases by 10% or more, reaching in some cases more than twice the firing rate during foveal fixation as during free gaze. Another group of cells shows a variation of less than 10% in the activity at peak between the two tasks. Finally, a few cells show activity that is higher during free gaze. More data will be needed in order to examine the distribution of these amplitude differences and determine whether there is a continuum of variation or whether these differences correspond to distinct cell classes.

Table 1. Peak response amplitude and receptive field size variation, expressed as number and proportion of cells that either did not change or increased or decreased (within a 10% confidence interval) from the foveal discrimination to the free gaze condition

	Decreased	No change	Increased
Peak response amplitude during free gaze			
Number of cells	12	9	5
% of cells	46.2	34.6	19.2
Width during free gaze			
Number of cells	5	7	14
% of cells	19.2	26.9	53.8

Neurons in area 7a show large and consistent increases in activity from a passive no-task to an active fixation condition (Mountcastle et al. 1981, 1987). Our findings in area LIP are partly similar, although the frequency and amount of enhancement was not as great as that reported for area 7a. This may reflect the fact that LIP neurons display relatively robust visual activity even in a "passive" viewing condition. However it is also possible that the apparent difference might be related, at least in part, to the fact that we compared each condition's peak, wherever those peaks were located, rather than the levels of activity at two identical retinal locations. As will be shown below, the location of the RF peak can vary between the two mapping conditions. Thus, a reduction in the rate of discharge at a given location may not necessarily indicate a global depression of neural activity, but could result instead from a shift of the peak of the RF to a different location.

Receptive Field Size. For about half of the cells, the RF was larger during free gaze than during fixation, although, as in the case of response amplitude, some cells may show little variation or even show the opposite trend (Table 1). In some cells, decrease in RF size and increase in peak amplitude from free gaze to fixation can covary as in the example shown in Fig. 2: during fixation, activity at the RF edges is damped, while sensitivity at the center is enhanced. However, in the population of cells tested, there seems to be no systematic correlation between the directions in which these two parameters varied.

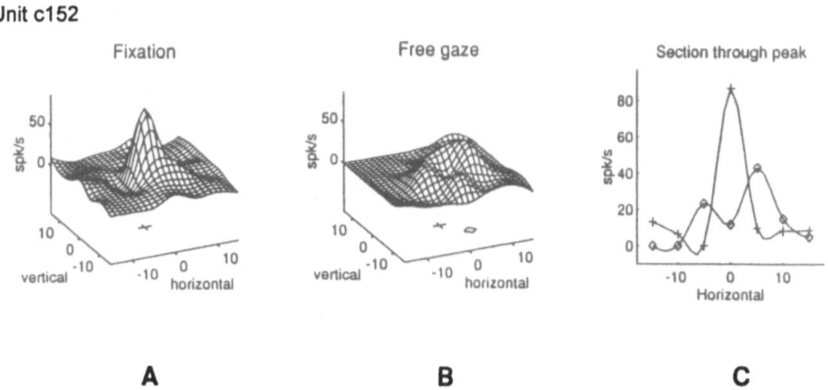

Fig. 2 A-C. RF of cell c152 calculated **A** during the simple discrimination task and **B** during the free gaze condition. The center of the receptive field is projected on the base of the 3D plots. In **B**, both RF centers are shown to visualize their relative locations. **C** Horizontal "cut" through the RFs' peaks. *Crosses* correspond to the RF during simple discrimination, *diamonds* to the RF during free gaze. Note the variation of the activity at peak and of the width. Note also the shift of the position of the peak as well as the activity depression at the fovea during the free gaze task

Location of Receptive Field Centers. There was a consistent tendency for RF centers (as defined by the coordinates of the peak in firing rate) to shift ipsilaterally toward the vertical midline from free gaze to fixation (24/26 cells). The magnitude of this shift varied among different neurons. The mean difference in the location of the RF peaks between the two conditions was 5.0° (SD=3.5°), a statistically significant difference ($p<0.04$). In about 40% of cells, this horizontal shift was also accompanied by a vertical shift toward the horizontal meridian, the net effect corresponding to a "foveotropic" displacement of the RF. For the remaining cells, only one of the two components of the shift, vertical or horizontal, was significant, and in a few cases the RF centers did not move at all. The main trend in the group of neurons tested under the two conditions thus reflects a representation of visual space which is biased towards the center of the visual field when attention emphasizes the fovea.

An RF shift can be achieved in at least two different ways: by a displacement of the entire RF (center and borders) across the visual field, or by a shift of the center of mass as a result of a change in the spatial distribution of activity within the receptive field itself. Both types of modifications were observed, either in isolation or in combination, and the overall shape of the RF could also be modified. Cell c127 (Fig. 3) exemplifies this: the size of the peaks, although shifted between fixation and free gaze remains constant. RF width at half peak is 2° during free gaze condition, and it increases to 4.5° during fixation. As can be seen, the increase in size during fixation is due to a "growth" of the RF towards the fovea. The cell is activated by stimulation of a part of the retina that was ineffective during free gaze. This is thus a case in which both the center of mass

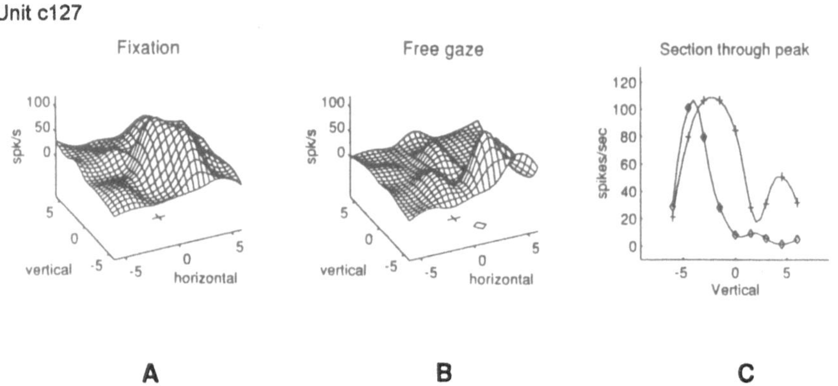

Fig. 3 A-C. RF of cell c127 calculated **A** during the simple discrimination task and **B** during free gaze. **C** Vertical cut through the RFs' peaks. Conventions as in Fig. 2. Note the extension of the RF towards the fovea during the simple discrimination task

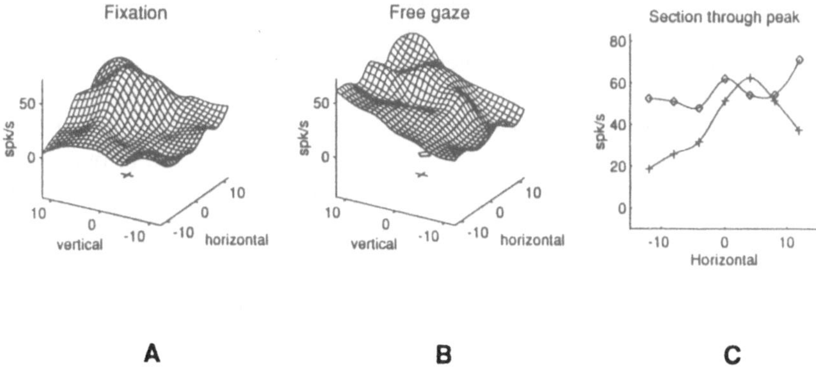

Fig. 4 A-C. RF of cell c184 calculated **A** during the simple discrimination task and **B** during free gaze. **C** Horizontal cut through the upper hemifield. Conventions as in Fig. 2. Note the extension of the RF to the ipsilateral hemifield during free gaze condition

and the inner border of the RF have shifted toward the fovea. Cell c152 (Fig. 2) shows all three types of changes that we have described so far. It is foveal during fixation, with a sharp peak of activity at 85 Hz and it is 6° wide. In contrast, it is more than 6° eccentric during free gaze, its peak activity drops to 40 Hz, and its width at half peak is broadened to 10°. In other words, through this combination of dynamic modifications, the neuron seems to undergo a functional change from a somewhat coarse detector during free gaze to a sharply tuned spatial filter during foveal fixation. The example of cell c184 (Fig. 4) is also instructive. During free gaze conditions, the RF extends to the ipsilateral hemifield, while it is restricted to the upper contralateral hemifield during the fixation task. Peak firing levels do not change between the two conditions, but the fixation peak is displaced toward the fovea relative to the free gaze peak.

Attention and Spatial Filtering of Visual Input

Our results suggest that the concept of the visual receptive field as a fixed characteristic of neurons should be revised. The present data indicate that receptive field properties change with the ongoing behavior, a finding consistent with results reported for other cortical areas (Sato 1989, Desimone and Duncan 1995, Motter 1993, DeAngelis et al. 1993a). The first issue which must be addressed is whether the changes observed can legitimately be attributed to attentional modulation or more simply reflect nonspecific variations in arousal level. It may be argued that cortical arousal is higher when the animal is actively engaged in the performance of a visual discrimination task as opposed to simply gazing at a screen with spots of light flashing on and off. The two RF mapping procedures employed were carried out as part of a testing protocol containing a number of other tasks that are routinely used to study the sensory, attentional and motor response characteristics of each of the recorded neurons. Since the monkey did not know when the task conditions would change and require switching to a new behavior (eye movement, fixation, bar press, etc), it needed to remain alert throughout the free-gaze mapping period. The fact that in the course of any given experiment, a spontaneous saccade 3° or larger in amplitude was made on average every 1.5 s is consistent with this assumption. Another argument against a simple arousal effect is the fact that the observed changes are broader and more complex than a simple variation of the overall strength of neuronal responses and seem to affect the spatial distribution of activity across a portion of the visual field. Two further questions need to be addressed: what are the critical variables controlling RF dynamics, and what are the implications for the analysis of visual space in parietal cortex?

The two situations studied can be construed as involving two distinct, but closely related components: an attentional one, which is related to the monitoring of foveal visual events and an oculomotor one, which is related to the maintenance of eye fixation. The neural processing underlying these two components could influence visual sensitivity of LIP neurons. Previous studies have amply demonstrated the convergence of extraretinal signals related to eye movements and attention upon neurons in this area (Andersen et al. 1987; Gnadt and Andersen 1988; Duhamel et al. 1992; Colby et al. 1993). We have specifically addressed the question of fixation-related activity in area LIP, using visually guided and memory-guided fixation tasks. In only a small proportion of cells was activity uniquely related to ocular fixation, e.g., to the behavioral act of maintaining the eyes at a given position in the orbit (Ben Hamed et al. 1995). In most cells studied neuronal activity recorded during fixation was dependent upon visual and attentional factors, such as the presence or absence of a foveal target, or the temporal proximity to a critical task event such as the change in color of the fixation target. Through its output to the superior colliculus, area LIP may well have a modulatory influence on fixation mechanisms (Munoz and Wurtz 1993; Dorris and Munoz 1995), but this may be mainly by relaying signals related to attentional disengagement and reengagement. Another argument for the attentional origin of these effects comes from experiments in which RFs were plotted during the delay interval of a memory-guided saccade task to peripheral targets, when attention can be assumed to be focal, but not foveal (Duhamel et al. 1995). In this situation RF shifts were found which differed according to the location in space of the memorized target. The effects of focal attention could be tested directly in further experiments by comparing the RF profiles obtained in conditions where the monkey needed to detect an event occuring consistently at the same spatial location (narrow focus) as opposed to detecting an event occurring unpredictably at one of several possible locations (wide focus), and irrespective of the presence of fixation target.

We also suggest that RF substructure plays an important role in spatial analysis. One particularly interesting class of neurons that was encountered had relatively low response rates along with large, peripheral RFs during free gaze, indicating that any stimulus falling in the contralateral visual field would be able to elicit at least a weak neural response, i.e., to be detected by the neuron. During foveal fixation, the responses were stronger, the area of excitability was more restricted, and the RF center was displaced towards the center of the visual field. In other words, sensitivity to a visual event around the point the monkey fixated - around the point on which his attention was locked- was increased whereas it was virtually non-existant elsewhere. Such a built-in capacity to increase signal-to-noise-ratio and spatial selectivity means that, under these conditions, a cell's RF can act as a very effective filter to increase stimulus saliency during demanding perceptual operations.

It is thus tempting to view single neurons as operators that switch from a detector function to a filter function, according to the behavioral conditions under which we are looking at them. These functions could be controlled by response strength variation, by RF size variation, or by both. Our cell sample does not suggest any covariance between these two parameters, and they may be considered as two solutions to a unique problem. The shift of RF centers can be viewed for its part as a means of controlling the spatial tuning of the filtering function, i.e., the location where visual processing resources must be directed. Usually, though not always necessarily, this point corresponds to the region of fixation that falls on the fovea.

It appears that the attentional modulation of visual processing by single neurons can be a powerful and highly plastic solution to the behavioral requirements of spatial analysis. More experimental evidence is required in order to explain the mechanisms that govern RF dynamics. In particular, specific tasks will have to be designed in order to dissociate oculomotor from attentional effects. Another important aspect is the time course of RF modulation. DeAngelis et al. (1993b) showed that the RF structure of V1 neurons varies with time, and that stimulation of a portion of space can induce either an excitatory or an inhibitory response, depending on the time at which we are viewing it. These authors suggest that spatiotemporal activity within the first 200 ms is crucial for analysis of stimulus features such as orientation, speed, direction of movement, and color. Longer term effects, allowing for attentional factors to come into play, could be important in understanding the physiology of neurons in higher-order areas such as the posterior parietal cortex.

Relevance to Neuropsychological Findings

The RF properties described here might account for some of the functional impairments observed following damage to parietal cortex. During free gaze, some cells were responsive to a rather large area, extending into the ipsilateral visual field, while during attentive fixation, spatial analysis was limited to the contralateral hemifield. The restriction of the attentional field to contralateral space generally correlates with clinical observations: parietal patients respond to contralesional stimuli when attention is disengaged. This performance could be mediated by the intact parietal lobe, since during this particular attentional state the RFs are expanded and span both hemifields. In contrast, when attention is engaged, manifestations of contralateral neglect dominate, because the intact parietal lobe is now dealing more exclusively with the ipsilesional field.

The syndrome described by Balint (1909) following bilateral parietal damage is quite rare, but a number of descriptions have recurred in the literature indicating that these patients have difficulty shifting attention from one object to another, even within the central visual field. Our finding of a relatively fine-grained representation of the foveal region is consistent with this clinical picture, since as a result of bilateral lesions both parietal representations of the central visual field would be damaged. Interestingly, patients with Balint's syndrome can often be shown to have intact visual fields, and, despite their subjective impression of tunnel vision, are able to detect peripheral objects. However, it appears as though stimuli from the environment enter and leave the patients' restricted window of visual awareness haphazardly: at times, gaze can wander aimlessly across the visual field, while at other times it remains locked onto the same object for such long periods that it will eventually fade away. It can be said that in these patients, attention can operate in one of two states, disengaged or engaged, but without any control over the process that allows it to shift between the two. The dynamic properties exhibited by the RFs of parietal neurons may play a key role in the balance between attentional engagement and disengagement required for an optimal allocation of processing resources across visual field representation throughout the cortical visual system.

Acknowledgements. This work was supported by Human Capital and Mobility grant n° ERBCHRXTC930267 from the European Commission.

References

Allman J, Miezin F, McGuinnes E (1985) Stimulus-specific responses from beyond the classical receptive field: neurophysiological mechanisms for local-global comparisons in visual neurons. Ann Rev Neurosci 8:407–430

Andersen RA, Asanuma C, Essick G, Siegel RM (1990) Corticocorical connections of anatomically and physiologically defined subdivisions within the inferior parietal lobule. J Comp Neurol 296:65–113

Andersen RA, Essick GK, Siegel RM (1987) Neurons of area 7 activated by both visual stimuli and oculomotor behavior. Exp Brain Res 67:316–322

Balint R (1909) Seelenlähmung des "Schauens", optische Ataxie, räumliche Störung der Aufmerksamkeit. Monatsschr Psychiatr Neurol 25:51–81

Ben Hamed S, Duhamel J-R, Bremmer F, Graf W (1995) Representation of the central visual field and fixation-related activity in the lateral intraparietal area of macaque monkeys. Soc Neurosci Abstr 21:665–268.16(Abstract)

Bowman EM, Brown VJ, Kertzman K, Schwartz U, Robinson DL (1993) Covert orienting of attention in macaques. I. Effects of behavioral context. J Neurophysiol 70:431–433

Brotchie PR, Andersen RA, Snyder LH, Goodman SJ (1995) Head position signals used by parietal neurons to encode locations of visual stimuli. Nature 375:232–235

Colby CL, Duhamel J-R (1991) Heterogeneity of extrastriate visual areas and multiple parietal areas in the macaque monkey. Neuropsychologia 6:517–537

Colby CL, Duhamel J-R, Goldberg ME (1993) The analysis of visual space by the lateral intraparietal area of the monkey: the role of extra-retinal signals. Prog Brain Res 95:307–316

DeAngelis GC, Ohzawa I, Freeman RD (1993a) Spatiotemporal organization of simple-cell receptive fields in the cat's striate cortex. I. General characteristics and postnatal development. J Neurophysiol 69:1091–1117

DeAngelis GC, Ohzawa I, Freeman RD (1993b) Spatiotemporal organization of simple-cell receptive fields in the cat's striate cortex. II. Linearity of temporal and spatial summation. J Neurophysiol 69:1118–1135

Desimone R, Duncan J (1995) Neural mechanisms of selective visual attention. Ann Rev Neurosci 18:193–222

Desimone R, Wessinger M, Thomas L, Schneider W (1991) Attentional control of visual perception: cortical and subcortical mechanisms. In: Cold Spring Harb Symp Quant Biol. pp 963–971

Dorris MC, Munoz DP (1995) A neural correlate for the gap effect on saccadic reaction times in monkey. J Neurophysiol 73:2558–2562

Duhamel J-R, Ben Hamed S, Bremmer F, Graf W (1995) The influence of attention and intended eye movements on the visual excitability of neurons in the lateral intraparietal area of macaque monkeys. Soc Neurosci Abstr 21:665–268.17(Abstract)

Duhamel J-R, Colby CL, Goldberg ME (1992) The updating of the representation of visual space in parietal cortex by intended eye movements. Science 255:90–92

Gnadt JW, Andersen RA (1988) Memory related motor planning activity in posterior parietal cortex of macaque. Exp Brain Res 70:216–220

Gnadt JW, Mays LE (1995) Neurons in monkey parietal area LIP are tuned for eye-movement parameters in three-dimensional space. J Neurophysiol 73:280–297

Lynch JC, Graybiel AM, Lobeck LJ (1985) The different projections of two cytoarchitectonic subregions of the inferior parietal lobule of macaque upon the deep layers of the superior colliculus. J Comp Neurol 235:241–254

Motter PC (1993) Focal attention produces spatially selective processing in visual cortical areas V1, V2, and V4 in the presence of competing stimuli. J Neurophysiol 70:909–919

Mountcastle VB, Andersen RA, Motter PC (1981) The influence of attentive fixation upon the excitability of light-sensitive neurons of the posterior parietal cortex. J Neurosci 1:1218–1225

Mountcastle VB, Motter PC, Steinmetz MA, Sestokas AK (1987) Common and differential effects of attentive fixation on the excitability of parietal and prestriate (V4) cortical visual neurons in the macaque monkey. J Neurosci 7:2239–2255

Munoz DP, Wurtz RH (1993) Fixation cells in the superior colliculus. J Neurophysiol 70:559–589

Robinson DL, Goldberg ME, Stanton GB (1978) Parietal association cortex in the primate: sensory mechanisms and behavioural modulations. J Neurophysiol 41:910–932

Sato T (1989) Interactions of visual stimuli in the receptive fields of inferior temporal neurons in awake macaques. Exp Brain Res 77:23–30

Steinmetz MA, Connor CE, Constantinidis C, McLaughlin JR (1994) Covert attention suppresses neuronal responses in area 7a of the posterior parietal cortex. J Neurophysiol 72:1020–1023

Trotter Y, Celebrini S, Stricanne B, Thorpe S, Imbert M (1992) Modulation of neural stereoscopic processing in primate area V1 by viewing distance. Science 257:1279–1281

Object-Based Attention in Visual Neglect: Conceptual and Empirical Distinctions

M. J. FARAH[1] and L. J. BUXBAUM[2]

[1]Department of Psychology, University of Pennsylvania, Philadelphia, USA
[2]Moss Rehabilitation Institute, Philadelphia, USA

Spatial and Object-Based Visual Attention

The earliest stages of visual perception operate in parallel, with equal processing accorded each location or object in the visual field. However, at later stages attention selects some parts of the visual field but not others for further processing. The winnowing of stimulus information occurs at several stages of visual perception, and the shape, orientation and identity of objects in the field may play a role at several of these stages.

In this chapter we will briefly review what is known about the roles of objects in normal attention, and ask how these roles might affect attention in the syndrome of visual neglect. Several hypotheses have been put forward to describe the role of objects in the perception of patients with neglect, and a variety of different findings have been interpreted in terms of object-based visual attention in neglect. We will attempt to distinguish between different hypotheses and to weigh the support for each in terms of empirical findings in patients with neglect. We begin with a basic distinction between spatial attention and object-based attention.

Spatial and Object-Based Attention

Two major forms of visual attention are spatial attention and object-based attention. Spatial attention has been the subject of much research in cognitive psychology. A common task used in studying spatial attention is the simple reaction time paradigm developed by Posner and colleagues, in which cues and targets appear to the left and right of fixation as shown in Fig. 1. The cue consists of a brightening of the box surrounding the target location and the target itself is a plus sign inside the box. The subject's task is to respond as quickly as possible once the target appears. The cues precede the targets and occur either at the same location as the target (a "valid" cue) or at the other location (an "invalid" cue). Subjects respond more quickly to validly than invalidly cued targets, and this

In: Parietal Lobe Contributions to Orientation in 3D Space (1997). P. Thier and H.-O. Karnath (eds). Springer-Verlag, Heidelberg.

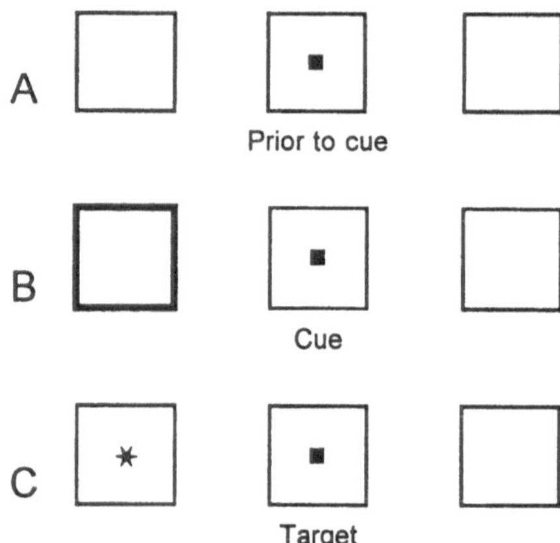

Fig. 1 A-C. Sequence of displays in a cued lateralized simple reaction time task used to demonstrate spatial attention

difference has been interpreted as an attentional effect. Specifically, when a target is invalidly cued, attention must be disengaged from the location of the cue, moved to the location of the target, and re-engaged there before the subject can complete a response. In contrast, when the target is validly cued, attention is already engaged at the target's location, and responses are therefore faster (see Posner and Cohen 1984).

More direct evidence that attentional selection is spatial in nature comes from the work of Shulman et al. (1979). In their experiment, subjects received cues and targets in far peripheral locations. However, a probe event could occur between the cue and target in an intermediate peripheral location. When subjects were cued to a far peripheral location, they were facilitated in processing the probe prior to the maximal facilitation of the far peripheral location. These results fit with the idea that if attention is moved from point A to point B through space, then it moves through the intermediate points in a spatial representation.

Another representative result supporting spatially allocated visual attention was reported by Hoffman and Nelson (1981). Their subjects were required to perform two tasks on each trial: A letter search task, in which subjects determined which of two target letters appeared in one of four spatial locations, and an orientation discrimination, in which subjects determined the orientation of a small U-shaped figure. Hoffman and Nelson found that when letters were correctly identified, the orientation of the U-shaped figure was better discriminated when it was adjacent to the target letter, compared to when the target letter and U-shape were not adjacent. These results support the hypothesis that attention is allocated to stimuli as a function of their location in visual space.

In recent years psychologists have also gathered evidence that attention operates on nonspatial representations of objects. That is, instead of selecting stimuli for further processing one location at a time, stimuli are selected one object at a time. Duncan (1984) presented subjects with brief presentations of superimposed boxes and lines, like the ones shown in Fig. 2. Each of the two objects could vary on two dimensions: The box could be either short or tall and have a gap on the left or right, and the line could be tilted clockwise or counterclockwise and be either dotted or dashed. The critical finding was that when subjects were required to make two decisions about the stimuli, they were more accurate when both decisions were about the same object. For example, subjects were more accurate at reporting the box's size and the side of the gap, compared to, say, reporting the box's height and the line's texture. This finding fits with the notion that attention is allocated to objects per se: It would be more efficient to attend to a single object representation rather than either attending to two object representations simultaneously, or attending to one representation and then the other.

Vecera and Farah (1994) repeated and extended the findings of Duncan by adding a condition to the experiment in which the box and line were spatially separated. If the object-based attention effects found by Duncan involved the cost of switching attention between purely object-based representations, from which spatial location information had truly been eliminated, then there should be no larger effect of switching attention between objects when they are separated than when they are superimposed. The findings in fact supported this entirely nonspatial object-based form of attention: There was a performance cost associated with switching between objects, but it did not depend upon the relative spatial locations of the two objects.

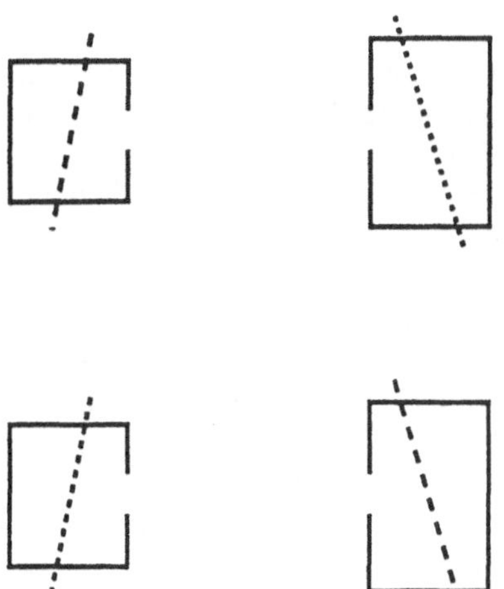

Fig. 2. Examples of stimuli used to demonstrate object-based attention

Frames of Reference

There is another sense of "object-based" attention that has recently attracted interest in neuropsychology which refers to the use of an object-centered frame of reference for allocating spatial attention. The relation between frames of reference and the issue of spatial versus object-based attention as discussed earlier is worth clarifying explicitly. Whatever frame of reference is used to allocate attention to locations in space, so long as attention is being allocated to locations and not objects, it is a form of spatial attention. However, objects may affect the allocation of spatial attention by affecting the frame of reference by which left, right, and other spatial directions are defined.

Any representation of spatial location must be specified relative to a frame of reference, in other words, an origin and axes. Visual stimuli are initially encoded in terms of a retinal frame of reference, that is, a frame of reference whose origin is at the subject's fixation point and whose axes measure displacement from the fixation point. In addition to the retinal frame of reference, there is evidence of head-centered frames (used in sound localization, among other processes), and trunk-, arm-, and hand-centered frames (used in planning reaching movements). These frames are all examples of viewer-centered frames, as their origins are tied to the viewer's body. With viewer-centered frames of reference, the representation of stimulus location changes whenever the viewer changes position. Viewer-centered representations are thus rather unstable, in contrast with two other kinds of frames of reference, environment-centered and object-centered.

An environment-centered frame represents location relative to the fixed environment, regardless of the viewer's location. Visual or gravitational cues or dead reckoning could potentially be used to determine the location of the environment as a viewer moves through it, for the purposes of representing the location of stimuli relative to an environment-centered frame.

An object-centered frame represents location relative to the intrinsic axes of an object, defined by the object's top, bottom, front, back, left and right. An example of a description of location in an object-centered frame is the description of the location of the bells on an old-fashion alarm clock as "above the clock face." The bells are *above* the clock face even when the clock is laying on its side or upside down. This example highlights an interesting feature of object-centered frames of reference: When they are used to represent the locations of different parts of an object, they result in representations that have location and orientation invariance. That is, the description of the parts' locations is the same regardless of the location or orientation of the object, because when the object moves the frame of reference moves too. For this reason, object-centered representations have sometimes been hypothesized to underlie the representation of shape for purposes of object recognition (Marr 1982). However, one difficulty with this approach to object recognition is that before the locations of an object's parts can be described relative to an object-centered frame, the top, bottom, left, right, front, and back of the object must be identified. This creates a chicken-and-egg problem, whereby the object-centered frame is needed for object recognition but object recognition appears to be required before the object-centered frame can be accessed.

Whatever the status of object-centered representations in object recognition, the idea that attention is allocated to spatial locations using an object-centered frame is clearly very different from the idea that attention is allocated to objects, in the sense that arose from the box and line experiments of Duncan and others.

Visual Attention in Neglect: Types of Object-Based Processing

Neglect is an impairment of visual attention the most pronounced characteristic of which is a spatial asymmetry of processing, such that stimuli on one side of a space receive less processing than stimuli on the other. However, recent observations suggest that objects may also influence the allocation of attention in neglect. In the remainder of this chapter we will review the different ways in which objects influence perception in visual neglect.

Is There Selection by Object As well As by Location in Neglect?

In one sense neglect might be said to be object-based if there were a tendency to attend to or neglect entire objects. This sense of object-based neglect corresponds most closely with Duncan's demonstrations of object-based neglect, in that selection is by object as opposed to by location. Farah et al. (1993) reported an experiment to test the possibility that an object-based component of attention contributes to neglect patients' performance, as well as the more obvious spatial or location-based component. Neglect patients were asked to name as many letters as they could from a scattered array of letters which were superimposed on one of two kinds of background: large, simple closed curves ("blobs") that either straddled the left and right sides of space or were contained entirely within either the left or right side of space.

Although the blobs are not relevant to the task of reading the letters, they are perceived by at least some levels of the visual system, and the question is what, if any effect, do they have on the distribution of attention over the stimulus field? If attention is object-based as well as location-based, then there will be a tendency for entire blobs to be either attended or nonattended, in addition to the tendency for the right to be attended and the left to be nonattended. This leads to different predictions for performance in the two conditions. When the blobs extend from the right to the left hemifield, there should be more attention allocated to the left than when the blobs are each contained within one side. This is what was found: Subjects named more letters on the left, and started their searches on the left more often in the straddling blobs condition. The results of this experiment suggest that objects do affect the distribution of attention in neglect, and imply that visual attention is object-based as well as location-based.

This conclusion is partly confirmed and expanded by the findings by Egly et al. (1994) with right and left parietal-damaged patients. Although these patients did not show overt neglect at the time of testing, they did exhibit the extinction-like reaction time pattern known as a "disengage deficit" whereby an ipsilesional stimulus preceding a contralesional stimulus greatly increases reaction time to the contralesional stimulus. Egly and colleagues used the stimulus display shown in Fig. 3, in which bars straddle either the upper and lower or left and right sides of the visual field. By cueing one end of one bar, and then presenting a target at a different location on the same or a different bar, they could hold spatial separation between the first and second stimulus constant and vary whether the disengagement from the first to the second stimulus involved disengaging from one bar object to another, or whether it involved disengaging from equally distant locations within a single object. They found that all subjects showed a disengage deficit, but that there was an additional difficulty disengaging from one object to attend to another after left, but not right, parietal damage.

In order to find out whether the damaged attention system in neglect takes knowledge about objects into account when allocating attention, or merely takes objects to be the products of low-level grouping principles, we followed up on a phenomenon first observed by Sieroff and Posner (1988). They observed that neglect patients are less likely to neglect the left half of a letter string if the string makes a word than if it makes a nonword. For example, patients are less likely to omit or misread the "t" in "table" than in "tifcl." Sieroff and Posner have assumed that the spatial distribution of attention is the same when reading words and nonwords, and that the difference in reading performance for words and nonwords is attributable to top-down support from word representations "filling in" the missing low-level information. However, there is another possible explanation for the superiority of word over nonword reading in neglect patients, in terms of object-based attentional processes. Just as a blob object straddling the two hemifields causes a reallocation of attention to the leftward extension of the blob,

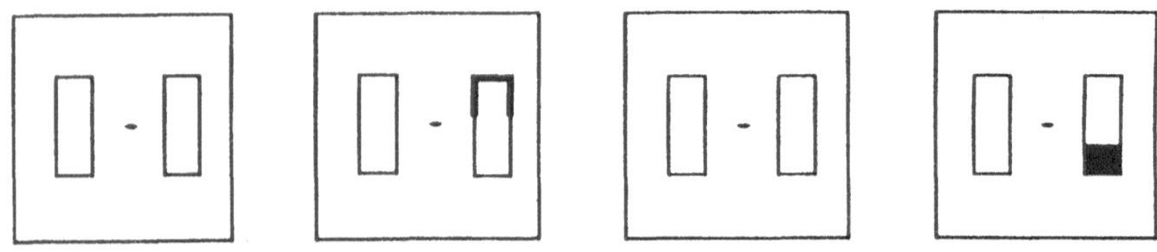

Fig. 3. Example sequence of displays in a cued lateralized simple reaction time task used to demonstrate spatial and object-based components of attention

because attention is being allocated to whole blobs, so, perhaps, might a lexical object straddling the two hemifields cause a reallocation of attention to the leftward extension of the word. Of course, this would imply that the "objects" that attention selects can be defined by very abstract properties such as familiarity of pattern, as well as low-level physical features.

In order to test this interpretation of Sieroff et al.'s observation, and thereby determine whether familiarity can be a determinant of objecthood for the visual attention system, we showed word and nonword letter strings printed with each letter a different color, and asked subjects to both read the letters and to name the colors (Brunn and Farah 1991). If there is a reallocation of attention to encompass entire objects, in this case lexical objects, then patients should be more accurate at naming the colors on the left sides of words than nonwords. This is what we found, providing further evidence that object-based selection affects the distribution of attention in neglect.

Is Attention Allocated Within an Object-Centered Frame of Reference in Neglect?

Patients with neglect may be impaired in processing the left sides of single objects as well as the left side of space more generally, and this has been interpreted in terms of the use of an object-centered frame for allocating attention to space. The copies shown in Fig. 4 illustrate this apparent form of object-based neglect. Information is included from both sides of space, but the left side of each individual object is omitted. Such observations appear, at least initially, to support the contention that visual processing in neglect may be disordered with respect to both object-based and spatial frames of reference. Evidence from copying tasks, however, is at best only suggestive. For example, neglect of the left sides of objects in Fig. 4 could reflect neglect in retinotopic coordinates if subjects make separate eye fixations on each object they copy. Copying a pattern of any complexity requires multiple fixations as different parts of the pattern are separately copied.

More direct tests of the hypothesis of object-centered neglect have been designed using more purely perceptual tasks such as visual search and same–different matching. The earliest study to address this issue was reported by Farah et al. (1990). We followed up on an earlier set of findings by Ladavas (1987) and Calvanio et al. (1987), who had assessed neglect in the four quadrants of the visual field when patients were upright and when they reclined on one side. When the patients reclined, the dividing line between their viewer-centered lefts and rights was orthogonal to the dividing line between the left and right of the environment and the stimulus objects. Decoupling these frames of reference revealed that neglect is defined relative to both a viewer-centered frame of reference and some external frame, but these original observations did not allow

Fig. 4. Copies suggesting object-based neglect

the environment-centered frame to be decoupled from the object-centered frame. In order to assess the contributions of the environment-centered and object-centered frames separately, we presented patients with scattered letters to read on pictures of objects, which were presented either upright or rotated by 90°, as shown in Fig. 5. We found evidence that attention was differentially allocated to the right sides of the viewer-centered and the environment-centered frames, but found no evidence of the use of an object-centered frame.

Behrmann and Moscovitch (1994) considered the possibility that objects did not affect distribution of attention because they were superfluous to the task. They redesigned Farah et al.'s task so that drawings were the object of the search, not merely a background, by outlining the drawings of objects with different colors to be reported. Even under such conditions they failed to find neglect of the intrinsic left of the objects. They next considered the possibility that object-centered frames would only be used for objects such as asymmetric letters whose identities depend on differences between the left and right. In keeping with this, subjects neglected the left sides of 90°-rotated asymmetric letters, consistent with object-centered neglect, but failed to show this effect with symmetric letters.

Another apparently positive finding of object-centered neglect comes from the reading errors of a neglect dyslexic studied by Caramazza and Hillis (1990). Their patient made more errors on the intrinsic right than on the left sides of words, and this was true whether the words were printed normally or mirror-reversed or even

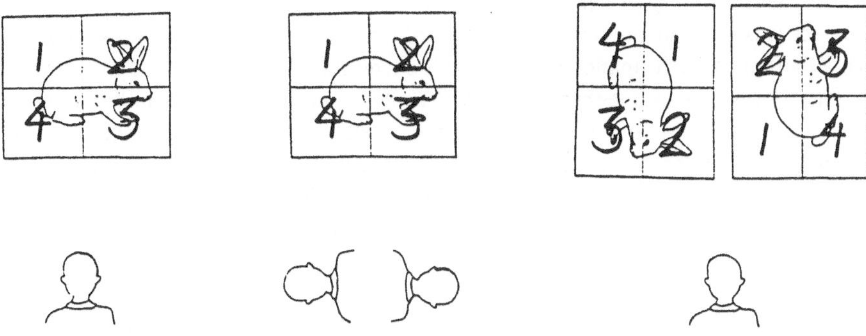

Fig. 5. Manipulations of stimulus and viewer position intended to disentagle contributions of viewer-centered, environment-centered, and object-centered frames of reference in neglect. For example, with the viewer rotated clockwise, quadrant 2 is on the left of the object and environment but on the right of the viewer. With the object rotated clockwise, quadrant 3 is on the left of the viewer and environment but the right of the object. See text for further explanation

printed vertically. That is, when the words were presented mirror-reversed, she neglected the letters on her left, and when they were presented vertically she neglected the bottom letters. Caramazza and Hillis interpreted these results as a demonstration that neglect may occur with respect to an orientation-invariant canonical representation of a word form.

Although the results of Behrmann and Moscovitch (1994) and Caramazza and Hillis (1990) are consistent with the use of an object-centered frame in neglect, a recent case study has led us to consider an alternative explanation of these results. According to several theorists (e.g., Jolicoeur 1985; Tarr and Pinker 1990), noncanonically oriented (e.g., tilted) objects may be recognized by mental rotation of the object to an upright position based on low-level properties of the object's shape, such as its axis of elongation. The rotated shape is subsequently matched to stored orientation-specific (e.g., upright) representations. If such mental rotation procedures occur prior to the processing stage at which neglect arises, then at least some cases of apparent object-centered neglect may not reflect neglect of the intrinsic left of the object. Instead, they could reflect neglect of the viewer- or environment-centered left of the object after it has been mentally rotated to its upright position. In other words, mental rotation procedures may allow viewer- or environment-centered neglect to masquerade as object-centered neglect.

Using the same color reporting procedure as Behrmann and Moscovitch (1994), we assessed whether this might be the case. We asked a subject with neglect to report colors outlining tilted asymmetric and symmetric objects and letters in two conditions (Buxbaum et al. 1996). In the first condition, he was asked to imagine that the tilted figures were upright and then name the colors. In the second condition, he was explicitly asked to refrain from mentally rotating the figures and instead simply to name the colors as he saw them. Apparent object-centered neglect was evident only in the first condition and not in the second condition. This pattern of performance held for both asymmetric and symmetric objects and letters. In other words, apparently object-centered neglect occured only when the left and right sides of the figures were mentally aligned with the viewer- and environment-centered left and right. The intrinsic characteristics of the objects (e.g., their symmetry) were irrelevant to this effect. This suggests that the neglect of the objects occurred with respect to the viewer- and/or environment-centered frames and not with respect to an orientation-invariant representation of the object.

It is possible that other cases of apparent object-centered neglect may be similarly attributable to viewer- or environment-centered neglect occuring after mental rotation. Indeed, the Caramazza and Hillis findings could be explained in this way, given Koriat and Norman's (1984) finding that normal subjects mentally rotate inverted words in order to read them. This possibility, coupled with our own failure to find object-centered neglect in symmetrical or asymmetrical objects or letters (except when mental rotation was explicitly used) leads us to doubt that object-centered frames of reference are used in allocating attention in neglect. However, this conclusion is drawn on the basis of a very small amount of data, and future research could prove it wrong.

Although we remain skeptical about the existance of object-centered neglect in the strong sense so far discussed, we believe there is good evidence of object-centered neglect in a weaker but still interesting sense. Driver et al. (1994) introduced the term "axis-based" neglect to describe the possiblity that an object's principle axis could determine the boundary between two sides of the object; the assignment of left and right to those sides is nevertheless relative to the viewer. For example, in the left part of Fig. 6, taken from the work of Driver and Halligan (1991), the object defines a dividing line between the space to the left and right of its axis of elongation. This contrasts with the stronger sense of object-centered neglect already discussed, in which stored knowledge about the left and right sides of objects determines the distribution of attention. For example, one can say that the capital B in Fig. 6 has the straight line on the left and the loops on the right, where left and right refer to intrinsic sides of the object. The axis-based type of object-centered frame is really only partly object-centered, as it is parasitic on the viewer-centered frame for the assignment of left and right. This point can be appreciated by looking at Fig. 7, in which the shapes from Fig. 6 have been rotated 180° from upright. The axis-based left of the novel shape is still on the viewer's left, whereas the left of the *B* is now on the viewer's right.

Driver and Halligan (1991) provided compelling evidence of axis-based neglect in a task requiring detection of differences between two elongated novel shapes in both upright and 45°-rotated orientations. In the rotated condition, portions of the left sides of the stimuli fell on the right side of viewer-centered and environment-centered space such that putative object-centered and viewer-centered frames were uncoupled. The subject was significantly less likely to detect left-sided as

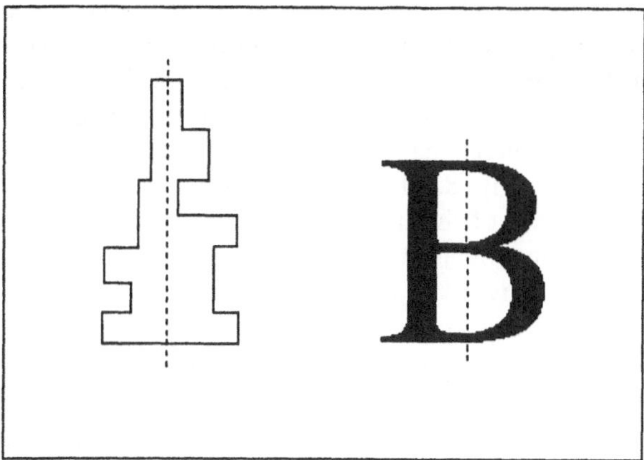

Fig. 6. Vertical axes defined by objects. For the novel shape, its elongation defines a vertical axis, relative to which left and right can be represented. On the right, stored knowledge about the shape also contributes to defining which axis is vertical and divides the shape into left and right halves

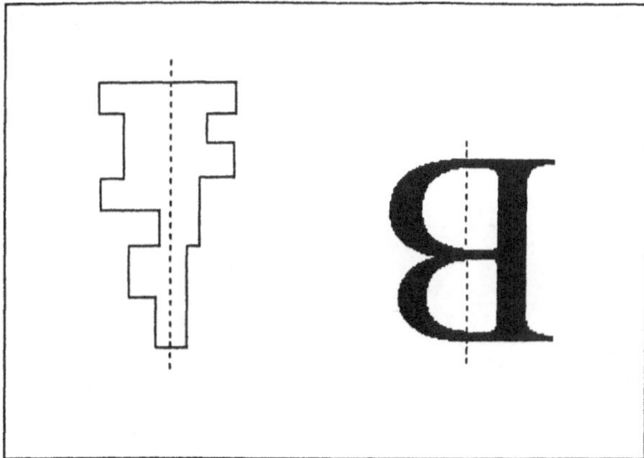

Fig. 7. Inverting the shapes shown in Fig. 6 demonstrates a difference between the two ways of defining axes: Whereas we would now say that different parts of the novel shape are on the left and right, the *B* has intrinsic left and right halves. The left side of the *B* is now on the right side of the page

compared to right-sided differences between the shapes in both the upright and rotated conditions. Driver and Halligan suggested that these data supported the presence of object-centered neglect defined with respect to the left and right of the objects' principal axes.

Subsequently, Driver et al. (1994) demonstrated that when egocentric location as well as object shape and orientation are held constant, the neglected regions of objects may be modified by a *perceived* major axis. Patients with neglect viewed equilateral triangles which, while identical in all cases, were biased to appear to be pointing in one of two directions by manipulation of the array in which they were presented, as shown in Fig. 8. The relative location of a gap with respect to the perceived principal axis of the triangle was varied, while holding constant the gap's egocentric location. Subjects neglected more gaps to the left than right of the perceived major axis. In response to these data, Driver et al. (1994) suggested that the principal axes of objects may indeed be used to divide objects into two regions. Critically, since triangles do not have intrinsic lefts and rights, the neglected side of the axis is determined by the viewer's left.

Further evidence that can be interpreted in terms of axis-based neglect comes from the study of Behrmann and Tipper (1994), in which subjects with neglect were asked to respond to targets appearing on the ends of a bar-bell-shaped stimulus. In the condition of interest, the bar-bell rotated 180° while the subjects watched, such that the left side of the bar-bell ended up on the right side of the subjects and vice versa. After the bar-bell stopped a target appeared on one end or the other. Response times to the targets in the moving condition were compared to those in a static condition in which the bar-bell remained stationary throughout the trial. Detection was faster for targets on subjects' lefts and slower for targets on their rights in the moving, relative to static, condition. Thus the neglect may be characterized as axis-based in the sense that the ends of the bar-bell were coded as left and right as a function of their original locations on either side of the bar-bell's

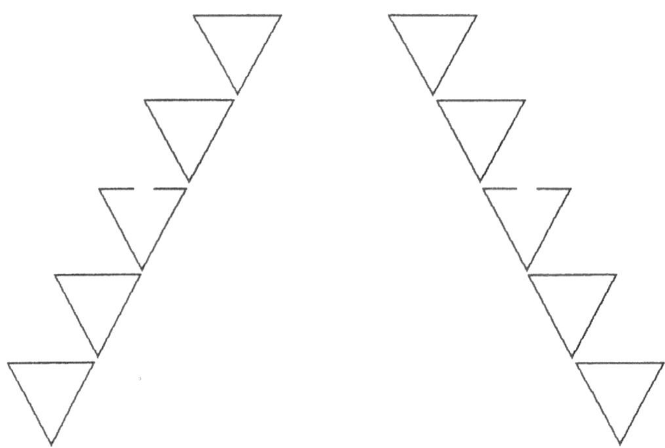

Fig. 8. A demonstration that the perceived major axis dividing a shape into left and right sides can be influenced by context

vertical axis, relative to the viewer. Critically, however, the neglect need not be characterized as object-centered in the strong sense because there is no evidence that stored knowledge concerning the left and right sides of bar-bells determined the allocation of attention to the left and right sides of the display.

Studies of the perception of chimeric figures, that is, figures composed of the left and right halves of two different objects as shown in the top part of Fig. 9, provide another demonstration that objects can influence the distribution of attention in neglect, without necessarily imposing an object-centered frame of reference specifying the object's intrinsic left and right. Some subjects with neglect are more likely to report the left sides of chimeric figures when the chimeric "object" is split down the middle by a small gap (Fig. 9, bottom) (Buxbaum and Coslett 1994; Young et al. 1992). In other words, for certain patients, when the two halves of the chimeric are joined to make a single object, the left side of the object is neglected. Although there seems to be some contribution of stored object knowledge to this effect, as the effect is stronger

Fig. 9. A chimeric stimulus with (*bottom*) and without (*top*) a gap

when the two half-objects are semantically related, the object effects are still axis-based rather than derived from a truly object-centered frame of reference, in that the object supplies the axis only, and the determination of left and right is viewer- or environment-centered. This conclusion follows from the observation that when the chimerics are inverted so that the object's intrinsic left is on the subject's right and vice versa, subjects with left neglect continue to neglect the half-object on their (viewer- or environment-centered) left.

A number of cases have been reported in which there is simultaneously right neglect for spatial search and constructional tasks, which involve perception of multiple objects in space, and left neglect for word reading and sometimes object naming (Bisiach et al. 1986; Costello and Warrington 1987; Humphreys and Riddoch 1995; Katz and Sevush 1989). These cases also demonstrate that objects affect the distribution of attention in neglect in a way that is dissociable from the more broadly distributed attention used for search and constructional tasks. Both these cases and the findings with chimeric objects can be interpreted in terms of axis-based neglect, with objects' vertical axes serving to divide the visual field into two lateral halves but the viewer- or environment-centered frame determining which half is left and which is right.

Conclusions

Early writings on neglect called the disorder "hemispatial neglect" and "spatial agnosia", emphasizing the most obvious feature of neglect which is that it varies according to spatial location. Only recently have the contributions of object structure to the attentional impairment of neglect been investigated. Given the young age of this research program, it is not surprising that many questions have yet to be answered decisively. Nevertheless, some preliminary conclusions can be drawn.

There is some evidence that in patients with neglect attention tends to select or fail to select entire objects at a time. This object-based selection is superimposed on the more pronounced spatial gradient of attention, such that objects extending into the attended hemifield are likelier to be perceived in their entirety than objects that do not. Left hemisphere-damaged patients show the strongest tendency to select whole objects at a time. Although right hemisphere-damaged neglect patients showed these object effects in a visual search task, only left hemisphere-damaged patients showed an object-based disengage deficit when tested after their neglect had resolved. Object-based selection in neglect appears to select objects defined not only by physical characteristics such as closure and continuity. Words behave more like objects than nonwords in the sense that entire words tend to be perceived, implying that long-term memory knowledge as well as physical form determines objecthood.

The distribution of attention in neglect can also be said to be "object-based" in the sense that objects' frames of reference have some, albeit limited, effect on the distribution of attention in the visual field. Specifically, the axis that divides an object into its left and right halves can influence the location and orientation of the dividing line between the left and right sides of the visual field for purposes of attentional allocation. However, which of these two halves, so divided, is considered left and which is considered right does not depend upon the object's intrinsic left and right, but on the viewer or environment's left and right. A number of findings of neglect of the left sides of objects can be interpreted in this way: Joined chimeric stimuli suffer more neglect than chimerics with small gaps between the two halves because they are more object-like and hence provide a stronger cue to major axis location; when the region of a stimulus to the (viewer- and environment-centered) left of its major axis is neglected, it will continue to be neglected to some degree after moving to the viewer- and environment-centered right; and there is some evidence that apparent neglect of objects' intrinsic left sides when presented sideways or upside-down results from mental rotation to the upright orientation and subsequent neglect of the viewer- or environment-centered left of the imagined object.

References

Behrmann M, Moscovitch M (1994) Object-centered neglect in patients with unilateral neglect: effects of left-right coordinates of objects. J Cognitive Neurosci 6:1–16

Behrmann M, Tipper SP (1994) Object-based attentional mechanisms: evidence from patients with unilateral neglect. In: Umilta C, Moscovitch M (eds) Attention and performance XV: Conscious and nonconscious information processing. MIT Press, Cambridge

Bisiach E, Perani D, Vallar G, Berti A (1986) Unilateral neglect: personal and extrapersonal. Neuropsychologia 24:759–767

Brunn JL, Farah MJ (1991) The relation between spatial attention and reading: evidence from the neglect syndrome. Cognitive Neuropsychol 8:59–75

Buxbaum LJ, Coslett HB (1994) Neglect of chimeric figures: two halves are better than a whole. Neuropsychologia 32:275–288

Buxbaum LJ, Coslett HB, Montgomery MW, Farah MJ (1996) Mental rotation may underlie apparent object-based neglect. Neuropsychologia 34:112–126

Calvanio R, Petrone PN, Levine DN (1987) Left visual spatial neglect is both environment-centered and body-centered. Neurology 37:1179–1183

Caramazza A, Hillis AE (1990) Spatial representation of words in the brain implied by studies of a unilateral neglect patient. Nature 346:19

Costello AL, Warrington EK (1987) The dissociation of visuospatial neglect and neglect dyslexia. J Neurol Neurosurg Psychiatry 50:1110–1116

Driver J, Halligan PW (1991) Can visual neglect operate in object-centered coordinates? An affirmative single-cased study. Cognitive Neuropsychol 8:475–496

Driver J, Baylis GC, Goodrich S, Rafal RD (1994) Axis-based neglect of visual shapes. Neuropsychologia 32:1353–1365

Duncan J (1984) Selective attention and the organization of visual information. J Exp Psychol Gen 113:501–517

Egly R, Driver J, Rafal R (1994) Shifting visual attention between objects and locations: evidence from normal and parietal lesion subjects. J Exp Psychol Gen 123:127–161

Farah MJ, Brunn JL, Wong AB, Wallace M, Carpenter PA (1990) Frames of reference for allocation of spatial attention: evidence from the neglect syndrome. Neuropsychologia 28:335–347

Farah MJ, Wallace MA, Vecera SP (1993) "What" and "where" in visual attention: evidence from the neglect syndrome. In: Robertson IH, Marshall JC (eds) Unilateral neglect: clinical and experimental studies. Lawrence Erlbaum, Hove, pp 123–137

Hoffman JE, Nelson B (1981) Spatial selectivity in visual search. Perception and Psychophysics 30:283–290

Humphreys GW, Riddoch MJ (1995) Separate coding of space within and between perceptual objects: evidence from unilateral visual neglect. Cogn Neuropsychol 12:283–311

Jolicoeur P (1985) The time to name disoriented natural objects. Memory and Cognition 13:289–303

Katz RB, Sevush S (1989) Positional dyslexia. Brain Lang 37:266–289

Koriat A, Norman J (1984) What is rotated in mental rotation? J Exp Psychol Learn Mem Cogn 10:421–434

Ladavas E (1987) Is the hemispatial deficit produced by right parietal lobe damage associated with retinal or gravitational coordinates. Brain 110:167–180

Marr D (1982) Vision. Freeman, San Francisco

Posner MI, Cohen Y (1984) Components of visual orienting. In: Bouma H, Bouwhuis DG (eds) Attention and performance X: Control of language processes. Lawrence Erlbaum, Hove

Shulman GL, Remington RW, McLean JP (1979) Moving attention through visual space. J Exp Psychol Hum Percept Perform 5:522–526

Sieroff E, Posner MI (1988) Cueing spatial attention during processing of words and letters strings in normals. Cogn Neuropsychol 5:451–472

Tarr MJ, Pinker S (1990) Human object recognition uses a viewer-centered reference frame. Psychol Sci 253–256

Vecera SP, Farah MJ (1994) Does visual attention operate on objects or locations? J Exp Psychol Gen 123:146–160

Young AW, Hellawell DJ, Welch J (1992) Neglect and visual recognition. Brain 115: 51–71

Extinction as a Paradigm Measure of Attentional Bias and Restricted Capacity Following Brain Injury

J. DRIVER[1], J. B. MATTINGLEY[2], C. RORDEN[2], and G. DAVIS[2]

[1]Department of Psychology, Birkbeck College, University of London, UK
[2]Department of Experimental Psychology, University of Cambridge, UK

Introduction

The present chapter focuses on just one aspect of the spatial derangement classically associated with unilateral parietal injury, namely the phenomenon of extinction. Patients with extinction respond appropriately for *isolated* stimuli, whether these fall on their contralesional or ipsilesional side. However, when presented with two *simultaneous* events, the more contralesional goes undetected (e.g., Anton 1899; Bender 1952; Critchley 1949; Oppenheim 1885; Wortis et al. 1948). Extinction can be found in vision, hearing and touch, and some dissociations between the modalities have been reported, as discussed later. In clinical practice, extinction is usually tested for by the informal method of 'confrontation', with the examiner's own fingers being used to stimulate the patient briefly on one or both sides. More recent studies use better-controlled computerized tests to elicit essentially the same phenomenon, by presenting brief stimuli on either or both sides of a VDU (e.g., Baylis et al. 1993; Di Pellegrino and De Renzi 1995; Ward et al. 1994).

As we shall see, extinction can actually be found after various unilateral lesions, which need not always directly involve the parietal lobe. Moreover, the disorder can apparently dissociate from other parietal functions. Nevertheless, we thought it timely to reconsider what is known about extinction, for the following reasons. First, measures of extinction have several methodological advantages over other indices of spatial abnormality following unilateral brain injury, such as the various measures of unilateral neglect. In particular, we shall argue that extinction offers a relatively pure measure of deficits in perceptual awareness following brain injury. Second, extinction has some theoretically interesting parallels with components of normal attention, as we will explain. Third, surprisingly little is known as yet about extinction, despite its textbook status. It has been studied much less than the more florid aspects of unilateral neglect, and many questions about extinction therefore remain unanswered. Finally, as noted above, extinction in some form can be seen after various unilateral lesions. While this fact might initially seem to suggest that the disorder is nonspecific, it may ultimately turn out to be an

In: Parietal Lobe Contributions to Orientation in 3D Space (1997). P. Thier and H.-O. Karnath (eds). Springer-Verlag, Heidelberg.

advantage for the researcher interested in relating brain and behavior, as we may succeed in identifying qualitatively different forms of extinction depending on the locus of brain damage.

Comparing Extinction with Other Aspects of Unilateral Neglect

Many textbooks discuss extinction in the context of the unilateral neglect syndrome, which has also traditionally been associated with parietal injury, especially to the right hemisphere. As discussed elsewhere in this volume (e.g., by Karnath, by Milner, and by Mattingley and Driver), the neglect syndrome can involve many florid abnormalities, that may dissociate on closer inspection. Thus, neglect is best regarded as a label for a range of spatial biases, rather than as one highly specific disorder. Patients labelled as having neglect will typically show disabling spatial biases towards the ipsilesional side in everyday life, and these may also be apparent in various simple clinical tests, such as cancellation or line-bisection.

Textbooks often present extinction as a particularly sensitive measure for the spatial bias in neglect, which can continue to pick up the bias even in chronic patients who have largely recovered from the florid symptoms of their acute neglect (e.g., Heilman et al. 1985). We would agree that this timecourse of events is often observed, and also that extinction can be a highly sensitive index of spatial bias when measured appropriately, as we elaborate below. However, in agreement with several recent authors (e.g., Bisiach 1991; Karnath, this volume; Milner, this volume), we suspect that extinction is not simply a mild form of neglect. Neglect is a multicomponent syndrome that can involve many deficits, only some of which will be present in any one case; and extinction is probably just one of these possible components.

Many conventional tests for neglect (e.g., cancellation) seem like the standard extinction procedure in relying on ipsilesional events to compete with contralesional events. For instance, in cancellation tasks, neglect patients typically fail to cancel lines toward the contralesional side of a page. However, their deficit can be reduced if every line has to be cancelled with an eraser, rather than crossed out (Mark et al. 1988). This reduced neglect for contralesional stimuli when ipsilesional competitors are removed has obvious parallels with the improved detection of contralesional events in extinction when they are presented on their own. However, most neglect tests also *differ* from extinction tests in several respects, as illustrated below.

In tasks like cancellation, the stimuli are usually present continuously, rather than only briefly as in extinction. They may be spread over a wide space, so that appropriate exploratory eye-, head- or body-movements are required for all the stimuli to be perceived. By contrast, exploration of this kind is ruled out in extinction by the use of brief displays. Cancellation tasks force each stimulus to be responded to serially, as an individuated item at local spatial scales. In extinction, the patient must instead take in as much information as possible rapidly and in parallel. Cancellation tasks require fine localization, both in perception, and also in the motor responses with the ipsilesional hand that are usually required. By contrast, no visually-guided movement is required in extinction tasks, and if perceptual localization is required at all, this is usually only at the crude level of reporting on which side an event occurred (though see Di Pellegrino and De Renzi 1995). Finally, cancellation tasks typically require considerable perseverance on each trial, whereas each extinction trial can be over within half a second.

The reader can doubtless think of many other differences between the tasks, but the above list probably suffices to indicate that many component processes will be required in cancellation tasks that are unnecessary for the extinction task. As such, extinction seems potentially to offer a purer measure, of a more specific disorder. For instance, standard extinction measures seem uncontaminated by exploratory or motoric biases, and thus should directly index just spatially-specific abnormalities in perceptual awareness. On the other hand, by virtue of this very purity, the extinction measure may fail to tap many of the processes that can potentially be disrupted in patients with neglect, such as distortions in the coding of egocentric space (Karnath, this volume) or motoric biases (Mattingley and Driver, this volume).

Several authors have now reported apparent double dissociations between clinical measures of extinction versus neglect (e.g., Barbieri and De Renzi 1989; Vallar et al. 1994). The single dissociation of extinction without neglect (e.g., Di Pellegrino and De Renzi 1995) is, on its own, entirely consistent with the traditional view that neglect is simply a more severe version of the extinction phenomenon. However, the opposite dissociation, of neglect with no apparent extinction (Liu et al. 1992; Vallar et al. 1994) seems to overturn this view, suggesting instead that extinction may be a quite separate disorder from the various components of neglect.

However, the apparent double dissociation must be interpreted with some care, for several reasons. First, clinical measures of extinction and neglect can be extremely crude. Though the confrontation method for testing extinction has proved useful, this standard procedure is less than ideal in many respects (e.g., the stimuli are produced by the examiner's fingers and so will doubtless vary in duration etc. from one trial to the next, from one patient to the next and from one examiner to another). Equally, while cancellation measures have also proved useful, they too are extraordinarily crude in some respects (e.g., a patient who requires considerable time and effort to cancel the contralesional items may be scored as having absolutely no neglect on some versions of the task). Given that such crude measures have been employed for extinction and neglect, it may come as no surprise that the correlation between these measures can be less than perfect across every patient. Some of the apparent 'double dissociations' may simply be due to inherent noise in the measurements, especially when only limited data are collected from each patient.

A second possible doubt about some of the reported cases of neglect without extinction is that they may be due to floor or ceiling effects in the extinction measure. When patients show "no extinction" on confrontation tests, their performance may simply be at ceiling. Extinction might still be found if briefer or less salient stimuli were used. Indeed, we have repeatedly observed, in recovering cases, that extinction can be found with brief computerized displays when it is no longer seen on confrontation (e.g., Baylis et al. 1993). Similarly, we have found that the duration of a computerized display often has to be reduced over blocks of trials to keep performance off ceiling, and thus maintain extinction, as a lengthy study continues (e.g., Ward et al. 1994). Floor effects can also present problems. For instance, in Vallar et al.'s (1994) anatomical study of visual and tactile extinction under confrontation, neglect patients were scored as having no extinction if they scored less than 80% correct for isolated contralesional events. However, one can imagine that these same patients might have performed better for the isolated contralesional events had different or longer stimuli been used; and they might then have gone on to show extinction. Future studies should perhaps always use precisely timed stimuli, so that these can be titrated to keep performance off the ceiling and floor for every condition that presents isolated stimuli, with extinction then being measured just by any worsening of contralesional performance in the presence of an ipsilesional competitior.

Despite the above methodological misgivings, we do suspect that some forms of neglect can indeed be observed without any extinction at all. In collaboration with Masud Husain (at Charing Cross Hospital, London), we recently saw a case with extensive right frontoparietal damage who showed very severe visual neglect in everyday life and in tests such as cancellation, yet showed absolutely no visual extinction on confrontation or in computerized tests presenting brief visual stimuli on either or both sides of a VDU. This applied even when the computerized displays were reduced in duration and salience to the point where detection of *isolated* events in either visual field was becoming unreliable.

Finally, we must note that computerized tests do not automatically provide a panacea for the various methodological problems raised above. Another recent case we saw with parietal damage showed reliable visual extinction on confrontation, yet none on brief computerized tests, even when the viewing angles were equated across these tasks and the computer displays titrated to bring performance off ceiling. Thus, while the standard confrontation method is flawed in many respects, it may unwittingly sensitize the extinction measure in other respects. In particular, the usual requirement that patients fixate the examiner's nose may enhance any attentional deficit by providing an interesting central stimulus for the patient to lock onto. Moreover, the movement of already visible objects (as with the examiner's fingers) is known to be less attention-capturing in normals (see Yantis 1993) than the abrupt onset of visual events that suddenly appear from nowhere (as in most computerized tests). Thus, movement detection may be a particularly sensitive measure of any visual extinction.

Whatever the ultimate relation of extinction to neglect (for further discussion see Karnath, this volume and Milner, this volume) it now seems clear that there are many potential components to the spatial disorder that follows parietal damage, with extinction being just one of these. The rest of our chapter focusses on just this particular component of the parietal syndrome.

Sensory versus Attentional Accounts for the Basic Facts of Extinction

Extinction is increasingly attributed to a pathological bias in attention towards the ipsilesional side (e.g., Baylis et al. 1993; Di Pellegrino and De Renzi 1995; Heilman et al. 1985; Kinsbourne 1993; Posner et al. 1984; Ward et al. 1994). On this view, the bias simply becomes more apparent when two events compete for attention (as on double-stimulation trials) than when an isolated event suddenly appears in an otherwise empty field (as on single trials), with nothing else to attract attention. However, many of the earliest investigators (e.g., Bender 1952) considered extinction to be a sensory rather than an attentional deficit. This may be partly attributable to the fact that attention did not become a scientifically respectable concept until the information-processing models of the late 1950s (e.g., Broadbent 1958). However, it should also be noted that the basic phenomenon of missing one target only under double-stimulation can be found (in touch) even after purely spinal injuries (Bender 1952) which presumably disrupt afferent pathways at the very earliest stages. This provides one case of extinction-like behavior that must surely be regarded as a sensory deficit rather than an attentional impairment.

Our own view is that the basic phenomenon of extinction can arise for different reasons in different cases. Some of these may indeed reflect a weakening of afferent input for the affected side (as in spinal injuries); but others reflect a disorder at much higher levels of processing, as we argue below. There are several standard arguments in favor of attentional accounts for extinction, but none of these seem entirely satisfactory to us on closer inspection. We first discuss these traditional arguments, and their possible shortcomings, before going on to cite our own reasons for approaching extinction as a possible deficit in attention.

The first traditional claim is that preserved detection of isolated contralesional events shows that the necessary afferent pathways must be fully intact. However, preserved single detection does not entail that afferent pathways for contralesional events are *entirely* normal; it merely entails that they are adequate to support a crude detection judgement in the absence of any distracting events. Detection of isolated contralesional events is usually 100% correct in standard confrontation measures of extinction (e.g., Di Pellegrino and De Renzi 1995). Since performance is at ceiling, a subtle impairment for single events on the contralesional versus ipsilesional side may go undetected. This could be avoided in future by titrating the duration or intensity of computerized stimuli to bring performance off ceiling in every condition.

If a subtle sensory impairment does exist, then why should it be more apparent on double-trials than on single trials? Ultimately, this probably reflects the fact that concurrent stimuli compete for limited-capacity attention (which applies even for normals, as we explain later). However, the important point for present purposes is that competition for attention can be biased by *sensory* factors, as a tactile example from everyday life illustrates. All pickpockets know that the tactile sensations of one's wallet being removed can escape awareness if a more powerful knock is delivered simultaneously to the other side of the body, even though the wallet removal would be detected in isolation. While the effect of the knock may be to distract attention, the knock wins out over the more subtle sensations on the other side in this example simply because it provides a stronger afferent input. In principle, the same might apply for extinction after unilateral brain injury; that is, stimuli on the ipsilesional side may simply produce a stronger "knock" than those on the contralesional side.

A second argument used to justify attentional accounts for extinction has been that the stimulated receptor locations that get extinguished need not be fixed. For instance, Kinsbourne (1993) noted that visual extinction can be found within the ipsilesional field, as well as between events in separate hemifields as on the standard test (see also Di Pellegrino and De Renzi 1995). Kinsbourne argued that the *relative* position of the two concurrent events is more critical than their absolute position, proposing that whichever event is further to the contralesional side should suffer under double-stimulation. He explains this in terms of his influential hemispheric-competition model, which posits that activation of each hemisphere induces a contralateral orienting tendency. The eventual direction of attention depends on the extent to which one hemisphere's orienting tendency is activated over the other's. The resulting competition between the hemispheres is often envisaged as being played out over inhibitory callosal connections (as in Cohen et al.'s (1994) toy connectionist simulation of extinction).

On Kinsbourne's model, the appropriate unilateral lesion will lead to a chronic orienting bias in the ipsilesional direction, due simply to the lesioned hemisphere being harder to activate than the intact hemisphere. This bias should be more apparent on double-stimulation than on single stimulation for the following reason. An isolated contralesional event may still activate the impaired hemisphere to some extent, and any degree of activation over that for the other hemisphere (which only has an empty field projecting directly to it) will produce some orienting in the contralesional direction. However, this residual activation of the lesioned hemisphere will be less than that for the intact hemisphere on double-stimulation trials, with the result that orienting will now strongly favor the ipsilesional event.

Kinsbourne's model provides a powerful account for extinction, and for several other aspects of unilateral neglect, and it has been incorporated into many subsequent accounts. However, several queries can be raised. First, while the model provides an elegantly simple mechanism for lateral shifts of attention, it says little about what attention actually then does, beyond the implicit premise that unless attention gets drawn to events, they may entirely escape awareness (as for contralesional events on double-stimulation). Second, if this implicit premise were strictly correct, so that events always go undetected unless directly attended, it might then become puzzling why *normal* subjects can detect both events on double-stimulation. On a literal version of Kinsbourne's approach, the two events should surely produce balanced orienting tendencies in normals, which would cancel out to leave them attending only centrally. (Alternatively, even normals might be expected to show left-sided extinction with sufficiently brief displays, if they have a somewhat stronger orienting tendency towards the right as Kinsbourne has argued). Thus, extinction patients do *not* seem to be exactly like normals with just an orienting bias added. We return to this point later, arguing that one probably has to posit a deficit in processing capacity in extinction patients, in addition to any lateral bias in attention.

A third query arises because, in the simple model outlined above, the hemispheres are characterized as homogenous structures competing with each other as wholes. This surely needs refining to take account of the fact that, within each hemisphere, different structures carry out different functions. Finally, the empirical point originally leading us into Kinsbourne's theory, namely that extinction can occur even within the ipsilesional hemifield, has as yet only been tested in a few cases. Moreover, the result might conceivably be explained by a gradient of *perceptual* impairment, extending with increasing severity in the contralesional direction, rather than requiring a purely attentional explanation.

Karnath (1988) has provided a third reason to suspect the involvement of attention in extinction. He reported that the failure to detect contralesional events on double-stimulation in unilateral parietal patients can be ameliorated if the patients are told to ignore all ipsilesional events, reporting only contralesional ones (see also Di Pellegrino and De Renzi 1995). This is an important finding, as it shows that extinction need not always be an inevitable, stimulus-driven outcome whenever two simultaneous events are presented. The result seems consistent with the view that extinction (at least, parietal extinction) only arises when the two events on double trials both compete for attention.

It would be interesting to test whether the same is found in extinction patients with different kinds of lesion. For instance, will those cases who show extinction after unilateral *frontal* damage (e.g., Heilman and Valenstein 1972; Stein and Volpe 1983) still be able to concentrate their covert attention deliberately on just the contralesional side when instructed to do so? Or will their performance be at the mercy of the stimulus, as for so many other frontal phenomena? Equally, will those cases who show extinction-like behavior after spinal damage to afferent pathways (Bender 1952) also be able to compensate by concentrating on the impaired side during double-stimulation trials? Or is this only possible in cases of higher-level extinction? For the moment, while Karnath's (1988) results show that parietal extinction can be modulated by deliberate attention, this alone does not prove that such extinction is caused by an attentional imbalance, rather than by a sensory imbalance. On the contrary, one might argue that a weakened contralesional input simply gets boosted when the patient concentrates on just the contralesional side, due to the quite normal operation of deliberate attention upon an abnormally weak perceptual input. This account could be tested by examining any effects of instructions for the spinal cases who undoubtedly do have weakened inputs.

The possible effects of instructions raise the point that there are both exogenous components to attention (which are stimulus-driven in a bottom-up manner), and endogenous components (which are strategy-driven in a top-down manner, as when instructions are given in Karnath's study). That is, normal attention can be drawn to a location either by a salient event suddenly appearing there (the exogenous case) or merely by an intention to attend there (the endogenous case). There is now reasonable evidence from normals to suspect that different neural mechanisms may be involved in these two forms of attention (see Klein et al. 1992; Rafal et al. 1991; Spence and Driver 1994, 1996, in press; Theeuwes 1994; Yantis 1993). A stimulus that produces a stronger input is obviously more likely to attract exogenous attention. For this reason, it will always be quite hard for neuropsychologists to tease sensory deficits away from attentional deficits when dealing with impaired exogenous mechanisms, as a weakened afferent input will inevitably lead to weaker exogenous orienting. A strict distinction between sensory and attentional factors may therefore be easier to maintain for the endogenous case. Nevertheless, we think that even the stimulus-driven aspects of extinction can still be usefully approached as potential deficits in attention, for the reasons given below.

New Reasons to Consider Extinction as a Possible Deficit in Attention

In our view (see also Driver 1994, 1995, 1996), it is only useful to consider extinction as a potential deficit in attention if doing so either leads directly to a better understanding of the phenomenon or leads to fruitful new experiments. The long-running debate over whether unilateral neglect is best considered as a deficit in attention, or in spatial representation, illustrates that merely labelling a defict as attentional does little to further its understanding unless one has a clear view of what the normal mechanisms of attention might do. Equally, until one gets specific about how the brain represents space (see Karnath, this volume), merely suggesting that neglect patients have some deficit in spatial representation seems about as useful as proposing that aphasic patients have some problem with language.

Fortunately, we think that work on normal attention has now reached a point where it raises specific experimental issues that can be usefully addressed in extinction patients. Thus, approaching extinction as a potential attentional deficit is at least a productive working hypothesis. The best justification for this hypothesis comes from a well-established fact about normal attention, that has striking parallels with extinction, yet remains little known by most neuropsychologists. We refer to the well-established finding (Duncan 1980, 1985; Eriksen and Spencer 1969; Shiffrin and Gardner 1972) that normal subjects can monitor two streams of possible target input as readily as one, provided that in fact they only ever receive a single target at a time (i.e. in signal detection terms, people can make a hit and a correct rejection simultaneously with no dual-task cost). By contrast, if two *targets* are presented simultaneously in the two separate streams, rather than a target plus a nontarget, then normals tend to report one target but to miss the other (i.e. they have difficulty in making two concurrent hits, tending to make one hit plus a miss).

Of course, this limit in performance only arises in normals when one uses rapid, masked displays to bring performance off the ceiling. Nevertheless, the normal two-target limit has many formal similarities to extinction. In extinction, a hit on either side can also be readily combined with a correct rejection (as on single trials), with the difficulty only arising when two simultaneous hits are required (as on double trials). Indeed, extinction seems to differ from the normal two-target limitation in only two main respects. First, it can be seen with quite long displays, under conditions in which normals would be at ceiling. This suggests some reduction in processing speed or capacity in the patients, in addition to their spatial bias. Second, in the extinction patients, but not in the normals, one can always predict which of the two targets will be missed on double trials (i.e. the more contralesional one). This is the spatially specific aspect of extinction, which is presumably due simply to the lateralisation of the brain damage.

Duncan (e.g., 1980, 1984, 1985) has argued that the two-target limitation in normals reflects a competition for limited-capacity attention, which he argues is necessary for stimuli to reach awareness, so that they enter short-term visual memory and arbitrary responses can then be made to them. This limited-capacity stage is contrasted with preattentive vision, which is thought to deal with multiple stimuli in parallel, albeit outside awareness. The distinction between attentive and preattentive vision is now commonplace in the normal literature and has been bolstered by several operational definitions, such as those from visual search tasks. In the specific context of Duncan's theory, preattentively derived properties are those that can guide access to the limited-capacity stage that produces the two-target limit. A further key aspect of Duncan's normal theory is his suggestion that the two-target limit only applies for target attributes from separate objects (Baylis and Driver 1993; Duncan 1984); but *not* for different attributes of the same object. That is, the competition for limited-capacity attention arises between distinct objects *after* preattentive segmentation of the scene has taken place.

We believe that this view of normal attention can be usefully applied to studies of extinction, on the hypothesis that extinction is a spatially specific exaggeration of the normal difficulty with concurrent targets. This approach does not merely provide a new perspective on the established phenomena, but also leads to novel testable predictions, which is where its usefulness lies. In cases of extinction that are due to an attentional deficit of the kind envisaged, preattentive vision should still proceed for extinguished items on double-stimulation trials. Moreover, extinction should be reduced if the two events on double trials are arranged so that preattentive segmentation processes will combine them both into a single common object. Our review of empirical work below shows that both these predictions have recently been confirmed for cases of parietal extinction, illustrating the potential for fruitful interplay between theories of normal attention and studies of extinction.

Parietal Visual Extinction and Object Segmentation

As noted above, the two-target limit in *normals* is avoided when the two target attributes belong to a common object (Baylis and Driver 1993; Duncan 1984). If a patient's extinction reflects a pathological exaggeration of the normal two-target limit, then it should similarly be ameliorated by linking the two events into a common object on double-stimulation trials. Driver et al. (submitted a) tested this using the most basic form of grouping between two visual events, namely uniform connectedness (Palmer and Rock 1994). Their displays comprised circles which could be connected by a line to form a dumbbell (see Fig. 1). A right-parietal patient with visual extinction had to detect circles presented briefly (to rule out saccades) on either the left or right side, on both sides, or not at all. In separate blocks, either no connecting line was ever present ('unlinked' displays in Fig. 1) or a horizontal line appeared between the potential target locations in all conditions ('linked' displays). The patient responded verbally to indicate the presence and location of any circles (i.e. he said "none", "left", "right" or "both"). Substantial extinction was observed for the unlinked displays (i.e. many erroneous "right" responses were made for displays with circles on both sides). Crucially, this extinction was drastically reduced for the linked displays, where the uninformative connecting bar (which also appeared in unilateral and catch trials) joined the two circles into a common object, so that they then became allies rather than competitors in the bid to attract attention.

This shows that parietal extinction can be reduced when the two events on double-stimulation form a common object, as predicted if such extinction is indeed a pathological exaggeration of the normal two-target attentional limit identified by Duncan (1980, 1984, 1985). It remains to be seen whether the same will be found for extinction resulting from different lesions. For now, the parietal result has been extended by a few studies that found it to hold for grouping principles that are considerably more sophisticated than literal connectedness. Ward et al. (1994) found that parietal visual extinction is reduced if the two events

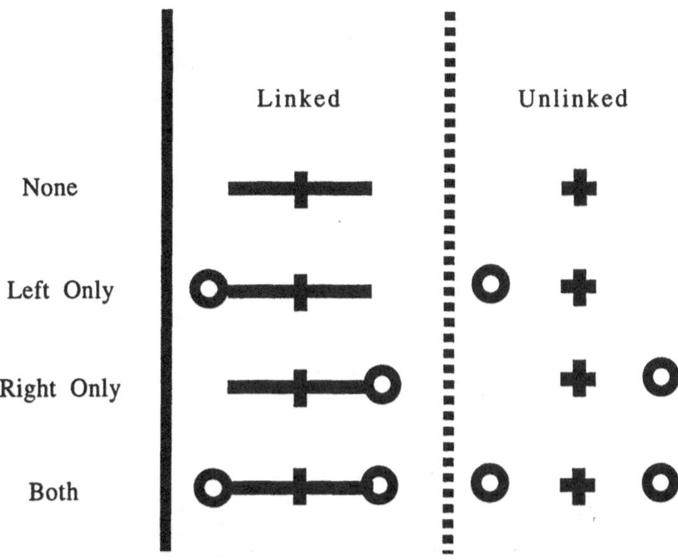

Fig. 1. Eight example displays from Driver et al. (submitted a). Their right-parietal patient had to detect the presence of *circles* which either appeared on both sides (*bottom row*), on just the right, just the left, or not at all (*top row*) on each trial. *Circles* were connected to a *horizontal bar* in some blocks (linked displays, examples on *left* of figure) but not in other blocks (unlinked displays shown at *right*). Substantial extinction was found for unlinked two-circle displays, but not for linked two-circle displays

on double stimulation are grouped by alignment. More recently, Mattingley et al. (submitted a) found that parietal extinction could be dramatically reduced by modal completion (which yields subjective figures such as those made famous by Kanizsa in 1979; see Fig. 2B); and also by amodal completion (which yields the impression of a partially occluded object; see Fig. 2D). In the modal completion task, their patient had to detect removal of quarter-segments from four circles, that could be taken away from just the left circles, just the right circles, not at all, or from circles on both sides (as illustrated in Fig. 2A). Extinction in the latter double-stimulation condition was greatly reduced if the offsets yielded a subjective Kanizsa rectangle (as shown in Fig. 2B), thus forming a single new object, as compared with trials where narrow arcs prevented any subjective figure between comparably aligned offsets on the two sides (Fig. 2C).

These results confirm that parietal extinction is reduced if the two events on double stimulation link to form a single common object, even when this is an entirely subjective object. Evidently, the visual processes that generate subjective figures can still be intact in parietal extinction patients. This accords with the notion that such extinction disrupts attentive vision but not preattentive vision, since recent evidence shows that Kanizsa figures arise preattentively in normals

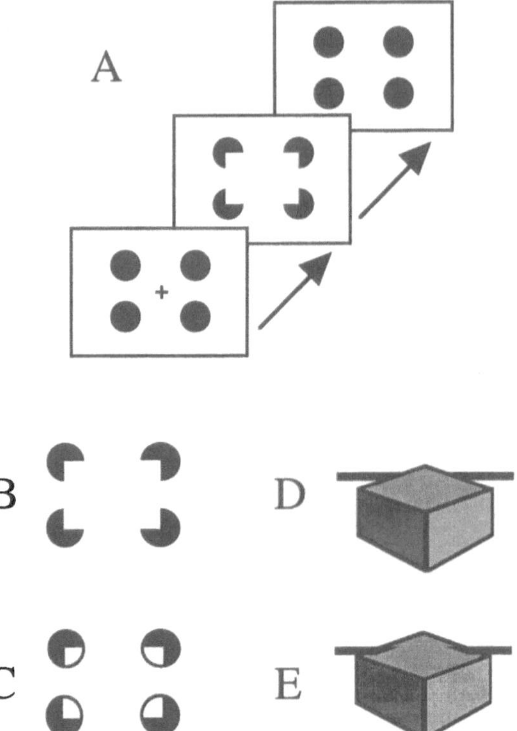

Fig. 2. A Example sequence of events in one trial of the extinction study by Mattingley et al. (submitted a), with the *arrows* depicting time. Each trial began with four *circles* around central fixation (*first frame*). Quarter-segments could then be briefly removed from all circles (as shown in the *middle frame*) or from just the two circles on the left, just the two on the right, or not at all. The patient had to detect these brief offsets. Little extinction was found when removals from both sides produced a Kanizsa subjective rectangle (as in **B**), but the addition of *small arcs* to prevent any subjective figure (as in **C**) reintroduced substantial extinction. In further conditions, the patient had to detect *horizontal bars* which could appear on either or both sides of a *central cube* (see **D** and **E**). Little extinction was found if the bars appeared as part of one continuous occluded object (**D**), but became apparent when the cube was moved up slightly to prevent any amodal completion (**E**)

(Davis and Driver 1994, in press a). Mattingley et al. (submitted a) went on to show that *amodal* completion of an object, under apparent occlusion (see Fig. 2D), can also reduce parietal extinction. Their patient now had to detect the onset of horizontal bars that could appear on either side of a central cube or on both sides. Extinction was greatly reduced when the bars abutted the cube so as to give the impression of a partially occluded, continuous rod on double-stimulation trials (Fig. 2D), rather than suggesting two separate rods (Fig. 2E). This suggests that processes of amodal completion behind occluders (Kanizsa 1979) can remain intact in parietal extinction patients, implying that their extinction arises at the level of a depth interpretation for the visual scene, rather than at the strictly 2D level that reflects the initial organisation of the afferent pathways. Once again, this result accords with the view that parietal extinction disrupts attentive vision but not preattentive vision, since recent normal studies have found that amodal completion, like modal completion, arises preattentively (He and Nakayama 1992; Enns and Rensink 1994), with attention then operating on the surfaces that arise in the resulting depth interpretation (Davis and Driver, in press b).

In summary, recent studies show that parietal extinction can be powerfully affected by sophisticated grouping processes, in keeping with the notion of a disruption to attentive rather than preattentive vision. Any attempt to explain these results in terms of weakened sensations for the contralesional side has to acknowledge that this weakening arises at a sufficiently late stage of processing for modal completion and depth interpretation to still take place in a normal fashion. Once this is acknowledged, then the essential point of attentional accounts for parietal extinction seems vindicated, namely that the deficit is *not* merely due to an afferent loss.

The above conclusions were based on the situations in which extinction can be prevented on double-stimulation trials, by appropriate grouping of a display. In the next section, we again conclude that relatively high levels of processing can be reached by contralesional stimuli in parietal extinction. However, this conclusion is now based on data from those trials in which the contralesional event *does* escape awareness due to extinction.

Implicit Measures of Extinguished Processing in Parietal Patients

The oldest question in the literature on normal attention concerns the level of processing for unattended stimuli (Broadbent 1958; Driver and Tipper 1989; Driver 1996). Various methodological problems arise when trying to answer this question, for the simple reason that directly asking a normal person about a supposedly unattended stimulus is very likely to turn it into an attended stimulus. Various sophisticated indirect measures have been developed in the normal literature to get around this problem. These all have the characteristic that the extent of processing for unattended stimuli is assessed without ever asking the subject to respond to them; in this sense, the methods provide *implicit* measures of unattended processing. Typical behavioral methods include interference measures, as with the classic Stroop task in which responses to a target stimulus are delayed by an irrelevant distractor if it is incompatible with the currently required response, or priming measures in which an unattended distractor can affect the speed of responses to a subsequent related target (e.g., Driver and Tipper 1989).

Further implicit techniques include measures of neural activity such as event related potentials (ERPs) which can be implemented for an unattended stimulus without ever requiring an overt response to it (e.g., Mangun et al. 1993). Doubtless other neuroimaging techniques (e.g., positron emission tomography (PET) or functional magnetic resonance imaging) will also soon be adapted to study the neural responses to ignored versus attended events.

These implicit measures of unattended processing in normals can all be used to study the extent of processing for contralesional items that remain undetected during double-stimulation trials in extinction. Several recent studies have done just this with right-parietal patients. Although some of the individual studies can be criticized on methodological grounds (see Driver 1996), when taken together the emerging picture is that considerable implicit processing can still take place for extinguished contralesional events, up to levels of representation that seem highly counterintuitive given the patient's lack of awareness. It remains to be seen how much this outcome depends on the exact lesion producing the extinction. Driver (1995, 1996) has argued that implicit recognition of extinguished contralesional events may only take place when the lesion is primarily dorsal, thus leaving inferotemporal systems relatively intact to support covert object recognition

Audet et al. (1991) first reported that the identity of an extinguished contralesional letter can influence the speed of response to a concurrent target letter on its ipsilesional side, a result since replicated and extended by Cohen et al. (1995) and by Di Pellegrino and De Renzi (1995). Baylis et al. (1993) found that parietal extinction greater between two identical letters than two different letters and argued that identity was extracted even for the extinguished letters; indeed, they suggested that in the case of repeated letters, contralesional items were often extinguished precisely *because* of their extracted identity.

Berti and Rizzolatti (1992) found that the speed of fruit/animal judgements for an ipsilesional line-drawing was affected by the category of an immediately preceding drawing on the contralesional side. McGlinchey-Berroth et al. (1993) found that an extinguished contralesional drawing could semantically prime lexical decision for a subsequent central letter string. Finally, Marzi et al. (1996) recently reported a case where simple detection responses were faster on double-stimulation trials than on trials with just an ipsilesional target, even when the patient could not consciously distinguish these conditions. In addition, these various findings of implicit processing for extinguished stimuli have been bolstered by similar findings for neglected stimuli, in patients who apparently fails to detect even isolated contralesional events (e.g., Ladavas et al. 1993).

These parietal results, showing considerable implicit processing for extinguished contralesional stimuli, accord with the view advocated earlier, namely that the deficit in parietal extinction primarily arises in attentive vision, with considerable processing still taking place preattentively. The finding of covert recognition for stimuli that the patient remains unaware of may seem highly counterintuitive but is perhaps less surprising when considered from an anatomical perspective. As noted by Driver (1995, 1996), the suspected substrate for object recognition lies in ventral visual areas which are relatively spared in most parietal extinction patients who have more dorsal lesions. This suggests that whether or not implicit recognition still takes place in extinction may depend on how far the unilateral lesion extends ventrally.

While the evidence shows that considerable perceptual processing can take place covertly for extinguished contralesional items, it should be acknowledged that this falls short of demonstrating that such processing is fully equivalent to that for an *un*extinguished item. Future neuroimaging studies might usefully compare the neural activity produced by extinguished versus unextinguished contralesional items on double stimulation trials (and could do so in various lesion groups). For the moment, as with the data on preattentive grouping that we reviewed in the previous section, the existing evidence on implicit processing suggests that if parietal extinction is due to a sensory impairment, then this impairment must operate at a very high level, which effectively concedes the central claim of attentional accounts.

Parietal Visual Extinction and Prior Entry

The preceding two sections illustrate that issues and methods from the normal attention literature (concerning segmentation processes and implicit measures of recognition) can be usefully applied to studies of extinction. A recent study by Rorden et al. (in press) further illustrates this point, and the results again argue against accounts of parietal extinction in terms of mere afferent loss. Rorden et al. applied new methods for studying normal visual attention to patients with parietal extinction. These new methods concern so-called prior entry measures of attention. Titchener (1908) originally asserted his law of prior entry on phenomenal grounds. This law claims that covertly attended stimuli are experienced sooner than comparable unattended stimuli, so that a more attended event will seem to precede another (i.e. will have "prior entry" to awareness) even if the two events are actually physically synchronous (see also Maylor 1985). This claim has recently been confirmed by Stelmach and Herdman (1991) who found, in normal subjects, that a visual event on the ignored side has to physically precede an event on the covertly attended side for the two to be perceived as synchronous.

This normal result has been interpreted as showing a speeding up of perceptual processing for attended events relative to unattended events. A seemingly related normal result was obtained by Hikosaka et al. (1993). They observed that a horizontal line presented instantaneously on a computer display appears to scroll from a covertly attended side towards the other side, as though the attended side provides input to a motion detector sooner. However, attentional interpretations of this "shooting line" phenomenon have recently been challenged in the normal literature (Downing and Treisman, submitted). At present, only the prior-entry phenomenon between two unconnected events (Stelmach and Herdman 1991; Rorden and Driver, submitted) remains unchallenged in the normal literature as an example of attentional influences on the speed of perceptual processing.

Rorden et al. (in press) applied the prior-entry measures of both Stelmach and Herdmann (1991) and Hikosaka et al. (1993) to two right-parietal patients with extinction. They reasoned that if these patients did indeed have an attentional bias towards the ipsilesional side, then events on that side should seem to appear sooner than physically synchronous events on the contralesional side. In their version of Stelmach and Herdmann's task, a small bar appeared in the left visual field and another small bar in the right visual field, with varied asynchronies of onset between them (see Fig. 3A). The patients simply had to judge which bar appeared first while fixating centrally. The results were clear and striking (Fig. 3B). Both patients judged that the ipsilesional right bar had appeared first unless the contralesional bar actually led by over 200 ms, suggesting a dramatic disruption to the normal time-course of visual awareness. This outcome is consistent with a powerful attentional bias towards the ipsilesional side, strongly favoring the earlier conscious detection of events on that side. However, these results alone might also be explained by a drastic *sensory* impairment for the contralesional side, such that afferent inputs on that side have an exceptionally sluggish rise time.

This alternative sensory account is rendered implausible by the performance of the same patients in the other task, the "shooting line" measure of Hikosaka et al. (1993). In this task, the two bars were connected into a continuous horizontal line, such that the various asynchronies in onset for left versus right visual fields now led to percepts of scrolling motion for normal observers (see Fig. 4A). The performance of the patients was indistinguishable from normal controls in this task (see Fig. 4B), in dramatic contrast to their highly abnormal performance when the two bars were separated. These results have several implications. The normal performance in the motion task suggests that afferent inputs were sufficiently intact to provide input to motion detectors without any temporal disruption. This contrasts with the highly abnormal time-course of visual awareness for the separated bars. The better performance with a single connected bar (Fig. 4A) than with two separated bars (Fig. 3A) is reminscent of the improvement in extinction observed in our earlier section on segmentation when the two events on double-stimulation trials were linked into a common object (Fig. 1).

Given the normal performance in the motion task, the difficulty in judging the temporal order of the two separate bars seems more plausibly attributed to an attentional bias than to a strictly sensory impairment. However, the result with separate bars does not accord with one specific account for the attentional abnormality in parietal extinction, namely Posner et al.'s (1984) suggestion of a particular difficulty in disengaging attention from ipsilesional events. Our patients

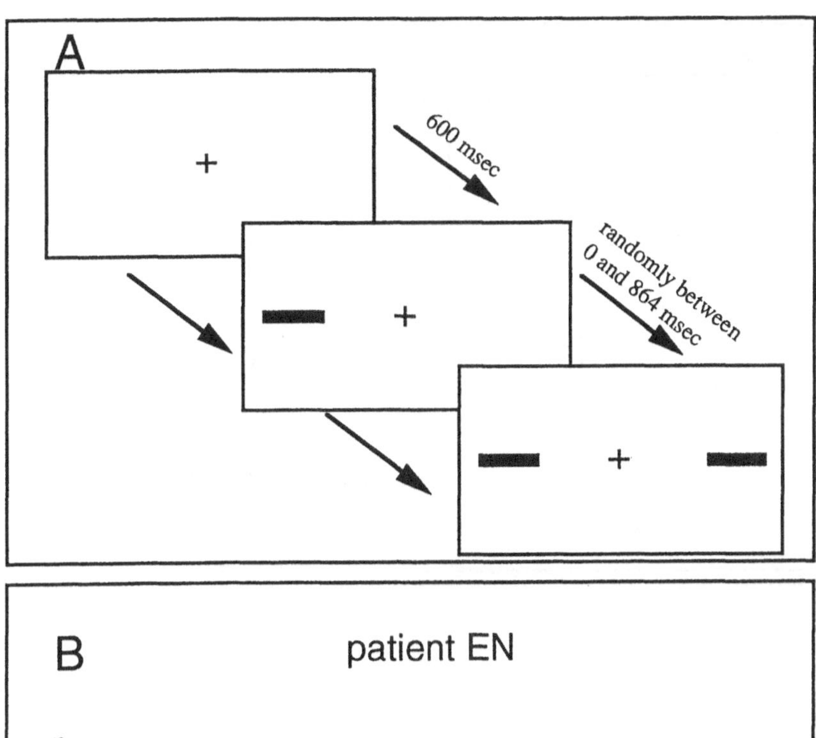

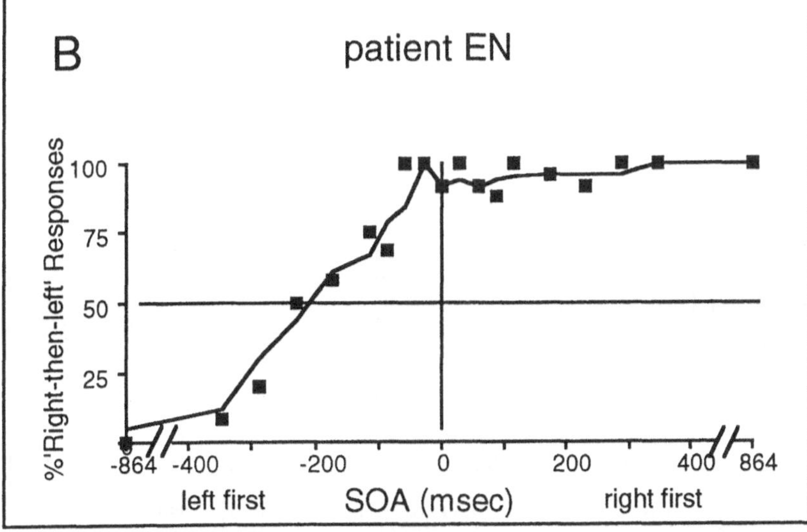

Fig. 3. A Example sequence of events in one trial of the study by Rorden et al. (in press), with *arrows* depicting time. The patients fixated centrally and judged whether the *bar* on the *left* or *right* appeared first (a *left-first* sequence is illustrated, but right-first sequences were as likely). Any temporal lead for one side was varied up to 864 ms. **B** Results for one right-parietal patient, showing the percentage of 'right-then-left responses' for each temporal lead, connected by a smoothed function. It can be seen that the patient reported the right item as first unless the left item actually led by over 200 ms

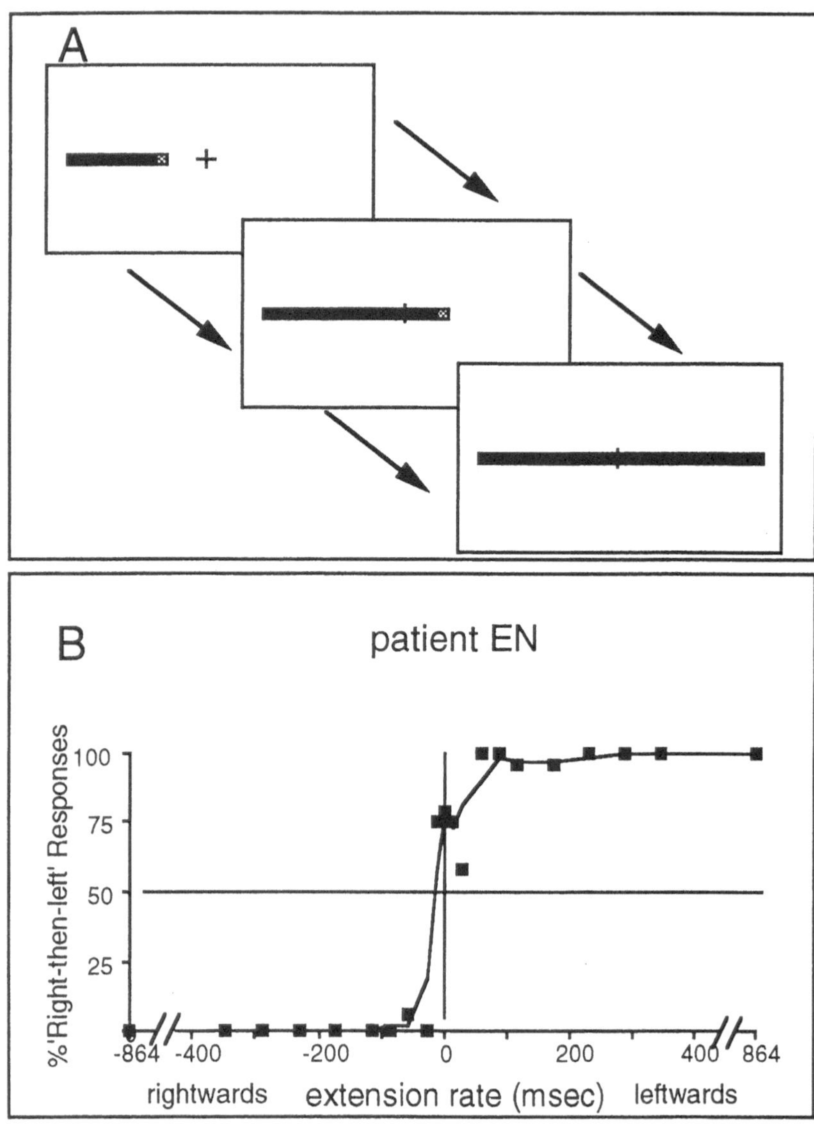

Fig. 4. A Example sequence of events from one trial in the 'shooting-line' condition of Rorden et al. (in press). This was similar to the task depicted in Fig. 3A, except that the *bars* on each side were now connected to yield a motion percept when one side led the other. **B** In contrast to the results in Fig. 3B, the patient's responses were now roughly symmetrical around true zero. These motion judgements were indistinguishable from normal

performed abnormally in the separated-bars task (Fig. 3) even when the contralesional bar preceded the ipsilesional bar by 200 ms, during which time there was no ipsilesional event to disengage from. In any case, while the proposal of a disengage deficit can nicely explain why unilateral parietal patients would miss the contralesional event once their attention has been drawn ipsilesionally on double-stimulation trials, in itself a disengage deficit does not explain why attention should invariably be drawn to the ipsilesional event first. A prior-entry advantage for that side, due to a chronic attention bias, could account for this.

The final implication of these results is that prior-entry tasks with two separated events (as in Fig. 3A) are more appropriate indices for the direction of attention than motion judgements, such as the 'shooting line' task (as in Fig. 4A). This point is also emerging in the normal attentional literature (Downing and Treisman, submitted; Rorden and Driver, submitted) again illustrating that studies of normal attention and parietal extinction can usefully converge on common conclusions.

Extinction Across Sensory Modalities

As our final illustration of the potential for interplay between studies of normal attention and of extinction, we turn to consider attentional issues for other modalities beyond just vision and, in particular, to the possibility of crossmodal links in attention. This is a relatively new topic in the normal literature, but substantial progress has recently been made (for review see Driver, in press). Two basic questions have now been addressed in normals. The first concerns whether shifts of attention in one modality have any implications for shifts of attention in another modality. This question needs to be considered separately for the endogenous case (of voluntary attention) and the exogenous case (of stimulus-driven attention). Spence and Driver (1996) recently reported that when normals deliberately direct their covert visual attention to one side because they expect a visual target there, then auditory attention tends to follow even if sound targets are actually more likely on the other side. Likewise, deliberate shifts of auditory attention tend to be accompanied by corresponding covert visual shifts. Crossmodal links can also be found in the exogenous case. For instance, a sudden sound will summon normal visual attention (Spence and Driver, in press), although sudden visual events apparently do not lead to auditory shifts of attention (Spence and Driver, in press) unless an eye-movement is made to the sound (Rorden and Driver, submitted). Sudden tactile events can also lead to exogenous shifts in visual attention (Grossenbacher and Driver, submitted).

The second crossmodal question that has been addressed for normal attention is whether integration between the modalities can affect the spatial coordinates in which attention operates. Two concrete examples may clarify this issue. Grossenbacher and Driver (submitted) found that the visual location which becomes attended following a sudden tactile event on one hand depends on where that hand is currently located, as signalled proprioceptively. This implies that tactile positions are recalibrated by proprioceptive information prior to the shift of visual attention being triggered. Driver and Grossenbacher (1996) studied deliberate tactile attention in normals, rather than exogenous responses to a sudden touch. They found that attending to vibrotaction on one hand, while ignoring distracting vibrotaction on the other, became easier if the two hands were placed further apart. This suggests that the representational space in which deliberate tactile attention operates can be modulated by proprioceptive information, about the current disposition of the tactile receptors in external space.

These various crossmodal issues can also be studied in patients with extinction. The issue of whether shifts of attention in one modality have consequences for another modality can be addressed by looking for extinction between events in separate modalities. Where extinction is due to early sensory competition in damaged afferent pathways (as, presumably, in spinal cases), one might expect extinction to be restricted to situations in which the two simultaneous events both depend on the same particular afferent pathway (e.g., for just tactile events in the spinal cases). Barbieri and De Renzi (1989) reported that even in unilateral parietal patients, extinction can be found within one modality, yet not within another in particular single cases (see also Vallar et al. 1994). However, these studies relied on the clinical confrontation method and as discussed earlier this can be somewhat insensitive, so it could be maintained that extinction might also have been found within the apparently intact modality had a more sensitive computerized method been employed using very brief stimuli.

In his classic treatise on extinction, Bender (1952) reported that it could be found between concurrent events in quite separate modalities (e.g., one visual and one tactile). However, he only reported anecdotal evidence for this claim. Inhoff et al. (1992) carried out a more systematic study of any extinction between concurrent tactile and visual events in unilateral parietal patients and found none. They therefore argued that the parietal cortex does not combine tactile and visual information. However, their conclusion seems somewhat surprising in two respects. From an anatomical perspective, it is well known that various regions of the parietal cortex have both tactile and visual inputs (e.g., Graziano and Gross 1996). And from a psychological perspective, Grossenbacher and Driver (submitted) have already shown that tactile events lead to shifts of visual attention in normals. Accordingly, Mattingley et al. (submitted b) reexamined any extinction between concurrent tactile and visual events in three patients with right middle cerebral artery infarctions. Crossmodal extinction was found in all cases, both from an ipsilesional visual event upon a contralesional tactile event and from an ipsilesional tactile event upon a contralesional visual event. It remains to be determined whether the difference in outcome from the Inhoff et al. (1992) study depends on the particular lesions involved in each patient or upon details of the procedure.

Extinction can also be used to address the more subtle crossmodal issue from the normal attention literature, namely whether integration between the modalities can affect the spatial coordinates in which attention operates. Recall Driver and Grossenbacher's (1996) finding that normal tactile attention is modulated by proprioceptive information about the current location of tactile receptors. Driver et al. (submitted b) examined this same issue for parietal patients showing tactile extinction. Extinction between the two hands was measured as a function of their current position. The unseen hands were either placed close together in the ipsilesional hemispace, close together in the contralesional hemispace, or far apart in separate hemispaces. More extinction was found when the unseen hands lay in separate hemispaces rather than together on the ipsilesional side, with intermediate performance when the hands were together on the contralesional side. This demonstrates that tactile extinction can be modulated by proprioceptive information about the current layout of body parts, just as for normal tactile attention. Moscovitch and Behrmann (1994) have similarly shown that tactile extinction in right-parietal patients can be modulated by proprioception. They measured extinction between two brief taps on either side of the ipsilesional wrist. The tap on the left side of space was extinguished, regardless of whether the hand was positioned palm up or palm down. Thus, the region of the skin surface that became extinguished could reverse, depending on proprioceptive information about the current posture. Note that this provides a further example of extinction that cannot be explained by mere afferent loss.

It remains to be determined whether these patterns of results will depend on the exact lesion producing the extinction. Nonetheless, it is already clear that studies of extinction can be revealing about interactions between the modalities and also about neural representations fot the spatial disposition of body parts (or "body-image"). Comparisons of different lesion groups should give us further information about the neural substrates involved in these mechanisms.

The Possible Anatomy of Extinction

In the preceding sections, we occasionally acknowledged that extinction can be produced by a range of unilateral lesions but concentrated primarily on data from unilateral parietal patients, in keeping with the theme of this book. Several caveats concerning the possible anatomy of extinction should now be made. First, the "unilateral parietal patients" referred to in the preceding sections typically had quite extensive lesions, usually produced by middle cerebral artery infarctions. These lesions would often include nonparietal regions (e.g., basal ganglia), and the specific parietal areas that were directly damaged will also have varied between cases. Future extinction studies should attempt to compare groups with more discrete lesions to distinct parietal areas and to different nonparietal areas. The existing data are insufficient to allow any firm conclusions on this issue, both because the details of the lesions are often inadequate and because the measure of extinction is usually clinical confrontation, which can be an unreliable index as discussed earlier. With the advent of widespread MRI scanning for lesion analysis and the increasing use of more sensitive computerized tests for extinction, this situation should hopefully improve in the near future.

Furthermore, in addition to merely testing for the presence or absence of extinction with computerized detection tasks, future studies could also apply the more specific behavioral measures reviewed in the above sections (e.g., concerning whether extinction is modified by instructions or by the grouping of the display; and also whether implicit recognition, prior-entry abnormalities and crossmodal interactions can be observed). By examining these issues in different lesion groups, we may discover qualitative differences between distinct forms of extinction that seem quite indistinguishable on simple detection tasks.

Several speculations can already be found in the literature concerning possible qualitative differences between extinction following distinct lesions. Vallar et al. (1994) suggested that subcortical lesions produce a sensory form of extinction, whereas cortical lesions produce a more attentional disorder. However, their claim was based primarily on arguments from the known anatomy, rather than on any firm behavioral difference between the lesion groups. Hopefully, their distinction might be confirmed by applying new behavioral measures, such as those discussed above, to the different lesion groups.

Milner (this volume) suggests that cortical extinction is primarily associated with superior parietal lesions. His suggestion is based on Posner et al.'s (1984) association of the so-called disengage deficit in cued attention with this region, plus recent PET data suggesting that the superior parietal cortex may be particularly involved in attention to peripheral targets. Two queries can be raised regarding the superior-parietal hypothesis. First, recent data on the disengage deficit suggest that it is found most strongly following more inferior lesions than originally claimed (Friedrich et al., submitted). Second, extinction in standard detection tasks can be found after various nonparietal cortical lesions, including frontal damage (Heilman and Valenstein 1972; Stein and Volpe 1983); moreover, we believe that extinction in some form can even be found after unilateral *temporal* damage, as described below, so the disorder may not be restricted to the dorsal stream as Milner proposes.

Desimone and Duncan (1995) as well as Duncan (1996) have recently suggested that extinction is a very general phenomenon, that will result in some form from virtually any sizeable lesion. They provide a neural perspective on Duncan's (1980, 1984, 1985) model of normal attention, which was described earlier. Their suggestion is that attentional capacitylimits arise when concurrent stimuli compete for control of a limited pool of neurons at very high levels of perceptual representation. Although distinct neural subsystems at these high levels may code different perceptual properties (e.g., color, motion or shape), Desimone and Duncan suggest that links between the subsystems ultimately lead to a single object coming to dominate activity across all of them at any one time. On this view, extinction will result from any lesion which gives an advantage to one class of object over another in the competition to control high-level neural activity. Thus, a unilateral lesion to a spatiotopic system will give a competitive advantage to stimuli from one side, as observed in standard extinction.The more general claim is that even lesions to nonspatiotopic systems (say, to a visual subsystem which represents faces) might give a competitive disadvantage to the impaired class of stimuli versus unimpaired classes (see Humphreys et al. 1994). Any mild impairment should become more apparent when impaired and unimpaired classes directly compete (as on double-stimulation trials) than when they are tested separately (as on single trials).

It is too early to judge the usefulness of this very general approach. However, we close this section with one example of atypical extinction that fits quite naturally with Desimone and Duncan's view. It was found following a unilateral lesion that would not traditionally be associated with the disorder, and the resulting 'extinction' seemed qualitatively different from that which follows parietal lesions. Berti et al. (1992) studied a patient with right unilateral temporal-lobe resection. When presented briefly with pictured objects in the left visual field, right visual field, or in both fields concurrently, the patient never failed to detect the contralesional event on double-stimulation trials. In this sense, the temporal-lobe patient did not show extinction. However, when asked instead to identify each picture, a new form of extinction became apparent. The patient could reliably identify the contralesional picture in isolation, but often failed to do so on double-stimulation trials. Thus, a form of extinction became apparent only when the task tapped an ability that is traditionally associated with the damaged region (namely inferotemporal object recognition); extinction was absent in the standard detection measure. Moreover, the extinction for object recognition seemed qualitatively different from that found in parietal patients in the following sense. Baylis et al. (1993) found that parietal extinction is more pronounced when two concurrent objects have the same identity rather than different identities. Exactly the opposite was found in Berti et al.'s temporal-lobe patient; she was better able to identify the contralesional picture on double-stimulation trials if it depicted an object with the same identity as the ipsilesional picture. Berti et al. attributed this to intact recognition of the ipsilesional object priming the partially impaired recognition of the contralesional object. This case provides a preliminary example that some form of extinction may be apparent after lesions which would not traditionally be associated with the disorder, as Duncan (1996) predicts. Moreover, the extinction became apparent only in a task that tapped into the usual presumed function of the damaged region (here, object recognition), and it seemed qualitatively different to that found after other lesions.

Extinction and Balint's Syndrome

In may be useful in closing to consider the possible relation of extinction following unilateral damage to aspects of Balint's (1909) syndrome following *bilateral* lesions. This syndrome is classically considered to involve several deficits, which may well dissociate in cases with more focal lesions. The traditional syndrome involves difficulties in spatial localization, in the control of reaching movements and saccades, and, of particular interest for us, an apparent inablity to be aware of more than one object at a time. Classic anecdotes include an inability to light cigarettes because at any given time only the match, or just the cigarette, could be seen. This difficulty in perceiving multiple objects simultaneously can readily be demonstrated with more formal measures (e.g., Humphreys et al. 1994).

This aspect of Balint's syndrome is highly reminiscent of extinction in that only one of two simultaneous events can be consciously perceived, just as on double-stimulation trials in extinction. The difference, of course, is that extinction is spatially specific, with the contralesional event invariably suffering on double trials; whereas in Balint cases, which event gets missed may not depend on spatial parameters (Humphreys et al. 1994). As noted earlier, the spatial specificity of extinction can be readily accounted for by the laterality of the lesion, for instance in terms of Kinsbourne's (1993) model of the opposing orienting tendencies produced by the two hemispheres. However, the restricted awareness for concurrent objects in bilateral Balint patients cannot be readily explained in these same terms. Since their damage is bilateral, no orienting bias in one particular direction should be expected.

Instead, it seems that the bilateral damage in Balint's syndrome may simply reduce the capacity to represent simultaneous objects, perhaps because coding the different locations of simultaneous objects is essential for producing distinct representations of them (Driver 1996; Friedman-Hill et al. 1995), and this coding of position is drastically disrupted in Balint patients. In any case, the essential point for now is that if bilateral lesions reduce the capacity to represent concurrent objects, unilateral lesions may well also do so, and this aspect of extinction may have been largely overlooked by researchers due to the more obvious spatial bias. At various points in our review, we have noted that deficits in the capacity to represent concurrent objects may be needed to explain parietal extinction, in addition to any spatial bias in attention. In collaboration with Humphreys and Bundesen, Duncan (1996) recently adapted normal measures of visual attention and capacity limits (Bundesen 1990) in order to separate any abnormal capacity restrictions from any spatial biases in unilateral extinction patients. The tasks involved unspeeded whole report and partial report for brief displays of multiple letters. The preliminary results suggest that when the appropriate brief displays are used, extinction patients with cortical damage show a drastically reduced capacity for reporting multiple concurrent items, even when all the items are arranged in a column rather than being spread over the left and right. Thus, there may be an element of Balint's syndrome even in unilateral cases, with the pronounced spatial bias tending to distract researchers from the reduced capacity. It is noteworthy in this respect that some right-hemisphere patients, who usually show left-sided extinction in standard tasks, can exhibit a paradoxical *right*-sided extinction instead if instructions of strategies lead them to report any left items first (Karnath

1988; Humphreys et al. 1996), suggesting a fundamental difficulty in becoming aware of two simultaneous objects in addition to any spatial bias.

Conclusions

We think the studies reviewed above show that there can be a fruitful interplay between studies of normal attention and studies of extinction after brain injury. Many of the same theoretical and methodological issues arise. Moreover, patients with extinction after parietal injury seem to have a deficit primarily in attentive vision, with relatively preserved preattentive vision. This relative preservation of early preattentive vision seems to make anatomical sense, given that the typical parietal lesions lie quite far into the visual system. Unilateral parietal patients also seem to have a reduced capacity for representing concurrent events, in addition to their spatial bias in attention. In a further parallel with normal attention, extinction can be observed between events in different modalities and can be used to study how the brain generates a body image. Finally, extinction can arise from various lesions. Instead of implying that the disorder is merely nonspecific, these different lesions may well produce distinct forms of extinction which should be particularly revealing about the functions of the damaged areas. In this respect, it is particularly exciting that a recent study has found that the basic phenomenon of extinction can be produced in *normal* subjects, by transient magnetic stimulation over parietal cortex (Pasqual-Leone et al. 1994). We may soon be in a position to compare different forms of extinction following different reversible functional 'lesions' within the same normal subject using variations on these new magnetic methods.

Acknowledgements. This work was supported by the Wellcome Trust (UK). Our thanks to Masud Husain, Hans-Otto Karnath and Bob Rafal for their collaboration and encouragement.

References

Anton G (1899) Über die Selbstwahrnehmung der Herderkrankungen des Gehirns durch den Kranken bei Rindenblindheit und Rindentaubheit. Arch Psychiatr Nervenkr 32:86–111

Audet T, Bub D, Lecours AR (1991) Visual neglect and left-sided context effects. Brain and Cogn 16:11–28

Balint R (1909) Seelenlähmung des "Schauens", optische Ataxie, räumliche Störung der Aufmerksamkeit. Monatsschr Psychiatr Neurol 25:51–81

Barbieri C, De Renzi E (1989) Patterns of neglect dissociation. Behav Neurol 2:13–24

Baylis GC, Driver J (1993) Visual attention and objects: evidence for hierarchical coding of location. J Exp Psychol: Hum Percept Perform 19:451–470

Baylis GC, Driver J, Rafal RD (1993) Visual extinction and stimulus repetition. J Cogn Neurosci 5:453–466

Bender MB (1952). Disorders in perception. Thomas, Springfield

Berti A, Rizzolatti G (1992) Visual processing without awareness: Evidence from unilateral neglect. J Cogn Neurosci 4:345–351

Berti A, Allport A, Driver J, Dienes Z, Oxbury J, Oxbury S (1992) Levels of processing for stimuli in an "extinguished" visual field. Neuropsychologia 30:403–415

Bisiach E (1991) Extinction and neglect: same or different? In: Paillard J (ed) Brain and space. Oxford University Press, Oxford

Broadbent DE (1958) Perception and communication. Oxford University Press, Oxford

Bundesen C (1990) A theory of visual attention. Psychol Rev 97:523–547

Cohen JC, Farah MJ, Romero RD, Servan-Schreiber D (1994) Mechanims of spatial attention: the relation of macrostructure to microstructure in parietal neglect. J Cogn Neurosci 6:377–387

Cohen A, Ivry RB, Rafal RD, Kohn C (1995) Activating response codes by stimuli in the neglected field. Neuropsychology 9:165–173

Critchley M (1949) The phenomenon of tactile inattention with special reference to parietal lesions Brain 72:538–561

Davis G, Driver J (1994) Parallel detection of Kanizsa subjective figures in the human visual system. Nature 371:791–793

Davis G, Driver J (in press a) Kanizsa subjective figures can act as obligatory occluders at parallel stages of visual search. J Exp Psychol: Hum Percept Perform

Davis G, Driver J (in press b). Spread of visual attention over modally- versus amodally-completed surfaces. Psychol Sci

Desimone R, Duncan J (1995) Neural mechanisms of selective visual attention Ann Rev Neurosci 18:193–222

Di Pellegrino G, De Renzi E (1995) An experimental investigation on the nature of extinction. Neuropsychologia 33:153–170

Downing PE, Treisman A (submitted). The line-motion illusion: Attention or impletion?

Driver J (1994) Unilateral neglect and normal attention. Neuropsychol Rehabil 4:123–126

Driver J (1995) Object segmentation and visual neglect. Behav Brain Res 71:135–146

Driver J (1996) What can visual neglect and extinction reveal about the extent of 'preattentive' processing? In: Kramer AF, Coles M,GH, Logan GD (eds) Convergent operations in the study of visual selective attention. APA Press, Washington DC, pp 193–224

Driver J (in press) Crossmodal links in attention: The 2nd EPS Prize lecture. Q J Exp Psychol

Driver J, Grossenbacher PG (1996) Multimodal constraints on tactile spatial attention. In: Innui T, McClelland JL (eds) Attention and Performance. XVI. MIT Press, Cambridge, pp 209–236

Driver J, Tipper SP (1989) On the nonselectivity of "selective" seeing: contrasts between interference and priming in selective attention. J Exp Psychol: Hum Percept Perform 15:304–314

Driver J, Goodrich S, Ward R, Rafal RD (submitted a) Grouping by uniform connectedness affects Balint's syndrome and unilateral visual extinction similarly

Driver J, Mattingley JB, Beschin N, Robertson I (submitted b). Tactile extinction as a function of unseen hand separation

Duncan J (1980) The locus of interference in the perception of simultaneous stimuli. Psychol Rev 87:272–300

Duncan J (1984) Selective attention and the organization of visual information. J Exp Psychol: Gen 113:501–517

Duncan J (1985) Visual search and visual attention. In: Posner MI, Marin OSM (eds) Attention and Performance. XI. Erlbaum, Hillsdale, pp 85–106

Duncan J (1996) Coordinated brain systems in selective perception and action. In: Innui T, McClelland JL (eds) Attention and Performance. XVI. MIT Press, Cambridge, pp 549–578

Enns JT, Rensink RA (1994) An object completion process in early vision. In: Visual search III: proceedings of the third international conference on visual search . Taylor and Francis, London

Eriksen CW, Spencer T (1969) Rate of information processing in visual percption: some methodological considerations. J Exp Psychol Monogr 79 (2, part 2)

Friedman-Hill SR, Robertson LC, Treisman A (1995) Parietal contributions to visual feature binding: evidence from a patient with bilateral lesions. Science 269:853–955

Friedrich FJ, Egly R, Beck D, Rafal RD (submitted) Covert orienting of attention: effects of parietal lobe and superior temporal gyrus lesions

Graziano MSA, Gross CJ (1996). Multiple pathways for processing visual space. In: Innui T, McClelland JL (eds) Attention and Performance.XVI. MIT Press, Cambridge, pp 181–208

Grossenbacher PG, Driver J (submitted). The coordinates of tactile-visual links in exogenous covert attention

He ZJ, Nakayama K (1992) Surfaces versus features in visual search. Nature 359:231–233

Heilman KM, Valenstein E (1972) Frontal lobe neglect in man. Neurology 22:660–664

Heilman KM, Watson RT, Valenstein E (1985) Neglect and related disorders. In: Heilman KM, Valenstein E (eds) Clinical Neuropsychology. Oxford University Press, Oxford, pp 279–336

Hikosaka O, Miyauchi S, Shimojo S (1993) Focal visual attention produces illusory temporal order and motion sensation. Vision Res 33:1219–1240

Humphreys GW, Romani C, Olson A, Riddoch MJ, Duncan J (1994) Non-spatial extinction following lesions of the parietal lobe in humans. Nature 372:357–359

Humphreys GW, Boucart M, Datar V, Riddoch MJ (1996) Processing of fragmented forms and strategic control of orienting in visual neglect. Cogn Neuropsychol 13:177–204

Inhoff AW, Rafal RD, Posner MI (1992) Bimodal extinction without crossmodal extinction. J Neurol, Neurosurg Psychiatry 55:36–39

Kanizsa G (1979) Organization in Vision. Praeger, New York

Karnath H-O (1988) Deficits of attention in acute and recovered visual hemi-neglect. Neuropsychologia 26:27–43

Kinsbourne M (1993) Orientational bias model of unilateral neglect: evidence from attentional gradients within hemispace. In: Robertson IH, Marshall JC (eds) Unilateral neglect: clinical and experimental findings. Erlbaum, Hove, pp 63–86

Klein RM, Kingstone A, Pontefract A (1992) Orienting of visual attention. In: Rayner K (ed) Eye movements and visual cognition: scene perception and reading. Springer-Verlag, New York, pp 46–65

Ladavas E, Paladini R, Cubelli R (1993) Implicit associative priming in a patient with left visual neglect. Neuropsychologia 31:1307–1320

Liu GT, Bolton AK, Price BH, Weintraub S (1992) Dissociated perceptual-sensory and exploratory-motor neglect. J Neurol Neurosurg Psychiatry 55:701–706

Mangun GR, Hillyard SA, Luck SJ (1993) Electrocortical substrates of visual selective attention. In: Meyer DE, Kornblum S (eds), Attention and Performance. XIV. MIT Press, Cambridge, pp 219–244

Mark VW, Kooistra CA, Heilman, KM (1988) Hemispatial neglect affected by non-neglected stimuli Neurology 38:1207–1211

Marzi CA, Smania N, Martini MC, Gambina G, Tomelleri G, Palamara A, Allesadrini F, Prior M (1996) Implicit redundant-targets effect in visual extinction. Neuropsychologia 34:9–22

Mattingley JB, Davis G, Driver J (submitted a). Visual filling-in of surfaces in unilateral parietal visual extinction.

Mattingley JB, Driver J, Beschin N, Robertson I (submitted b) Crossmodal extinction between vision and touch.

Maylor EA (1985). Facilitory and inhibitory components of orienting in visual space. In: Posner MI, Marin OSM (eds) Attention and Performance. XI. Erlbaum, Hillsdale, pp 189–204

McGlinchey-Berroth R, Milberg WP, Verfaiellie M, Alexander M, Kilduff PT (1993) Semantic processing in the neglected visual field: evidence from a lexical decision task. Cogn Neuropsychol 10:79–108

Moscovitch M, Behrmann M (1994) Coding of spatial information in the somatosensory system: Evidence from patients with neglect following parietal lobe damage. J Cogn Neurosci 6:151–155

Oppenheim H (1885) Über eine durch eine klinisch bisher nicht verwerthete Untersuchungsmethode ermittelte Form der Sensibilitatsstörung bei einseitigen Erkrankungen des Grosshirns. Neurologisches Centralblatt 4:529–533

Palmer S, Rock I (1994) Rethinking perceptual organization: The role of uniform connectedness. Psychonom Bull Rev 1:29–55

Pasqual-Leone A, Gomez-Tortosa E, Grafman J, Alway D, Nichelli P, Hallet M (1994) Induction of visual extinction by rapid-rate transcranial magnetic stimulation of parietal lobe. Neurology 44:494–498

Posner MI, Walker JA, Friedrich FJ, Rafal RD (1984) Effects of parietal injury on covert orienting of attention. J Neurosci 4:1863–1874

Rafal R, Henik A, Smith J (1991) Extrageniculate contributions to reflex visual orienting in normal humans: A temporal hemifield advantage. J Cogn Neurosci 3:322–328

Rorden C, Driver J (submitted) Saccade preparation and visual prior entry revisited.

Rorden C, Mattingley JB, Karnath H-O, Driver J (in press). Visual extinction and prior entry: impaired perception of temporal order with intact motion perception after parietal injury. Neuropsychologia

Shiffrin RM, Gardner GT (1972) Visual processing capacity and attentional control. J Exp Psychol 93:72–83

Spence CJ, Driver J (1994) Covert spatial orienting in audition: exogenous and endogenous mechanisms facilitate sound localization. J Exp Psychol: Hum Percepti Perform 20:555–574

Spence CJ, Driver J (1996) Audiovisual links in endogenous covert orienting. J Exp Psychol: Hum Percept Perform 22:1005–1030

Spence, CJ, Driver J (in press) Audiovisual links in exogenous covert orienting. Percept Psychophys

Stein S, Volpe BT (1983) Classical "parietal" neglect syndrome after subcortical right frontal lobe infarction. Neurology 33:797–799

Stelmach L B, Herdman CM (1991) Directed attention and perception of temporal order. J Exp Psychol: Hum Percept Perform 17:539–550

Theeuwes J (1994) Endogenous and exogenous control of visual selection Perception 23:429–440

Titchener EB (1908) Lectures on the elementary psychology of feeling and attention. MacMillan, New York

Vallar G, Rusconi ML, Bignamini L, Germiani G, Perani D (1994) Anatomical correlate of visual and tactile extinction in humans: a clinical CT scan study. J Neurol Neurosurg Psychiatry 57:464–470

Ward R, Goodrich S, Driver J (1994) Grouping reduces visual extinction: neuropsychological evidence for weight-linkage in visual selection. Vis Cogn 1:101–129

Wortis SB, Bender MB, Teuber HL (1948) The significance of the phenomenon of extinction. J Nerv Ment Dis 107:382–387

Yantis S (1993) Stimulus-driven attentional capture. Curr Dir Psych Sci 5:156–161

Is Extinction Following Parietal Damage an Interhemispheric Disconnection Phenomenon?

C. A. MARZI[1], A. FANINI[1], M. GIRELLI[1], A. E. IPATA[1], C. MINIUSSI[1], M. PRIOR[2], and N. SMANIA[3]

[1]Department of Neurological and Visual Sciences, University of Verona, Verona, Italy
[2]Department of General Psychology, University of Padua, Padova, Italy
[3]Rehabilitation Center, Ospedale Policlinico Borgo Roma, Verona, Italy

Contralateral extinction of sensory stimuli is a relatively frequent consequence of unilateral damage to the parietal lobe both in nonhuman and human primates (for a review see Heilman et al. 1993). In the visual modality it consists of the inability to report a stimulus following simultaneous presentation of another similar stimulus in the opposite hemifield, while targets presented individually are correctly detected. A broadly similar picture is present in the tactile and auditory modality. Unilateral damage to the right rather than left parietal cortex, as well as to other cortical and subcortical sites, is more often implicated in extinction phenomena in humans (Vallar et al. 1994), but the mechanisms underlying this phenomenon and its laterality are still debated. Various theories have attempted to explain extinction on the basis of a single factor, but it is very likely that its neural mechanisms are multiform. Several accounts have been proposed: attentional, sensory, interference and interhemispheric inhibition (for a review see Heilman et al. 1993), and it is beyond the aim of this article to examine them thoroughly. What we would like to do instead is to discuss the possibility that a frequent, albeit not exclusive, cause of extinction following unilateral parietal-temporal damage is callosal disconnection.

Interhemispheric disconnection may occur because of a degeneration of callosal receiving and projecting neurons resulting from cortical damage. Such neurons are particulary widespread in the parietal and temporal cortex as opposed to the occipital cortex where only areas with a visuotopic representation of the vertical meridian and adjoining visual field zones are interhemispherically connected (for a review see Cusick and Kaas 1986; Marzi 1986; Kennedy et al. 1991; Kaas 1995). Obviously, such a hypothesis relies upon two fundamental bodies of evidence: first, that patients with a complete section of the corpus callosum (CC) show extinction-like phenomena, and, second, that extinction patients with a parietal-temporal lesion have a callosal disconnection syndrome. Furthermore, a callosal disconnection hypothesis of extinction must specify the nature of the interhemispheric integration of information required by the task used to tap extinction phenomena. This is a particulary important point because, in principle, extinction could result from the lack of interhemispheric integration of different sensory, attentional or response-related operations (for a discussion see Smania et

In: Parietal Lobe Contributions to Orientation in 3D Space (1997). P. Thier and H.-O. Karnath (eds). Springer-Verlag, Heidelberg.

al. in press). One could hypothesize that in a typical task designed to assess extinction the response is subserved by the left hemisphere (LH) and therefore the information presented to the right hemisphere (RH) must be transferred to the LH. Such a transfer may be impaired in extinction patients. However, typically, extinction patients can detect unilateral contralesional stimuli, and therefore a simple failure of callosal transfer of low-level sensory information is unlikely. An impairment may occur only with bilateral presentations because the weak signal transcommissuarally relayed from the RH may be difficult to detect when the LH is simultaneously activated by the more efficient direct pathway. This might explain why RH extinction patients are unimpaired in giving a verbal response to single contralesional stimuli while they fail with bihemispheric stimulus presentations. Moreover, interhemispheric integration presumably requires a higher-quality callosal transmission in comparison to detecting a single stimulus. Therefore, the same callosal transmission channel may fail in interhemispheric comparison but not with single stimuli. That the normal CC may entail different morphofunctional channels with different properties and transmission time has been suggested recently by a study of Iacoboni and Zaidel (1995). They found that increasing the complexity of the motor response in a Poffenberger (1912) paradigm increased the difference between crossed and uncrossed visuomotor responses, i.e., a measure that is considered an estimate of callosal transmission time (for a review see Marzi et al. 1991). Moreover, laterality effects concerning the hemifield of stimulus entry and the hand used for response also changed as a function of the task. Importantly, these effects were present only in normal but not in split-brain subjects, and this suggests that they were related to a shift in the callosal channel used for interhemispheric transmission.

One potential problem concerning a hemispheric disconnection theory of extinction following unilateral parietal damage is posed by the finding (see Di Pellegrino and De Renzi 1995; Smania et al., in press) that extinction occurs also for two stimuli presented to one hemisphere only. In both the above studies only presentations to the contralesional (left) hemifield resulted in extinction of the leftmost stimulus while no effect was found for presentations to the ipsilesional hemifield.

Such a differential hemifield effect suggests that stimuli presented to the RH cannot be properly evaluated and have to be transferred to the LH for a decision to be made as to their number and/or for verbal response. Recent experiments in our laboratory (Smania et al., in press) have shown that when a verbal response is not required, extinction rate decreases substantially in RH lesioned patients. This suggests that during an extinction test, access to LH is crucially important for carrying out the verbal response rather than for mastering the task.

Extinction Phenomena in Callosotomized Patients

Evidence that callosal section may result in hemineglect has been provided in the literature by various authors (Goldenberg 1986; Corballis 1995; Kumral et al. 1995) although the presence or frequency of true hemineglect in split-brain subjects has been questioned (Joynt 1977; Plourde and Sperry 1984). In this report, we will focus on extinction-like phenomena in commissurotomized patients rather than on neglect because the two syndromes are dissociable anatomically (Vallar et al. 1994) as well as neuropsychologically (Bisiach 1991), with extinction apparently a more specific consequence of parietal damage than neglect.

Before concentrating on visual extinction, which will be the major focus of this report, it is interesting to briefly mention other modalities as well to show that the phenomenon of extinction is not modality-specific.

A clear instance of *tactile* extinction in a patient with a hematoma localized at the trunk of the CC and sparing the cerebral hemispheres has been described by Mayer et al. (1988). The two hands were perceptually impaired only in dichaptic conditions of stimulus presentation when interhemispheric integration was required. Perception of unimanual stimuli, though, was unimpaired.

In the *auditory* modality, following the pioneering experiments of Milner et al. (1968) using dichotic listening in split-brain patients, many reports have shown extinction of the ear ipsilateral to the language-dominant hemisphere (usually the left ear) when speech stimuli are presented (for a review see Sugishita et al. 1995).

As to the *visual* modality, an interesting early study casting light on the effect of simultaneous bilateral stimulus presentation in patients lacking the CC is that of Teng and Sperry (1974). Six split-brain patients were tested on a dot counting task with one to five dots flashed either to the right (RVF) or left visual field (LVF) or simultaneously to both fields. The patients were supposed to indicate the number of dots seen in a given hemifield by extending the same number of fingers of the ipsilateral hand. Frequent unilateral extinction was observed during bilateral testing in both the RVF and LVF, with the laterality of the effect most probably related to the presence of some extracallosal brain damage. Extinction occurred in the weaker hemisphere but sometimes also occurred when both hemispheres were equally proficient in processing the unilateral input. The authors point out that the pattern of extinction of their split-brain patients is remarkably similar to that seen in brain-damaged patients (Benton and Levin 1972; Bender 1977). Interestingly, following bilateral presentation, no concurrent increase in dot counting errors for nonextinguished presentations was observed in spite of frequent extinction. In other words, following bilateral presentations, one hemisphere either fails to perform at all or performs as well as it would following unilateral presentations. It is also important to mention that there was no correlation between poor performance in unilateral and bilateral trials. These results, according to the authors, suggest that extinction is caused by an "all-or-none rivalry in some gating mechanisms" related to the output stage of information processing. This is an important issue that deserves further investigation. The study of Teng and Sperry (1974) shows that commissurotomy patients are impaired during simultaneous bilateral stimulus presentation. This suggests that the forebrain commissures are necessary for carrying out the numerosity judgment task used to tap extinction.

A series of experiments by Corballis and collaborators (Corballis and Sergent 1992; Corballis and Trudel 1993; Corballis 1994) has provided further extensive evidence of the limited interhemispheric exchange of information available to fully commissurotomized patients when they have to compare simultaneously presented visual digits or forms in the two hemispheres. Both shape and numerical information show little, if any, interhemispheric integration, which is in partial disagreement with evidence provided on the same patients by Sergent (1987, 1990). All these studies have not specified whether the patients showed some degree of unilateral extinction during bilateral presentations. However, the pattern of errors made by Corballis' patients suggests that some information was relayed from the weaker hemisphere, and therefore complete extinction of the stimulus in one hemifield is very unlikely to have taken place.

Further evidence of a broadly similar nature comes from another split-brain patient tested on a simple visual reaction time (RT) paradigm, who showed electrophysiological as well as behavioral evidence of a rightward bias toward the right hemifield (Proverbio et al. 1994). A roughly similar finding has been obtained in the same fully commissurotomized patient (case M.E., see below) discussed in the present report who has been studied by Berlucchi et al. (in press a, b) on a cue-target light detection task using a RT paradigm.

Other reports on the ability of split-brain subjects to perform simple perceptual interhemispheric integrations have yielded somewhat contrasting results. Ramachandran et al. (1986) described one of three split-brain patients (N.G.) who was impaired in judging the apparent motion of lights presented simultaneously across the vertical meridian, while having no trouble with within-hemifield presentations. By the same token, Gazzaniga (1987) described another split-brain subject (J.W.) who was unable to distinguish successive pairs of lights, one in each hemifield, from the presentation of a single unilateral light. Both of these examples may be interpreted as a kind of extinction even though it does not seem as if one stimulus in the pair was entirely missed, as occurs in extinction patients. In contrast to the above evidence, Naikar and Corballis (1996) have recently reported that another split-brain patient (L.B.) had little difficulty discriminating single from double bilateral flashes.

A more direct comparison of commissurotomy and extinction patients with RH lesions can be found in a recent study by Reuter-Lorenz et al. (1995) on the same split-brain patient (J.W.) studied by Gazzaniga (1987), who was tested on a task similar to that used to tap extinction phenomena in patients with unilateral RH lesions. The patient was supposed to say if he saw one or two flashes generated by two horizontally aligned red light emitting diodes (LEDs) positioned 7° to the left and right of fixation. Following stimulus presentation, the patient was supposed to indicate whether he saw a stimulus on the left, on the right or on both sides. The results showed good accuracy for unilateral detection, but extinction of the left stimulus on about half of the bilateral trials. Extinction was present, albeit to a lesser extent, even when a verbal report was not required.

As in the study of Teng and Sperry (1974), the authors say that the extinction shown by J.W. resembles that seen in patients with unilateral parietal damage. Interestingly, such impairment is not present in another patient (S.C.) with only a partial anterior callosotomy.

On the whole, the studies reviewed here show that there are numerous indications that callosotomy may result in extinction-like symptoms. Further evidence that full commissurotomy may result in unilateral extinction has been provided recently in our laboratory.

Case M.E.

Patient M.E. is a right-handed 26 year old man who in 1989 underwent a two-stage commissurotomy for treatment of a severe form of post-traumatic epilepsy. The completeness of the callosal resection and the sparing of the anterior commissure have been verified by magnetic resonance imaging (MRI) (Aglioti et al. 1993). It is important to note that, as a consequence of a head trauma suffered when he was 8 years old, M.E. underwent partial right frontal polectomy to remove a subdural hematoma shortly after the accident. MRI showed an involvement of area 8 with a substantial sparing of area 6 (see Aglioti et al. 1993; Berlucchi et al. in press a, b). We tested the above patient on a simple reaction time (RT) paradigm similar to that used in other experiments carried out recently in our laboratory (Marzi et al. 1996). Briefly, the subject was supposed to press a key as fast as possible following tachistoscopic presentation on a PC screen of a 2° x 2° (°= degree of visual angle) luminous square 7° either to the RVF or LVF or located bilaterally along the horizontal meridian. The response was performed with one or the other hand in different blocks of trials. Immediately after completing the response, the subject was to report verbally whether he saw one or two stimuli. On trials in which he reported seeing one stimulus only, the patient was asked to indicate whether the stimulus had appeared on the LVF or RVF. Such a procedure allows testing for the presence of a redundant target effect (RTE), i.e., a decrease in RT when responding to two vs. one stimulus only, as well as assessing extinction on a trial-to-trial basis. In a previous study (Marzi et al. 1996), we found that some extinction patients show a RTE for bilateral stimuli presented simultaneously in the two hemifields even in trials in which they report having seen one stimulus only. This can be considered evidence of an implicit processing of extinguished stimuli.

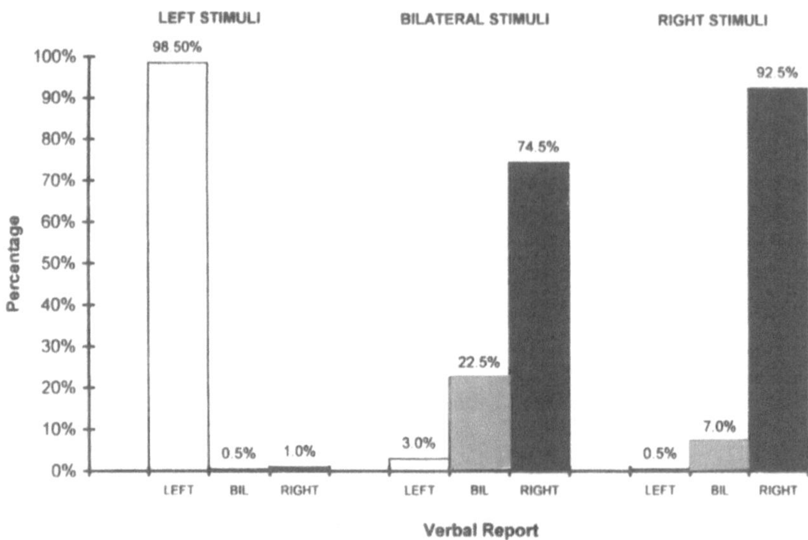

Fig. 1. Percentage of verbal responses of patient M.E. (complete commissurotomy) following presentations of left visual field (*LVF*), bilateral (*BIL*) or right visual field stimuli (*RVF*). The great majority of unilateral *LVF* or *RVF* stimuli were responded to correctly while in only 22.5% of bilateral stimuli M.E. gave the correct response "bilateral" and in 74.5% of these trials he responded "right hemifield"

What we found in M.E. was that, confirming the results of Reuter-Lorenz et al. (1995) mentioned above, during bilateral presentations most left stimuli were extinguished while unilateral stimuli were correctly detected (see Fig.1). In spite of that, there was a clearcut RTE for responses to bilateral stimuli that were reliably faster than responses to the hemifield yielding the fastest responses to unilateral stimuli (see Fig. 2). This advantage of bilateral vs. single stimuli was present for either hand and either hemifield and was reliable as assessed by a one-way ANOVA showing a highly significant effect of stimulation type (unilateral LVF, unilateral RVF and bilateral; $F_{2,566}=46.9$, $p<.0001$). Post-hoc Tukey tests showed that all differences, and in particular the crucial difference between the fastest unilateral condition (RVF) and the bilateral condition, were highly significant.

Mathematical analysis using Miller's method (Miller 1982; Reuter-Lorenz et al. 1995; Marzi et al. 1996) of the RTs' cumulative frequency distributions (see Fig. 3) showed that in M.E. the RTE cannot be attributed to probability summation and is likely to be related to neural summation. This result is in good accord with the study of Reuter-Lorenz et al. (1995) showing a large RTE in a fully commissurotomized patient in spite of his extinguishing many of the stimuli presented to the LVF. That M.E. indeed suffered from hemispheric disconnection

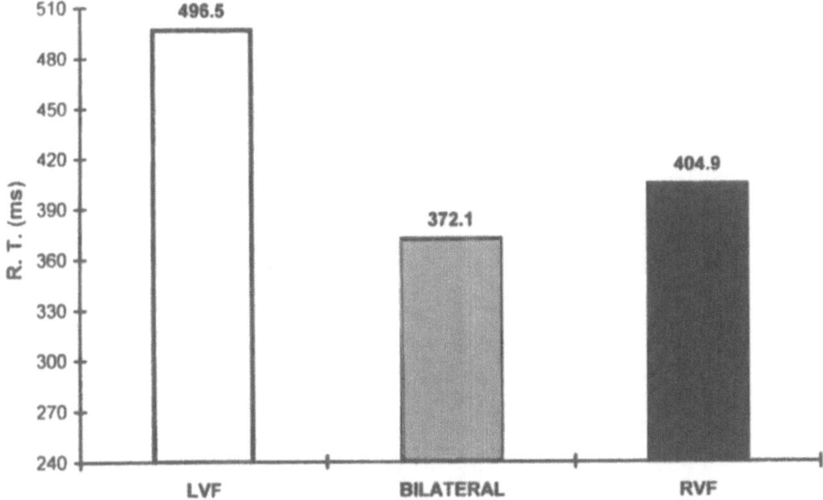

Fig. 2. Simple reaction time paradigm. Mean speed of response of patient M.E. to unilateral left visual field (*LVF*), right visual field (*RVF*) or to bilateral stimuli. Responses to bilateral stimuli were clearly faster than those to *RVF*, i.e., than those to the faster unilateral responses. Such a difference was statistically reliable (see text). Note the markedly slower mean reaction time (*RT*) to *LVF* stimuli than to *RVF* stimuli. This effect might be related to the presence of right hemisphere (*RH*) damage in the patient

was assessed both behaviorally and electrophysiologically in a further study carried out in our laboratory. Confirmimg previous findings (Aglioti et al. 1993; for a review see Marzi et al. 1991), our patient had a prolonged interhemispheric transmission (IT) time as witnessed by a larger than normal crossed uncrossed difference (CUD) between hemifield-hand combinations with the former condition yielding a slower mean RT (458.0 ms) than the latter (443.6 ms). The CUD of about 15 ms is small in comparison with other studies on split-brain subjects but still represents an almost threefold increase in IT time in comparison to normal subjects (Marzi et al. 1991). The cumulative frequency distribution (see Fig.4) of RTs using the hand contralateral to the stimulated hemifield (crossed condition, requiring an IT) or ipsilateral to the stimulated hemifield (uncrossed condition, requiring no IT) shows very little overall overlap. However, it is interesting to note that the speed advantage of the uncrossed vs. crossed condition is much more clearcut for fast than slow RTs. This hints to the interesting possibility that slow responses make use of channels not requiring the integrity of the cerebral commissures in the crossed condition. Moreover, in agreement with the observation of Mangun et al. (1991; see also Tramo et al. 1995) in other split-brain patients, we found that in M.E. there was no clear electrophysiological sign of callosally transmitted visual signals.

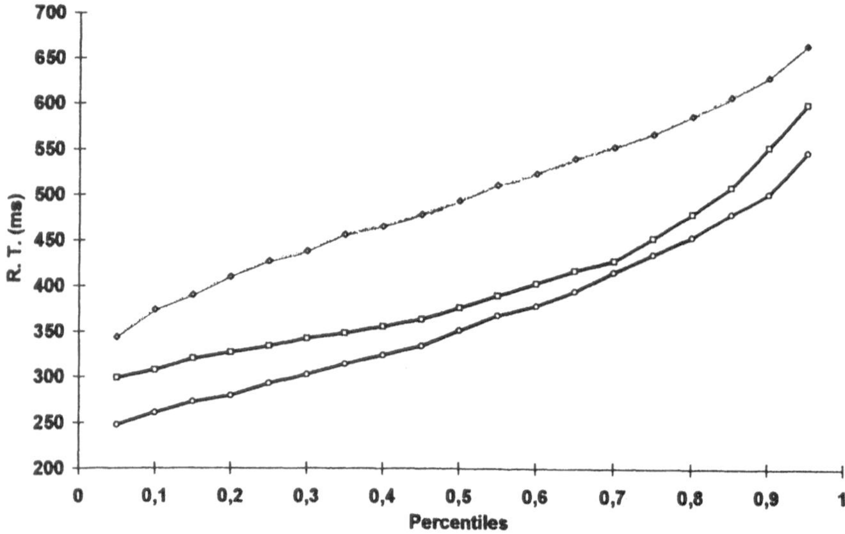

Fig. 3. Case M.E.: Cumulative frequency distributions for response times (*RTs*) in the three conditions of stimulus presentation. Note that there is no overlap between the three conditions. *LVF* (*diamonds*), left visual field; *RVF* (*squares*), right visual field; *BIL* (*circles*), bilateral presentations

We believe that the extinction results are important for two reasons: First, they extend to commissurotomized patients' evidence of a dissociation between explicit and implicit processing similar to that found in some hemianopic (Marzi et al. 1986) and extinction patients (Marzi et al. 1996).

Second, and more relevant to the present topic, the extinction results show that complete section of the CC results in unilateral extinction of the left stimuli during bilateral presentations. The concordance as to the side of extinction between the study of Reuter-Lorenz et al. (1995) and our own study may be attributed simply to chance, to the laterality of extracallosal damage, or, more interestingly, to a rightward attentional bias caused by hemispheric disconnection as has been suggested by various studies (Berlucchi et al. in press a, b; Mangun et al. 1994; Proverbio et al. 1994). Our patient M.E. has a right prefontal lesion which might be responsible for his left extinction although such lesions are not a frequent cause of extinction or of attentional bias toward the ipsilesional field (Henik et al. 1994). Theoretically, it is possible that such a lesion alone is not sufficient to yield an extinction effect but that it becomes effective in combination with a callosal section as is the case in monkeys (Crowne et al. 1981; Watson et al. 1984). This issue can be resolved by studying patients with a prefrontal lesion comparable to that of M.E. but without callosal surgery. It should be mentioned that the patient studied by Reuter-Lorenz et al. (1995) also shows evidence of RH damage as assessed by EEG signs of paroxisms that, following the second stage callosotomy,

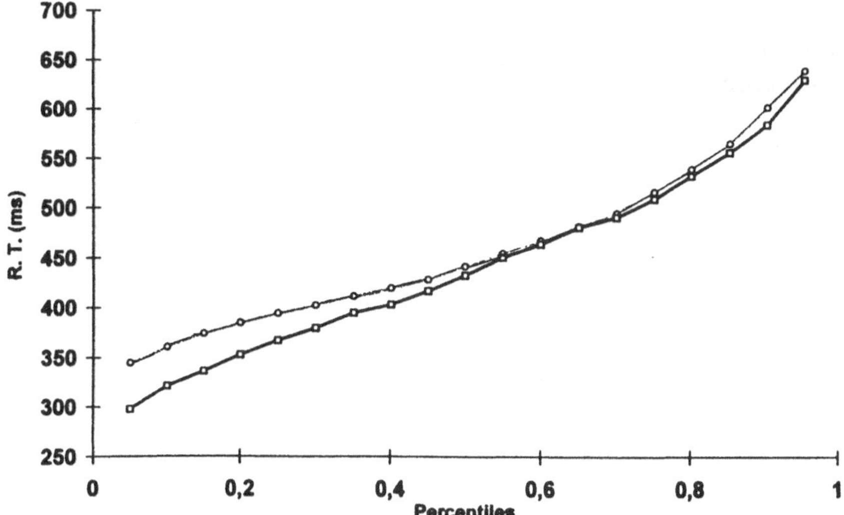

Fig. 4. Case M.E.: Cumulative frequency distributions for response times (*RTs*) in the crossed (*circles*) and uncrossed (*squares*) conditions of visual presentation in relation to the hand used for response. The circles represent the mean *RTs* of the two crossed conditions (left visual field right hand and right visual field left hand). The squares represent the mean *RTs* of the two uncrossed conditions (left visual field left hand and right visual field right hand). Crossed conditions require an interhemispheric transfer while such is not the case for uncrossed conditions

were lateralized to the RH (for more details see Tramo and Bharucha 1991). Therefore, the possibility of left extinction in both our patients and those of Reuter-Lorenz et al. (1995) being dependent on RH extracallosal damage cannot be disproved without studying a larger number of patients without RH pathology. In spite of the EEG abnormalities, the main thrust of the study of Tramo and Bharucha (1991) was that the RH but not the LH of patient J.W. showed a normal perception of harmony in music. Therefore, the RH of this patient did not seem to suffer from an overall functional impairment in comparison with the LH.

Finally, it is important to point out that in our study one patient with callosal agenesis and three others with partial surgical or ischemic lesion of the CC did not show any unilateral extinction when tested in the same apparatus as described above. Therefore, it seems justified to conclude that only a complete commissurotomy, although sparing the anterior commissure, is sufficient for extinction to show up.

One important question that we intend to tackle in a forthcoming study is whether or not M.E. shows the phenomenon of extinction when both stimuli are presented to one and the same hemisphere. We know that such is the case when two stimuli are presented to the damaged RH of extinction patients (Di Pellegrino and De Renzi 1995; Smania et al. in press) while no extinction is present for pairs of stimuli presented to the intact LH.

Callosal Disconnection in Extinction Patients: Electrophysiological Evidence

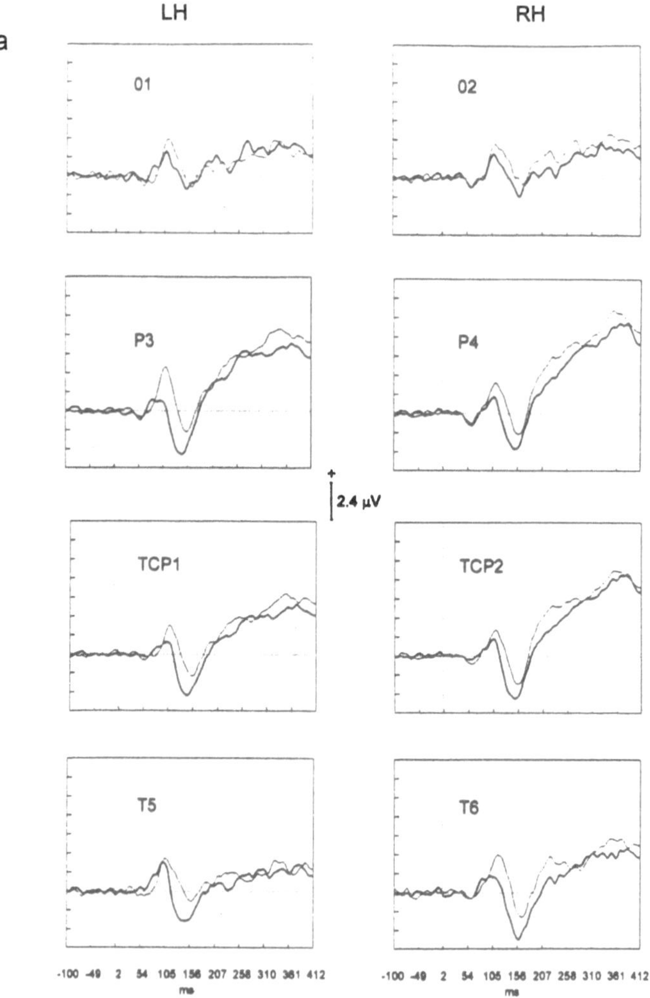

Fig. 5 a, b. Average visually evoked potentials (*VEPs*) obtained from various electrode sites (*O1-O2*, *P3-P4*, *TCP1-TCP2*, *T5-T6*) in the left hemisphere (*LH*) and right hemisphere (*RH*) of four normal subjects (**a**) and four patients (**b**) with contralateral extinction as a result of parietal-temporal damage of the *RH*. The stimuli were small white rectangles presented tachistoscopically at 7° to either the *LVF* or the *RVF*. In the normal subjects (**a**), both stimulation of the ipsilateral hemifield (interhemispheric response; *thin line*) and that of the contralateral hemifield (direct response; *thick line*) yielded a clear *P1* (positive waveform peaking around 100 ms after stimulus onset) and a large *N1* (negative waveform peaking

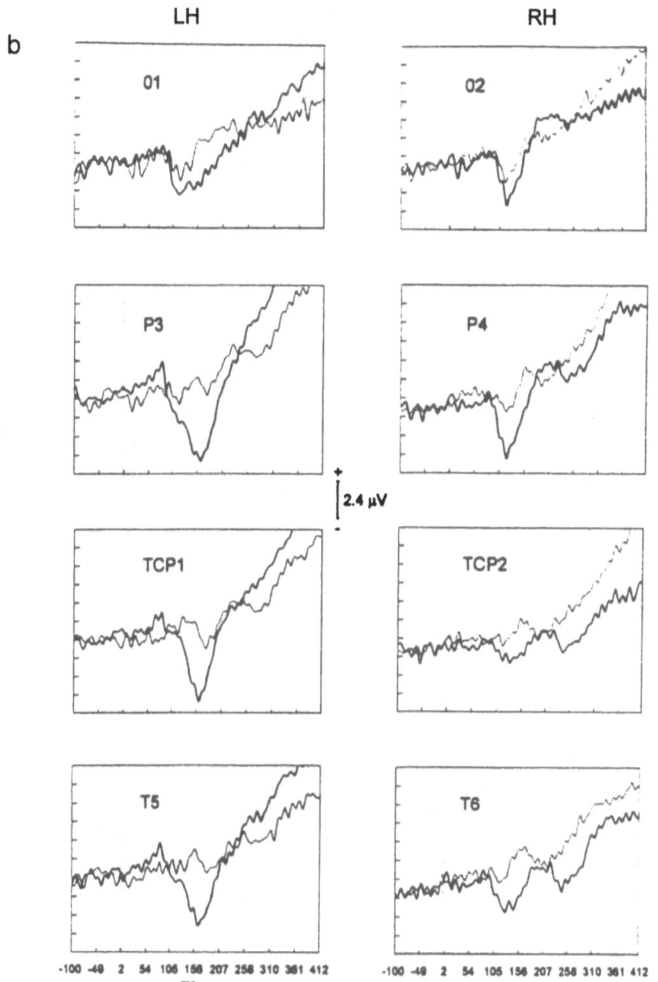

around 150 ms after stimulus onset). In patients (**b**), at all electrode sites, but more markedly at *P3*, *P4*, *TCP1* and *T5*, the direct input from the contralateral visual hemifield (*thick line*) evoked a *P1* and a large *N1*, as in normals. In contrast, the indirect (interhemispheric; *thin line*) input from the ipsilateral hemifield yielded practically no response barring a small *N1* at the occipital leads of the two hemispheres (*O1* and *O2*). Notice that the direct responses recorded in patients at *TCP1* and *TCP2* yielded a clear laterality effect with the latter site (*RH*) showing a much smaller *N1* (and an absent *P1*) than the former (*LH*). This probably represents a direct effect of the vascular damage to the *RH*. *LVF*, left visual field; *RVF*, right visual field

Recently, we (Miniussi et. al 1995) studied four extinction patients with a right unilateral ischemic lesion centered on the parietal lobe in a task identical to that performed by M.E. During performance of the task we recorded visual event-related potentials and found that the transcallosal responses to visual stimuli presented to the hemifield ipsilateral to the recording hemispheric sites were severely disturbed and in some cases totally lacking. This was true for both the ipsi- and the contralesional hemisphere and was in sharp contrast to what we found in the same experiment on four normal subjects. In these subjects the indirect (callosal) responses obtained from stimulation of the hemifield ipsilateral to the recording hemisphere were clearly present and broadly similar to the direct responses from the contralateral hemifield (compare Fig. 5a [normals] with Fig. 5b [patients]).

In contrast to callosal responses, in patients, the direct responses to stimuli presented to the hemifield contralateral to the recording hemisphere were normal in the intact LH and somewhat reduced but still present in the damaged portions of the RH. Thus, only the transcallosal response was selectively altered in these patients, and this suggests that IT of visual information in both directions is impaired. Such impairment is particulary evident in the parietal leads (e.g., see P3-P4 in Fig. 5b) and is broadly similar to that found in our callosotomized patient M.E. and in those of Mangun et al. (1991) and Tramo et al. (1995).

Therefore, it is reasonable to assume that in these patients the task used to tap extinction, namely, deciding whether stimuli have been presented unilaterally or bilaterally, might become extremely difficult or impossible because of the impaired capacity to integrate information across the CC. We are currently undertaking an experiment on a larger number of extinction patients as well as controls to try to contrast IT in trials in which there was extinction and in those trials in which the patients reported the bilateral stimuli correctly. In addition to such a within-subject comparison, we also plan to compare callosal transmission among patients with different unilateral lesions and with or without extinction.

All in all, we believe that patients with large lesions centered on the parietal lobe are likely to suffer from a degeneration of the callosal fibers necessary for integrating visual and, possibly, tactile and auditory information presented simultaneously to the two hemispheres. Therefore, it is conceivable to attribute to a callosal disconnection syndrome some form of extinction following parietal-temporal damage while extinction following subcortical lesions (Vallar et al. 1994) or cortical lesions in areas with sparse callosal connections may be attributable to different mechanisms.

Acknowledgements. We wish to thank Giovanni Berlucchi, Edoardo Bisiach, Michael Corballis and Tim Shallice for their comments on an earlier draft and Marco Veronese for his skill in the preparation of the figures. This work was supported by Contributo CNR n. 94.02765.CT04 awarded to C.A.M.

References

Aglioti S, Berlucchi G, Pallini R, Rossi GF, Tassinari G (1993) Hemispheric control of unilateral and bilateral responses to lateralized light stimuli after callosotomy and in callosal agenesis. Exp Brain Res 95:151–165

Bender MB (1977) Extinction and other patterns of sensory interaction. In: Weinstein EA, Friedland RP (eds) Advances in neurology, vol 18. Raven, New York, pp 107–110

Benton AL and Levin HS (1972) An experimental study of "obscuration". Neurology 22:1176–1181

Berlucchi G, Mangun GR and Gazzaniga MS (in press a) Visuospatial attention and the split-brain. News in Physiol Sci

Berlucchi G, Aglioti S, Tassinari G (in press b) Rightward attentional bias and left hemisphere dominance in a cue-target light detection task in a callosotomy patient. Neuropsychologia

Bisiach E (1991) Extinction and neglect: same or different? In: Paillard J (ed) Brain and space. Oxford University Press, Oxford, pp 251–257

Corballis MC (1994) Can commissurotomized subjects compare digits between the two visual fields? Neuropsychologia 32:1475–1486

Corballis MC (1995) Line bisection in a man with complete forebrain commissurotomy. Neuropsychology 9:147–156

Corballis MC, Sergent J (1992) Judgements about numerosity by a commissurotomized subject. Neuropsychologia 30:865–876

Corballis MC, Trudel CI (1993) The role of the forebrain commissures in interhemispheric integration. Neuropsychology 7:306–324

Crowne DP, Yeo CH, Steele Russell IS (1981) The effects of unilateral frontal eye field lesion in the monkey: visual motor guidance and avoidance behaviour. Behav Brain Res 2:165–187

Cusick CG and Kaas JH (1986) Interhemispheric connections of cortical sensory and motor representations in primates. In: Lepore F, Ptito M, Jasper HH (eds) Two hemispheres-one brain: functions of the corpus callosum. Liss, New York, pp 83–102

Di Pellegrino G, De Renzi E (1995) An experimental investigation on the nature of extinction. Neuropsychologia 33:153–170

Gazzaniga MS (1987) Perceptual and attentional processes following callosal section in humans. Neuropsychologia 25:119–133

Goldenberg G (1986) Neglect in a patient with partial callosal disconnection. Neuropsychologia 24:397–403

Heilman KM, Watson RT, Valenstein E (1993) Neglect and related disorders. In: Heilman KM and Valenstein E (eds) Clinical neuropsychology. 3rd ed. Oxford University Press, New York, pp 279–336

Henik A, Rafal R, Rhodes D (1994) Endogenously generated and visually guided saccades after lesions of the human frontal eye fields. J. Cogn Neurosci 6:400–411

Iacoboni M, Zaidel E (1995) Channels of the corpus callosum. Evidence from simple reaction times to lateralized flashes in the normal and the split brain. Brain 118:779–788

Joynt RT (1977) Inattention syndromes in split-brain man. In: Weinstein EA, Friedland RP (eds) Advances in neurology, vol 18. Raven, New York, pp 33–39

Kaas JH (1995) The organization of callosal connections in primates. In: Reeves AG, Roberts DW (eds) Epilepsy and the corpus callosum II. Plenum, New York, pp 15–27

Kennedy H, Meissirel C, Dehay C (1991) Callosal pathways and their compliancy to general rules governing the organization of corticocortical connectivity. In: Dreher B, Robinson SR (eds) Neuroanatomy of the visual pathways and their development. Vision and visual dysfunction, vol 3. Macmillan, London, pp 324–359

Kumral E, Kocaer T, Sagduyu A, Sirin H, Togyar A, Evyapan D, Vuilleumier P (1995) Infarctus calleux après occlusion bilatérale des artères carotides internes avec syndrome d' héminégligence et astasie-abasie. Rev Neurol (Paris) 151:202–205

Mangun GR, Luck SJ, Gazzaniga MS, Hillyard SA (1991) Electrophysiological measures of interhemispheric transfer of visual information: studies in split-brain patients. Soc Neurosci Abstr 17:866

Mangun GR, Luck SJ, Plager R, Loftus W, Hillyard SA, Handy T, Clark VP and Gazzaniga MS (1994) Monitoring the Visual World: Hemispheric Asymmetries and Subcortical Processes in Attention. J Cogn Neurosci 6:267–275

Marzi CA (1986) Transfer of visual information after unilateral input to the brain. Brain Cogn. 5:163–173

Marzi CA, Tassinari G, Aglioti S, Lutzemberger L (1986) Spatial summation across the vertical meridian in hemianopics: a test of blindsight. Neuropsychologia 24:749–758

Marzi CA, Bisiacchi P, Nicoletti R (1991) Is interhemispheric transfer of visuomotor information asymmetric? Evidence from a meta-analysis. Neuropsychologia 29:1163–1177

Marzi CA, Smania N, Martini MC, Gambina G, Tomelleri G, Palamara A, Alessandrini F, Prior M (1996) Implicit redundant-targets effect in visual extinction. Neuropsychologia 34:9–22

Mayer E, Koenig O, Panchaud A (1988) Tactual extinction without anomia: evidence of attentional factors in a patient with a partial callosal disconnection. Neuropsychologia 26:851–868

Miller J. (1982) Divided attention: Evidence for coactivation with redundant signals. Cogn Psychol 14:247–279

Milner B, Taylor L, Sperry RW (1968) Lateralized suppression of dichotically presented digits after commissural section in man. Science 161:184–186

Miniussi C, Girelli M, Ipata AE, Smania N, Marzi CA (1995) Impaired interhemispheric transmission in patients with unilateral visual extinction. Eur J Neurosci [Suppl] 8:136

Naikar N, Corballis MC (1996) Perception of apparent motion across the retinal midline following commissurotomy. Neuropsychologia 34:297–309

Plourde G, Sperry RW (1984) Left hemisphere involvement in left spatial neglect from right-sided lesions. Brain 107:95–106

Poffenberger AT (1912) Reaction time to retinal stimulation, with special reference to the time lost through nerve centers. Arch Psychol 23:1–73

Proverbio AM, Zani A, Gazzaniga, MS, Mangun GR. (1994) ERP and RT signs of a rightward bias for spatial orienting in a split-brain patient. Neuroreport 5:2457–2461

Ramachandran VS, Cronin-Golomb A, Myers JJ (1986) Perception of apparent motion by commissurotomy patients. Nature 320:358–359

Reuter-Lorenz PA, Nozawa G, Gazzaniga MS, Hughes HC (1995) Fate of neglected targets: a chronometric analysis of redundant target effects in the bisected brain. J. Exp Psychol Hum Percept Perform 21:211–230

Sergent J (1987) A new look at the human split brain. Brain 110:1375–1392

Sergent J (1990) Furtive incursions into bicameral minds: integrating and coordinating role of subcortical structures. Brain 113:537–568

Smania N, Martini MC, Prior M, Marzi CA (in press) Input and response determinants of visual extinction: A case study. Cortex

Sparks R, Geschwind N (1968) Dichotic listening in man after section of the neocortical commissures. Cortex 4:3–16

Sugishita M, Otomo K, Yamazaki K, Shimizu H, Yoshioka M, Shinohara A (1995) Dichotic listening in patients with partial section of the corpus callosum. Brain 118:417–427

Teng EL, Sperry RW (1974) Interhemispheric rivalry during simultaneous bilateral task presentation in commissurotomized patients. Cortex 10:111–120

Tramo MJ, Bharucha JJ (1991) Musical priming by the right hemisphere post-callosotomy. Neuropsychologia 29: 313–325

Tramo MJ, Baynes K, Fendrich R, Mangun GR, Phelps EA, Reuter-Lorenz PA, Gazzaniga MS (1995) Hemispheric specialization and interhemispheric integration: insights from experiments with commissurotomy patients. In: Reeves AG, Roberts DW (eds) Epilepsy and the corpus callosum II. Plenum, New York, pp 263–295

Vallar G, Rusconi ML, Bignamini L, Geminiani G, Perani D (1994) The anatomical correlates of visual and tactile extinction in humans. A clinical CT-scan study. J Neurol Neurosurg Psychiatry 57:464–470

Watson RT, Valenstein E, Day AL, Heilman KL (1984) The effects of corpus callosum lesions on unilateral neglect in monkeys. Neurology 34:812–815

Action and Perception in the Visual Periphery

M. A. GOODALE and K. J. MURPHY

Department of Psychology, University of Western Ontario, London, Canada

Introduction

The visual identification of objects depends primarily on central vision. We tend to perceive best what we are looking at. Why then does the visual field extend well beyond the few degrees of the fovea and parafoveal regions? Part of the answer is related to the need to detect biologically relevant stimuli across a wide region of visual space[1]. Indeed, humans (and other highly 'visual' vertebrates) typically orient their gaze towards objects and events that suddenly appear in the visual periphery. But this is not the only reason why our visual fields are so large. Many of the movements that humans make through the world require processing of information far into the periphery. Thus, when we walk from one place to another in our immediate environment we use the optic flow that is generated by our movement to control both our heading and our rate of locomotion, and much of this flow occurs in the peripheral visual field (see Lappe, this volume). Discontinuities in that flow also mark the presence of obstacles that we might wish to avoid. Yet typically we are quite unaware that we are using such information. Indeed, as the evidence reviewed in this chapter will show, even when we direct our attention to stimuli in our visual periphery, our ability to describe or discriminate them is not as accurate as one would expect from the way we use those stimuli to control skilled movements of our limbs and body. We just can't see things as well in the visual periphery as we can in the fovea even though, paradoxically, our motor control systems remain remarkably sensitive to visual cues even in the far periphery.

---

[1] Theoretically, it would be possible to have the equivalent of foveal vision over the entire retina. Such an adaptation, however, would require that an enormous amount of the brain be devoted to processing all the objects and events projected on the retina. Eric Schwartz (cited in Aloimonos 1994) has calculated that if this were the case, then humans would have a brain weighing nearly 30.000 pounds.

In: Parietal Lobe Contributions to Orientation in 3D Space (1997). P. Thier and H.-O. Karnath (eds). Springer-Verlag, Heidelberg.

Perception in the Visual Periphery

The fact that visual perception is poorer in the periphery than in the fovea has been related to a number of different factors. At the level of the retina, the density of photoreceptors and ganglion cells has been shown to decrease as a function of eccentricity (Curcio and Allen 1990; Curcio et al. 1987; Østerberg 1935; Rolls and Cowey 1970). The average human retina, for example, contains approximately 1.07 million ganglion cells and roughly half of these are located within 16° of the foveal center, an area that encompasses less than 8% of the total retina (Curcio and Allen 1990). In addition to the increased spacing between photoreceptors in the periphery, there are irregularities in their position (Hirsch and Curcio 1989; Hirsch and Miller 1987; Yellot 1982, 1984). Moreover, the increased convergence of photoreceptors on ganglion cells in the periphery (e.g., Perry and Cowey 1985; Wässle et al. 1989, 1990) is followed by more convergence of peripheral ganglion cells on cells in the dorsal lateral geniculate nucleus (Connolly and Van Essen 1984), and again by convergence of cells in the dorsal lateral geniculate nucleus on cells in primary visual cortex (Daniel and Whitteridge 1961; Dow et al. 1981; Van Essen et al. 1984). Cortical magnification, the fact that proportionally more primary visual cortex is devoted to the fovea than to the peripheral retina, is thought to be directly related to the decline in ganglion cell density from central to peripheral retina and to the corresponding increase in convergence of peripheral information between retina and cortex (Azzopardi and Cowey 1993; Curcio and Allen 1990; Schein and De Monasterio 1987). It has been suggested therefore that the drop in visual sensitivity as viewing shifts from the fovea to the far periphery is a simple consequence of the reduction in sampling of the image (e.g., Van Doorn et al. 1972).

There is evidence, however, that the decline in visual sensitivity in the periphery cannot be entirely explained in terms of undersampling of the retinal image or peripheral irregularities. While it is true that some psychophysical functions, such as grating acuity at different eccentricities scale in a manner consistent with cortical magnification (e.g., Wertheim 1894; Cowey and Rolls 1974), other functions, such as stereo acuity (Fendick and Westheimer 1983) and numerosity judgments (Parth and Rentschler 1984), do not. Pattern recognition is particularly poor – much poorer than one would expect on the basis of cortical magnification alone (Strasburger et al. 1994).

How then can one explain the sharp decline in psychophysical performance with peripheral viewing? A number of factors are probably at work. For one thing, the two prominent input channels to primary visual cortex, the so-called magno and parvo channels, may differ in their representation of the visual field, with the parvo system declining much more rapidly with eccentricity than the magno system (Connolly and Van Essen 1984; Schein and De Monasterio 1987). This difference is apparent first at the level of the retina where the density of midget ganglion cells, which give rise to the parvo pathway, declines much more rapidly between the fovea and the far periphery than the density of parasol ganglion cells, which give rise to the magno pathway (De Monasterio and Gouras 1975; Dacey and Peterson 1992). At the level of V1, the afferent density of parvo cells has been found to increase only slightly with eccentricity while the afferent density of magno cells rises steeply (Schein and De Monasterio 1987). As a consequence, the number of parvo cells per point-image area declines with eccentricity, while the number of magno cells per point-image area remains quite constant. Although not all studies have found such large differences between the two channels as a function of eccentricity (e.g., Livingstone and Hubel 1988; Perry et al.1984), it is still possible that cortical magnification may be more of a parvo than a magno channel phenomenon. As a consequence, those psychophysical functions which are mediated largely by parvo-dependent mechanisms might be expected to degrade much more with eccentric viewing than those mediated by magno-dependent mechanisms.

Perception Versus Action in the Visual Periphery

There is another explanation for the decline in performance in the visual periphery, however, that is related not so much to differences in the distribution of the various input channels as to differences in the nature of the response that is demanded of the subject. In the typical experiment, conventional psychophysical testing is carried out and subjects are required to indicate whether or not they can discriminate between different visual displays in which the stimulus parameters are systematically manipulated. In short, subjects are asked what they can *see* in their visual periphery. But as we suggested earlier, what people report seeing and what actually happens to their behavior in the presence of particular visual stimuli do not necessarily coincide. In many situations, stimuli in the peripheral visual field can exert considerable control over the movements that people make even though they remain quite unaware of the controlling stimuli. Putting it more concretely, the neural mechanisms subserving the visual control of actions may have access to visual information from the peripheral fields that is not shared by the neural mechanisms mediating the experiential perception of objects and events in the visual world. Support for this idea comes from studies that have examined the pattern of inputs from the peripheral and central visual fields to different regions of extrastriate visual cortex. In the section below, we briefly review some of this evidence.

The caudal half of the monkey's cerebral cortex contains a complex mosaic of interconnected visual areas (for a review see Zeki 1993). Despite the complexities of the interconnections between these cortical areas, Ungerleider and Mishkin (1982) were able to identify two broad "streams" of projections arising from V1 and projecting to higher visual areas: a ventral stream arising from V1 and projecting to inferotemporal cortex, and a dorsal stream also arising from V1 but projecting instead to the posterior parietal cortex. Although one must always be cautious when drawing homologies from monkey to human neuroanatomy (Crick and Jones 1993), it seems likely that the visual projections from primary visual cortex to the temporal and parietal lobes in the human brain may involve a separation into dorsal and ventral streams similar to that seen in the monkey.

Ungerleider and Mishkin (1982) originally proposed that the ventral stream plays a special role in the identification of objects, whereas the dorsal stream is responsible for localizing objects in visual space. In a recent re-interpretation of the functions of the two streams, Goodale and Milner (1992; Milner and Goodale 1995) place less emphasis on the differences in the visual information that is received by the two streams (object features versus spatial location) than they do on the differences in the transformations that the streams perform upon that information. According to Goodale and Milner's account, both streams process information about object features and about their spatial relations, but each stream uses this visual information in different ways. In the ventral stream, the transformations focus on the enduring characteristics of objects and their relations, permitting the formation of long-term perceptual representations. Such representations play an essential role in the identification of objects and enable us to classify objects and events, attach meaning and significance to them, and establish their causal relations. Such operations are essential for accumulating a knowledge-base about the world. In contrast, the transformations carried out by the dorsal stream deal with the moment-by-moment information about the location and spatial disposition of objects in particular egocentric coordinates and thereby mediate the visual control of skilled actions, such as manual prehension, directed at those objects. As such, the dorsal stream can be regarded as a cortical extension of the dedicated visuomotor modules that mediate visually guided movements in all vertebrates. To summarize the Goodale and Milner proposal: the perceptual representations constructed by the ventral stream interact with various high-level cognitive mechanisms and enable an organism to select a particular course of action with respect to objects in the world, while the visuomotor networks in the dorsal stream (and associated cortical and subcortical pathways) are responsible for the programming and on-line control of the particular movements that action entails.

What is most interesting in the context of the arguments about sensitivity to stimuli in the visual periphery is the fact that the patterns of inputs from the peripheral and central visual fields are quite different in the two streams. Many areas in the posterior parietal cortex have an extensive representation of the visual periphery (Baizer et al. 1991; Colby et al. 1988; Motter and Mountcastle 1981; Motter et al. 1987). Area PO, for example, which is located in the parietooccipital sulcus and sends projections to a number of areas in posterior parietal cortex, appears to be largely dominated by the visual periphery; unlike area V1, there is no evidence in area PO of cortical magnification of foveal vision (Colby et al. 1988; Gattass et al. 1985). It is not surprising therefore that projections to PO and other dorsal stream areas have been found to arise largely from those portions of areas V2 and V3 that subserve the visual periphery (Baizer et al. 1991). A very different pattern of projections can be seen in the ventral stream. Projections from areas V2, V3, and V4 to inferotemporal cortex tend to arise from regions in these areas that represent the central visual fields (Baizer et al. 1991). Moreover, neurons in area V4 and inferotemporal cortex have receptive fields that are restricted largely to central vision, particularly the fovea (Gross et al. 1972; Desimone et al. 1984). In summary, the dorsal stream has an extensive representation of the entire visual field, including the far periphery, while the ventral stream is dominated by central vision.

The difference in the representation of the visual field in the dorsal and ventral streams conforms rather well to the proposed division of labor between the two streams put forward by Goodale and Milner (1992; Milner and Goodale 1995). The ventral stream's concentration on central vision is exactly what one might expect to see in a system devoted to constructing our percepts of the world and the objects within it. The extensive representation of the peripheral visual fields in the dorsal stream is exactly what is needed for the efficient visual control of many skilled actions. The emphasis on different regions of the visual field in the two streams may also help to explain the differential representation of the parvo and magno inputs across the retinal representation in V1. As was discussed in the preceding section, cortical magnification in V1 may be more related to parvo than to magno mechanisms (Schein and De Monasterio 1987). Although the ventral stream receives roughly the same number of inputs from the parvo and magno channels (Ferrera et al. 1992), the dorsal stream is dominated by the magno channel with only a few inputs coming from the parvo channel (e.g., Maunsell et al. 1990). Thus, the over-representation of central parvo inputs at the level of V1 may reflect the fact that the information coded by these cells may be conveyed largely to the ventral rather than the dorsal stream.

We can now see why conventional psychophysical assessment of sensitivity to stimuli in the visual periphery might miss much of the visual processing that occurs in this region of the visual field. Verbal reports of what one can see would tap only ventral stream mechanisms. The accuracy of such reports would of necessity fall off rapidly as the stimuli were presented more and more peripherally. But if subjects were tested in a variety of visuomotor tasks, then a rather different story might emerge. As the controlling stimuli are moved into the visual periphery, dorsal stream mechanisms would continue to be engaged and the subjects' motor performance might not deteriorate nearly as rapidly as one might expect on the basis of their verbal report of what they see. It was this possibility that motivated us to compare visuomotor performance and verbal report to stimuli presented in the visual periphery.

Grasping Without Perceiving in the Visual Periphery

Humans are often able to grasp objects quite well that they are not looking at directly. It is possible that the dorsal stream mechanisms that mediate the programming and on-line control of such movements can utilize information about the size and shape of the goal object to calibrate the grasp even though experiential perception of these object dimensions might be rather poor.

To test this idea, we carried out an experiment in which we looked at the ability of human subjects to use visual information about the dimensions of five different rectangular blocks which were presented for 100 ms at various positions in the peripheral visual field along the horizontal meridian from 5° to 70°. Although the

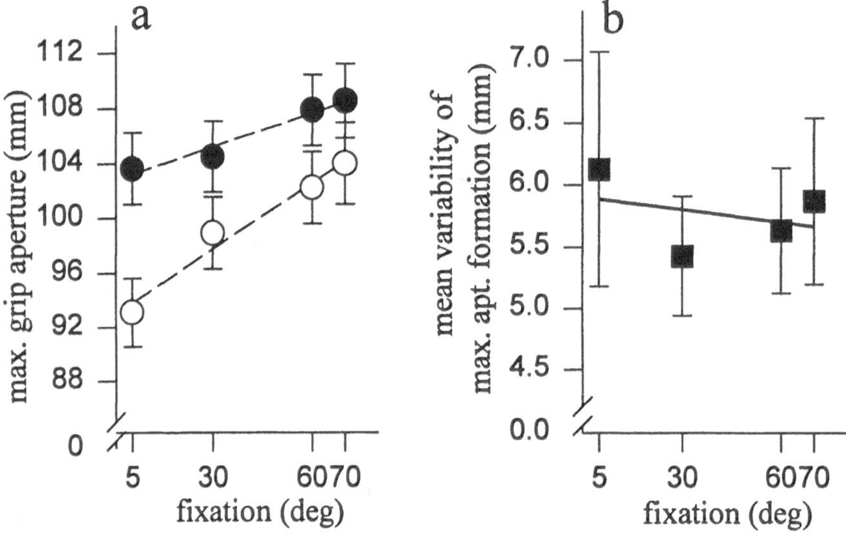

Fig. 1a, b. The effect of retinal eccentricity on the scaling of the grasp in 14 right-handed students (seven women; seven men; aged 22-29). **a** Grip scaling to two of the five different sized objects that were presented at different retinal eccentricities. *Filled circles* are responses to object 4. *Open circles* are responses to object 2. Sensitivity to object width was measured as the maximum opening between thumb and forefinger during the grasping movement using standard opto-electronic recording with infrared-light-emitting diodes (WATSMART, Northern Digital Inc., Waterloo, Canada). Maximum grip aperture increased with retinal eccentricity ($p<.001$) but remained scaled to object width at each peripheral view ($p<.001$). The magnitude of the increase in grip aperture, as a function of peripheral viewing angle, was greater for object 2 compared to object 4 ($p<.001$). **b** Variability of grip formation as a function of retinal eccentricity. There was no significant change in the variability of grip scaling performance with retinal eccentricity. *Error bars* indicate standard error of the mean

dimensions of the blocks varied, their overall surface area did not. Sensitivity to the width of these objects was measured under two response conditions: a *visuomotor* condition in which subjects were required to reach out and grasp the object, and a *visuoperceptual* condition in which subjects were required to categorize the different objects on the basis of their width.

In the visuomotor task, which was carried out first, we measured the maximum aperture between the thumb and forefinger as subjects reached out to pick up one of the five blocks. The change in eccentric viewing was accomplished by having the subjects direct their gaze increasingly leftward at fixation points located at different distances from the target object along the same horizontal plane. Thus, although the retinal location of the target block varied from trial to trial, the location of the block with respect to the subject's hand and body remained constant (centered at the body midline at a viewing distance of about 50 cm). On each trial, the subject began a grasp with the right hand placed at the body midline 30 cm from the object.

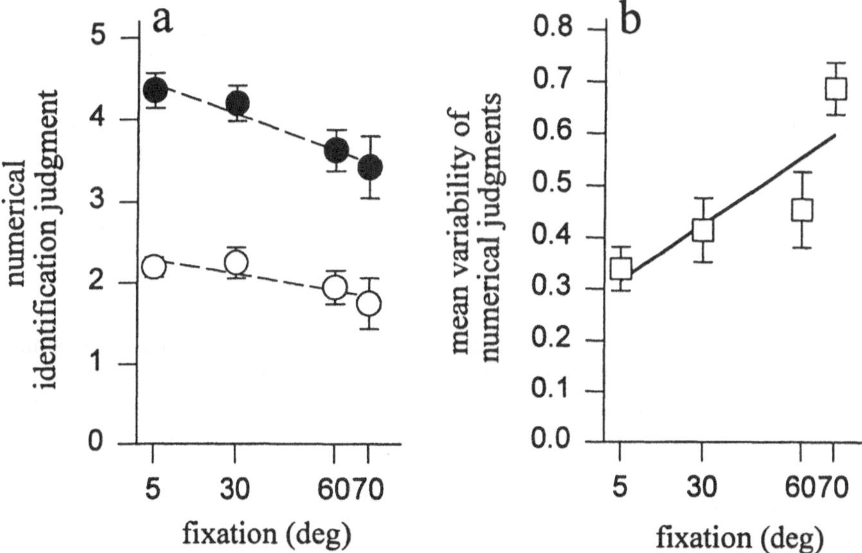

Fig. 2a, b. The effect of retinal eccentricity on visuoperceptual identification of object width using a numerical label in the same 14 subjects described in Fig.1. **a** Visuoperceptual judgments of object width for two of the same five objects viewed at different retinal eccentricities. *Filled circles* are responses to object 4. *Open circles* are responses to object 2. Subjects were asked to assign a numerical label from 1 (smallest width) to 5 (largest width). Although they underestimated object width as retinal eccentricity increased ($p<.001$), verbal judgments remained correlated with the target width at all eccentricities ($p<.001$). **b** Variability of visuoperceptual responses. In contrast to the visuomotor results, the variability in assigning numerical labels increased with retinal eccentricity ($p<.001$). *Error bars* indicate standard error of the mean

Previous work in our laboratory and others has shown that maximum grip aperture, which is achieved well before contact, is highly correlated with object width when the objects are presented in central vision (Jakobson and Goodale 1991; Jeannerod 1988). Subjects in our experiment behaved similarly. In fact, as Fig. 1a shows, the correlation between maximum aperture and object width was maintained even when the object was presented in the far periphery of the visual field. Moreover, there was no change in the variability of the grasp calibration as viewing angle increased (Fig. 1b). What did change was the amplitude of the grasp, which increased as a function of retinal eccentricity (Fig. 1a). Does this mean that subjects saw the objects as being larger when they were presented at more and more eccentric views? Research in visual psychophysics, dating from the pioneering work of Helmholtz, would suggest exactly the opposite: objects are typically reported to be diminished in size when viewed in the visual periphery (e.g., Helmholtz 1962; Collier 1931; James 1890; Newsome 1972; Schneider et al. 1978). It has been suggested that peripherally viewed objects often appear smaller because of optical effects; that is, because the peripheral retina is closer to the nodal point of the lens than the fovea, the same object will subtend a smaller angle on the peripheral retina than the fovea. Spatial "undersampling" of the sort described earlier could also be at work. The reduction in size with peripheral viewing could also arise from the fact that the ventral stream, which mediates perceptual report, is largely dominated by central vision. But whatever the explanation(s) might be, the phenomenon appears to be quite robust and we too obtained a similar result in our visuoperceptual task.

In our visuoperceptual task, subjects were first trained to categorize the five different blocks according to their dimensions from 1 (smallest width) to 5 (largest width). Their view of the objects was quite unrestricted at this stage and they could explore them in central vision. Following this training, they were presented with each of the five blocks in random order in different positions in the visual periphery. The same 100 ms viewing time that had been used in the visuomotor task was also used here. Consistent with earlier studies, we found that subjects progressively underestimated object width as retinal eccentricity increased (Fig. 2a). Moreover, as Fig. 2a,b shows, even though the subjects continued to discriminate between blocks in the peripheral visual field, the variability of their numerical category judgments increased significantly the further away from central vision the object was viewed. Thus, their visuoperceptual performance differed from their visuomotor performance in two important respects. First, perceptual estimates of object width decreased with eccentricity while grip aperture increased. Second, the variability of the perceptual estimates also increased with eccentricity while the variability in grip aperture did not change.

The question remains, of course, as to why grip aperture increased as a function of retinal eccentricity. As we have just seen, this effect could not have arisen from

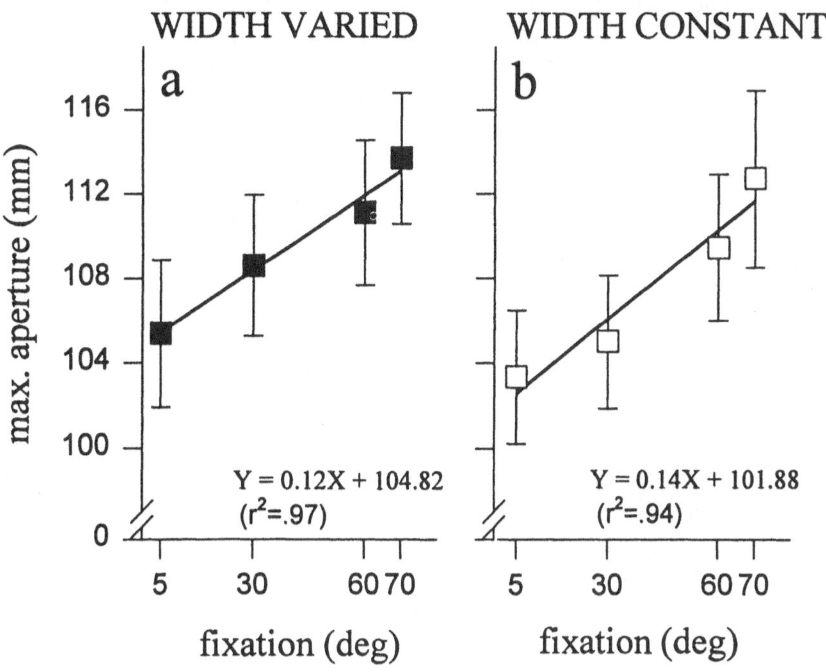

Fig. 3a,b. The effect of retinal eccentricity on the scaling of the grasp in eight right-handed students (five men; three women; aged 18–22) under two conditions. **a** Object-varied condition. The same set of five objects used earlier was presented. Grasps to only two of the five objects were analyzed. Each point represents the maximum aperture of both objects combined. Consistent with the results of the experiment shown in Fig. 1a, grip aperture increased with retinal eccentricity ($p<.001$) when the object dimensions were varied from trial to trial. **b** Object-constant condition. The same two objects that were selected for analysis in the object-varied condition were presented separately in two blocks of trials. Thus, in any one block of trials, the same object was presented over and over again. Moreover, subjects were given some foveal experience with each target object before beginning the run of repeated trials. Even though subjects were picking up the same object from trial to trial, they continued to open their hand wider as retinal eccentricity increased ($p<.001$). There was no effect of retinal eccentricity on the variability of grasp formation in either the object-varied or the object-constant conditions. Moreover, as can be seen from the *graphs*, there was no difference in either grasp aperture or variability between the two conditions. Grip aperture was well-scaled to the width of the two objects in both the object-varied (object 4, M=114.07, SE=3.4; object 2, M=105.3, SE 3.3; $p<.001$) and the object-constant condition (object 4, M=113.44, SE=3.9; object 2, M=101.83, SE=3.4; $p<.001$) and there was no difference in the scaling between the two conditions. In all cases, *error bars* indicate standard error of the mean

a systematic change in the subjects' perception with peripheral viewing, since subjects, rather than perceiving the blocks as larger than they really were, perceived them as smaller. Although it is true that there was an increase in the variability of their perceptual judgments with eccentricity, this uncertainty about object width cannot explain the increase in grip aperture either, since grip aperture

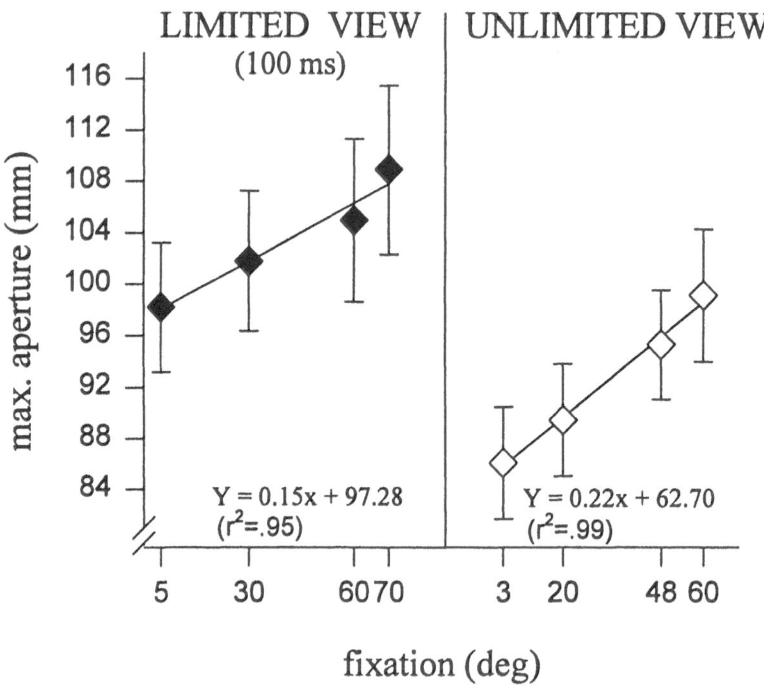

Fig. 4. The effect of retinal eccentricity on grip aperture under limited (100 ms; open loop) and unlimited (closed loop) viewing conditions in eight right-handed students (four women; four men; aged 18-22). The same target object was used throughout both conditions. Grip aperture increased with retinal eccentricity under both limited ($p<.001$) and unlimited viewing conditions ($p<.001$). There was no significant difference in the variability of grasp formation between the two conditions. Consistent with previous research (e.g., Jakobson and Goodale 1991), however, grip scaling was smaller and more tightly coupled to object width under the unlimited view condition ($p<.001$). The latter result can be attributed to the fact that under unlimited viewing, visual information can be used 'on-line' to fine tune the precision of the grasp during the closing phase of the reaching movement. *Error bars* indicate standard error of the mean. (The peripheral viewing angles were different in closed loop reaching because a chin rest was used to permit more sophisticated monitoring of eye movements in this condition.)

was no more variable in the far periphery than it was in the near periphery. In fact, in a second experiment, we demonstrated that previous perception-based knowledge of the dimensions of a goal object exerts little influence over the calibration of the grasp when reaching to that object in the peripheral visual field. Using the same experimental set-up employed in the original visuomotor task, we compared grip aperture under two conditions: in one condition, both the object dimensions and retinal eccentricity varied (as in the first experiment); in the other condition, retinal eccentricity varied but the object dimensions remained constant. Thus, in the latter condition, the same object was presented on every trial and the same grip aperture was required on every trial. As Fig. 3a,b shows, the slope of the function describing the relationship between grip aperture and retinal eccentricity was virtually identical under both 'object varied' and 'object constant' conditions. Grip aperture increased with eccentricity in the same way, independent of whether or not the size of the block was predictable. In addition, the variability in grip aperture did not change as a function of retinal eccentricity in either condition and there was no difference in the degree of variability between the two conditions. Finally, in a third experiment we demonstrated that the effect of retinal eccentricity on grip aperture could not be attributed to the limited (100 ms) view subjects were permitted. When subjects were allowed an unlimited view of the goal object (and their reaching hand) throughout the grasp, they continued to open their hand wider as retinal eccentricity increased – even when the same object was presented trial after trial (Fig. 4).

The results of the last two experiments undercut any argument that perceptual judgments of object dimensions were driving the formation of the grasp. Grip formation was virtually identical under the 'object-varied' and 'constant' conditions. In addition, grip aperture also changed with eccentricity under unlimited viewing conditions when object dimensions were held constant. These findings support the earlier suggestion that subjects do not use perceptual knowledge of object dimensions to program their grasping movements. These results are also consistent with other work in our laboratory showing that the visuomotor system relies very little on the stored representations of a familiar goal object and instead programs each grasping movement de novo according to the visual information currently available (Goodale et al. 1994). A final question remains to be answered: why did subjects increase their grip aperture with increased eccentricity of viewing? We have seen that, even when subjects are highly familiar with the goal object, they still open their hand wider the more eccentric the view. Perhaps the wider opening is an obligatory strategy that the visuomotor system employs to handle the final approach phase of the grasp when information is less than optimal. Certainly there is a good deal of evidence to suggest that the hand opens wider when subjects reach in visual 'open loop', in which they have no opportunity to monitor the relative position of their hand and the goal object during the closing phase (Jakobson and Goodale 1991; Jeannerod 1988). As Figs. 3b and 4 show, this strategy is employed even when subjects 'know' the dimensions of the objects they are reaching to. Thus, a similar strategy might be employed for reaching to objects in the visual periphery where some reduction in information resolution occurs[2].

In our study, of course, the difference in width between different target objects was quite large – and thus, even in the visuoperceptual task, subjects are still able to see a difference in the dimensions of the targets from trial to trial. Their performance, however, was still more reliable in the visuomotor task (and the functions describing the relationship between the response and retinal eccentricity in the two tasks were quite different). Nevertheless, in future experiments, it might be useful to test the resolution of both systems by using objects the dimensions were much more similar. One might expect that the discrimination thresholds for grip calibration might be more sensitive in the far periphery than the discrimination thresholds for perceptual judgments.

Conclusion

The apparent dissociation between the effect of eccentric viewing on the visual control of grasping and its effect on visual identification is consistent with the idea that the neural mechanisms subserving visuomotor control are quite separate from those subserving our perception of objects and events in the world (Goodale and Milner 1992; Milner and Goodale 1995). Moreover, as we reviewed earlier, the dorsal stream, which is thought to play a critical role in the visual control of skilled actions like prehension, receives extensive inputs from the peripheral visual fields. Inputs to the ventral stream, which is presumed to play a major role in object perception, originate largely from the central visual field. As we discussed earlier, this striking difference in the way in which the visual field is represented in the two streams is exactly what one might expect to see in systems that play such different roles in the visual life of the organism. Indeed, it is likely that this difference in representation might explain why the visual control of grasping movements directed at targets in the peripheral visual field is so much more reliable than perceptual judgments about those same objects. Finally, the difference in visual field representation in the two streams might also explain the puzzling observation that many psychophysical functions do not scale appropriately with either ganglion cell density or cortical magnification in primary visual cortex.

---

[2] As Fig. 1a shows, peripheral viewing had a relatively greater effect on the amplitude of grasps directed to the smaller of the two objects. This difference probably reflects the fact that grip aperture was approaching a maximum in the case of grasps directed to the large object in the far periphery. Clearly, an upper limit would have to be reached as object size increases and viewing angle becomes more eccentric. Nevertheless, these large grasps showed the same inter-trial variability as smaller grasps; in fact, variability did not vary with either viewing angle or object size.

Acknowledgements. This research was supported by a grant from the Medical Research Council of Canada to M.A. Goodale. K.J. Murphy was a recipient of a postgraduate studentship from the Natural Sciences and Engineering Research Council of Canada.

References

Aloimonos Y (1994) What I have learned. Computer Vision and Image Understanding 60:74–85

Azzopardi P, Cowey A (1993) Preferential representation of the fovea in the primary visual cortex. Nature 361:719–720

Baizer JS, Ungerleider LG, Desimone R (1991) Organization of visual inputs to the inferior temporal and posterior parietal cortex in macaques. J Neurosci 11:168–190

Colby CL, Gattass R, Olson CR, Gross CG (1988) Topographical organization of cortical afferents to extrastriate visual area PO in the macaque: a dual tracer study. J Comp Neurol 269:392–413

Collier RM (1931) An experimental study of form perception in peripheral vision. J Comp Psychol 1:281–289

Connolly M, Van Essen D (1984) The representation of the visual field in parvicellular and magnocellular layers of the lateral geniculate nucleus in the macaque monkey. J Comp Neurol 226:544–564

Crick F, Jones E (1993) Backwardness of human neuroanatomy. Nature 361:109–110

Cowey A, Rolls E T (1974) Human cortical magnification factor and its relation to visual acuity. Exp Brain Res 21:447–454

Curcio CA, Allen KA (1990) Topography of ganglion cells in human retina. J Comp Neurol 300:5–25

Curcio CA, Sloan KR, Packer O, Hendrickson AE, Kalina RE (1987) Distribution of cones in human and monkey retina: individual variability and radial asymmetry. Science 236:579–582

Dacey DM, Petersen MR (1992) Dendritic field size and morphology of midget and parasol ganglion cells of the human retina. Proc Natl Acad Sci USA 89:9666–9670

Daniel PM, Whitteridge D (1961) The representation of the visual field on the cerebral cortex in monkeys. J Physiol 159:203–221

De Monasterio FM, Gouras P (1975) Functional properties of ganglion cells of the rhesus monkey retina. J Physiol 251:167–196

Desimone R, Albright TD, Gross CG, Bruce C (1984) Stimulus selective properties of inferior temporal neurons in the macaque. J Neurosci 4:2051–2062

Dow BM, Synder AZ, Vautin RG, Bauer R (1981) Magnification factor and receptive field size in foveal striate cortex of the monkey. Exp Brain Res 44:213–228

Fendick M, Westheimer G (1983) Effects of practice and the separation of test targets on foveal and peripheral stereoacuity. Vision Res 23:145–150

Ferrera VP, Nealey TA, Maunsell JHR (1992) Mixed parvocellular and magnocellular geniculate signals in visual area V4. Nature 358:756–758

Gattass R, Sousa APB, Covey E (1985) Cortical visual areas of the macaque: possible substrates for pattern recognition mechanisms. In: Chagas C, Gattass R, Gross C (eds) Pattern recognition mechanisms. Springer-Verlag, New York, pp 1–20

Goodale MA, Milner AD (1992) Separate visual pathways for perception and action. Trends Neurosci 15:20–25

Goodale MA, Jakobson LS, Keillor J (1994) Differences in the visual control of pantomimed and natural grasping movements. Neuropsychologia 32:1159–1178

Gross CG, Rocha-Miranda CE, Bender DB (1972) Visual properties of neurons in inferotemporal cortex of the macaque. J Neurophysiol 35:96–111

Helmholtz H von (1962) Helmholtz's treatise on physiological optics, vol III. (translated from the 3rd German edition) Dover, New York

Hirsch J, Curcio CA (1989) The spatial resolution capacity of human foveal retina. Vis Res 29:1095–1101

Hirsch J, Miller WH (1987) Does cone positional disorder limit resolution? J Opt Soc Am A 4:1481–1492

Jakobson LS, Goodale MA (1991) Factors affecting higher order movement. Exp Brian Res 86:199–208

James W (1890) Principles of psychology, vol 2. Macmillan, London

Jeannerod M (1988) The neural and behavioral organization of goal-directed movements. Oxford University Press, Oxford

Livingstone MS, Hubel DH (1988) Do the relative mapping densities of the magno- and parvocellular systems vary with eccentricity? J Neurosci 8:4334–4339

Maunsell JHR, Nealey TA, DePriest DD (1990) Magnocellular and pavocellular contributions to responses in the middle temporal visua area (MT). J Neurosci 10:3323–3334

Milner AD, Goodale MA (1995) The visual brain in action. Oxford University Press, Oxford

Motter BC, Mountcastle VB (1981) The functional properties of light-sensitive neurons of the posterior parietal cortex studied in waking monkeys: foveal sparing and opponent vector organization. J Neurosci 1:3–26

Motter BC, Steinmetz MA, Duffy CJ, Mountcastle VB (1987) Functional properties of parietal visual neurons: mechanisms of directionality along a single axis. J Neurosci 7:154–176

Newsome LR (1972) Visual angle and apparent size of objects in peripheral vision. Percept Psychophys 12:300–304

Østerberg G (1935) Topography of the layer of rods and cones in the human retina. Acta Ophthalmol Suppl 6:1–102

Parth P, Rentschler I (1984) Numerosity judgments in peripheral vision: limitations of the cortical magnification hypothesis. Behav Brain Res 11:241–248

Perry VH, Cowey A (1985) The ganglion cell and cone distributions in the monkey's retina: implications for central magnification factors. Vision Res 25:1795–1810

Perry VH, Oehler R, Cowey A (1984) Retinal ganglion cells that project to the dorsal lateral geniculate nucleus in the macaque monkey. Neuroscience 12:1101–1123

Rolls ET, Cowey A (1970) Topography of the retina and striate cortex and its relationship to visual acuity in rhesus monkeys and squirrel monkeys. Exp Brain Res 10:298–310

Schein SJ, De Monasterio FM (1987) Mapping of retinal and geniculate neurons onto striate cortex of macaque. J Neurosci 7:996–1009

Schneider B, Ehrlich DJ, Stein R, Flaum M, Mangel, S (1978) Changes in the apparent lengths of lines as a function of degree of retinal eccentricity. Perception 7:215–223.

Strasburger H, Rentschler I, Lewis O, Harvey L O Jr (1994) Cortical magnification theory fails to predict visual recognition. Eur J Neurosci 6:1583–1588

Ungerleider LG, Mishkin M (1982) Two cortical visual systems. In: Ingle DJ, Goodale MA, Mansfield RJW (eds) Analysis of visual behaviour. MIT Press, Cambridge, pp 549–586

Van Doorn AJ, Koenderink JJ, Bouman MA (1972) The influence of retinal inhomogeneity on the perception of spatial patterns. Kybernetik 10:223–230

Van Essen DC, Newsome WT, Maunsell JHR (1984) The visual field representation in striate cortex of the macaque monkey: asymmetries, anisotropies and individual variability. Vision Res 24:429–448

Wässle H, Grünert U, Röhrenbeck J, Boycott BB (1989) Cortical magnification factor and the ganglion cell density in the primate retina. Nature 341:643–646

Wässle H, Grünert U, Röhrenbeck J, Boycott BB (1990) Retinal ganglion cell density and cortical magnification factor in the primate. Vision Res 11:1897–1911

Wertheim, T (1894) Über die indirekte Sehschärfe. Z Psychol Physiol Sinnesorg 7:172–187

Yellot JI (1982) Spectral analysis of spatial sampling by photoreceptors: topological disorder prevents aliasing. Vision Res 22:1205–1210

Yellot JI (1984) Image sampling properties of photoreceptors: a reply to Miller and Bernard. Vision Res 24:281–282

Zeki S (1993) A vision of the brain. Blackwell, Oxford.

Representation of Space

The Spatial Features of Unilateral Neglect

E. BISIACH

Dipartimento di Psicologia, Università di Torino, Torino, Italy

Map me no maps, sir, my head is a map, a map of the whole world.
– Henry Fielding: Rape upon Rape

... the unitary space of common sense is a construction, though not a deliberate one. It is part of the business of psychology to make us aware of the steps in this construction.
– Bertrand Russell: Human Knowledge. Its Scope and Limits

The Spatial Frames of Unilateral Neglect

Figure 1 shows the performance of a left-neglect patient on Albert's (1973) line cancellation task. On first consideration, to state that all *left* targets are ignored, while no target is missed on the *right* side may seem to be an adequate description of the patient's performance. That description, however, would miss the point as to the exact reference, in the present case, of the terms 'left' and 'right'. Indeed, regardless of whether it is relatively continuous, or (as *prima facie* suggested by Fig. 1) characterized by a sharp change between two adjacent regions (see Section 4), the directional gradient of neglect has to be defined with respect to a *system* of frames of reference with egocentric and allocentric components.

Egocentric Components

The articulate organism to which the term 'egocentric space' is referred can not be treated as a single central point or plane in a coordinate system. It is therefore amazing that the implications of this obvious fact were realized so late in the study of unilateral neglect. With reference to which organismic axes or planes (or

In: Parietal Lobe Contributions to Orientation in 3D Space (1997). P. Thier and H.-O. Karnath (eds). Springer-Verlag, Heidelberg.

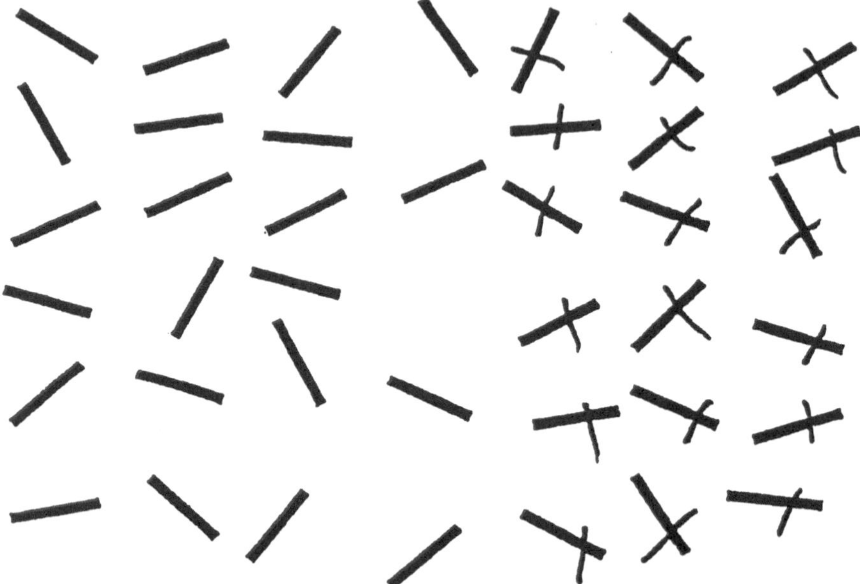

Fig. 1. Performance of a left-neglect patient on a line cancellation task

combinations of axes and planes) does the disorder manifest itself? Consider a task in which a neglect patient has to pick up, one after another, pegs from a pegboard hidden from sight (Fig. 2). Would the boundary between the area from which the patient collects pegs and the area from which he or she fails to collect them be related to any of the following: (a) the line of sight over the hidden pegboard; (b) the sagittal midplane of the head; (c) the sagittal midplane of the trunk[1]; or (d) the radial axis relative to the limb outstretched over the pegboard? (In usual testing conditions these axes and planes are in register, but imagine, for example, the case of a patient looking straight ahead while collecting pegs from a pegboard placed 60° to the right.)

The first step in trying to unravel this problem was undertaken by Heilman and Valenstein (1979). They asked right brain-damaged patients with left visual and somatosensory extinction on double simultaneous stimulation to bisect horizontal lines placed in three different locations with respect to the sagittal midplane of their trunk: to the left, in the middle, and to the right. Rightward bisection errors were large with lines to the left, small with lines to the right and intermediate with lines in the middle. Similar results were later obtained (e.g., Butter et al. 1988; Nichelli et al. 1989; Tegnér et al. 1990). Other investigators failed to demonstrate a significant effect of line location (Riddoch and Humphreys 1983; Ishiai et al.

[1] For the sake of simplicity, I will treat thorax and pelvis as a single rigid part of the body.

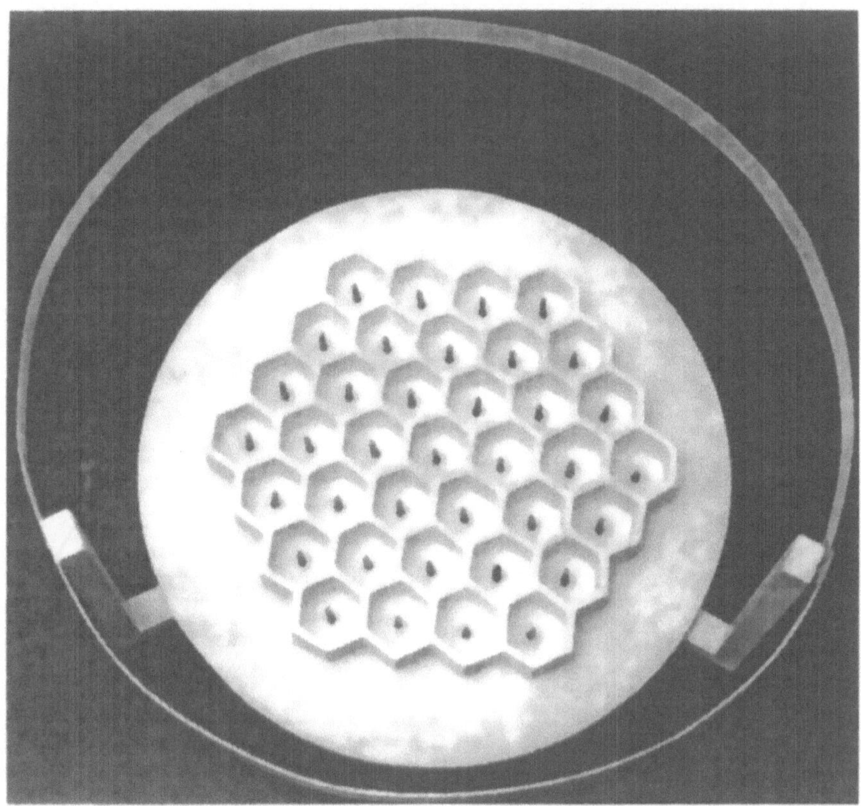

Fig. 2. Apparatus for the investigation of tactile exploration. (From Bisiach et al. 1985)

1989; Reuter-Lorenz and Posner 1990), although a trend in agreement with Heilman and Valenstein's findings was evident in the only paper in which the results were reported *in extenso* (Riddoch and Humphreys 1983).

Karnath et al. (1991) investigated, in four right brain-damaged patients with left neglect, the effects of rotation of head or trunk around their vertical axes upon the RTs of saccadic eye-movements to targets 7° to the left or right of a central fixation point. They found reaction times (RTs) to left targets to be significantly longer in the condition in which both head and trunk faced the VDU screen than in the condition in which the trunk alone was rotated leftwards by 15°. Since the retinal projection of visual stimuli did not change from the first to the second condition, the difference in RTs seems to be due to the fact that left stimuli appeared to the left of the trunk's sagittal midplane in the first condition but to its right in the second. No significant differences emerged from other comparisons, including the condition in which the head was turned to the left or right while eyes and trunk did not move. Karnath et al. (1993) also found improved detection and

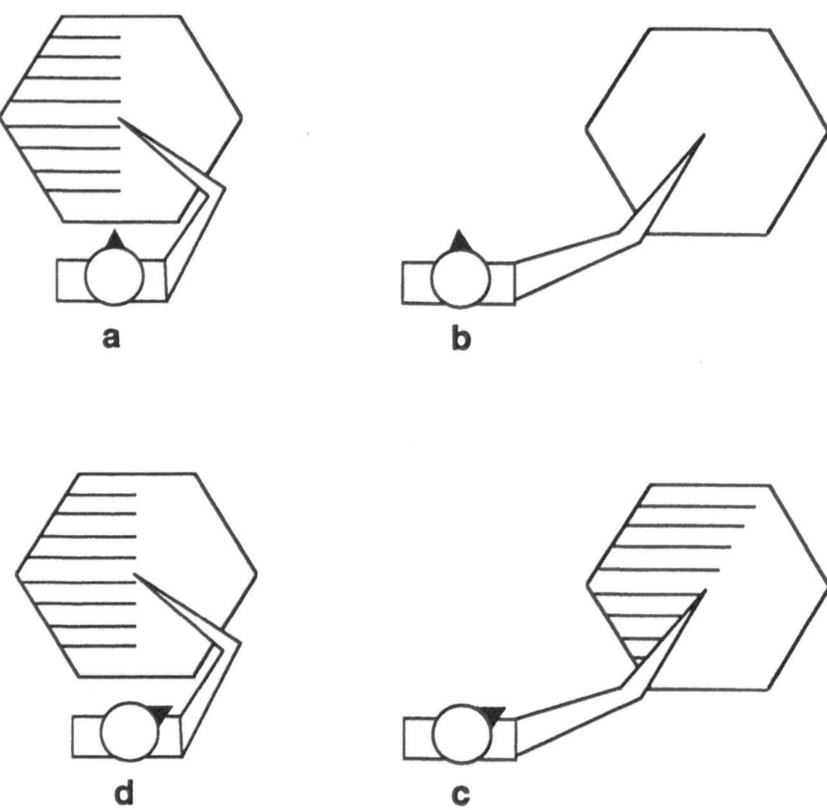

Fig. 3 a-d. Left neglect related to the line of sight and the trunk's sagittal midplane

identification of stimuli in the left visual field of three left-neglect patients who kept facing the screen with eyes and head after leftward trunk rotation. The same effect was obtained through left neck muscle vibration. The opposite manoeuvres did not cause a worsening of neglect, a fact that, as the authors rightly noted, is still in need of clarification. They also noted that the remission of neglect could not be explained only as being due to the real or illusory change of a head-on-trunk signal caused by trunk rotation or neck muscle vibration, respectively. If that were the case, indeed, the same effect would have been obtained by rightward head rotation with eyes and trunk fixed. The authors therefore suggested an interaction between eye-in-head and head-on-trunk signals as responsible for the observed phenomenon. Chokron and Imbert (1995) asked a left neglect patient to point straight ahead while blindfolded; the rightward deviation recorded when the sagittal midplanes of head and trunk coincided was increased or reversed by rotating the trunk, but not the head, rightwards and leftwards, respectively.

Altogether, these data suggest that neglect may be related to a frame centred on the sagittal midplane of the patient's trunk. They, however, leave unanswered the important question as to whether this is the only egocentric spatial frame within which the disorder may manifest itself.

This question was addressed in a study using a tactile exploration task (Bisiach et al. 1985). Fifteen patients with left neglect were asked to extract pegs from holes in which they were inserted, without visual control. The apparatus (Fig. 2) was placed on a table and covered by a cloth. In each of the four conditions of the experiment, patients had to remove the pegs leaving each of them on the bottom of the hexagonal cell to which they it belonged. The starting point was in all conditions the central cell, over which the patients' hands were passively positioned. The apparatus lay in front of the patient (0°) in conditions *a* and *d* (see Fig. 3), whereas it was placed 60° to the right in conditions *b* and *c*; patients were required to look straight ahead in conditions *a* and *b*, but 60° to the right in conditions *c* and *d*. Left- and right-side omissions were recorded in each condition with reference to the diameter of the hexagonal honeycomb structure aligned with the radial axis relative to the patients' right upper limbs in the starting position of that condition. As expected, left-side omissions were more frequent than right-side omissions in condition *a*. No left/right differences were found in condition *b*, which shows that neglect was not sensitive, in our experiment, to the direction of the limb in its workspace. However, left neglect was present in condition *c*, in which the tactual display was to the right of the trunk's sagittal midplane but astride the line of sight, and in *d*, in which it was to the left of the line of sight but astride the trunk's sagittal midplane. Both factors – line of sight and the trunk's sagittal midplane – were significant; their interaction was not[2].

Our experiment seems therefore to have individuated two components of the ego-centred spatial frame that constrains neglect phenomena. Such components are likely to correspond to different neural circuits; if so, the potential for a double dissociation should arise. In our group of patients there was a weak and statistically nonsignificant positive correlation between the amount of left neglect in conditions *c* and *d*. This implies that neglect relative to the line of sight and neglect relative to the trunk's sagittal midplane were not mutually exclusive. However, a double dissociation was evident from the fact that two patients, one with a parietal and one with a thalamic lesion, showed left neglect related to the trunk's sagittal midplane (condition *d*) but unrelated to the line of sight (condition *c*), whereas another patient, the only one with frontal lesion, showed left neglect related to the line of sight (condition *c*) but unrelated to the trunk's sagittal midplane (condition *d*).

In our experiment, patients were free to move their head, so that their line of sight always lay on the head's sagittal midplane. This did not allow distinguishing between a retinotopic frame and a frame centred on the head's sagittal midplane. Kooistra and Heilman (1989), however, reported the case of a patient whose

---

[2] Condition *c* was consistently mislabelled as *b* on p 143 of the original article.

neglect was framed by a combination of *retinotopic*, and body-centred coordinates. Her visual field was examined under three different eye-rotation conditions: 0°, 30° to the left, 30° and to the right. In all conditions her head was fixed, so that the sagittal midplanes of head and trunk coincided. The patient behaved as if she had left hemianopia when tested with eyes at 0° or 30° to the left, whereas no visual field defect was apparent when she deviated her gaze 30° to the right. That study, therefore, individuated a retinotopic component though failing to distinguish between head-centred and trunk-centred neglect.

To summarize, four egocentric frames of unilateral neglect may be envisaged, in which 'left' and 'right' are respectively defined with reference to: (1) the trunk's sagittal midplane, (2) the head's sagittal midplane, (3) the vertical meridian of the visual field, and (4) the longitudinal axis of a limb ready to carry out an action (or a series of actions) in its working space. Frame (1), trunk-centred, seems to be unequivocally demonstrated by the results of bisection experiments with lines placed on either side of the trunk's sagittal midplane or astride it, as well as by data such as those reported by Bisiach et al. (1985), Karnath et al. (1991, 1993) and Chokron and Imbert (1995). Frame (2), head-centred, remains putative since, with the exception of Karnath et al. (1991), no experiment has been carried out, so far, to disentangle it from frames (1) and (3)[3]. Frame (3), retinotopic, has been individuated in the patient described by Kooistra and Heilman (1989). Incidentally, it is worth remembering that a *true* retinotopic frame should be definitely unconfounded, as in the study by Kooistra and Heilman (1989), from effects of hemiamblyopia, no matter how subtle these might be. As for frame (4), limb-centred, we have so far only negative evidence (Bisiach et al. 1985), although neurophysiological data such as those reported by Graziano and Gross (1993) are in strong support of spatial frames individually centred on parts of the monkey's body and probably implemented in the putamen and inferior area 6.

Allocentric Components

Egocentric frames do not exhaustively describe the 'left' and 'right' of unilateral neglect. They interplay with allocentric frames specified with respect to the environment and the ways its contents are perceptually grouped.

Early evidence of allocentric determinants of neglect was independently provided in 1962 by Apfeldorf and by Kinsbourne and Warrington. Apfeldorf noted that neglect may affect the contralesional side of single components of a complex drawing made by the patient, rather than the contralesional side of the

[3] Although Karnath et al. (1991) did not find a significant effect of head rotation by averaging data from these four patients, the contribution of a head-centred frame could be suspected in patients 1 and 3, whose performance was better with leftward than rightward head rotation (see their Fig. 4).

Fig. 4. Neglect of the left side of single objects in a series (From Gainotti, Messerli and Tissot 1972)

drawing as a whole. Kinsbourne and Warrington found neglect or pathological completion of the left side of words even when these were tachistoscopically projected within the right visual field. We know that patients may, when reading a text, neglect – or pathologically complete – the contralesional side of individual words rather than one side of the page (Schott et al. 1966). As first remarked by Gainotti et al. (1972), the same may happen when patients are asked to copy a series of drawings aligned from left to right (see Fig. 4).

Caramazza and Hillis (1990) reported the case of left brain-damaged patient NG, who also neglected or miscompleted the last segment of words when they were written vertically or mirror-reversed (Fig. 5). Driver and Halligan (1991) described a patient who neglected the left side of each of the two parts resulting from deletion of the two middle columns in the visual array of Albert's (1973) line cancellation task. The patient was also asked to give same/different judgements to pairs of vertically elongated random shapes drawn one above the other (Fig. 6). The shapes were presented upright or rotated 45° clockwise or counterclockwise,

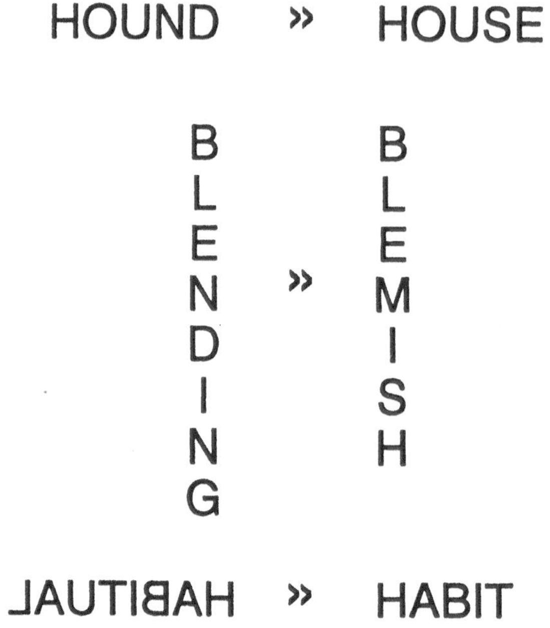

Fig. 5. Neglect of the last segment of a word in patient NG

so that the side of the differing detail was independently defined in terms of viewer- and object-centred frames. The patient's performance was worse when the critical detail was to the left of the principal axis of the shapes, irrespective of its location in egocentric coordinates. Performance was also worse when the critical detail fell to the left of the midline in egocentric coordinates, irrespective of its side in the object-centred frame. Interestingly, the effect of the egocentric frame was found to be less strong than that of the allocentric frame. Halligan and Marshall (1994) reported a similar finding: their patient BS, when asked to copy the drawing of a tower, completely omitted the left side even when the model was rotated 45° clockwise (Fig. 7).

Halligan and Marshall (1993; see also Marshall and Halligan 1993) asked left-neglect patients to copy a drawing with two flowers, one on the left and the other on the right, in two different conditions: in the first, the two flowers originated from a common stem; in the second, the lower portion of the model was deleted, so that the two flowers appeared as two independent objects. Figure 8 is a nice illustration of their findings: in the first condition, the patient's copy omitted the flower on the left and some left details of the flower on the right: in the second, both flowers were present in the patient's copy, but with details missing on the left side of each of them.

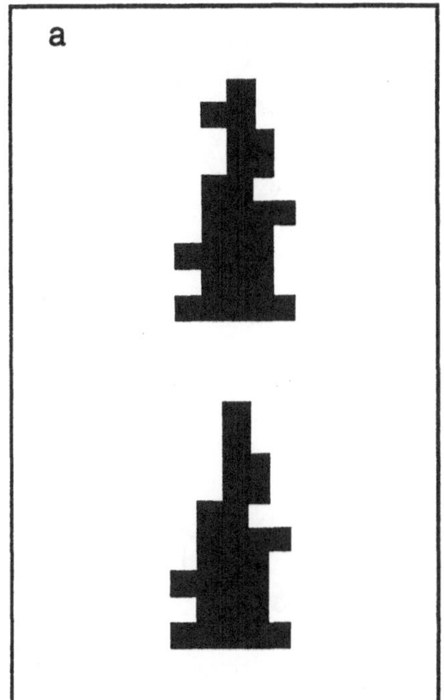

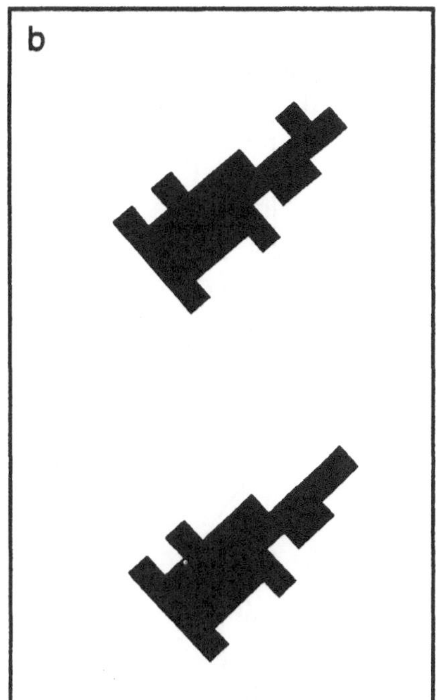

Fig. 6 a,b. Random shapes used in Driver and Halligan's (1991) experiment. (From Driver and Halligan 1991)

Parsing a perceptual array might even reverse the side of neglect in particular patients. This is suggested by findings of Humphreys and Riddoch (1994) in cases JR, a right-handed subject suffering from bilateral (left occipitoparietal, right cerebellar and frontoparietal) vascular lesions, and EL, a left-hander with radiologically less defined brain damage (multiple low-density areas in the left capsular region and generalized cortical atrophy). Both patients seemed to neglect the right side of a visual array as a whole, but the left side of single objects within that array.

The distinction between array-related and object-related neglect was not found by Chatterjee (1994) to correlate with presence/absence of visual-field defect, nor with the location of the lesion. His patients were instructed to take photographs of horizontal lines or objects in such a way that they would appear exactly in the centre of the frame of the picture: the assumption was that array-related and object-related neglect would result in a rightward and leftward displacement of the object, respectively. The former was found in four and the latter in three patients, while another patient showed variable performance.

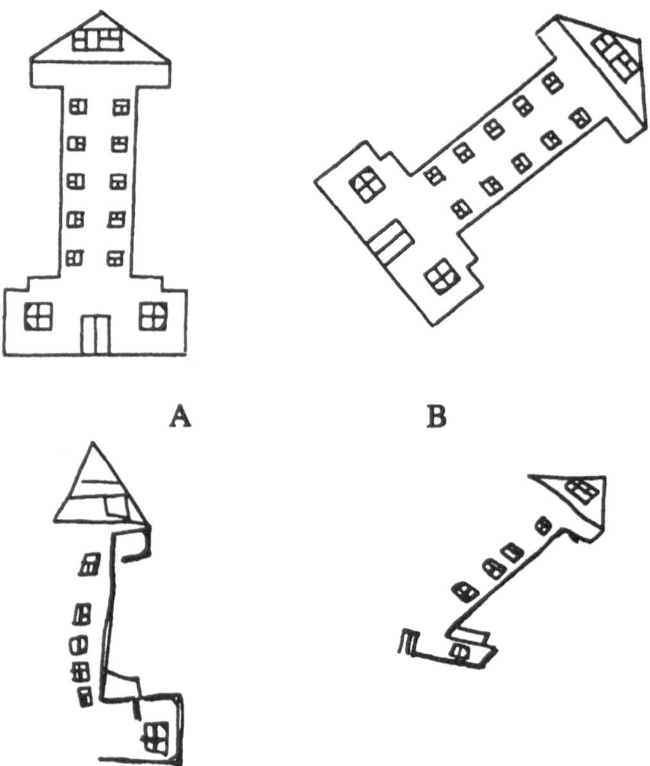

Fig. 7 A, B. Example of object-related neglect unaffected by rotation. (From Halligan and Marshall 1994)

These observations confirm that global neglect of one side of a perceptual array may multiply through segmentation of that array into perceptual units and be substituted by multicentric neglect: while moving from one perceptual unit to the next, either leftwards or rightwards, the patient skips, as it were, his/her own neglect. As demonstrated by Kinsbourne and Warrington's tachistoscopic experiment, by the reading errors of Caramazza and Hillis's patient, and by the experiments with tilted shapes mentioned above, this multiplication is not just the consequence of a reestablishment of retinotopic or head-centred neglect on the contralesional side of whatever perceptual unit happens to capture visual fixation through 'preattentive' processes.

The most dramatic indication of neglect secondary to the segregation of individual units within the perceptual array and unrelated to the line of sight, however, is the following. Patients unable to detect flagrant differences on the contralesional side of two (otherwise identical) drawings placed one above the

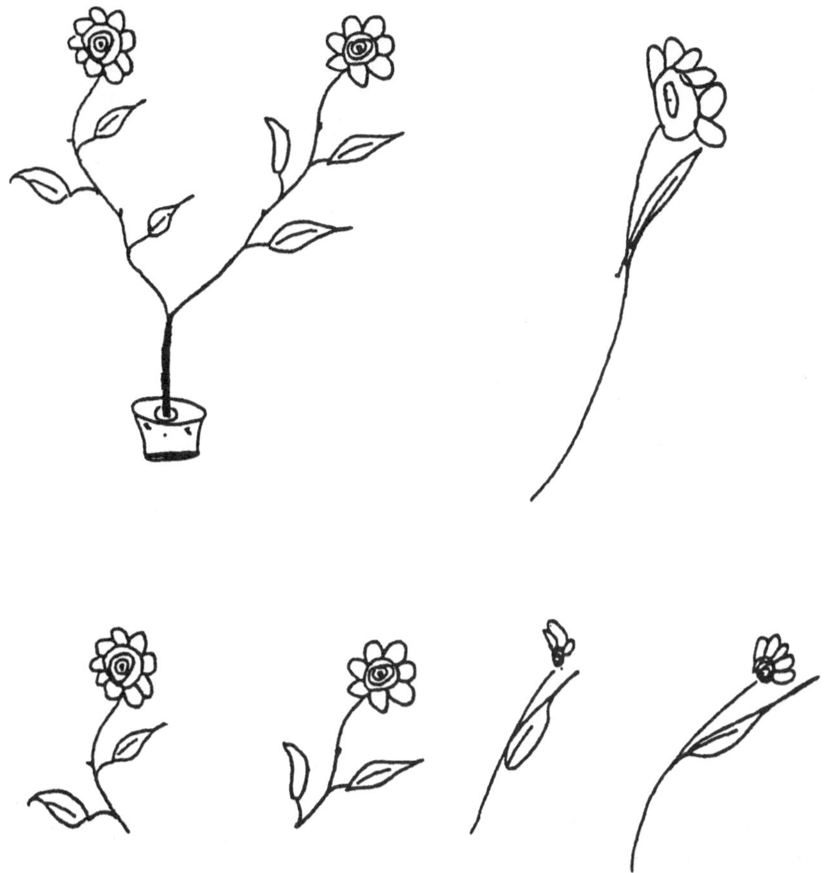

Fig. 8. Left neglect modulated by perceptual parsing. (From Halligan and Marshall 1993)

other may fail to detect the differences even if they accurately slide their fingertip across the differing details or along the differing edges while tracing the contour of each drawing (Bisiach and Rusconi 1990; see also Young et al. 1992; Vallar et al. 1994).

It is necessary to note that, whatever the contribution of the environment in determining a frame for neglect phenomena might be, such a frame cannot be conceived other than in patient-relative terms, as resulting, that is, from an interplay of egocentric and allocentric coordinates. There is in fact no intrinsic left (or right) side of an object that undergoes neglect independently of the perspective under which that object is perceived or mentally represented by the patient. This point leads to another set of data that have greatly enriched our understanding of neglect and, consequently, of the structure of the mental representation of spatial properties.

Two papers published in 1987, one by Làdavas and the other by Calvanio et al., extended the analysis of the spatial frame of contralesional neglect by exploring the effects of rotations of the patient's head and body in the coronal plane.

In her experiment 3, Làdavas asked five right brain-damaged patients with left visual extinction to respond as fast as possible to visual stimuli appearing on either side of the fixation point and above it, while their head was tilted 90° either to the left or to the right. In the condition in which the patient's head was tilted to the left, both stimuli appeared therefore in the (ipsilesional) right visual hemifield; nonetheless, mean RTs were 535 ms to the left and 475 ms to the right stimulus. Conversely, when the head was tilted to the right, stimuli appeared in the (contralesional) left visual hemifield; mean RTs, in this condition, were 651 msec and 546 msec to left and right stimuli, respectively. Both effects – left vs. right stimuli and left vs. right visual hemifield – were statistically significant[4]. Similar results were later obtained in four right parietal and, less definitely, in four left parietal patients, but not in four right temporal and four left temporal patients (Làdavas et al. 1989).

Calvanio et al. (1987) asked ten patients with left neglect to name objects or read words arranged in 5 x 5 matrices while sitting upright or reclining on either side (stimulus orientation being constant). The mean number of reports for each quadrant is shown in Fig. 9 (maximum number=4; responses relative to the middle column and row of each matrix were not considered). Shaded cells refer to the left side with respect to the sagittal midplane of the head and trunk. It is evident from Fig. 9 that neglect was relative to the left side of the body in its three different orientations but also to the left side of the visual array *as if* viewed while sitting upright. Indeed, the worst performance was found, for objects as well for with words, in the lower left quadrant under the 'recline left' condition and in the upper left quadrant under the 'recline right' condition[5]. Calvanio et al. wondered whether the first frame was typical of parietal and the second of frontal neglect; unfortunately, this could not be settled on the basis of their anatomical data[6].

Results similar to those obtained by Calvanio et al. (1987) were later reported by Farah et al. (1990). They used as stimuli 16 letters, four for each quadrant of a picture representing an animal (or object) facing left or right. In addition to the

[4] In this experiment subjects responded to left and right stimuli by pressing a left and a right key with the index and the middle fingers of their right hands, respectively. This may give rise to some problems in the interpretation of the results, whose meaning, however, is clarified by the results obtained by Calvanio et al. (1987), and by Làdavas and associates in their replication of her first study (Làdavas et al. 1989), in which they recorded simple RTs.

[5] Calvanio et al. argued, on the basis of data from literature, that their findings were not an artifact due to ocular counter-rotation in the body-tilt condition.

[6] The hypothesis was suggested by Teuber and Mishkin's (1954) study on 'field dependence' in brain-damaged people.

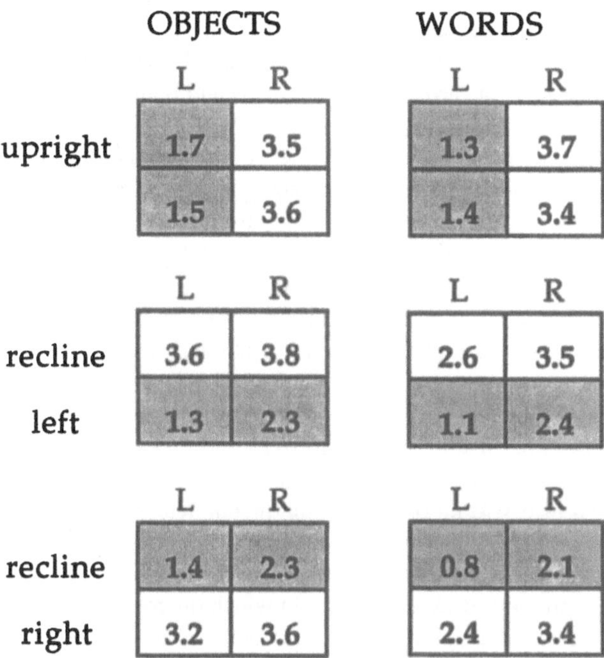

Fig. 9. Influence of head and body rotation on neglect (From Calvanio et al. 1987). *L*, environment-centered left; *R*, environment-centered right; *shaded cells*, body centered left; *unshaded cells*, body centered right

conditions employed in the study by Calvanio et al., the authors included in their investigation two further conditions: while sitting upright, patients looked to visual arrays where the figure enclosing the letters to be reported (e.g. a left- or a right-facing rabbit) was rotated in such a way as to point either upwards or downwards. Under these two additional conditions, the ten patients who participated in the experiment showed neglect relative to the left side of the visual display but not relative to the area of the figure that, with respect to the viewer, would have been the left side if that figure were canonically oriented. Farah et al. concluded that their data offered no evidence of a further, 'object-centered' frame to which neglect phenomena are sensitive. Evidence of the latter was however provided by Behrmann and Moscovitch (1994) with a similar experiment. Neglect patients were asked to name the colors (four for each quadrant) forming the outline of common objects or letters that could be either symmetrical or asymmetrical about their vertical axis. Stimuli were presented upright or tilted 90° clockwise or counterclockwise. With asymmetrical letters (though not with objects and symmetrical letters) neglect was found to be present in both viewer-centred and object-centred coordinates.

A most impressive instance of object-centred neglect unaffected by object rotation was found by Tegnér (1994) in patients who had been asked to paint toy animals with patches or dots. After having painted one side, the patient had to rotate the animal by 180° around a vertical axis and paint the opposite side. Most patients painted the animals only on the right side from each perspective. A few of them, however, after having painted the right side on the first perspective, irrespective of whether the animal was facing right or left, turned the animal as required and only painted the *left* side as seen from their viewpoint.

A similar result was obtained by Behrmann and Tipper (1994) and Tipper and Behrmann (in press). In their experiment, left-neglect patients had to detect a target appearing in either the left or the right of two circles joined by a horizontal bar to form the shape of a dumb-bell. In a stationary condition, patients showed worse detection of left-side targets. In a condition in which the target appeared after the dumb-bell had undergone a 180° clockwise or counterclockwise rotation in 1.7 s, the detection of left targets improved and that of right targets worsened. The side of neglect was thus completely reversed in some patients (those in whom neglect on other tests was more severe), but incompletely reversed in others (those with less severe neglect). The authors concluded that their data implied the existence of two different frames: one object-related and one ('location-based') relative to the stationary background within which the rotating objects appeared.

The most likely explanation of results such as those reported by Làdavas, Calvanio et al., Làdavas et al., Farah et al., Tegnér, and Behrmann and Moscovitch seems to involve a process of mental rotation of the perceiver or the percept, leading to a canonical representation of the visual array as it would appear to an upright viewer. This explanation could also apply to the findings by Caramazza and Hillis (1990), Driver and Halligan (1991) and Halligan and Marshall (1994) mentioned above. Neglect phenomena might thus relate to the rotated representation in the same way as left/right asymmetries arise in the processing of letters that probably cannot be explained only in terms of direct hemispheric access. In fact, Robertson and Lamb (1988) found that normal subjects, asked to judge whether letters were normal or mirror-reversed, were – as expected – faster with right than with left visual field presentation when letters presented on either side of the fixation point were upright; when the slides were rotated clockwise or counterclockwise by 90°, RTs were shortest with lower and upper visual field presentation, respectively. If a process of mental rotation in the coronal plane were responsible for the results of Calvanio et al., one would expect neglect of the lower and upper side of the stimulus array in the recline-left and recline-right conditions, respectively, to be more pronounced with words than with objects, since it is plausible to assume that the former would induce mental rotation more than the latter. These are indeed the findings of Calvanio et al. The data reported by Tegnér and by Tipper and Behrmann, on the other hand, are a clear demonstration of a relative invariance of allocentric neglect with respect to external (as opposed to mental) rotation of the object towards which the patient's attention is directed.

The Construction of Unitary Space

To conclude, 'left' and 'right' are qualified in unilateral neglect with reference to a multiplicity of frames. Although all frames *concern* – in a sense strikingly conforming to the etymology of this term – both the subject of perception or representation and the object that is perceived or represented, some of them are relatively egocentric while others are relatively allocentric. We have seen that in the case of egocentric neglect, reference to one frame does not necessarily exclude reference to another (Bisiach et al. 1985; Kooistra and Heilman 1989). The same is likely to be true for allocentric frames, as suggested by Tipper and Behrmann's findings and by Fig. 1, in which left neglect is evident with regard to the whole display as well as with regard to single lines, marked by the patient with rightwardly displaced strokes. Furthermore, combined reliance on ego- and allocentric frames is suggested, as we have seen, by data such as those reported by Driver and Halligan (1991).

What kind of synthesis should follow our analysis? Suppose that left neglect patients who were asked to draw a daisy with their right upper limb outstretched and abducted by 90°, while keeping eyes and head aligned with the trunk's sagittal midplane, omitted details on the left side with respect to that limb's longitudinal axis. Should this be regarded as a demonstration of an egocentric frame related to that axis, or as an instance of object-centred neglect? Maybe there is no contradiction in giving an affirmative answer to both questions. In fact, the different ego- and allocentric frames with respect to which the spatial aspects of neglect can be described are nothing but projections of the interplay between two structures: the perceiver-representer's body itself, throughout all changes in the reciprocal articulations of receptor and effector divisions, and the perceived or imagined environment. The kaleidoscopic features of spatial neglect resulting from various experimental manipulations are therefore the expression of different task-related kinds of parsing of these two structures, as well as of the different ways in which any elements resulting from a particular kind of parsing are put in (or out of) register with each other through routines such as mental rotation, translation, etc. These are the mental operations to an understanding of which the investigation of unilateral neglect and related disorders is, in the end, meant to contribute.

There are probably two unifying drives acting upon the construction of space representation. One is gravitation, by which a canonical orientation is established for the subject as well as for the object(s) of representation. The other one is the alignment of the manifold of ego- and allocentric frames of reference in a *virtual* common frame (see Stein 1992, and the following commentaries). By stating that 'the unitary space of common sense is a construction', Bertrand Russell (1948, p 234) gives the term 'common sense' the Latin meaning of *sensus communis*, i.e., common acceptance of a belief, rather than the Aristotelian meaning of supramodal sensory faculty. We do in fact experience space as something unitary.

The phenomenology of unilateral neglect, however, makes us aware of some of the steps leading to the construction of the system of functional interactions that corresponds to such an experience; a system for which we have no reasons to posit a neural substrate separate from, and superordinate to, the circuits in which the different frames of space representation are separately implemented.

Altitudinal and Radial Neglect

As noted by Halligan and Marshall (1989), there are sparse reports relative to the behavior of left neglect patients on cancellation tasks in which omissions are more frequent in the lower than in the upper quadrants of the stimulus array. Although overlooked by their authors, these observations captured Halligan and Marshall's attention, so that a systematic investigation was carried out by them in 23 left and three right neglect patients using a modified version of Albert's (1973) cancellation task. The rate of omissions they found in the four quadrants of the stimulus array had the following distribution (percentage values):

Left neglect		*Right neglect*	
39	5	3.3	43.4
48	8	3.3	50

The difference between top and bottom quadrants, although not as pronounced as the difference between ipsi- and contralesional quadrants, was statistically significant and could also be observed in patients free of visual field defects on clinical assessment.

These results are in agreement with those reported a few years earlier in abstract form by Morris et al. (1985). Their interpretation and, therefore, their theoretical relevance are still uncertain. It is still unknown whether or not they are also typical of neglect patients with exclusively frontal lesions, that is, of patients with a lesion that does not directly involve the visual pathways. Therefore, the fact that unilateral neglect is frequently due to impairment of the inferior parietal lobule, in proximity to the optic fibers related to the lower quadrant of the opposite visual hemifield, may suggest that the worst performance in this quadrant is due to clinically undetectable amblyopia. In order to rule out this hypothesis a control experiment is needed in which neglect patients tactually explore a stimulus array while blindfolded. An alternative explanation could be suggested in terms of fatigue, since the usual scanpath in neglect patients starts from the upper ipsilesional corner of the visual array and proceeds with variable (often very irregular) strategy towards the lower contralesional corner. This hypothesis is weakened, however, by Mark and Heilman's (1988) finding that neglect of the lower contralesional quadrant may still be observed when patients are required to initiate cancellation tasks from the bottom of the visual array.

Much more interesting are manifestations of altitudinal neglect due to midbrain or bilateral brain lesions. Regarding the former, neglect of upper or lower visual space was found in the cat following lesion of different sectors of the mesencephalic commissural system (Matelli et al. 1983), and in patients suffering from progressive supranuclear ophthalmoplegia (Posner et al. 1982). The latter deserve a more detailed account.

Rapcsak et al. (1988) observed neglect phenomena relative to lower space in a patient with a lesion involving areas 39 (angular gyrus) and 19 (lateral occipital gyrus) on both sides of her brain, as well as area 37 (middle temporal gyrus) in the right hemisphere. She had no visual field defects but showed visual extinction in the lower quadrants when two stimuli were simultaneously presented, one above and the other below the horizontal meridian. The patient bisected above the objective midpoint vertical lines shown at eye level, 30 cm in front of her, in each of three modalities: visual, tactual, and visuo- tactual. Further data were collected from a subsequent study on the same patient (Mennemeier et al. 1992), in which visual and tactual bisection were tested with horizontal lines (centred on the trunk's midsagittal plane or located on either side of it), vertical lines (above, at, or below eye level) and radial lines (at three different distances from the patient's body). Horizontal and radial lines were presented on a table whose level relative to the patient's trunk was not specified. Earlier findings concerning the vertical dimension were confirmed. Left and proximal neglect were also found, though less pronounced, in both the visual and tactile modality. Errors were independent of the hand used and the position of the line along each dimension. On the basis of additional findings the authors concluded that the patient's behavior was not due to her limbs' directional hypokinesia, i.e., reluctance to move towards the contralesional side.

Neglect of lower space was also found by Butter et al. (1989) in a patient in whom the CT scan showed bilateral damage in the dorsal portions of the occipital lobe, extending to cuneal and precuneal regions. Left inferior quadrantanopia had been initially diagnosed; later, she showed extinction in the lower visual half-field when stimuli were simultaneously presented one above and the other below the horizontal meridian. She bisected above the objective midpoint vertical lines presented in the frontal plane above, at, and below eye level. Tactual bisection was more impaired than visual bisection; visuotactual performance was at an intermediate level. Bisection of lines below eye level was more impaired than at, or above, eye level.

Upper-space neglect was instead reported by Shelton et al. (1990) in a patient suffering from a bilateral inferior temporal and deep occipital infarction with no evidence of parietal damage. Visual and tactual bisection were examined in conditions similar to those used in the study by Mennemeier et al.. Downward deviation of the subjective midpoint was recorded in both the visual and tactual modality with vertical lines; in the visual modality, errors were greater with lines above than below eye level. (Only the eye-level location was tested in the tactual modality.) In the radial condition, distal neglect (i.e. bisection errors towards the patient's body) was found, both visual and tactual, and was more severe with distal than with proximal lines.

Upper-space neglect was also found by Henaff and Michel (1992) following a mirror-symmetrical lesion in the vicinity of the parietoccipital sulcus that also involved the right optic radiations and caused left hemianopia. The patient neglected the upper segment of vertical words projected within her right visual field, independent of the level at which words appeared with respect to the horizontal meridian. Upper extinction was found with stimuli presented, one above the other, within her right visual field. Phenomena of left visual neglect were also present.

These observations are very important because they provide further evidence of the topological relationships between brain anatomy and spatial attributes of mental representation: they suggest that, in the same way as left and right neglect are respectively caused by right and left brain damage, bilateral lesions in the parietoccipitotemporal areas may cause neglect of lower or upper space depending, respectively, on whether they are located more dorsally or more ventrally. They also suggest that, whatever its ultimate explanation might be, spatial neglect does not necessarily depend on the disruption of the normal interaction between structures located in *opposite hemispheres*.

By contrast, these observations are somewhat equivocal with regard to neglect along the radial dimension. In the conditions of the studies reviewed above, this dimension was indeed confounded, at least in part, with the retinotopic vertical dimension. In the tactual modality, this was less obvious but still plausible as a result of intersensory integration (i.e. in the sense of a mental translation from a somatosensory-motor into a visual frame of reference). The question as to whether or not neglect may also affect the anteroposterior radial dimension is therefore still open.

True radial neglect was ruled out by Adair et al. (1995) in a patient who showed upper altitudinal neglect as a consequence of bilateral infarction within the inferior occipital and temporal regions. While sitting upright, she misbisected towards the near end radial lines (perpendicular with respect to the frontal plane of her body) located *below* eye level, but misbisected towards the far end radial lines located *above* eye level. Bisection errors in the radial dimension were therefore simply due to neglect of upper space in a retinotopic frame. (Upper retinotopic space, in fact, is located nearer to the body when gaze is directed above eye level, but farther from it when gaze is directed below eye level.) While sitting upright, the patient also misbisected downwards vertical lines. When lying on either side, she still bisected below the objective midpoint lines that were vertically oriented with respect to retinotopic coordinates (i.e., horizontally oriented with respect to the gravitational axis). The patient also showed mild left neglect. The size of the bisection error with retinotopically vertical lines was greater when she was lying on the left side (so that environment-related left neglect and retinotopic upper neglect coincided) than when she was lying on her right side (so that environment-related left neglect and retinotopic upper neglect were in opposition). This could suggest an interplay of egocentric and allocentric frames of reference; however, her performance was mainly dependent on the former.

The converse was found in a study in which Mennemeier et al. (1994) were able to disentangle ego- and allocentric determinants of neglect in patients CH (lower altitudinal neglect, formerly reported by Rapczak et al. 1988, and Mennemeier et al. 1992) and FT (upper altitudinal neglect, formerly reported by Shelton et al. 1990). These two patients were asked to bisect horizontal, vertical, and radial lines in different body positions. In some combinations of line placement and body orientation, line bisection errors in opposite directions were expected depending on the spatial frame to which neglect was related. The results were decisive. For example, in the upright position patient CH bisected vertical lines above and radial lines beyond the objective midpoint. With CH lying prone above them, these lines reversed their orientation in terms of bodily, but not in terms of environmental coordinates. In this position, the patient misbisected towards her body lines with a radial orientation with respect to her body, but a vertical orientation with respect to the environment. This outcome was consistent with the hypothesis according to which neglect was related to a spatial frame anchored to the environment rather than to the patient's body. From this and similar results the authors concluded that whenever environment- and body-related frames were unconfounded neglect consistently showed in these two patients in terms of the former. Since in normal subjects changes in body orientation did not significantly modify (minimal) directional biases in line bisection, the authors also suggested that neglect patients' reliance on the environmental frame in line bisection might be due to a release phenomenon and/or inability to compute body-related spatial information. Furthermore, since the interplay of ego- and allocentric frames was lesser in patient CH (parietotemporal lesion) than in patient FT (mesial occipitotemporal lesion), the authors tentatively speculated that the temporo-parietal areas could be responsible for the interplay of different reference frames; they also pointed out that this conjecture would agree with the view according to which temporoparietal areas are the locus of polymodal sensory convergence. Further research is required to assess the generality of these phenomena and, as stated by the authors themselves, to assess the relative involvement of somatosensory and vestibular input in their causation.

Proximal and Distal Frames

'In the effects of right parietal lesion in a right-handed person, amorphosynthesis is found in terms of neglect and lack of awareness of left half of person, neglect and lack of awareness of the left extracorporeal field. The relative accentuation of person or extrapersonal space correlates with lesions in anterior parietal and in parietooccipital areas respectively, with a large overlap in the intraparietal sulcus' (Denny-Brown 1962, pp 244-245).

The above statement, unfortunately, left out any mention of the evidence on the basis of which it was made. Earlier suggestions regarding the possible fractionation of unilateral neglect with respect to proximal and distal spatial frames (Brain 1941; Paterson and Zangwill 1944) are scanty and, in retrospect, vague. As yet, however, no great progress has been made, at least regarding the study of brain-damaged people.

A double dissociation between contralesional neglect of peripersonal and extrapersonal space was reported by Rizzolatti and coworkers following lesion of different cortical areas in the monkey (Rizzolatti et al. 1983; Rizzolatti et al. 1985). Lesion of postarcuate cortex (area 6) or inferior parietal lobule (area 7b, an area projecting to area 6) was found to cause bimodal neglect of peribuccal tactile and visual stimuli. The normal response to such stimuli, i.e., opening of the mouth and attempting to bite (usually occurring independent of eye fixation), was delayed or absent. No attempt to grasp such stimuli with the ipsilesional hand was reported. Some monkeys also showed visual neglect for more distal stimuli within hand-reach (at a 30 cm distance). By contrast, all animals oriented correctly to visual stimuli presented further away (at a 60 cm distance). Conversely, lesion of area 8 did not abolish responses to peripersonal stimuli though giving rise to neglect for stimuli beyond reaching distance.

Personal and extrapersonal neglect were assessed in a series of 97 right brain-damaged patients (Bisiach et al. 1986). Personal neglect was tested by asking patients to reach for the left hand with the right (ipsilesional) one. Extrapersonal neglect was tested by asking patients to cross out 13 circles drawn on an A4 sheet of paper. Medium and severe personal neglect were found to be much less frequent than extrapersonal neglect (six cases vs. 35), though no firm conclusion can be drawn from this finding because of the dissimilarity of the two tests. Medium and severe extrapersonal neglect were found in many patients in whom personal neglect was minimal or absent altogether. Only a single (though clear-cut) observation of a patient, who showed the highest degree of personal neglect while correctly executing the cancellation task, was indicative of a double dissociation between the two types of neglect. Unlike extrapersonal neglect, personal neglect was found to be present only in patients in whom motor, somatosensory and visual functions were jointly impaired to a marked degree. With regard to anatomo-clinical correlations, both types of neglect were found to be associated with lesions involving the inferoposterior parietal region or subcortical structures such as the thalamus and basal ganglia (for further details, see also Vallar and Perani 1987). The comparison with the data obtained by lesion studies in the monkey may suggest a dissimilarity of the neural circuits responsive for proximal and distal space representation in the two species.

Further evidence of selective impairment within a proximal (personal or extrapersonal) frame of reference comes from single-case studies. Halligan and

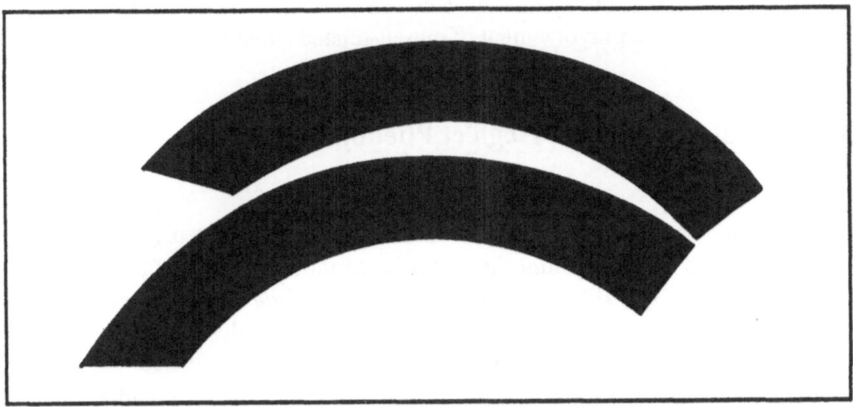

Fig. 10. Wundt-Jastrow illusion: the lower shape looks longer than the upper one. (From Pizzamiglio et al. 1989)

Marshall (1991; see also Marshall and Halligan 1991) reported the case of a patient who manifested severe left neglect in line bisection when he had to mark with a pen the midpoint of lines placed 45 cm away from his body, while showing correct performance when requested to bisect by means of a light-pointer lines presented at a distance of 244 cm. No indications of personal neglect were observed. The extent of the lesion (a frontotemporoparietal infarct) precluded precise conclusions regarding the neural circuits whose damage was responsible for his impairment. Selective personal neglect was instead observed by Guariglia and Antonucci (1992) in a patient who had undergone surgery for right parietal lobe haemorrhage. The patient had no manifestation of neglect in extrapersonal space, but failed to comb his hair on the left side and shave his left cheek. Neglect of the left side of his body was also evident in every-day behavior. Once again, clinico-anatomical correlations were limited by the presence of an extremely large area of brain dysfunction: repeated single photon emission computed tomography examinations demonstrated hypoperfusion of the whole right hemisphere, most severe in the frontoparietal areas and involving both cortical and subcortical structures.

Particularly interesting are the considerations suggested by the result of a study in which neglect in near and far extrapersonal space was assessed by means of a test that did not require motor responses of any kind (Pizzamiglio et al. 1989). Configurations inducing the Wundt-Jastrow illusion were exposed at two different distances (35 and 151 cm) to 28 left-neglect patients. When looking at a configuration such as that shown in Fig. 10, normal subjects misjudge the lower shape as being longer while left-neglect patients make the opposite error. The authors found a very strict positive correlation between the results obtained in the two viewing conditions. Since the task did not require motor responses, the authors suggested that the distinction between proximal and distal visual neglect might only concern the involvement of output-related attentional systems.

Directional Gradient of Neglect Phenomena

A sharp boundary between normal (ipsilesional) and neglected (contralesional) field is predicted by *hemispatial* theories such as those of Heilman and associates (e.g., Heilman et al. 1987), whereas Kinsbourne's *directional* theory (1970) predicts a continuous gradient along the ipsilesional-contralesional axis. Truth, of course, might lie in between; so far however, there is no objective certainty of any sharp demarcation between an area enjoying functional integrity and an area of homogeneous neglect. On the contrary, there are several *prima facie* demonstrations of a more or less continuous gradient of neglect even within the ipsilesional hemispace, that is, in the area that should be uniformly spared according to radical hemispatial theories of unilateral neglect. To be sure, a neglected area with a sharp boundary may be observed, for example, in the patients' drawings or execution of cancellation tasks, but this might depend on a critical point at which the progressive decline of a certain function gives rise to an abrupt change (Fig. 11).

Clinical findings suggestive of a continuous gradient of neglect are the (probably infrequent) extinction of the left of two stimuli simultaneously presented in the right (ipsilesional) visual field (Rapcsak et al. 1987; Bisiach and Geminiani 1991, p 33) and the phenomenon described by Cohn (1972), consisting of the magnetic attraction of a patient's fixation by an object appearing in the ipsilesional periphery of the visual field unaffected by the lesion despite the simultaneous appearance (within the same visual field) of another object in more central position (see also Weinstein and Friedland 1977, p 51). There is however further evidence from more formal investigations to be considered before drawing conclusions.

Kinsbourne (1977) mentioned the results of an experiment in which brain-injured people had to report strings of letters tachistoscopically presented across the fixation point or within either visual hemifield. With right visual field exposure, right brain-damaged patients showed better performance with rightmost letters. The opposite is true of normal subjects. Posner et al. (1984) found that patients with right parietal lesions could not easily move attention from one target to another located to its left, even if both targets were within the right visual hemifield. De Renzi et al. (1989) measured choice RTs to a target letter in a row of four equally spaced letters presented in the right visual field of left neglect patients. The leftmost letter in the row corresponded to the central fixation point; the rightmost letter was 96 mm to the right of it. RTs were found to be shortest when the target was in the rightmost position and longest when it was in the leftmost; intermediate RTs were found for the targets appearing in intermediate

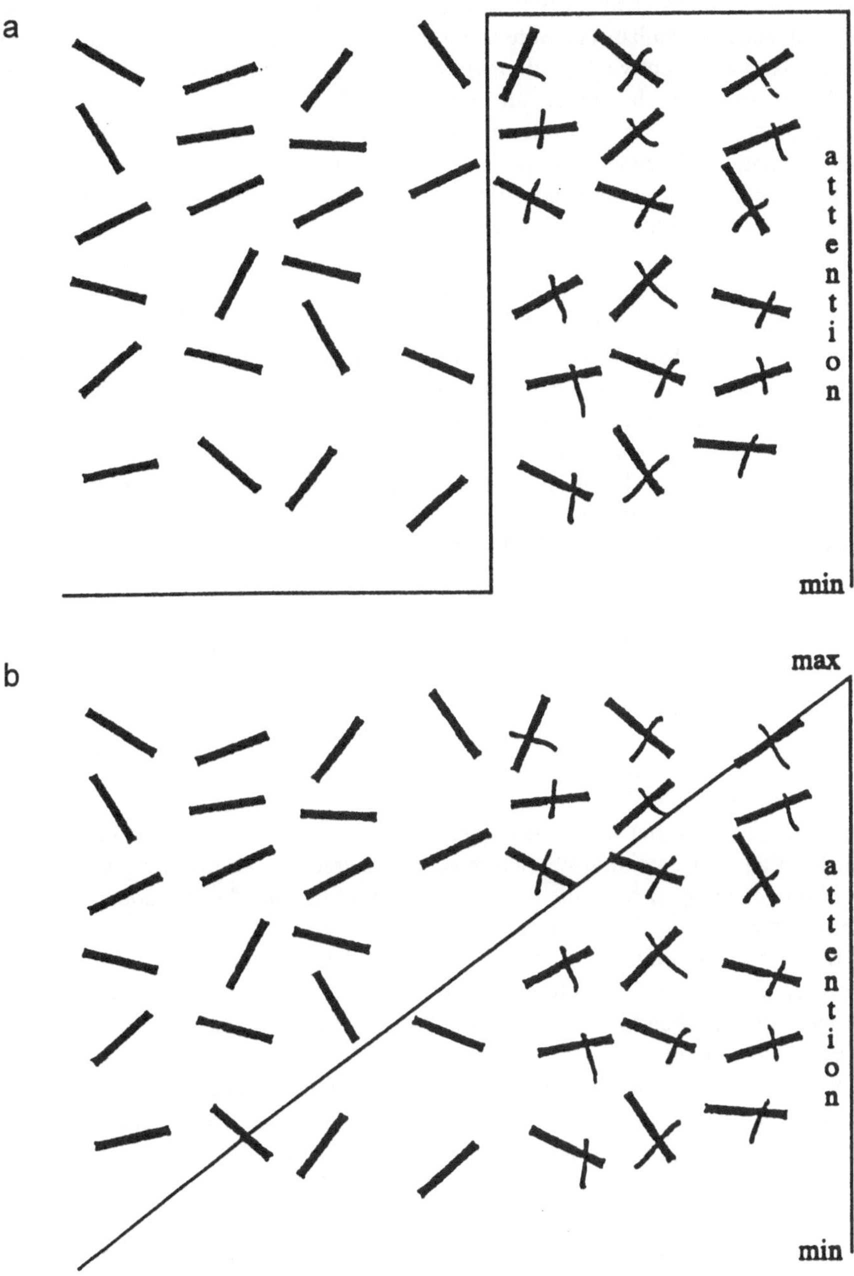

Fig. 11 a, b. The sharp boundary of neglect on a cancellation task could either be due to sudden loss of attentional resources on the contralesional side (**a**), or progressive decline (**b**) with a critical point at which neglect manifests itself on one side of the stimulus array

positions. Làdavas et al. (1990) found that in left-neglect patients, contrary to normal subjects, choice RTs were faster to visual stimuli appearing in the right of two boxes permanently exposed to the right of the fixation point.

All these data, however, are equivocal. Each of them, indeed, could alternatively be interpreted in terms of allocentric neglect relative to one side of the (no matter how composite) perceptual unit to which the patient is attending on that particular task. This suggestion may appear somewhat stretched regarding extinction on double stimulation within the ipsilesional visual hemifield. It must, however, be noted that extinction is usually observed when the patient is given a sequence of events constituted by irregular alternation of single and double stimuli, so that the latter could be processed as a unitary, spatially distributed pattern of stimulation over an invariant background. It must be noted that contralesional stimuli may be detected despite the large amount of heterogeneous and sustained background stimulation on the ipsilesional side when they are single, though they are extinguished by cooccurrent discrete (often minute and ipsimodal) ipsilesional stimuli; this supports the hypothesis according to which recurring stimulus pairs are treated by the nervous system, as it were, as single *Gestalten* disembedded from background stimulation and therefore spread from side to side over a perceptual-unit-centred frame. Data such as those reported by De Renzi et al., on the other hand, may easily be explained as being due to right-to-left scanning of the perceptual array – which is the rule with left-neglect patients – rather than in terms of a continuous, side-to-side gradient of attention.

Stronger indication of a pathological gradient equally affecting each point of space representation along the horizontal dimension is the finding that right brain-damaged patients, especially those showing neglect, systematically mislocate rightward, in the coronal plane, the acoustic image resulting from fusion of dichotic tones of differing intensity (Bisiach et al. 1984). As shown by Fig. 12, the mislocation is a continuous function of the degree of eccentricity given to the acoustic image by varying the relative intensity of monaural stimulations, with no abrupt change along the ear-to-ear axis.

More recently, Mattingley et al. (1992) gave left-neglect patients a sequential, visually guided, right-to-left motor task. They found a smooth increase of movement execution times with no discontinuity between the ipsilesional and the contralesional field.

A continuous side-to-side gradient of neglect would accord with the way some spatial relations are coded at the neuronal level. Galletti and coworkers (1993), for example, have shown that in the monkey's area V3A the responses of some gaze-sensitive visual neurons are *progressively* facilitated as gaze deviates from one side to the opposite (see Figs. 1 and 2 in their paper). In the intact brain, the uniform representation of space along the left-right dimension could itself depend on the balance between similarly calibrated space-coding neurons with opposite continuous gradients of directional responsiveness. Discrete lesions within space-coding networks could thus be such as to selectively impair representation of one side of egocentric space with no sharp boundary between any two adjacent areas.

The same could apply, *mutatis mutandis*, to disorders of space representation along the vertical dimension giving rise to altitudinal neglect.

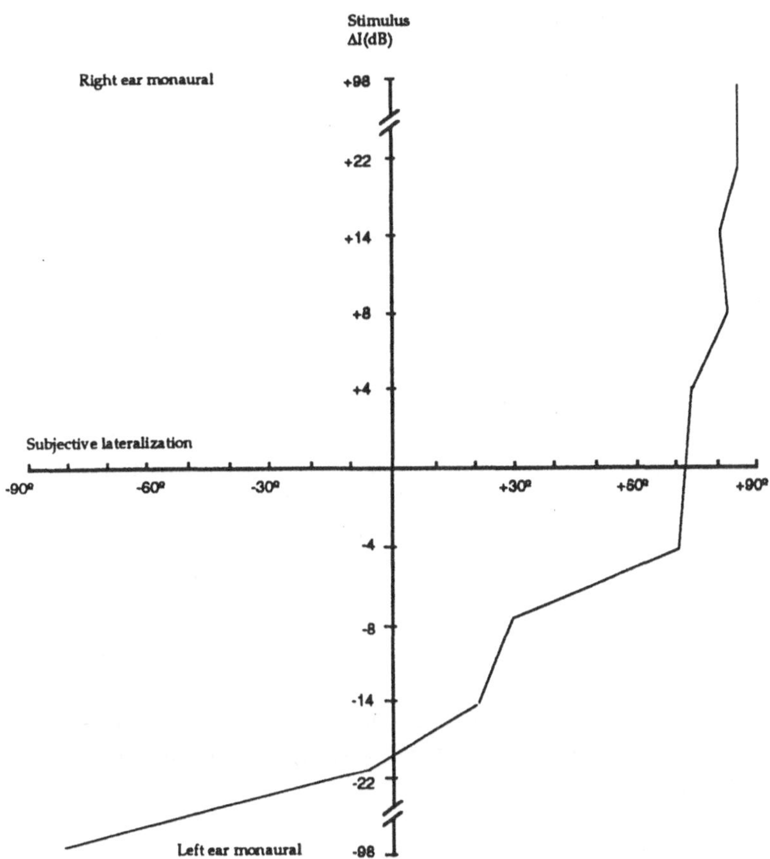

Fig. 12. Rightward mislocation of dichotic sounds in left-neglect patients. *Ordinate*: interaural intensity difference; positive and negative values correspond to higher right or left ear intensity, respectively. *Abscissa*: perceived sound lateralization; 0° corresponds to the head's sagittal midplane. (From Bisiach et al. 1984)

Misrepresentation of Rear Space

Until recently, phenomena of space misrepresentation due to unilateral brain lesion have only been investigated with respect to the left-right axis of a frontal plane facing the patient. This yielded foreshortened descriptions insufficient for the reconstruction of the underlying dysfunction in full relief. Some of these phenomena, such as the ipsilesional deviation of the subjective *straight ahead* (e.g., Heilman et al. 1983; Mark and Heilman 1990; Chokron and Imbert 1995), can be described as a lateral shift of an egocentric frame of reference. Recording these phenomena in a frontal plane facing the patient, however, is not sufficient to settle the question as to whether the ipsilesional shift is due to an illusory *rotation* of the patient's sagittal midplane, or to an illusory *translation* of that plane towards the side of the lesion. It could be argued that the former does not accord with the type of mislocation of dichotic sounds found in right brain-damaged patients (Bisiach et al. 1984). Indeed, it would predict *no* lateral shift of the subjective location of dichotic sounds with no interaural intensity difference, since an illusory rotation of the patient's sagittal midplane around the vertical axis of (complanar) head and body would leave this axis unaffected. Contrary to that, Fig. 12 shows a striking ipsilesional mislocation of the fused auditory image even when the sounds delivered to each ear were of equal intensity. The question, however, can be more directly addressed by ascertaining the presence and the direction, in neglect patients, of a shift in the subjective *straight behind*. This was done by means of a free-field auditory localization task in which ten left neglect patients were asked to report verbally whether stimuli given in the front or back half-spaces were to the left or to the right of their body's sagittal midplane (Vallar et al. 1995). Normal subjects and right brain-damaged patients without unilateral neglect made slight errors, with a prevailing leftward bias in rear space. Neglect patients, by contrast, showed an average well-defined rightward (i.e., ipsilesional) bias both in front and rear half-spaces. The procedure of the experiment was such as to make unlikely front/rear confusions in sound localization. No instances of such a confusion, however, were found in a study in which ten right brain-damaged patients with left neglect had to localize, by pointing, free-field auditory stimuli in the front half-space, although a rightward pointing bias was present (Pinek et al. 1989).

In the study by Vallar et al. a dissociation was only found in one neglect patient who showed a rightward bias in back, but no bias in front space. However, three right brain-damaged patients who did not manifest neglect on the screening (*visuo*spatial) tests showed a rightward shift confined to the subjective straight ahead in one case and to the subjective straight behind in two. While dissociations of neglect between and within sensory modalities are well-known and amenable to relatively well defined courses of speculation, the dissociation with respect to front and rear spaces would add, if confirmed, a further riddle to the study of the mechanisms underlying spatial attention.

Conclusions

I have summarized in this chapter clinical data showing the highly composite structure of the system of ego- and allocentric frames of spatial reference. I have also reported, in the section above, findings suggesting that lesions of the right hemisphere may give rise to an ipsilesional distortion of that system, affecting both the space in front and behind the patient. I passed over the still very poorly understood issue of the different frequency, severity, duration and, possibly, kind of lateral distortion of space representation after left and right hemisphere lesion, because it was not strictly pertinent to the subject of this chapter. This chapter, indeed, was mainly concerned with the multiplicity of frames of spatial reference as revealed by the study of neglect and related phenomena. For this reason, I have also left out any attempt at defining the nature of the distortion of space representation within individual frames, although this point is, in all likelihood, strictly related to the issue of the gradient of neglect. Such an attempt has been made elsewhere on the basis of recently acquired empirical data (Bisiach et al. 1994; Bisiach et al. 1996). Very briefly, the distortion underlying neglect and related phenomena has been likened to a pathological remapping of an Euclidean onto a logarithmic scale, with spatial expansion on the contralesional and compression on the ipsilesional side, giving rise to something similar to the Oppel-Kundt illusion (Watt 1994). This hypothesis fits well with phenomena such as the ipsilesional bias in line bisection, pointing to an imaginary point in space, and localization of acoustic stimuli; it is also perfectly compatible with a continuous side-to-side gradient of distortion. Whether it may also fit neglect *strictu sensu* (i.e. in the sense, for instance, of omission of contralesional targets on cancellation tasks) is left to future research and speculation to ascertain.

Acknowledgements. I am indebted to Marlene Behrmann, Ken Heilman, and Otto Karnath for helpful comments on the first version of this chapter. Figures 1 and 2 have been reproduced from Halligan, Marshall (1993) *When two is one...* first published in *Perception* 22, pp 309–312, with permission of Pion Limited, London. Figure 12 redrawn from Bisiach et al. 1984 by permission of Oxford University Press.

References

Adair JC, Williamson DJ, Jacobs DH, Na DL, Heilman KM (1995) Neglect of radial and vertical space: importance of the retinotopic reference frame. J Neurol Neurosurg Psychiatry 58:724–728

Albert ML (1973) A simple test of visual neglect. Neurology 23:658–664

Apfeldorf M (1962) Perceptual and conceptual processes in a case of left-sided spatial inattention. Percept Mot Skills 14:419–423

Behrmann M, Moscovitch M (1994) Object centered neglect in patients with unilateral neglect: effects of left-right coordinates of objects. J Cogn Neurosci 6:1–16

Behrmann M, Tipper SP (1994) Object–based attentional mechanisms: Evidence from unilateral neglect. In: Umiltà C, Moscovitch M (eds) Attention and Performance XV. MIT Press, Cambridge, pp 351–376

Bisiach E, Rusconi ML (1990) Break-down of perceptual awareness in unilateral neglect. Cortex 26:643–649

Bisiach E, Geminiani G (1991) Anosognosia related to hemiplegia and hemianopia. In: Prigatano GP, Schacter DL (eds) Awareness of deficit after brain injury. Oxford University Press, New York, pp 17–39

Bisiach E, Cornacchia L, Sterzi R, Vallar G (1984) Disorders of perceived auditory lateralization after lesions of the right hemisphere. Brain 107:37–52

Bisiach E, Capitani E, Porta E (1985) Two basic properties of space representation. J Neurol Neurosurg Psychiatry 19:543–551

Bisiach E, Perani D, Vallar G, Berti A (1986) Unilateral neglect: personal and extrapersonal. Neuropsychologia 24:759–767

Bisiach E, Rusconi ML, Peretti VA, Vallar G (1994) Challenging current accounts of unilateral neglect. Neuropsychologia 32:1431– 1434

Bisiach E, Pizzamiglio L, Nico D, Antonucci G (1996) Beyond hemineglect. Brain 119:851–857

Brain WR (1941) Visual disorientation with special reference to lesions of the right cerebral hemisphere. Brain 64:244–272.

Butter CM, Mark VW, Heilman KM (1988) An experimental analysis of factors underlying neglect in line bisection. J Neurol Neurosurg Psychiatry 51:1581–1583

Butter CM, Evans J, Kirsch N, Kewman D (1989) Altitudinal neglect following traumatic brain injury: a case report. Cortex 25:135–146

Calvanio R, Petrone PN, Levine DN (1987) Left visual spatial neglect is both environment-centered and body-centered. Neurology 37:1179–1183

Caramazza A, Hillis AE (1990) Spatial representation of words in the brain implied by studies of a unilateral neglect patient. Nature 346:267–269

Chatterjee A (1994) Picturing unilateral spatial neglect: viewer versus object centred reference frames. J Neurol Neurosurg Psychiatry 57:1236–1240

Chokron S, Imbert M (1995) Variations of the egocentric reference among normal subjects and a patient with unilateral neglect. Neuropsychologia 33:703–711

Cohn R (1972) Eyeball movements in homonymous hemianopia following simultaneous bitemporal object presentation. Neurology 22:12–14

Denny-Brown D (1962) Discussion fourth session, A. In: Mountcastle VB (ed) Interhemispheric relations and cerebral dominance. Johns Hopkins Press, Baltimore, pp 244–252

De Renzi E, Gentilini M, Faglioni P, Barbieri C (1989) Attentional shift towards the rightmost stimuli in patients with left visual neglect. Cortex 25:231–237

Driver J, Halligan PW (1991) Can visual neglect operate in object-centred co-ordinates? An affirmative single-case study. Cogn Neuropsychol 8:475–496

Farah MJ, Brunn JL, Wong AB, Wallace MA, Carpenter P (1990) Frames of reference for allocating attention to space: evidence from the neglect syndrome. Neuropsychologia 28:335–347

Gainotti G, Messerli P, Tissot R (1972) Qualitative analysis of unilateral spatial neglect in relation to laterality of cerebral lesions. J Neurol Neurosurg Psychiatry 35:545:550

Galletti C, Battaglini PP, Fattori P (1993) Cortical mechanisms of visual space representation. Biomed Res 14 (Suppl 4):47–54

Graziano MSA, Gross CG (1993) A bimodal map of space: somatosensory receptive fields in the macaque putamen with corresponding visual fields. Exp Brain Res 97:96–109

Guariglia C, Antonucci G (1992) Personal and extrapersonal space: a case of neglect dissociation. Neuropsychologia 30:1001–1009

Halligan PW, Marshall JC (1989) Is neglect (only) lateral? A quadrant analysis of line cancellation. J Clin Exp Neuropsychol 11:793–798

Halligan PW, Marshall JC (1991) Left neglect for near but not far space in man. Nature 350:498–500

Halligan PW, Marshall JC (1993) When two is one: a case study of spatial parsing in visual neglect. Perception 22:309–312

Halligan PW, Marshall JC (1994) Toward a principled explanation of unilateral neglect. Cogn Neuropsychol 11:167–206

Heilman KM, Valenstein E (1979) Mechanisms underlying hemispatial neglect. Ann Neurol 5:166–170

Heilman KM, Bowers D, Watson RT (1983) Performance on hemispatial pointing task by patients with neglect syndrome. Neurology 33:661–664

Heilman KM, Bowers D, Valenstein E, Watson RT (1987) Hemispace and hemispatial neglect. In: Jeannerod M (ed) Neurophysiological and neuropsychological aspects of spatial neglect. Elsevier, Amsterdam, pp 115–150

Henaff MA, Michel F (1992) A peculiar top/left visual neglect following a prestriate lesion. 10th European Workshop on Neuropsychology, Bressanone, 26–31 January.

Humphreys GW, Riddoch MJ (1994) Attention to within-object and between-object spatial representations: multiple sites for visual selection. Cogn Neuropsychol 11:207–241

Ishiai S, Furukawa T, Tsukagoshi H (1989) Visuospatial processes of line bisection and the mechanisms underlying unilateral spatial neglect. Brain 112:1485–1502

Karnath H-O, Schenkel P, Fischer B (1991) Trunk orientation as the determining factor of the 'contralateral' deficit in the neglect syndrome and as the physical anchor of the internal representation of body orientation in space. Brain 114:1997–2014

Karnath H-O, Christ K, Hartje W (1993) Decrease of contralateral neglect by neck muscle vibration and spatial orientation of trunk midline. Brain 116:383–396

Kinsbourne M (1970) A model for the mechanism of unilateral neglect of space. Trans Amer Neurol Assoc 95:143–146

Kinsbourne M (1977) Hemineglect and hemisphere rivalry. In: Weinstein EA, Friedland RP (eds) Hemi-inattention and hemisphere specialization. Raven, New York, pp 41–49

Kinsbourne M, Warrington EK (1962) A variety of reading disability associated with right hemisphere lesions. J Neurol Neurosurg Psychiatry 25:339–344

Kooistra CA, Heilman KM (1989) Hemispatial visual inattention masquerading as hemianopia. Neurology 39:1125–1127

Làdavas E (1987) Is the hemispatial deficit produced by right parietal lobe damage associated with retinal or gravitational coordinates? Brain 110:167–180

Làdavas E, Del Pesce M, Provinciali M (1989) Unilateral attention deficits and hemispheric asymmetries in the control of visual attention. Neuropsychologia 27:353–366

Làdavas E, Petronio A, Umiltà C (1990) The deployment of visual attention in the intact field of hemineglect patients. Cortex 26:307–317

Mark VW, Heilman KM (1988) Does fatigue account for left peripersonal neglect? J Clin Exp Neuropsychol 10:335

Mark VW, Heilman KM (1990) Bodily neglect and orientational biases in unilateral neglect syndrome and normal subjects. Neurology 40:640–3

Marshall JC, Halligan PW (1991) Spatial maps. Nature 352:673–674

Marshall JC, Halligan PW (1993) Visuo-spatial neglect: a new copying test to assess perceptual parsing. J Neurol 240:37–40

Matelli M, Olivieri MF, Saccani A, Rizzolatti (1983) Upper visual space neglect and motor deficits after section of the midbrain commissures in the cat. Behav Brain Res 10:263–285

Mattingley JB, Bradshaw JL, Phillips JG (1992) Impairments of movement initiation and execution in unilateral neglect: directional hypokinesia and bradykinesia. Brain 115:1849–1874

Mennemeier M, Wertman E, Heilman KM (1992) Neglect of near peripersonal space. Brain 115:37–50

Mennemeier M, Chatterjee A, Heilman KM (1994) A comparison of the influences of body and environment centered reference frames on neglect. Brain 117:1013–1022

Morris S, Mickel S, Brooks M, Swavely S, Heilman KM (1985) Recovery from neglect. J Clin Exp Neuropsychol 7:609

Nichelli P, Rinaldi M, Cubelli R (1989) Selective spatial attention and length representation in normal subjects and in patients with unilateral spatial neglect. Brain and Cognition 9:57–70

Paterson A, Zangwill OL (1944) Disorders of visual space perception associated with lesions of the right cerebral hemisphere. Brain 67:331–358.

Pinek B, Duhamel JR, Cavé C, Brouchon M (1989) Audio-spatial deficits in humans: differential effects associated with left versus right hemisphere parietal damage. Cortex 25:175–186

Pizzamiglio L, Cappa S, Vallar G, Zoccolotti P, Bottini G, Ciurli P, Guariglia C, Antonucci G (1989) Visual neglect for far and near extra-personal space in humans. Cortex 25:471–477

Posner MI, Cohen J, Rafal RD (1982) Neural systems control of spatial orienting. Philos Trans R Soc Lond B Biol Sci 298:187–198

Posner MI, Walker JA, Friedrich FJ, Rafal RD (1984) Effects of parietal injury on covert orienting of attention. J Neurosci 4:1863–1874

Rapcsak SZ, Watson RT, Heilman KM (1987) Hemispace-visual field interactions in visual extinction. J Neurol Neurosurg Psychiatry 50:1117–1124

Rapcsak SZ, Cimino CR, Heilman KM (1988) Altitudinal neglect. Neurology 38:277–281

Reuter-Lorenz PA, Posner MI (1990) Components of neglect from right-hemisphere damage: an analysis of line bisection. Neuropsychologia 28:327–333

Riddoch MJ, Humphreys GW (1983) The effect of cueing on unilateral neglect. Neuropsychologia 21:589–599

Rizzolatti G, Matelli M, Pavesi G (1983) Deficit in attention and movement following the removal of postarcuate (area 6) and prearcuate (area 8) cortex in macaque monkeys. Brain 106:655–673

Rizzolatti G, Gentilucci M, Matelli M (1985) Selective spatial attention: one center, one circuit, or many circuits? In: Posner MI, Marin OSM (eds) Attention and performance XI. Erlbaum, Hillsdale, pp 251–265

Robertson LC, Lamb MR (1988) The role of perceptual reference frames in visual field asymmetries Neuropsychologia 26:145–152

Russell B (1948) Human knowledge. Its scope and limits. Allen and Unwin, London

Shelton PA, Bowers, D Heilman KM (1990) Peripersonal and vertical neglect. Brain 113:191–205

Schott B, Jeannerod M, Zahin MZ (1966) L'agnosie spatiale unilatérale: perturbation en secteur des méchanismes d'exploration et de fixation du regard. J Méd Lyon 47:169–195

Stein JF (1992) The representation of egocentric space in the posterior parietal cortex. Behav Brain Sci 15:691–700

Tegnér R (1994) Playing with toy animals: examples of object-based neglect. 12th European Workshop on Cognitive Neuropsychology, Bressanone, 23–28 January.

Tegnér R, Levander M, Caneman G (1990) Apparent right neglect in patients with left visual neglect. Cortex 26:455–458

Teuber HL, Mishkin M (1954) Judgement of visual and postural vertical after brain injury. J Psychol 38:161–175

Tipper SP, Behrmann M (in press) Object-centered not scene-based visual neglect. J Exp Psychol: Hum Percept Perform

Vallar G, Perani D (1987) The anatomy of spatial neglect in humans. In: Jeannerod M (ed) Neurophysiological and neuropsychological aspects of spatial neglect. Elsevier, Amsterdam, pp 235–258

Vallar G, Rusconi ML, Bisiach E (1994) Awareness of contralesional information in unilateral neglect: effects of verbal cueing, tracing and vestibular stimulation. In: Umiltà C, Moscovitch M (eds) Attention and Performance XV. MIT Press, Cambridge, pp 377–391

Vallar G, Guariglia C, Nico D, Bisiach E (1995) Spatial hemineglect in back space. Brain 118:467–472

Watt R (1994) Some points about human vision and visual neglect. Neuropsychol Rehabil 4:213–219

Weinstein EA, Friedland RP (1977) Behavioral disorders associated with hemi-inattention. In: Weinstein EA, Friedland RP (eds) Hemi-inattention and hemisphere specialization. Raven Press, New York, pp 51–62

Young AW, Hellawell DJ, Welch J (1992) Neglect and visual recognition. Brain 115:51–71

Neural Encoding of Space in Egocentric Coordinates ? –

Evidence for and Limits of a Hypothesis Derived from Patients with Parietal Lesions and Neglect

H.-O. KARNATH

Department of Neurology, University of Tübingen, Tübingen, Germany

Patients suffering from neglect after (predominantly) right parietal brain lesions show a characteristic disturbance of visuospatial behavior. They demonstrate a deficient response to stimuli located contralaterally to the lesion and fail to explore the contralesional part of space by eye or limb movements. On the ward, it can be observed that these patients collide with objects positioned on the contralesional side. When working on a visual search field, they only mark those targets located on the ipsilesional side while neglecting those on the contralesional side (Fig. 1). Correspondingly, during the spontaneous ocular exploration of a

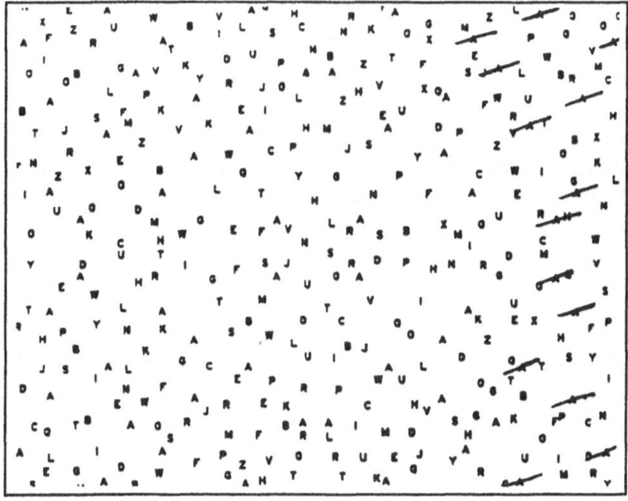

Fig. 1. Cancellation performance of a patient with a right parietal lesion and spatial neglect. In this task (Mesulam 1985), subjects have to mark all target letters "*A*" (n=30 each on the left and right side of the test sheet) randomly distributed with other letters in a search field (DIN A4 size). The patient marked only 14 targets on the extreme right side and likewise neglected the other 46 target letters

In: Parietal Lobe Contributions to Orientation in 3D Space (1997). P. Thier and H.-O. Karnath (eds). Springer-Verlag, Heidelberg.

stimulus array, eye movements are almost exclusively confined to the right, ipsilesional side (Karnath 1994b). Even in darkness, i.e., without any visual stimulus, neglect patients show this bias toward the right side in their exploratory eye movements (Hornak 1992).

To explain this symptomatology, we argued for an altered central representation of egocentric space due to a disturbance of the multisensory coordinate transformation process involved in neural space encoding (Karnath et al. 1993; Karnath 1994a). In accordance with this hypothesis, several recent studies have reported compensatory effects on neglect symptomatology using vestibular, optokinetic, and neck-proprioceptive stimulation that directly affect the egocentric coding of space.

Compensation of Neglect Symptomatology by Vestibular, Optokinetic and Neck-Proprioceptive Stimulation

Silberpfennig (1941) first described a case of temporary remission of visual neglect by peripheral vestibular activation on the ipsilesional side. Rubens (1985) replicated this finding, demonstrating in 17 right brain-damaged patients with left-sided hemineglect that caloric stimulation of the contralesional ear with cold water could transiently reduce the patients' neglect in word reading, line cancellation (Fig. 2), and in pointing to people standing on the left side of their bed. Using the same paradigm, Cappa et al. (1987) confirmed the results. They observed an improvement in cancellation, in patients' ability to point to parts of their upper left extremity, and a regression of anosognosia (the latter in two of their four neglect patients). More recently, Bisiach et al. (1991) reported a patient suffering from a right hemispheric lesion, left sided hemineglect, and somatoparaphrenic delusion. The patient's belief that her arm did not belong to her own body could be temporarily suppressed by vestibular cold water stimulation in the contralesional ear. Rode and Perenin (1994) investigated the effect of vestibular stimulation upon representational neglect with regard to the imagination. Eight neglect patients with right hemispheric lesions were asked to evoke mentally the map of France and to name as many towns as possible. After caloric stimulation with cold water on the left, their poor performance concerning the left side of France, i.e., the western region of France, improved immediately, while their naming of towns on the right side remained unchanged.

Recent studies reported an influence of optokinetic stimulation on deficient task performance in neglect patients. Pizzamiglio et al. (1990) asked non-brain-damaged subjects, and right brain-damaged patients with and without hemineglect, to bisect lines in front of a stationary as well as a moving background. When a moving background was used, a displacement of the subjective midpoint in the direction of movement was observed in all three groups of subjects. Concerning the group of neglect patients, optokinetic stimulation by

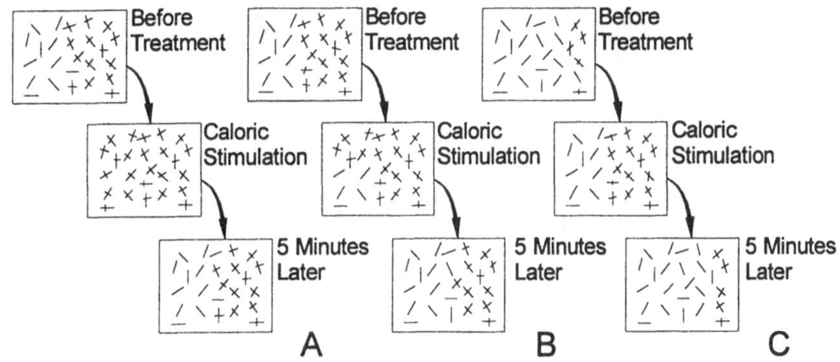

Fig. 2 A-C. Cancellation performance of patients with right hemispheric lesions and spatial neglect. The patients were asked to cross out lines slanted in various directions and spaced randomly on a 9 x 12-inch card. There are three different patterns of line-crossing before, during, and after left ice- or right warm-water vestibular stimulation. With stimulation, five patients crossed out all lines **(A)**, seven improved primarily in the left upper quadrant **(B)**, and five others improved only up to or near the midline **(C)**. (From Rubens 1985, p 1021)

leftward movement of the background could reduce their pathological displacements of bisection marks, whereas optokinetic stimulation to the right markedly increased displacement.

Vallar et al. (1993) found positive effects of optokinetic stimulation on the deficient sense of forearm position in neglect patients with right-sided brain damage. The forearm (not visible for the subject) was moved by the examiners to one of four different locations in the horizontal plane. The patients had to determine the position of the forearm. Compared with right brain-damaged and left brain-damaged patients without neglect, and normal controls, position judgements by the neglect patients were more severely impaired with regard to both the contralateral and the ipsilateral arm. Optokinetic stimulation by a leftward movement reduced the deficit in the position sense of the forearm, while stimulation by a rightward movement led to a worsening of the patients' performance.

Karnath et al. (1993) investigated the effect of neck muscle vibration in neglect patients. Vibration of neck muscles leads to kinaesthetic illusions. In normal subjects, the illusion involves a spatial displacement of trunk or head position (Biguer et al. 1988; Taylor and McCloskey 1991). The illusion can be explained as a (false) central interpretation of the altered proprioceptive input from the periphery; the discharge is centrally coded as lengthening of the vibrated muscles.

Three patients with right-hemisphere lesions and left-sided spatial neglect without visual field defects had to detect and identify stimuli which were tachistoscopically presented in the left visual field (LVF) or right visual field

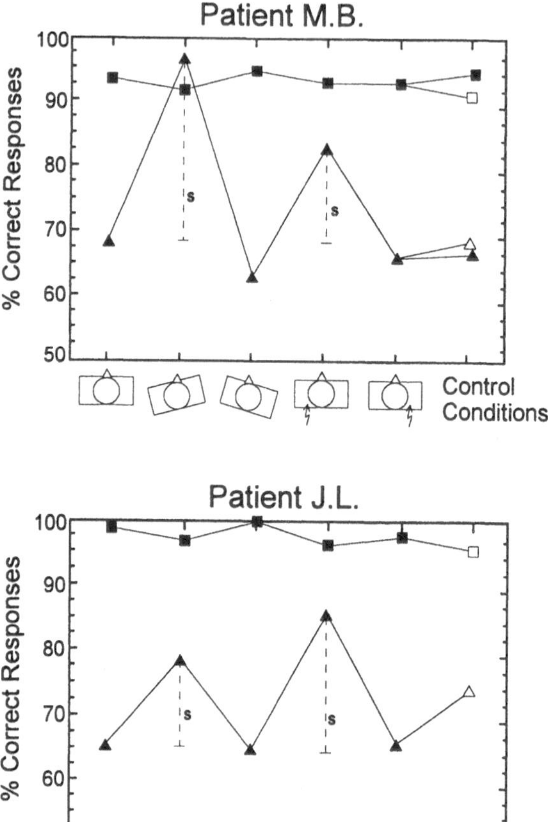

Fig. 3. Percentage of correct responses for unilateral left (*triangles*) and unilateral right (*squares*) visual field stimuli given by two patients with a right hemispheric lesion and spatial neglect under five different test conditions and two control conditions. The spatial relation between the orientation of head and trunk is illustrated as seen from above. The trunk is represented by a *rectangle*, the head by a *circle*. The *arrow* indicates the side of neck muscle vibration. In the first condition, orientation of the head and trunk midline was aligned with the middle of the projection screen and the location of the fixation point. The trunk was then turned 15° to the left (condition II) and to the right (condition III) while the head was facing straight ahead. In the following two conditions, the left (condition IV) and right (condition V) posterior neck muscles were vibrated while trunk and head were facing straight ahead. In the control conditions the head was turned to the left (*filled square* and *triangle*) and/or the left hand muscles were vibrated (*open square* and *triangle*) (*s*, $p \leq 0.05$). (From Karnath et al. 1993, p 390)

(RVF). Neglect of stimuli presented in the contralesional visual field was observed when the patient's body was in a normal upright position with trunk, head, and gaze oriented straight ahead to the middle of the projection screen. The contralateral neglect could be reduced by vibrating the left posterior neck muscles as well as by turning the trunk 15 degrees to the left (Fig. 3). The results show that the proprioceptive information given by real (= turning the trunk) or apparent (= vibration) lengthening of the left posterior neck muscles leads to a remission of contralateral neglect.

Can the Compensatory Effects be Explained by Nonspecific Orientation of General Attentional Processes to the Impaired Side ?

Positive effects on neglect symptoms by simply directing the patients' attention to the neglected side using visual or acoustic cues have frequently been described (e.g., Riddoch and Humphreys 1983; Karnath 1988). The compensatory effects on spatial neglect by vestibular and neck-proprioceptive stimulation, however, cannot be explained by general arousal and activation, due to some salient sensory stimulation on the contralesional side of the body, which might have led to an orientation of general attentional processes to that side.

First, caloric stimulation of the *ipsilesional* external auditory ear canal with *warm* water can reduce the patients' contralateral neglect symptoms, just as stimulation of the contralesional ear canal with cold water can (Rubens 1985). It is thus the direction of apparent ego-motion, as determined by vestibular induction, rather than the stimulated side that correlates with the compensatory effect. Second, a specific effect was also seen when proprioceptive stimulation of different sites on the contralesional side of the body (e.g., the left neck region and the left hand) was compared (Karnath et al. 1993). Vibrating the left posterior neck muscles induced an illusory displacement of body orientation and led to a significant remission of neglect, while vibrating the hand muscles on the same side of the body had no comparable effect (Fig. 3), even though this should presumably attract any general spatial attentional system.

To further examine whether or not the compensatory effect of neck muscle vibration on spatial neglect has a specific origin, Karnath (1995) compared *at the same anatomical site* mechanical muscle vibration versus transcutaneous electrical stimulation; the latter activates afferent nerve fibres rather unselectively. Four neglect patients without visual field defects, one with a lesion of the right basal ganglia and three with a right-sided, predominantly parietal lesion, were examined using a cancellation task and a copying task before, during and after neck muscle vibration, during transcutaneous electrical stimulation of neck muscles and during vibration of hand muscles on the left side of the body. In all patients, neck muscle vibration clearly improved task performance, while transcutaneous electrical stimulation and hand vibration had little or no effect (Fig. 4). The results demonstrate that the effect of neck muscle vibration cannot be explained as arousal and activation due to unspecific sensory stimulation on the contralesional side of the body. Instead, the results support arguments for a specific effect of neck muscle vibration on neglect symptomatology, namely, an influence on the central representation of egocentric space. This conclusion is strengthened by the observation that in man neck muscle vibration induces kinaesthetic illusions and a displacement of the egocentric midline (Biguer et al. 1988; Karnath 1994c), while transcutaneous electrical stimulation produces no such illusions (Karnath 1995; Vallar et al. 1995).

Vallar et al. (1995) also investigated the effect of transcutaneous electrical stimulation on visuo-spatial neglect using a similar cancellation task. In a larger sample of neglect patients, they found a slight, but significant increase in successfully cancelled letters using transcutaneous electrical stimulation on the left side of the neck. The authors, however, observed the same positive effect (about 11% increase) also on transcutaneous electrical stimulation of the left hand. Their finding that stimulation of both the left hand and the left side of the neck results in the same increase in correct target letters strongly supports the assumption that the effect induced by transcutaneous electrical stimulation in

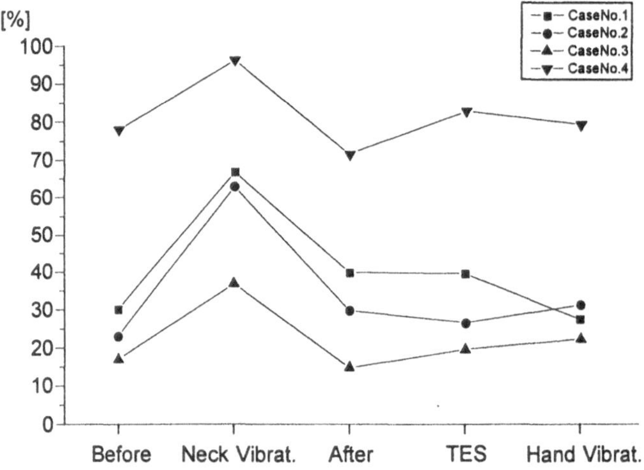

Fig. 4. Cancellation performance of four patients with a right hemispheric lesion and spatial neglect. The cancellation task was identical with that described and shown in Fig. 1. Illustrated is the percentage of correct target letters before, during, and after neck muscle vibration, during transcutaneous electrical stimulation (*TES*) of the left posterior neck muscles, and during vibration of hand muscles on the left. (From Karnath 1995, p 323)

neglect is based on an unspecific sensory activation leading to orientation of general attentional processes to the stimulated side. Such a positive effect on task performance, induced by nonspecifically directing the patients' attention to the neglected side, has previously been found using contralaterally located visual cues (Riddoch and Humphreys 1983), acoustic signals such as a noisy buzzer (Seron et al. 1989), or simply verbal instructions to attend to that side (Karnath 1988). It may thus be concluded that both orientation of spatial attention and encoding of egocentric space seem to play a role in spatial neglect.

To explain the different effects on neglect symptomatology by muscle vibration versus transcutaneous electrical stimulation of the neck, it can be assumed that the two types of stimulation activate nerve fibres differentially. It is broadly accepted that the vibration effect is mediated via muscle spindle activation, while transcutaneous electrical stimulation activates afferent nerve fibres rather unselectively. Although we know that percutaneous vibration in man also activates spindle secondaries and other receptors (Burke et al. 1976a, b), evidence has been reported that muscle spindle primary endings (Ia fibres) are most sensitive to mechanical vibration (Roll et al. 1989). Thus, it can be assumed that the compensatory effect of neck vibration on neglect is induced by the predominant activation of afferent Ia nerve fibres which seem to contribute specifically to elaborating the egocentric spatial reference frame.

Spatial Neglect as the Clinical Consequence of Disturbed Coordinate Transformation in the Central Representation of Egocentric Space

Recent neurophysiological research strongly supports the hypothesis that the brain uses neural representations of space organized in nonretinal, body-centered and/or world-centered coordinates (Andersen et al. 1993; Battaglini et al., this volume; Thier and Andersen 1996 and this volume). To obtain such reference frames, the information coded in coordinates of the peripheral sensory organs must be transformed and integrated. The input from the retina has to be combined with eye position signals as well as head position information (Fig. 5). The eye position information is derived from eye-muscle proprioception as well as the efference copy. One source of information about head position is the proprioceptive input derived from the neck region. Another source of this information is the afferent input from the vestibular system. Clear evidence has been reported that these different input channels directly interact in generating a spatial reference frame (Mergner et al. 1992; Karnath 1994c). Further, it must be assumed that the space around our body is not only perceived by visual, vestibular, and proprioceptive signals but also by tactile and auditory signals. These sensory modalities need to be integrated into a neural representation of space in order to accurately orient the body and guide limb or eye movements within space (Fig. 5). The posterior parietal cortex seems to be the most prominent area of the brain involved in such multimodal transformations (Grüsser et al. 1990a, b; Andersen 1995).

Interpreting their findings in normal humans and cats, Ventre et al. (1984) and Jeannerod and Biguer (1987) speculated that one cause of asymmetrical spatial behavior following a brain lesion in higher vertebrates might be an alteration of the representation of body-centered space. They further speculated that any region of space where the position of objects cannot be encoded and represented in a proper system of coordinates and where orienting movements (e.g., reaching for visual objects) can no longer be generated will be disregarded and neglected.

The compensatory effects on spatial neglect of vestibular, optokinetic, and neck-proprioceptive stimulation seem to confirm this speculation. The findings support the hypothesis that the brain uses a form of neural representation organized in nonretinal coordinates to encode spatial information. They demonstrate that the input from these different information channels directly contributes to computation. The central representation of egocentric space allows us to determine our body position in space and is the basis of explorative motor behavior in space. As illustrated in Fig. 5, spatial neglect seems to result from a disturbance of those cortical structures that are crucial to transforming the sensory input coordinates of the peripheral organs into the egocentric coordinate system.

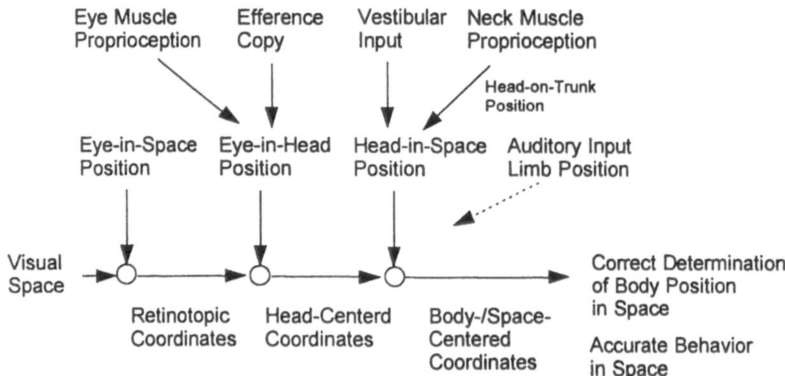

Fig. 5. Coordinate transformations required for the central representation of (objects in) visual space. To correctly determine body orientation and accurate behavior in space, the retinotopic coordinates of visual space are determined and then combined at least with the information of eye-in-head, head-in-space, and head-on-trunk position. Further, it can be assumed that tactile and auditory signals are integrated. In patients with spatial neglect the coordinate transformation seems to work with a systematic error and distortion of the egocentric spatial reference frame leading to an ipsilesional displacement of subjective localization of body orientation and of explorative motor behavior in space. Coordinate transformation is shown to consist of separate steps only for didactic purposes without meaning to rule out other forms of processing. (From Karnath 1994a, p 149)

The disturbed multisensory coordinate transformation leads to (a) an altered perception of body orientation, i.e., to a disparity of the subjective and objective body orientation with a displacement of the subjectively perceived position to the ipsilesional side and (b) to a bias of space exploration in the same direction. Compensating spatial neglect symptomatology by manipulating vestibular, optokinetic and neck-proprioceptive input is interpreted as a central "correction" of neural coordinate transformation, leading to a reorientation of the distorted or deviated egocentric spatial reference frame.

In accordance with the first of these two claims is the observation of an altered perception of body orientation in neglect patients (Heilman et al. 1983). The authors asked five patients with right hemispheric lesions and left-sided neglect to point to an imaginary spot in space perpendicular to the midline of the chest with their eyes closed and found a deviation to the ipsilesional side. Performing a similar task, neglect patients did not benefit from seeing their bodies' orientation (Mark and Heilman 1990). The pointing errors toward the ipsilesional side of a horizontal slit positioned at the center of a board were the same as when their bodies were shielded from view.

Both studies used a motoric pointing task to determine the subjective perception of body orientation. The authors thus could not exclude that a bias in the patients' motor response was responsible for their findings. To prevent any possible interactive effects between the perception of body orientation and motor behavior, Karnath (1994c) asked three patients with a right-sided parietal lesion and neglect but no visual field defects to direct a laser point to the position which they felt to lie exactly "straight ahead" of their bodies' orientation. The laser point, however, was then directed by the experimenter according to the verbal instructions given by the patients.

Whereas in light as well as in darkness the subjective body orientation of the control groups was close to the objective body position, the three neglect patients localized their bodies' sagittal midplane about 15 degrees to the right of the objective orientation. No relevant differences in "straight ahead" were found between the neglect patients and controls in the vertical plane.

The neglect patients' horizontal displacement of sagittal midplane to the right could be compensated for either by neck muscle vibration or by caloric vestibular stimulation of the left side (Fig. 6). When vestibular stimulation was combined with neck muscle vibration, in the control groups as well as in the neglect patients, horizontal deviation was linearly combined by adding or neutralizing the effects observed when either type of stimulation was applied exclusively (Fig. 6). Data analysis further revealed that the neglect patients' ipsilesionally displaced subjective body orientation did not result from disturbed primary perception or disrupted transmission of the vestibular or proprioceptive input from the periphery.

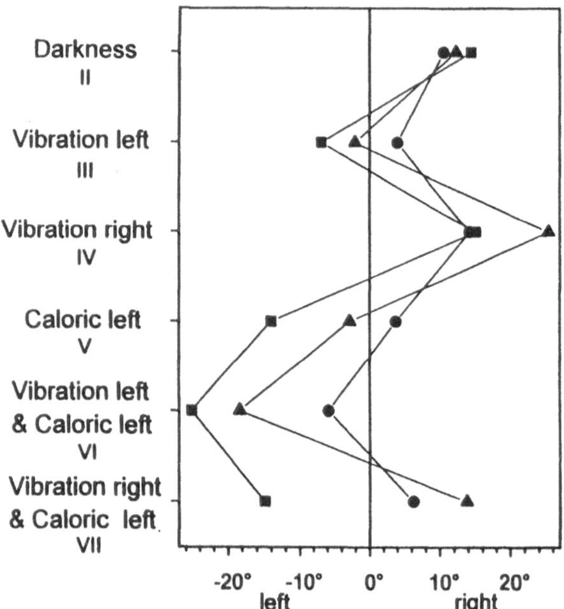

Fig. 6. The influence of neck muscle vibration and vestibular stimulation on the subjective perception of "straight ahead" body orientation was investigated in three patients with right parietal lesions and spatial neglect (*circle*, *triangle*, *square*) in six different test conditions. Condition II (baseline condition): darkness, no additional stimulation; condition III: vibration of the left posterior neck muscles; condition IV: vibration of the right posterior neck muscles; condition V: caloric stimulation (cold-water irrigation) of the left external auditory canal; condition VI: left-sided caloric stimulation (cold-water irrigation) plus vibration of the left posterior neck muscles; condition VII: left-sided caloric stimulation plus vibration of the right posterior neck muscles. Illustrated are the patients' judgements of the subjective "straight ahead" body orientation (in degree of visual angle) in the horizontal plane. (From Karnath 1994c, p 1005)

The effect of optokinetic stimulation on the disturbed perception of body orientation was examined in three additional patients with right-sided brain damage and spatial neglect, using a similar experimental procedure as above (Karnath 1996). Without stimulation these three neglect patients again localized their bodies' sagittal midplane markedly to the right of the objective orientation (by about 16 degrees of visual angle). Optokinetic stimulation was then evoked by a constant linear moving pattern of randomly distributed white dots of different sizes. The patients' horizontal displacement of sagittal midplane was reduced by movement of the surround to the left, and worsened by movement to the right.

The second claim made about the behavioral consequences of disturbed coordinate transformation (see (b) above) is strengthened by a recent study investigating the relationship between the perception of body orientation and ocular exploration of space in patients with parietal lesions and neglect (Karnath and Fetter 1995). The authors investigated whether or not the ipsilesional bias of ocular space exploration seen in neglect patients with respect to the *objective* orientation of the sagittal midplane also exists in relation to the patients' *subjective* localization of body orientation. As expected, ocular exploration of neglect patients was biased toward the ipsilesional right side of the sagittal midplane. However, in relation to the patients' *subjective* localization of the sagittal midplane, exploratory eye movements were symmetrically distributed on the subjective "left" and "right" sides as observed in non-brain-damaged controls. It was concluded that the spatial reference frame, and thus the spatial area in which motor behavior is executed, deviates to the objective right in neglect patients due to disturbed central representation of egocentric space. Consequences of this distortion are a displacement of subjective localization of body orientation and, to the same degree, a displacement of the spatial area in which exploratory motor behavior is executed.

To summarize, several observations have been reported that favor the above model of spatial neglect. They suggest that the information of visual input, together with vestibular and neck-proprioceptive input, is used for computing a unitary central representation of egocentric space. Integration of the contributing input channels is used for object motion detection, dynamic spatial orientation, space exploration, and determination of body position in space.

Does Optokinetic, Neck-Proprioceptive and Vestibular Stimulation Influence the Basic Disorder (*Grundstörung*) Leading to Spatial Neglect?

It is the asymmetric stimulation of the peripheral sensory organs of an input channel that induces apparent ego-motion and the feeling of a displacement of body orientation in space. Thus, one would expect that the same stimulation that improved the disturbed perception of body orientation in patients with brain damage and neglect should induce a transient disparity of subjective and objective body orientation in normal subjects. In fact, this has been observed with vestibular (Fischer and Kornmüller 1931), optokinetic (Brecher et al. 1972) and neck-proprioceptive stimulation (Jeannerod and Biguer 1987). A displaced localization of the body's orientation should further result when the asymmetric input is induced by disturbed primary perception of stimuli via these information channels. Indeed, this has been found in patients with acute unilateral peripheral vestibular disorder (Hörnsten 1979), in patients with hemianopia (Fuchs 1920) and in normal subjects following a prolonged exposure to prismatic displacement of the visual scene (Held and Bossom 1961). However, no neglect was observed in any of these situations.

The findings thus demonstrate that deviation of the egocentric spatial reference frame in normal subjects and in subjects with primary perceptual disorders does not consequently lead to clinical symptomatology identical with that seen in patients with brain damage and spatial neglect. In other words, whereas stimulation reduced the neglect of contralateral stimuli in brain damaged patients, stimulation in normal subjects did not lead to a typical, clinically manifest neglect of stimuli on one side of space. The question may thus arise whether these observations contradict the above explanation of spatial neglect. In this context, Bisiach et al. (1996) discussed the possibility of compensatory vestibular, optokinetic, and neck-proprioceptive stimulation in patients with spatial neglect not affecting the basic disorder, i.e., the *Grundstörung* of the disease, but only modulating the patients' behavior *on top* of it.

However, it has to be taken into consideration that the specific manipulation of sensory input in normal subjects, which obviously induces a behavioral and perceptual disturbance that is also found in patients with spatial neglect, is not a realistic model of clinical neglect and cannot become one, either. It is simply an illustration of "neglect-like" behavior in normal subjects, evoked by certain experimental conditions. The cortical structures that are associated with the occurrence of spatial neglect in case the patients have lesions are, of course, intact in these normal subjects. In contrast, in neglect patients the central structures involved in the multisensory coordinate transformation process, and thus the central representation of egocentric space *itself*, are affected. The brain lesion leads to disturbance of the body-centered reference frame involved in spatial cognition rather than to disturbance of the sensory input, or parts thereof, hence contributing to its elaboration.

The difference between healthy subjects being stimulated, patients with primary perceptual disorders and patients with brain lesions leading to spatial neglect is also illustrated by the following. When one input channel is manipulated in non-brain-damaged subjects, e.g., by vibrating the posterior neck muscles, the subjects still have access to the correct information transmitted by the other contributing channels. Thus, a frequent consequence is a central suppression of the conflicting information. For example, in non-brain-damaged subjects neck muscle vibration induces an illusion of body displacement only in complete darkness. With the light on, no comparable effects are achieved due to the additional information about body position available from the visual system. This compensation for misleading information is, however, not observed in patients with brain damage and spatial neglect. In light as well as darkness, neglect patients localize the body's sagittal midplane to the right of objective orientation (Karnath 1994c).

Besides patients with perceptual disorders leading to asymmetrical sensory deafferentation, a displacement of subjective "straight ahead" is also found in patients with optic ataxia (Perenin, this volume), i.e., in patients with parietal lesions who show a specific visuomotor disorder (misreaching) but no clinical signs of neglect. Obviously, the subjective perception of body orientation does not accompany the same type of visuomotor deficits in patients with optic ataxia and in patients with spatial neglect. Based on additional observations speaking for optic ataxia and neglect being two clinical entities with different anatomical correlates, Perenin (this volume) thus argues for two types of spatial representation and functions represented in the parietal lobe. The upper part of parietal cortex (injured in optic ataxia) is supposed to be involved mainly in direct coding of space for action by means of several effector-specific representations. The inferior part (predominantly associated with the occurrence of spatial neglect) is responsible for more enduring and conscious representations underlying spatial cognition and awareness.

Where are the Limits of Disturbed Representation of Egocentric Space in Accounting for Clinical Neglect Symptoms?

Object-Based Neglect

Recent studies demonstrate that the failure of neglect patients to respond to contralateral stimuli can operate in an object-centered (Driver and Halligan 1991; Driver et al. 1994) as well as a body-centered frame of reference. Clinical bedside-testing reveals this differing behavior when patients are asked to copy a scene consisting of several objects situated on the left and right sides of the test sheet (Gainotti et al. 1972). While some patients only draw the right side of all objects irrespective of whether they are located on the right or on the left side, others copy the objects on the right side correctly but fail to draw those on the left side completely. Both types of behavior can also be observed when analyzing eye movements during exploration of visual stimuli. Rizzo and Hurtig (1992) and Karnath (1994b) described two patients with pure neglect, i.e., without additional visual field defects, who performed eye movements that were almost exclusively confined to the right side of a complex visual scene. On that side, single objects were scanned by eye movements and correctly described. Karnath (1994b) varied the cognitive impact of the drawing's centrally located elements in "pointing" towards supplementary information in the picture's left half. Irrespective of whether there were weak or strong connective elements, exploration of this half did not take place.

Walker et al. (1996) recently reported a patient who made left saccades to locate objects to the left of the display's midline, but restricted most of his fixations to the right side of individual objects. They interpret their findings as their neglect patient having a deficit in orienting his attention within an object-based frame of reference, while being less impaired at orienting between objects. Humphreys and Riddoch (1994) even reported two neglect patients with bilateral lesions who showed neglect for one side of individual objects in the left hemifield when items in the display were coded as parts of a single object, but for entire objects in the right hemifield when items were coded as separate perceptual objects. The authors interpreted their findings as evidence of two forms of spatial representation. One form was thought to be concerned with the relations between parts of single objects, the other one with the relations between separate objects. In their patients, the authors assumed impairments of both representations.

Disturbed coordinate transformation leading to an altered central representation of egocentric space can readily explain spatial neglect, i.e., the patients' deficient response to stimuli located contralaterally to the lesion and their lack of exploration of the contralesional part of space. At first glance, the same mechanism does not account for object-based deficits found in neglect patients. An altered perception of body orientation with subjective displacement of body position to the ipsilesional side and a bias of space exploration in the same direction seem to be of minor relevance for deficits concerning the relation between parts of single objects. However, the existence of object-based neglect phenomena is not inconsistent with the above explanation. As Baylis and Driver (1993) and Humphreys and Riddoch (1994) proposed, there may be distinct cognitive processes of visual selection and perception which can be selectively disturbed. Bisiach (this volume) argues that the different neglect phenomena result from an interplay of egocentric and allocentric coordinates, and thus have to be defined with respect to a system of frames of reference that have both egocentric and allocentric components (for further discussion of this topic see also Farah and Buxbaum as well as Walker and Findlay, this volume).

Visual Extinction

Another issue of current debate is the question of whether or not neglect and extinction are manifestations of the same or different underlying mechanisms (Bisiach 1991; Driver et al., this volume). Following unilateral brain lesions both phenomena can be observed together as well as separately. "Extinction" describes the patients' deficit to report a contralaterally located visual, tactile or auditory stimulus when a second stimulus is simultaneously presented on the ipsilesional side. The same contralesional stimulus can, however, be detected and correctly reported when presented alone. Analysis of CT scans in a series of patients with right-sided brain damage suggested that neglect and extinction might be associated with different anatomical correlates (Vallar et al. 1994). Milner (1995 and this volume) even speculated on whether both phenomena might be associated with a disturbance of two different functional systems, namely, the dorsal and ventral projections from the primary visual cortex to the posterior parietal and inferior temporal cortex, i.e., the dorsal and ventral streams according to Ungerleider and Mishkin (1982).

Thus, it would be interesting to study whether the compensatory effects found in spatial neglect with regard to vestibular, optokinetic and neck-proprioceptive stimulation can also be achieved in patients with extinction. Such investigations would address the question of whether or not the clinical phenomenon of extinction is due to the same mechanism as assumed for spatial neglect, i.e., whether extinction can be regarded as another consequence of disturbed neural coordinate transformation leading to an altered central representation of egocentric space.

An experiment directly concerning this question was recently carried out (Karnath et al., in preparation). The study investigated the effect of manipulating the proprioceptive head-on-trunk signal either by turning the trunk (relative to the stationary head) or by vibrating the posterior neck muscles under conditions of bilateral simultaneous presentation of visual stimuli in patients with clinically manifest extinction and spatial neglect. If extinction is a disturbance with its own pathogenetic origin, then it is expected to occur independently of, and should not be based upon, the afferent signals from the periphery contributing to the elaboration of egocentric spatial coordinates. The patient's real or apparent body orientation in space then should not influence the extinction of contralateral stimuli.

Four patients with unilateral right hemispheric lesions and normal binocular visual fields were investigated. In all patients the lesions were due to an infarct in the territory of the right middle cerebral artery. Three patients showed a lesion centered on the right basal ganglia. The CT scan of the fourth patient revealed a hypodense area in the temporo-parietal cortex. The four patients showed pronounced extinction of contralateral stimuli. Clinical confrontation testing with visual, auditory, and tactile stimuli showed a normal reaction to unilaterally presented right or left stimuli, but in most of the trials there was no reaction to the left stimuli in all three modalities with bilateral presentation. Neuropsychological examination including copying, line bisection, cancellation, and picture comparison further revealed left-sided spatial neglect in each of these subjects. Five patients with unilateral left-sided brain damage served as a control group. None of these patients showed any signs of extinction, hemineglect or visual field defects. As an additional control group, 15 non-brain-damaged dermatological patients were examined.

Stimuli were presented tachistoscopically in the patients' LVF and RVF with an exposure time of 180 ms. The stimuli were all located in an area (4° x 8.6°) beginning 4° to the left and right of the central fixation point. The stimuli were color photographs of geometrical figures. The patients were asked to name the features of the stimuli, i.e., number, color, and shape, on both sides.

Presenting these stimuli simultaneously in the left and right visual field, Karnath (1988) observed that neglect patients spontaneously and stereotypically oriented their attention primarily to the ipsilesional side and always began naming with the stimulus on the side ipsilateral to the lesion. In most of the trials the contralesional stimuli were either completely neglected or only single aspects (e.g., the color of the stimulus) were reported. Recently, Baylis et al. (1993) made a similar observation examining five patients with visual extinction, briefly presenting them with bilateral visual stimuli in the LVF and RVF. Although the order of report for the two sides was not constrained, all extinction subjects spontaneously responded to the ipsilesional side first on each trial. Thus, it seems that regarding bilateral stimulation the patients' attention is automatically attracted by the stimulus located on the ipsilesional side.

The possible influence that spatial trunk orientation and neck muscle vibration might have on patients' extinction and on their spontaneously chosen order of naming bilateral stimuli was investigated in four experimental conditions. In *Condition I*, the baseline condition, patients sat in a normal upright body position in front of the projection screen. Orientation of the head and trunk midline was aligned with the middle of the projection screen and the location of the fixation point. In *Condition II* the trunk was turned 15° to the left while the head was facing straight ahead. This deviation of the trunk midline resulted in both the LVF and RVF stimuli being situated to the right of the trunk midline. In *Condition III* the left posterior neck muscles were vibrated (100 Hz) while trunk and head were facing straight ahead. A *Control Condition* was added to examine the possibility of any decrease in visual extinction under the conditions of neck muscle vibrating and trunk turning being caused by an arousal effect due to nonspecific sensory or proprioceptive stimulation on that side. Therefore, the muscles of the left hand of the patients were vibrated (100 Hz) while the body was in the normal upright position.

In each of the four experimental conditions, 20 bilateral stimuli were presented. In half of the instances, the presentation of a unilateral stimulus in the RVF preceded a bilateral stimulus; in the other half of the instances, a unilateral LVF stimulus preceded the presentation of a bilateral stimulus. This manipulation was used to check the possibility that the spontaneous order of naming the bilaterally presented figures might be influenced mainly by an attentional bias towards that side due to the stimulus location that was relevant for stimulus naming on the preceding trial. In addition, ten stimuli were unilaterally presented in the LVF, ten stimuli unilaterally in the RVF, and five presentations of blank fields on both sides of fixation were used. These stimuli were added and presented with the other stimuli in a pseudo-randomized order so that the subjects could not know ahead of presentation whether or not a stimulus would appear in the RVF alone, in the LVF alone, simultaneously in both visual half-fields, or would not appear at all.

Two of the four patients who clinically showed extinction and spatial neglect, M.B. and J.L., had also been subjects in a previous study (Karnath et al. 1993). In these two patients, the present and previous experiments were conducted within a two-day interval. As illustrated above in Fig. 3, M.B.'s and J.L.'s spatial neglect of unilaterally presented LVF stimuli could be reduced by vibrating the left posterior neck muscles as well as by turning the trunk 15 degrees to the left. The present experiment thus allows a direct comparison of the effect of neck-proprioceptive stimulation on visual extinction with its effect on spatial neglect. Such an *intra*individual comparison of both clinical phenomena, using the same experimental manipulations in the same patients, is of course of special relevance for the theoretical interpretation of the results (compare Figs. 3 and 7).

The present results demonstrate, first, that when two stimuli were presented simultaneously in the right and left visual half-field, the four right brain-damaged patients neglected the contralesional LVF stimulus (Fig. 7). In contrast, these LVF stimuli could be detected and correctly reported when presented alone. In cases M.B. and J.L., the latter finding might first surprise because the previous experiment (see Fig. 3) revealed deficient naming of unilateral LVF stimuli in these patients. The difference, however, is explained by the different duration of stimulus presentation in both experiments. As shown earlier (Karnath 1988; Berti et al. 1992), exposure time, as one parameter that determines task difficulty, influences the degree of the patients' neglect of unilaterally presented stimuli on the contralesional side. The ability to correctly identify and name unilaterally

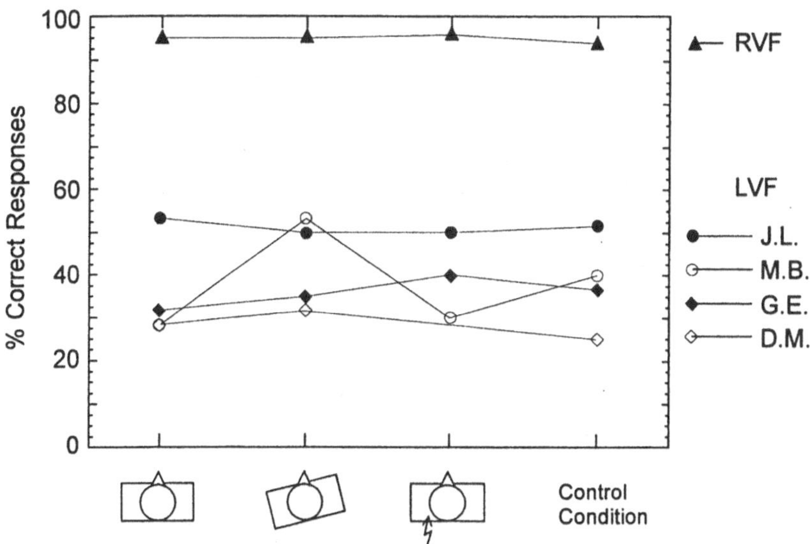

Fig. 7. Percentage of correct responses to bilateral, simultaneously presented left-visual-field (*LVF*) and right-visual-field (*RVF*) stimuli by four right brain-damaged patients with clinically manifest extinction and spatial neglect (J.L., M.B., G.E., D.M.) under three different test conditions and one control condition. The spatial orientation of head and trunk midline is illustrated as described in Fig. 3. The trunk is represented by a *rectangle*, the head by a *circle*. The *arrow* indicates the side of neck muscle vibration. In the first condition, orientation of the head and trunk midline was aligned with the middle of the projection screen and the location of the fixation point. The trunk was then turned 15° to the left (condition II) while the head was facing straight ahead. In the following condition III, the left posterior neck muscles were vibrated while trunk and head were facing straight ahead. In the control condition, the left hand muscles were vibrated. *Filled triangles* represent average RVF performance of the four patients

presented stimuli decreases with brief exposure times as were used in the previous study.

The present results further show that regarding bilateral stimulus presentation the four patients stereotypically oriented their attention primarily in the ipsilesional direction. In nearly all of these trials, they spontaneously began with the figures presented in the ipsilesional RVF when naming stimuli (Table 1). It seems that the patients' attention was automatically attracted by the stimulus located on the ipsilesional side. This fits similar previous observations in patients with neglect and extinction, which were obtained using various experimental techniques (Mark et al. 1988; De Renzi et al. 1989; Gainotti et al. 1991; D'Erme et al. 1992; Baylis et al. 1993).

Table 1. Average number of the spontaneously chosen order of naming bilaterally presented visual stimuli

Condition	Preceding unilateral stimulus (10 trials in each condition)	Onset of naming with the stimulus in the	
		LVF	RVF
I	LVF	0.2	9.8
II	LVF	0.7	9.3
III	LVF	2.0	8.0
CC	LVF	1.2	8.8
I	RVF	0.5	9.5
II	RVF	0.5	9.5
III	RVF	0.0	10.0
CC	RVF	0.0	10.0

In the three experimental conditions and one control condition (CC), four right brain-damaged patients with clinically manifest extinction and spatial neglect were asked to identify and name the stimuli on both sides (n=20 bilateral stimuli each). In each condition, the presentation of a unilateral stimulus in the right visual field (RVF) preceded a bilateral stimulus in ten trials; in ten further trials, a unilateral left visual field (LVF) stimulus preceded the presentation of a bilateral stimulus.

Two distinct types of behavior were found in the group of non-brain-damaged controls. One subgroup almost always began with the figures presented in the LVF when naming bilateral stimuli. This was also observed in the group of left brain-damaged controls. In contrast, a second subgroup of non-brain-damaged controls seemed to show no preference regarding their primary orientation of attention. Further analysis revealed that this balanced distribution of initial orientation of attention can be regarded as the consequence of a priming effect triggered by the preceding unilateral stimulus in the sense of summoning and covertly directing focal attention to that location.

With one exception in one test condition, none of the three groups of subjects showed any reliable change of performance in the four different experimental conditions. The patients' extinction and their spontaneous tendency to orient attention primarily to the ipsilesional side with bilateral stimulation did not change throughout the whole investigation; this behavior occurred when the patients were sitting in a normal upright position and could be influenced neither by turning the trunk to the left, nor by vibrating the left posterior neck muscles. Extinction and the ipsilesional orienting bias occurred irrespective of the patients' real or apparent body orientation with respect to the spatial location of the visual stimuli.

These findings are a contrast to previous observations that showed a remission of spatial neglect with the same manipulations of neck-proprioceptive input (Karnath et al. 1993). In two of the four patients with clinically manifest extinction and spatial neglect, M.B. and J.L., this discrepancy between both phenomena was even observed intraindividually within a two-day interval of testing.

One way of interpreting the discrepancy would be to assume that the experimental manipulation chosen to influence the egocentric spatial reference was simply not strong enough to override the spontaneous rightward orientation of attention. Thus, it appears necessary to conduct further investigations to clarify whether strengthening the effect of reorienting the egocentric spatial reference frame by, for example, combining vibration of neck muscles with caloric vestibular stimulation (Karnath 1994c) and/or modifying the spatial arrangement of the bilateral stimuli, will have any influence on contralateral extinction of stimuli.

A second interpretation would be that extinction and the ipsilesional bias of primary orienting of attention are not based upon a disturbance of the central representation of egocentric space as has been assumed for spatial neglect, but rather have their own pathogenetic origin which is independent of a disrupted body-centered reference frame. With regard to the latter interpretation the question arises as to the mechanism underlying the ipsilesional orienting bias. A model that could readily explain this bias is that proposed by Kinsbourne (1977, 1987). It describes an imbalance in lateral orienting tendencies, with excessive orienting towards the ipsilesional side in neglect patients. Kinsbourne argued that attention is directed along the vector resultant from the interaction of paired opponent processors that are controlled by the right and left hemispheres respectively, each of which directs attention towards the opposite end of a visual display. Activation imbalance in neglect patients biases the vector of attentional orienting and, therefore, elicits ipsilesional shifts of attention and gaze.

According to this model, it is possible that neck-proprioceptive stimulation is sufficient to correct neglect for a single stimulus presented on the contralesional side; however, when an additional stimulus pulls attention rightward, proprioceptive stimulation is not effective to overcome the states of the visual opponent system. Moreover, it could be that the deviated or distorted representation of egocentric space itself is influenced by additional visual stimulation pulling attention rightward. If such a stimulus comes up, as in the extinction paradigm, the deviation might be magnified. Subsequent studies have to test these possibilities.

A severe attentional bias to the right might also affect the time-course of visual awareness, which could explain visual extinction in these patients. Rorden et al. (in press) recently tested the perception of temporal order for separate visual events in two patients with left-sided extinction following right parietal damage. The patients were given two "prior entry" tasks. The first task presented two unconnected bars, one in each visual hemifield, with the patients asked to judge which appeared sooner. Both patients reported that the right bar preceded the left unless the latter led by over 200 ms. The second task presented one continuous line in a scrolling format across the same spatial extent, with the patients asked to judge in which direction the line moved. The patients now performed normally. Thus, the perception of temporal order for separate events was impaired by the lesions, but motion perception within single events was without disruption. These findings strongly suggest a genuine abnormality in the times at which discrete visual events were experienced as occurring by the extinction patients. The contralesional event had to be given a temporal lead of over 200 ms before it was reliably judged to have occurred first. The findings favor the argumentation of Ward et al. (1994) that extinction might be a pathological exaggeration of the attentional competition among simultaneous visual targets that is well established for normal subjects (Duncan 1980). Ward and colleagues thus predicted that visual extinction should be reduced if the two events of double stimulation could be linked to form a single global object. The results of Rorden and coworkers appear to confirm this prediction (a further discussion of this topic is given by Driver et al., this volume).

With respect to the question of whether neglect and extinction are manifestations of the same or different underlying mechanisms, it would be interesting to study whether the delay of visual awareness of contralesional events could be significantly manipulated by sensory input that is assumed to contribute to and thus predominantly influence the egocentric representation of space rather than evoke nonspecific orienting of attention. It should be tested whether or not the delay could be diminished by vestibular, optokinetic or neck-proprioceptive stimulation.

A further conclusion that could be drawn from the present results is that neglect of stimuli presented on the contralesional side follows a *combination* of different components of disturbance (Karnath 1988), i.e., at least, (a) an imbalance of lateral orienting tendencies with spontaneous primary orienting towards the ipsilesional side, and (b) a deficit in reorienting attention from this primary focus to the contralesional side. A similar line of reasoning was put forward by Gainotti and coworkers (1989, 1991). The authors likewise explained contralateral neglect of stimuli by a combination of different pathogenetic components in the context of a multicomponent concept of disturbance.

Based on the present results it may be cautiously speculated whether the first component, the spontaneous primary orienting towards the ipsilesional side, might be the consequence of an attentional bias to the ipsilesional side, while the second component, i.e., the deficit in voluntarily orienting attention from this primary focus toward locations on the contralesional side, may result from a distortion or deviation of the egocentric spatial reference frame.

Acknowledgements. This work was supported by grants from the Deutsche Forschungsgemeinschaft (KA 1258/1-1) and the Bundesministerium für Bildung, Wissenschaft, Forschung und Technologie (01KO9501/III6) awarded to the author. I am grateful to Edoardo Bisiach, Jon Driver, Johannes Dichgans, and Wolfgang Hartje for their helpful discussion and comments on the manuscript. Moreover, I would like to thank Antje Waddington for her language assistance.

References

Andersen RA (1995) Encoding of intention and spatial location in the posterior parietal cortex. Cereb Cortex 5:457–469

Andersen RA, Snyder LH, Li C-S, Stricanne B (1993) Coordinate transformations in the representation of spatial information. Curr Opin Neurobiol 3:171–176

Baylis GC, Driver J (1993) Visual attention and objects: evidence for hierarchical coding of locations. J Exp Psychol 19:451–470

Baylis GC, Driver J, Rafal RD (1993) Visual extinction and stimulus repetition. J Cogn Neurosci 5:453–466

Berti A, Allport A, Driver J, Dienes Z, Oxbury J, Oxbury S (1992) Levels of processing for visual stimuli in an "extinguished" field. Neuropsychologia 30:403–415

Biguer B, Donaldson IML, Hein A, Jeannerod M (1988) Neck muscle vibration modifies the representation of visual motion and direction in man. Brain 111:1405–1424

Bisiach E (1991) Extinction and neglect: same or different? In: Paillard J (ed) Brain and space. Oxford University Press, Oxford, pp 251–257

Bisiach E, Rusconi ML, Vallar G (1991) Remission of somatoparaphrenic delusion through vestibular stimulation. Neuropsychologia 29:1029–1031

Bisiach E, Pizzamiglio L, Nico D, Antonucci G (1996) Beyond unilateral neglect. Brain 119:851–857

Brecher GA, Brecher MH, Kommerell G, Sauter FA, Sellerbeck J (1972) Relation of optical and labyrinthean orientation. Opt Acta 19:467–471

Burke D, Hagbarth KE, Lofstedt L, Wallin BG (1976a) The response of human muscle spindle endings to vibration of non–contracting muscles. J Physiol (Lond) 261:673–693

Burke D, Hagbarth KE, Lofstedt L, Wallin BG (1976b) The response of human muscle spindle endings to vibration during isometric contraction. J Physiol (Lond) 261:695–711

Cappa S, Sterzi R, Vallar G, Bisiach E (1987) Remission of hemineglect and anosognosia during vestibular stimulation. Neuropsychologia 25:775–782

De Renzi E, Gentilini M, Faglioni P, Barbieri C (1989) Attentional shift towards the rightmost stimuli in patients with left visual neglect. Cortex 25:231–237

D'Erme P, Robertson I, Bartolomeo P, Daniele A, Gainotti G (1992) Early rightwards orienting of attention on simple reaction time performance in patients with left–sided neglect. Neuropsychologia 30:989–1000

Driver J, Halligan PW (1991) Can visual neglect operate in object–centred co–ordinates? An affirmative single–case study. Cogn Neuropsychol 8:475–496

Driver J, Baylis GC, Goodrich SJ, Rafal RD (1994) Axis–based neglect of visual shapes. Neuropsychologia 32:1353–1365

Duncan J (1980) The locus of interference in the perception of simultaneous stimuli. Psychol Rev 87:272–300

Fischer MH, Kornmüller AE (1931) Egozentrische Lokalisation. 2. Mitteilung (optische Richtungslosigkeit beim vestibulären Nystagmus). J Psychol Neurol 41:383–420

Fuchs W (1920) Untersuchung über das Sehen der Hemianopiker und Hemiamblyopiker. Z Psychol Physiol Sinnesorg 84:67–169

Gainotti G, Messerli P, Tissot R (1972) Quantitative analysis of unilateral spatial neglect in relation to lateralisation of cerebral lesions. J Neurol Neurosurg Psychiatry 35:545–550

Gainotti G, D'Erme P, De Bonis C (1989) Aspects cliniques et mécanismes de la négligence visuo–spatiale. Rev Neurol (Paris) 145:626–634

Gainotti G, D'Erme P, Bartolomeo P (1991) Early orientation of attention toward the half space ipsilateral to the lesion in patients with unilateral brain damage. J Neurol Neurosurg Psychiatry 54:1082–1089

Grüsser O-J, Pause M, Schreiter U (1990a) Localization and responses of neurones in the parieto–insular vestibular cortex of awake monkeys (Macaca fascicularis). J Physiol (Lond) 430:537–557

Grüsser O-J, Pause M, Schreiter U (1990b) Vestibular neurones in the parieto-insular cortex of monkeys (Macaca fascicularis): visual and neck receptor responses. J Physiol (Lond) 430:559–583

Heilman KM, Bowers D, Watson RT (1983) Performance on hemispatial pointing task by patients with neglect syndrome. Neurology 33:661–664

Held R, Bossom J (1961) Neonatal deprivation and adult rearrangement: complementary techniques for analyzing plastic sensory-motor coordinations. J Comp Physiol Psychol 54:33–37

Hörnsten G (1979) Constant error of visual egocentric orientation in patients with acute vestibular disorder. Brain 102:685–700

Hornak J (1992) Ocular exploration in the dark by patients with visual neglect. Neuropsychologia 30:547–552

Humphreys GW, Riddoch MJ (1994) Attention to within-object and between-object spatial representations: multiple sites for visual selection. Cogn Neuropsychol 11:207–241

Jeannerod M, Biguer B (1987) The directional coding of reaching movements. A visuomotor conception of spatial neglect. In: Jeannerod M (ed) Neurophysiological and neuropsychological aspects of spatial neglect. Elsevier, Amsterdam, pp 87–113

Karnath H-O (1988) Deficits of attention in acute and recovered visual hemi-neglect. Neuropsychologia 26:27–43

Karnath H-O (1994a) Disturbed coordinate transformation in the neural representation of space as the crucial mechanism leading to neglect. In: Halligan PW, Marshall JC (eds) Spatial neglect: position papers on theory and practice. Erlbaum, Hillsdale, pp 147–150

Karnath H-O (1994b) Spatial limitation of eye movements during ocular exploration of simple line drawings in neglect syndrome. Cortex 30:319–330

Karnath H-O (1994c) Subjective body orientation in neglect and the interactive contribution of neck muscle proprioception and vestibular stimulation. Brain 117: 1001–1012

Karnath H-O (1995) Transcutaneous electrical stimulation and vibration of neck muscles in neglect. Exp Brain Res 105:321–324

Karnath H-O (1996) Optokinetic stimulation influences the disturbed perception of body orientation in spatial neglect. J Neurol Neurosurg Psychiatry 60:217–220

Karnath H-O, Fetter M (1995) Ocular space exploration in the dark and its relation to subjective and objective body orientation in neglect patients with parietal lesions. Neuropsychologia 33:371–377

Karnath H-O, Christ K, Hartje W (1993) Decrease of contralateral neglect by neck muscle vibration and spatial orientation of trunk midline. Brain 116:383–396

Kinsbourne M (1977) Hemi-neglect and hemisphere rivalry. Adv Neurol 18:41–49.

Kinsbourne M (1987) Mechanisms of unilateral neglect. In: Jeannerod M (ed) Neurophysiological and neuropsychological aspects of spatial neglect. Elsevier, Amsterdam, pp 69–86

Mark VW, Heilman KM (1990) Bodily neglect and orientational biases in unilateral neglect syndrome and normal subjects. Neurology 40:640–643

Mark VW, Kooistra CA, Heilman KM (1988) Hemispatial neglect affected by non-neglected stimuli. Neurology 38:1207–1211

Mergner T, Rottler G, Kimmig H, Becker W (1992) Role of vestibular and neck inputs for the perception of object motion in space. Exp Brain Res 89:655–668

Mesulam M-M (1985) Attention, confusional states, and neglect. In: Mesulam M-M (ed) Principles of behavioral neurology. Davis, Philadelphia, pp 125–168

Milner AD (1995) Cerebral correlates of visual awareness. Neuropsychologia 33: 1117–1130

Pizzamiglio L, Frasca R, Guariglia C, Incoccia C, Antonucci G (1990) Effect of optokinetic stimulation in patients with visual neglect. Cortex 26:535–540

Riddoch MJ, Humphreys GW (1983) The effect of cueing on unilateral neglect. Neuropsychologia 21:589–599

Rizzo M, Hurtig R (1992) Visual search in hemineglect: what stirs idle eyes? Clin Vision Sci 7:39–52

Rode G, Perenin M-T (1994) Temporary remission of representational hemineglect through vestibular stimulation. Neuroreport 5:869–872

Roll JP, Vedel JP, Ribot E (1989) Alteration of proprioceptive messages induced by tendon vibration in man: a microneurographic study. Exp Brain Res 76:213–222

Rorden C, Mattingley JB, Karnath H-O, Driver J (in press) Visual extinction and prior entry: impaired perception of temporal order with intact motion perception after unilateral parietal damage. Neuropsychologia

Rubens AB (1985) Caloric stimulation and unilateral visual neglect. Neurology 35: 1019–1024

Seron X, Deloche G, Coyette F (1989) A retrospective analysis of a single case neglect therapy: a point of theory. In: Seron X, Deloche G (eds) Cognitive approaches in neuropsychological rehabilitation. Erlbaum, Hillsdale, pp 289–316

Silberpfennig J (1941) Contributions to the problem of eye movements. III. Disturbances of ocular movements with pseudohemianopsia in frontal lobe tumors. Confin Neurol 4:1–13

Taylor JL, McCloskey DI (1991) Illusions of head and visual target displacement induced by vibration of neck muscles. Brain 114:755–759

Thier P, Andersen RA (1996) Electrical microstimulation suggests two different forms of representation of head-centered space in the intraparietal sulcus of rhesus monkeys. Proc Natl Acad Sci USA 93:4962–4967

Ungerleider LG, Mishkin M (1982) Two cortical visual systems. In: Ingle DJ, Goodale MA, Mansfield RJW (eds) The analysis of visual behavior. MIT Press, Cambridge, pp 549–586

Vallar G, Antonucci G, Guariglia C, Pizzamiglio L (1993) Deficits of position sense, unilateral neglect and optokinetic stimulation. Neuropsychologia 31:1191–1200

Vallar G, Rusconi ML, Bignamini L, Geminiani G, Perani D (1994) Anatomical correlates of visual and tactile extinction in humans: a clinical CT scan study. J Neurol Neurosurg Psychiatry 57:464–470

Vallar G, Rusconi ML, Barozzi S, Bernardini B, Ovadia D, Papagno C, Cesarani A (1995) Improvement of left visuo-spatial hemineglect by left-sided transcutaneous electrical stimulation. Neuropsychologia 33:73–82

Ventre J, Flandrin JM, Jeannerod M (1984) In search for the egocentric reference. A neurophysiological hypothesis. Neuropsychologia 22:797–806

Walker R, Findlay JM, Young AW, Lincoln NB (1996) Saccadic eye movements in object-based neglect. Cogn Neuropsychol 13:569–615

Ward R, Goodrich S, Driver J (1994) Grouping reduces visual extinction: neuropsychological evidence for weight-linkage in visual selection. Vis Cogn 1: 101–129

Lesion in a Basis Function Model of Parietal Cortex: Comparison with Hemineglect

A. POUGET[1] and T. J. SEJNOWSKI[2]

[1] Institute for Cognitive and Computational Sciences, Georgetown University, Washington D.C., USA

[2] Howard Hughes Medical Institute, The Salk Institute for Biological Studies, La Jolla, USA and Department of Biology University of California, San Diego, La Jolla, USA

According to current theories of spatial representation, the positions of objects are represented in multiple processing systems throughout the brain, each system specialized for a particular sensorimotor transformation and using its own frame of reference (Stein 1992; Goldberg et al. 1990). The lateral intraparietal area (LIP), for example, appears to encode the locations of objects in oculocentric coordinates, presumably for the control of saccadic eye movements (Colby et al. 1995). The ventral intraparietal cortex (VIP; Colby and Duhamel 1993) and the premotor cortex (Fogassi et al. 1992; Graziano et al. 1994), on the other hand, seem to use head-centered coordinates and might be involved in the control of hand movements toward the face.

This modular theory of spatial representations is not fully consistent with the behavior of patients with parietal or frontal lesions. Such lesions cause hemineglect, a syndrome characterized by a lack of response to sensory stimuli appearing in the hemispace contralateral to the lesion (Heilman et al. 1985). According to the modular view, the deficit should depend on behavior (e.g., oculocentric for eye movements, head-centered for reaching). However, experimental and clinical studies show a more complex pattern. Instead, neglect affects multiple frames of reference simultaneously, and to a first approximation, independently of the task.

This point is particularly clear in an experiment by Karnath et al. (1993; Fig. 1A). Subjects were asked to identify a stimulus that can appear on either side of the fixation point. In order to test whether the position of the stimuli with respect to the body affects performance, two conditions were tested: a control condition with head straight ahead (C1) and a second condition with head rotated 15° on the right (where right is defined with respect to the trunk) or, equivalently, with the trunk rotated 15° on the left (where left is defined with respect to the head) (see Fig. 1A, C2). In C2, both stimuli occurred further to the right of the trunk than in C1, though at the same location with respect to the head and retina. Moreover, the trunk-centered position of the left stimulus in C2 was the same as the trunk-centered position of the right stimulus in C1.

In: Parietal Lobe Contributions to Orientation in 3D Space (1997). P. Thier and H.-O. Karnath (eds). Springer-Verlag, Heidelberg.

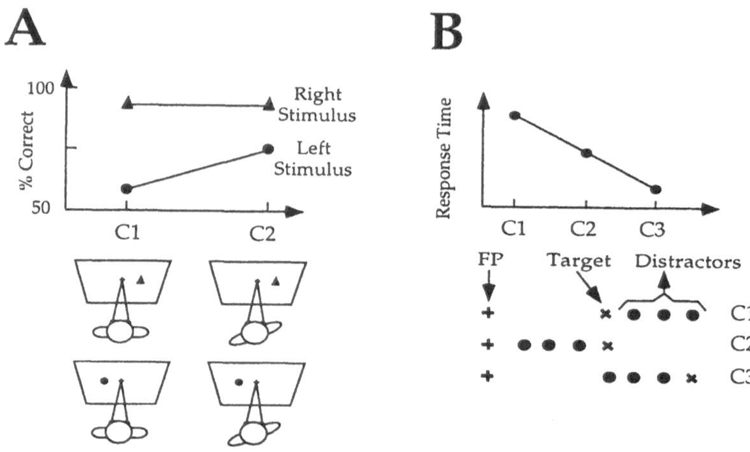

Fig. 1. A Percentage of correct identification in Karnath et al. experiment (1993). In condition 1 (*C1*), subjects were seated with eyes, head, and trunk lined up whereas in condition 2 (*C2*) the trunk was rotated by 15° to the left. The overall pattern of performance is not consistent with pure retinal or pure head-centered neglect and suggests a deficit affecting a mixture of these two frames of reference. **B** Response times for Arguin and Bub (1993) experiment for the three experimental conditions illustrated below the graph (*FP*, fixation point). The decrease from condition 1 (*C1*) to condition 2 (*C2*) is consistent with object centered neglect, i.e., subjects are faster when the target is on the right of the distractors then when it is on the left, even though the retinal position of the target is the same. The further decrease in reaction time in condition 3 (*C3*) shows that the deficit is also retinotopic

As expected, subjects with right parietal lesions performed better on the right stimulus in the control condition (C1), a result consistent with both retinotopic and trunk-centered neglect. To distinguish between the two frames of reference, performance should be compared across conditions.

If the deficit is purely retinocentric, the results should be identical in both conditions since the retinotopic locations of the stimuli do not vary. On the other hand, if the deficit is purely trunk-centered, the performance on the left stimulus should improve when the head is turned right since the stimulus now appears further toward the right of the trunk-centered hemispace. Furthermore, performance on the right stimulus in the control condition should be the same as performance on the left stimulus in the rotated condition since they share the same trunk-centered position in both cases.

Neither of these hypotheses can fully account for the data. As expected from retinotopic neglect, subjects always performed better on the right stimulus in both conditions. However, performance on the left stimulus improved when the head was turned right (C2), though not sufficiently to match the level of performance on the right stimulus in the control condition (C1, Fig. 1A). Therefore, these results suggest a retinotopic neglect modulated by trunk-centered factors.

In addition, Karnath et al. (1991) tested patients on a similar experiment in which subjects were asked to generate a saccade toward the target. The analysis of reaction time revealed the same type of results as the one found in the identification task, thereby demonstrating that the spatial deficit is, to a first approximation, independent of the task. Several other experiments have found that neglect affects a mixture of frames of reference in a variety of tasks (Ladavas 1987; Ladavas et al. 1989; Calvanio et al. 1987; Farah et al. 1990; Bisiach et al. 1985; Behrmann and Moscovitch 1994).

An experiment by Arguin and Bub (1993) suggests that neglect can be object-centered as well. As shown in Fig. 1B, they found that reaction times were faster when a target (cross "x" in Fig. 1B) appeared on the right of a set of distractors (C2) instead of on the left side (C1), even though the target is at the same retinotopic location in both conditions. Interestingly, moving the target further to the right led to even faster reaction times (C3), showing that hemineglect is not only object-centered but retinotopic as well in this task. Several other experiments have led to similar conclusions (Bisiach et al. 1979; Driver and Halligan 1991; Driver et al. 1994; Halligan and Marshall 1994; Husain 1995).

These results strongly support the existence of spatial representations using multiple frames of reference simultaneously shared by several behaviors. We recently developed a theory with these properties (Pouget and Sejnowski 1995, in press); we examine here whether a simulated lesion leads to a deficit similar to hemineglect. Our theory posits that parietal neurons compute basis functions (BFs) of sensory signals, such as visual or auditory inputs, and posture signals, such as eye or head position. The resulting representation, which we called a basis function map, can be used for performing nonlinear transformations of the sensory inputs – the type of transformations required for sensorimotor coordination.

The basis function hypothesis is briefly summarized in the first section of this chapter. In the second section, we describe the network architecture and the various methods used to assess the network performance in behavioral tests. In the third section, we compare the behavior of parietal patients with the network after a unilateral lesion of the basis function representation.

Basis Function Representation

The receptive field of most parietal cells is fixed on the retina, as for V1 neurons. The amplitude of their response to a light, however, is modulated by eye position (Andersen et al. 1985): typically, the response to a visual stimulus in the center of the receptive field increases monotonically as the eye moves along a particular direction in space, specific to each cell.

We have shown in a previous study that these response properties are consistent with the hypothesis that parietal neurons compute basis functions of their inputs

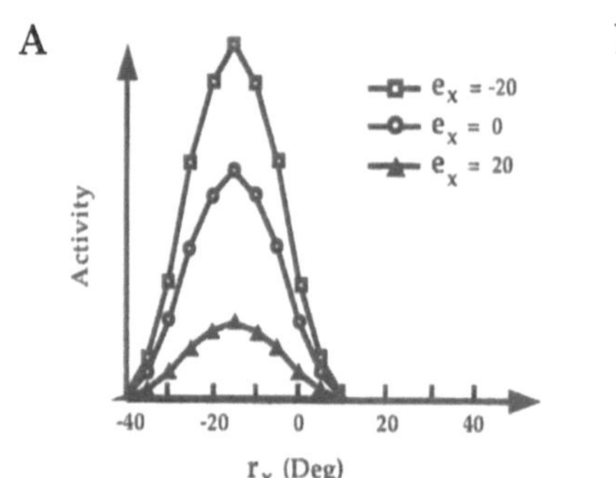

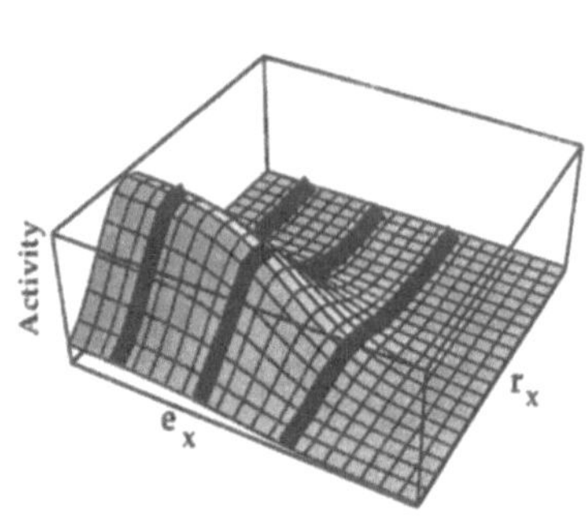

Fig. 2. A Idealization of a retinotopic visual receptive field of a typical parietal neuron for three different gaze angles (e_x). Note that eye position modulates the amplitude of the response but does not affect the retinotopic position of the receptive field (adapted from Andersen et al. 1985). **B** 3D plot showing the response function of an idealized parietal neuron for all possible eye and retinotopic positions, e_x and r_x. The plot in *A* was obtained by mapping the visual receptive field of this idealized parietal neuron for three different eye positions as indicated by the bold lines

(Pouget and Sejnowski 1995, in press). Their response can be described as the product of a gaussian function of retinal location multiplied by a sigmoid function of eye position (Fig. 2B). Sets of both Gaussians and sigmoids are basis functions, and the set of all products of these two basis functions also forms basis functions over the joint space.

A set of basis functions has the property that any *nonlinear* function can be approximated by a *linear* combination of the basis functions. Therefore, basis functions reduce the computation of *nonlinear* mappings to a *linear* transformation – a simpler computation. Most sensorimotor transformations are *nonlinear* mappings of the sensory and posture signals into motor coordinates; hence, given a set of basis functions, the motor command can be obtained by a *linear* combination of these functions. Basis functions are precisely what parietal neurons appear to compute. This formalization entails that the parietal cortex recodes the sensory inputs in a format that facilitates the computation of motor commands. This perspective is consistent with the Goodale and Milner suggestion that the dorsal pathway mediates object manipulation (the "How" pathway), as opposed to simply localizing objects as Mishkin et al. previously suggested (the "Where" pathway) (Goodale and Milner 1990; Mishkin et al. 1983).

It is important to realize that not all models of parietal cells have the properties of simplifying the computation of nonlinear motor commands. For example, Goodman and Andersen (1990) as well as Mazzoni and Andersen (1995) have proposed that parietal cells simply add the retinal and eye position signals. The output of this linear model does not reduce the computation of motor commands to linear combinations because linear units cannot provide a basis set. By contrast, the hidden units of the Zipser and Andersen model (1988) have response properties closer to the basis function units, and the basis function hypothesis can be seen as a formalization of this previous model (for a detailed discussion see Pouget and Sejnowski, in press).

One particularly interesting property of basis functions is the fact that they represent the positions of objects in multiple frames of reference simultaneously. Thus, one can recover simultaneously the position of an object in retinotopic *and* head-centered coordinates from the response of a group of basis function units similar to the one shown in Fig. 2B (Pouget and Sejnowski 1995, in press). As shown in the next section, this property allows the same set of units to be used to perform multiple spatial transformations in parallel.

This approach can be readily extended to other sensory and posture signals and to other parts of the brain where similar gain modulations have been reported (Trotter et al. 1992; Field and Olson 1994; Boussaoud et al. 1993; Bremmer and Hoffmann 1993; Brotchie et al. 1995). When generalized to other posture signals such as vestibular inputs of head position, the resulting representation encodes simultaneously the retinal, head-centered, body-centered, and world centered coordinates of objects.

The study presented here explores the effects of a lesion in a spatial representation using basis functions. It is an attempt to bridge the gap between our current understanding of spatial representations at the neurophysiological and neuropschological levels.

Model Organization

The model contains two distinct parts: a network for performing sensorimotor transformations and a selection mechanism. The selection mechanism is used when there is more than one object present in the visual field at the same time.

Network Architecture

We implemented a network using basis function units in the intermediate layer to perform a transformation from a visual retinotopic map to two motor maps in head-centered and oculocentric coordinates respectively (Fig. 3). The input contains a retinotopic visual map analogous to the one found in the early stages of visual processing and a set of units encoding eye position, similar to the neurons found in the intralaminar nucleus of the thalamus (Schlag-Rey and Schlag 1984). These input units project to a set of intermediate units that contribute to both output transformations. Each intermediate unit computes a Gaussian of the retinal location of the object, r_x, multiplied by a sigmoid of eye position, e_x:

$$o_{ij} = \frac{e^{-\frac{(r_x - r_{xi})^2}{2\sigma^2}}}{1 + e^{-\beta(e_x - e_{xj})}} \tag{1}$$

We consider horizontal positions only because the vertical axis is irrelevant for hemineglect. These units are organized in two two-dimensional maps covering all possible combinations of retinal and eye position selectivities. The only difference between the two maps is the sign of the parameter β which controls whether the units increase or decrease activity with eye position. β was set to 8° for one map and -8° for the other map. The indices (i, j) refer to the position of the units on the maps. Each location is characterized by a position for the peak of the retinal receptive field, r_{xi}, and the midpoint of the sigmoid of eye position, e_{xj}. These quantities are systematically varied along the two dimensions of the maps in such a way that in the upper right corner r_{xi} and e_{xj} correspond to right retinal and right eye positions whereas in the lower left they correspond to left retinal and left eye positions.

As emphasized previously, this type of response function is consistent with the responses of single parietal neurons found in area 7a. The resulting population of units forms basis function maps encoding the locations of objects in head-centered and retinotopic coordinates simultaneously.

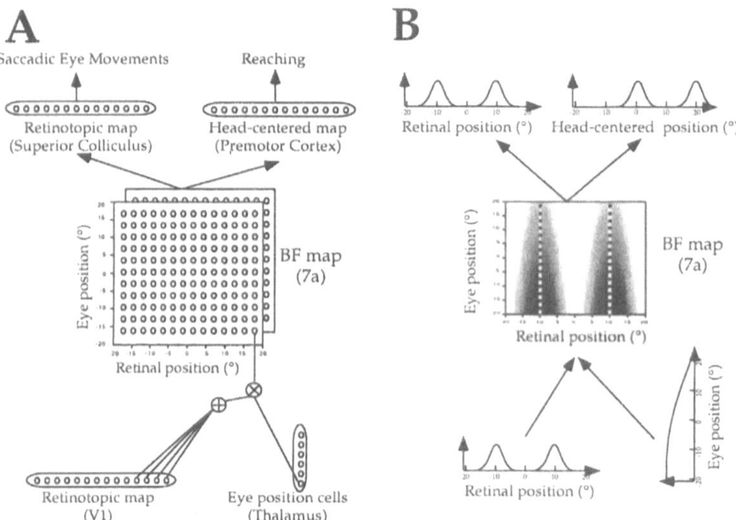

Fig. 3. A Network architecture. Each unit in the intermediate layers is a basis function unit with a gaussian retinal receptive field modulated by a sigmoid function of eye position. This type of modulation is characteristic of the response of parietal neurons. **B** Pattern of activity for two visual stimuli presented at +10° and 10° on the retina with the eye pointing at +10°

The activities of the units in the output maps are computed by a simple linear combination of the BF unit activities. Appropriate values of the weights were found by using linear regression to achieve the least mean square error.

This architecture mimics the pattern of projections of the parietal area 7a, which projects to both the superior colliculus and the premotor cortex (via the ventral parietal area VIP; Andersen et al. 1990; Colby and Duhamel 1993), where neurons have retinotopic and head-centered visual receptive fields respectively. Figure 3B shows a typical pattern of activity in the network when two stimuli are presented simultaneously while the eye is fixated 10° toward the right (only the BF map with positive $\beta = +8°$ is shown).

Hemispheric Biases and Lesion Model

Neurophysiological data indicate that although the parietal cortices in both hemispheres contain neurons with all possible combinations of retinal and eye position selectivities, most cells tend to have their retinal receptive field on the contralateral side (Andersen et al. 1990). Whether a similar contralateral bias exist for the eye position in the parietal cortex remains to be determined although several authors have reported such bias for eye position selectivities in other parts of the brain (Schlag-Rey and Schlag 1984; Galletti and Battaglini 1989; van Opstal et al. 1995).

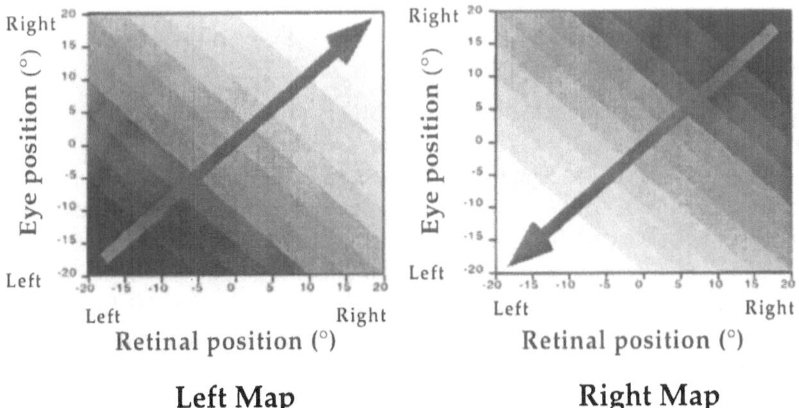

Fig. 4. Neuronal gradients in *left* and *right* basis function maps for which the parameter β is positive, i.e, the units increase activity with eye position. The right map contains more neurons for left retinal and left eye positions while the left map has the opposite gradient

In the model, we divide the two BF maps into two sets of two maps, one set for each hemisphere (again the two maps in each hemisphere correspond to two possible values for the parameter β). Units are distributed across each hemisphere to create *neuronal* gradients. These *neuronal* gradients induce contralateral *activity* gradients such that there is more activity overall in the left maps than in the right maps when an object appears on the right of the retina and the eyes are turned to the right, with the opposite being true in the right maps.

Several types of *neuronal* gradients can lead to such activity gradients. The gradients we use for the simulations presented here affect only the maps with positive β, i.e, maps with units whose activity increases as the eyes turn right. In both the right and left map, the number of units for a given pair of (r_{xi}, e_{xi}) increases for contralateral values of eye and retinal location as indicated in Fig. 4 which is consistent with the experimental observation that hemispheres overrepresent contralateral positions.

To model a right parietal lesion, we removed the right parietal maps and studied the network behavior with the left maps alone. The effect of the lesion is therefore to induce a *neuronal* gradient such that there is more activity in the network for right *retinal* and right *eye* positions.

We found that the exact profile of the *neuronal* gradient across the basis function maps did not matter as long as it induces a monotonically increasing *activity* gradient as objects are moved further to the right of the retina and the eyes fixate further to the right. The results presented in this chapter were obtained with linear *neuronal* gradients.

Selection Model

We adapted a selection mechanism from Burgess (1995) which itself was inspired by the visual search theory of Treisman and Gelade (1980) and the saliency map mechanism proposed by Koch and Ullman (1985). It was used to model the behavior of patients when presented with several stimuli simultaneously and it operates on what we call the *saliency* value associated with each stimulus.

The simultaneous presentation of stimuli induces multiple hills of activity in the network (see for instance the pattern of activity shown in Fig. 1B for two visual stimuli). We defined the stimulus *saliency*, s_i, as being the sum of the activities of all the basis function units whose receptive field is centered exactly on the retinal position of the stimulus (it is the sum of activities along the dotted line shown on the basis function map in Fig. 3B). The index i varies from 1 to n, with n as the number of stimuli in view at a given time. This method is mathematically equivalent to looking at the profile of activity in the superior colliculus output map and defining the saliency of the stimulus as the peak value of activity. Consequently, one need only consider the profiles of activity in the colliculus output map to determine the network behavior. Qualitatively similar values could also be obtained by looking at the profile of activation in the head-centered map.

At the first time step, the stimulus with the highest saliency is selected by winner-take-all, and its corresponding saliency is set to zero to implement inhibition of return. At the next time step, the second highest stimuli is selected, and inhibited, while the previously selected item is allowed to recover slowly. These operations are repeated for the duration of the trial. This procedure ensures that the most salient items are not selected twice in a row, but because of the recovery process, the stimuli with the highest saliencies might be selected again if displayed long enough.

This mechanism is such that the probability of selecting an item is proportional to two factors: the absolute saliency associated with the item and the saliency relative to the ones of competing items.

Evaluating Network Performance

We used this model to simulate several experiments in which patient performance was evaluated according to reaction time or percent of correct response.

In reaction time experiments, we assumed that processing involves two sequential steps: target selection and target processing. Target selection time was assumed to be proportional to the number of iterations, *n*, required by the selection network to select the stimulus using the mechanism described above. Each iteration was arbitrarily chosen to be 50 ms long. This duration matters only when more than one stimulus is present, so that distractors could delay the detection of the target by winning the competition.

The time (RT) for target processing (that is to say, target recognition, target naming, etc.) was assumed to be inversely proportional to stimulus saliency, s_i.

$$RT = 100 + 50n + \frac{500}{1000 s_i}. \tag{2}$$

We determined the percentage of correct responses to a stimulus to be a sigmoid of the stimulus saliency:

$$p = \frac{0.5}{(1 + exp(-(s_i - s_O)/t)} + 0.5, \tag{3}$$

where s_o and t_o are constants.

This is a standard method in signal detection theory when assuming gaussian noise of equal variance for signal and noise (Green and Swets 1966). This is equivalent to assuming that the rate of correct detection (hit rate) is the integral of the probability distribution of the signal from the decision threshold to infinity.

In line bisection experiments, subjects are asked to judge the midpoint of a line segment. Our network estimated the midpoint of a line, $\vec{m}$, by computing the center of mass of the activity induced by the line in the BF map.

$$\vec{m} = \frac{\sum_{all\ units} a_i r_{xi}}{\sum_{all\ units} a_i} \quad (4)$$

where r_{xi} is the retinal position of the peak of the visual receptive field of unit i.

Results

All the results concern the lesioned model only, i.e., the model in which the right BF maps have been removed.

Line Cancellation

We first tested the network on the line cancellation test, a test in which patients are asked to cross out short line segments uniformly spread over a page. To simulate this test, we presented the display shown in Fig. 5A and ran the selection mechanism to determine which lines get selected by the network. As illustrated in Fig. 5A, the network crossed out only the lines located in the right half of the display, just as left neglect patients do in the same task (Heilman et al. 1985). The rightward gradient introduced by the lesion makes the right lines more salient than the left lines. As a result, the rightmost lines always win the competition, preventing the network from selecting the left lines.

We computed the probability that the line was crossed out as a function of its position in the display, where position is defined with respect to the frame of the display (Fig. 5A). We found that there is a sharp jump in the probability function such that lines on the right of this break have a probability near 1 of being selected whereas lines on the left of the break have a probability near zero (Fig. 5B).

The sharp jump in the probability of selection stands in contrast to the smooth and monotonic profile of the neuronal gradient. Whereas the sharp boundary in the pattern of line crossing may suggest that the model "sees" only one half of the display, the linear profile of the neuronal gradient shows that this is not the case. The sharp jump is mostly the consequence of the dynamics of the selection process: because right bars are associated with higher saliencies, they consistently win the competition to the detriment of left bars. Consequently, the network starts by selecting the bar the furthest on the right and due to inhibition of return moves its way toward the left. Eventually, however, previously inhibited items recover

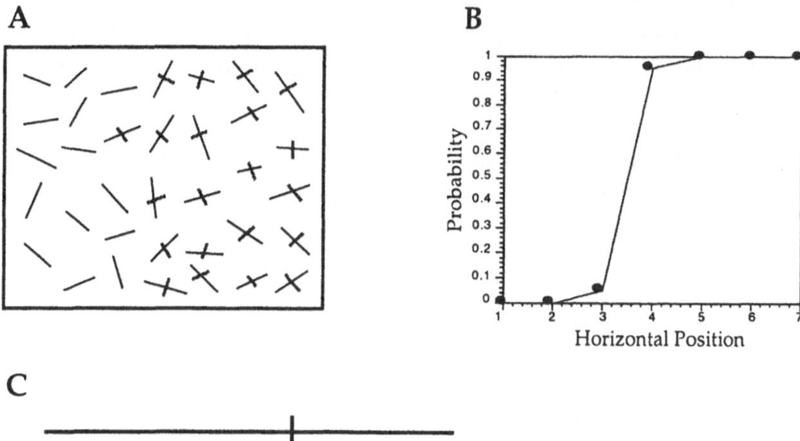

Fig. 5. A Network behavior in *line cancellation* task. As with right parietal patients, the network fails to cross out the line segments on the left of the page. **B** Probability of crossing a line as a function of its horizontal position in the display. The network behaves as if it had no representation of the left side of the display, i.e, as if the neuronal gradient introduced by the lesion were a step function. The gradient however is smooth, and the sudden change in behavior in the middle of the display is the result of the dynamics of the selection mechanism. **C** Network behavior in *line bisection* task. The midpoint is estimated too far to the right due to the overrepresentation of the right side of space

and win the competition again, preventing the network from selecting the leftmost bars. The point at which the network stops selecting bars toward the left depends on the exact recovery rate and the total number of items displayed.

The pattern of line crossing by the network is not the result of a deficiency in the selection mechanism. It is the result of a selection mechanism operating on a lesioned spatial representation. The network had trouble detecting stimuli on the left side of space not because it was unable to orient toward that side of space – it would orient to the left if only one stimulus were presented in the left hemifield – but because the bias in the representation favored the rightmost bars in the competition.

Line Bisection

In the line bisection task, the network estimated the line midpoint to be slightly to the right of the actual midpoint (Fig. 5C) as reported in patients with left neglect (Heilman et al. 1985). In contrast, the performance of an intact network was perfect (not shown).

The error was not because the lesioned network did not "see" the left side of the line. On the contrary, the whole line was represented in the lesioned network but due to the neuronal gradient, more neurons respond to the right side of the line than the left side. As a result, the center of mass calculation used to estimate the middle of the line leads to a rightward error.

Therefore, as assessed by the line cancellation and line bisection tests, a lesioned network exhibited a behavior consistent with the neglect syndrome observed in humans following unilateral parietal lesions.

Mixture of Frames of Reference

Next, we sought to determine the frame of reference of neglect in the model. Since Karnath et al. (1993) manipulated head position, we simulated their experiment by using a BF map integrating visual inputs with head position, rather than eye position. We show in Fig. 6B the pattern of activity obtained in the retinotopic output layer of the network in the various experimental conditions. In both conditions, head straight ahead (dotted lines) or turned on the side (solid lines), the right stimulus is associated with more activity than the left stimulus. This is the consequence of the larger number of cells in the basis function map for

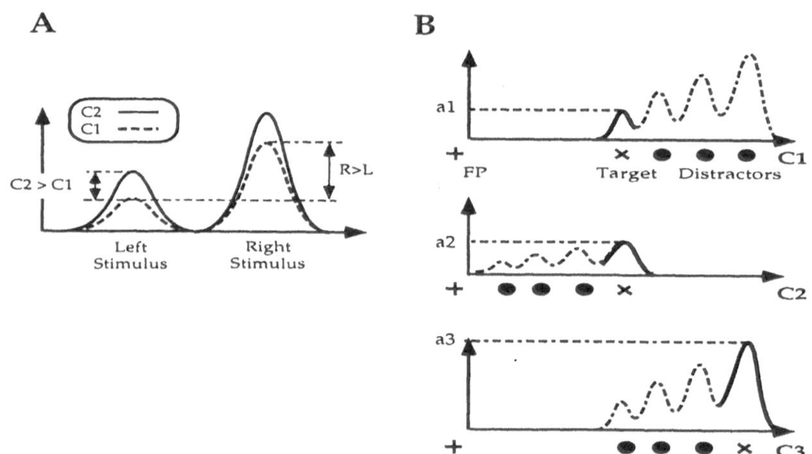

Fig. 6 **A, B.** Activity patterns in the retinotopic output layer when simulating the experiments by **A** Karnath et al. (1993) and **B** Arguin and Bub (1993). **A** Performance on the left stimulus improves from condition 1 (*C1*) to condition 2 (*C2*) because the stimulus saliency increases across conditions. **B** Reaction time between conditions 1 and 2 decreases due to the change in the relative saliency of the target with respect to the distractors, even though the absolute saliency of the target is the same in these two conditions (*a1=a2*). *FP*, Fixation point; *C3*, condition 3

rightward position. In addition, the activity for the left stimulus increases when the head is turned to the right. This effect is related to the larger number of cells in the basis function maps tuned to right head positions.

Since network performance is proportional to activity strength, the overall pattern of performance was found to be similar to what has been reported in human patients (Fig. 1A): the right stimulus was better processed than the left stimulus, and performance on the left stimulus increases when the head is rotated toward the right, although not sufficiently to match the peformance on the right stimulus in condition 1. Therefore, as in humans, neglect in the model is neither retinocentric nor trunk-centered alone but both at the same time.

Similar principles can be used to account for the behavior of patients in many other experiments dealing with frames of reference (Ladavas 1987; Ladavas et al. 1989; Calvanio et al. 1987; Farah et al. 1990; Bisiach et al. 1985; Behrmann and Moscovitch 1994).

Object-Centered Effect

The network reaction times in simulations of the Arguin and Bub (1993) experiments followed the same trends reported in human patients (Fig. 1B). Figure 6B illustrates the patterns of activity in the retinotopic output layer of the network for the three conditions in those experiments. Although the absolute levels of activity associated with the target (solid lines) in conditions 1 and 2 are the same, the activity of the distractors (dotted lines) differed in the two conditions. In condition 1, they had relatively higher activity and thereby strongly delayed the detection of the target by the selection mechanism. In condition 2, the distractors were less active than the target and did not delay target processing as much as they did in condition 1. The reaction time decreased even more in condition 3 because the absolute activity associated with the target was higher. Therefore, the network exhibited retinocentric and object-centered neglect, with the same pattern observed in parietal patients (Arguin and Bub 1993).

The object-centered effect might not have been expected since there was no *explicit* object-centered representation in the model. This result demonstrates that object-based neglect does not necessarily imply that an explicit object-based representation has been lesioned in neglect patients. The form of neglect found in the Arguin and Bub (1993) experiment could be a consequence of *relative* neglect since the apparent object-based effect could be explained by the relative saliency of the subparts of the object.

Other results in neglect can be explained with the same principles if we assume that the basis function map does not simply reflect the retinal image but receives instead a preprocessed version of the image. If the parietal cortex represents only the object that is attended, it is then possible to account for the interaction that has been recently reported between scene segmentation and neglect (Driver et al.

1992; Halligan and Marshall 1994). We predict in particular that subjects with right parietal lesions will neglect the left side of the attended object – the one that has been segmented and selected – regardless of its position in space. This is indeed consistent with results reported from patients (Driver et al. 1992; Halligan and Marshall 1994).

If preprocessing of the image involves a normalization of the image for rotation – a form of mental rotation – then neglect of the left side of rotated objects can also be explained (Driver and Halligan 1991; Driver et al. 1994; Buxbaum et al. 1995). Similarly, if the basis function representation is the "screen" used for visual imagery, then the model can replicate the well-known inability of neglect patients to imagine the left side of a visual scene (Bisiach and Luzzatti 1978).

Discussion

The model of the parietal cortex presented here was originally developed by considering the response properties of parietal neurons and the computational constraints inherent in sensorimotor transformations. It was not designed to model neglect, so its ability to account for a wide range of deficits is additional evidence in favor of the basis function hypothesis.

As we have shown, our model captures three essential aspects of the neglect syndrome: (1) It reproduces the pattern of line crossing of parietal patients in line cancellation and line bisection experiments; (2) the deficit coexists in multiple frames of reference simultaneously; and (3) the model accounts for some of the object-based effects. These results rely in part on the existence of monotonic gradients along the retinal and eye position axis of the basis function map. As we have seen, the retinal gradient is supported by neurophysiological recordings in the parietal cortex (Andersen et al. 1990), but gradients for the postural signals remain to be demonstrated. The retinal gradient hypothesis is also at the heart of Kinsbourne's theory of hemineglect (Kinsbourne 1987) and some models of neglect dyslexia and line bisection are based on a similar idea (Mozer and Behrmann 1990; Mozer et al. in press).

Recent lesion experiments in monkeys suggest that, contrary to what was widely assumed, area 7 in the monkey may not be the homologue of the inferior parietal areas 39 and 40 in humans, the ones that are typically lesioned in the neglect syndrome (Watson et al. 1994). Instead, it would appear that the areas found in the superior temporal sulcus (STS) of the monkey cortex are the analogues to areas 30 and 40. If this report is confirmed, then we predict that the responses of cells in the STS should have gain fields to integrate sensory and posture signals, as in the parietal cortex.

Our approach can account for many studies beyond the ones considered here by using similar computational principles. It can reproduce, in particular, the behavior of patients in line-bisection experiments (Halligan and Marshall 1989;

Burnett-Stuart et al. 1991; Bisiach et al. 1994) and a variety of experiments dealing with frames of reference, whether in retinotopic, trunk-centered (Bisiach et al. 1985; Moscovitch and Behrmann 1994), environment-centered (Ladavas 1987; Farah et al. 1990) (i.e., with respect to gravity), or object-centered coordinates (Driver and Halligan 1991; Halligan and Marshall 1994; Husain 1995). It is also possible to account for the inability of parietal patients to imagine the contralesional side of a visual scene if visual imagery uses a basis function map as its "projection screen" (Bisiach and Luzzatti 1978). Finally, a model with a basis function map integrating sensory signals with vestibular inputs would also exhibit a temporary recovery after strong vestibular stimulation, as reported in humans following caloric stimulation of the inner ear. The mechanisms at play would be identical to the ones involved in the performance improvement on left targets in Karnath et al. (1993) experiments when subjects turn their head to the right (Figs. 1A and 6A).

The results presented in this chapter have been obtained without using explicit representations of the various cartesian frames of reference (except for the retinotopy of the BF map). In fact, it is precisely because the lesion affected noncartesian representations that the model was able to reproduce these results. The lesion affects the functional space in which the basis functions are defined, which shares common dimensions with cartesian spaces, but cannot be reduced to them. Hence, a basis function map integrating retinal location and head position is retinotopic, but not solely retinotopic. Consequently, any attempt to determine the cartesian space in which hemineglect operates is bound to lead to inconclusive results in which cartesian frames of reference appear to be mixed.

It would be interesting to see if the basis function hypothesis could also account for sensorimotor adaptation, such as learning to reach accurately while wearing visual prisms. We predict that adaptation takes place in several frames of reference simultaneously, a prediction that is testable and would provide further support for the basis function framework.

Acknowledgements. This research was supported in part by a fellowship from the McDonnell-Pew Center for Cognitive Neuroscience to A.P. and grants from the Office of Naval Research and the Howard Hughes Medical Institute to T.J.S. We thank Daphne Bavelier for stimulating and inspiring discussions.

References

Andersen R, Essick G, Siegel R (1985) Encoding of spatial location by posterior parietal neurons. Science 230:456–458

Andersen R, Asanuma C, Essick G, Siegel R (1990) Corticocortical connections of anatomically and physiologically defined subdivisions within the inferior parietal lobule. J Comp Neurol 296:65–113

Arguin M, Bub D (1993) Evidence for an independent stimulus-centered reference frame from a case of visual hemineglect. Cortex 29:349–357

Behrmann M, Moscovitch M (1994) Objects-centered neglect in patients with unilateral neglect: effects of left-right coordinates of objects. J Cogn Neurosci 6:151–155

Bisiach E, Luzzatti C (1978) Unilateral neglect of representational space. Cortex 14:129–133

Bisiach E, Luzzatti C, Perani D (1979) Unilateral neglect, representational schema and conciousness. Brain 102:609–618

Bisiach E, Capitani E, Porta E (1985) Two basic properties of space representation in the brain: evidence from unilateral neglect. J Neurol Neurosurg Psychiatry 48:141–144

Bisiach E, Rusconi M, Peretti V, Vallar G (1994) Challenging current accounts of unilateral neglect. Neuropsychologia 32:1431–1434

Boussaoud D, Barth T, Wise S (1993) Effects of gaze on apparent visual responses of frontal cortex neurons. Exp Brain Res 93:423–434

Bremmer F, Hoffmann K (1993) Pursuit related activity in macaque visual cortical areas MST and LIP is modulated by eye position. Soc Neurosci Abstr: 1283

Brotchie P, Andersen R, Snyder L, Goodman S (1995) Head position signals used by parietal neurons to encode locations of visual stimuli. Nature 375:232–235

Burgess N (1995) A solvable connectionist model of immediate recall of ordered lists. In: Tesauro G, Touretzky D, Leen T (eds) Advances in neural information processing Systems, vol 7. MIT Press, Cambridge

Burnett-Stuart G, Halligan P, Marshall J (1991) A newtonian model of perceptual distortion in visuo-spatial neglect. Neuroreport 2:255–257

Buxbaum LJ, Coslett HB, Montgomery MW, Farah MJ (1996) Mental rotation may underlie apparent object-based neglect. Neuropsychologia 34:112–126

Calvanio R, Petrone P, Levine D (1987) Left visual spatial neglect is both environment-centered and body-centered. Neurology 37:1179–1181

Colby C, Duhamel J (1993) Ventral intraparietal area of the macaque: anatomic location and visual response properties. J Neurophysiol 69:902–914

Colby C, Duhamel J, Goldberg M (1995) Oculocentric spatial representation in parietal cortex. Cereb Cortex 5:470–481

Driver J, Halligan P (1991) Can visual neglect operate in object-centered coordinates? An affirmative single case study. Cogn Neuropsychol 8:475–496

Driver J, Baylis G, Rafal R (1992) Preserved figure-ground segregation and symmetry perception in visual neglect. Nature 360:73–75

Driver J, Baylis G, Goodrich S, Rafal R (1994) Axis-based neglect of visual shapes. Neuropsychologia 32:1353–1365

Farah M, Brunn J, Wong A, Wallace M, Carpenter P (1990) Frames of reference for allocating attention to space: evidence from the neglect syndrome. Neuropsychologia 28:335–47

Field P, Olson C (1994) Spatial analysis of somatosensory and visual stimuli by single neurons in macaque area 7B. Soc Neurosci Abstr 20:317.12

Fogassi L, Gallese V, di Pellegrino G, Fadiga L, Gentilucci M, Luppino G, Matelli M, Pedotti A, Rizzolatti G (1992) Space coding by premotor cortex. Exp Brain Res 89:686–690

Galletti C, Battaglini P (1989) Gaze-dependent visual neurons in area V3a of monkey prestriate cortex. J Neurosci 9:1112–1125

Goldberg M, Colby C, Duhamel J (1990) Representation of visuomotor space in the parietal lobe of the monkey. Cold Spring Harb Symp Quant Biol 55:729–739

Goodale M, Milner A (1990) Separate visual pathways for perception and action. Trends Neurosci 15:20–25

Goodman S, Andersen R (1990) Algorithm programmed by a neural model for coordinate transformation. International Joint Conference on Neural Networks, San Diego

Graziano M, Yap G, Gross C (1994) Coding of visual space by premotor neurons. Science 266:1054–1057

Green D, Swets J (1966) Signal detection theory and psychophysics. Wiley, New York

Halligan P, Marshall J (1989) Line bisection in visuo-spatial neglect: disproof of a conjecture. Cortex 25:517–521

Halligan P, Marshall J (1994) Figural perception and parsing in visuospatial neglect. Neuroreport 5:537–539

Heilman K, Watson R, Valenstein E (1985) Neglect and related disorders. In: Heilman K, Valenstein E (eds) Clinical neuropsychology. Oxford University Press, New York, pp 243–294

Husain M (1995) Is visual neglect body-centric? J Neurol Neurosurg Psychiatry 58:262–263

Karnath H-O, Schenkel P, Fischer B (1991) Trunk orientation as the determining factor of the 'contralateral' deficit in the neglect syndrome and as the physical anchor of the internal representation of body orientation in space. Brain 114:1997–2014

Karnath H-O, Christ K, Hartje W (1993) Decrease of contralateral neglect by neck muscle vibration and spatial orientation of trunk midline. Brain 116:383–396

Kinsbourne M (1987) Mechanisms of unilateral neglect. In: Jeannerod M (ed) Neurophysiological and Neuropsychological aspects of spatial neglect. North-Holland, pp 69–86

Koch C, Ullman S (1985) Shifts in selective visual attention: towards the underlying neural circuitry. Human Neurobiol 4:219–227

Ladavas E (1987) Is the hemispatial deficit produced by right parietal lobe damage associated with retinal or gravitational coordinates? Brain 110:167–180

Ladavas E, Pesce M, Provinciali L (1989) Unilateral attention deficits and hemispheric asymmetries in the control of visual attention. Neuropsychologia 27:353–366

Mazzoni P, Andersen R (1995) Gaze coding in the posterior parietal cortex. In: Arbib M (ed) The handbook of brain theory and neural networks. MIT Press, Cambridge, pp 423–426

Mishkin M, Ungerleider L, Macko K (1983) Object vision and spatial vision: two cortical pathways. Trends Neurosci Oct: 414–417

Moscovitch M, Behrmann M (1994) Coding of spatial information in the somatosensory system: evidence from patients with neglect following parietal lobe damage. J Cogn Neurosci 6:151–155

Mozer M, Behrmann M (1990) On the interaction of selective attention and lexical knowledge: a connectionist account of neglect dyslexia. J Cogn Neurosci 2:96–123

Mozer M, Halligan P, Marshall J (in press) The end of the line for a brain-damaged model of hemispatial neglect. J Cogn Neurosci

Pouget A, Sejnowski T (1995) Spatial representations in the parietal cortex may use basis functions. In: Tesauro G, Touretzky D, Leen T (eds) Advances in neural information processing systems, vol 7. MIT Press, Cambridge

Pouget A, Sejnowski T (in press) Spatial transformations in the parietal cortex using basis functions. J Cogn Neurosci

Schlag-Rey M, Schlag J (1984) Visuomotor functions of central thalamus in monkey. I. unit activity related to spontaneous eye movements. J Neurophysiol 51:1149–1174

Stein J (1992) The representation of egocentric space in the posterior parietal cortex. Behav Brain Sci 15:691–700

Treisman A, Gelade G (1980) A feature integration theory of attention. Cogn Psych 12:97–136

Trotter Y, Celebrini S, Stricanne B, Thorpe S, Imbert M (1992) Modulation of neural stereoscopic processing in primate area V1 by the viewing distance. Science 257:1279–1281

van Opstal A, Hepp K, Suzuki Y, Henn V (1995) Influence of eye position on activity in monkey superior colliculus. J Neurophysiol 74:1593–1610

Watson R, Valenstein E, Day A, Heilman K (1994) Posterior neocortical systems subserving awareness and neglect. Arch Neurol 51:1014–1021

Zipser D, Andersen R (1988) A back-propagation programmed network that stimulates reponse properties of a subset of posterior parietal neurons. Nature 331:679–684

Neuronal Coding of Visual Space in the Posterior Parietal Cortex

P. P. BATTAGLINI[1], C. GALLETTI[2], and P. FATTORI[2]

[1]Institute of Physiology, University of Trieste, Trieste, Italy
[2]Department of Human and General Physiology, University of Bologna, Bologna, Italy

If the eyes were always still, the problem of a proper evaluation of object position in the field of view would be relatively simple: stationary objects in different spatial locations of the visual field would project their images onto different points of the retina, and the detection of their spatial location would simply be based on the retinotopic organization of the brain structures involved in the analysis of visual information. Since, however, eyes move continuously, the image of one and the same object can be projected anywhere onto the retina. For example, an object to the left of an observer looking straight ahead projects its image onto the right hemiretinas, but when the eyes move to the left, the image of the same object moves on the retina, occupying a different location. In spite of this retinal displacement, the observer always perceives the object in the same position in space, hence the retinotopic organization of the visual structures of the brain does not account for a proper evaluation of object position in space. Of course, things are much more complicated if the observer also moves his/her head or the entire body. In every circumstance, even in the simplest one, the visual system must take into account any modification of gaze direction so as to compensate for the image displacements due to ocular movement and to correctly locate an object in visual space.

More than 10 years ago, Sparks and Mays (1983) hypothesized that in order to correctly direct the gaze towards a visual target in space, three sets of information are needed: the position of the target image on the retina, the target position in visual space with respect to the head of the observer, and the difference between current and desired eye position. According to these authors, the three sets of information are probably provided by three different types of neurons: (a) neurons with receptive fields organized in retinotopic coordinates, as the great majority of neurons in the visual system are, (b) neurons discharging whenever a stimulus appears in a specific region of the visual environment regardless of gaze direction, and (c) neurons that alter their discharge rate when there is a specific difference between current and desired eye position. While neurons of the first and third type had already been found at the time of their suggestion, neurons of the second type had not been. According to the hypothesis of Sparks and Mays, these cells would directly encode visual space in craniotopic, instead of retinotopic, coordinates.

In: Parietal Lobe Contributions to Orientation in 3D Space (1997). P. Thier and H.-O. Karnath (eds). Springer-Verlag, Heidelberg.

According to Sparks (1991), "These neurons are responsive to stimuli occupying a particular region of visual space, regardless of the retinal locus of the target image and regardless of the position of the eye in the orbit. The spatial properties of these neurons would not be detected in an acute recording study nor in a chronic recording experiment in which the receptive field was plotted with a simple fixation point."

In the same way the oculomotor system needs spatial information about visual targets in order to direct the gaze towards them, the skeletomotor system needs spatial information about objects in the visual field in order to direct correct reaching movements towards those objects. In 1983, the group of Rizzolatti (Gentilucci et al. 1983) reported having found, in the monkey premotor cortex, neurons possibly involved in guiding reaching movements, with visual receptive fields organized in craniotopic coordinates. In their experiments the animal faced a visual stimulus which was rotated in a vertical plane describing a circular trajectory in front of its face. The spike activity of the recorded neuron was plotted along the stimulus trajectory. The discharge rate increased whenever the stimulus was in the lower part of the visual field, independent of the animal's eye movements. Unfortunately, the receptive fields of these neurons were not mapped, hence it was not possible to demonstrate that their position in space remained unchanged, as the position of receptive fields of neurons coding for absolute spatial locations must.

Another, perhaps less known description of neurons possibly encoding absolute positions in space has been provided by Pigarev and Rodianova (1988), who published their study in Russian. By recording from the parietal cortex of behaving cats, they found a population of visual neurons responding to visual stimuli with spatial constancy. The whole extent of the receptive fields was not mapped in this study either, hence the finding of neurons encoding the absolute location of visual stimuli in space was not proved.

As a matter of fact, receptive fields in the monkey premotor and cat parietal cortices are often very large and their excitability is not uniform throughout their areas, thus making it very difficult to clearly define their borders and demonstrate that they do not move in space along with eye movements. In addition, a different population of neurons has been described which is reportedly able to code visual space in spatial coordinates. This finding was so convincing that the hypothesis of Sparks and Mays of a coding of visual space at the single neuron level actually faded out.

These so-called gaze-sensitive visual neurons were first reported by Andersen and Mountcastle (1983) in area 7a and then described in detail by Andersen and coworkers in 1985 (Andersen et al. 1985). According to their description, each of these neurons has a large receptive field, with a strong gradient of excitability. The authors demonstrated that the region of higher excitability inside the receptive field was strictly retinotopic, but its excitability was modulated by the direction of gaze in such a way that the same retinotopic stimulation could elicit different responses according to gaze direction. This behavior is schematically illustrated

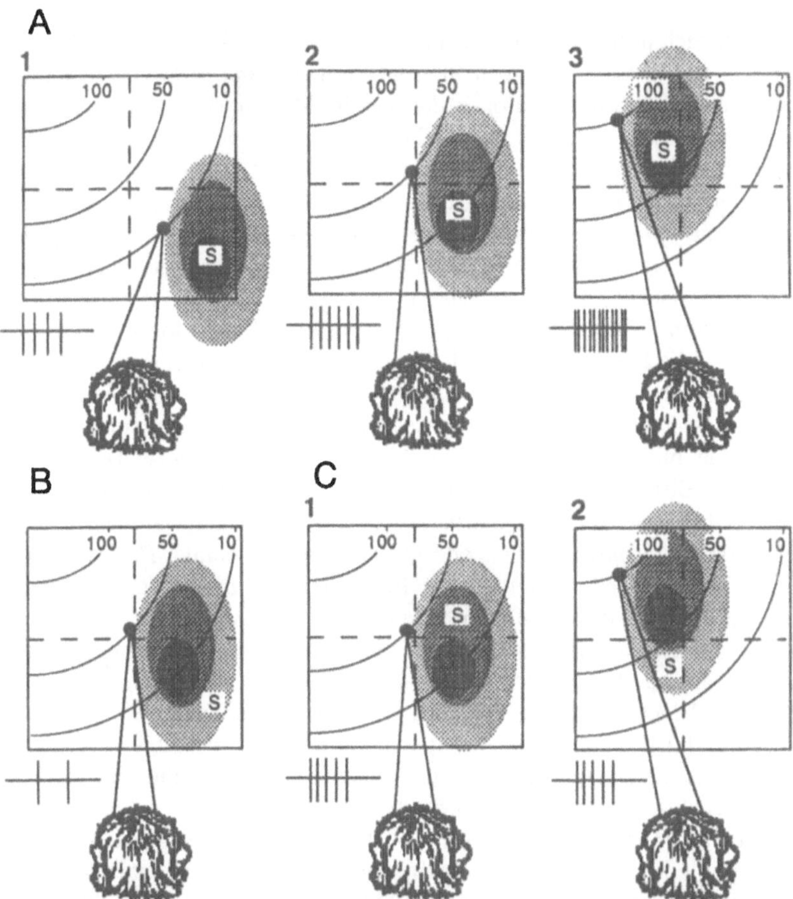

Fig. 1 A-C. Behavior of a gaze-sensitive visual neuron of area 7a (**A**) and of its impossibility to univocally code single spatial locations due to the gradient of excitability inside the receptive field (**B-C**). In each insert, the square represents the screen which the animal is fixating. The curves marked "100", "50" and "10" are isoresponsiveness lines which show the gradient of excitability to visual stimulation due to the influence of the direction of gaze on the neuronal responsiveness. The gradient is such that the best responses to the retinotopic stimulation of the receptive field can be obtained when the animal looks upwards and to the left (100% of best response). The *grey ovals* indicate the extent of the receptive field. Different shades of grey indicate the gradient of excitability inside the receptive field itself. The darker oval is the region with the greatest excitability to visual stimulation, independently of the direction of gaze. The *square* marked "S" indicates the location of a visual stimulus. **A**, *1-3*: behavior of a gaze-dependent visual neuron of area 7a, as described by Andersen et al. (1985). **B** The stimulation of the same spatial location as in **A**, *1* causes a different firing rate as a consequence of the combination of the gradient of excitability inside the receptive field with the influence exerted by the direction of gaze. **C**, *1-2* Different combinations of gaze modulation and gradient of excitability inside the receptive field produce equal patterns of discharge for the stimulation of different regions of the screen

in Fig. 1. In each panel of the figure, the head of a monkey is shown while gazing at different positions on a screen in front of it. On the screen, a hypothetical receptive field of a parietal neuron is sketched in different shades of gray, with the darkest area the zone of higher excitability, whereas the pale one is the peripheral region of lower excitability to visual stimulation. Curved numbered lines indicate the gain field of the neuron, that is, the extent and impact of gaze modulation on cell excitability. When the monkey fixates to the right (Fig. 1A, 1), the excitability of the receptive field is very low (10% of maximum), hence a visual stimulus (S) applied to its visually most responsive region evokes a weak response from the recorded neuron. When the fixation point is positioned close to the center of the screen (Fig. 1A, 2) or upwards and to the left (Fig. 1A, 3), the receptive field moves coherently with gaze, and stimulation of the same retinotopic position elicits more and more vigorous responses from the neuron, as a consequence of the favorable modulation exerted by these directions of gaze. Because it is stimulated in different screen locations, this neuron is able to encode different spatial locations with different discharge rates. The discharge of such a neuron, however, does not univocally encode the location of stimuli. The same spatial location can be coded with different frequencies of discharge, as shown in Fig. 1B. In this case the stimulation is applied to the same spatial location as in Fig. 1A, 1, but in a different gaze direction. Conversely, the same pattern of discharge can be equivocally obtained for different combinations of gaze directions and locations of the stimulus within the receptive field, as shown in Fig. 1C, 1 and 2. Here the animal is looking at different locations on the screen, and visual stimulation is applied to different regions within the receptive field, hence the neuronal activity is the same in both situations.

Andersen and coworkers were aware of the problem and suggested that, even though a single neuron cannot univocally code for locations in space, a group of neurons actually can, as later demonstrated by Zipser and Andersen (1988). These authors succeeded in training a neuronal network based on gaze-sensitive visual neurons to map visual targets in head-centered coordinates.

Soon after the publication of this model, another population of gaze-dependent visual neurons was described in area V3A, a prestriate area of the occipital lobe located at the base of the lunate sulcus (Galletti and Battaglini 1989). The behavior of these neurons was similar to that of those previously described in area 7a, but differed in an important property: the neurons in area V3A have smaller receptive fields. Moreover, the stimulation of only a part of the receptive field did not a evoke a response which was appreciably different from that obtained from any other part of the receptive field itself. So these neurons do not seem to have a strong gradient of excitability inside their receptive fields. Receptive fields of V3A neurons are retinotopically organized, so they move coherently with gaze like those of area 7a neurons; but due to their small size and the absence of a clear gradient of internal excitability, they encode new, small screen locations at each new direction of gaze, as shown in Fig. 2A.

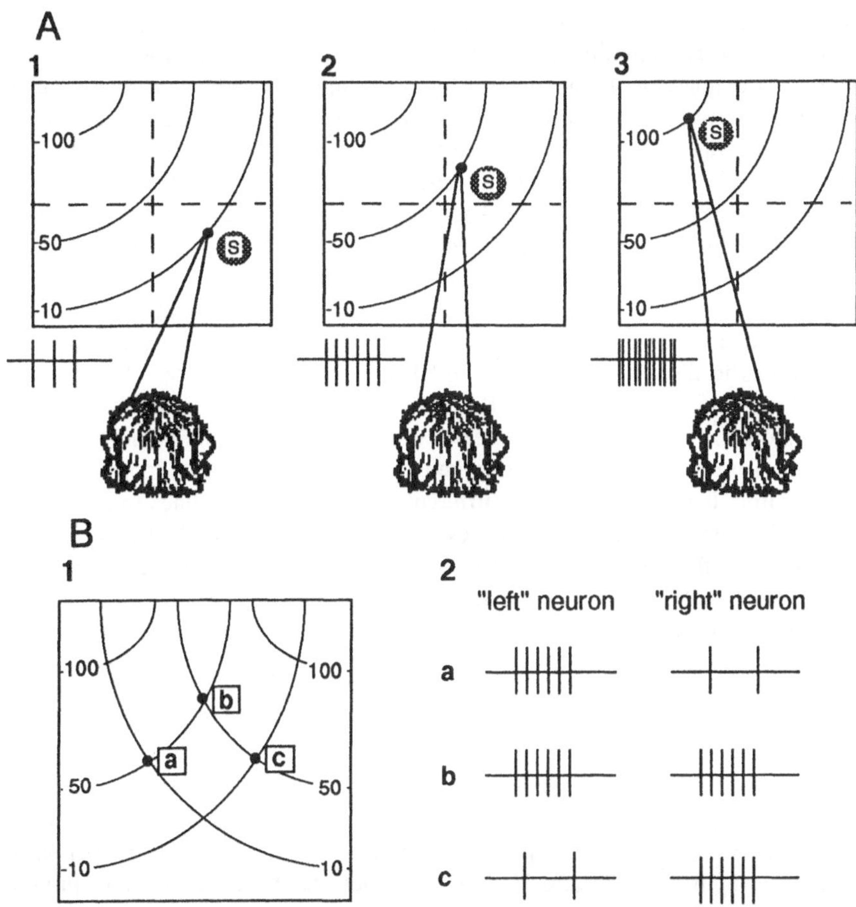

Fig. 2 A, B. Behavior of a gaze-dependent visual neuron of area V3A (**A**) and proposed model of space coding by small samples of such neurons (**B**). **A**, *1-3* Conventions as in Fig. 1. Note the small size of the receptive field and the absence of a gradient of excitability inside it. **B**, *1* Gaze-related distribution of visual responsiveness of two hypothetical neurons, one facilitated when the animal looks leftwards, the other when it looks rightwards. *Curved lines* represent isoresponsiveness lines, as in **A**. *Squares* marked "a," "b" and "c" indicate the locations of a visual stimulus for three different fixation points (*solid circles*). **B**, *2* Responses of the two neurons shown in **B**, *1* to the stimulation of locations a, b and c

For the sake of completeness, however, one has to mention another complication applying to gaze-sensitive visual neurons of both area 7a and V3A. The direction of gaze along which neurons become more and more reactive to visual stimulation often has a planar distribution, or an even more complex one.

When the animal looks anywhere along an isoresponsiveness line, a properly positioned visual stimulus will elicit almost the same response. Thus, these neurons are not able to unequivocally code single locations in space either, although relatively small groups of V3A neurons might do so in a simpler and more direct way than in area 7a, since the complication of a large receptive field with a strong gradient of excitability inside it was not present. Figure 2B shows the behavior of two hypothetical V3A gaze-dependent visual neurons: one neuron (the "left" neuron) is facilitated when the animal looks towards the left upper corner of the screen, the other neuron (the "right" one) is facilitated when the animal looks towards the right upper corner. The animal looks at three different locations, and visual stimulation is consequently applied to the receptive fields of the neurons in three different positions of the screen (marked "a," "b," and "c," respectively). As shown in Fig. 2B, 2, the stimulation of positions "a" and "b" yields the same response from the left neuron, with the fixation points located on the same isoresponsiveness line (marked "50"), but different responses from the right neuron, because they lie on different isoresponsiveness lines (marked "10" and "50"); the same is true, but reversed, for positions "b" and "c," where responses are the same when recorded for the right neuron, but different when recorded for the left neuron. Thus, the spatial locations "a," "b," and "c" cannot be distinguished by one of these neurons alone: instead they are codified differently by the two neurons together, the simultaneous activity of which unambiguously codes the three spatial locations on the screen. For instance, if the left neuron discharges at 50% of its higher rate and the right neuron is almost silent, the visual stimulus is in position "a."

The discovery of gaze dependency outside area 7a allowed Pouget et al. (1993; see also Pouget and Sejnowski this volume) to postulate a new model for the egocentric representation of space, which took into account the properties of neurons in both areas 7a and V3A. The model was tested in order to produce saccades towards targets in space, and it succeeded in proving that an egocentric representation of space may also be present in early stages of visual processing in which retinotopy is a dominant feature.

It is important to emphasize that gaze-sensitive visual neurons in these models are confined to the inner layers of the multilayer neuronal network whose output layer might be represented by neurons which directly encode visual space in egocentric coordinates.

Recently, a further population of gaze-sensitive visual neurons has been found in the anterior bank of the parietooccipital sulcus (Galletti et al. 1995). The visual receptive fields of these neurons were uniform in excitability, larger in size than those of area V3A, but smaller than those of area 7a. Quite frequently the gain fields were not planar and were also very restricted in extent, so that the neurons were visually responsive only when the animal looked towards certain parts of the visual field. The functional role of this population of gaze-sensitive visual neurons might possibly be the same as the one attributed to similar neurons discovered in

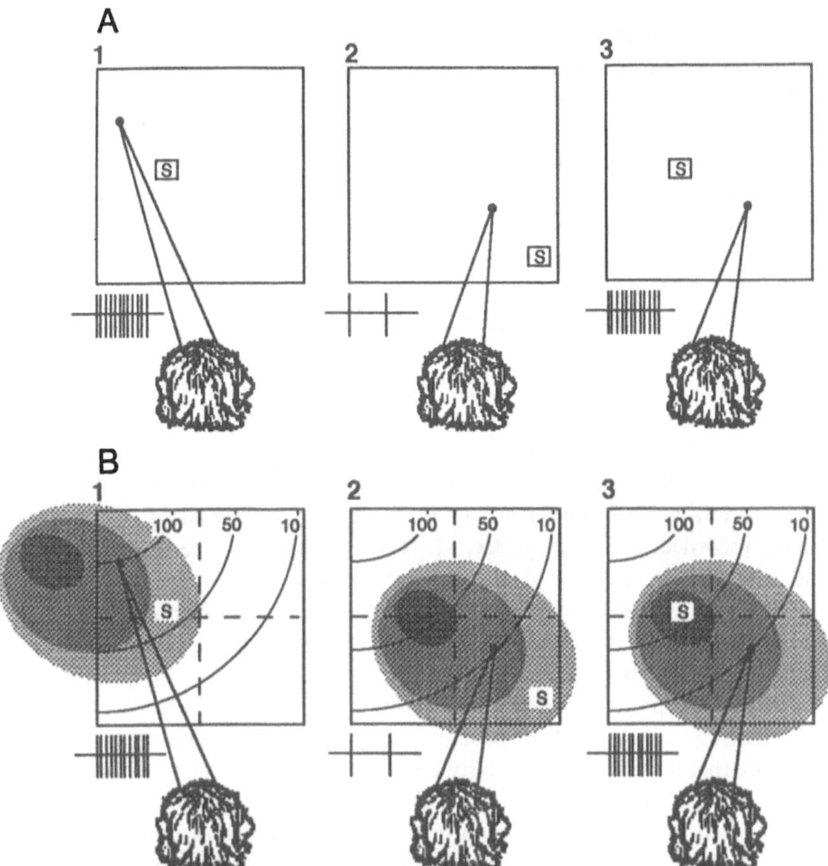

Fig. 3 A, B. Behavior of a putative neuron encoding spatial locations independently of the direction of gaze (**A**) and interpretation of its pattern of response assuming that it is a gaze-sensitive visual neuron (**B**). Conventions as in Fig. 1A, *1-3*: The neuron gives a strong response only when the stimulus is in a precise location on the screen (**A**, *1* and **A**, *3*), irrespective of the direction of gaze. In **A**, *2* stimulation of the same retinotopic location as in **A**, *1*, at a different screen location gives no response. **B**, *1-3* Explanation of the finding illustrated in **A** assuming that the neuron was a gaze-sensitive visual neuron with a gradient of excitability within the receptive field. Each panel has to be compared to the one with the corresponding number in **A**

other cortical regions, or these neurons might contribute in a different way to space-location coding, as will be proposed later.

So far, two main considerations emerge from the experimental data reported above: (1) the only way in which the brain can encode visual space purposefully seems to be based on the activity of large groups of gaze-sensitive visual neurons;

their distributed activity can supply the information necessary to direct the subject's eyes and/or behavior towards surrounding objects; (2) this way of encoding visual space is not unique to the parietal cortex, since other regions of the brain can use it for their own purposes. There is no reason to reject these conclusions, and actually, the reported models can describe at least one of the neuronal mechanisms used by the brain to interact with extrapersonal space.

The detailed description of gaze-sensitive visual neurons, the possibility of describing with computational models their possible role in space constancy, and their discovery in several cortical regions (actually more than just those mentioned above) could explain, at least in part, why the experimental data of the groups of both Rizzolatti and Pigarev was not taken into proper consideration. It was argued, in fact, that the behavior of those neurons could be easily explained by the fact that they were gaze-dependent visual neurons with a strong gradient of excitability within their receptive fields, as shown in Fig. 3A illustrating a schematic representation of the observations made by the two research groups. In Fig. 3A, the phenomenon of spatial constancy seems evident. When the animal fixates the left part of the screen, a visual stimulus yields an optimal response from the recorded neuron (Fig. 3A, 1). If the animal looks to the right and the visual stimulus is repositioned accordingly, the response of the neuron is very poor (Fig. 3A, 2); however, if, at this same angle of gaze, the visual stimulus is positioned where it initially was, the neuron again gives an optimal response (Fig. 3A, 3). This shows that the neuron gives a strong response only if the visual stimulus is in the proper position in space, irrespective of the direction of gaze of the animal. Figure 3B shows the possible explanation of this finding, based on the assumption that those neurons actually were gaze-dependent visual neurons with a gradient of visual responsiveness within the receptive field. In this case the neuron is supposed to have a large, nonuniform receptive field that is affected by the direction of gaze, so the neuron is facilitated when the animal looks upwards and to the left. If a weak, peripheral part of the receptive field is stimulated while the animal is looking to the left, the cell will give a strong response owing to gaze facilitation (Fig. 3B, 1). When the animal looks to the right, the visual stimulation of the same retinotopic location evokes a very weak response, or no response at all (Fig. 3B, 2) because the gaze effect is less favorable than before. On the other hand, in this new direction of gaze, visual stimulation of the same spatial position as the initial one evokes a strong response in spite of the unfavorable gradient of excitability, since the hot spot of the receptive field is stimulated (Fig. 3B, 3). If it is not possible to map the whole extent of the receptive field, it is not possible to demonstrate that the borders of the receptive fields remain in the same spatial position in spite of gaze displacements.

Recently, neurons with receptive fields organised in craniotopic coordinates have been described in the dorsal part of the anterior bank of the parietooccipital sulcus (Battaglini et al. 1990; Galletti et al. 1993), in an area later defined as V6A (Galletti et al. 1996). The behavior of these neurons appears to be similar to that shown in Fig. 3A, but in these studies the entire extent of the receptive field has

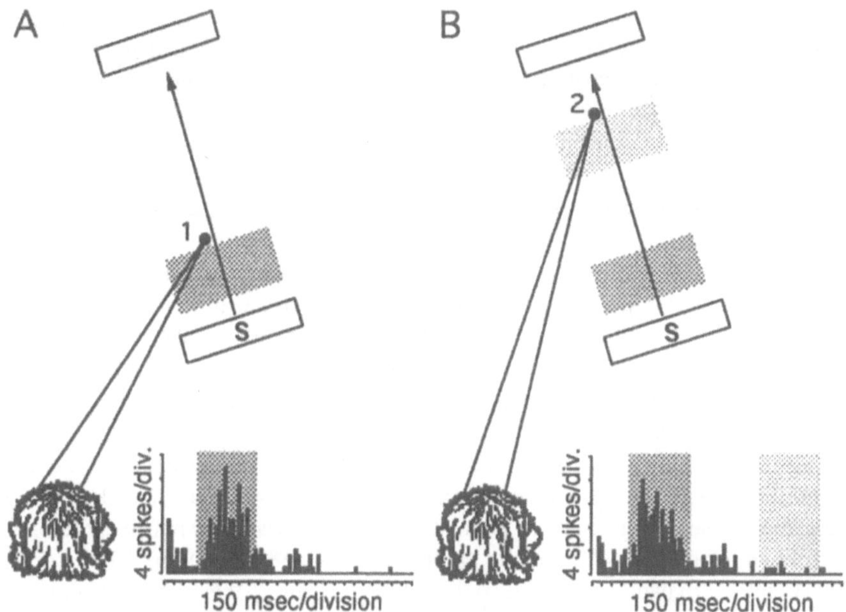

Fig. 4 A, B. Spatial constancy of receptive field borders in a real-position cell of area V6A. **A** A visual stimulus (*S*) is repeatedly moved in the direction shown by the *arrow* accross the receptive field of the neuron (*gray rectangle*) when the animal is fixating in position 1. The location and size of the receptive field was determined during manual mapping with optimal stimuli. Spike activity of the neuron was compiled and plotted as a peri-stimulus time histogram (PSTH, *bottom*). Thirteen repetitions of the stimulation were used for this display. Data from single trials was aligned at the onset of the stimulus movement. The *dark gray area* in the PSTH indicates the time during which the visual stimulus swept across the receptive field. **B** As in **A**, but in this case the animal is looking at position 2, about 30° away from the previous one. Visual stimulation is as before. The *pale gray rectangle* shows the position that the receptive field should have assumed at this different fixation point if it were linked to retinotopic coordinates. In the PSTH, the *dark* and *pale gray areas* indicate the time during which the stimulus swept across the *dark* and *pale gray areas* shown above. (Redrawn from Galletti et al. 1993)

been mapped, making it possible to show that it did not move when the eyes moved. The receptive fields were in fact remapped each time the fixation point was displaced. Whenever possible, moving stimuli sweeping across the receptive field, instead of stationary stimuli flashed onto it, were used to visually activate the neurons. By using moving stimuli it is possible to detect and document the spatial location of the borders of the receptive field, as shown in Fig. 4 for two of them. In Fig. 4A the animal gazes at a certain location, while a moving stimulus crosses the receptive field of the recorded neuron. The onset and offset of the neuronal response are clearly detectable in the peri-stimulus time histogram, indicating the time the stimulus enters and exits the receptive field. In Fig. 4B the fixation point is placed on a different location of the screen, but the stimulus starts its trajectory from the same position as before. The neuronal response has the same time course and clear onset and offset as in Fig. 4A and it is not shifted proportionally in the course of time to the distance between the actual and the previous fixation point, as a retinotopic receptive field would have required. This clearly means that the two borders remained in the same spatial location regardless of the direction of gaze, hence the receptive field was organized in spatial instead of retinotopic coordinates.

Figure 5 schematically shows the peculiarity of such a real-position cell with respect to a gaze-dependent visual neuron and a classical visual cell. In each panel, the location of the receptive field of the same neuron is shown with respect to different locations of the fixation point. While the receptive field of a classical visual neuron (left) can be equally driven in every direction of gaze, in a gaze-sensitive visual neuron (center) the receptive field can be driven only in a more or less restricted region of space linked to the gain field of the neuron. The term "gain field" is used here with the definition suggested by Andersen et al. (1990) to indicate that part of the field of view that influences visual responsiveness of the neuron while being gazed at by the animal. Galletti and co-workers (Galletti et al. 1995) called "dynamic receptive field" (Fig. 5, center) the restricted region of space from which the neuron is able to receive visual information during changes

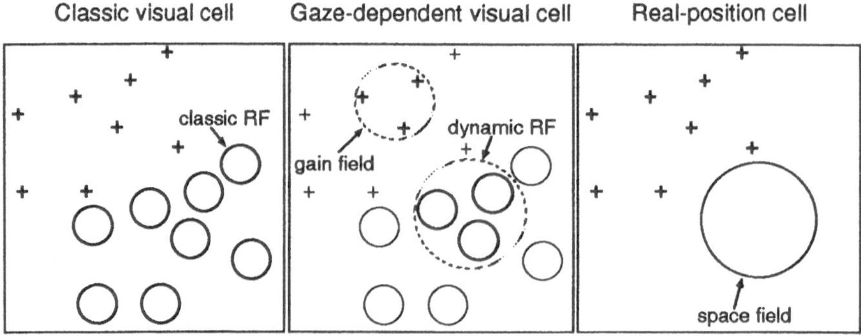

Fig. 5. Behavior of three different types of visual neurons. *Left*: In a classic visual cell, the receptive field (*RF*, *circle*) moves coherently with every change in fixation point location (*cross*). Eight different fixation points are shown with their respective receptive-field positions. *Center*: In a gaze-dependent visual neuron, the receptive field is still retinotopic and moves with the gaze, but it responds best at certain directions of gaze (*bold crosses* and *circles*) and not at others (*thin crosses* and *circles*). The region where fixation points are more effective in facilitating the neuronal response is called "gain field;" the corresponding region occupied by the receptive fields is called "dynamic receptive field." *Right*: In a real-position cell, at every fixation point it is possible to evoke a strong response from the neuron, if the stimulation is carried out in a precise location on the screen called "space field," that does not change with eye movements. (Redrawn from Galletti et al. 1995)

in gaze direction. Finally, the term "space field" (Fig. 5, right) is used to indicate the region of space the stimulation of which always activates a real-position cell, irrespective of the actual direction of gaze (cf. Galletti et al. 1995).

Soon after the first report on real-position cells in the parietooccipital cortex, the group of Rizzolatti (Fogassi et al. 1992) reported having found both gaze-sensitive visual neurons with retinotopic receptive fields and real-position cells in the monkey premotor cortex (inferior area 6). The real-position neurons of inferior area 6 have three-dimensional visual receptive fields anchored to spatial locations that remain unchanged in spite of the animal's eye movements. The fact that these neurons were strongly reactive to visual stimuli moved in depth near the face of the animal seems to be a crucial property. As a matter of fact, Boussaud et al. (1993) were able to find in the same cortical region a high percentage of cells modulated by the direction of gaze, but could not find any real-position behavior by presenting stationary stimuli in a bidimensional plane and not near to the animal's face.

The discovery of real-position cells in different cortical areas and their description by different laboratories support the evidence of their existence. In addition, the presence of real-position cells mixed with gaze-sensitive visual neurons in the same areas (both inferior area 6 and area V6A) suggests the possibility of an interaction between these two types of cells. Area V6A is, from this point of view, the most likely region where this interaction may occur. The cortical region containing area V6A has very recently been described in detail by Galletti et al. (1996), reporting that it can be functionally and anatomically subdivided in two adjacent areas: V6A dorsally and V6 ventrally. Area V6 corresponds to the latest definition of area PO (Colby et al. 1988), while V6A is a cortical region located dorsally to PO itself.

Area V6A contains neurons with various properties, and some of them seem not to be visual at all (Galletti et al. 1991). Among visual neurons, several peculiarities were found which distinguish this area from the underlying area V6 (PO): most cells have larger receptive fields, some of them have complex visual selectivity, respond to the expansion of bidimensional stimuli or prefer very low stimulus speed or are even modulated by the animal's attentive state. In area V6A, gaze-sensitive visual neurons were found in a high percentage. Interestingly, this area also contains gaze-sensitive visual neurons so strongly modulated by gaze that their receptive fields can be activated only when receiving visual information from a very restricted part of the field of view. This occurs often in area V6A, but not in V3A (Galletti and Battaglini 1989), nor in 7a or LIP (Andersen et al. 1990), where a high percentage of gaze-sensitive visual neurons has also been found. Finally, V6A but not V6 contains real-position cells (Galletti et al. 1996).

As mentioned previously, the presence of gaze-sensitive visual neurons in area V6A with very restricted dynamic receptive fields and of real-position cells suggests that gaze-dependent visual neurons might represent a step along a hierarchical chain possibly leading to real-position behavior. This suggestion is supported by the fact that in some microelectrode penetrations into V6A, gaze-sensitive visual neurons were found which had dynamic receptive fields whose space location could predict the behavior of a real-position cell, which was actually found in the same penetration (Galletti et al. 1995). Figure 6 schematically reports data from a single penetration in which this congruent behavior was observed.

Figure 6 shows receptive-field locations and preferred gaze directions of five gaze-dependent visual neurons clustered around a real-position cell in a range of about 200 µm. Neuron 1 had its receptive field below the fovea and it was optimally responsive when the animal looked upwards (arrow in 6A, 1), bringing

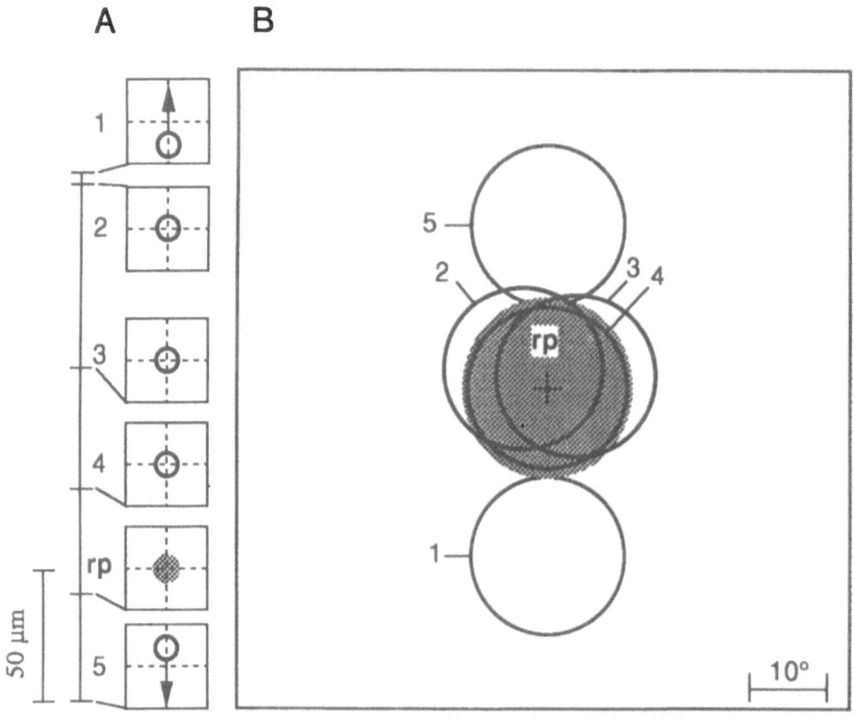

Fig. 6 A, B. Receptive field locations and depths along a single electrode penetration of five gaze-dependent visual neurons (*1-5*) and one real-position cell (*rp*). **A** Relative position of the neurons along a penetration in area V6A. The *squares* represent the screen in front of the animal and the *circles* show the position of the receptive field when the animal was looking at the center of the screen, indicated by the *crossing of the dashed lines*. *Arrows* indicate the preferred direction of gaze. **B** All the receptive fields are plotted in a larger scale on the screen. The receptive fields of the five gaze-dependent visual neurons are plotted with respect to the fovea (*cross*), while the space field of the real-position cell (*gray circle*) is plotted in screen coordinates. (Redrawn from Galletti et al. 1995)

the receptive field to the center of the screen; neurons 2, 3 and 4 had their receptive fields around the fovea, and they were best activated when the animal looked straight ahead; finally, neuron 5 had its receptive field above the fovea and was best activated when the animal looked downwards (arrow in 6A, 5), bringing the receptive field to the center of the screen. Figure 6B shows, on a larger scale, the screen locations of the receptive fields of the different neurons when the animal looked at the center of the screen. The real-position cell ("rp" in the figure) was found at an intermediate depth between cells 4 and 5, a few tenths of microns away from them. Its space field was located on the center of the screen and remained equally responsive in the same spatial position wherever the animal was looking. The spatial coordinates of the space field were the same as those of the dynamic receptive fields of the gaze-sensitive visual neurons encountered along the penetration, suggesting a role for gaze-dependent visual neurons in building up real-position behavior.

The existence of real-position cells suggests a simpler way of building an internal representation of visual space than the one based on gaze-sensitive visual neurons. Each part of the visual field could be directly encoded by the space fields of real-position cells, so that the real-position cell population as a whole might actually "hold" an objective representation of the visual space that does not change with the changes of the eye position in the orbit, as is the case for the spatial map constructed by gaze-sensitive visual neurons. The activity of these neurons would change only if the visual stimulus changes, thus signalling that something has happened in that particular region of space.

Since gaze-dependent visual neurons have so far been found in several brain structures where, in turn, real-position cells were not present, it is reasonable to assume that there are at least two ways to encode visual space: one based on gaze-sensitive visual neurons and the other on real-position cells. They might be independently used by different areas subserving different or complementary functions. Gaze-sensitive visual neurons probably encode visual space dynamically, updating its internal representation at every change in the direction of gaze. This dynamic representation might be used for the multiple purposes of perceptual processes, as suggested by the wide distribution of these neurons in the brain. Real-position cells, in turn, possibly encode visual space in a static way, because their excitability is independent of the direction of gaze and only relies on the spatial location of the visual stimulus. This property might be particularly useful for motor purposes, allowing a direct interaction with the visual environment, as suggested by the presence of real-position cells in both the parietal and the premotor cortices. This view seems to be supported by the effects of lesions to these cerebral regions, which drastically reduce the awareness of objects in space (neglect) or impair the visuospatial ability to perform reaching movements (cf. Bisiach and Vallar 1988 and Rizzolatti and Gallese 1988 for reviews on this topic).

Acknowledgements. Work in the authors' laboratory has been supported by EC grant CHRX-CT93-0267 (DG 12 COMA) and by grants from MURST and CNR, Italy. P.P. B. is indebted to the C. & D. Callerio Foundation, Trieste, Italy, for the facilities generously provided.

References

Andersen RA, Mountcastle VB (1983) The influence of the angle of gaze upon the excitability of the light-sensitive neurons of the posterior parietal cortex. J Neurosci 3:532–548

Andersen RA, Essik GK, Siegel RM (1985) Encoding of spatial location by posterior parietal neurons. Science 230:456–458

Andersen RA, Bracewell RM, Barash S, Gnadt JW, Fogassi L (1990) Eye-position effect on visual, memory and saccade-related activity in areas LIP and Ta of macaque. J Neurosci 10:1176–1196

Battaglini PP, Fattori P, Galletti C, Zeki S (1990) The physiology of area V6 in the awake, behaving monkey. J Physiol (Lond) 423:100P

Bisiach E, Vallar G (1988) Hemineglect in humans. In: Boller F, Grafman J (eds) Handbook of neuropsychology. Elsevier, Amsterdam, vol 1, pp 195–222

Boussaud D, Barth TM, Wise SP (1993) Effects of gaze on apparent visual responses of frontal cortex neurons. Exp Brain Res 93:423–434

Colby CL, Gattas R, Olson CR, Gross CG (1988) Topographical organization of afferents to extrastriate visual area PO in the macaque: a dual tracer study. J Comp Neurol 269:392–413

Fogassi L, Gallese V, di Pellegrino G, Fadiga L, Gentilucci M, Luppino G, Matelli M, Pedotti A, Rizzolatti G (1992) Space coding by premotor cortex. Exp Brain Res 89: 686–690

Galletti C, Battaglini PP (1989) Gaze-dependent visual neurons in area V3A of monkey prestriate cortex. J Neurosci 9:1112–1125

Galletti C, Battaglini PP, Fattori P (1991) Functional properties of neurons in the anterior bank of the parieto-occipital sulcus of the macaque monkey. Eur J Neurosci 3:452–461

Galletti C, Battaglini PP, Fattori P (1993) Parietal neurons encoding spatial locations in craniotopic coordinates. Exp Brain Res 96:221–229

Galletti C, Battaglini PP, Fattori P (1995) Eye position influence on parieto-occipital area PO (V6) of the macaque monkey. Eur J Neurosci 7:2486–2501

Galletti C, Fattori P, Battaglini PP, Shipp S, Zeki S (1996) Functional demarcation of a border between areas V6 and V6A in the superior parietal gyrus of the macaque monkey. Eur J Neurosci 8:30–52

Gentilucci M, Scandolara C, Pigarev IN, Rizzolatti G (1983) Visual responses in the postarcuate cortex (area 6) of the monkey that are independent of eye position. Exp Brain Res 50:464–468

Pigarev IN, Rodionova EI (1988) Neurons with visual receptive fields independent of the position of the eyes in cat parietal cortex. Sensory Systems (Moscow) 2:245–254

Pouget A, Fisher SA, Sejnowski TJ (1993) Egocentric spatial representation in early vision. J Cogn Neurosci 5:150–161

Rizzolatti G, Gallese V (1988) Mechanisms and theories of spatial neglect. In: Boller F, Grafman J (eds) Handbook of neuropsychology. Elsevier, Amsterdam, vol 1, pp 223–246

Sparks DL (1991) The neural control of orienting eye and head movements. In: Humphrey DR, Freund HJ (eds) Motor control: concepts and issues. Wiley, London, pp 263–275

Sparks DL, Mays LE (1983) The role of the monkey superior colliculus in the spatial localization of saccade targets. In: Hein A, Jeannerod M (eds) Spatially oriented behaviour. Springer, Berlin Heidelberg New York, pp 63–86

Zipser D, Andersen RA (1988) A back-propagation programmed network that simulates response properties of a subset of posterior parietal neurons. Nature 331:679–684

Modulation of the Neglect Syndrome by Sensory Stimulation

G. VALLAR[1,2], C. GUARIGLIA[1,2], and M. L. RUSCONI[3]

[1]Dipartimento di Psicologia, Universita' di Roma "La Sapienza", Roma, Italy
[2]IRCCS Clinica "S. Lucia", Roma, Italy
[3]Dipartimento di Psicologia Generale, Universita' di Padova, Padova, Italy

Early Observations

The general notion that specific sensory stimulations may affect spatial representations is not novel. With reference to patients who suffered from peripheral disorders of the vestibular system, Pierre Bonnier (1905) introduced the term *aschématie*, to denote a disorder of body schema defined as *l'anesthésie limitée à la notion topographique*. The reported symptoms included a variety of abnormalities, such as global disorders of localization, de-personalization, perceiving the whole body or body parts as pathologically large in size, feeling that the patients' selves were divided into two, or their bodies no longer existent. Paul Schilder (1933) described a number of patients whose visual and somatosensory phenomenal experiences were modulated by the activity of the vestibular system. For instance, after vestibular stimulation a patient suffering from hallucinations produced by belladonna intoxication "... saw the Titanic sinking with many passengers moving on the deck ... The face of a woman shows a swollen lip. The jaw disappears" (p. 138). Another patient, a 51 year-old woman studied in 1927, had putative vestibular attacks of a few seconds' duration, with specific somatosensory and visual unilateral symptoms: She reported a variety of abnormalities concerning her left side, which was "lame", "burning", "heavier"; objects in the left visual field were blurred, and she heard the ticking of a watch less on the left side; a one-sided depersonalization was also present, since "she can't think on the left side" (p. 152-153). Jean Lhermitte (1952), in a brief review concerning the building up and modulation of the body image by afferent sensory inputs, concluded that the vestibular nerve has a most relevant role, deserving to be termed "*nerf de l'espace*".

These clinical observations, of which some illustrative examples have been given, are not controlled experiments. Their basic message, however, is that the vestibular system is substantially involved in the building up and ongoing operation of internal representations of the body and of objects in extrapersonal space. A disordered vestibular function may produce deficits ranging from

In: Parietal Lobe Contributions to Orientation in 3D Space (1997). P. Thier and H.-O. Karnath (eds). Springer-Verlag, Heidelberg.

different varieties of *asomatognosia*, to a sort of unilateral personal and extrapersonal *spatial hemineglect*, as in one of Schilder's patients.

Effects of Vestibular, and Other Sensory Stimulations, on the Syndrome of Spatial Hemineglect

One direct implication of the clinical findings mentioned in the previous section is that vestibular stimulation may affect disorders of spatial processing produced by cerebral lesions. In two patients with tumors in the right frontal lobe, Silberpfennig (1941) reported temporary recovery from left hemineglect after vestibular stimulation: The gaze to the left became prompt, the head ceased to lean towards the extreme right, and reaching to the left and neglect dyslexia were improved. Marshall and Maynard, 40 years later (1983), showed that left cold caloric stimulation improved the ability of a right brain-damaged patient to compensate for left homonymous hemianopia by scanning the left visual half-field. Rubens (1985) undertook a systematical investigation of the effects of vestibular stimulation on left visuospatial hemineglect in a series of 18 right brain-damaged patients. He replicated the earlier observations using the standard clinical tests of line cancellation and reading.

In the following years, two relevant developments took place: (a) the effects of vestibular stimulation on other aspects of the neglect syndrome were explored; and (b) other sensory stimulations were used. Table 1 summarizes the results of these studies. The common trait of the deficits listed in Table 1 is *spatial*: In any given domain (extrapersonal space, body space, internally generated images) they concern the side contralateral to the lesion (left in right brain-damaged patients) with reference to a given coordinate system (Vallar 1994). Table 2 illustrates the experimental paradigms used to investigate the more extensively assessed disorder, visuospatial hemineglect. This clinical term refers to the defective ability of patients with unilateral cerebral damage to explore the side of space contralateral to the lesion, and to report stimuli presented in this portion of space (Vallar 1994). The differential clinical diagnosis of hemineglect as opposed to global disorders of space exploration and perception is based on the presence of a contra/ipsilateral gradient, whereby performance is comparatively well preserved in the ipsilateral side. The studies cited in Tables 1 and 2 do not include altitudinal and radial neglect (Butter et al. 1989; Mennemeier et al. 1992; Rapcsak et al. 1988; Shelton et al. 1990), in which the effects of sensory stimulations have not been assessed.

The inspection of Tables 1 and 2 shows a highly convergent pattern of results, even though the effects of all stimulations have not been assessed on each disorder and task. Vestibular and optokinetic stimulations producing a nystagmus with a slow phase towards the left contralesional side, and left-sided transcutaneous mechanical vibration and electrical nervous stimulation improve a number of

Table 1. Effects of sensory stimulations on component deficits of the neglect syndrome (N)

	Stimulation							
	V		OPK		TMV		TENS	
Deficit	L	R	L	R	L	R	L	R
Extrapersonal N								
– Visuospatial [a]	+	–	+	–	+	–	+§	–
– Nonvisual [b]	+							
Imaginal N [c]	+							
Personal N [d]	+							
Anosognosia for hemiplegia [e]	+	–						
Somatoparaphrenia [f]	+							
Somatosensory deficits								
– Touch [g]	+						+	–
– Position sense [h]			+	–				
– Extinction [i]	+							
Motor deficits [j]	+							

Stimulation: V= vestibular; OPK= optokinetic; TMV= transcutaneous mechanical vibration; TENS: transcutaneous electrical nervous stimulation. L= left; R= right. +/– = positive/negative or undetectable effects on the assessed deficit. L and R refer to the direction of the slow phase of the nystagmus produced by V and OPK, to the stimulated side in the case of TMV and TENS.

[a] **V** (Cappa et al. 1987; Karnath 1994; Marshall and Maynard 1983; Rubens 1985; Silberpfennig 1941; Vallar et al. 1990, 1991a, 1993b, 1994); **OPK** (Karnath 1996; Pizzamiglio et al. 1990); **TMV** (Karnath 1994, 1995; Karnath et al. 1993); **TENS** (Vallar et al., in press b), (§no effects: Karnath 1995)

[b] **V** (Vallar et al. 1990)

[c] **V** (Geminiani and Bottini 1992; Rode and Perenin 1994)

[d] **V** (Cappa et al. 1987; Rode et al. 1992)

[e] **V** (Cappa et al. 1987; Ramachandran 1995; Rode et al. 1992; Vallar et al. 1990)

[f] **V** (Bisiach et al. 1991; Rode et al. 1992)

[g] **V** (Vallar et al. 1990, 1991a, 1993b) ; **TENS** (Vallar et al., in press b)

[h] **OPK** (Vallar et al. 1993a, 1995a)

[i] **V** (Vallar et al. 1993b)

[j] **V** (Rode and Perenin 1994)

Table 2. Effects of sensory stimulations on visuo-spatial hemineglect, by task used to assess the deficit

	Stimulation							
	V		OPK		TMV		TENS	
Task	L	R	L	R	L	R	L	R
Reading [a]	+	–						
Visuomotor exploration [b]	+	–			+	–	+§	–
Tactile-motor exploration [c]	+							
Line bisection [d]			+	–				
Detection [e]					+	–		
Object identification [f]	+				+	–		
Midsagittal plane [g]	+		+	–	+	–		

Tasks requiring visuomotor exploration include target cancellation and copying tests. Stimulation: V= vestibular; OPK= optokinetic; TMV= transcutaneous mechanical vibration; TENS: transcutaneous electrical nervous stimulation. L= left; R= right. +/– = positive/negative or undetectable effects on the assessed deficit. L and R refer to the direction of the slow phase of the nystagmus produced by V and OPK, to the stimulated side in the case of TMV and TENS.

[a] **V** (Rubens 1985; Silberpfennig 1941)

[b] **V** (Cappa et al. 1987; Rubens 1985; Vallar et al. 1990, 1991a, 1993b, 1994); **TMV** (Karnath 1995); **TENS** (Vallar et al. 1995d), (§no effects: Karnath 1995)

[c] **V** (Vallar et al. 1990)

[d] **OPK** (Pizzamiglio et al. 1990)

[e] **TMV** (Karnath et al. 1993), related evidence in Karnath et al. (1991)

[f] **V** (Vallar et al. 1994); **TMV** (Karnath et al. 1993)

[g] **V and TMV** (Karnath 1994); **OPK** (Karnath 1996)

component deficits of the syndrome of spatial hemineglect[1]. By contrast, vestibular and optokinetic stimulations producing a nystagmus with a slow phase towards the right ipsilesional side and right-sided transcutaneous stimulations

[1] The term syndrome is used here in a probabilistic sense, to indicate a set of frequently cooccurring deficits, with no implications concerning the mechanisms involved (Vallar 1991).

worsen the disorders[2]. The only discrepancy concerns the effects of transcutaneous electrical stimulation of the left neck on visuospatial hemineglect. Vallar et al. (1995d) reported positive effects, which Karnath (1995) failed to replicate. Methodological differences, however, prevent a direct comparison of the two studies (e.g., duration of the stimulation, sequence of the experimental conditions).

The studies summarized in Tables 1 and 2 show that different direction-specific sensory stimulations affect the three main manifestations of hemineglect: (a) for objects in extra-personal space; (b) for objects in internal visuo-spatial representations; and (c) for personal, bodily space. These stimulations also influence, in a similar fashion, other deficits that fit under the rubric of the neglect syndrome (Heilman et al. 1993). These include extinction to double simultaneous stimulation, anosognosia for hemiplegia, and productive phenomena, such as delusional thoughts concerning the left side of the body (somatoparaphrenia, see Gerstmann 1942).

There is, however, a productive deficit that is not improved by direction-specific sensory stimulations. Bisiach et al. (1994) have recently shown that patients with left hemineglect, when required to set the endpoints of an imaginary horizontal line of a given length, can make a displacement error towards the left side (related evidence in Ishiai et al. 1994a, b). This paradoxical behavior contrasts sharply with the rightward directional bias observed in tasks such as line bisection, and setting the subjective straight ahead. Bisiach et al. (1996) have also found that the leftward displacement of the left endpoint is further increased by the optokinetic stimulation (a leftward movement of luminous dots), that improves other components of the neglect syndrome (see Tables 1 and 2), both defective (e.g., deficits of position sense, see Vallar et al. 1993a, 1995a) and productive (somatoparaphrenia). The precise relationships of the leftward pathological extension found in the task of setting the endpoints of an imaginary segment with the other aspects of hemineglect, however, remain unclear.

Anatomical Basis

The exploration of the neural correlates of the effects of sensory stimulations on the neglect syndrome has so far been confined to the vestibular system. In normal subjects Bottini et al. (1994) investigated its central projections by a Positron emission tomography activation method. The irrigation of the left external ear

---

[2] Hemineglect is more frequent and severe after lesions in the right hemisphere (review in Vallar 1993; Vallar et al. 1993b). Henceforth, the terms *left/right* will be used for *contralesional/ipsilesional*. The effects of sensory stimulations have been investigated in a few left brain-damaged patients with right hemineglect, in whom similar, mirror-reversed effects have been observed (Vallar et al. 1993b, 1995c, in press b).

canal with cold water (the stimulation which temporarily improves spatial hemineglect) activated a number of cerebral areas in the contralateral right hemisphere, including the posterior insula, the putamen, the transverse and superior temporal gyri, the inferior parietal lobule, the anterior mesial cingulate gyrus, and the primary sensory cortex. The irrigation of the right external ear canal with cold water produced a similar pattern of activation in the left hemisphere. These results agree with data from studies in the monkey, in which an extensive field with neurons exhibiting vestibular responses has been found in the parietoinsular vestibular cortex, located in the parietal operculum of the sulcus lateralis and in the retro-insular region. Neurons activated by vestibular stimulation have also been found in the parietal areas 7a and 7b, 2v and 3a ("neck region"; review in Grüsser et al. 1992). The distribution of the cortical direct efferent projections to the brainstem vestibular nuclei is in line with these findings. In the study by Akbarian et al. (1994) these regions included the parietoinsular vestibular cortex, the parietal regions 2v and 3a, a temporal region at the fundus of the lateral sulcus, the premotor cortex, and the anterior cingulate region. The results of Faugier-Grimaud et al. (this volume) were similar, with the addition of a projection from the frontal eye fields.

A number of cortical and subcortical regions therefore constitute the central vestibular projections, with a main insular-posterior parietal component. Lesions in this region, which is substantially involved in the internal representation of space (Andersen 1995a; Sakata and Kusunoki 1992), are the more frequent anatomical correlate of spatial hemineglect in humans (review in Vallar 1993). The parietal-insular region may then be a main, though not exclusive, locus where the modulation of spatial processes by afferent sensory systems takes place.

The neural correlates of the temporary recovery from left hemianesthesia produced by vestibular stimulation (cold water in the left ear) were investigated by Bottini et al. (1995) in a patient with a lesion involving the right somatosensory cortex and the supramarginal gyrus. Perceptual awareness of tactile stimuli delivered to the left hand was associated with activation of three undamaged regions in the right hemisphere: the putamen and the insular and inferior frontal cortices. In normal subjects, the right putamen and the insular region are activated both by left cold vestibular stimulation and by conjoint left cold vestibular and left tactile stimulations. These data, albeit concerning a single patient, indicate that the activation of a subset of undamaged regions of the right hemisphere may constitute the anatomical correlate of the temporary recovery from left hemianesthesia. A similar pattern of right hemisphere activation may take place during the improvement of other aspects of the neglect syndrome, after vestibular stimulation.

Anatomical and physiological studies in the cat (Kornhuber and Da Fonseca 1964; Mergner 1979) and in the monkey (Büttner and Lang 1979; Fredrickson et al. 1974) have however shown that the cortical projections from the vestibular nuclei are *bilateral,* although the contralateral component seems to be more relevant. Similarly, the cortical efferent projection to the brainstem vestibular complex is bilateral (Faugier-Grimaud et al., this volume). In patients with extensive lesions in the right hemisphere (see Cappa et al. 1987; Vallar et al. 1993b), the improvement of hemineglect through vestibular stimulation might also reflect the activation of regions in the undamaged left hemisphere, via ipsilateral pathways.

In the case of other sensory stimulations which improve hemineglect, no direct anatomo-physiological evidence is available in humans. Vestibular neurons, however, may also respond to visual optokinetic and proprioceptive inputs (Büttner and Lang 1979; Mergner 1979). In the monkey Grüsser et al. (1992) have described cortical neurons, termed *polymodal vestibular cells*, which respond to direction-specific vestibular and optokinetic stimulations and to somatosensory/proprioceptive stimulations of the neck and other body parts. These units may be a neural basis of the modulatory effects on hemineglect, summarized in Tables 1 and 2.

These polymodal vestibular cells may also constitute a neural substrate of the improvement of hemineglect through movements of the left hand. Limb activation reduces the rightward bias in visuomotor exploratory tasks, line bisection, and walking trajectories (Halligan et al. 1991; Robertson and North 1992, 1993; Robertson et al. 1992, 1994). Limb activation is a complex maneuver, however. Different components may contribute to improve hemineglect: somatosensory/proprioceptive inputs, which may specifically activate the vestibular cells, spatiomotor cueing towards the left side, on which attention is focused, and non-specific hemispheric activation (see Halligan et al. 1991; Robertson et al. 1994). The interpretation of these effects is therefore far from straightforward.

Interpretation

Nonspecific Hemispheric Activation

A plausible explanation of the effects of sensory stimulations on hemineglect is in terms of nonspecific general cerebral activation. This is compatible with the finding, reviewed in the previous section, that the vestibular stimulation which improves the deficit activates a number of regions in the right hemisphere. Vallar et al. (1995c) investigated this problem assessing the effects of vestibular stimulation in a patient who had a lesion in the left parietooccipital paraventricular white matter and suffered from both right visuo-spatial hemineglect, and fluent dysphasia. The irrigation of the right ear with cold water temporarily improved right hemineglect, as assessed by a visuospatial exploratory task, but did not affect the language disorder. The deficits of speech production and comprehension, of repetition and of immediate auditory-verbal memory span did not improve after the stimulation. Minimal improvement was observed in a tasks with a relevant visuospatial component (i.e., confrontation naming of pictures and black-and-white photographs of objects), which may be affected by hemineglect and, therefore, by vestibular stimulation.

A second argument concerns the hemispheric asymmetry of the effects of vestibular and optokinetic stimulations on somatosensory deficits. The hypothesis that they reflect a nonspecific activation of the undamaged regions of the contralateral hemisphere predicts comparable patterns of recovery in patients with both left- and right-sided hemispheric lesions. In the study of Vallar et al. (1993b), however, vestibular stimulation improved left hemianesthesia in 15 out of 17 (88%) right brain-damaged patients. Right somatosensory deficits temporarily improved only in two out of 11 (18%) left brain-damaged patients; these two cases also had right visuospatial hemineglect. Similarly, transcutaneous electrical stimulation improved left hemianesthesia in all ten right brain-damaged patients; it had no detectable effects, however, on the right somatosensory deficit of three out of four left brain-damaged patients; the patient who showed a temporary improvement also had right visuospatial hemineglect (Vallar et al., in press b). Finally, in the two experiments of Vallar et al. (1993a, 1995a) optokinetic stimulation modulated the severity of the deficit of position sense only in right brain-damaged patients with left visuospatial hemineglect. No effects were observed in right and left brain-damaged patients without hemineglect. These data indicate that vestibular, optokinetic, and transcutaneous electrical nervous stimulations affect the operation of specific neural systems, which are more frequently disrupted by lesions in the right hemisphere.

Improvement of Defective Primary Sensory Processes

The activation effects of sensory stimulations may enhance the sensory processing (somatotopic and retinotopic levels of representation) of stimuli presented in the left side, and therefore improve hemineglect. This interpretation is physiologically plausible. Both in the monkey (Grüsser et al. 1992) and in humans (Bottini et al. 1994) vestibular stimulation activates portions of the somatosensory cortex. Furthermore, cells in the primary visual cortex may respond to vestibular stimulation (e.g., Vanni-Mercier and Magnin 1982), even though secondary mechanisms are likely to be involved (Grüsser et al. 1992).

A number of different observations, however, militate against this conclusion. First, it has long been known that left visuo-spatial hemineglect may occur in the absence of both somatosensory and visual half-field deficits (left hemianesthesia and hemianopia, review in Bisiach and Vallar 1988). Second, spatial hemineglect is associated with damage to the right inferior-posterior parietal regions (supramarginal gyrus), and, less frequently, damage to the right premotor cortex, but is not associated with lesions confined to the primary somatosensory and visual cortices (review in Vallar 1993). Third, right brain-damaged patients with left hemianesthesia and hemianopia may show preserved early somatosensory (Vallar et al. 1991b) and visual (Spinelli et al. 1994; Vallar et al. 1991b; Viggiano et al. 1995) evoked potentials. Similarly, one right brain-damaged patient with a left hemianesthesia improved by vestibular stimulation showed conductance skin responses to unreported left tactile stimuli (Vallar et al. 1991a). These data suggest that primary sensory processing in both the visual and somatosensory systems may be largely spared in patients with left hemineglect and defective awareness of visual and tactile stimulations. Accordingly, the possibility that the main effects of sensory stimulations take place at the level of dysfunctional primary sensory processes is unlikely.

Also, the hemispheric asymmetry discussed in the previous section does not support an interpretation in terms of improvement of defective primary sensory processes. In line with this view, unilateral lesions confined to the primary somatosensory (Corkin et al. 1970; Pause et al. 1989) and visual (e.g., Holmes 1918; Weiskrantz 1986) cortices do not bring about sensory deficits, which differ according to the side of the lesion.

The conclusion that a nonsensory spatial factor provides a substantial contribution to the left visual half-field deficit of patients with right hemisphere lesions and hemineglect is in line with the results of a single case study by Kooistra and Heilman (1989). The left hemianopia of their patient, who had a right thalamic and temporo-occipital lesion, improved when her eyes were directed 30° towards the right side. In this condition, in which left visual half-field testing fell into the right half-space, the patient's left hemianopia improved significantly. Applying the same logic to left hemianesthesia, Smania and Aglioti (1995) found that the detection of left-sided tactile stimuli by patients with lesions in the right hemisphere improved when the left hand was held in the right half-space. This effect was present during both single and double (two stimuli simultaneously delivered to the left and the right hand) stimulation. That left tactile extinction to double simultaneous stimulation has a spatial non-sensory component is also suggested by its temporary improvement after vestibular stimulation (related evidence in Moscovitch and Behrmann 1994; three right brain-damaged patients in Vallar et al. 1993b).

Leftward Reorienting of Spatial Attention

It is well-known that visual (Butter and Kirsch 1992; Mattingley et al. 1995; Riddoch and Humphreys 1983) or verbal (Karnath 1988) stimulations which direct the patients' attention towards the neglected side may improve the disorder. In the case of vestibular input, however, an interpretation in terms of the orientation of spatial attention towards the *side* which has received the sensory stimulation cannot account for the whole range of observed phenomena. The irrigation of the *left* ear with warm water, which produces a nystagmus with a slow phase towards the right side, worsens hemineglect, whereas the douching of the *right* ear with warm water improves the disorder (Rubens 1985; Vallar et al. 1990). Under these conditions, a specific left-sided vestibular stimulation, which might attract the patient's attention, actually worsens the disorder, while a right-sided stimulus has positive effects. There is no direct relationship, therefore, between the stimulated side and improvement or worsening of the deficit.

The relevant factor could be, however, the direction of the induced orientation of attention (see Gainotti 1993; Làdavas 1993), rather than the side of the stimulation. A number of studies exploring the attentional deficit of patients with left hemineglect found a pathological shift towards the right side (De Renzi et al. 1989; Làdavas 1990; Làdavas et al. 1990), which has been termed *magnetic attraction* (De Renzi et al. 1989) towards objects in the extreme right end of space, and *hyperattention* (Làdavas 1993) to right-sided stimuli. According to these attentional views the relevant parameter of the vestibular stimulations that improve hemineglect is not the stimulated side (cold water in the left ear, warm water in the right ear). Rather, the nystagmus with a slow phase towards the left side, and the leftward postural and kinetic deviations (Cohen et al. 1964; Shanzer and Bender 1959) indicate a leftward orientation of attention. By contrast, the stimulations that worsen hemineglect (warm water in the left ear, cold water in the right ear) have rightward directional effects. A similar line of reasoning applies to optokinetic inputs, by which the whole visual field is stimulated, but the relevant factor is the direction of the movement of the luminous dots, that produces an optokinetic nystagmus with a leftward or rightward slow phase (Howard and Templeton 1966). An attentional interpretation of this sort is also compatible with the observation, discussed in the previous section, that primary sensory processes may be preserved in patients with hemineglect. The pathological shift of attention towards the right side may prevent the access of left-sided information to conscious perceptual experience.

This attentional account is not consistent, however, with an observation by Vallar et al. (1993a, 1995a) . Right brain-damaged patients with left visuospatial hemineglect have a deficit of position sense, which involves both the left and the right arm, although in the latter the impairment is much less severe. In these patients, optokinetic stimulation with a leftward direction of the movement improved the disorder of position sense not only in the left, but also in the right arm, opposite to the direction of optokinetic stimulation, and of the putative leftward orientation of attention. Conversely, a stimulation with a rightward direction worsened performance level not only in the left, but also in the right arm, namely, the side towards which attention was putatively oriented. The view that horizontal optokinetic stimulation produces an orientation of spatial attention towards the direction of the movement of the luminous dots does not therefore explain the pattern of results observed in the right arm. According to this attentional hypothesis, a stimulation with a leftward direction of the movement should worsen (or be ineffective), but not improve performance level. Conversely, stimulation with a rightward direction should increase or have no effect on, but not diminish, position sense in the right arm.

To summarize, the range of effects produced by sensory stimulations in patients with hemineglect cannot be entirely explained in terms of the modulation of the lateral orientation of spatial attention, which is pathologically shifted towards the left side. Even though this may be a component deficit of hemineglect, the findings discussed above suggest that optokinetic, and other sensory stimulations, bring about a reorganization of spatial processes, which involves the whole representation of space and the body.

Modulation of Spatial Representations by Sensory Input

The processing of sensory input comprises multiple levels of representation, with different coordinate frames. In the process of perceiving and localizing objects in the visual environment, distinctions can be made between retinotopic coordinate systems on the one hand, and head-, body-, and object-centered (etc.) spatial frames on the other. The computation of the latter frames involves the integration of visual input with information from other sensory sources (vestibular, proprioceptive/somatosensory) concerning the position of the eyes, of the head, and of the body. A similar distinction can be drawn between somatotopic representations of the body and higher order spatial frames (see the time-honored concept of body image or schema, "the tridimensional inner diagram each person has of its own body", Roth 1949, p. 89; Schilder 1950). The basic distinction between primarily sensory and spatial representations has an anatomical counterpart. The primary visual and somatosensory cortices are the neural correlates of the retinotopic and somatotopic levels of representation, while a network including the posterior parietal region and the premotor cortex constitutes the neural basis of the spatial co-ordinate systems (reviews in Andersen 1995a, b; Grüsser and Landis 1991, ch. 21 and 22; Lacquaniti, in press). In line with these observations, the more frequent anatomical correlate of the manifold manifestations of hemineglect in humans is a lesion involving the inferior parietal lobule (supramarginal gyrus, review in Vallar 1993). An alternative account of the encoding of the spatial location of the stimulus in terms of oculocentric representations, which do not require further coordinate transformations, may be found in the work of Goldberg and coworkers (Colby et al. 1995; Duhamel et al. 1992; Goldberg and Colby 1989).

Neurons that code information in visual extrapersonal space with reference to non-retinotopical coordinates (head-, body-centered) have been found in the premotor cortex (area 6, Fogassi et al. 1992), and in the parietooccipital sulcus (Galletti et al. 1993, 1995a, b). Neurophysiological evidence of a modulation of visual processing by sensory information has been recently provided by Brotchie et al. (1995), who found that the position of the head, through proprioceptive inputs from the neck muscles and vestibular signals, affects the visual and saccadic activities of parietal neurons.

As noted above, the building up and updating of spatial representations of both objects in extrapersonal spaces (e.g., grasping space, distant action space, see Grüsser and Landis 1991) and of the body in personal space involve a multisensory integration, in which gravitational and positional cues provided by the vestibular and the proprioceptive/somatosensory systems play a relevant role. This feature, shared by extrapersonal (objects) and personal (body) spatial representations, may account for the similarity of the effects of direction-specific sensory stimulations on the extrapersonal and personal deficits of the neglect syndrome (see Table 1).

Suggestions have been repeatedly made that the mechanism underlying the manifold manifestations of hemineglect involves the unbalanced activity of a *bilateral* set of cerebral structures concerned, as discussed above, with the building up and operating of spatial representations, through the processing and integration of information from different sensory modalities (Kinsbourne 1993, for a related functional account; Vallar et al. 1993b; Ventre et al. 1984). This notion of *unbalance* can be traced back to a study by Sprague (1966), who showed in the cat that hemianopia and neglect, produced by extensive temporooccipital damage, improved after either a subsequent lesion in the contralateral superior colliculus or splitting the collicular commissure.

One effect of the unbalanced activity produced by a unilateral lesion is the comparatively minor weight given to left-sided sensory information. This results in impoverished representations, with a left-right gradient, or a *leftward distortion*. The unilateral or direction-specific stimulations listed in Tables 1 and 2 provide supplementary information which modulates (reducing or worsening) the distortion towards the left side. The present view has three implications.

1. The stimulations that modulate the manifold manifestations of hemineglect produce in normal subjects directional effects similar to those found in patients. In patients with left hemineglect, and in both normal subjects and patients without hemineglect, optokinetic stimulation modulates in a similar direction the subjective midpoint in line bisection (Pizzamiglio et al. 1990), and the subjective "straight ahead" of body orientation (Karnath 1996). In normal subjects vibration of the left posterior neck muscles produces a rightward deviation of pointing at midsagittal targets (Jeannerod and Biguer 1987) and displacement of visual targets viewed in the dark (Biguer et al. 1988; Taylor and McCloskey 1991). Unilateral vibration of the neck muscles and vestibular stimulation of the left ear have similar directional effects on the subjective straight ahead in patients with hemineglect and in both left brain-damaged patients and normal controls (Karnath 1994; Karnath et al. 1994). Studies in normal subjects concerning the effects of visual (optokinetic) and nonvisual (e.g., labyrinthine) factors on the apparent straight ahead and on postural stability are in line with these results (Howard and Templeton 1966; Lackner 1978). With respect to their perceptual and motor effects, these

stimulations may be considered equivalent signals (see a discussion of the optovestibular equivalence in Bischof 1974).

2. Unilateral peripheral disorders bring about directional effects similar to those observed in hemineglect. This is the case for the shift of the subjective visual midline towards the side of the ocular deviation, produced by unilateral vestibular disorders (Hörnsten 1979; Kanner and Schilder 1930, for early related evidence). The early clinical observations by Bonnier (1905) and Schilder (1933) extend the effects of peripheral disorders to the induction and modulation of illusory perceptions or delusions concerning the subject's body (see also Schilder 1950, p. 117). Proprioceptive stimulations may affect the perceived shape and orientation of body parts (Lackner 1988; Roll et al. 1991, p. 113: "Experimental Asomatognosia"). These clinical and experimental data provide evidence converging with the results of vestibular stimulation on somatoparaphrenia (Table 1) to the effect that the processing levels at which such delusions arise are modulated by afferent sensory inputs.
3. The substantial bilateral reduction of one source of sensory information may attenuate the rightward processing bias, improving hemineglect. When subjects are placed in the supine position, the input from the otolith organs of both sides is greatly reduced (Howard 1982, p. 433; Morrow and Sharpe 1993), and, therefore, one source of rightward processing bias is abolished. Pizzamiglio et al. (1995) verified this prediction, showing that the rightward error in line bisection was reduced when right brain-damaged patients were supine, as opposed to the standard upright position.

The Nature of the Rightward Distortion of Spatial Representations in Hemineglect

The many instances of directional error towards the side of the hemispheric lesion have been revealed by tasks investigating the patients' performance in the *front* half-space (see references concerning the tasks assessing line bisection and the subjective midsagittal plane in Table 2 and Heilman et al. 1983; Mark and Heilman 1990). The occurrence of rightward directional errors in the front half-space is however compatible with two different disorders of the egocentric coordinate system: a rightward *translation* of the whole (front and back) frame, and a *clockwise rotation* around the vertical axis of the body (see a discussion of hemineglect in terms of "rotation" in Ventre et al. 1984). Both hypotheses predict a rightward shift of the subjective midsagittal plane in the front half of egocentric space. In the back half of space, however, a clockwise rotation would involve a leftward displacement and directional error; translation, by contrast, would still produce a rightward displacement, and directional error. A comparison of the performance of right brain-damaged patients with hemineglect in a free-field auditory localization task in the *front* and in the *back* half-spaces thus adjudicates

A

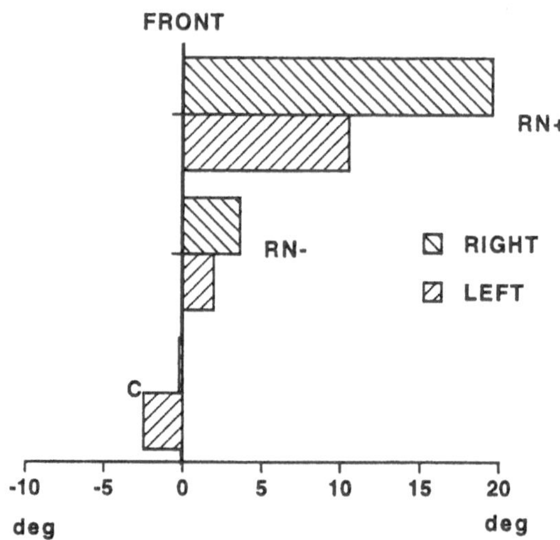

B

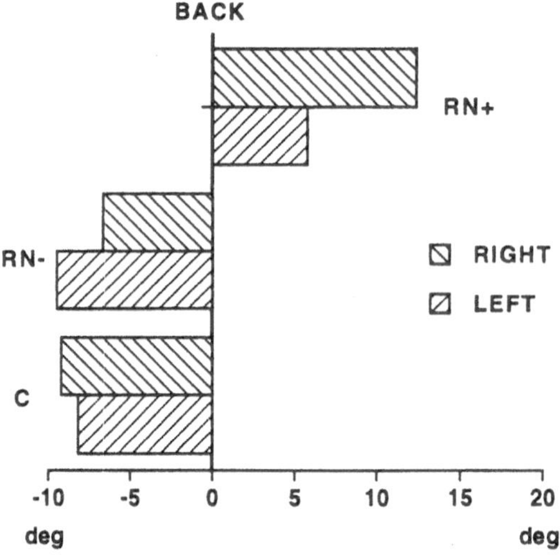

Fig. 1 A, B. Subjective localization of the auditory midline in the front (**A**) and back (**B**) half-spaces by group [right brain-damaged patients with/without visuospatial hemineglect (*RN+*/*RN-*), control subjects (*C*)], and starting position of the stimulation sequence. (From Vallar et al. 1995b)

the two views (Vallar et al. 1995b). In line with the hypothesis of a pathological translation, patients with left visuospatial hemineglect displaced the subjective auditory mid-sagittal plane towards the right side in both the front and the back half-spaces (Fig. 1). The translation hypothesis is also consistent with both an early clinical observation (Wortis and Pfeffer 1948) and the results of studies using auditory lateralization paradigms, in which subjects localize the perceived position of fused sound images generated by dichotic stimuli in the frontal plane passing through the ears. Right brain-damaged patients made a rightward systematical directional error for all stimuli, even when the interaural intensity or time differences were zero, while normal subjects localized the fused sound images in the intersection between the midfrontal and the midsagittal planes (the midbody axis; Altman et al. 1979; Bisiach et al. 1984; Teuber 1962). The hypothesis of a rightward rotation of the egocentric co-ordinate system would predict, by contrast, a normal localization of dichotic stimuli when the time or intensity differences are zero, since the vertical axis of the body would be unaffected.

The pathological distortion of egocentric space cannot be confined, however, to the *translation* towards the side of the hemispheric lesion of spatial coordinate systems. Were this the case, the only type of error made by patients with hemineglect would be a rightward displacement. This indeed occurs in tasks assessing the subjective midpoint of a horizontal line, the subjective straight ahead (see Tables 1 and 2), and in the phenomenon of *allochiria*, frequently associated with right brain damage and neglect (Critchley 1953; Heilman 1979; Joanette and Brouchon 1984; Kawamura et al. 1987)[3]. A main feature of hemineglect is, however, the patients' *defective awareness* of left-sided information, even though a considerable amount of nonconscious processing may take place (e.g., Berti and Rizzolatti 1992; Bisiach and Rusconi 1990; Marshall and Halligan 1988; Vallar et al. 1994, in press a). The distortion towards the right side, modulated by sensory systems, thus involves a more profound, and largely unexplored, anisotropic derangement of the spatial representations, disrupting the patients' awareness.

That an interpretation in terms of mere translation is insufficient is illustrated also by the pattern of the deficit of position sense investigated by Vallar et al. (1993a, 1995c). Right brain-damaged patients with left visuospatial hemineglect showed a defective perception of the spatial positions of the forearms both in the horizontal plane, where positions varied along a left-right dimension, and in the vertical plane, where the up-down dimension was the relevant parameter. In both planes, the deficit was modulated by horizontal optokinetic stimulation in a similar fashion, consistent with the hypothesis of a rightward distortion of the

[3] The term *allochiria* denotes a disorder whereby sensory stimuli are dislocated to the opposite side of the body or space (Meador et al. 1991). Patients with lesions in the right hemisphere usually displace the perceived stimulus from the left to the right side (Critchley 1953).

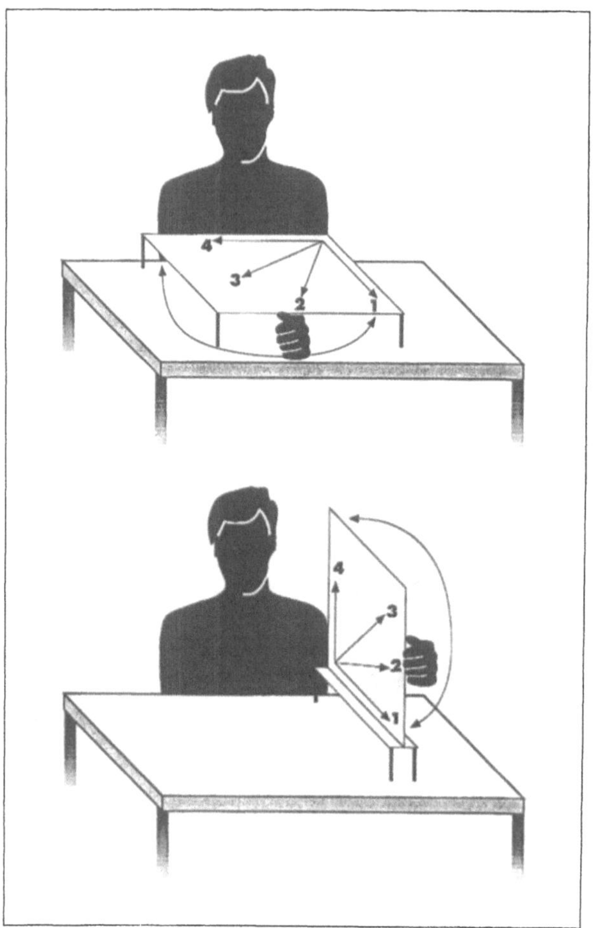

Fig. 2. Schematic drawing (front view of the left forearm's assessment) of the four assessed positions (1–4) in the horizontal (*top*) and vertical (*bottom*) plane. (From Vallar et al. 1995a)

coordinate frame in which the spatial position of the forearms was represented (Figs. 2 and 3). A simple rightward translation, however, does not readily predict a deficit in a task assessing the perception of the spatial position of the forearm relative to the arm in the vertical plane. In this condition, not only the relationships between the two relevant body segments should be preserved, but the relevant vertical dimension should be unaffected by the lateral displacement. Perenin (this volume) reports that the perceived subjective midline is displaced towards the side of the lesion also in patients with different disorders, such as optic ataxia and homonymous hemianopia (see related evidence in Corin and

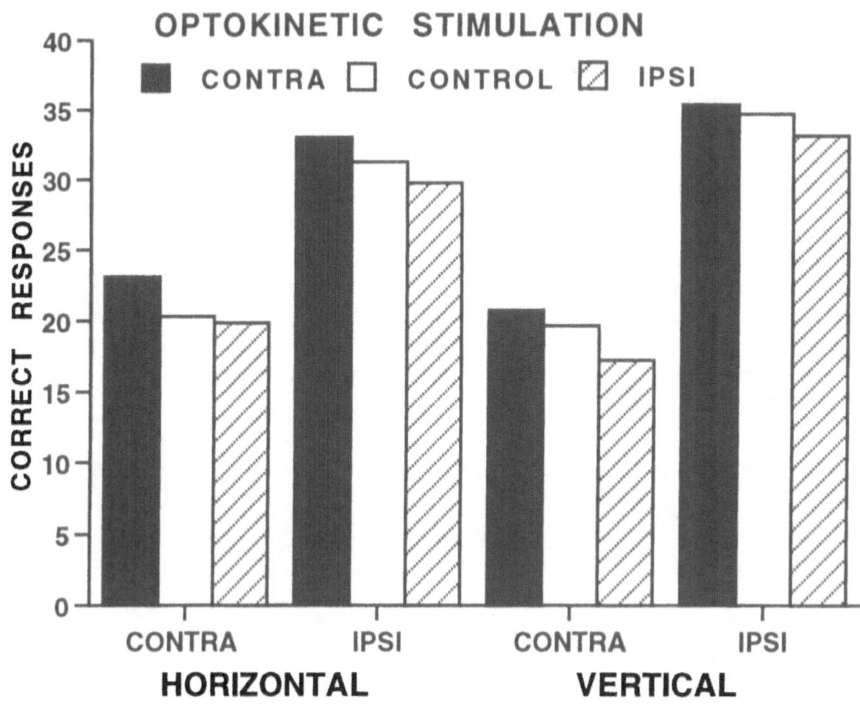

Fig. 3. Mean correct responses of right brain-damaged patients with left visuospatial hemineglect in the horizontal and vertical planes by forearm [contralateral (left) and ipsilateral (right) to the side of the lesion] and by stimulation [control: stationary dots; stimulation with a contralateral (left) or ipsilateral (right) direction of the movement]. *CONTRA*, contralateral; *IPSI*, ipsilateral. (From Vallar et al. 1995a)

Bender 1972), and concludes that interpretations of hemineglect in terms of the mere rightward translation of spatial coordinates are not sufficient to account for defective spatial awareness.

A final implication of the investigation of the effects of sensory stimulations concerns the interpretation of the dissociations among neglect-related disorders (discussion in Halligan and Marshall 1992; Vallar 1994). It is now widely accepted that deficits such as extrapersonal visuospatial and personal neglect (Bisiach et al. 1986a, b), neglect hemianaesthesia (Vallar et al. 1993b, in press b), and imaginal neglect (Guariglia et al. 1993) do not necessarily co-occur. Selective patterns of impairment suggest the existence of discrete representations, concerning different sectors of space and the body, and are in line with neurophysiological observations (e.g., Graziano and Gross 1995; Rizzolatti et al. 1983). A number of sensory stimulations, however, modulate the manifold components of the neglect syndrome in a similar directional fashion. This indicates that the different representations involved (and their pathological counterparts) share a basic *spatial* feature over and above the specific sector of space that may be selectively impaired. Sensory inputs contribute to the building up and ongoing operation of this spatial medium.

Acknowledgements. This work was supported in part by grants from CNR, MURST, Ministero della Sanita', and EC (BMH1-CT94-1133). We are grateful to Marc Jeannerod, Hans-Otto Karnath, Francesco Lacquaniti, and Luigi Pizzamiglio for their helpful suggestions. Usual disclaimers apply. Figure 1 reprinted by permission of Oxford University Press. Figure 2 reprinted from Cortex by permission of Masson.

References

Akbarian S, Grüsser OJ, Guldin WO (1994) Corticofugal connections between the cerebral cortex and brain-stem vestibular nuclei in the macaque monkey. J Comp Neurol 339:421–437

Altman JA, Balonov LJ, Deglin VL (1979) Effects of unilateral disorder of the brain hemisphere function in man on directional hearing. Neuropsychologia 17:295–301

Andersen RA (1995a) Coordinate transformations and motor planning in posterior parietal cortex. In: Gazzaniga MS (ed) The cognitive neurosciences. MIT Press, Cambridge, pp 519–548

Andersen RA (1995b) Encoding of intention and spatial location in the posterior parietal cortex. Cereb Cortex 5:457–469

Berti A, Rizzolatti G (1992) Visual processing without awareness: Evidence from unilateral neglect. J Cogn Neurosci 4:345–351

Biguer B, Donaldson IML, Hein A, Jeannerod M (1988) Neck muscle vibration modifies the representation of visual motion and direction in man. Brain 111:1405–1424

Bischof N (1974) Optic-vestibular orientation to the vertical. In: Kornhuber HH (ed) Handbook of sensory physiology, vol. 6. Springer, Berlin Heidelberg New York, pp 155–190

Bisiach E, Rusconi ML (1990) Break-down of perceptual awareness in unilateral neglect. Cortex 26:643–649

Bisiach E, Vallar G (1988) Hemineglect in humans. In: Boller F, Grafman J (eds) Handbook of neuropsychology, vol.1. Elsevier, Amsterdam, pp 195–222

Bisiach E, Cornacchia L, Sterzi R, Vallar G (1984) Disorders of perceived auditory lateralization after lesions of the right hemisphere. Brain 107:37–52

Bisiach E, Perani D, Vallar G, Berti A (1986a) Unilateral neglect: personal and extrapersonal. Neuropsychologia 24:759–767

Bisiach E, Vallar G, Perani D, Papagno C, Berti A (1986b) Unawareness of disease following lesions of the right hemisphere: anosognosia for hemiplegia and anosognosia for hemianopia. Neuropsychologia 24:471–482

Bisiach E, Rusconi ML, Vallar G (1991) Remission of somatoparaphrenic delusion through vestibular stimulation. Neuropsychologia 29:1029–1031

Bisiach E, Rusconi ML, Peretti V, Vallar G (1994) Challenging current accounts of unilateral neglect. Neuropsychologia 32:1431–1434

Bisiach E, Pizzamiglio L, Nico D, Antonucci G (1996) Beyond unilateral neglect. Brain 119:851–857

Bonnier P (1905) L'aschématie. Rev Neurol 12:605–609

Bottini G, Sterzi R, Paulesu E, Vallar G, Cappa SF, Erminio F, Passingham RE, Frith CD, Frackowiak RSJ (1994) Identification of the central vestibular projections in man: a positron emission tomography activation study. Exp Brain Res 99:164–169

Bottini G, Paulesu E, Sterzi R, Warburton E, Wise RJS, Vallar G, Frackowiak RSJ, Frith CD (1995) Modulation of conscious experience by peripheral stimuli. Nature 376:778–781

Brotchie PR, Andersen RA, Snyder LH, Goodman SJ (1995) Head position signals used by parietal neurons to encode locations of visual stimuli. Nature 375:232–235

Butter CM, Kirsch N (1992) Combined and separate effects of eye patching and visual stimulation on unilateral neglect following stroke. Arch Phys Med Rehabil 73:1133–1139

Butter CM, Evans J, Kirsch N, Kewman D (1989) Altitudinal neglect following traumatic brain injury: a case report. Cortex 25:135–146

Büttner U, Lang W (1979) The vestibulocortical pathway: neurophysiological and anatomical studies in the monkey. In: Granit R, Pompeiano O (eds) Progress in brain research, vol. 50. Reflex control of posture and movement. Elsevier, Amsterdam, pp 581–588

Cappa S, Sterzi R, Vallar G, Bisiach E (1987) Remission of hemineglect after vestibular stimulation. Neuropsychologia 25:775–782

Cohen B, Suzuki J-I, Shanzer S, Bender MB (1964) Semicircular canal control of eye movements. In: Bender MB (ed) The oculomotor system. Harper and Row, New York, pp 163–172

Colby CL, Duhamel J-R, Goldberg ME (1995) Oculocentric spatial representation in parietal cortex. Cereb Cortex 5:470–481

Corin MS, Bender MB (1972) Mislocalization in visual space. With reference to the midline at the boundary of a homonymous hemianopia. Arch Neurol 27:252–262

Corkin S, Milner B, Rasmussen T (1970) Somatosensory thresholds. Arch Neurol 23:41–58

Critchley M (1953) The parietal lobes. Hafner, New York

De Renzi E, Faglioni P, Gentilini M, Barbieri C (1989) Attentional shift towards the rightmost stimuli in patients with left visual neglect. Cortex 25:231–237

Duhamel J-R, Colby CL, Goldberg ME (1992) The updating of the representation of visual space in parietal cortex by intended eye movements. Science 255:90–92

Fogassi L, Gallese V, Di Pellegrino G, Fadiga L, Gentilucci M, Luppino G, Matelli M, Pedotti A, Rizzolatti G (1992) Space coding by premotor cortex. Exp Brain Res 89:686–690

Fredrickson JM, Kornhuber HH, Schwarz DWF (1974) Cortical projections of the vestibular nerve. In: Kornhuber HH (ed) Handbook of sensory physiology, vol. 6. Vestibular system. Springer, Berlin Heidelberg New York, pp 565–582

Gainotti G (1993) The role of spontaneous eye movements in orienting attention and in unilateral neglect. In: Robertson IH, Marshall JC (eds) Unilateral neglect: clinical and experimental studies. Erlbaum, Hove, pp 107–122

Galletti C, Battaglini PP, Fattori P (1993) Parietal neurons encoding spatial locations in craniotopic coordinates. Exp Brain Res 96:221–229

Galletti C, Battaglini PP, Fattori P (1995a) Eye position influence on the parieto-occipital area PO (V6) of the macaque monkey. Eur J Neurosci 7:2486–2501

Galletti C, Fattori P, Battaglini PP, Shipp S, Zeki S (1995b) Functional demarcation of a border between area V6 and V6A in the superior parietal gyrus of the macaque monkey. Eur J Neurosci 8:30–52

Geminiani G, Bottini G (1992) Mental representation and temporary recovery from unilateral neglect after vestibular stimulation. J Neurol Neurosurg Psychiatry 55:332–333

Gerstmann J (1942) Problem of imperception of disease and of impaired body territories with organic lesions. Arch Neurol Psychiatr 48:890–913

Goldberg ME, Colby CL (1989) The neurophysiology of spatial vision. In: Boller F, Grafman J (eds) Handbook of neuropsychology, vol. 2. Elsevier, Amsterdam, pp 301–315

Graziano MS, Gross CG (1995) The representation of extrapersonal space: a possible role for bimodal, visual-tactile neurons. In: Gazzaniga MS (ed) The cognitive neurosciences. MIT Press, Cambridge, pp 1021–1034

Grüsser O-J, Landis T (1991) Visual agnosias and other disturbances of visual perception and cognition. Macmillan, Houndmills

Grüsser O-J, Guldin W, Harris J-C, Lefèbre J-C, Pause M (1992) Cortical representation of head-in-space movement and same psychophysical experiments on head movement. In: Berthoz A, Graf W, Vidal PP (eds) The head-neck sensory motor system. Oxford University Press, New York, pp 497–509

Guariglia C, Padovani A, Pantano P, Pizzamiglio L (1993) Unilateral neglect restricted to visual imagery. Nature 364:235–237

Halligan PW, Marshall JC (1992) Left visuo-spatial neglect: a meaningless entity? Cortex 28:525–535

Halligan PW, Manning L, Marshall JC (1991) Hemispheric activation vs spatio-motor cueing in visual neglect: a case study. Neuropsychologia 29:165–176

Heilman KM (1979) Neglect and related disorders. In: Heilman KM, Valenstein E (eds) Clinical neuropsychology. Oxford University Press, New York, pp 268–307

Heilman KM, Bowers D, Watson RT (1983) Performance on hemispatial pointing task by patients with neglect syndrome. Neurology 33:661–664

Heilman KM, Watson RT, Valenstein E (1993) Neglect and related disorders. In: Heilman KM, Valenstein E (eds) Clinical neuropsychology. Oxford University Press, New York, pp 279–336

Holmes G (1918) Disturbances of vision by cerebral lesions. Br J Ophthalmol 2:353–384

Hörnsten G (1979) Constant error of visual egocentric orientation in patients with acute vestibular disorder. Brain 102:685–700

Howard IP (1982) Human visual orientation. Wiley, Chichester

Howard IP, Templeton WB (1966) Human spatial orientation. Wiley, London

Ishiai S, Sugishita M, Watabiki S, Nakayama T, Kotera M, Gono S (1994a) Improvement of left unilateral spatial neglect in a line extension task. Neurology 44:294–298

Ishiai S, Watabiki S, Lee E, Kanouchi T, Odajima N (1994b) Preserved leftward movement in left unilateral spatial neglect due to frontal lesions. J Neurol Neurosurg Psychiatry 57:1085–1090

Jeannerod M, Biguer B (1987) The directional coding of reaching movements. A visuomotor conception of spatial neglect. In: Jeannerod M (ed) Neurophysiological and neuropsychological aspects of spatial neglect. Elsevier, Amsterdam, pp 87–113

Joanette Y, Brouchon M (1984) Visual allesthesia in manual pointing: some evidence for a sensorimotor cerebral organization. Brain Cogn 3:152–165

Kanner L, Schilder P (1930) Movements in optic images and the optic imagination of movements. J Nerv Ment Dis 72:489–517

Karnath H-O (1988) Deficits of attention in acute and recovered visual hemi-neglect. Neuropsychologia 26:27–43

Karnath H-O (1994) Subjective body orientation in neglect and the interactive contribution of neck muscle proprioception and vestibular stimulation. Brain 117:1001–1012

Karnath H-O (1995) Transcutaneous electrical stimulation and vibration of neck muscles in neglect. Exp Brain Res 105:321–324

Karnath H-O (1996) Optokinetic stimulation influences the disturbed perception of body orientation in spatial neglect. J Neurol Neurosurg Psychiatry 60:217–220

Karnath H-O, Schenkel P, Fischer B (1991) Trunk orientation as the determining factor of the 'contralateral' deficit in the neglect syndrome and as the physical anchor of the internal representation of body orientation in space. Brain 114:1997–2014

Karnath H-O, Christ K, Hartje W (1993) Decrease of contralateral neglect by neck muscle vibration and spatial orientation of trunk midline. Brain 116:383–396

Karnath H-O, Sievering D, Fetter M (1994) The interactive contribution of neck muscle proprioception and vestibular stimulation to subjective "straight ahead" orientation in man. Exp Brain Res 101:140–146

Kawamura M, Hirayama K, Shinohara Y, Watanabe Y, Sugishita M (1987) Alloaesthesia. Brain 110:225–236

Kinsbourne M (1993) Orientational bias model of unilateral neglect: evidence from attentional gradients within hemispace. In: Robertson IH, Marshall JC (eds) Unilateral neglect: clinical and experimental studies. Erlbaum, Hove, pp 63–86

Kooistra CA, Heilman KM (1989) Hemispatial visual inattention masquerading as hemianopia. Neurology 39:1125–1127

Kornhuber HH, Da Fonseca JS (1964) Optovestibular integration in the cat's cortex: a study of sensory convergence on cortical neurons. In: Bender MB (ed) The oculomotor system. Harper and Row, New York, pp 239–277

Lackner JR (1978) Some mechanisms underlying sensory and postural stability in man. In: Held R, Leibowitz HW, Teuber H-L (eds) Handbook of sensory physiology, vol. 8. Perception. Springer, Berlin Heidelberg New York, pp 805–845

Lackner JR (1988) Some proprioceptive influences on the perceptual representation of body shape and orientation. Brain 111:281–297

Lacquaniti F (in press) Frames of reference in sensorimotor coordination. In: Boller F, Grafman J (eds) Handbook of neuropsychology. Elsevier, Amsterdam

Làdavas E (1990) Selective spatial attention in patients with visual extinction. Brain 113:1527–1538

Làdavas E (1993) Spatial dimensions of automatic and voluntary orienting components of attention. In: Robertson IH, Marshall JC (eds) Erlbaum, Hove, pp 193–209

Làdavas E, Petronio A, Umiltà C (1990) The deployment of visual attention in the intact field of hemineglect patients. Cortex 26:307–317

Lhermitte J (1952) L'image corporelle en neurologie. Schweiz Arch Neurol Psychiatr 69:214–236

Mark VW, Heilman KM (1990) Bodily neglect and orientational biases in unilateral neglect syndrome and normal subjects. Neurology 40:640–643

Marshall CR, Maynard FM (1983) Vestibular stimulation for supranuclear gaze palsy: a case report. Arch Phys Med Rehabil 64:134–136

Marshall JC, Halligan P (1988) Blindsight and insight in visuo-spatial neglect. Nature 336:766–767

Mattingley JB, Bradshaw JL, Bradshaw JA (1995) Horizontal visual motion modulates focal attention in left unilateral spatial neglect. J Neurol Neurosurg Psychiatry 57:1228–1235

Meador KJ, Allen ME, Adams RJ, Loring DW (1991) *Allochiria* vs *Allesthesia*. Is there a misperception? Arch Neurol 48:546–549

Mennemeier M, Wertman E, Heilman KM (1992) Neglect of near peripersonal space. Brain 115:37–50

Mergner T (1979) Vestibular influences on the cat's cerebral cortex. In: Granit R, Pompeiano O (eds) Progress in brain research, vol. 50. Reflex control of posture and movement. Elsevier, Amsterdam, pp 567–579

Morrow MJ, Sharpe JA (1993) The effects of head and trunk position on torsional vestibular and optokinetic eye movements in humans. Exp Brain Res 95:144–150

Moscovitch M, Behrmann M (1994) Coding of spatial information in the somatosensory system: evidence from patients with neglect following parietal lobe damage. J Cogn Neurosci 6:151–155

Pause M, Kunesch E, Binkofsky F, Freund HJ (1989) Sensorimotor disturbances in patients with lesions of the parietal cortex. Brain 112:1599–1625

Pizzamiglio L, Frasca R, Guariglia C, Incoccia C, Antonucci G (1990) Effect of optokinetic stimulation in patients with visual neglect. Cortex 26:535–540

Pizzamiglio L, Vallar G, Doricchi F (1995) Gravity and hemineglect. Neuroreport 7:370–371

Ramachandran VS (1995) Anosognosia in parietal lobe syndrome. Consc Cogn 4:22–51

Rapcsak SZ, Cimino CR, Heilman KM (1988) Altitudinal neglect. Neurology 38:277–281

Riddoch MJ, Humphreys GW (1983) The effect of cueing on unilateral neglect. Neuropsychologia 21:589–599

Rizzolatti G, Matelli M, Pavesi G (1983) Deficits in attention and movement following the removal of postarcuate (area 6) and prearcuate (area 8) cortex in macaque monkeys. Brain 106:655–673

Robertson IH, North N (1992) Spatio-motor cueing in unilateral left neglect: the role of hemispace, hand and motor activation. Neuropsychologia 30:553–563

Robertson IH, North N (1993) Active and passive activation of left limbs: influence on visual and sensory neglect. Neuropsychologia 31:293–300

Robertson IH, North NT, Geggie C (1992) Spatiomotor cueing in unilateral left neglect: three case studies of its therapeutic effects. J Neurol Neurosurg Psychiatry 55:799–805

Robertson IH, Tegnér R, Goodrich SJ, Wilson C (1994) Walking trajectory and hand movements in unilateral left neglect: a vestibular hypothesis. Neuropsychologia 32:1495–1502

Rode G, Perenin MT (1994) Temporary remission of representational hemineglect through vestibular stimulation. Neuroreport 5:869–872

Rode G, Charles N, Perenin MT, Vighetto A, Trillet M, Aimard G (1992) Partial remission of hemiplegia and somatoparaphrenia through vestibular stimulation in a case of unilateral neglect. Cortex 28:203–208

Roll JP, Roll R, Velay J-L (1991) Proprioception as a link between body space and extra-personal space. In: Paillard J (ed) Brain and space. Oxford University Press, Oxford, pp 112–132

Roth M (1949) Disorders of the body image caused by lesions of the right parietal lobe. Brain 72:89–111

Rubens AB (1985) Caloric stimulation and unilateral visual neglect. Neurology 35:1019–1024

Sakata H, Kusunoki M (1992) Organization of space perception: neural representation of three-dimensional space in the posterior parietal cortex. Curr Opin Neurobiol 2:170–174

Schilder P (1933) The vestibular apparatus in neurosis and psychosis. J Nerv Ment Dis 78:137–164

Schilder P (1950) The image and appearance of the human body. International Universities Press, New York

Shanzer S, Bender MB (1959) Oculomotor responses on vestibular stimulation of monkeys with lesions of the brain-stem. Brain 82:669–682

Shelton PA, Bowers D, Heilman KM (1990) Peripersonal and vertical neglect. Brain 113:191–205

Silberpfennig J (1941) Contributions to the problem of eye movements. III. Disturbances of ocular movements with pseudohemianopsia in frontal lobe tumors. Confin Neurol 4:1–13

Smania N, Aglioti S (1995) Sensory and spatial components of somaesthetic deficits following right brain damage. Neurology 45:1725–1730

Spinelli D, Burr DC, Morrone MC (1994) Spatial neglect is associated with increased latencies of visual evoked potentials. Vis Neurosci 11:909–918

Sprague J (1966) Interaction of cortex and superior colliculus in mediation of visually guided behavior in the cat. Science 153:1544–1547

Taylor IL, McCloskey DI (1991) Illusions of head and visual target displacement induced by vibration of neck muscles. Brain 114:755–759

Teuber H-L (1962) Effects of brain wounds implicating right or left hemisphere in man: hemisphere differences in vision, audition and somesthesis. In: Mountcastle VB (ed) Interhemispheric relations and cerebral dominance. John Hopkins Press, Baltimore, pp 131–157

Vallar G (1991) Current methodological issues in human neuropsychology. In: Boller F, Grafman J (eds) Handbook of neuropsychology, vol. 5. Elsevier, Amsterdam, pp 343–378

Vallar G (1993) The anatomical basis of spatial hemineglect in humans. In: Robertson IH, Marshall JC (eds) Unilateral neglect: clinical and experimental studies. Erlbaum, Hove, pp 27–59

Vallar G (1994) Left spatial hemineglect: An unmanageable explosion of dissociations? No. Neuropsychol Rehabil 4:209–212

Vallar G, Sterzi R, Bottini G, Cappa S, Rusconi ML (1990) Temporary remission of left hemianaesthesia after vestibular stimulation. Cortex 26:123–131

Vallar G, Bottini G, Sterzi R, Passerini D, Rusconi ML (1991a) Hemianesthesia, sensory neglect and defective access to conscious experience. Neurology 41:650–652

Vallar G, Sandroni P, Rusconi ML, Barbieri S (1991b) Hemianopia, hemianesthesia and spatial neglect. A study with evoked potentials. Neurology 41:1918–1922

Vallar G, Antonucci G, Guariglia C, Pizzamiglio L (1993a) Deficits of position sense, unilateral neglect, and optokinetic stimulation. Neuropsychologia 31:1191–1200

Vallar G, Bottini G, Rusconi ML, Sterzi R (1993b) Exploring somatosensory hemineglect by vestibular stimulation. Brain 116:71–86

Vallar G, Rusconi ML, Bisiach E (1994) Awareness of contralesional information in unilateral neglect: effects of verbal cueing, tracing and vestibular stimulation. In: Umiltà CA, Moscovitch M (eds) Attention and performance XV. Conscious and nonconscious information processing. MIT Press, Cambridge, pp 377–391

Vallar G, Guariglia C, Magnotti L, Pizzamiglio L (1995a) Optokinetic stimulation affects both vertical and horizontal deficits of position sense in unilateral neglect. Cortex 31:669–683

Vallar G, Guariglia C, Nico D, Bisiach E (1995b) Spatial hemineglect in back space. Brain 118:467–472

Vallar G, Papagno C, Rusconi ML, Bisiach E (1995c) Vestibular stimulation, spatial hemineglect and dysphasia. Selective effects? Cortex 31:589–593

Vallar G, Rusconi ML, Barozzi S, Bernardini B, Ovadia D, Papagno C, Cesarani A (1995d) Improvement of left visuo-spatial hemineglect by left-sided transcutaneous electrical stimulation. Neuropsychologia 33:73–82

Vallar G, Guariglia C, Nico D, Tabossi P (in press a) Left neglect dyslexia and the processing of *neglected* information. J Clin Exp Neuropsychol

Vallar G, Rusconi ML, Bernardini B (in press b) Modulation of neglect hemianaesthesia by transcutaneous electrical stimulation. J Exp Neuropsychol Soc

Vanni-Mercier G, Magnin M (1982) Single neuron activity related to natural vestibular stimulation in the cat's visual cortex. Exp Brain Res 45:451–455

Ventre J, Flandrin JM, Jeannerod M (1984) In search for the egocentric reference. A neurophysiological hypothesis. Neuropsychologia 22:797–806

Viggiano MP, Spinelli D, Mecacci L (1995) Pattern reversal visual evoked potentials in patients with hemineglect syndrome. Brain Cogn 27:17–35

Weiskrantz L (1986) Blindsight. A case study and implications. Clarendon, Oxford

Wortis SB, Pfeffer AZ (1948) Unilateral auditory-spatial agnosia. J Nerv Ment Dis 108:181–186

Ipsilesional Displacement of Egocentric Midline in Neglect Patients with, but Not in Those Without, Extensive Right Parietal Damage

M. HASSELBACH and C. M. BUTTER

Department of Psychology, The University of Michigan, Ann Arbor, USA

Introduction

Unilateral spatial neglect is a disorder characterized by a deficiency in responding or orienting to stimuli located on the side of space contralateral to a cerebral lesion. Neglect is more frequently found after right than after left hemispheric lesions, especially those involving the inferior parietal lobule. However, it also is present after damage restricted to the frontal lobes, cingulate cortex, basal ganglia, or thalamus (Vallar and Perani 1987). Several dissociations in the clinical manifestation of neglect have been described. Patients may display neglect in one modality and not another (visual, auditory, tactile), or for one region of space and not another (personal, peripersonal, extrapersonal; Halligan and Marshall 1992). In addition, space can be represented in environment-centered, viewer-centered, and even object-centered coordinates (for a review see Farah et al. 1990), and neglect patients can show deficits in any one, or more, of these coordinates (Bisiach et al. 1985; Chatterjee 1994; Calvanio et al. 1987; Driver et al. 1994). Unfortunately, the mechanisms involved in these different manifestations of neglect are not well understood. A dissociation that has been predicted is that between "perceptual" and "motor" neglect. According to Mesulam's cortical network theory of directed attention (Mesulam 1990), anterior lesions should result in a deficit in scanning and exploring contralateral space (motor neglect or directional hypokinesia), and posterior lesions should result in diminished awareness of contralateral space (perceptual neglect). In fact, several studies have confirmed this dissociation (Daffner et al. 1990; Tegnér and Levander 1991; Liu et al. 1992; Làdavas et al. 1993). In the remainder of this paper we focus on perceptual neglect.

Several interpretations of the disturbance underlying perceptual neglect have been suggested; among the most prominent of these are disorders of attention and of central representations. Attentional theories of neglect have taken three main forms. According to one, neglect is due to an impairment in directing attention to the contralesional side of space (Riddoch and Humphreys 1983). This view is supported by studies showing that neglect patients improve their performance on a

In: Parietal Lobe Contributions to Orientation in 3D Space (1997). P. Thier and H.-O. Karnath (eds). Springer-Verlag, Heidelberg.

line bisection task when cued to the neglected end of the line (Riddoch and Humphreys 1983; Butter et al. 1990). The second type of attentional theory suggests that neglect is due to hyperattention to stimuli on the ipsilesional side of space (Làdavas et al. 1990; Kinsbourne 1993). Evidence for this theory is provided by findings that patients with left-sided neglect, unlike normal subjects, show faster reaction times to stimuli to the right of fixation than to central stimuli (Làdavas et al. 1990). Finally, some investigators attribute neglect to an impairment in disengaging attention from ipsilesional stimuli (Posner et al. 1987; Marshall and Halligan 1994; Mark et al. 1988). This interpretation is supported by the finding that neglect patients show fewer left-sided omissions in a line cancellation task when they are instructed to erase the targets instead of making a mark through them (Mark et al. 1988).

An alternative to attentional accounts of neglect proposes that the disorder is due to an alteration in the representation of egocentric space (Bisiach and Berti 1987). This view is based upon the finding that patients with left-sided neglect fail to report the left side of an imagined scene regardless of the direction from which they are instructed to view it (Bisiach et al. 1981). Additional evidence comes from a study by Bisiach and colleagues (1979) in which neglect patients were asked to judge whether pairs of moving forms viewed successively were the same or different in shape. Since the forms were viewed through a central slit, the judgments required the mental reconstruction of the forms. The patients displayed a deficit in detecting the left-sided but not right-sided differences in these mentally reconstructed forms. Although these results suggest an impairment in internal representations, it has been argued that they could also be explained by an attentional loss. Kinsbourne's (1993) orientational bias model of neglect proposes that attention can be deployed both at the level of direct perception and the level of internal representation through the interaction of lateralized and opposed orienting mechanisms. This model would predict that after right-hemispheric damage, scanning of the left side of a mental image as well as of the left half of the environment would be impaired. By this view, the above mentioned results of Bisiach and his colleagues could be interpreted as a deficit in an internal attentional scanning mechanism, rather than as a permanent distortion in internal representations. It is difficult to see, however, how the orientational bias model of neglect could account for Bisiach and colleagues' (1984) finding that right-hemispheric patients with neglect show alterations in localizing auditory stimuli; their perceived position was shifted to the (unneglected) right side of space. The authors interpret this finding as consistent with the theory that neglect is due to a severe distortion in the internal representation of egocentric space.

Although the findings of Bisiach and his colleagues (summarized above) support the claim that at least some forms of neglect are due to a distortion in the internal representation of egocentric space, they do not identify which egocentric coordinate system, or systems, are altered in neglect. The reference frame for egocentric coordinates can be defined with respect to the retina, the head, or the body. Because these three frames were aligned with one another in the above cited studies, it is impossible to determine which was contributing the most to "defining" the neglected side of space. In contrast, studies of neglect patients in which the body and head were aligned and eye movements monitored have shown that retinal coordinates do not determine the border between the neglected and the unneglected side of space (Johnston and Diller 1986; Posner et al. 1987). In addition, Karnath and colleagues (1991) conducted a study in which the relation between the head and trunk coordinates were varied while the retinal position of the stimuli was held constant. They found that left-sided neglect is reduced when the trunk is turned to the left with respect to the head, but not when the head is turned to the left with respect to the trunk. This points to an important role of body-centered coordinates in neglect. A later study showed that left-sided neglect is also reduced when the left posterior neck muscles are mechanically vibrated (Karnath et al. 1993). This procedure signals muscle stretch to the brain (Roll et al. 1989) and induces the illusion that the trunk has been turned to the left (Taylor and McCloskey 1991; Biguer et al. 1988; Karnath 1994; Karnath et al. 1994).

Additional evidence that body-centered coordinates are disrupted in neglect comes from studies of midline localization (i.e., the point that is perceived as straight ahead of the body) in neglect patients with right parietal lobe damage. Mark and Heilman (1990) reported that when asked to point straight ahead, these patients, unlike control subjects, pointed approximately 4° to the right on average. However, it was unclear from this study whether the neglect patients' deviated pointing was due to directional hypokinesia, a disorder shown by patients with left-sided neglect (Heilman et al. 1985), or due to a disorder in the representation of egocentric space. More recently, Karnath (1994) tested patients with left-sided neglect due to lesions involving the right parietal lobe in a midline localization task that did not require pointing. Instead, the patients instructed the examiner to adjust a point of light so that it was aligned with the center of their trunk. Karnath concluded that unilateral spatial neglect is due to an alteration in the central representation of space, resulting in a systematic ipsilesional deviation of the perceived body midline.

The present study investigates the relationship between lesion location and the disruption of body-centered, egocentric coordinates. Lesion location is a factor whose influence on distortions of midline perception is poorly understood. The research on the relation between body midline perception and unilateral spatial neglect has exclusively employed patients with extensive damage to the right parietal lobe. Because neglect can occur after damage to other cortical areas as well as after damage restricted to subcortical structures, it is unclear whether the body midline distortions found in the above mentioned studies are directly related to the neglect phenomenon, or whether they are related specifically to parietal lobe damage only. In the present study, we raised the question whether neglect due to right hemispheric lesions located primarily outside the parietal lobe is also accompanied by rightward shifts in the perceived body midline. In order to investigate this, we used a psychophysical midline localization task that has no directional motor component and permits estimation of the just noticeable difference (JND) associated with the perceived midline. In addition to using a commonly employed search task to screen for neglect, we also employed a test of visual search to determine whether neglect was present in the same situation in which we tested for midline localization.

Materials and Methods

Subjects

Three groups of subjects were tested. The normal control group consisted of ten subjects (five males and five females) ranging in age from 51 to 72 years (mean = 61 yrs). The right-hemispheric control group consisted of five males and two females with right-hemispheric lesions (see Table 1) who did not display neglect (more than one omission of a target stimulus) in the Star Cancellation test, a test shown to be more sensitive to neglect than a number of other commonly used tests (Halligan et al. 1989). The third group consisted of five patients with right-hemispheric lesions who displayed neglect in the Star Cancellation test. Their case histories are briefly described below.

Table 1. Characteristics of right-hemispheric control subjects

Patient	Age	Sex	Visual field defect	Symptom onset-test	Lesion location[a]
ML	75	F	Yes	9 Mos.	O
JH	65	M	Yes	3 Yrs.	IC
JE	58	M	Yes	2 Yrs.	OT
HB	61	M	No	4 Yrs.	T
HW	54	M	No	25 Days	F, BG, IC
CA	51	F	No	18 Days	F, T, P
PB	50	M	No	3 Yrs.	O

[a] O, occipital lobe; IC, internal capsule; OT, optic tract; T, temporal lobe; F, frontal lobe; BG, basal ganglia; P, parietal lobe

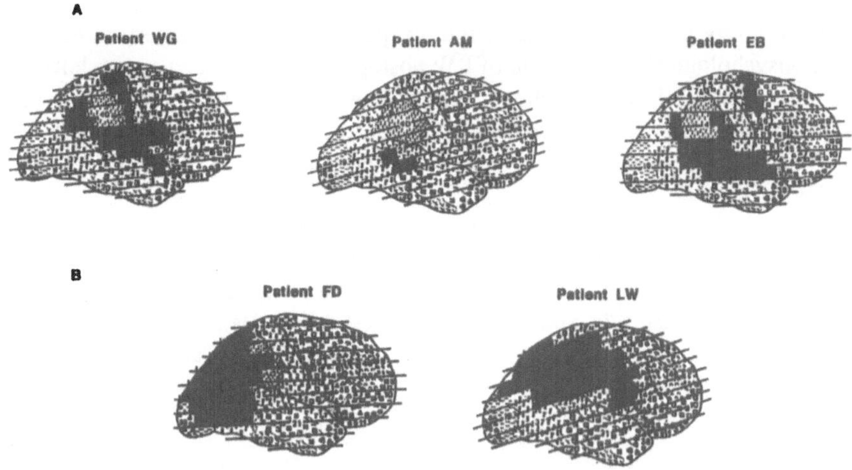

Fig. 1 A, B. A Lateral view of lesions of neglect patients without extensive parietal lobe damage and **B** neglect patients with extensive parietal lobe damage. Lesions were plotted from computed tomography scans using the method described by Damasio and Damasio (1989)

WG is a 50-year-old woman who suffered a right-hemispheric stroke 45 days before testing. CT scans showed a hypodensity involving the frontal, temporal, and parietal lobes, but not extensive inferior parietal lobule damage (see Fig. 1A). Her visual fields were intact to confrontation, and she showed no visual extinction. She showed mild left-sided neglect on the Star Cancellation test, in which she omitted three targets on the left side of the sheet.

AM is a 53-year-old man who suffered a right-hemispheric stroke 4 years before testing. CT scans soon after his stroke revealed an acute cortical infarct involving the temporal and occipital lobes but not the parietal lobe (see Fig. 1A). Visual fields were intact to confrontation, and he showed no visual extinction. In the Star Cancellation test he showed moderate left-sided neglect, omitting seven targets on the left side.

EB is a 65-year-old man who suffered a right-hemispheric stroke 45 days before testing. CT scans revealed an acute cortical infarct involving the right temporal lobe and extending into the frontal and parietal lobes and the internal capsule, but involving only slight damage to the inferior parietal lobule (see Fig. 1A). He displayed a left homonymous hemianopia to confrontation and showed severe left-sided neglect in the Star Cancellation test: He marked only 16 out of 50 targets, and those were on the far right side of the sheet.

LW is a 69-year-old woman who underwent a craniotomy for tumor resection in 1965. Four years before testing, the patient underwent a second craniotomy and resection of a benign meningioma of the right parietal area. Two days postoperatively, CT scans indicated a CVA in the distribution of the right middle cerebral artery involving extensive inferior parietal lobule damage (see Fig. 1B). Neuropsychological examination of LW postoperatively revealed severe left-sided neglect and a left homonymous hemianopia. When tested four years later, immediately prior to participating in the current experiment, LW omitted 19 targets on the Star Cancellation test. Her left homonymous hemianopia, however, had resolved.

FD, a 62-year-old man, suffered a right-hemispheric stroke 60 days before testing. CT scans demonstrated that the lesion involved an extensive portion of the inferior parietal lobule, as well as the temporal and occipital lobes, with extension into the internal capsule and basal ganglia (see Fig. 1B). He displayed a left homonymous hemianopia to confrontation. In the Star Cancellation test, he showed severe left-sided neglect, marking only 12 targets on the far right side of the sheet.

Procedures

Midline Localization Test. The stimulus used to assess egocentric midline localization was a black circle 0.5° in diameter. It was affixed to one end of a clear plastic rod that could be moved to any position on the board. The stimulus was presented at eye level on an opaque white, featureless 122 cm wide by 61 cm high Plexiglas board which rested on a table in the subject's frontal plane. Results of a pilot study indicate that at this distance, subjects did not use the edges of the board as cues when judging whether the stimulus is to the left or right of the middle of their body.

The circle was presented ten times directly in front of the subjects' actual midline and ten times in at least four positions (at 1.1° increments) to each side of this point. A scale on the back of the board, out of the subjects' view, was used by the examiner to measure 1.1° increments of the circle to the left and right of the subjects' midline, up to 13.2° in either direction.

Subjects were seated 51 cm from the board with the midline of their body and head aligned with the middle of the board. The investigator told the subjects that it was important to keep their head and body still during testing. Another investigator stood behind each neglect patient in order to hold their head straight and watch for any change in body posture. If the subjects' body posture did change, testing was suspended until their body returned to the original position. Testing began with 11 trials of the search task (described below), followed by 45 trials of the midline localization task. At this point subjects rested for several minutes. Testing then continued with 45 more trials of the midline localization task followed by ten trials of the search task.

Prior to testing, subjects' were shown the stimulus and told that it would be covered, moved to a different position on the board, and then uncovered. Subjects was instructed to indicate whether the stimulus appeared to the left or the right of the middle of the subject's body each time it was uncovered. A white 30 x 50 cm board was used to cover the stimulus while it was moved to the different positions between trials. The investigator illustrated what was meant by "the middle of the body" by pointing from the subjects' jugular notch of the sternum to their navel. All patients were instructed to respond both verbally and by pointing with the right hand in the direction indicated by their verbal response in order to eliminate errors due to left-right confusion (no patients showed left-right confusion using this method).

The normal control subjects and the right-hemisphere control subjects were tested in one session of 90 trials. The trials were presented in pseudorandom order with the stimulus appearing ten times in each of nine positions. The positions included the subjects' actual midline and four positions to either side of the midline in 1.1° increments. The order was determined before testing and remained the same for each subject.

The neglect patients were tested in either one session (WG, AM, LW) or two sessions (EB, FD), depending on the consistency of their responses. Before starting the test trials, they were given a series of preliminary trials to ensure that they understood the task and to determine the range within which to present the stimulus. The preliminary trials started in the same manner for each patient: The experimenter presented the stimulus three times 4.4° to the left and three times 4.4° to the right of the patients' actual midline (the same range used for the control groups) in random order. If the patients responded correctly on each trial ("right" each time the stimulus was to the right and "left" each time the stimulus was to the left), testing proceeded as with the control groups described above. If the patients responded inconsistently within this range, however, the range was adjusted until the subjects made three consistent responses to stimuli at the limits of the range. Each 1.1° increment within that range was then sampled at least ten times in pseudorandom order. However, FD's range was so broad that increments of 2.2° were each sampled at least ten times, and each increment between these values was sampled three to nine times. Otherwise, the procedures employed with FD were the same as those described above. All patients were encouraged to rest between trials when they needed to.

The estimate of perceived egocentric midline was achieved using a probit model of maximum likelihood estimation (SYSTAT 1992). This analysis gives slope and intercept estimations, which were used to plot each subject's psychophysical function and estimate the 50% point. Ninety-five percent confidence intervals for the midline were calculated using the procedure described by Deming (1950). A midline outside the 95% confidence interval of all the control subjects combined was considered abnormally displaced for neglect patients. Positive values indicate perceived midlines to the right of the physical midline; negative values indicate perceived midlines to the left of the physical midline. The JND was defined as the region within which subjects judged the stimulus to be to the right of the midline 25%-75% of the time.

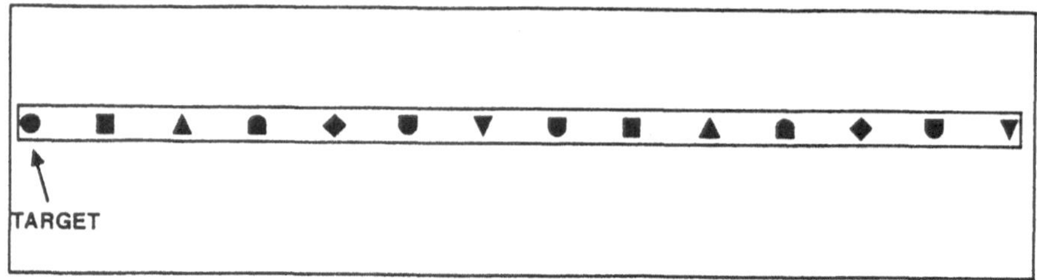

Fig. 2. Example of stimulus array used in the search task

Search Task. The stimuli used in the search task were 21 horizontal arrays of solid shapes, each approximately 0.5° in lateral extent mounted on a white poster board (see Fig. 2). Each array extended 35.2° laterally and contained 14 shapes. Fourteen of the 21 arrays included a single circle that appeared in one of the 14 possible positions (these arrays contained six other shapes, five of which appeared in two positions and one that appeared in three). The circle appeared on the left and right sides of the array equally often. The remaining seven arrays contained no circle (these arrays contained six shapes, four of which appeared in two positions and two of which appeared in three positions). The arrays were affixed, one at a time, in the center of the same board used for midline localization.

Prior to testing, subjects were told that they would see a number of arrays of shapes one after another and that each would contain either one circle or no circle. They were instructed to search the array for a circle and to say "yes" as soon as they saw one, or "no" if they did not see one. Time to search each array was recorded by the examiner with a stopwatch accurate to 10 ms.

Two measures were used to determine impairment on this task. One was a lateralized search-time score, defined as average search time for target stimuli on the right minus average search time for target stimuli on the left, divided by the total search time, and expressed as a percentage. A score outside the range of controls was considered abnormal. The second measure was the number of omissions, i.e., a "no" response when the target stimulus was actually present. Because the control subjects made no omissions on this task, one or more omissions were considered abnormal.

Results

Midline Localization Test

Results for all groups are summarized in Table 2. The mean egocentric midline judgment for the normal control group was 0.51° (SD=0.45; range=0.1° to the left

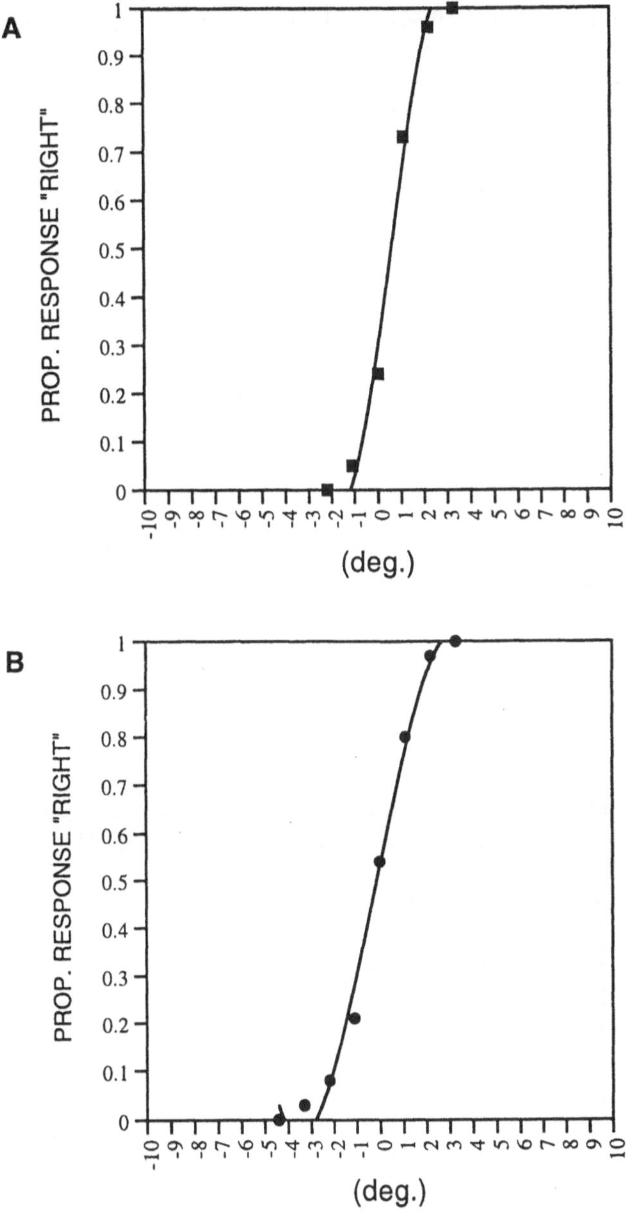

Fig. 3 A, B. Perceived egocentric midline of **A** normal control subjects and **B** right-hemispheric control subjects

Table 2. Results of midline localization and search tests

Subjects	Extensive parietal damage	Midline (degrees)[d]	95% confidence limits (degrees)[d]	JND (degrees)	Slope[a]	r-squared[a]	Percent lateralized search-time score[b]	Search omissions[d]
Controls[c] (n=17)	No	0.2R	1.2L–1.7R	3.3	0.152	0.902	-6.6	0
WG	No	0.1R	0.5L–0.8R	2.6	0.189	0.952	-32.6	0
AM	No	0.7R	0.3R–1.1R	2.0	0.270	0.902	-51.6	1L
EB	No	0.2L	1.5L–1.0R	8.5	0.059	0.930	-38.3	7L
FD	Yes	5.5R	3.6R–7.4R	13.8	0.036	0.871	–	7L, 5R
LW	Yes	5.9R	5.3R–6.6R	3.2	0.174	0.904	-9.6	1L

[a] Applies to linear functions shown in Fig. 4.
[b] Negative scores = longer search times for left- than for right-sided targets.
[c] Includes normal intact subjects and patients with right-sided lesions but without neglect.
[d] R = right of midline; L = left of midline.

to 1.3° to the right) to the right of their objective midline (see Fig. 3), a value that was significantly different from 0 (t=3.54; p=0.006; df=9). The range of their JNDs was 1.8°–2.5°; the range of their slopes was 0.15–0.50. The right-hemispheric control subjects (those without neglect) displayed an egocentric midline of 0.15° (SD=0.92; range=1.8° to the left to 1.2° to the right) to the left of their objective midline (see Fig. 3), a value not significantly different from 0. The range of their JNDs was 1°–3.5°; the range of their slopes was 0.17–0.50. Since the normal and right-hemispheric groups did not differ significantly in their perceived egocentric midline location, their midline judgment scores and 95% confidence limits were combined, resulting in an average midline location of 0.23° (SD=0.74) to the right of the objective midline. The 95% confidence limits for the combined data (1.2° left to 1.7° right) were used for comparison with neglect patients.

The midline-localization functions of neglect patients without extensive parietal damage are shown in Fig. 4A; the parameters of these functions are shown in Table 2. Patient WG responded consistently within the same range as the controls did; her perceived egocentric midline was within the 95% confidence limits of the

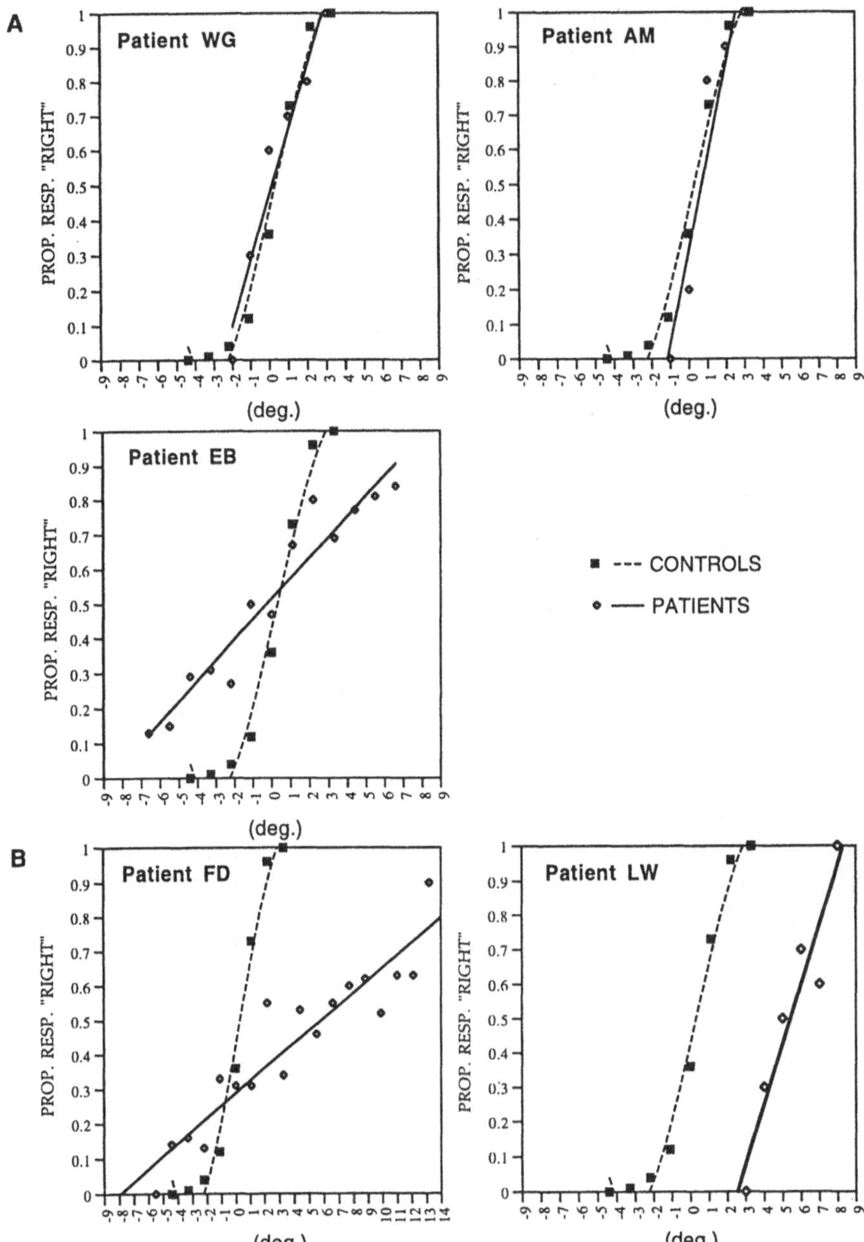

Fig. 4 A, B. Perceived egocentric midline of **A** neglect patients without extensive parietal lobe damage and **B** neglect patients with extensive parietal lobe damage, compared to combined egocentric midline results for both control groups (*CONTROLS*)

control subjects. Her JND and slope also were within the normal range. Patient AM also showed a perceived egocentric midline within the 95% confidence limits and a JND and slope within the normal range. Likewise, patient EB's perceived egocentric midline was within the 95% confidence limits. His JND (8.5°) and slope (0.059), however, were outside the normal range.

The midline-localization scores of the neglect patients with extensive parietal damage are shown in Fig. 4B; the parameters of these functions are shown in Table 2. Patient FD responded consistently within the range of 5.5° to the left (where he responded "left" 100% of the time) to 13.2° to the right (where he responded "right" 90% of the time). FD displayed a perceived egocentric midline displacement of 5.5° to the ipsilesional side (95% confidence limits = 3.61° to 7.39° right) and a JND (13.8°) and slope (0.036) outside the normal range. Patient LW responded consistently within the range of 3.3° to the right (where she responded "left" 100% of the time) and 8.8° to the right (where she responded "right" 100% of the time). LW displayed a perceived egocentric midline displacement of 5.9° to the ipsilesional side (95% confidence limits = 5.31° to 6.57° right) and a JND and slope within the normal range.

Search Task

The results of search testing are summarized in Table 2. The normal control group had a mean lateralized search-time score of -4.61% (SD=13.35); a negative score indicates a longer search time for targets on the left than for targets on the right of the arrays. The right-hemispheric control subjects had a mean lateralized search-time score of -9.36% (SD=8.52), which was not different from the normal control subjects' score (t=.827; *p*>.4; df=15). Therefore, the combined score (mean= -6.6%; SD=11.5; range= -21% – +26%) was used for comparison with those of neglect patients.

Patient WG had a lateralized search-time score (-32.6%) which fell outside the normal range. However, she made no omissions. Patient AM had a lateralized search-time score (-51.6%) which also fell outside the normal range; he made one omission when the stimulus was on the left. Patient EB had a search-time score (-38.3%) which was outside the normal range and omitted all seven stimuli on the left. FD's search-time score could not be calculated because he made 12 omissions. He reported seeing the stimulus only when it was in the two positions on the far right of the array. Patient LW had a search-time score which was within the normal range. She made an omission, however, when the stimulus was on the left and was therefore considered to be impaired in this task.

In summary, the two neglect patients with extensive parietal damage (FD and LW) displayed egocentric midlines that were shifted to the right of those of the control subjects. Furthermore, the 95% confidence intervals for the midline of these two patients did not overlap with those of the control subjects. In contrast, the neglect patients without extensive parietal lobe damage (WG, AM, and EB) did not show an impairment in egocentric midline. In addition, two patients had JNDs and slopes outside the range of the controls – one with extensive parietal damage (FD) and one without (EB). Finally, all neglect patients were impaired on the search task.

Discussion

Our finding that the perceived midline of neglect patients with extensive parietal lesions is displaced to the ipsilesional side is consistent with previous results (Heilman et al. 1983; Mark and Heilman 1990; Karnath 1994; Perenin, this volume). The midline shifts that we found cannot be accounted for by directional hypokinesia or problems with pursuit eye movements, for the psychophysical procedure that we used did not require manual pointing or visual following of moving stimuli. It is possible that the midline displacements reported here were smaller than those described by Karnath (1994) because of differences in size or location of lesions or in testing methods in the two studies.

Furthermore, we found in the present research that two patients with right-hemispheric lesions showed abnormally large JNDs for the perceived midline. One of these patients (EB) did not show displacement of the perceived midline, suggesting that the two abnormalities are dissociated. Whether the abnormally large JND for the perceived midline that our patients showed reflects a specific disorder (i.e., in variability of midline judgments) or a more general variability in perceptual thresholds remains to be determined.

The major finding of this study is the dissociation between unilateral spatial neglect and shifts in the perceived egocentric midline in patients with right-hemispheric infarcts located largely outside of the right parietal lobe. One might argue, however, that the absence of a midline shift in these patients may be due to some factor other than lesion location. Because the number of neglect patients in the two subgroups was small, lesion size or severity of neglect cannot be entirely ruled out as factors determining midline displacement. However, lesion size is unlikely to be a critical factor: EB, one of the three neglect patients without a displaced midline, had a lesion that was similar in size to that of FD, one of the two neglect patients whose midline was displaced. Furthermore, CA, one of the right-hemispheric control subjects (none of whom showed a midline shift) had an infarct that was at least as large as FD's. Nor can the absence of a midline shift in neglect patients readily be attributed to severity of neglect in the Star Cancellation test, for the two subgroups with neglect did not differ in this regard. In particular, EB, who did not display a displacement in midline perception, showed severe neglect in the Star Cancellation test, whereas LW, who did display a midline displacement, had less severe neglect in this test. Furthermore, of the two neglect patients most severely impaired on the Star Cancellation test, one showed a midline displacement (FD) and one did not (EB). Also, it should be pointed out that the neglect patients without midline displacements were more impaired in searching for left-side visual targets in the Circle Search task than one neglect patient with a midline displacement (LW). The visual displays used in this test were placed on the same screen and covered the same range of stimuli used in the midline localization test. Thus, the patients' search deficits were displayed in the same spatial dimensions and distance used to measure the perceived midline. Consequently, it is unlikely that neglect shown by patients with normal midlines was unrelated to the spatial parameters of the midline localization test.

Another possible confounding factor in the present study was visual field defects, which were present in two of the neglect patients we tested (EB and FD). Taken together, our findings suggest that hemianopia and a shift in the perceived midline are dissociable. Specifically, LW, a neglect patient with a displaced midline, had no visual field defect at the time of testing; EB, a neglect patient without a displaced midline, had a homonymous hemianopia; and two right-hemispheric control subjects (ML and JH) and one subject with a right optic-tract lesion (JE) had homonymous hemianopias but no midline shift. Fuchs (1920), however, described hemianopic patients with open-head brain injury whose perception of straight ahead was displaced first *contralesionally*, and then ipsilesionally. This difference was explained in terms of recovery from a complete hemianopia to an incomplete hemianopia, respectively. Thus, it is possible that the effect of a visual field loss and of neglect in patient EB (who showed no midline displacement) interacted in such a way that opposing biases in midline judgments were canceled. Furthermore, FD's displaced midline may have been due to a combination of neglect and hemianopia, or perhaps even to hemianopia alone. Unfortunately, it is difficult to compare Fuchs' findings to our own because the lesion sites in his patients were not described. They may have involved large amounts of parietal as well as occipital lobe tissue. This possibility is consistent with their clinical descriptions, which included definite signs of contralesional spatial neglect.

It is unlikely that low sensitivity of our midline localization test was responsible for failure to find a midline shift in the hemianopic control subjects and the hemianopic neglect patient without an extensive parietal lesion. Our test repeatedly sampled subjects' judgments of a range of target positions. Moreover, it was sufficiently sensitive to detect a small but significant rightward shift in control subjects' judgments of the midline, a finding previously reported (Werner et al. 1953; Mark and Heilman 1990).

With these considerations in mind, it appears that the crucial factor determining the ipsilesional midline shift in our patients was the presence of extensive parietal lobe damage. The proposal of a direct relationship between the parietal lobe and perception of visual straight ahead is strengthened by evidence from physiological and anatomical studies suggesting that the posterior parietal cortex is crucially involved in sensory, sensory-motor, and attentional processes contributing to the construction of a spatial coordinate system for localization in extrapersonal space (see review by Andersen 1989). Specifically, Andersen and colleagues (1993) propose that representations of space in body-centered coordinates may be achieved by combining information about head, eye, and retinal position.

The anatomical and physiological evidence is consistent with the proposal that lesions of the parietal lobe disrupt the representation of egocentrically defined space. In particular, our findings and those of others (Karnath et al. 1991; Karnath 1994; Perenin, this volume) suggest that the right parietal lobe is critical for normal perception of the egocentric straight ahead, which is an important reference plane for spatial orientation. However, the present findings are still open to two possible interpretations. The first is that some, but not all, manifestations of left-sided neglect are related to disorders of egocentric midline representation – in particular, that neglect following stroke involving primarily the frontal or temporal lobes (or both) is not due to this kind of disorder. The second possibility is that a disruption in egocentric midline is specific to parietal lobe damage but is not causally related to neglect. Evidence supporting this second view comes from the report by Bisiach and colleagues (1984) that localization of sounds was shifted rightward in patients with right cerebral lesions (involving the parietal lobe) who did not show visual neglect as well as in those patients who did show neglect. These findings, together with the present finding of no midline displacement in some neglect patients, suggest that displacement of the egocentric midline and spatial neglect are completely unrelated. However, this conclusion is difficult to reconcile with reports that left-sided neglect due to right parietal lesions is reduced by trunk rotation relative to the head and by vibration of left posterior neck muscles (Karnath et al. 1991, 1993). An alternative explanation of Bisiach and colleagues' findings is that their "nonneglect" patients had auditory neglect that was not detected in the visual task used to screen for neglect. In fact, the authors suggested that their findings may indicate that auditory localization is a more sensitive tool to detect neglect than the visual cancellation task. Taken together, the findings of Karnath and colleagues, Bisiach and colleagues, and the present findings do not allow us to determine whether the relationship between a disruption in egocentric midline and parietal-based neglect is a causal one.

Acknowledgements. The research reported in this article was supported by a grant from the National Institutes of Health (NS 28330). The authors thank Birdie Goynes and Christina Marshuetz for their assistance in testing subjects, J.E. Keith Smith for statistical advice and Hans-Otto Karnath for his helpful comments.

References

Andersen RA (1989) Visual and eye movement functions of the posterior parietal cortex. Annu Rev Neurosci 12:377–403

Andersen RA, Snyder LH, Li C-S, Stricanne B (1993) Coordinate transformations in the representation of spatial information. Curr Opin Neurobiol 3:171–176

Biguer B, Donaldson ML, Hein A, Jeannerod M (1988) Neck muscle vibration modifies the representation of visual motion and direction in man. Brain 111:1405–1424

Bisiach E, Berti A (1987) Dyschiria. An attempt at its systematic explanation. In: Jeannerod M (ed) Neurophysiological and neuropsychological aspects of spatial attention. Elsevier, Amsterdam, pp 183–201

Bisiach E, Luzzatti C, Perani D (1979) Unilateral neglect, representational schema and consciousness. Brain 102:609–618

Bisiach E, Capitani E, Luzzatti C, Perani D (1981) Brain and conscious representation of outside reality. Neuropsychologia 19:543–551

Bisiach E, Cornacchia L, Sterzi R, Vallar G (1984) Disorders of perceived auditory lateralization after lesions of the right hemisphere. Brain 107:37–52

Bisiach E, Capitani E, Porta E (1985) Two basic properties of space representation in the brain. J Neurol Neurosurg Psychiatry 48:141–144

Butter CM, Kirsch NL, Reeves G (1990) The effect of lateralized dynamic stimuli on unilateral spatial neglect following right hemisphere lesions. Restorative Neurol Neurosci 2:39–46

Calvanio R, Petrone PN, Levine DN (1987) Left visual spatial neglect is both environment-centered and body-centered. Neurology 37:1179–1183

Chatterjee A (1994) Picturing unilateral spatial neglect: viewer versus object centered reference frames. J Neurol Neurosurg Psychiatry 57:1236–1240

Daffner K, Ahern G, Weintraub S, Mesulam M-M (1990) Dissociated neglect behavior following sequential strokes to the right hemisphere. Ann Neurol 28:97–101

Damasio H, Damasio AR (1989) Lesion analysis in neuropsychology. Oxford University Press, New York

Deming WE (1950) Some theories of sampling. Wiley, New York

Driver J, Baylis GC, Goodrich SJ, Rafal RD (1994) Axis-based neglect of visual shapes. Neuropsychologia 32:1363–1365

Farah MJ, Brunn JL, Wong AB, Wallace MA, Carpenter PA (1990) Frames of reference for allocating attention to space: evidence from the neglect syndrome. Neuropsychologia 28:335–347

Fuchs W (1920) Untersuchung über das Sehen der Hemianopiker und Hemiamblyopiker: I. Verlagerungserscheinungen. Z Psychol Physiol Sinnesorg 84:67–169

Halligan PW, Marshall JC (1992) Left visuo-spatial neglect: a meaningless entity? Cortex 28:525–535

Halligan P, Marshall JC, Wade DT (1989) Visuospatial neglect. Lancet 2:908–911

Heilman KM, Bowers D, Watson RT (1983) Performance on hemispatial pointing task by patients with neglect syndrome. Neurology 33:661–664

Heilman KM, Bowers D, Coslett HB, Whelam H, Watson RT (1985) Directional hypokinesia: prolonged reaction times for leftward movements in patients with right hemisphere lesions and neglect. Neurology 35:855–859

Johnston CW, Diller L (1986) Exploratory eye movements and visual hemi-neglect. J Clin Exp Neuropsychol 8:93–101

Karnath H-O (1994) Subjective body orientation in neglect and the interactive contribution of neck muscle proprioception and vestibular stimulation. Brain 117:1001–1012

Karnath H-O, Schenkel P, Fischer B (1991) Trunk orientation as the determining factor of the 'contralateral' deficit in the neglect syndrome and as the physical anchor of the internal representation of body orientation in space. Brain 114:1997–2014

Karnath H-O, Christ K, Hartje W (1993) Decrease of contralateral neglect by neck muscle vibration and spatial orientation of trunk midline. Brain 116:383–396

Karnath H-O, Sievering D, Fetter M (1994) The interactive contribution of neck muscle proprioception and vestibular stimulation to subjective "straight ahead" orientation in man. Exp Brain Res 101:140–146

Kinsbourne M (1993) Orientational bias model of unilateral neglect: evidence from attentional gradients within hemispace. In: Robertson IH, Marshall JC (eds) Unilateral neglect: clinical and experimental studies. Erlbaum, Hove, pp 63–86

Làdavas E, Petronio AI, Umiltà C (1990) The development of visual attention in the intact field of hemineglect patients. Cortex 26:307–317

Làdavas E, Umiltà C, Ziani P, Brogi A, Minarina M (1993) The role of right side objects in left side neglect: a dissociation between perceptual and directional motor neglect. Neuropsychologia 31:761–773

Liu GT, Bolton AK, Price BH, Weintraub S (1992) Dissociated perceptual-sensory and exploratory-motor neglect. J Neurol Neurosurg Psychiatry 55:701–706

Mark VW, Heilman KM (1990) Bodily neglect and orientational biases in unilateral neglect syndrome and normal subjects. Neurology 40:640–643

Mark VW, Kooistra CA, Heilman KM (1988) Hemispatial neglect affected by non-neglected stimuli. Neurology 38:1207–1211

Marshall JC, Halligan PW (1994) Independent properties of normal hemispheric specialization predict some characteristics of visuo-spatial neglect. Cortex 30:509–517

Mesulam M-M (1990) Large-scale neurocognitive networks and distributed processing for attention, language, and memory. Ann Neurol 28:597–613

Posner MI, Walker JA, Friedrich FF, Rafal RD (1987) How do the parietal lobes direct covert attention? Neuropsychologia 25:135–145

Riddoch MJ, Humphreys GW (1983) The effect of cueing on unilateral neglect. Neuropsychologia 21:589–599

Roll JP, Vedel JP, Ribot E (1989) Alteration of proprioceptive messages induced by tendon vibration in man: a microneurographic study. Exp Brain Res 76:213–222

SYSTAT for Windows (1992) Statistics version 5th ed. SYSTAT Inc, Evanston, pp 453–455

Taylor JL, McCloskey DI (1991) Illusions of head and visual target displacement induced by vibration of neck muscles. Brain 114:755–759

Tegnér R, Levander M (1991) Through a looking glass. A new technique to demonstrate directional hypokinesia in unilateral neglect. Brain 114:1943–1951

Vallar G, Perani D (1987) The anatomy of spatial neglect in humans. In: Jeannerod M (ed) Neurophysiological and neuropsychology aspects of spatial neglect. Elsevier, Amsterdam, pp 235–258

Werner H, Wapner S, Bruell JH (1953) Experiments on sensory-tonic field theory of perception: VI. Effect of position of head, eyes, and of object on position of the apparent median plane. J Exp Psychol 46(4):293–299

Analysis of Self-Motion by Parietal Neurons

M. LAPPE

Department of Zoology and Neurobiology, Ruhr University Bochum, Bochum, Germany

Most animals, including primates and man, spend much of their time moving around in their environment. Sensing and controlling self-motion is an important requirement for a normally functioning and behaving animal. The last few years have seen primate neurophysiology starting to investigate the cortical mechanisms of self-motion processing. This paper presents a current view on how the determination and representation of self-motion in primate cortex is organized. It focuses on the visual information, i.e., the optic flow field. The literature on the processing of optic flow in the primate visual system is reviewed, and neurophysiological results are discussed in relation to computational requirements. Later, the integration of visual with nonvisual information supporting self-motion analysis is considered.

Multiple Sensory Signals Are Available During Self-Motion

For a complete description of self-motion, a number of different motions have to be considered. While the body translates and rotates in space, the head can move on the body and the eyes can move in the head. The central nervous system has a multitude of signals available for the determination of self-motion. These signals operate in various frames of reference. Otolith and labyrinth organs signal translation and rotation of the head in space. Somatosensory proprioception and sensorimotor feedback from the legs, trunk, and neck might be used to signal the direction of body motion during walking and the position of the head on the body. Eye muscle proprioception and motor efference can be used to signal the occurrence and velocity of eye movements. Nevertheless, in primates vision probably provides the most important signal for self-motion determination. Visual input often dominates other sensory signals. It can elicit postural responses and induce the sensation of circular, linear, or curvilinear self-motion (Dichgans and Brandt 1978; Lee 1980; Lestienne et al. 1977; Sauvan and Bonnet 1993). Visual input can also be used for the accurate estimation of self-motion parameters such

In: Parietal Lobe Contributions to Orientation in 3D Space (1997). P. Thier and H.-O. Karnath (eds). Springer-Verlag, Heidelberg.

as the time-to-contact with an obstacle (Regan and Hamstra 1993) or the direction of heading (Warren et al. 1991; Warren and Hannon 1990).

The importance of the visual input for the estimation of self-motion has first been pointed out by Gibson (1950). He introduced the term optic flow for the motion in the optic array surrounding a moving observer. This optic flow field in Gibson's sense thus describes the self-induced motion of the visual world in a body-centered reference frame. The optic flow carries information about self-motion parameters and also about the spatial relationship of objects in the three-dimensional world. The latter is conveyed by motion parallax. Motion parallax is the difference in the apparent motion of two objects positioned at different distances from the observer. Near objects appear to move faster than objects located further away. With respect to self-motion, Gibson noted that for a translating observer the motion in the optic array contains a singularity he termed the "focus of expansion" that lies in the direction of heading. The term "singularity", or "singular point", refers to an idealized point in the flow field at which visual motion is zero, i.e., a point that remains stationary in the optic array. For an observer moving on a straight line, the destination of travel is such a point, because all visual motion seems to expand radially from this point. Gibson's suggestion that the visual system might directly use this singularity to determine the direction of heading by analyzing the global optic flow structure has started a long line of research on optic flow perception that still continues to pose new questions to psychophysicists, computer scientists, and, as of recently, neurophysiologists.

Retinal Optic Flow Is a Complicated Self-Motion Signal

The analysis of motion in the optic array is complicated by the fact that the visual system senses movement in the optic array by virtue of the retinae, which are movable sensors themselves. Thus, movements of the retinae, i.e., eye movements, superimpose onto movements in the optic array. The result is a confounded flow field in which the focus of expansion is often obscured (Koenderink and van Doorn 1981; Nakayama and Loomis 1974; Regan and Beverly 1982). The visual system has to work with this retinal flow (Warren and Hannon 1990) as input and develop methods to deal with the influences of eye movements.

The potential complexity of the retinal flow field is illustrated in Figure 1. Several examples all describe essentially the same body motion. An imaginary observer moves on top of a ground plane covered with visible features schematized as black dots. Moving forward, he gazes at some point in space (Fig. 1a). The distinction between the direction of gaze and the direction of movement is very important. The direction of gaze defines the center of the retinal coordinate system in which the retinal flow is eventually represented, while the direction of

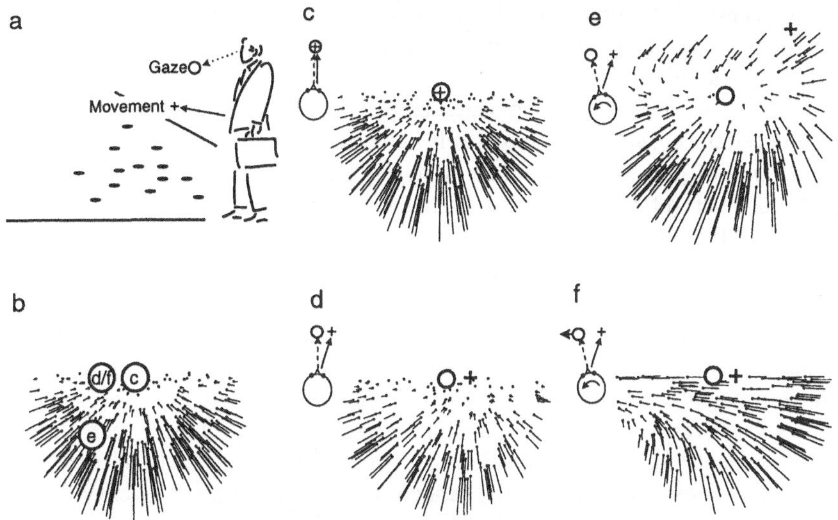

Fig. 1 a-f. Various examples of the dependence of the retinal flow field of a moving observer on gaze directions and eye movements. The observer moves on top of a ground plane. Direction of heading is indicated by a plus sign, direction of gaze by a circle

movement is what the visual system has to figure out. Figure 1b depicts the flow field in a body-centered coordinate system equivalent to the optic array in Gibson's terminology. All motion is directed away from the focus of expansion that coincides with the direction of heading. When this flow field gets projected onto the retina, the projection depends on the direction of gaze. Three gaze points indicated by circles in Fig. 1b will serve for the following retinal flow examples.

Figure 1c shows the retinal flow when the direction of gaze and the direction of movement coincide, i.e., when the observer looks straight ahead into the direction of movement. In this case the focus of expansion is visible and centered on the retina. In Fig. 1d the observer looks in a direction distinct from the direction of movement. Gazing at some fixed point on the horizon allows him to keep his eyes stationary, i.e., no eye movements occur. Again, the focus of expansion is visible, but now it is displaced from the center of the visual field. Still, it immediately indicates the direction of heading.

Figure 1e shows a situation in which the observer's gaze is directed not at the horizon but at a some element of the ground plane located below the horizon. For illustrative purposes, this element is marked by an 'e' in Fig. 1b. There are two consequences of this change in gaze direction. The first one is related to eye position (as in Fig. 1d), the second to eye velocity. The immediate effect of a change of eye position on the retinal flow field is an opposite displacement of the retinal image. The horizon has moved up in the visual field. The second, more important difference is that, unlike the case of Fig. 1d, the point at which gaze is directed is now in motion. To stabilize the image of that point on the fovea would thus require some kind of tracking eye movement. In case the observer wants to actively track this point, he would perform a voluntary smooth pursuit in the direction of the motion of the foveated target. However, large field visual motion experienced during self-motion by itself induces reflectory eye movements for the stabilization of gaze of the optokinetic nystagmus type, accompanied by a linear vestibuloocular reflex (Miles and Busettini 1992). However, optic flow fields stimulate different parts of the retina with different directions of motion. Since optokinetic responses in foveated animals show greater sensitivity of the fovea than of the periphery, one might assume that the optokinetic eye movement is also linked to that part of optic flow that falls on the fovea. Indeed, observations in monkeys (Pekel et al. 1995) and humans have shown that optic flow fields can induce nystagmic eye movements. The direction of these eye movements closely follows the foveal motion direction, similar to the case of voluntary pursuit. Both eye movements try to null foveal flow. Thus, in both cases, the direction and speed of the eye movement are related to the observer's movement. Since direction is determined by the direction of the flow on the fovea, it is always directed away from the retinal projection of the direction of heading. Eye speed, however, might be less well-defined, depending on the gain of the eye movement.

As a consequence of eye movement, the retinal flow becomes confounded and the focus of expansion becomes obscured. The result somewhat resembles a distorted spiraling motion around the fovea (Fig. 1e). This flow field also contains a singular point. Yet this singular point is different from a focus of expansion and related to the stabilizing eye movement. Perfect stabilization of gaze (unity gain) would result in a singular point located exactly on the fovea. The visual system might capitalize on the resulting restrictions on the structure of the retinal flow in this situation (Lappe and Rauschecker 1995).

Eye movements completely unrelated to the observer's movement might also occur. The optokinetic and vestibuloocular reflexes can be suppressed during smooth pursuit of a moving target. In Fig. 1f the observer again directs his gaze to the horizon as in Fig. 1d. Yet here it is assumed that he performs a smooth eye movement in order to track an object that moves leftward along the horizon, independent of the observer's movement. In this case, the retinal flow has a very different structure, resembling a curved movement. No singular point is visible. In summary, the visual signal available to a moving observer can change fundamentally during eye movements although self-motion remains unchanged.

Humans Can Use Optic Flow for Self-Motion Estimation

What cues does the visual system use to determine self-motion from retinal flow fields in the presence of eye movements? Undoubtedly, extraretinal information is used to cope with the visual effects of eye movements. In most studies that compared judgements of the direction of heading from optic flow in the presence and absence of extraretinal input, performance was better when extraretinal input was available (Banks et al. 1996; Royden et al. 1994; van den Berg 1992; Warren and Hannon 1990). But several of these studies have also shown that humans are quite successful even in the absence of extraretinal information, thus using only visual cues. Human subjects are highly accurate in locating the focus of expansion, in flow fields such as Fig. 1b and Fig. 1c. Errors range from less than 1° to 2.3° of visual angle, depending on the eccentricity of the focus of expansion (Warren and Kurtz 1992). In this situation, the primary cue is the radial arrangement of the motion vectors around the focus of expansion, while the distribution of flow field speeds (motion parallax) is less relevant (Warren et al. 1991).

When pursuit eye movements are performed during the presentation of optic flow stimuli, performance is similarly accurate (Royden et al. 1994; Warren and Hannon 1990). Extraretinal input can be removed by requiring subjects to fixate and then simulate eye movements along with observer movement. In this case, the results obtained in different studies were mixed. Warren and Hannon (1990) found good performance in many experimental conditions, but with an exception in the absence of motion parallax. These findings were challenged by Royden et al. (1994). Using a stimulus as in Fig. 1f, they claimed that the purely visual estimation of the direction of heading can only be performed for very slow simulated eye movements. At eye speeds greater than 1°/s extraretinal information would be required. However, eye speed proved not to be the essential parameter. Van den Berg (1992,1993) found good performance with simulated eye speeds up to 6°/s using a stimulus as in Fig. 1e. More likely, the difficulties observed with the flow field in Fig. 1f arise from its difference in structure from the other flow fields (Lappe and Rauschecker 1995) and its resemblance to a flow field produced by movement on a curved path (Royden 1994).

The results can thus be summarized as follows: The human visual system is able to accurately determine the direction of heading from retinal flow fields. Extraretinal signals are often not required, but usually improve performance. In situations in which the flow field is inherently ambiguous, extraretinal information can disambiguate the visual signal.

What Neural Mechanisms Support the Visual Estimation of Heading in Primate Cortex?

Gibson's original suggestion that the pattern of motion vectors determines the direction of heading does not specify any concrete method or procedure. A large number of computational schemes have been suggested since, but only few would be considered relevant from a neurobiological point of view. One of the most influential ideas was presented by Koenderink and van Doorn (1976), who observed that mathematically any flow field can be locally approximated by a set of four differential invariants: divergence, curl, and two components of deformation. This has led many neurophysiological studies on optic flow to use these basic components as stimuli in order to specify neuronal properties of optic flow processing.

Middle temporal Area Computes the Flow Field

The question of the cortical mechanisms of optic flow processing naturally has to start with the determination of the flow field from the retinal afferents. In the cerebral cortex of the monkey, the processing of visual motion is attributed to a successive series of areas within the dorsal stream, which is believed to be specialized in the analysis of motion and spatial relationships (Mishkin et al. 1983). Motion information proceeds from the primary visual cortex (V1) to the middle temporal area (MT) and the medial superior temporal area (MST) and from there to several higher areas in the parietal cortex (see for example Distler et al. (1993) for a recent schematic summary).

Area MT is the first area in the visual cortex that is dedicated specifically to the processing of motion. First of all, it contains a high proportion of direction selective neurons (Maunsell and Van Essen 1993). Furthermore, it has been linked to behavioral responses to motion stimuli in lesion (Newsome and Pare 1988) and microstimulation (Salzman et al. 1990) studies. The restricted size of the receptive fields in area MT would not allow individual MT neurons to perform a global analysis of the optic flow field. Rather, the properties of MT neurons are well suited for establishing a cortical representation of the retinal flow field, solving problems occurring in early estimation of visual motion (Adelson and Bergen 1985; Hildreth and Koch 1987; Movshon et al. 1985).

When the global organization of the representation of motion in the visual field in MT is considered, a relation to optic flow is already apparent. Preferred speeds increase with eccentricity (Maunsell and Van Essen 1983) similar to the way optic flow speeds naturally do. Moreover, in the peripheral visual field representation in area MT the number of direction sensitive neurons preferring motion away from the fovea is significantly higher than the number of neurons preferring motion towards the fovea (Albright 1989), a property well suited for the centrifugal structure of the flow field under natural self-motion conditions (Lappe and Rauschecker 1995).

Cells in Area MST Respond to Optic Flow Patterns

Area MST was deemed a likelier candidate for global analysis of optic flow fields because of its much larger receptive fields (Desimone and Ungerleider 1986). Early studies could occasionally identify neurons in the superior temporal sulcus that responded selectively to rotational motion, either in the frontoparallel plane or in depth, or to optical expansions (Bruce et al. 1981; Saito et al. 1986; Sakata et al. 1986). Tanaka and Saito (1989a,b) were the first to systematically investigate the response properties of MST neurons to random dot optic flow patterns. They used a set of stimuli that contained unidirectional motion, rotation, expansion, and contraction. For each of these elementary flow fields, they found neurons that responded selectively to only a single one. The selective responses of these neurons depended mainly on the spatial arrangement of motion vectors and on the size of the stimulated area. Other parameters such as shape and size of the individual motion elements, contrast, or speed gradients did not have much influence on neuronal responses.

One major concern of Tanaka and Saito was to differentiate genuine optic flow selectivity from allegedly different simple direction selectivity. This is a difficult problem, since an absolute criterion for a distinction cannot be easily formulated. In order to guarantee that some neurons possess a genuine selectivity for pattern motion (expansion, contraction, rotation) Tanaka and Saito required a neuron to be completely unselective for unidirectional motion. Only then would it be classified as expansion/contraction or rotation selective. While this is certainly a criterion to differentiate pattern motion selectivity from direction selectivity it discounted a large proportion of neurons in MST. Direction selectivity is a fundamental property of MST neurons that has led to the physiological characterization and determination of this area in the first place.

The need to differentiate optic flow processing and directional selectivity was also felt by Orban and coworkers (Lagae et al. 1994; Orban et al. 1992). They suggested a criterion based on the following assumption: Suppose a neuron has a receptive field or an area within its receptive field containing exclusive selectivity for unidirectional motion towards the right. A rotational optic flow pattern is presented to the neuron. When the center of rotation is located below the receptive field, clockwise rotation will result at first approximation in unidirectional motion towards the right in the neuron's receptive field. Counterclockwise rotation will result in leftward motion in the neuron's receptive field. If this neuron responds during clockwise but not during counterclockwise rotation, its response might be due to its direction selectivity. Now suppose that the center of rotation is moved to a position above the receptive field of the neuron. Clockwise rotation will now result in leftward, counterclockwise rotation in rightward motion in the receptive field. Then, if the neuron favors counterclockwise rotation, its response is again consistent with its direction selectivity. If, however, the neuron would respond to clockwise rotations in both positions, then another mechanism than simple directional selectivity would be required.

Elaborate application of this procedure indeed yielded a proportion of MST neurons fulfilling this position invariance criterion, whereas cells in area MT, although responsive to optic flow components, did not (Graziano et al. 1994; Lagae et al. 1994; Orban et al. 1992). However, there are problems with this criterion, too. First of all, complete invariance towards the visual field position of an expansion stimulus would render such a neuron probably useless for a task such as heading detection, in which the position of the focus of expansion can be one of the major sources of information (Graziano et al. 1994). However, this argument would only apply to neurons that respond with exactly equal firing rates at different stimulus positions, and not if some orderly variation occurs. Second, while it is obvious that simple direction selective neurons could not pass the position invariance test, the reverse assumption, namely, that failing the position invariance test discounts a neuron from being optic flow selective, is not valid. In fact, computational considerations have shown that reversals of selectivity can be quite consistent with neuronal specification for optic flow analysis (Lappe and Rauschecker 1993a,b). Thus, the position invariance criterion is also an ad hoc requirement that would be sufficient but not necessary. The conclusions so far are mixed. There are neurons in area MST that genuinely respond to optic flow stimuli, but whether or not these neurons form a separate subpopulation differing from other MST neurons is not clear.

A more wholistic account of the complexity of optic flow sensitivity in MST was presented by Duffy and Wurtz (1991a,b), who found that the majority of MST neurons responded to several different stimuli. They suggested a classification of neurons according to the number of different flow components they respond to. Triple component neurons respond simultaneously to unidirectional motion in a preferred direction, rotation in one of the two principle directions (clockwise, counterclockwise), and either expansion or contraction. Double component cells respond to unidirectional motion and to either rotation or expansion/contraction. These two classes made up the majority of cells. Very few neurons respond to rotation and expansion/contraction but lack direction selectivity. The most selective, but least populous group consisted of single component neurons that responded only to one of the stimuli: 9% of all neurons to respond expansion or contraction, 4% to rotation. However, Duffy and Wurtz also immediately stressed that this was a rather arbitrary classification. The neuronal responses instead represented a continuum of selectivities covering not only these classes but also all stages in between.

This raises the question of whether neurons in MST perform a mathematical decomposition of the flow field into a set of basic components such as expansion and rotation. Several lines of evidence in addition to the observations of Duffy and Wurtz suggest that this is not the case. A cell that extracts expansion (divergence) from an optic flow stimulus in a mathematical sense would have to respond with equal activity to pure expansion and to a stimulus where the same expansion is disturbed by adding some other flow component, for instance rotation. This is not true for MST cells (Graziano et al. 1994; Orban et al. 1992). Instead, some MST cells even prefer vectorial combinations of rotation and expansion/contraction over the two individual flow patterns and display a selectivity for spiral motion (Graziano et al. 1994). Furthermore, for a complete mathematical decomposition, two components of deformation are neccessary in addition to divergence (expansion/contraction) and curl (rotation) (Koenderink and van Doorn 1976). Cells responding exclusively to deformation were practically absent in MST (Lagae et al. 1994). There also are theoretical considerations reminding us to consider the decomposition into basic flow components with care. Decomposition is a local linearization of the flow field (Koenderink and van Doorn 1976). This means that mathematically at any point in a dense flow field, decomposition is valid only within a small vicinity of that point. This creates two problems with respect to the properties of MST neurons. First, the receptive fields are too large to perform a local analysis. Second, the observed position invariance would render a proposed attribution of the response to a localized position in the visual field inappropriate.

In summary, these results suggest that MST does not use a set of channels selective for just expansion, contraction and rotation, but rather a graded continuum of response properties in which single neurons might respond to several different flow components and to combinations of them.

Can Area MST Compute Self-Motion?

How else, then, would self-motion be represented or computed in MST? A mechanism well established for other complex brain functions is the distributed encoding of external, behavior-relevant parameters by whole populations of neurons (Andersen et al. 1993; Georgopoulos et al. 1986; Lee et al. 1988; Van Gisbergen et al. 1987). In this view, the representation of self-motion would be assumed to be spread across a large number of neurons in MST. Each individual neuron would make only a small contribution to the global self-motion representation by modulating its response strength depending on whether the flow field input is consistent with a preferred self-motion. A computational model of such a distributed representation of self-motion has been proposed by Lappe and Rauschecker (1993b). This model implements a minimization procedure for visual heading detection in two layers of neuron-like elements. The first layer contains a representation of the flow field modeled on area MT. The second layer contains cells that individually test the compatibility of their flow field input with a certain direction of heading. The overall direction of heading specified by the input flow field is finally computed by the full population of second layer neurons. This model predicts a computational map of heading directions in area MST. This map could either be represented implicitly by the population, or alternatively read out by a set of dedicated neurons either within area MST or in subsequent brain areas that require self-motion information.

When it comes to the question of whether optic flow sensitive neurons in MST possess the capability to estimate self-motion from optic flow fields, there are two problems with the paradigms used in most studies. The first concerns the size of the stimuli. During self-motion, optic flow normally covers the full visual field of the observer, and the global structure of the flow is of importance. The second problem concerns the placement of the stimuli. An important parameter for the determination of heading in the absence of eye movements is the position of the focus of expansion in the visual field. However, most studies in area MST that tested the positional modulation of optic flow responses used small stimuli (diameters ranging from 10° in Lagae et al. 1994 to 20° in Graziano et al. 1994 and 33° in Duffy and Wurtz 1991b), placed the stimuli with respect to the receptive field of the neuron (with the exception of Duffy and Wurtz 1991b), and tested positional variation by moving the full stimulus instead of only the singular point. These tests are adequate for the basic response properties of individual neurons. However, they confound stimulus extent in the visual field with position of the focus of expansion.

To test a possible relation of the neuronal response properties to self-motion, a variation of the location of the singular point in full-field stimuli would be more suitable. This has been acknowledged only recently (Duffy and Wurtz 1995; Pekel et al. 1996). Lappe et al. (1996) recently presented experiments explicitly carried out in analogy to computer simulations of model neurons. In these experiments, a monkey was required to fixate a central spot of light on a projection screen. Full-field (90° by 90°) computer generated optic flow stimuli simulating approaching (expansion) and subsequently receding (contraction) self-motion with respect to a random cloud of dots in three-dimensional space were projected. They displayed entirely realistic self-motion flow fields. Dots accelerated with eccentricity, grew larger in size as they approached the monkey, and moved with motion parallax according to their distance in depth. To determine the dependence of neuronal responses on the singular point in the optic flow stimulus, 17 different movie sequences were used. In each of these sequences the singular point was located in a different position, either in the center of the screen, or at one of 16 peripheral locations.

The responses of a single examplary neuron are shown in Fig. 2a,b. The modulations of response strength with the position of the singular point on the projection screen are displayed by three-dimensional surface plots. In these plots, the (x,y)-plane represents the positions of the singular point on the screen, while the z-axis indicates the response rate in spikes per second. The neuronal responses to expansion (a) and contraction (b) vary smoothly with the position of the singular point. Such a sigmoidal response profile is observed in many MST neurons (see also Duffy and Wurtz 1995, their Fig. 13B). Responses to both expansion and contraction can be elicited by proper placement of the singular point. Therefore, the neuron's response is not position invariant when the singular

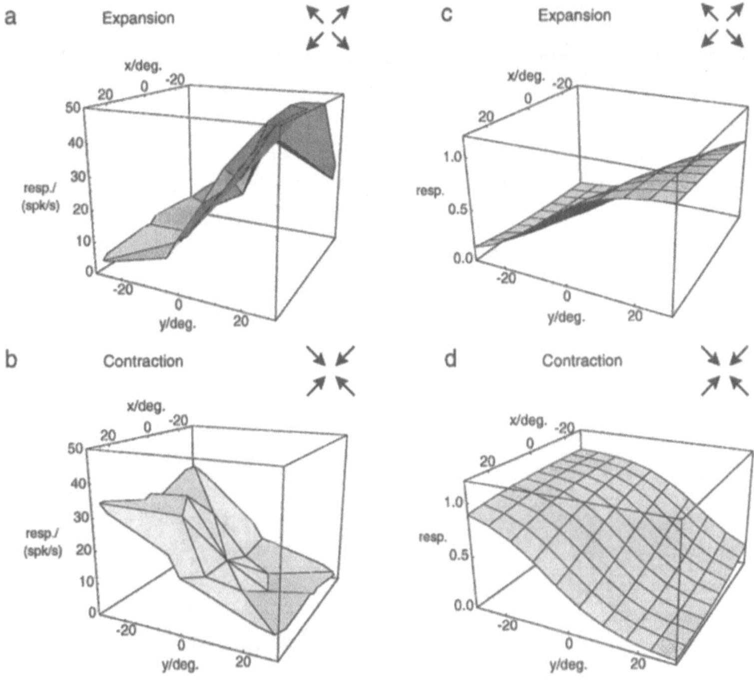

Fig. 2 a-d. Single-neuron activity as a function of the position of the focus of expansion/contraction in the visual field. *Left*: Data from a recording in area MST of an awake macaque monkey. *Right*: Computer simulation of a single neuron from the model of Lappe and Rauschecker (1993b)

point is moved over the full visual field. However, if the singular point were to be moved within a restricted area of the size used in most previous studies, the observed responses could be restricted to only one stimulus type. For instance, in all of the lower hemifield ($y<0$, left half of the diagrams), the neuron responds exclusively to contraction.

These experimental findings can also be recreated in computer simulations of the model of Lappe and Rauschecker (1993b). The characteristics of a single model neuron (Fig. 2c,d) are quite similar to the recorded data in Fig. 2a,b with respect to the sigmoidal shape of response modulation. In simulations, it has been found that single neurons in this model can capture a wide range of the properties of MST neurons. Single model neurons can respond to unidirectional translation, expansion/contraction, and rotation rather like triple component cells in area MST (Lappe and Rauschecker 1993b). They might also respond to only a subset of these stimuli (Lappe and Rauschecker 1993a) or to linear combinations (spiral patterns). Thus, the model successfully captures the observed continuum of response selectivities. Model neurons display position invariant responses within medium-sized areas of the visual field and position-dependent responses for large shifts of the singular point (Lappe and Rauschecker 1993b). As in MST (Duffy and Wurtz 1991b; Lagae et al. 1994), position invariance is weakest for neurons responding to several elementary flow field stimuli and stronger for more selective neurons (Lappe and Rauschecker 1993a) The model suggests a way to determine the position of the focus of expansion despite large areas of position invariant responses. Information is conveyed by the flanks of the responses, not by the maximum response. This is illustrated in Fig.3a. Simple summation of the responses of several neurons with differently oriented sigmoidal response curves results in population activity that peaks where the individual response curves overlap.

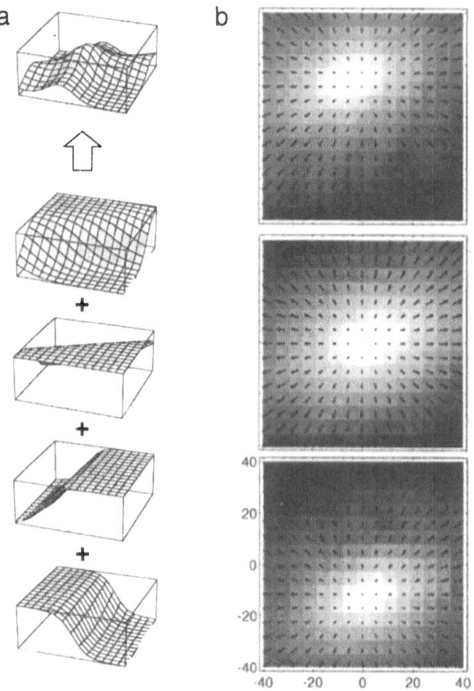

Fig. 3 a,b. Population encoding of the direction of heading. *Left*: Schematic illustration showing how sigmoidal response curves of individual neurons can be combined to give a maximum in population activity. *Right*: *Grayscale maps* indicating the location of the focus of expansion as retrieved from the recorded neuronal activities of 31 MST cells for three different true locations. Flow stimuli are shown superimposed

It would be interesting to know whether the tuning properties of MST neurons are indeed sufficient for precisely locating the focus of expansion. To test this, a least-mean-square minimization scheme can be used to derive the position of the focus of expansion of an optic flow pattern from neuronal activities. This procedure determines the capabilities of the MST population in the sense of an optimal observer theory. It proceeds as follows. For each neuron, a sigmoid response curve is fitted to the recorded activities. Then the difference between the actual recorded activity of a given position of the focus of expansion and the fitted curve is used as a constraint for the location of the focus of expansion. The constraints from the individual neurons are squared and summed to give a map of the least-mean-squared errors as a function of the location of the singular point. The resulting maps for three positions of the focus of expansion are shown in Fig.3b. Thirty-one recorded neurons contributed. In these maps, the least-square error obtained for a specific direction of heading is coded by the gray value at that map location. The most likely direction of heading is given by the brightest square in the map. For comparison with the true direction of heading, the optic flow stimuli are plotted schematically on top of the gray-scale maps. From these maps, it is evident that the potential to locate the focus of expansion is present in the neuronal population. The mean error over nine positions presented within the central 15° of the visual field shows that the location of the focus of expansion can be retrieved from neuronal activities with an average precision of 4.3°. This is quite close to the mean error of 2.3° obtained in a human psychophysical study with comparable stimuli (Warren and Kurtz 1992).

Some MST neurons have a peak-shaped response dependence on the position of the singular point rather than a sigmoidal one (Duffy and Wurtz 1995). These neurons might individually prefer a single direction of heading. A hypothesis already put forward by Lappe and Rauschecker (1993b) proposes that such neurons might already read out the activity of MST subpopulations provided by the more basic sigmoidal shaped response curves. The response properties of these neurons would then be similar to the top panel in Fig. 3a.

Eye Movements

Areas MT and MST are related to eye movements in several ways. Both areas are involved in the generation of smooth pursuit and optokinetic eye movements (Erickson and Dow 1989; Kawano et al. 1994; Komatsu and Wurtz 1988a). Lesions in MT/MST produce pursuit and optokinetic deficits (Dürsteler and Wurtz 1988). A subset of MT neurons selectively projects to subcortical structures that control optokinetic eye movements (Hoffmann et al. 1992; Ilg and Hoffmann 1993).

But area MST also contains information about the occurrence of eye movement. A subpopulation of MST neurons discharges during ongoing smooth pursuit eye movements (Erickson and Dow 1989; Komatsu and Wurtz 1988a). Part of the response of these neurons originates from the visual signal induced by eye movement, but some neurons retain their firing rate even in the absence of visual stimulation (Newsome et al. 1988).

Other MST cells show a striking ability to distinguish between active, eye movement-induced motion and passive, externally-induced object motion (Erickson and Thier 1991). This behavior was found in experiments using a paradigm quite similar to that used in psychophysical investigations of extraretinal signals in heading detection. Erickson and Thier compared two situations with identical visual, but differing extraretinal input. In the passive condition, a moving bar was swept across the receptive field of a neuron, while the monkey was required to keep its eyes stationary. In the active condition, the bar was stationary on the screen, and the monkey actively performed an eye movement that induced identical visual movement of the bar on the retina. Some neurons responded preferentially in the passive condition, thus revealing selectivity only to motion in the real world. Just as the extraretinal input to the pursuit neurons (Newsome et al. 1988) this phenomenon was not found earlier in the visual motion pathway, neither in area MT (Erickson and Thier 1991), nor in area V1 (Ilg and Thier 1996).

Could these extraretinal signals be useful for self-motion detection? A dependence of some of the optic flow processing MST neurons on pursuit signals has been reported by Duffy and Wurtz (1994). Lappe et al. (1994) have argued that pursuit neurons in MST could provide a signal of present eye velocity that might allow optic flow processing neurons to compensate to some degree for the visual effects of this eye movement. In this model, the pursuit signal is used to subtract a purely eye movement induced flow field component from the full retinal flow. However, this process is likely to be incomplete. For instance, the speed of eye movement seems to be less well represented by the pursuit neurons than its direction. Precise heading detection thus would still have to rely heavily on the visual signal, which would be merely augmented by extraretinal compensation. Such a hybrid model can account for the most prevalent conditions in which human heading detection has been shown to rely on extraretinal input, and it also leads to neurons that implement the active/passive distinction observed in MST (Lappe et al. 1994).

If optic flow processing neurons use pursuit signals in such a way, it should be reflected by neural responses to optic flow stimuli during ongoing pursuit eye movement. Model simulations predict that the responses to optic flow stimuli presented in the screen center are altered in a direction dependent manner during pursuit. Figure 4a,b shows the results of such a simulation. A neuron was probed with flow field stimuli displaying a centered expansion or contraction, accompanied by simulated rotations in different directions and an extraretinal signal either switched on or off. The circle indicates the response to the flow stimulus alone. The dark bar shows the responses during additional simulated eye

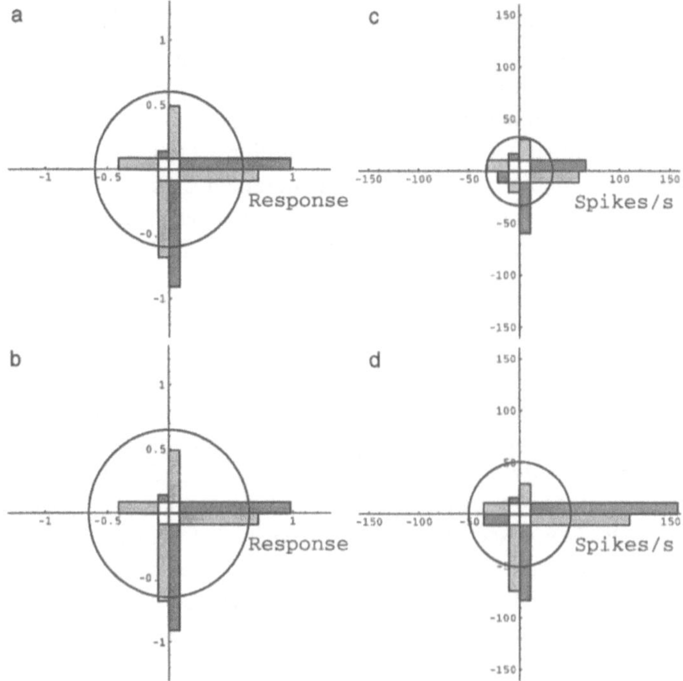

Fig. 4 a-d. Directional modulation of optic flow responses by real (light bars) or simulated (dark bars) pursuit eye movements. The polar plots show the neuronal response strengths for eye movements in four directions. *Left*: Model prediction. *Right*: Data from a single-neuron recording

movements in the four cardinal directions. The light bar shows the responses when an extraretinal signal allows partial remedy of eye movement effects. For both flow stimuli, the response is altered by the addition of eye rotation in the same way. Simulated downward pursuit enhances the response; simulated upward pursuit leads to a response weaker than the response to pure optic flow. In the presence of a simulated extraretinal signal, response modulation is diminished, and the response becomes more similar to the pure optic flow response.

When the same paradigm is employed in single unit recordings in monkey area MST, some neurons show similar behavior. An example of the responses of a single neuron is shown in Fig. 4c,d. The experiment compared three different conditions. In the first condition (eye stationary, circle) the monkey fixated a stationary target and was presented with an expanding or contracting optic flow stimulus. In the second condition (simulated pursuit, dark bars) the monkey again fixated a stationary target, but the stimulus contained expansion/contraction superimposed with a rotation around one of four axes. In the third condition (real pursuit, light bars) the stimulus was pure expansion or contraction, but the target moved, and the monkey followed the movement of the target with its eyes. For the neuron in Fig. 4c,d, optic flow response is modulated by the presence of eye rotation (simulated or real pursuit), but the modulation is weaker for real than for simulated pursuit. Such behavior was found for some of the neurons while others responded with similar strength in both conditions. Only a subset of the flow sensitive neurons might have access to extraretinal signals (Duffy and Wurtz 1994). Judging from the similarity to model simulations, these neurons seem well suited for providing a basis for the integration of visual and extraretinal signals for analyzing self-motion.

At this point it is also important to mention again that the relationship between retinal flow and eye movements is a mutual one. Not only do eye movements influence retinal flow, retinal flow also induces eye movements (Pekel et al. 1995). Thus it is conceivable that optic flow processing neurons might also contribute to the generation of eye movements in a process not yet investigated.

This might help in solving a potential problem in evaluating the function of area MST, since voluntary eye movements such as smooth pursuit of a small moving object by a stationary observer are also driven in part by area MST. Area MST contains at least two representations of the visual field, a dorsal one (MSTd) and a ventrolateral one (MSTl or MSTv). Some evidence exists that these two representations might serve two different purposes, MSTd the analysis of self-motion, MSTl the generation of smooth pursuit and the analysis of object motion. The original report of Tanaka and Saito (1989a) found expansion, contraction, and rotation cells to be clustered within the dorsal part of MST. In contrast, many cells in the ventrolateral part of MST respond well to small moving stimuli or to differential motion between a small point and a large background (Komatsu and Wurtz 1988b; Tanaka et al. 1993). Such a response would be appropriate for the control of pursuit eye movements and the analysis of object motion. However, there are some problems with such a strict differentiation. Neuronal response properties in MSTl are quite heterogeneous in several aspects. A sizable portion of the MSTl cell population prefers large stimuli instead of small spots and their visual and pursuit related functions resemble those of cells in MSTd (Komatsu and Wurtz 1988a,b). This is also true for extraretinal influences (Erickson and Thier 1991; Newsome et al. 1988) and for visual responses related to ocular following and optokinetic nystagmus (Kawano et al. 1994). Responses to optic flow stimuli have also been reported outside MSTd near or in the fundus of the superior temporal sulcus (Lagae et al. 1994; Pekel et al. 1996). Also, Tanaka and Saito (1989a) as well as Tanaka et al. (1993) found optic flow selective cells in the ventrolateral part of MST, albeit in a much smaller percentage than in MSTd. On the other hand, Graziano et al. (1994) showed that neurons in MSTd respond well to small optic flow stimuli. They suggested that such neurons could also contribute to the analysis of object motion. However, the preferred stimulus size might not be a decisive criterion for distinguishing between object- and self-motion, since psychophysical effects clearly related to self-motion can also be observed with rather small optic flow stimuli (Andersen and Braunstein 1985; Warren and Kurtz 1992). In summary, the dorsal part of MST seems to be devoted mainly to the processing of self-motion and possibly to the analysis of the visual consequences of eye movements in this task. The ventrolateral part is more heterogeneous and contains cells involved in the generation and maintenance of eye movements. However, since any self-motion immediately poses a challenge to the stability of the retinal image or at least to parts of it, the finding that neuronal functions of self-motion processing and of eye movement generation are located in immediate vicinity in the cerebral cortex should come as no surprise.

Multimodal Representation of Self-Motion

For a full representation of self-motion, other sensory signals need to be incorporated, and the derived self-motion parameters finally need to be represented in a reference frame suitable to controlling body movements (Telford et al. 1995). One signal, which only recently came to be appreciated for its supporting role in heading detection, is retinal disparity. In principle, since optic flow depends on the depth structure of the visual environment, depth cues obtained from retinal disparities can be helpful in determination of the direction of heading. However, a role of stereo vision in heading detection has usually been dismissed on the grounds that disparity only operates successfully in near vision while heading detection is essentially a far-vision process. Only recently it was shown that humans can use depth cues derived from retinal disparities to improve their performance in noisy flow fields (van den Berg and Brenner 1994).

Like cells in many other visual areas, MST cells carry a disparity signal. However, some neurons in area MST show a distinct disparity dependence that suggests that stereoscopic depth is integrated with visual motion in a manner suitable for self-motion analysis. These neurons reverse their preferred direction depending on whether unidirectional motion is presented in a plane nearer or farther than fixation distance (Roy and Wurtz 1992). Such an arrangement of motion directions is experienced during self motion when one fixates a stationary object towards the side of the movement path.

Also, vestibular and proprioceptive signals might contribute to heading detection. However, their primary value lies in the transformation of visually determined heading into a representation suitable to controlling body movements (Telford et al. 1995). A vestibular input to the pursuit neurons in MST is organized to signal the direction of combined eye and head movements (Thier and Erickson 1992). Also, eye-position-dependent modulation of neuronal responses in MST (Bremmer et al., in press) might be used to shift from a retinal to a head centered coordinate system using a mechanism proposed for other parietal areas (Andersen et al. 1993). Evidence that area MST is indeed involved in the visual control of body movements was obtained in a recent lesion study (Duffy and

Wurtz 1996). Lesions of area MST resulted in postural instabilities and disruption of the ability to perform postural responses to optic flow stimuli.

There is also some recent evidence that another area in parietal cortex, namely, the ventral intraparietal (VIP) area, might be involved in self-motion processing from multiple sensory input. Activity during optic flow stimulation was recorded not only in areas MT and MST but also in a region corresponding to area VIP. In physiological experiments by Bremmer et al. (this volume), neurons in area VIP were shown to respond to optic flow stimuli, but also to vestibular stimulation and during eye movements.

Thus, it is fair to conclude that many sensory signals arising during self-motion converge in parietal cortex. The visual determination of heading most likely takes place in area MST. However, from the data available so far, it seems too early to conclude how and in which particular area the multiple signals supporting self-motion analysis are transformed into a representation of body motion in space.

Conclusion

From the described physiological, psychophysical, and computational studies, a unified account of self-motion processing in the dorsal stream of the primate cortex emerges. It starts with local motion sensitive cells in primary visual cortex that are integrated in area MT to form the exquisite directional selectivity of MT neurons and compute local two-dimensional motion. Area MT contains a representation of the optic flow field that is well suited to serve as a basis for flow field analysis. Area MST uses the flow field representation provided by area MT to extract self-motion parameters. Cells responding to a variety of flow patterns could form a population encoding of the direction of heading. At the level of area MST, a number of additional signals that support heading detection are combined with visual information. These are eye movement signals, retinal disparity and, possibly, vestibular signals. The result is a multimodal, vision-dominated representation of self-motion in area MST or even higher parietal areas.

References

Adelson EH, Bergen JR (1985) Spatiotemporal energy models for the perception of motion. J Opt Soc Am A 2:284–298

Albright TD (1989) Centrifugal directionality bias in the middle temporal visual area (MT) of the macaque. Vis Neurosci 2:177–188

Andersen RA, Snyder LH, Li CS, Stricanne B (1993) Coordinate transformations in the representation of spatial information. Curr Opin Neurobiol 3:171–176

Anderson GJ, Braunstein ML (1985) Induced self-motion in central vision. J Exp Psychol: Hum Percept Perform 11:122–132

Banks MS, Ehrlich SM, Backus BT, Crowell JA (1996) Estimating heading during real an simulated eye movements. Vision Res 36:431–443

Bremmer F, Ilg UJ, Thiele A, Distler C, Hoffmann KP (in press) Eye position effects in monkey cortex I: Visual and pursuit related activity in extrastriate areas MT and MST. J Neurophysiol

Bruce C, Desimone R, Gross CG (1981) Visual properties of neurons in a polysensory area in superior temporal sulcus of the macaque. J Neurophysiol 46:369–384

Desimone R, Ungerleider LG (1986) Multiple visual areas in the caudal superior temporal sulcus of the macaque. J Comp Neurol 248:164–189

Dichgans J, Brandt T (1978) Visual vestibular interaction: effects on self-motion perception and postural control. In: Held R, Leibowitz HW, Teuber HL (eds) Perception. Handbook of sensory physiology, vol 8. Springer, Berlin Heidelberg New York, pp 755–804

Distler C, Boussaoud D, Desimone R, Ungerleider LG (1993) Cortical connections of inferior temporal area TEO in macaque monkeys. J Comp Neurol 334:125–150

Duffy CJ, Wurtz RH (1991a) Sensitivity of MST neurons to optic flow stimuli. I. A continuum of response selectivity to large-field stimuli. J Neurophysiol 65:1329–1345

Duffy CJ, Wurtz RH (1991b) Sensitivity of MST neurons to optic flow stimuli. II. Mechanisms of response selectivity revealed by small-field stimuli. J Neurophysiol 65:1346–1359

Duffy CJ, Wurtz RH (1994) Optic flow responses of MST neurons during pursuit eye movements. Soc Neurosci Abstr 20:1279

Duffy CJ, Wurtz RH (1995) Response of monkey MST neurons to optic flow stimuli with shifted centers of motion. J Neurosci 15:5192–5208

Duffy CJ, Wurtz RH (1996) Optic flow, posture, and the dorsal visual pathway. In: Ono, McNaughton, Molotchnikoff, Rolls, Nishijo (eds) Perception memory and emotion: frontier in neuroscience. Pergamon, Oxford

Dürsteler MR, Wurtz RH (1988) Pursuit and optokinetic deficits following chemical lesions of cortical areas MT and MST. J Neurophysiol 60:940–965

Erickson RG, Dow BM (1989) Foveal tracking cells in the superior temporal sulcus of the macaque monkey. Exp Brain Res 78:113–131

Erickson RG, Thier P (1991) A neuronal correlate of spatial stability during periods of self-induced visual motion. Exp Brain Res 86:608–616

Georgopoulos AP, Schwartz AB, Kettner RE (1986) Neural population coding of movement direction. Science 233:1416–1419

Gibson JJ (1950). The perception of the visual world. Houghton Mifflin, Boston

Graziano M SA, Andersen RA, Snowden R (1994) Tuning of MST neurons to spiral motions. J Neurosci 14:54–67

Hildreth EC, Koch C (1987) The analysis of visual motion: From computational theory to neuronal mechanisms. Annu Rev Neurosci 10:477–533

Hoffmann KP, Distler C, Ilg U (1992) Callosal and superior temporal sulcus contributions to receptive field properties in the macaque monkey's nucleus of the optic tract and dorsal terminal nucleus of the accessory optic tract. J Comp Neurol 321:150–162

Ilg UJ, Hoffmann KP (1993) Functional grouping of the cortico-pretectal projection. J Neurophysiol 70:867–869

Ilg UJ, Thier P (1996) Inability of rhesus monkey area V1 to discriminate between self-induced and externally induced retinal image slip. Eur J Neurosci 8:1156–1166

Kawano K, Shidara M, Watanabe Y, Yamane S (1994) Neural activity in cortical area MST of alert monkey during ocular following responses. J Neurophysiol 71:2305–2324

Koenderink JJ, van Doorn AJ (1976) Local structure of movement parallax of the plane. J Opt Soc Am 66:717–723

Koenderink JJ, van Doorn AJ (1981) Exterospecific component of the motion parallax field. J Opt Soc Am 71:953–957

Komatsu H, Wurtz RH (1988a) Relation of cortical areas MT and MST to pursuit eye movements. I. Localization and visual properties of neurons. J Neurophysiol 60:580–603

Komatsu H, Wurtz RH (1988b) Relation of cortical areas MT and MST to pursuit eye movements. III. Interaction with full-field visual stimulation. J Neurophysiol 60:621–644

Lagae L, Maes H, Raiguel S, Xiao DK, Orban GA (1994) Responses of macaque STS neurons to optic flow components: a comparison of areas MT and MST. J Neurophysiol 71:1597–1626

Lappe M, Rauschecker JP (1993a) Computation of heading direction from optic flow in visual cortex. In: Giles CL, Hanson SJ, Cowan JD (eds) Advances in Neural Information Processing Systems 5, Morgan Kaufmann, San Mateo CA, pp 433–440

Lappe M, Rauschecker JP (1993b) A neural network for the processing of optic flow from ego-motion in higher mammals. Neural Comp 5:374–391

Lappe M, Rauschecker JP (1995) Motion anisotropies and heading detection. Biol Cybern 72:261–277

Lappe M, Bremmer F, Hoffmann KP (1994) How to use non-visual information for optic flow processing in monkey visual cortical area MSTd. In Marinaro M, Morasso PG (eds) ICANN 94 - Proceedings of the international conference on artificial neural networks, 26-29 May 1994, Sorrento. Springer, Berlin Heidelberg New York, pp 46-49

Lappe M, Bremmer F, Pekel M, Thiele A, Hoffmann KP (1996) Optic flow processing in monkey STS: a theoretical and experimental approach. J Neurosci 16:6265–6285

Lee DN (1980) The optic flow field: the foundation of vision. Philos Trans R Soc Lond Biol Sci 290:169–179

Lee C, Rohrer WH, Sparks DL (1988) Population coding of saccadic eye movements by neurons in the superior colliculus. Nature 332:357–360

Lestienne F, Soechting J, Berthoz A (1977) Postural readjustments induced by linear motion of visual scenes. Exp Brain Res 28:363–384

Maunsell JHR, Van Essen DC (1983) Functional properties of neurons in middle temporal visual area of the macaque monkey I. Selectivity for stimulus direction speed and orientation. J Neurophysiol 49:1127–1147

Miles FA, Busettini C (1992) Ocular compensation for self-motion. Ann NY Acad Sci 656:220–232

Mishkin M, Ungerleider LG, Macko KA (1983) Object vision and spatial vision: two cortical pathways. Trends Neurosci 6:414–417

Movshon JA, Adelson EH, Gizzi MS, Newsome WT (1985) The analysis of moving visual patterns. In Chagas C, Gattass R, Gross C (eds) Pattern recognition mechanisms. Springer, Berlin Heidelberg New York

Nakayama K, Loomis JM (1974) Optical velocity patterns, velocity sensitive neurons, and space perception: a hypothesis. Perception 3:63–80

Newsome WT, Pare, EB (1988) A selective impairment of motion perception following lesions of the middle temporal visual area (MT). J Neurosci 8:2201–2211

Newsome WT, Wurtz RH, Komatsu H (1988) Relation of cortical areas MT and MST to pursuit eye movements. II. Differentiation of retinal from extraretinal inputs. J Neurophysiol 60:604–620

Orban GA, Lagae L, Verri A, Raiguel S, Xiao D, Maes H, Torre V (1992) First-order analysis of optical flow in monkey brain. Proc Nat Acad Sci USA 89:2595–2599

Pekel M, Lappe M, Hoffmann KP (1995) Eye movements during optic flow stimulation in monkeys. Eur J Neurosci 8 (suppl):97

Pekel M, Lappe M, Bremmer F, Thiele A, Hoffmann KP (1996) Neuronal responses in the motion pathway of the macaque monkey to natural optic flow stimuli. Neuroreport 7:884–888

Regan D, Beverly KI (1982) How do we avoid confounding the direction we are looking and the direction we are moving? Science 215:194–196

Regan D, Hamstra SJ (1993) Dissociation of discrimination thresholds for time to contact and for rate of angular expansion. Vision Res 33:447–462

Roy JP, Wurtz RH (1992) Disparity sensitivity of neurons in monkey extrastriate area MST. J Neurosci 12:2478–2492

Royden CS (1994) Analysis of misperceived observer motion during simulated eye rotations. Vision Res 34:3215–3222

Royden CS, Crowell JA, Banks MS (1994) Estimating heading during eye movements. Vision Res 34:3197–3214

Saito HA, Yukie M, Tanaka K, Hikosaka K, Fukada Y, Iwai E (1986) Integration of direction signals of image motion in the superior temporal sulcus of the macaque monkey. J Neurosci 6:145–157

Sakata H, Shibutani H, Ito Y, Tsurugai K (1986) Parietal cortical neurons responding to rotatory movement of visual stimulus in space. Exp Brain Res 61:658–663

Salzman CD, Britten KH, Newsome WT (1990) Cortical microstimulation influences perceptual judgements of motion direction. Nature 346:174–177

Sauvan XM, Bonnet C (1993) Properties of curvilinear vection. Percep Psychophys 53:429–435

Tanaka K, Saito HA (1989a) Analysis of motion of the visual field by direction, expansion/contraction and rotation cells clustered in the dorsal part of the medial superior temporal area of the macaque monkey. J Neurophysiol 62(3):626–641

Tanaka K, Saito HA (1989b) Underlying mechanisms of the response specificity of expansion/contraction and rotation cells in the dorsal part of the medial superior temporal area of the macaque monkey. J Neurophysiol 62:642–656

Tanaka K, Sugita Y, Moriya M, Saito HA (1993) Analysis of object motion in the ventral part of the medial superior temporal area of the macaque visual cortex. J Neurophysiol 69:128–142

Telford L, Howard IP, Ohmi M (1995) Heading judgements during active and passive self-motion. Exp Brain Res 104:502–510

Thier P, Erickson RG (1992) Resonses of visual-tracking neurons from cortical area MST-l to visual eye and head motion. Eur J Neurosci 4:539–553

van den Berg AV (1992) Robustness of perception of heading from optic flow. Vision Res 32:1285–1296

van den Berg AV (1993) Perception of heading. Nature 365:497–498

van den Berg AV, Brenner E (1994) Why two eyes are better than one for judgements of heading. Nature 371:700–702

Van Gisbergen JAM, Van Opstal AJ, Tax AAM (1987) Collicular ensemble coding of saccades based on vector summation. Neurosci 21:541–555

Warren WH, Blackwell AW, Kurtz KJ, Hatsopoulos NG, Kalish ML (1991) On the sufficiency of the velocity field for perception of heading. Biol Cybern 65:311–320

Warren WH, Hannon DJ (1990) Eye movements and optical flow. J Opt Soc Am A 7:160–169

Warren WH, Kurtz KJ (1992) The role of central and peripheral vision in perceiving the direction of self-motion. Percep Psychophys 51:443–454

The Representation of Movement in Near Extra-Personal Space in the Macaque Ventral Intraparietal Area (VIP)

F. BREMMER, J.-R. DUHAMEL, S. BEN HAMED, and W. GRAF

CNRS - Collège de France, Paris, France

Perception and control of self-motion require the processing of a number of different incoming sensory signals, such as visual, vestibular, somatosensory and even auditory information. These signals are combined to generate an unambiguous internal representation of movement in space. Studies to investigate the neuronal basis underlying the analysis of self-motion information have focused a great deal on the visual system, especially on single cell recordings in monkey medial temporal (MT) and medial superior temporal (MST) areas in the superior temporal sulcus (STS) (Duffy and Wurtz 1991a,b, 1995; Graziano et al. 1994; Lagae et al. 1994; Lappe et al. 1994; Treue and Maunsell, this volume; Ilg and Thier, this volume). Neurons in these areas respond selectively to the direction of moving stimuli. Those of area MT have relatively small receptive fields and respond preferably to small visual stimuli (Maunsell and van Essen 1983a,b; Albright 1984). Area MST receives massive projections from area MT (Maunsell and van Essen 1983c; Desimone and Ungerleider 1986). Usually, MST neurons have large receptive fields and respond selectively to whole field pattern motion (Saito et al. 1986; Tanaka et al. 1986, 1989; Tanaka and Saito 1989; Komatsu and Wurtz 1988a,b; Roy and Wurtz 1990; Roy et al. 1992; Kawano et al., this volume).

Another MT projection zone is the ventral intraparietal (VIP) area, located in the fundus of the intraparietal sulcus (Maunsell and van Essen 1983c). Neurons there also respond selectively to the direction and speed of a moving visual stimulus (Duhamel et al. 1991; Colby et al. 1993). By contrast to the motion areas in the STS, many neurons in area VIP in addition also have tactile sensitivity. In such cases, location and response properties of visual and somatosensory receptive fields are congruent.

The existence of multiple representations of motion in the dorsal stream of the macaque visual cortical system raises the question about the specific function of these different areas. In order to contrast the functional context of area VIP as compared to areas MT and MST, we tested further the involvement of area VIP in the analysis of self-motion and the representation of extrapersonal space (c.f., Duhamel et al. 1991; Colby et al. 1993).

In: Parietal Lobe Contributions to Orientation in 3D Space (1997). P. Thier and H.-O. Karnath (eds). Springer-Verlag, Heidelberg.

To this end, single cells were recorded extracellularly in one surgically prepared, awake rhesus monkey (*Macaca mulatta*) who performed several behavioral tasks. A total of 96 cells was recorded. Since at the present time, the animal is still employed in ongoing experiments, neurons were classified as being located in area VIP on the basis of their response properties and with respect to the recording sites within the intraparietal sulcus. Most importantly, neurons in area VIP differ from those in neighboring areas MIP (in the medial bank) and LIP (in the lateral bank of the intraparietal sulcus) with respect to their strong preference for the direction and speed of a moving visual stimulus.

Directional Selectivity for Visual and Tactile Stimuli

Visual stimuli as well as a central fixation target were back-projected onto a translucent tangent screen covering an area of 90° x 80° at a viewing distance of 57 cm. The animal was given liquid rewards for keeping the eyes within an electronically defined (2° x 2°) window, centered on a fixation target. Visual receptive field (RF) contours were delineated manually using a hand held projector displaying bars and spots. Quantitative mapping of the full RF was accomplished for some neurons using computer generated stimuli (Duhamel et al. 1995). Sensitivity to large field motion was assessed by presenting a moving random dot pattern (240 dots) covering the whole tangent screen.

Directional selectivity was assessed by moving the random dot pattern along a circular pathway at either 40°/s or 27°/s. The speed of the pattern was kept constant throughout a stimulus trial (cycle), but stimulus direction was changed continuously (0° to 360°) within a complete stimulus cycle. With this kind of stimulation, the full two-dimensional projection space could be covered during a single trial without the need to test a critical number of unidirectional pattern movements (c.f., Schoppmann and Hoffmann 1976). Speed tuning was determined by moving the random dot pattern into the cell's preferred direction with six different speeds (5°/s, 10°/s, 20°/s, 40°/s, 80°/s, 160°/s). Finally, somatosensory responsiveness was tested qualitatively with cotton tip applicators touching different parts of the head, body, and limbs, and by passive rotation of single joints of arms and hands.

Visual receptive fields were usually large, covering sometimes more than half of the animal's visual field (see also Duhamel et al. 1991; Colby et al. 1993). About 80% of the tested cells revealed a significant speed tuning (p<0.05, distribution-free ANOVA). On average, higher speeds (40°/s - 160°/s) evoked higher neuronal discharges than speeds in the lower range (5°/s - 20°/S). The vast majority of cells was strongly selective for the direction of a moving stimulus. An example is shown in Fig. 1. This cell preferred stimulus motion down and to the right, whereas movement up and to the left caused inhibition.

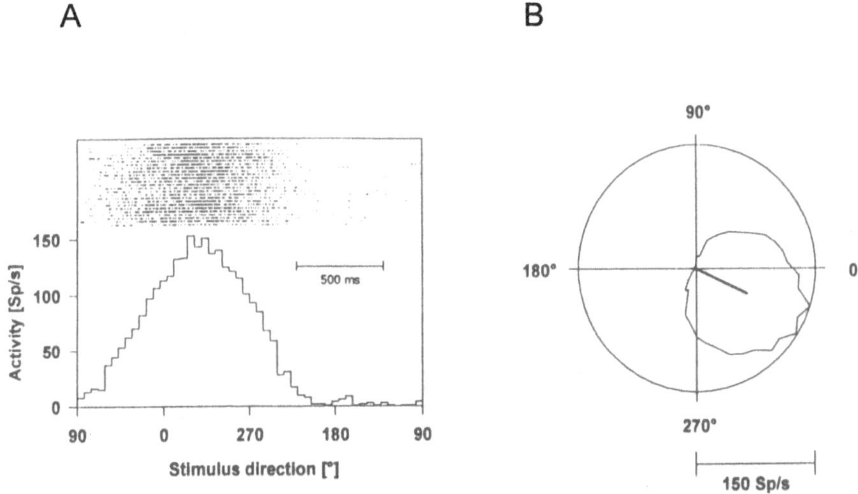

Fig. 1 A, B. Directional tuning to large field visual stimulus motion. **A** Average response of a cell to 25 consecutive stimulus trials. **B** Representation of the same data as a polar plot. Polar axes depict direction of stimulus motion. The preferred direction (polar vector) is computed quantitatively by weighted vector summation

Somatosensory responses were tested in 41 neurons. Of these, 20 neurons revealed a response with RFs located in the monkey's face region. About two thirds of these neurons (13 of 20) had directionally selective tactile responses, i.e., they only responded to movement of a tactile stimulus in one direction but not in the opposite direction. All neurons responsive to tactile stimulation also responded directionally selective to visual motion. In case of a directionally selective tactile response, preferred directions for visual and tactile stimuli were identical.

Optic Flow: Responsiveness to Simulated Movement in Three-Dimensional Space

Neuronal responsiveness to simulated forward or backward motion was assessed by presenting radially expanding or contracting optic flow stimuli (see Fig. 2). Dot speed in these patterns varied according to a dot's distance to the singularity (focus) of the pattern. The average dot speed was 40°/s.

The majority of tested neurons (47 of 70) responded in a significantly stronger manner to either expanding or contracting stimuli with the focus at the screen center ($p<0.05$, distribution-free ANOVA). An example of two different cells is given in Fig. 3. Visual pattern onset was often accompanied by a phasic response

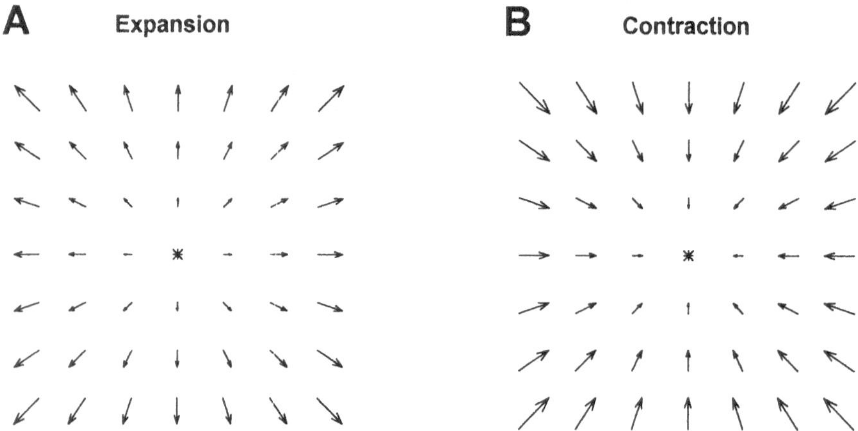

Fig. 2 A, B. Schematic drawing of two of the employed optic flow stimuli. Radially expanding (**A**) and contracting (**B**) pattern simulating a linear forward and backward movement

burst as was onset of stimulus motion for both expansion and contraction. Neurons could be classified into either expansion or contraction cells depending on whether presentation of an expansion or contraction stimulus elicited a tonic neuronal discharge. Presentation of the antagonistic stimulus, i.e., a contraction stimulus for an expansion cell and vice versa, often resulted in inhibition of the cell's discharge.

At this point, it should be mentioned that the above-described visual response characteristics of area VIP neurons closely resembled that of MST neurons. Neurons in the macaque MST area are also selectively responsive to optic flow fields, and an involvement of area MST in the perception of the direction of heading during locomotion was suggested (Duffy and Wurtz 1991a; Lagae et al. 1994; Lappe, this volume). In light of these similarities, the potential differential functional involvement regarding motion perception of these two areas needs to be considered. One key difference between both areas is the tactile responsiveness of many VIP area neurons.

In our study, 34 neurons, which were tested with optic flow stimuli, were also tested for tactile responsiveness. A large subpopulation (17 of 34) was responsive to both types of stimuli, while 14 of 34 of the neurons responded to either expansion or contraction but not to tactile stimuli. A small subpopulation (3 of 34) responded only to somatosensory stimulation, however two of these showed only a tactile on or off response but no directional selectivity. These findings might suggest that VIP neurons do not specifically code information concerning the far

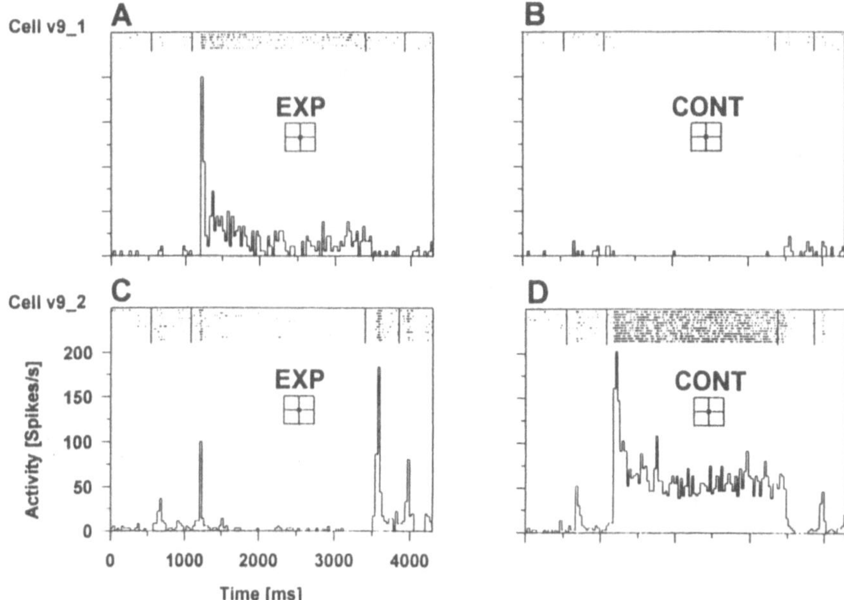

Fig. 3 A-D. Optic flow responses in the ventral intraparietal area (VIP). Responses of two different neurons to radially expanding and contracting optic flow stimuli. **A, C** Response to an expanding stimulus. **B, D** Response to a contracting stimulus. Vertical lines within each post-stimulus time histogram (PSTH) depict (first) the onset of the visual stimulus, (second) the beginning of stimulus motion, (third) the end of stimulus motion, (fourth) the extinction of the visual stimulus. In all cases, the location of the optic flow singularity was the center of the visual field, as indicated by insets within each PSTH. The expansion cell (**A, B**) responds to an expanding pattern with excitation (**A**), and to a contracting pattern with inhibition (**B**). The opposite result was obtained for a contraction cell (**C, D**). Both cells were located next to each other in a single electrode penetration

extrapersonal space including the direction of heading rather than information concerning self-motion in the near and ultra-near space. Information processed in area VIP could then be used for instance to avoid approaching objects.

Direction-Selective Responses for Vestibular Stimulation

Self-motion not only includes linear displacement into one direction with fixed gaze as simulated by the two-dimensional and three-dimensional visual stimuli but usually comprises also movement of the eyes and especially of the head when tracking objects. We thus also tested vestibular responsiveness of the recorded neurons. In this paradigm, the animal was rotated sinusoidally on a vertical-axis

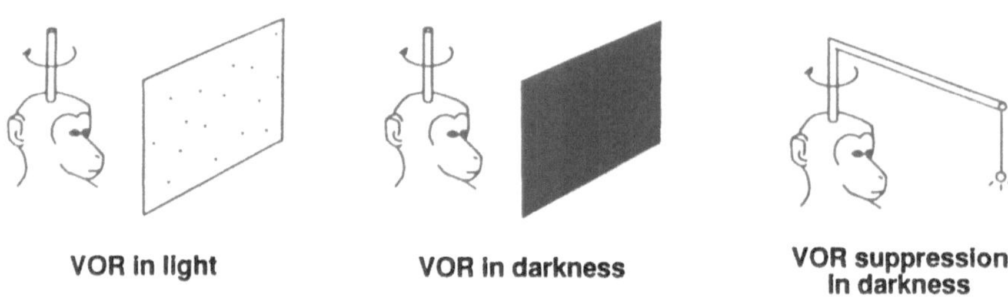

Fig. 4. Schematic representation of the employed vestibular stimulation paradigms. Yaw-axis rotation of the animal occurred either in light, in darkness, or, in some cases, in darkness during vestibulo-ocular reflex (VOR) suppression (not illustrated here).

turntable either in light or in darkness (0.25 Hz; ±30°) (Fig. 4). The animal was either left free to make compensatory eye movements (vestibulo-ocular reflex, VOR), or had to suppress the VOR by fixating a chair mounted light-emitting diode (LED).

About half of the tested neurons (25 of 53) showed significant vestibular responsiveness to yaw-axis stimulation ($p<0.05$, distribution-free ANOVA). Two neurons were biphasic, i.e., they responded with activation during rotation in both directions. One neuron's activity seemed to be more related to the position of the turntable than to the direction of its movement. The majority of neurons (22 of 25), however, revealed a clear preference for movement of the turntable either to the right or to the left. A typical result is shown in Fig. 5. Interestingly, the preferred directions for visual and vestibular stimulation of these neurons were non-complementary, i.e., the neuron preferred visual stimulus motion and also rotation of the head into the same direction.

This response pattern is quite in contrast to the characteristics of neurons in brain-stem or even higher order visual-vestibular processing centers (Büttner and Buettner 1978; Grüsser et al. 1990), where visual and vestibular on-directions would always be complementary, i.e., the visual on-direction would always be in the opposite direction to the respective head movement. The non-complementary on-directions of the recorded visual and vestibular responses in the VIP neurons, in essence, should generate a visual-vestibular conflict situation. In light of this latter consideration, it is even more interesting that, in the majority of the recorded neurons, the discharge was stronger during vestibular stimulation in light than in darkness. Although the animal made compensatory eye movements (VOR), the gain of the VOR was typically less than 1.0. Thus, for instance, the resulting retinal slip signal during head rotation to the right was visual motion to the left, i.e., the non-preferred direction of such a neuron. The incoming vestibular signal in this condition thus clearly overrules the incoming visual signal.

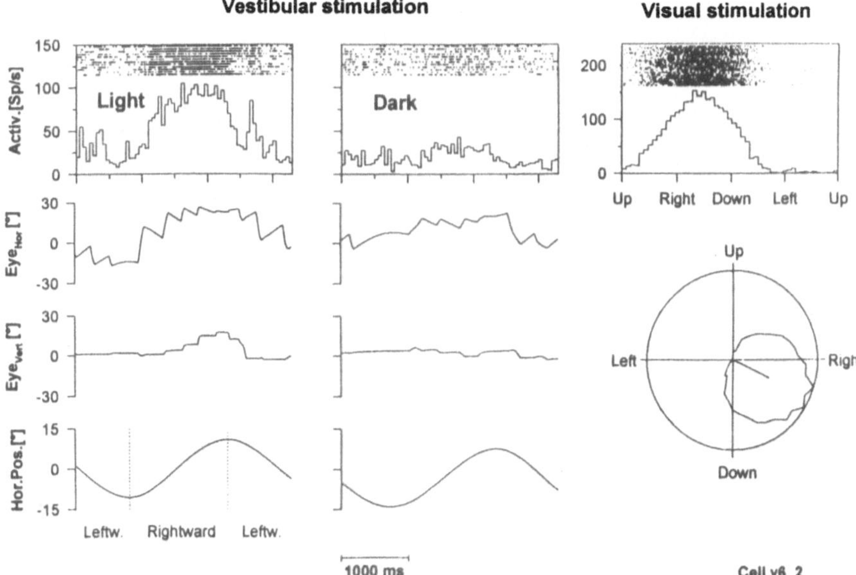

Fig. 5. Responses of a single neuron to vestibular and visual stimulation. The first two columns represent the neuronal responses (*first row*), sample horizontal (*second row*) and vertical (*third row*) eye position traces as well as horizontal turntable position (*fourth row*) during vestibular stimulation in light (*first column*) and in darkness (*second column*). The last column depicts the responses of this neuron to a large field stimulus moving along a circular pathway in the fronto-parallel plane (same neuron as in Fig. 1). The polar-plot shows the same data as the above PSTH, clearly indicating the preferred direction of this neuron for stimulus motion down and to the right. The preferred direction for vestibular stimulation was non-complementary (rightward) to the cell's preferred directions for visual motion (*first two columns*). Discharge was stronger for vestibular stimulation in light than in darkness

The non-complementary response characteristic of visual and vestibular preferred directions was obtained in all but one of the tested neurons (21 of 22). In this one instance, the preferred visual on-direction was about 90° off the vestibular on-direction. In all other cases, visual preferred directions were distributed about the vestibular ipsilateral and contralateral horizontal on-directions (Fig. 6A).

The limitation to only yaw axis stimulation most likely led to an underestimation of the proportion of neurons responsive to vestibular stimulation. Within the neuron population unresponsive to vertical axis rotation (28 of 53), no clear distribution of visual preferred directions could be discerned (Fig. 6B). These neurons could have either not received vestibular input at all, or they may have been connected with vertical canal or otolith systems.

All possibilities of unimodal, bimodal or multisensory convergence were present. However, vestibular-sensitive neurons were always also visually direction

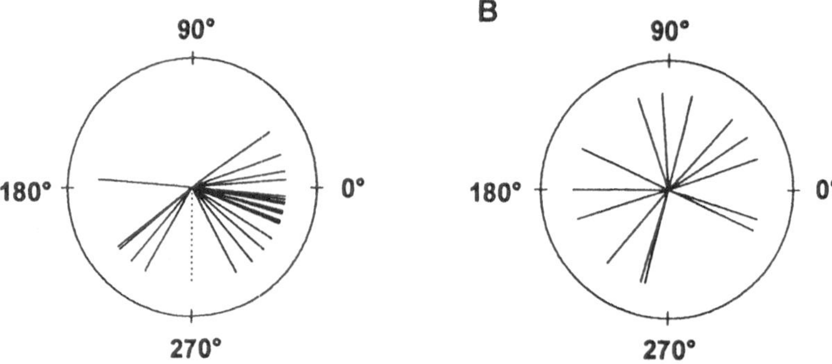

Fig. 6 A, B. Preferred visual directional tuning of ventral intraparietal area neurons with (**A**) and without (**B**) vertical axis rotational sensitivity. In neurons responding to yaw-axis stimulation (**A**), tuning directions form two clusters, representing association with ipsilateral (about 0°) and contralateral (about 180°) vestibular on-direction (type 1 and type 2, respectively). The *dotted line* pointing towards 270° is the only example of a neuron whose visual tuning direction is 90° off the vestibular on-direction. Visual tuning directions of neurons without vertical axis rotational sensitivity (**B**) do not show a preferred directional distribution

selective, including sensitivity to optic flow stimulation (n=27). In a similar fashion, somatosensory directional selective responsiveness was always accompanied by visual directional selectivity (n=13). As already mentioned, when tactile, visual and vestibular responses were combined (n=14), their directional tuning was always aligned (except in one case), although visual and tactile on-directions were non-complementary with regard to the vestibular response (13 of 14). In the sole instance of non-alignment of visual-tactile versus the vestibular response, the former sensory qualities remained co-directional, but about 90° off the vestibular on-direction.

Responses to vestibular stimulation so far have been reported in different regions of the macaque parietotemporal cortex. Ödkvist et al. (1974) described such responses in the "neck region" of somatosensory cortical area 3a. Two other regions considered "vestibular" are area 2v at the anterior tip of the intraparietal sulcus (Büttner and Buettner 1978; Graf et al. 1995), and the parietoinsular vestibular cortex (PIVC), located in the parietal operculum of the sulcus lateralis and the retroinsular region (Grüsser et al. 1990). Sakata et al. (1994) reported vestibular responses in some neurons "localized in the posterolateral part of area PG (area 7a of Vogt), on the anterior bank of the caudal superior temporal sulcus (STS), in the region partly overlapping the medial superior temporal (MST) area" (p 183). These latter neurons shared the non-complementary response characteristics of area VIP neurons regarding the preferred directions for visual and vestibular stimulation. The majority of neurons in the other areas respond in a complementary fashion. Such behavior is also typical for all vestibular brain stem neurons.

Further evidence of vestibular input to cortical areas of the dorsal stream of the macaque visual system comes from two other studies. Thier and Erickson (1992) showed a direction selective increase in activity during VOR suppression in neurons of area MST. Brotchie et al. (1995) found an influence of head position on the activity of parietal neurons (areas 7 and LIP). In the latter case, an integrated vestibular head velocity signal could be the origin for this head position signal.

Area VIP is reciprocally connected to areas 7, LIP, and MST. Area 7 in turn is connected to area PIVC which itself is part of a cortical network of vestibular sensitive areas (2v and 3a). The observation of vestibular responses in area VIP thus was not entirely unexpected. More unexpected was the finding that the preferred directions for visual and vestibular stimuli were non-complementary for essentially all investigated neurons. Regarding this aspect, area VIP differs remarkably from most other vestibular cortical areas.

Conclusion

The macaque area VIP is clearly involved in the processing of motion signals. The majority of neurons responds to optic flow pattern, simulating forward or backward motion. About one third of the tested neurons had tactile directional selectivity, and almost half of the investigated neurons responded to vertical axis vestibular stimulation. While preferred directions for visual and tactile stimuli were always identical, they were always non-complementary vis-à-vis the vestibular on-directions, i.e., non-synergistic. This peculiarity raises the question about the functional role of the vestibular responses in light of the hypothesis of a unified multimodal representation of movement in space in area VIP. Envisioning self-motion of an animal through a natural environment might give an answer to this question. Self-motion usually not only includes linear displacement into one direction with fixed gaze as simulated by the employed two-dimensional and three-dimensional visual stimuli. Movement through the environment also comprises orienting and tracking of objects, i.e., movement of the eyes and of the head. Since we did not test systematically optokinetic or pursuit eye movements, we will consider only head movements here. This is a reasonable assumption since, in everyday life, most tracking movements largely consist of head movements. If we now consider the visual information arriving at the retina during linear forward self-motion and a simultaneous rotational head movement e.g., to track some object, it becomes clear that the direction of motion on the retina induced by objects in the environments depends on the location of these objects with respect to the fixation plane. Objects located beyond the fixation plane induce visual motion on the retina which is opposite to the direction of head movement. However, objects with a distance up to the fixation plane induce visual motion into the same direction as the head movement. Objects in question could even approach the body surface such as to produce tactile flow, for instance across the face. Thus, the functional role of motion sensitive neurons in area VIP could be to encode self-motion in the *near* extrapersonal space. Alternatively, these neurons could also play a role in the analysis of object trajectory in this portion of space. Both hypotheses could imply a key difference between movement processing in areas VIP and MST. The latter is widely considered to be involved in heading detection analysis, i.e., in the encoding of visual information in the *far* extrapersonal space. However, further experiments in head-free animals during linear forward and backward motion need to be conducted to verify this hypothesis.

Acknowledgements. This research was supported by HCM grant ERBCHRXCT9930267 from the European Community and by Human Frontier Science Program grant RG71/96B. The authors wish to thank Dr. M. Lappe for critical reading of the manuscript and helpful suggestions with the interpretation of the results.

References

Albright T D (1984) Direction and orientation selectivity in visual area MT of the macaque. J Neurophysiol 52:1106–1130

Brotchie PR, Andersen RA, Snyder LH, Goodman SJ (1995) Head position signals used by parietal neurons to encode locations of visual stimuli. Nature 375:232–235

Büttner U, Buettner UW (1978) Parietal cortex (2v) neuronal activity in the alert monkey during natural vestibular and optokinetic stimulation. Brain Res 153:392–397

Colby CL, Duhamel J-R, Goldberg ME (1993) The ventral intraparietal Area (VIP) of the macaque: anatomical location and visual properties. J Neurophysiol 69:902–914

Desimone R, Ungerleider LG (1986) Multiple visual areas in the caudal superior temporal sulcus of the macaque. J Comp Neurol 248:164–189

Duffy CJ, Wurtz RH (1991a) Sensitivity of MST neurons to optic flow stimuli. I. A continuum of response selectivity to large-field stimuli. J Neurophysiol 65:1329–1345

Duffy CJ, Wurtz RH (1991b) Sensitivity of MST neurons to optic flow stimuli. I. Mechanisms of response selectivity revealed by small-field stimuli. J Neurophysiol 65:1346–1359

Duffy CJ, Wurtz RH (1995) Response of monkey MST neurons to optic flow stimuli with shifted centers of motion. J Neurosci 15:5192–5208

Duhamel J-R ,Colby CL, Goldberg ME (1991) Congruent representations of visual and somatosensory space in single neurons of monkey ventral intra-parietal cortex (area VIP). In: Paillard J (ed) Brain and Space. Oxford University Press, Oxford, pp 223–236

Duhamel J-R, Ben Hamed S, Bremmer F, Graf W (1995) The influence of attention and intended eye movements upon the excitability of neurons in the lateral intraparietal area of macaque monkeys. Soc Neurosci (Abstr) 21:268–17

Graf W, Bremmer F, Sammaritano M, Ben Hamed S, Duhamel J-R (1995) Oculomotor, vestibular, and visual response properties of neurons in the anterior inferior parietal lobule of macaque monkeys. Soc Neurosci (Abstr) 21:268–16

Graziano MSA, Andersen RA, Snowden R (1994) Tuning of MST neurons to spiral motions. J Neurosci 14:54–67

Grüsser O-J, Pause JM, Schreiter U (1990) Localization and responses of neurons in the parieto-insular vestibular cortex of awake monkeys (*Macaca fascicularis*). J Physiol (Lond) 430:537–557

Komatsu H, Wurtz RH (1988a) Relation of cortical areas MT and MST to pursuit eye movements. I. Localization and visual properties of neurons. J Neurophysiol 60:580–603

Komatsu H, Wurtz RH (1988b) Relation of cortical areas MT and MST to pursuit eye movements. III. Interaction with full field stimulation. J Neurophysiol 60:621–644

Lagae L, Maes H, Raiguel S, Xiao D-K, Orban GA (1994) Responses of macaque STS neurons to optic flow stimuli: a comparison of areas MT and MST. J Neurophysiol 71:1597–1626

Lappe M, Bremmer F, Hoffmann K-P (1994) How to use non-visual information for optic flow processing in monkey visual cortical area MSTd. In: Marinaro M, Morasso PG (eds) ICANN 94. Springer Verlag, Berlin, pp 46–49

Maunsell JHR, Van Essen DC (1983a) Functional properties of neurons in middle temporal visual area of the macaque monkey. I. Selectivity for stimulus direction, speed, and orientation. J Neurophysiol 49:1127–1147

Maunsell JHR, Van Essen DC (1983b) Functional properties of neurons in middle temporal visual area of the macaque monkey. II. Binocular interactions and sensitivity to binocular disparity. J Neurophysiol 49:1148–1167

Maunsell JHR, Van Essen DC (1983c) The connections of the middle temporal visual area (MT) and their relationship to a cortical hierarchy in the macaque monkey. J Neurosci 3:2563–2580

Ödkvist LM, Schwarz DWF, Fredrickson JM, Hassler R (1974) Projection of the vestibular nerve to the area 3a arm field in the squirrel monkey (Saimiri sciureus). Exp Brain Res 21:97–105

Roy J-P, Wurtz RH (1990) The role of disparity-sensitive cortical neurons in signalling the direction of self-motion. Nature 348:160–162

Roy J-P, Komatsu H, Wurtz RH (1992) Disparity sensitivity of neurons in monkey extrastriate area MST. J Neurosci 12:2478–2492

Saito H, Yukie M, Tanaka K, Hikosaka K, Fukada Y, Iwai E (1986) Integration of direction signals of image motion in the superior temporal sulcus of the macaque monkey. J Neurosci 6:145–157

Sakata H, Shibutani H, Ito Y, Tsurugai K, Mine S, Kusunoki M (1994) Functional properties of rotation-sensitive neurons in the posterior parietal association cortex of the monkey. Exp Brain Res 101:183–202

Schoppmann A, Hoffmann K-P (1976) Continuous mapping of direction selectivity in the cat's visual cortex. Neurosci Lett 2:177–181

Tanaka K, Hikosaka K, Saito H, Yukie M, Fukada Y, Iwai E (1986) Analysis of local and wide-field movements in the superior temporal visual areas of the macaque monkey. J Neurosci 6:134–144

Tanaka K, Fukada Y, Saito H (1989) Underlying mechanisms of the response specificity of expansion/contraction and rotation cells in the dorsal part of the medial superior temporal area of the macaque monkey. J Neurophysiol 62:642–656

Tanaka K, Saito H (1989) Analysis of motion of the visual field by direction, expansion/contraction, and rotation cells clustered in the dorsal part of the medial superior temporal area of the macaque monkey. J Neurophysiol 62:626–641

Thier P, Erickson RG (1992) Responses of visual-tracking neurons from cortical area MST-I to visual, eye and head motion. Eur J Neurosci 4:539–553

Subject Index